# FORTSCHRITTE DER CHEMIE ORGANISCHER NATURSTOFFE

---

# PROGRESS IN THE CHEMISTRY OF ORGANIC NATURAL PRODUCTS

---

# PROGRÈS DANS LA CHIMIE DES SUBSTANCES ORGANIQUES NATURELLES

HERAUSGEGEBEN VON   EDITED BY   RÉDIGÉ PAR

## L. ZECHMEISTER
CALIFORNIA INSTITUTE OF TECHNOLOGY, PASADENA

SIEBZEHNTER BAND
SEVENTEENTH VOLUME   DIX-SEPTIÈME VOLUME

VERFASSER   AUTHORS   AUTEURS
P. H. ABELSON · H. BARKEMEYER · K. BERNAUER · A. E. DIMOND
H. H. INHOFFEN · K. IRMSCHER · F. KORTE · I. KORTE · H. KUHN
W. A. SCHROEDER · B. B. STOWE · K. VENKATARAMAN

MIT 57 ABBILDUNGEN   WITH 57 FIGURES   AVEC 57 ILLUSTRATIONS

SPRINGER-VERLAG WIEN GMBH

ISBN 978-3-7091-8054-9      ISBN 978-3-7091-8052-5
DOI 10.1007/978-3-7091-8052-5

# Inhaltsverzeichnis.
# Contents. — Table des matières.

## Neuere Ergebnisse der Chemie pflanzlicher Bitterstoffe. Von F. KORTE, H. BARKEMEYER und I. KORTE, Chemisches Institut der Universität Bonn

## Alkaloide aus Calebassencurare und südamerikanischen Strychnosarten.

Von K. BERNAUER, Chemisches Institut der Universität Zürich...... 183

## Occurrence and Metabolism of Simple Indoles in Plants. By BRUCE B. STOWE, The Biological Laboratories, Harvard University, Cambridge, Massachusetts

## Some Biochemical Aspects of Disease in Plants. By A. E. DIMOND,

## The Chemical Structure of the Normal Human Hemoglobins.

# Flavones and Isoflavones.

## By K. VENKATARAMAN, Poona, India.

### With 5 Figures.

### Contents.

# I. Introduction.

Among the important plant coloring matters used for dyeing and printing in the middle ages or earlier were weld *(Reseda luteola)*, young fustic (wood of *Rhus cotinus*), old fustic (wood of *Chlorophora tinctoria*), quercitron bark *(Quercus tinctoria)*, and Persian berries (from various species of *Rhamnus*), which gave yellow, orange, brown and olive shades on aluminium, tin, chromium and iron mordants. According to Colour Index (1956) they find considerable use even at the present time, especially old fustic, osage orange (from the wood of *Maclura pomifera* which also contains morin), and quercitron (Flavine) in the U. S. A., for dyeing silk, wool, nylon and leather, for calico printing, and for shading logwood blacks.

The first flavone to be isolated in the pure state was chrysin from poplar buds (PICCARD, 1864). Morin from old fustic, luteolin from weld, fisetin ("fustin") from young fustic, and quercitrin from quercitron bark were obtained in the crystalline state, but probably not pure, by CHEVREUL in 1814 or 1815. From quercitrin LIEBERMANN and HAMBURGER in 1879 prepared quercetin, but assigned to it an erroneous formula.

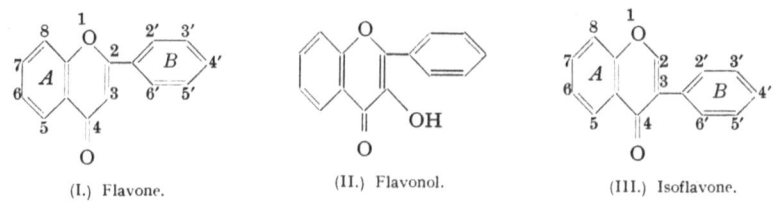

(I.) Flavone.          (II.) Flavonol.          (III.) Isoflavone.

The structures of fisetin, the coloring matter of young fustic, and of quercetin were elucidated by HERZIG (1884–1891). Shortly thereafter KOSTANECKI proved the constitution of chrysin, and he also gave the names flavone (from Latin *flavus*, yellow) and flavonol to the parent ring system (I) and its 3-hydroxy derivative (II). In the present review the term flavone normally includes a flavonol.

HERZIG and KOSTANECKI showed that the flavones and their alkyl ethers could be hydrolysed by alkali to mixtures of aromatic acids and phenolic ketones or phenols from which the structures of the flavones could be deduced. In 1898 KOSTANECKI, who made flavones his life work, achieved the first synthesis of a flavone, chrysin, and he subsequently developed several methods for the synthesis of flavones and flavonols. Numerous plant materials containing flavones were investigated by A. G. PERKIN during the period 1895 to 1920. In 1924 ROBINSON described a general reaction, which in its original form or with minor modifications has been widely used for synthesis in this field. More recently, a major contribution to our knowledge of natural flavones has been made by SESHADRI. Over 70 flavones, excluding different glycosides of the same flavone, have so far been isolated from plants, and nearly all have been synthesized.

The distribution of isoflavones in nature is very much more limited, and only 13 have so far been isolated, again excluding the glycosides. A few constituents of the soya bean first described as new isoflavones were later shown to be impure specimens of known isoflavones. An isoflavone structure was first assigned to a natural product in 1910 by FINNEMORE, who suggested that prunetin, isolated as its glucoside prunetrin from a *Prunus* bark, was probably 5,4'-dihydroxy-7-methoxy-isoflavone. Prunetin contained one methoxyl group, and it yielded phloroglucinol and *p*-hydroxyphenylacetic acid by alkali fusion. The production of a phenol and a phenylacetic acid (or a desoxybenzoin under milder conditions), together with formic acid, is now recognized as a clear indication of an isoflavone. Demethylation of prunetin gave prunetol, which proved to be identical with genistein isolated by PERKIN and NEWBURY from dyers' broom in 1899. The correctness of FINNEMORE's suggestion was proved in 1925 by BAKER and ROBINSON, who also succeeded in synthesizing the first isoflavone type. The parent iso-flavone (III) was synthesized in 1934 (*107*).

During the last few years there has been a revival of interest in the chemistry, biogenesis, and physiological properties of flavones, isoflavones and other flavonoids. The term flavonoid covers a large group of naturally occurring compounds in which two benzene rings are linked by a propane bridge ($C_6$—C—C—C—$C_6$, except in the isoflavones in which the

$$\begin{array}{c} C_6 \\ \diagup \end{array}$$

arrangement is $C_6$—C—C—C). The flavonoids include chalcones, dihydro-chalcones, aurones, flavanones, flavones and isoflavones, flavonols, 2,3-dihydroflavonols (flavanonols), flavan-3,4-diols (leucoanthocyanidins), anthocyanidins, and catechins. SESHADRI (*197*) has discussed their biochemistry, and in two reviews of the flavonoids and related plant pigments (*69, 67*) GEISSMAN has given a detailed account of their distribution in plants, methods of isolation and analysis, and possible mechanisms of biogenesis; SESHADRI and GEISSMAN have also listed the natural flavonoids known in 1950.

## II. Structural Relations and Interconversions.

The flavonoids vary from two (in the catechins) to five (in the flavonols) in the state of oxidation (or oxidation number) of the propane chain, calculated as the number of hydroxyl groups attached to the three carbon atoms, including the hydroxyls obtained by hydration of the double bonds and by hydrolytic fission of the pyran ring. Many interconversions of the flavonoids are possible in the laboratory as shown in *Chart 1*; there is no evidence that they represent biosynthetic pathways, although two or more flavonoid types (flavonols and flavanonols for instance) often occur together in the same plant. Several of the transformations, some of which are discussed later in connection with the synthesis of flavones, can be effected by reagents and routes other than those indicated in Chart 1. Some of the reactions by which chalcones and the isomeric flavanones can be converted to flavones and flavonols can take an alternative course leading to aurones (2-benzylidene coumaranones), which are isomeric with flavones and have been found to accompany chalcones and flavanones in a few plants.

Flavonols represent the highest oxidation state among the flavonoids, and they can be reduced, directly or by stages, to flavonoids at lower oxidation levels. Reduction of flavonols in sodium carbonate solution with sodium hydrosulphite yields flavanonols; taxifolin (XXV, p. 18) can be thus prepared from quercetin (IV, p. 7), a by-product being the corresponding 2-benzylcoumaranone (*175, 70*). However, datiscetin and morin (XIX, p. 17) resist such reduction, probably because of a steric effect of the 2'-hydroxyl group (*41*). WILLSTÄTTER and MALLISON in 1914 reduced quercetin to cyanidin by mercury, magnesium and hydrochloric acid in 2% yield. A conversion of flavonols to anthocyanidins by reduction with lithium aluminium hydride in ether followed by treatment with hydrochloric acid was reported by MIRZA and ROBINSON in 1950 (*156*). A useful general method for the production of anthocyanidins

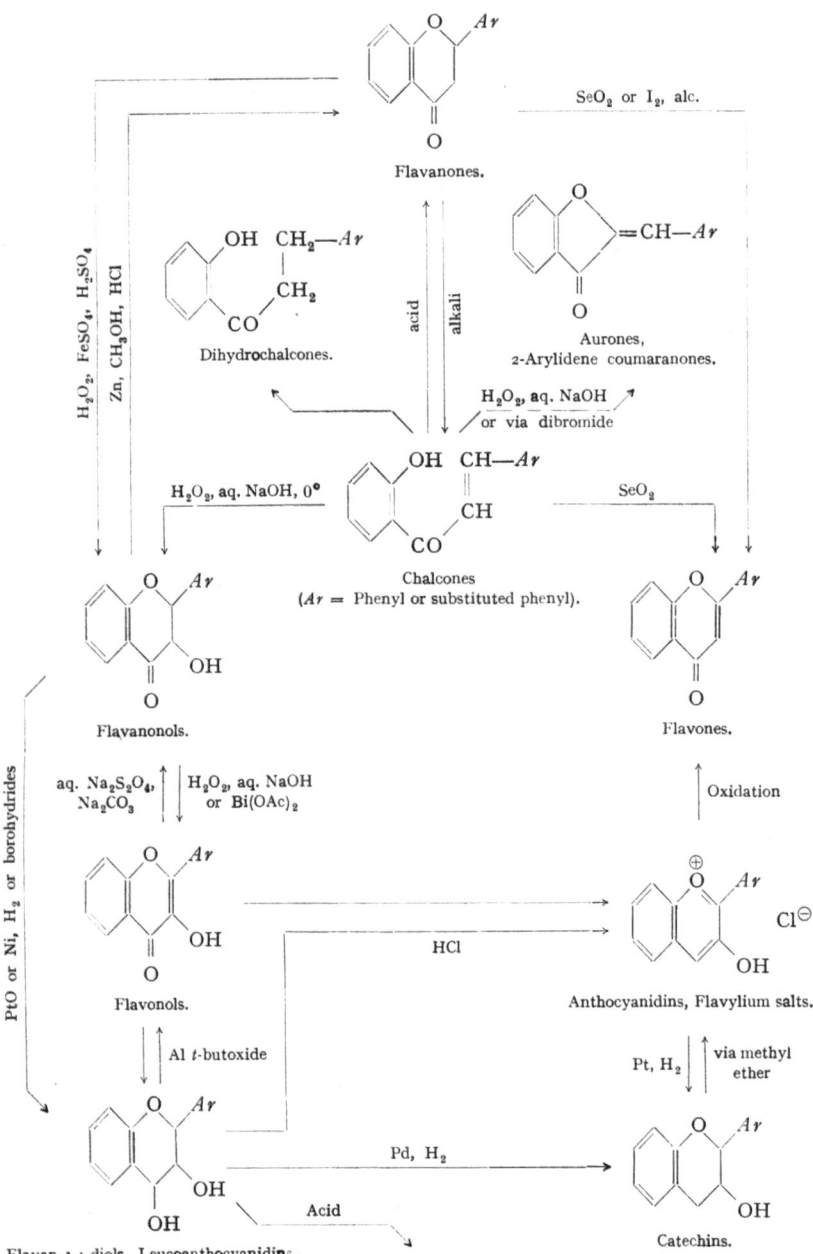

Chart *1*. Interconversions of Flavonoids.

in good yield from flavonols, described by KING and WHITE in 1957, is to submit them to reductive acetylation with zinc dust, sodium acetate and acetic anhydride, and treat the acetylated leucoanthocyanidins thus obtained with hot mineral acid (*126*). Flavan-3,4-diols or leucoanthocyanidins can be prepared (*106*) by catalytic hydrogenation of flavonols or dihydroflavonols or by reduction with lithium aluminium hydride or sodium borohydride.

## III. Occurrence in Plants.

Flavones, isoflavones, and the flavonoids in general are synthesized by higher plants; and they have not been found in lichens, as mould metabolites, or in the animal organism, with the exception that flavones have been detected in butterfly wings (*61*), the likely source being the plant food of the larva. Considering the very wide distribution of flavones in plants, it is reasonable to assume that they fulfil some purpose in plant life. SZENT-GYÖRGYI and others have suggested (cf. *197*) that they are involved in oxidation-reduction systems, but no real evidence of any physiological function of flavones or other flavonoids has yet been obtained. They play no part in photosynthesis except as end products.

KLEIN and WERNER (1925) associated the occurrence and localisation of flavones with the relationship of plants in families, genera and species (*127*). Flavones and flavanones are among the heartwood constituents of conifers which have been examined by ERDTMAN in relation to their taxonomy. He is of the view that flavone derivatives are formed in the cambium and transported to the heartwood via the rays in the sapwood; the flavone and other constituents of the heartwood and bark are relatively stable "end products" which may sometimes function as fungicides, insecticides or antioxidants. Taken in conjunction with the pinosylvins (3,5-dihydroxystilbene and its methyl ethers), the distribution of flavones and flavanones can be broadly related to species differences in the pines (*53*). After an extensive study of flavones occurring in different plant families, HÖRHAMMER (*88*) was unable to trace any phytochemical regularity. BATE-SMITH (*19*) has correlated the distribution of leucoanthocyanins, flavones and other phenolics with woody and herbaceous habit and with systematic position.

### Glycosides.

Flavones and isoflavones are present in plants both in the free state and as glycosides. Unlike the anthocyanins in which the sugar residue is invariably attached to the 3-hydroxyl (and in addition often to the 5-hydroxyl group), the sugar residue in flavones may be attached to a hydroxyl group in any position except the 6 and the 2' in which it has not so far been encountered; the most frequent attachment is to the 3- or 7-hydroxyl. Rhamnosides are common, although the occurrence of free rhamnose in plants appears to be doubtful; the best source for the preparation of *L*-rhamnose is quercitrin (commercial Flavine). There are glycosides containing two sugar residues, which may be identical or different, attached to two hydroxyl groups. Among the limited number of isoflavones known at present the sugar residue is in the 7- or 4'-position, and the isoflavones appear to exist in plants as the aglycones more commonly than the flavones. Flavones and isoflavones occur in all parts of plants—roots, rhizomes, inner and outer bark, heartwood and sapwood, leaves, flowers, fruits, seeds, and the resinous exudates of wood. The same plant often contains more than one flavone or isoflavone; there are plants which contain both flavones and isoflavones. The same flavone may be found in different

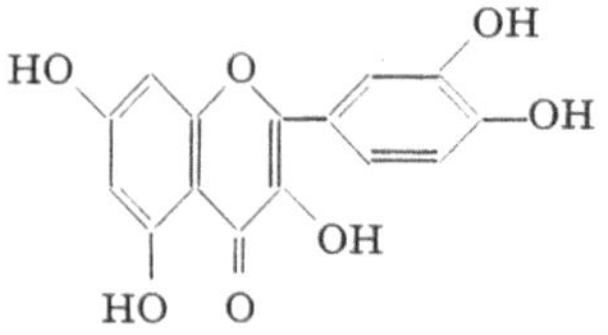

IV.) Quercetin. 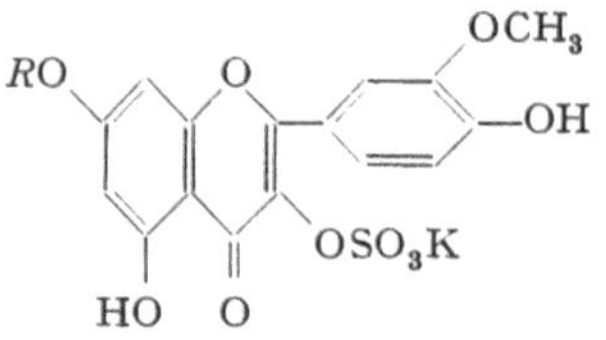  (V.) (R = H) Persicarin.

plants and in combination with a variety of sugars; quercetin (IV) in particular is wide-spread in plants in the form of numerous glycosides and ethers, and among its sugar components are arabinose, xylose, glucose, galactose, rhamnose, and rutinose. The leaves of etiolated beans contain quercituron, a glucuronide of quercetin; chrysin, baicalein, and scutellarein also occur as the glucosiduronic acids (*151*).

Isorhamnetin and rhamnazin have been found to occur as the 3-potassium sulphates (V; $R = H$ or $CH_3$). These appear to be the only two sulphuric esters of flavones occurring in nature; but other sulphuric esters are known, e. g. sinigrin in the black mustard and other mustard oil glycosides. -

### Hydroxylation Patterns.

The degree of hydroxylation in natural flavones varies from none in the parent compound to seven in hibiscetin, erianthin and gardenin, the two latter being pentamethyl and hexamethyl ethers of isomeric heptahydroxyflavones. A fact which may have some biogenetic significance is that at lower levels of hydroxylation (in which methoxylation is included) flavones unsubstituted in the 3-position predominate, while the highly hydroxylated compounds are mostly flavonols. Flavonol and monohydroxyflavonols do not occur in nature; among ten trihydroxy-flavones only three are flavonols; at the tetrahydroxy stage flavones and flavonols are about equal in number (six and eight); among forty penta- and hexahydroxy derivatives thirty-six are flavonols, and the three heptahydroxy compounds are all flavonols.

The $A$-ring is most frequently a resorcinol or phloroglucinol derivative, and in some flavones a derivative of 1,2,3,5-tetrahydroxybenzene. Auranetin (VI) is the only flavone in which the $A$-ring is derived from 1,2,3,4-tetrahydroxybenzene, and gardenin (LXXXVII, p. 40) the only flavone in which it is derived from 1,2,4,5-tetrahydroxybenzene. In tangeretin (VII), nobiletin, and calycopterin (LI, p. 30) the $A$-ring is derived from 1,2,3,4,5-pentahydroxybenzene and therefore substituted in all the six positions.

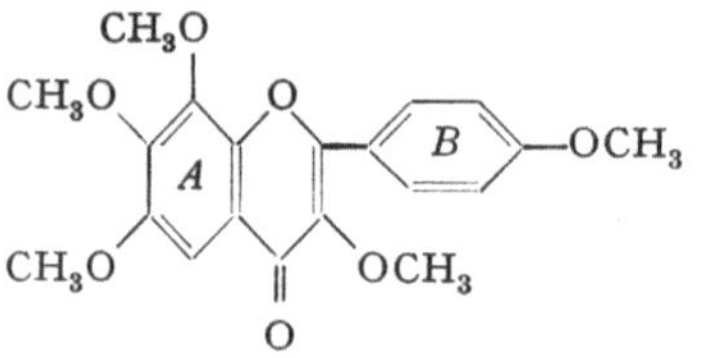

(VI.) Auranetin. 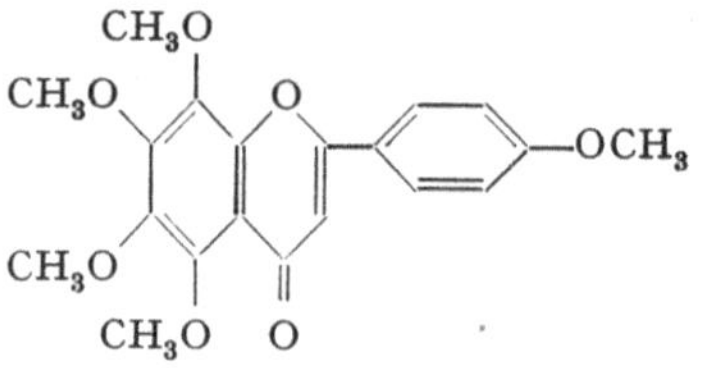          (VII.) Tangeretin.

The $B$-ring is a derivative of protocatechuic acid in thirty-one flavones, $p$-hydroxybenzoic acid in eighteen, benzoic acid in thirteen, gallic acid in seven, $\beta$-resorcylic acid in three (including artocarpin, XVIII, p. 17), salicylic acid in two, and 2,4,5-trihydroxybenzoic acid in one (oxyayanin-A; LII, p. 30). The three flavones (morin, artocarpetin and artocarpin) having the resorcinol pattern in the $B$-ring occur in the same plant, *Artocarpus integrifolia*.

Many methyl ethers and seven methylene ethers of polyhydroxy-flavones have been isolated. Among the di-, tri- and tetrahydroxyflavones the ethers are derived from the naturally occurring hydroxyflavones, exceptions being 5,6-dimethoxyflavone, wogonin and artocarpetin (XVII, p. 17); but in the more highly substituted flavones several ethers are known corresponding to polyhydroxyflavones which have not yet been encountered in plants. Further, a hydroxyflavone and its ether may occur in plants belonging to totally different families. Monomethyl ethers are the commonest, but 5,6-dimethoxyflavone, tangeretin, auranetin and nobiletin are flavones in which all the hydroxyl groups are found to be methylated. There are a few other flavones in which all the hydroxyl groups are protected, partly by methylation and partly by methylenation. Demethoxykanugin and kanugin contain one methylenedioxy group, together with two and three methoxyl groups, respectively. In meliternin (X, p. 12) and melisimplexin (VIII) four of the six hydroxyl groups are methylated and two are in the form of the methylene ether. Meliternatin (IX), another flavone in which there are no free hydroxyl groups, is the only example of a flavone containing two methylenedioxy groups. Five of the seven flavones containing methylenedioxy groups occur in *Melicope ternata* and *M. simplex*, and the other two in *Pongamia glabra*.

The isoflavone, pseudobaptigenin, contains one hydroxyl and one methylenedioxy group.

The plants in which seven flavones and one isoflavone containing $CH_2O_2$ groups occur belong to families (*Leguminosae* and *Rutaceae*) included in a list of eleven to which SIMMONDS and STEVENS (206) have drawn attention as families in which methylenated phenols are distributed. The recently isolated tlatlancuayin (XXXIX, p. 23) belongs to a twelfth family, *Amarantaceae*. SIMMONDS and STEVENS have observed that the methylenated phenols mentioned in GEISSMAN and HINREINER's review (69) are highly methylated and that none contains a free hydroxyl group in the ring carrying the methylenedioxy bridge; the subsequently described wharangin (XI, p. 12) is an exception.

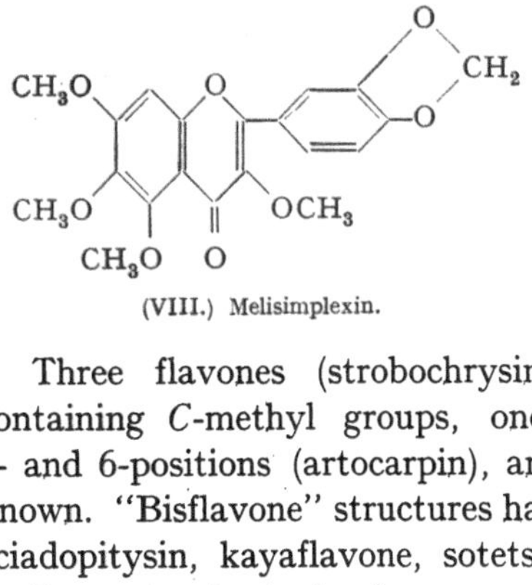

(VIII.) Melisimplexin. 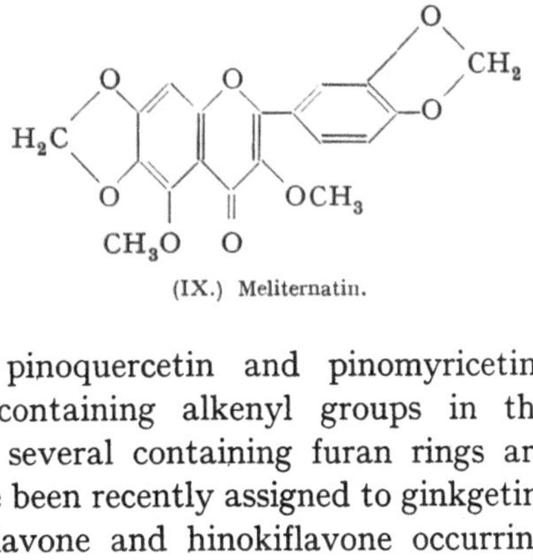        (IX.) Meliternatin.

Three flavones (strobochrysin, pinoquercetin and pinomyricetin) containing C-methyl groups, one containing alkenyl groups in the 3- and 6-positions (artocarpin), and several containing furan rings are known. "Bisflavone" structures have been recently assigned to ginkgetin, sciadopitysin, kayaflavone, sotetsuflavone and hinokiflavone occurring in Japanese plants (112).

Since only thirteen natural isoflavones are known, it is obvious that the variations in the number and orientation of hydroxyl and methoxyl groups are also limited; the lowest degree of hydroxylation (in which methoxylation is included) is two and the highest six, but no pentahydroxyisoflavone has yet been isolated. There are several more complex natural products, such as jamaicin, osajin, pomiferin, and munetone, which contain the isoflavone ring system.

## IV. Isolation.

General methods for the isolation of plant constituents are well known.

It is usual to carry out a preliminary extraction with a series of solvents of increasing polarity, separation into acid, basic and neutral constituents, and qualitative tests for the presence of common types of natural products such as carotenoids, anthocyanins, alkaloids, tannins and resin acids. A procedure for the isolation of flavones and isoflavones, which may often be detected by color reactions, may then be evolved. Mere solvent extraction may sometimes yield a flavone or isoflavone constituent. Examples are the flavone artocarpin and the isoflavone muningin, both of which are constituents of heartwoods. The wood of *Artocarpus integrifolia* from which PERKIN isolated the flavonol morin, together with cyanomaclurin which is probably a flavan derivative (*154*), readily yields a new flavone, artocarpin, on extraction in a Soxhlet with *n*-hexane (*48*). Muningin likewise is isolable by extraction of the powdered heartwood of the African tree, *Pterocarpus angolensis*, with boiling ethanol and crystallising the brittle resin thus obtained from dioxan (*123*).

As in other fields, paper chromatography has now provided a rapid and invaluable method for determining if more than one flavone or isoflavone is present and suitably modifying the isolation procedure, for following the progress of the large-scale separations, and for examining flower petals and other plant materials available in small quantities.

The use of borate-impregnated paper facilitates the separation if one of the constituents is reactive towards boric acid because of the presence of 5-hydroxyl or vicinal hydroxyl groups (*215*). The flavones and isoflavones present in a plant may often be identified by comparison with the $R_F$ values of the known compounds, as shown by BATE-SMITH and SWAIN, ERDTMAN, SESHADRI and others. An extensive survey of anthoxanthins in Indian plants has been made by SESHADRI (*171*), employing circular paper chromatography and phenol saturated with water.

Some caution must be exercised in drawing final conclusions, for which it is desirable to isolate the constituents in the pure state and make a direct comparison with authentic samples. Complete dependence on paper chromatography, particularly on the basis of $R_F$ values recorded in the literature without running a chromatogram of known flavones and isoflavones simultaneously with the plant extract under examination, may result in the oversight of new constituents very closely related to known compounds in structure and therefore in chromatographic behaviour. Examples are pinoquercetin (6-methylquercetin) and pinomyricetin (6-methylmyricetin) (*132*) which for this reason the earlier workers failed to isolate.

Reference should be made to standard books on paper chromatography and the original papers for details of the procedure as applied to flavones. Glycosides may be distinguished from aglycones by the movement of the former with water as the developing solvent, while the latter remain more or less stationary.

Among the solvent systems that have been used for the separation of flavones are: (a) *n*-butanol-acetic acid-water (BAW); (b) acetic acid-water; (c) water-saturated isopropanol, *sec.*-butanol, ethyl acetate, chloroform, phenol or cresols; (d) *n*-heptane-*n*-butanol-water; and (e) water-saturated benzene-ligroin containing a trace of methanol. Further separation may be effected as usual by two-dimensional paper chromatography, for instance by using chloroform-ligroin-methanol after water-saturated benzene-ligroin (*21, 42, 64, 138, 139, 172, 190*). The flavone constituents are then detected by their characteristic fluorescence in ultraviolet light or by spraying with reagents such as ammonia, alcoholic ferric chloride, aluminium chloride, neutral and basic lead acetate, or diazonium salts. BAW and water-saturated ethyl acetate have been used for the separation of isoflavones.

Similar paper chromatographic technique and identification by chromogenic sprays are also useful for the separation and characterisation of chalcones, flavanones and other phenolic pigments which often accompany the flavones. Some flavonoids have been separated by paper electrophoresis (*83*).

Substantially pure flavonoids have been isolated from water extracts by adsorption on a weakly acid cation-exchange resin and elution with an organic solvent (*219*). Magnesol (hydrated magnesium trisilicate) is a useful adsorbent for the chromatographic separation of flavonoids (*97*); using acetone and water-saturated ethyl acetate as solvents quercetin and five of its glycosides have been separated (*99*). Chalcones, flavanones and flavonols have been separated by chromatography on polyamides such as powdered Perlon; methanol is used instead of complicated solvent systems and the procedure is suitable for large-scale separations (*94, 169*).

Flavone and isoflavone glycosides are usually water-soluble and ether-insoluble, and the aglycones soluble in ether; if the aglycones are highly hydroxylated, they may have considerable solubility in water, and if they are highly or fully methylated, they are water-insoluble. Both the glycosides and aglycones are generally alcohol-soluble. Some flavone glycosides and aglycones have been separated by the CRAIG countercurrent distribution technique, using ethyl acetate-butanol-water and other solvent systems (*92, 93*).

Selective precipitation of compounds containing vicinal hydroxyl groups by neutral lead acetate, followed by precipitation with basic lead acetate, is an old and very useful step in the isolation and purification of phenolic coloring matters (*174*).

PICCARD treated a hot alcoholic extract of poplar buds with lead acetate, filtered the yellow precipitate, removed lead salts from the filtrate by hydrogen sulphide, evaporated the clear filtrate to dryness, and crystallised the product from alcohol. Quercitrin can be prepared similarly from quercitron bark, except that a little acetic acid is used with lead acetate. Alternatively, after the acetic acid-lead acetate treatment in which the quantity of lead acetate is just enough to remove the impurities as a dark colored precipitate, more lead acetate is added to precipitate quercitrin as the bright orange-yellow lead salt, from which quercitrin is then recovered by treating an alcoholic suspension with hydrogen sulphide.

BRIGGS and LOCKER (*38 a*) found that the bark of *Melicope ternata* contained two basic constituents which gave precipitates with alkaloidal reagents and crystalline salts with acids, but they proved to be fully alkylated hydroxyflavones free from nitrogen. Four new flavones were then isolated by the following procedure. The solvent was removed from the acetone extract and the residue exhausted with trichloroethylene. Concentrated hydrochloric acid removed the completely alkylated flavones, which were precipitated on dilution with water and separated by fractional crystallisation into meliternatin (IX, p. 9) and meliternin (X). Extraction of the trichloroethylene solution with aqueous sodium carbonate led to wharangin (XI), and subsequent extraction with caustic soda solution led to ternatin (XII, p. 12).

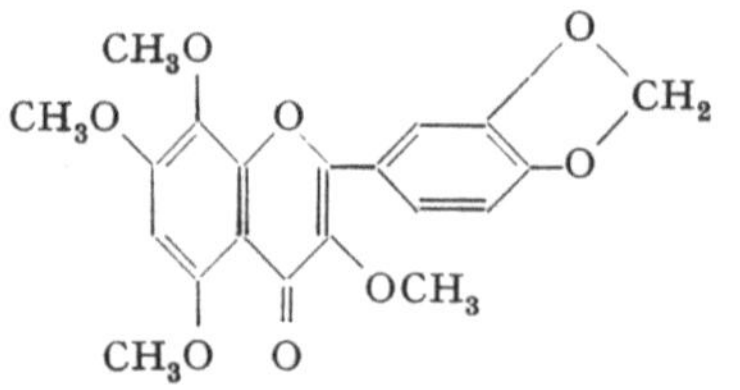

(X.) Meliternin.

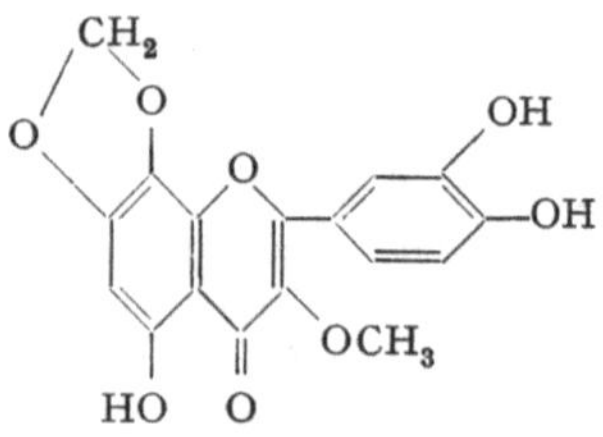

(XI.) Wharangin.

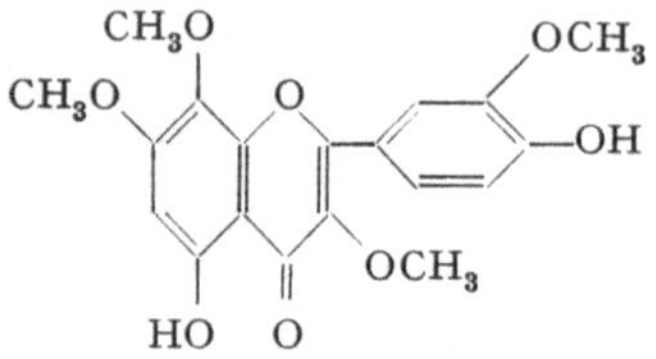

(XII.) Ternatin.

Shimizu (205) has shown the usefulness of borax solution in the isolation of flavones. Vicinal hydroxyl groups and a hydroxyl in the 5-position take part in the formation of soluble borax complexes. By this method myricitrin can be extracted from *Myrica rubra* bark in a yield of 5.5%, while only 0.19% is obtained by methanol extraction.

## V. Color Reactions.

Three reagents useful for detecting the presence of flavones and related substances in plant materials, which have been freed from anthocyanins and chlorophyll, are (a) aqueous ammonium or sodium hydroxide, (b) magnesium and hydrochloric acid, and (c) sodium amalgam, followed after a suitable interval by hydrochloric acid.

(a) A yellow to orange-red color with alkali indicates a phenolic coloring matter of the chalcone, flavanone, aurone, flavone or isoflavone type; the depth of the color, particularly on heating, can give some additional information, a flavanone for instance becoming a deeper yellow, orange or red as the result of ring fission to a chalcone. Flavonols in alkaline solution are unstable to air oxidation, and this is a valuable test for distinguishing between flavonols and flavones not containing a free hydroxyl group in the 3-position. Herzig (1891) observed that the deep yellow solution of fisetin in aqueous sodium hydroxide gradually turned brown on exposure to air (and rapidly by bubbling air or oxygen through the solution), breaking down to resorcinol and protocatechuic acid. Three hydroxyl groups in vicinal positions (5,6,7 or 3',4',5', but not 6,7,8) are indicated by the Bargellini test, a green coloration and the separation of dark green or greenish blue flocks when an aqueous or alcoholic solution is treated with cold dilute sodium hydroxide solution.

Color changes in alkaline buffer solutions can give useful indications of the constitution of a flavone (*1*). 5,7,8-Trihydroxyflavonols (gossypetin type) exhibit a series of color changes; somewhat similar colorations are given by myricetin and other flavonols with three vicinal hydroxyl groups in the *B*-ring.

(b) When a colorless or yellow alcoholic extract of a plant material is treated with magnesium and a few drops of concentrated hydrochloric acid, an orange, red, magenta or violet color develops almost immediately if a flavone, flavanone, flavonol, flavanonol or xanthone is present (SHINODA, 1928); some flavanones have been observed to give blue and green colors. The depth of the coloration partly depends on the concentration of the flavonoid in the extract under test, and for obtaining specific information it is desirable to compare the color directly with the colors produced by known flavonoids at a series of concentrations. In general, flavones and xanthones give orange to red, flavonols (including their ethers) red to magenta, and flavanones (including flavanonols) more intense red to magenta colorations. The influence of the number and position of the hydroxyl groups on the visible color and wave length of maximum absorption of a methanolic solution of a few polyhydroxy-flavones after reduction by magnesium and hydrochloric acid is shown in *Table 1* (p. 51).

When magnesium is replaced by zinc in the SHINODA test, flavanonols can be distinguished from flavanones (*175*), and flavonol 3-glycosides from the aglycones (*204*), since only the former give deep colors.

(c) The appearance of a red to magenta color, when an alcoholic solution is treated with sodium amalgam and after an interval (which has to be varied from a few minutes to several hours) with acid, usually indicates a flavone, flavonol 3-methyl ether or 3-glycoside, or flavanone [ASAHINA and INUBUSE (cf. *38a*)]. MARINI-BETTOLO and BALLIO (*149*) have drawn attention to the limitations and the non-specific character of the SHINODA and ASAHINA tests; and they have described a test with antimony pentachloride in carbon tetrachloride which sharply distinguishes chalcones (intense red or violet precipitate) from flavanones, flavones and flavonols (yellow or orange precipitate).

Flavones (and isoflavones with less intensity) usually exhibit visible fluorescence in concentrated sulphuric acid; the parent flavone fluoresces violet; a hydroxyl group in the 4'-position greatly increases the intensity of the fluorescence. Flavonols are fluorescent in organic solvents, and flavonols such as fisetin without a 5-hydroxyl group adsorbed on filter paper exhibit an intense fluorescence in ultraviolet light. Morin forms a green-fluorescing complex with aluminium salts used for the detection of aluminium. Flavonols in which the 3-hydroxyl is free are distinguished from those in which the 3-hydroxyl is protected by etherification or glycosidation by the production of a distinct yellow color on treatment with methanolic zirconium oxychloride and citric acid (*90*).

Like *o*-hydroxycarbonyl compounds, 5-hydroxyflavones react with the DIMROTH reagent (boric acid dissolved in hot acetic anhydride) to form orange to red solutions, from which a crystalline boroacetate sometimes separates.

When 5-hydroxyflavones or 5-hydroxyflavonols or their methyl ethers are treated in acetone solution with a solution of boric and citric (or oxalic) acid in acetone, a yellow color (*224*) with a yellowish green fluorescence (*212*) develops. 2'-Hydroxychalcones and other compounds containing a CO group attached to ·a benzene ring, an *o*- or *peri*-hydroxyl, and a double bond conjugated with the CO group respond to the test. Thus the test is negative with *o*-hydroxyaceto-phenones and 5-hydroxyflavanones. A variation of the test is to add boric acid to a solution of the compound in concentrated sulphuric acid (*167*).

There appears to be a relation between the boron and flavonol contents of plants and of individual plant organs, and the possible physiological significance of this relationship has been discussed (*212, 24*).

5-Hydroxyflavones and 5-hydroxyisoflavones in which the 8-position is unoccupied are shown by the GIBBS test: the development of a blue or blue-green color (characteristic absorption in the 500–700 m$\mu$ region) by the addition of 2,6-dichlorobenzoquinone chloroimide in borate buffer to a phenol in pyridine solution (*119*).

The orientation of two hydroxyl groups in *o*- or *p*-positions is shown by a violet color when alcoholic *o*-dinitrobenzene is added to a solution in aqueous sodium carbonate or hydroxide (*32*). Hydroxyl groups in the 5,8-positions in a flavone are identified by the "gossypetone reaction"— the development of a red-brown color or precipitate when an alcoholic solution is treated with *p*-benzoquinone. A methylenedioxy group is indicated by a green color when alcoholic gallic acid is added to a solution of the substance in sulphuric acid.

Flavones and isoflavones containing a free hydroxyl group in the 3-, 5- or 8-position and those containing vicinal dihydroxy and tri-hydroxy groups give characteristic colors with aqueous or alcoholic ferric chloride (*38g*).

## VI. Survey of Natural Flavones.*

### Flavone, 5-Hydroxyflavone and Dihydroxyflavones.

Flavone (I, p. 2), the parent member of the group, occurs as a mealy deposit on the leaves, stalks and seed-capsules of several *Primulae*. From the incrustation on *P. imperialis* var. *gracilis* KARRER and

---

* For complete references to the occurrence of flavones, see W. KARRER (*115*); for the literature before 1941 see (*154*) and (*174*).

SCHWAB (*114*) isolated, in addition to flavone which was the major constituent, a small amount of 5-hydroxyflavone (primuletin), readily separable by chromatography on alumina. This is the only monohydroxy-flavone encountered in nature so far.

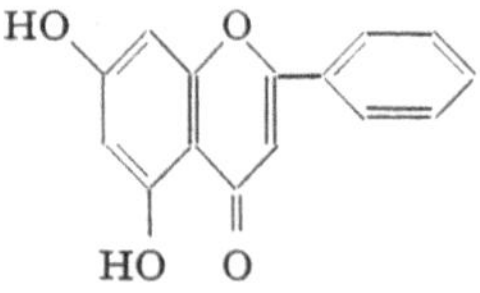

(XIII.) Chrysin.

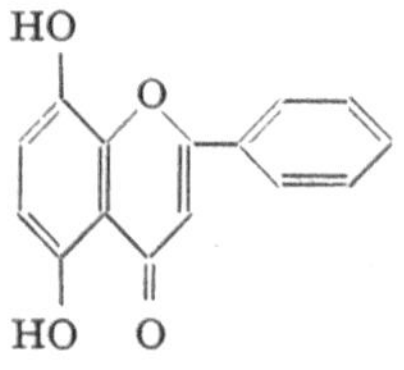

(XIV.) Primetin.

Only two dihydroxyflavones have been found to occur in nature: chrysin (XIII) in poplar buds and pine heartwood (*53, 138*) and primetin (XIV) which accompanies flavone in the deposit on the leaf of the Japanese *Primula modesta* (*154*); but a third (5,6-dihydroxy-flavone) occurs as the dimethyl ether in the bark of *Casimiroa edulis* (*101*), and it is one of several examples of flavones which were synthesized before they were identified as natural products. In the bark of *Pinus toringo* chrysin occurs as the 7-glucoside, toringin. Both in poplar buds and in pine heartwoods chrysin is accompanied by its 7-methyl ether, tectochrysin, obtained synthetically by the methylation of chrysin with diazomethane in ether. Some pine heartwoods also contain strobo-chrysin (6-methylchrysin), the flavanones, pinocembrin and pinostrobin, corresponding to chrysin and tectochrysin, and the flavanones, crypto-strobin and strobopinin, which are the 6 (or 8)- and 8 (or 6)-methyl derivatives of pinocembrin. (*139, 153*). Liquiritigenin, occurring as the glucoside liquiritin in *Glycyrrhiza glabra*, is 7,4'-dihydroxyflavanone (*154*), and the corresponding flavone has not yet been found in nature. The 4'-methyl ether was suspected to be identical with pratol in the blossom of the red clover, *Trifolium pratense*, but pratol has recently been shown to be the similarly substituted isoflavone, formononetin (*20*). Thus, in the one monohydroxyflavone and the five dihydroxyflavone derivatives (chrysin, tectochrysin, strobochrysin, primetin, and 5,6-dimethoxy-flavone), substitution occurs only in the *A*-ring.

An unidentified dihydroxyflavone, orange-yellow monoclinic crystals, m. p. 228°, accompanies flavone in the secretion of *Primula denticulata* (*26*).

### Trihydroxyflavones.

Three trihydroxyflavones have been isolated: 5,6,7 (baicalein), 5,7,4' (apigenin), and 3,5,7 (galangin); all three also occur as ethers (*Table 2*, p. 51). According to Colour Index (1956) chamomile flowers (*Anthemis nobilis*), which contain apigenin and its 7-glucoside, are much used as a pale yellow dye in hair washes. 5,7,8-Trihydroxyflavone has not been found to occur in nature, but the 8-methyl ether (wogonin), accompanies baicalein in *Scutellaria baicalensis.*

Buddleoflavonol, occurring as a glycoside (buddleoflavonoloside) in the flowers and leaves of *Buddleia variabilis*, was stated to be 3-acetylacacetin (*154*), but has now been shown to be identical with acacetin (*11*); buddleoflavonoloside is identical with linarin, the 7-rutinoside of acacetin. Vitexin, isolated (*174*) from *Vitex lucens* (formerly *V. littoralis*), has been tentatively assigned the structure (XV) (*54*).

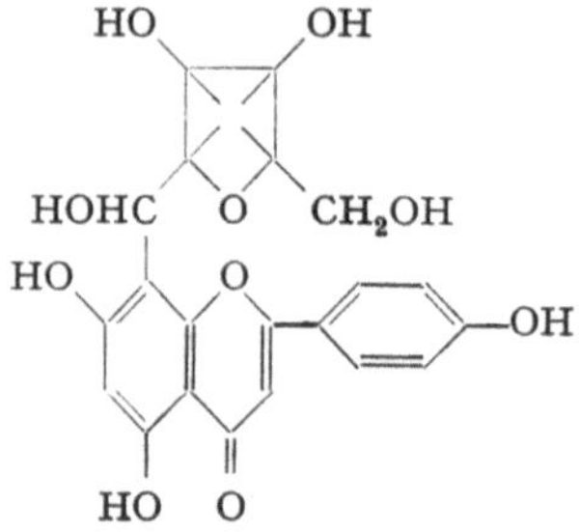

(XV.) Vitexin. 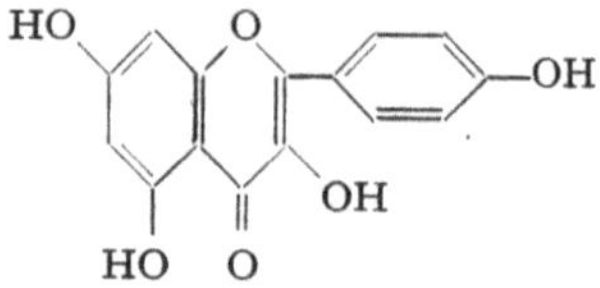     (XVI.) Kaempferol.

Vitexin and closely related substances have also been isolated from barley and several other plants (*196*). Homovitexin (*174*), for which the name "isovitexin" is preferred, is dextrorotatory and isomeric with vitexin which is laevorotatory (*37*). Saponaretin, which occurs in barley and in *Saponaria officinalis* as the glucoside, saponarin, is probably an analogue of vitexin in which the 8-position of apigenin is occupied by the open-chain group —(CHOH)$_5$CH$_2$OH (*196*).

### Tetrahydroxyflavones.

Tetrahydroxyflavones and their methyl ethers are listed in *Table 3* (p. 52). There are six tetrahydroxyflavones: scutellarein, luteolin, kaempferol (XVI), datiscetin, fisetin, and pratoletin.

Kaempferol occurs in numerous plants, being second in this respect only to quercetin. Several flavones described in the older literature as new flavones (nimbicetin, populnetin, robigenin, trifolitin, swartziol) have been subsequently shown to be identical with kaempferol. Two monomethyl ethers of luteolin (chrysoeriol and diosmetin) and of kaempferol (kaempferide and rhamnocitrin), a dimethyl ether of scutellarein (pectolinarigenin), a methylene-dimethyl ether of fisetin (desmethoxy-kanugin), and a monomethyl ether of datiscetin (ptaeroxylol) are known.

By hydrolysis of a glycoside, lotusin, occurring in *Lotus arabicus*, DUNSTAN and HENRY (cf. *174*) obtained maltose, hydrocyanic acid, and a flavone, "lotoflavin", to which they assigned the structure 5,7,2',4'-tetrahydroxyflavone. WHEELER (*50*) has recently shown that "lotoflavin" consists mainly of quercetin with a trace of a second flavonol, probably kaempferol.

Two derivatives of 5,7,2',4'-tetrahydroxyflavone have been isolated more recently (*47*, *48*) from the heartwood of *Artocarpus integrifolia*; these are the 7-methyl ether, artocarpetin (XVII), and the 3,6-diisopentenyl derivative, artocarpin (XVIII). This wood also contains morin (XIX) and cyanomaclurin, the latter compound probably being constituted as the cyclic hemiketal of 5,7,2',4'-tetrahydroxy-3-keto-flavan (*154*); all the four constituents therefore have the unique resorcinol pattern in the *B*-ring.

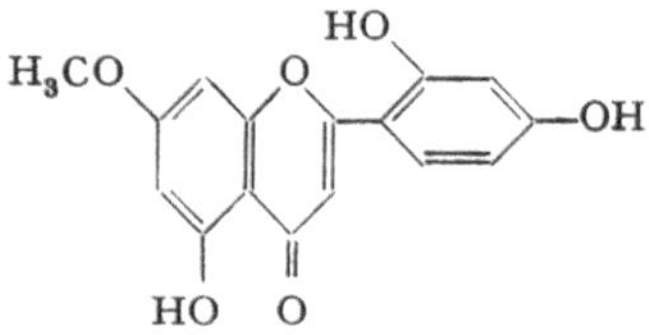

(XVII.) Artocarpetin.

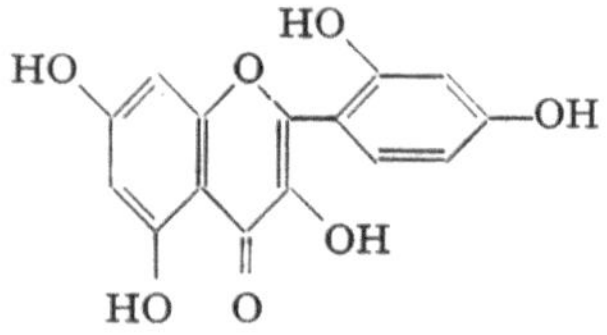

(XIX.) Morin.

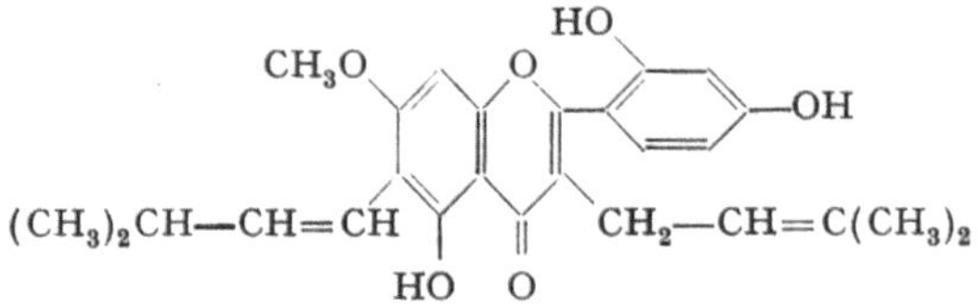

(XVIII.) Artocarpin.

Icariin (XX), found in *Epimedium macranthum*, has been shown (*154*) to be a rhamnoside-glucoside of icaritin (XX), the 8-(3-methyl-3-hydroxy-butyl) derivative of kaempferide. Icaritin readily loses a molecule of water, forming anhydroicaritin (XXI). The trimethyl ether of icaritin and the dimethyl ether of anhydroicaritin have been synthesized. Icariin

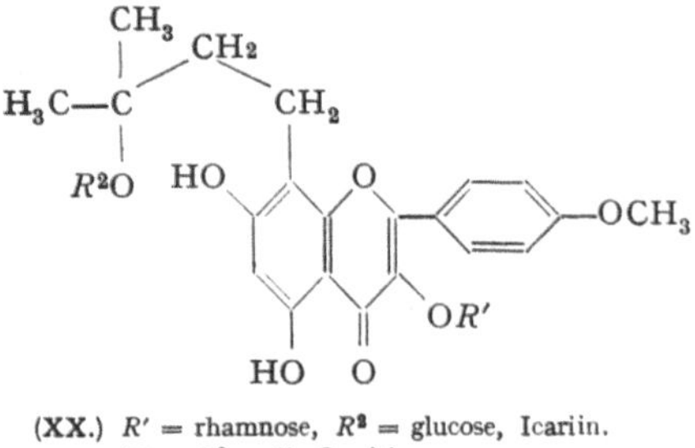

(XX.) $R'$ = rhamnose, $R^2$ = glucose, Icariin.
$R' = R^2$ = H, Icaritin.

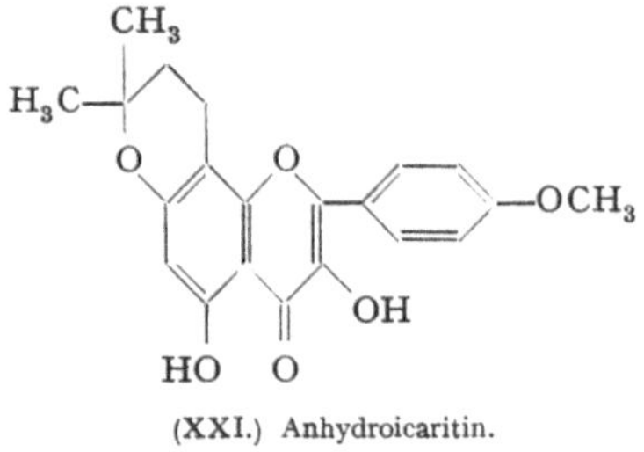

(XXI.) Anhydroicaritin.

is accompanied by nor-icariin, the analogous derivative of kaempferol. Amurensin, the 7-glucoside of nor-icaritin, and the corresponding flavanonol (phellamurin) have been isolated from the leaves of *Phellodendron amurense* (*82*).

Another unusually substituted flavone is fukugetin (garcinin), isolated from *Garcinia spicata*, for which the structures (**XXII**) and (**XXIII**) have been suggested (*129, 154*).

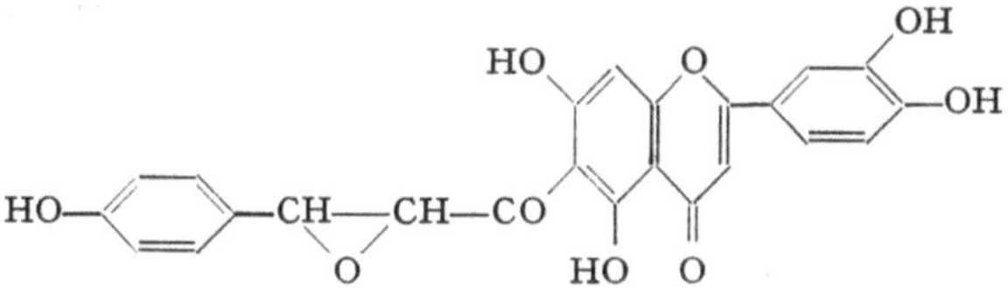

(XXII.)

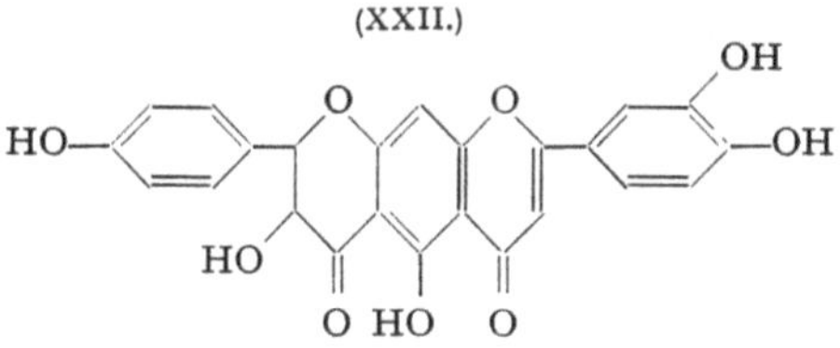

(XXXIII.)

The wood of *Zelkowa serrata* contains two pigments, keyakinin and keyakinol; keyakinin has the probable constitution of rhamnocitrin carrying a —$(CHOH)_4CH_2OH$ side-chain in the 8-position, and keyakinol is the corresponding flavanonol (*63*).

## Pentahydroxyflavones.

Five pentahydroxyflavones have so far been isolated: quercetin, morin, robinetin, herbacetin, and a constituent of the heartwood of *Acacia melanoxylon* for which the trivial name melanoxetin is suggested. Melanoxetin accompanies the corresponding flavan-3,4-diol (melacacidin, **XXIV**) and can be prepared from the latter by OPPENAUER oxidation (*117*).

Quercetin (**IV**, p. 7) is the most widely distributed of the flavones, and perhaps of all the plant pigments with the exception of chlorophyll and some of the carotenoids; it occurs both in the free form and as glycosides, the sugar residue being usually attached in the 3-position (see *Table 4*, p. 54).

According to a recent report, quercetin is now being produced in pilot plant quantities by Weyerhaeuser Timber in the U. S. The raw material is Douglas fir bark which contains taxifolin (**XXV**), dehydrogenated to quercetin by air oxidation under slightly alkaline conditions. Among the suggested uses for quercetin

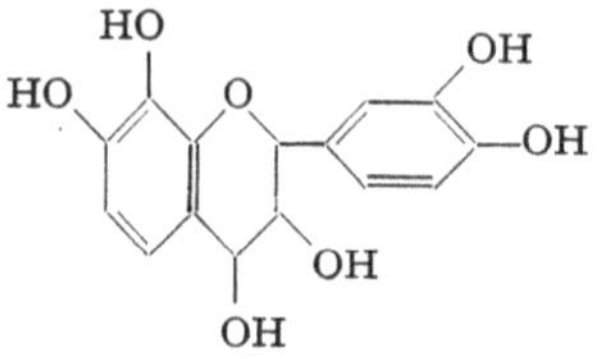

(XXIV.) Melacacidin.

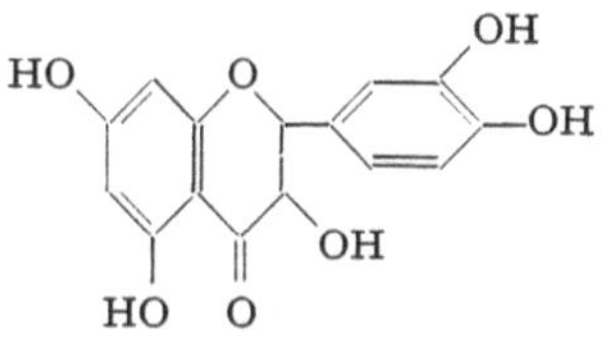

(XXV.) Taxifolin.

are as an anti-oxidant in foods, an ultraviolet absorber, and a chemical intermediate from which phloroglucinol and protocatechuic acid can be prepared.

Rutin, the 3-rutinoside of quercetin, is manufactured from buckwheat, *Sophora japonica* and Eucalyptus leaves for use as "vitamin P" in the treatment of capillary fragility; quercetin itself, recovered as a by-product from rutin manufacture or the lumber industry, is used for the same purpose.

Seven methyl ethers of quercetin are known: four monomethyl ethers (azaleatin, rhamnetin, tamarixetin and isorhamnetin), two dimethyl ethers (ombuin and rhamnazin), and a trimethyl ether, ayanin.

Azaleatin (quercetin 5-methyl ether; LIX; $R = H$, p. 32) appears to be the only exception among flavones to Bose's generalization (34) that in partially methylated polyhydroxyflavones occurring in nature a methoxyl group is never found in the 5-position. No biogenetic consideration, however, supports Bose's suggestion, and nonconforming examples are known among flavanones (isopedicin) and isoflavones (muningin). Azaleatin also contravenes the statement of Briggs and Locker (38e) that "no naturally occurring 3,5-dihydroxyflavone exists which is methylated at $C_{(5)}$ but not at $C_{(3)}$".

*Table 5* (p. 54) lists the pentahydroxyflavones other than quercetin and their methyl or methylene ethers. No ether of morin is known. Kanugin is a derivative of robinetin in which all the five hydroxyl groups are protected. Tambuletin, tambulin and flindulatin are ethers of herbacetin; the pentahydroxyflavones corresponding to the ethers, tricin, tangeretin (VII, p. 8), penduletin and auranetin (VI, p. 8), have not been isolated. There is no flavone other than auranetin in which hydroxyl or methoxyl groups occur in the 6,7,8-positions, although such substitution has been found among the coumarins (fraxin, fraxidin and isofraxidin).

### Hexahydroxyflavones.

*Table 6* (p. 55) lists nineteen naturally occurring hexahydroxyflavone derivatives. There are three hexahydroxyflavones: myricetin (XXVI), gossypetin (XXVII) and quercetagetin (XXVIII); in addition there is a C-methyl derivative of myricetin (pinomyricetin). No myricetin ether is known. A dimethyl ether of gossypetin occurs in lemon peel.

Ternatin (XII, p. 12), wharangin (XI), meliternin (X), melisimplin (XXIX), melisimplexin (VIII, p. 9), and meliternatin (IX), all of which occur in the bark of two New Zealand trees of the genus *Melicope*, are a remarkable series; three are ethers of gossypetin (the pigment of cotton flowers) and three are ethers of quercetagetin (the pigment of the flowers of the marigold). Quercetagetin is accompanied by a monomethyl ether, patuletin, in the African marigold. Three other quercetagetin ethers have been isolated: chrysosplenetin, oxyayanin-B and artemisetin. The hexahydroxyflavones corresponding to nobiletin,

2*

acrammerin, calycopterin (LI, p. 30) and oxyayanin-A have not been isolated from plants. Oxyayanin-A (LII, p. 30) is unique in the orientation of substituents in the B-ring, being the only derivative of hydroxyhydroquinone.

The structure assigned (66) to acrammerin has been questioned by SESHADRI (15), who synthesized 5,7,8,3',4',5'-hexamethoxyflavone and found that its m. p. differed from that of acrammerin pentamethyl ether.

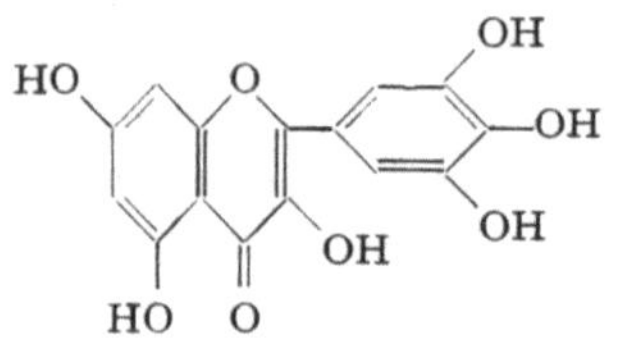

(XXVI.) Myricetin.

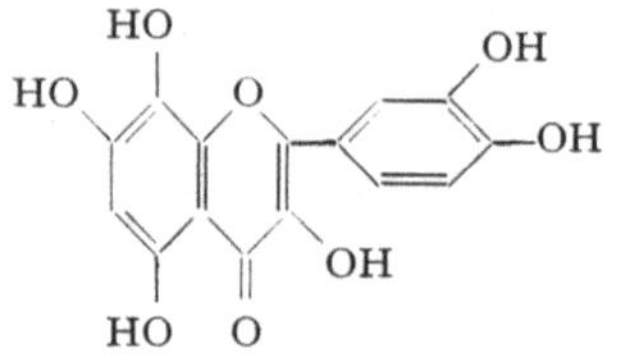

(XXVII.) Gossypetin.

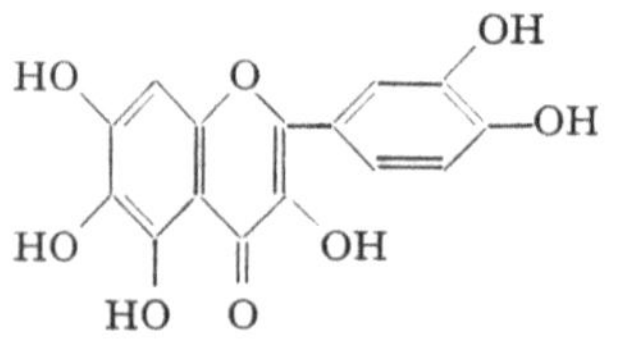

(XXVIII.) Quercetagetin.

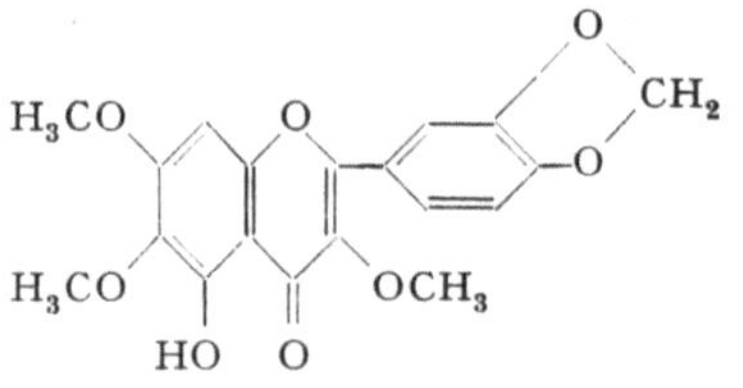

(XXIX.) Melisimplin.

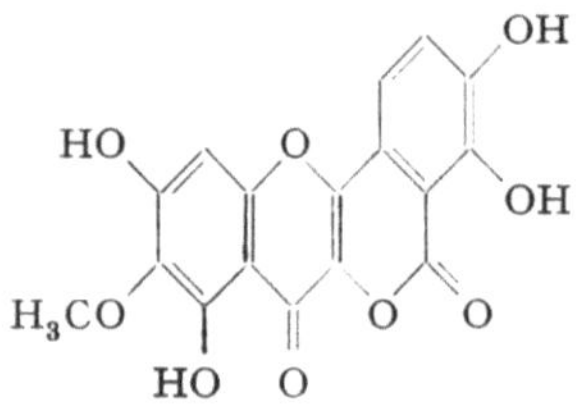

(XXX.) Distemonanthin.

A quercetagetin derivative, which represents a new type of flavone pigment containing a lactone (isocoumarin) ring, is distemonanthin (XXX) which accompanies ayanin and the oxyayanins A and B in *Distemonanthus Benthamianus* (122).

## Heptahydroxyflavones.

Only one heptahydroxyflavone, hibiscetin (XXXI), occurring as the 3-glucoside (hibiscitrin) in the flowers of *Hibiscus sabdariffa* and *H. cannabinus*, has so far been isolated (185); but two isomeric hepta-hydroxyflavones occur in the form of pentamethyl and hexamethyl ethers (erianthin and gardenin). The constitution of hibiscetin has been confirmed by synthesis: the ALLAN-ROBINSON reaction on 2,4-dihydroxy-ω-3,6-trimethoxy-acetophenone and trimethylgallic anhydride, followed by demethylation (*184*).

Erianthin, found in the flowers of *Blumea eriantha*, was assigned the structure (XXXII) by Bose and Dutt (cf. *154*). By heating the ketone (XXXIII), obtained by hydrolysis of calycopterin dimethyl ether, with veratric anhydride Seshadri obtained a heptamethoxyflavone together with a hydroxyhexamethoxy-flavone; the latter had the properties of a 5-hydroxyflavone and therefore the structure (XXXIV), but it was not identical with erianthin monomethyl ether (*200*).

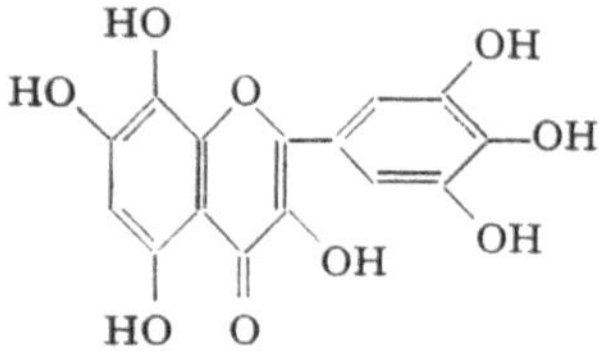

(XXXI.) Hibiscetin.

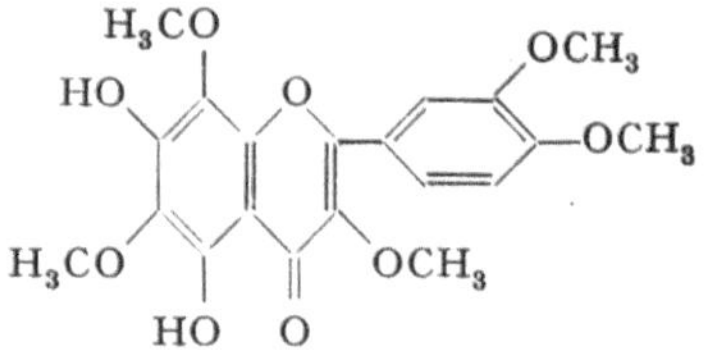

(XXXII.) Erianthin.

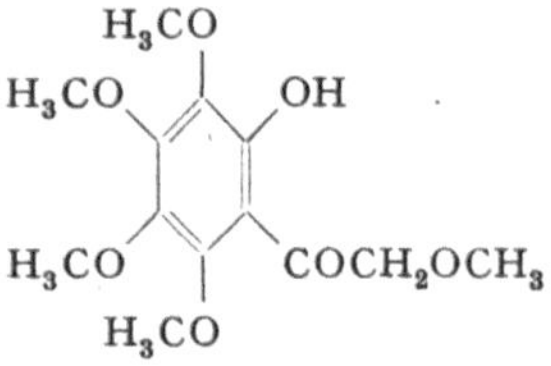

(XXXIII.)

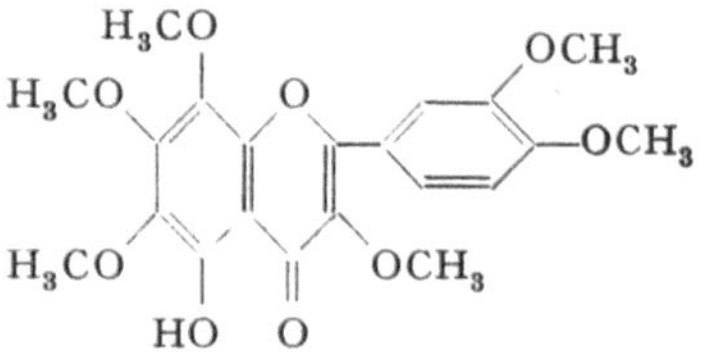

(XXXIV.)

Gardenin in the resin of *Gardenia lucida* is 5-hydroxy-3,6,8,3',4',5'-hexamethoxyflavone (LXXXVII, p. 40) (*33*, *34*), confirmed by synthesis (*2*); and it is unique among flavones in the orientation of hydroxyl and methoxyl groups in the 5,6,8-positions.

### Furanoflavones.

From the oil of *Pongamia glabra* Limaye (*135*) isolated karanjin and demonstrated its constitution as (XXXV), confirmed by a synthesis due to Manjunath (*146*). The parent furanoflavone (XXXV), lanceolatin B, has been found to occur in the root bark of *Tephrosia lanceolata* Grab. (*180*), together with the diketone (pongamol, XXXVI), which is also a constituent of Pongamia oil (*166*) and is of great interest as

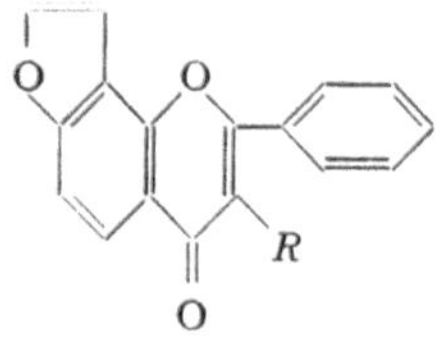

(XXXV.) R = OCH₃, Karanjin.
R = H, Lanceolatin B.

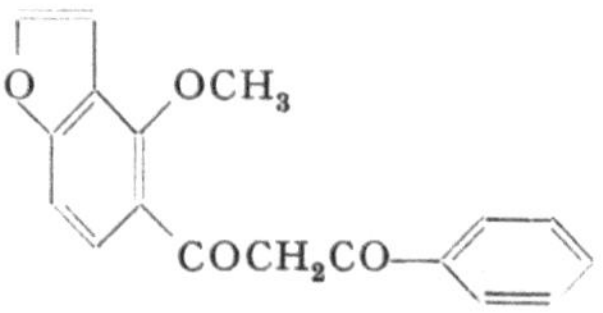

(XXXVI.) Pongamol.

the only naturally occurring diketone related to a flavone. Demethylation of pongamol yields lanceolatin B.

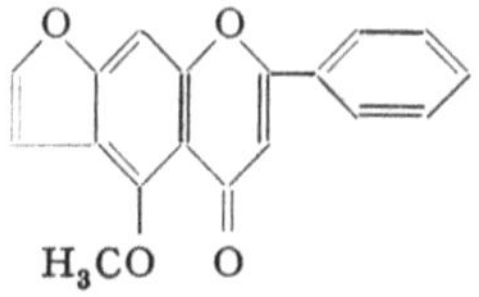

(XXXVII.) Pinnatin.

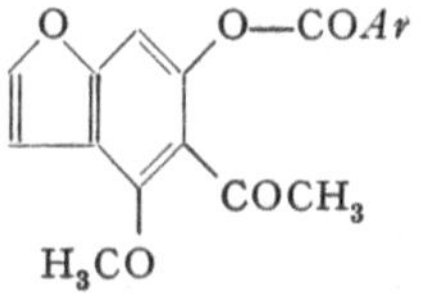

(XXXVIII.)

The root of *Pongamia pinnata* contains four furanoflavones: karanjin, pongapin (3′,4′-methylenedioxykaranjin), pinnatin (XXXVII) and gamatin (3′,4′-methylenedioxypinnatin). Pinnatin and gamatin have been synthesized by carrying out the BAKER-VENKATARAMAN transformation on the O-aroyl derivative (XXXVIII; *Ar* = phenyl or 3,4-methylenedioxyphenyl) of visnaginone (5-acetyl-6-hydroxy-4-methoxy-coumarone), obtained by alkaline degradation of visnagin (*173*).

## VII. Survey of Natural Isoflavones.

Only thirteen isoflavones (*Table 7*, p. 56), excluding those containing additional ring systems, have so far been isolated from plants, although these belong to widely different families. They vary in the degree of hydroxylation from daidzein (7,4′-dihydroxyisoflavone) to irigenin which contains three hydroxyl and three methoxyl groups; the parent isoflavone, monohydroxyisoflavones and pentahydroxyisoflavones have not been noticed in plants. No isoflavone unsubstituted in the *B*-ring is known. Of the thirteen isoflavones, three are polyhydroxy compounds and the others contain one or more ether groups; the polyhydroxy compounds corresponding to pseudobaptigenin and maximin, tectorigenin and muningin, tlatlancuayin, and irigenin do not occur in plants. The 7-position is always occupied; eight of the thirteen isoflavones contain a hydroxyl, three a methoxyl, and one an isopentenyloxy group; in one it is part of a methylenedioxy group. The *A*-ring is substituted in the 7 or 5,7 or 5,6,7-positions; and the *B*-ring in the 4′ or 3′,4′, or 3′,4′,5′-positions.

Unique in several ways is the recently isolated (and unpronounceable) isoflavone, tlatlancuayin (XXXIX); the first flavonoid isolated from a plant of the *Amarantaceae* family, it is the only isoflavone in which all the hydroxyl groups are protected by methylation or methylenation; it has a methylenedioxy group in the *A*-ring and a 2′-methoxyl group in the *B*-ring; even among the much more numerous flavones meliternatin is the only one with a methylenedioxy group in the *A*-ring; and datiscetin and ptaeroxylol (its 2′-methyl ether) are the only flavones which carry a 2′-hydroxyl or methoxyl as the only substituent in the *B*-ring.

A 2'-hydroxylated isoflavone ("isogenistein") and three C-methylisoflavones, reported to be present in soya bean, have now been shown to be different from the authentic compounds obtained by synthesis and to consist very probably of impure daidzein and genistein. Biochanin-B is identical with formononetin, and pratensol, isolated by Power and Salway (cf. *154*) from red clover, is probably identical with biochanin-A (*30*). Olmelin, isolated from *Gleditschia triacanthos*, has been assigned the same structure as biochanin-A (*65*), but the melting points of the two isoflavones and of their derivatives do not agree.

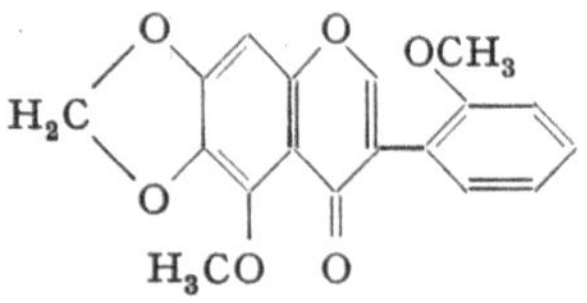

(XXXIX.) Tlatlancuayin.

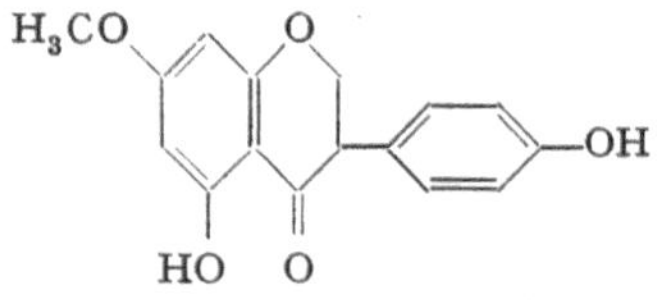

(XL.) Padmakastein.

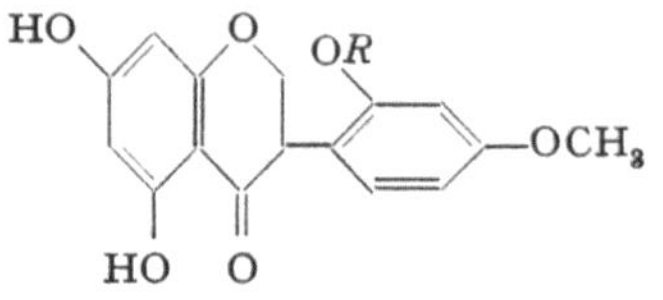

(XLI.) R = H, Ferreirin.
R = CH₃, Homoferreirin.

Over thirty flavanones (including the 3-hydroxy derivatives or flavanonols) have been isolated from plants, but only three isoflavanones: padmakastein (XL), ferreirin and homoferreirin (XLI). Padmakastein (free and as a glycoside) accompanies the corresponding isoflavone, prunetin, in the bark of *Prunus puddum* (*179*). The heartwood of *Ferreirea spectabilis* contains pigments of four types: an isoflavone (biochanin-A), two isoflavanones (ferreirin and homoferreirin), a flavanone (naringenin), and an anthrone (chrysophanol-anthrone) (*118*). By the selenium dioxide dehydrogenation method (*141*) O-trimethylferreirin gives 5,7,2',4'-tetramethoxyisoflavone (*125*); conversely, homoferreirin has been synthesized by hydrogenation of 5,7-dihydroxy-2',4'-dimethoxyisoflavone in the presence of platinum-charcoal (*168*).

### Polycyclic Compounds Containing the Isoflavone or Isoflavanone Nucleus.

Isoflavones and isoflavanones containing additional furan or pyran rings are relatively wide-spread in plants. The rotenoids, fish-poisons and insecticides occurring in many species of *Derris*, *Tephrosia* and other leguminous plants (*56*), possess a 5-ring system built round an isoflavanone core, and examples are rotenone (XLII) and toxicarol (XLIII). The structures assigned to rotenone and several other rotenoids are based on sound degradative evidence, but final confirmation by total synthesis has not yet been achieved (*192*).

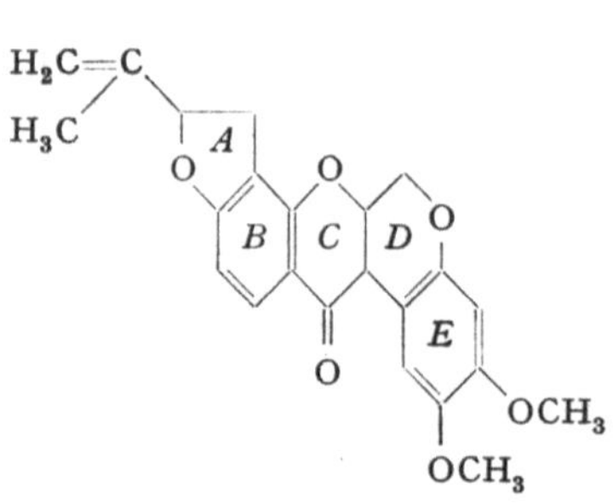

(XLII.) Rotenone.

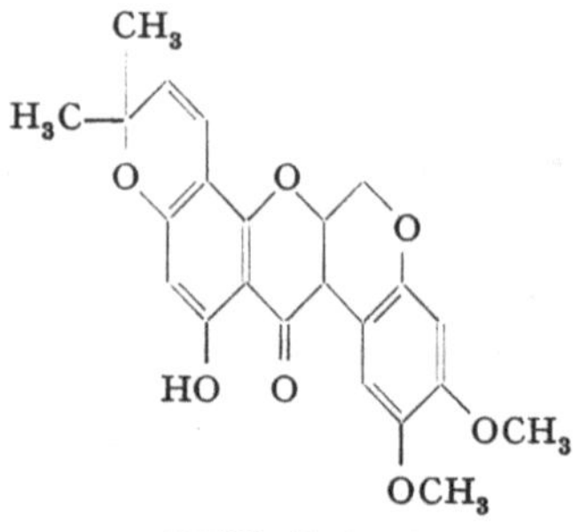

(XLIII.) Toxicarol.

*Mundulea suberosa* and *Piscidia erythrina* are rotenone-bearing plants from which two isoflavone derivatives, munetone (XLIV) and jamaicin (XLV), have been isolated (*52, 210*). In these two compounds the *D*-ring of rotenone and toxicarol is absent; but both contain a methoxyl group in the 2'-position of the isoflavone nucleus indicating the close relationship to the pentacyclic rotenone and toxicarol types.

If the methylenedioxy group in jamaicin is replaced by two methoxyl groups and a hydroxyl group is introduced in the 5-position of the chromone nucleus, the structure becomes identical with that proposed by HARPER (*80*) for a substance ("toxicarol-isoflavone") which accompanied α-toxicarol, rotenone and other rotenoids in the resin from *Derris malaccensis*.

Osajin and pomiferin are two yellow pigments similar to munetone and jamaicin in structure, but they were isolated from the fruit of the

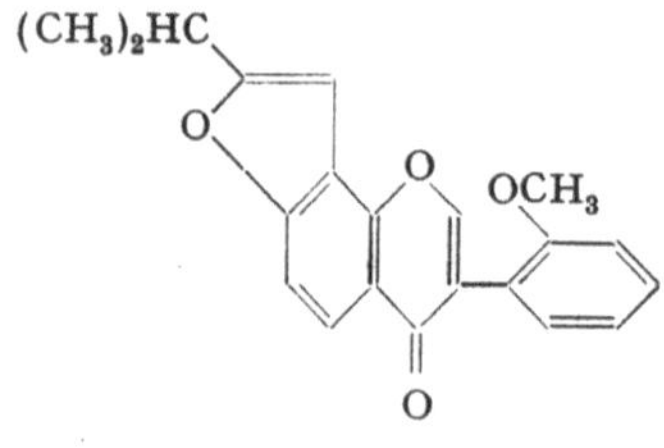

(XLIV.) Munetone.

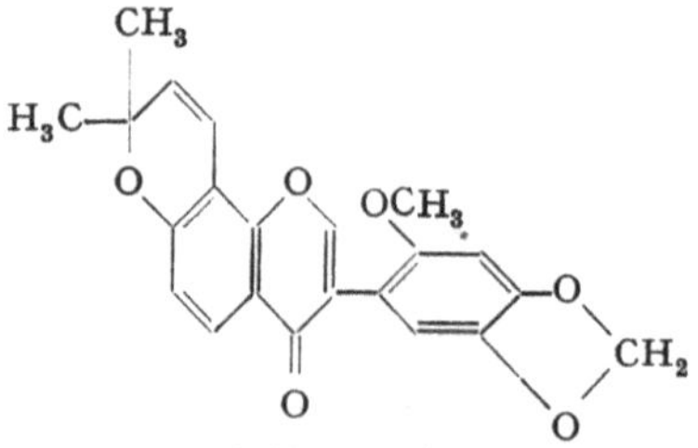

(XLV.) Jamaicin.

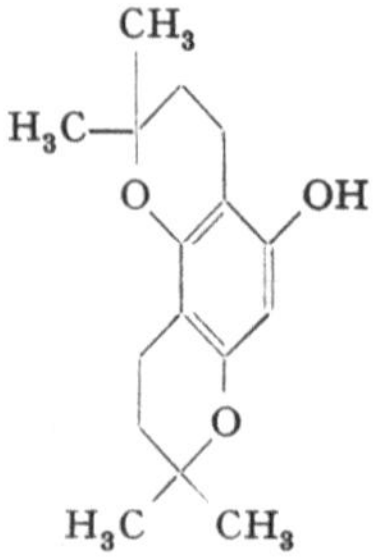

(XLVIII.)

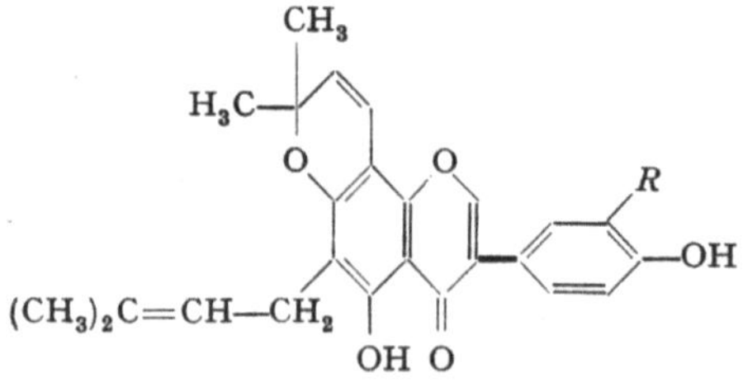

(XLVI.) *R* = H, Osajin.
        *R* = OH, Pomiferin.

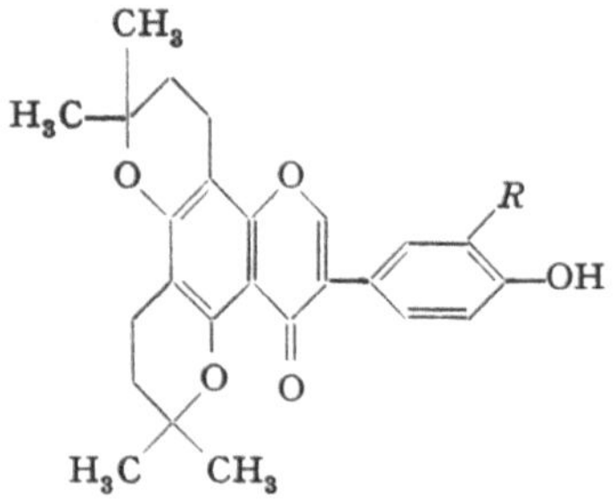

(XLVII.) R = H, Dihydro-iso-osajin.
         R = OH, Dihydro-iso-pomiferin.

osage orange *(Maclura pomifera)*, and therefore very different in botanical origin. WOLFROM *(226)* has demonstrated the structures of osajin and pomiferin to be (XLVI); unlike jamaicin and munetone, the 2'-position in osajin and pomiferin is free. Osajin and pomiferin possess two ethylenic bonds which are saturated readily, but in two stages, by catalytic hydrogenation. When osajin and pomiferin are treated with sulphuric acid in glacial acetic acid, they cyclize to the isomeric iso-osajin and iso-pomiferin; dihydro-osajin and dihydro-pomiferin behave similarly, yielding dihydro-iso-osajin and dihydro-iso-pomiferin (XLVII), the structures of which have been confirmed by synthesis starting from dihydro-iso-osajinol (XLVIII, p. 24) *(227)*.

## VIII. Absorption Spectra of Flavones and Isoflavones.

*Ultraviolet Spectra.* Flavones exhibit two absorption bands in the 250 mμ and 300–380 mμ regions *(105, 209)*; in addition, substituted flavones show another maximum between 210 and 220 mμ. Flavone itself absorbs at 250 mμ and 297.5 mμ; flavonol absorbs at 239, 305 and 347.5 mμ; the long wave-length band in flavone therefore undergoes a bathochromic shift, and an additional band, associated with the additional possibility of tautomerism *(209)*, appears at 305 mμ, which is also present in the spectra of many flavonols. Etherification of the 3-hydroxyl restores the character of the flavone spectrum *(209)*. A hydroxyl group in the 4'-position produces a large bathochromic effect. Hydroxyls in the 5 and 7-positions also produce bathochromic shifts, and the effect of further hydroxylation is cumulative *(44)*. O-methylation of hydroxyl groups in the 5, 6, 7, 2' or 3'-positions produces little or no change in the absorption spectrum; methylation of a 4'-hydroxyl usually produces a hypsochromic shift of the long wavelength band *(209)*. Data for a series of naturally occurring flavones and flavonols collected from the literature *(38g, 44, 209)* are presented in *Table 8* (p. 57) and *Figures 1* and *2* (pp. 26, 27).

Shifts in the flavone spectrum by hydroxyl substitution have been explained in terms of resonance structures involving the carbonyl group (7), chromone as the main chromophore (209), and polarizations along two perpendicular axes involving electronic transitions in the chalcone and $\gamma$-pyrone components of the molecule (44).

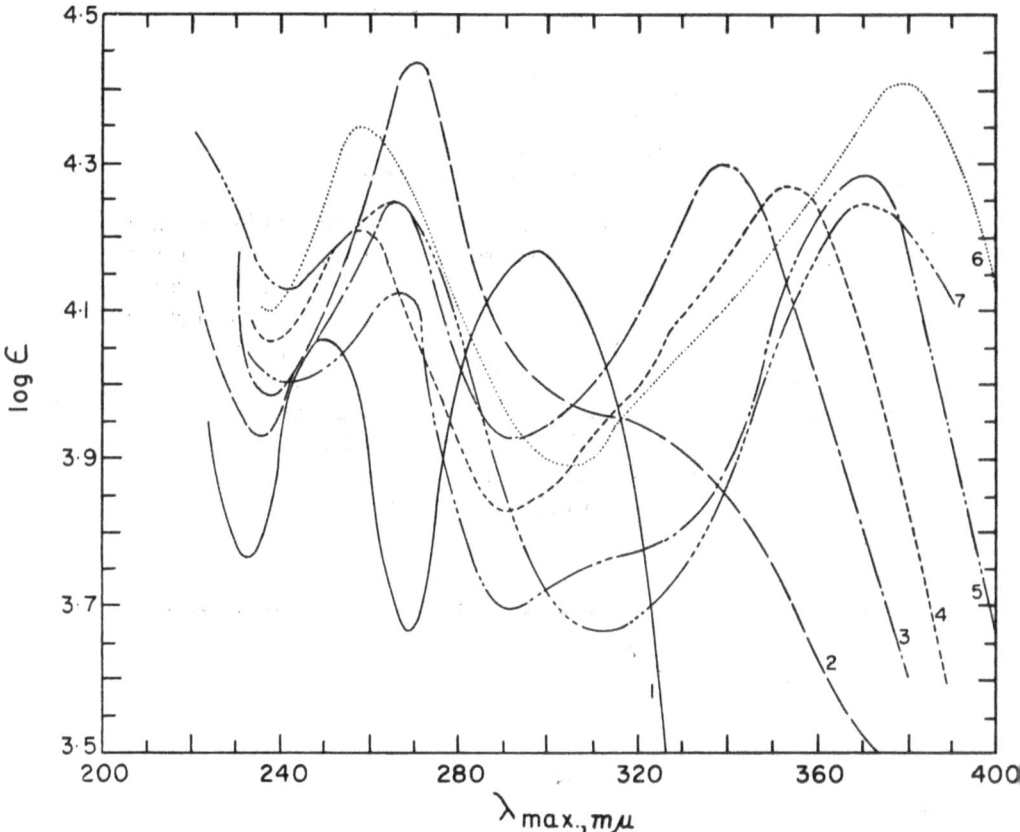

Fig. 1. (1) ———— Flavone, (2) — — — Chrysin, (3) —·—·— Apigenin, (4) ········ Luteolin, (5) —··— Kaempferol, (6) ·········· Quercetin, (7) —····— Morin.

Isoflavones (*Table 9*, p. 58; *Fig. 3*, p. 28) (36, 46, 79) have a high intensity band at about 260 m$\mu$ (more intense than the flavone band in the same region) and a very weak band (often a shoulder or inflection) in the 300 m$\mu$ region. The chalcone chromophore, responsible for the intense long wave-length band in flavone spectra, is absent in isoflavones, and there is instead a $C_6H_5$—CO—C=C chromophore with a phenyl group in cross conjugation.

The bathochromic shift obtained when sodium ethoxide is added to a solution of a flavone or isoflavone in absolute ethanol can be used for locating a free hydroxyl group; the isomeric methyl ethers and glucosides of apigenin and genistein can thus be distinguished (*147*). The shift in $\lambda_{max}$ of 50–75 m$\mu$ when aluminium chloride is added to an ethanolic solution is useful for characterizing 5-hydroxyflavones (*211*).

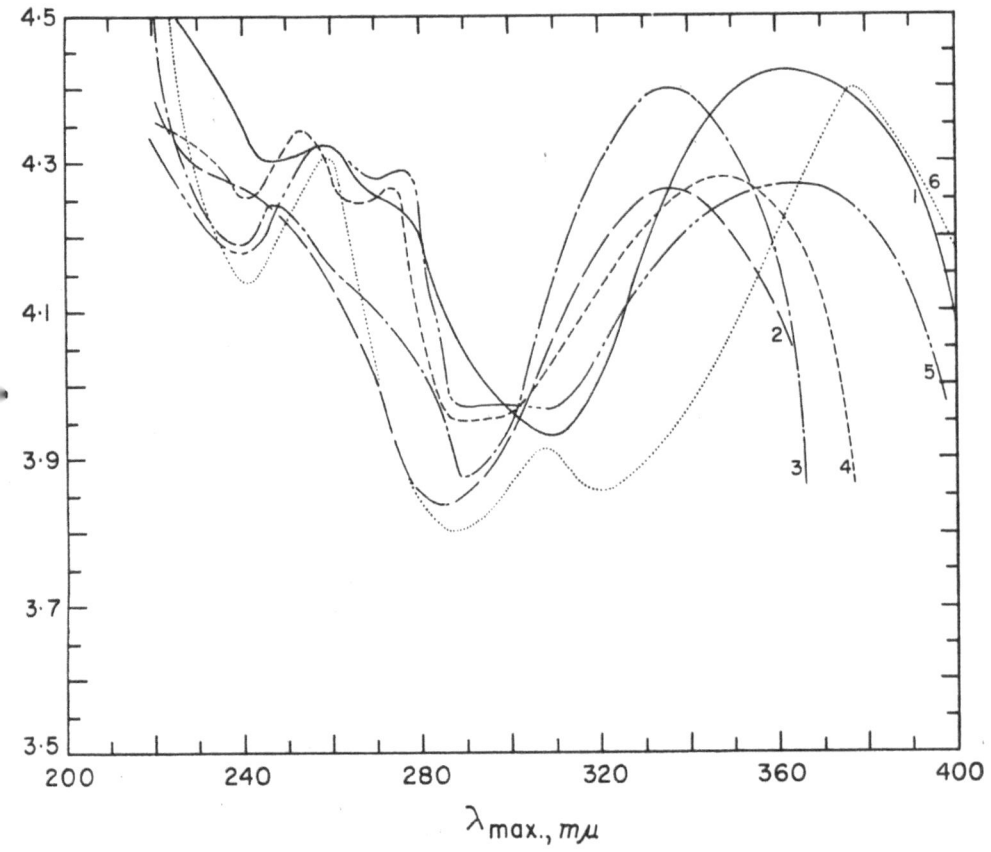

Fig. 2. *(1)* ———— Quercetagetin, *(2)* — — — Melisimplexin, *(3)* — · — · Meliternatin, *(4)* — — — — — Meliternin, *(5)* — — — Ternatin, *(6)* ·········· Myricetin.

Some of the hydroxyl groups in a flavonol may be located by determining the absorption spectra successively in ethanol, ethanolic sodium acetate, ethanolic boric acid-sodium acetate, and sodium ethylate. A bathochromic shift (8–20 m$\mu$) of the short wave-length band in ethanolic sodium acetate indicates a 7-hydroxyl. A bathochromic shift (15–30 m$\mu$) of the long wave-length band in ethanolic boric acid-sodium acetate shows the presence of *o*-dihydroxyl groups. Disappearance of the long wave-length band in ethanolic sodium ethylate indicates decomposition and the presence of hydroxyls in the 3,4'-positions; stability in sodium ethylate shows that at least one of these is alkylated (*108, 110*). *Figs. 4* and *5* (pp. 32, 33) demonstrate the application of these principles to the constitution of azalein (5-*O*-methylquercetin 3-rhamnoside) (LIX) and its aglycone, azaleatin (LIX) (*110*), to be discussed on p. 32.

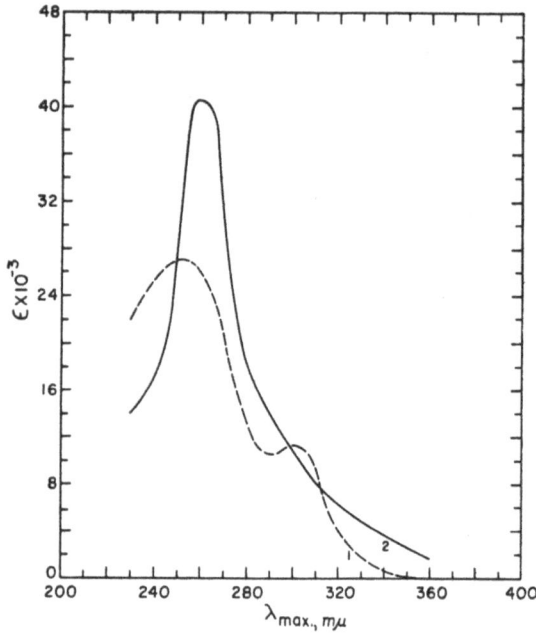

Fig. 3. *(1)* – – – – Formononetin, *(2* ,———— Genistein.

*Infrared Spectra.* In connection with a study of the occurrence of flavonoids in allergic pollens Inglett (*100*) has recorded the infrared spectra of several naturally occurring flavones and flavanones (*Table 10*, p. 58). The carbonyl frequency of flavanone undergoes a large shift to a lower frequency by hydroxyl substitution in the 5-position; but the carbonyl frequency of flavone is only slightly influenced by a 5-hydroxyl, although hydrogen bonding is no less effective in 5-hydroxy-flavones than in 5-hydroxyflavanones.

## IX. Determination of the Structure of Flavones and Isoflavones.

Hydrolytic fission is the main mode of degradation of flavones and isoflavones, and in conjunction with color reactions and absorption spectra, is usually adequate for determining their structure. Flavones unsubstituted in the 3-position require more drastic conditions of alkali treatment than flavonols for their degradation. By digesting chrysin (XIII) with boiling concentrated potassium hydroxide PICCARD (1877) obtained phloroglucinol, benzoic acid, acetic acid and acetophenone (the last in very small amount); KOSTANECKI (1893) deduced from this result the constitution of chrysin and also the mechanism of hydrolysis involving the formation of a 1,3-diketone as an intermediate.

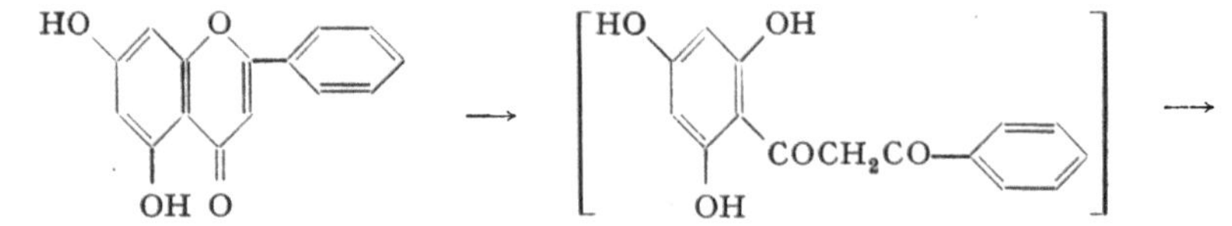

(XIII.) Chrysin.

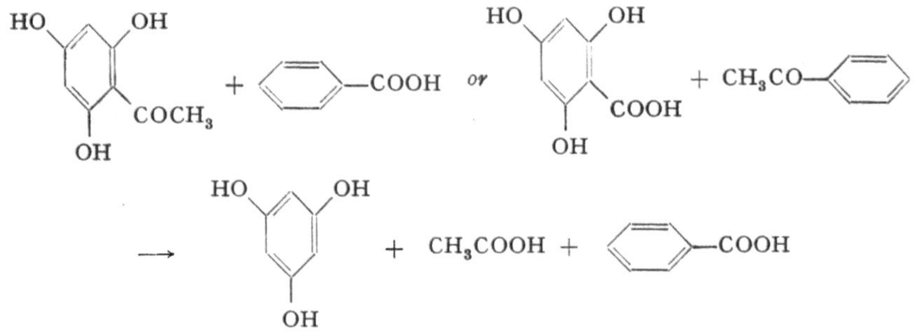

Fisetin tetraethyl ether (XLIX) breaks down to fisetol diethyl ether (L) and 3,4-diethoxybenzoic acid (HERZIG, 1891) by heating on a water-bath with about 10% alcoholic KOH; such mild hydrolysis after complete O-alkylation, followed by identification of the ω-alkoxy-acetophenone and aromatic acid, continues to be a routine step in the investigation of flavonols.

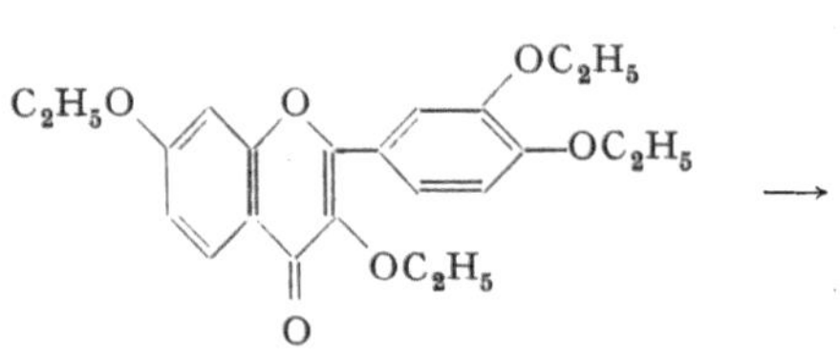

(XLIX.) Fisetin tetraethyl ether.

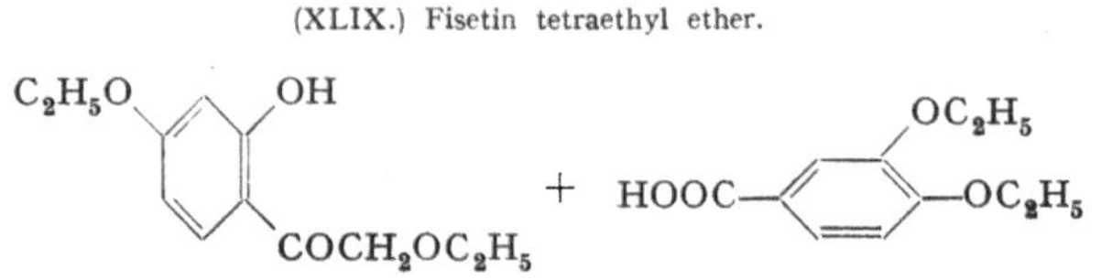

(L.) Fisetol diethyl ether.

The aromatic acid obtained by alkaline fission of a flavone or its methyl ether is usually easy to identify. Color reactions can be used to detect the phenol, confirmed by the preparation of derivatives such as the azo dye obtained by coupling with diazotized aniline. Alkaline degradation of flavones may be carried out on a micro scale and the products examined by paper chromatography (*139*).

For characterizing phenols (including hydroxyflavones) on a paper chromatogram tetrazotized benzidine has been used as a spray reagent (*138*), but it is preferable to use a fast-coupling diazonium salt from a monoamine such as diazotized *p*-nitroaniline or 4-nitro-2-methoxyaniline (commercially available as a stabilized diazonium salt, Fast Red B Salt).

Color reactions have been mentioned on p. 12 which indicate the orientation of hydroxyl groups in a flavone. Additional evidence of a 5-hydroxyl, used for instance in investigating the constitution of calycopterin (*202*), is its relative resistance to methylation. Calycopterin (LI), a dihydroxy-tetramethoxyflavone, gave *p*-hydroxybenzoic acid, from which it followed that calycopterin was a flavone of a type unknown at the time in which all the 3, 5, 6, 7 and 8-positions were occupied by hydroxyl or methoxyl groups. Stability of an alkaline solution to air oxidation located a methoxyl in the 3-position. Since a monomethyl ether was obtained by treatment with diazomethane in ether, the 5-position carried a hydroxyl group, and calycopterin was therefore constituted as (LI).

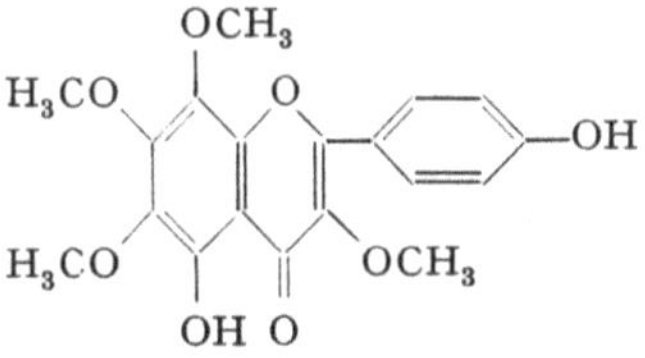

(LI.) Calycopterin

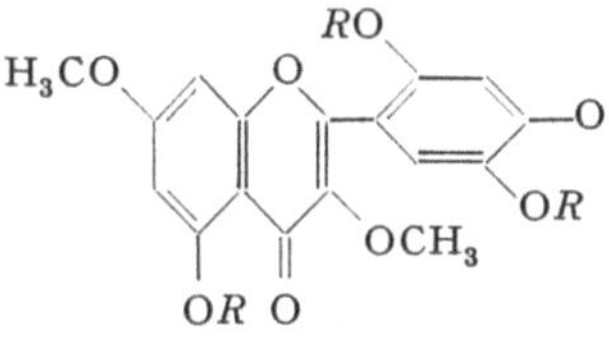

(LII.) R = H, Oxyayanin-A.
R = C$_2$H$_5$, Oxyayanin-A triethyl ether.

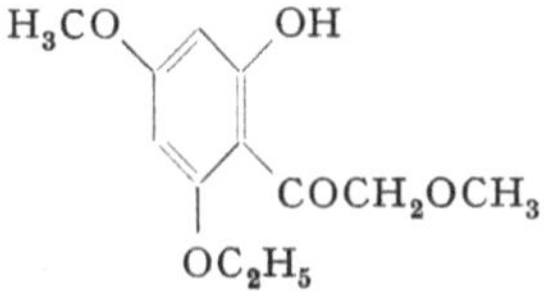

(LIII.) 2-Ethoxy-6-hydroxy-ω,4-dimethoxyacetophenone. 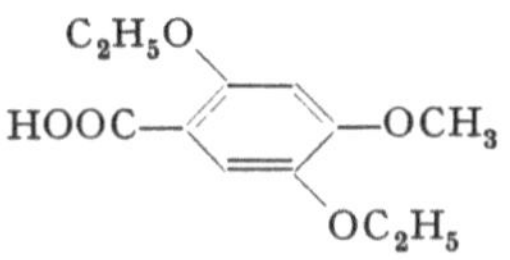   (LIV.) 2,5-Diethoxy-4-methoxybenzoic acid.

A method for determining the position of hydroxyl and methoxyl groups in a partially methylated polyhydroxyflavone is to ethylate the hydroxyl groups, hydrolyse the product with alkali, and identify the resultant ethoxymethoxy compounds. Thus the triethyl ether of oxyayanin-A (LII) gave 2-ethoxy-6-hydroxy-ω,4-dimethoxyacetophenone (LIII) and 2,5-diethoxy-4-methoxybenzoic acid (LIV) (*121*).

BOSE's work on gardenin (*33, 34*) provides a good example of methods by which the structure of a complex flavone can be deduced. Gardenin was found to be a flavone containing one hydroxyl and six methoxyl groups, being the first heptahydroxyflavone derivative to be isolated. The hydroxyl was placed in the 5-position in view of its sparing solubility in aqueous alkali, ferric color, resistance to methylation, and the chlorine content of the product obtained with stannic chloride which corresponded to a complex produced by replacement of H by $SnCl_3$ and not to a double salt with stannic chloride. Alkali fusion yielded the readily identifiable trimethylgallic acid and a chocolate-colored ketone, whose properties showed that it was a quinone constituted as (LV) or (LVI). Since the product of demethylation of gardenin with hydriodic acid did not give the BARGELLINI test, two methoxyls were not in the 6,7-positions, and probably also not in the 7,8-positions because the WESSELY-MOSER change (*220*) would result in a 5,6,7-trihydroxyflavone derivative. The structure (LVII) for gardenin was ruled out by its inability to couple with diazonium salts, and gardenin therefore was constituted as (LVIII).

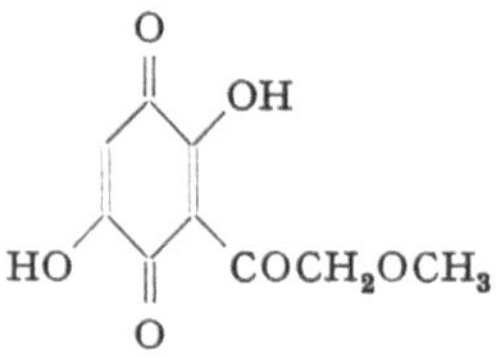

(LV.) 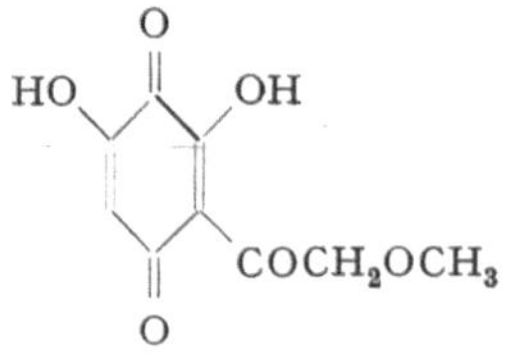          (LVI.)

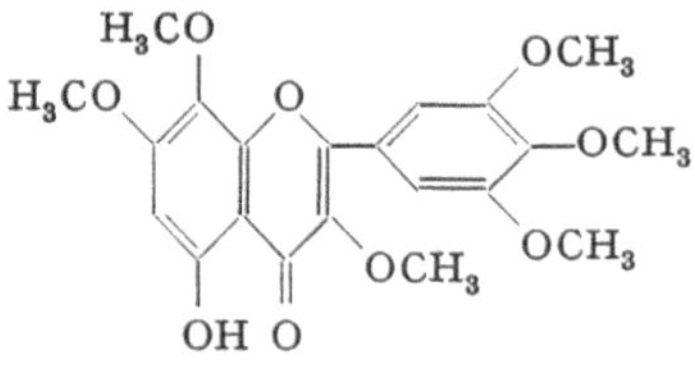

(LVII.) 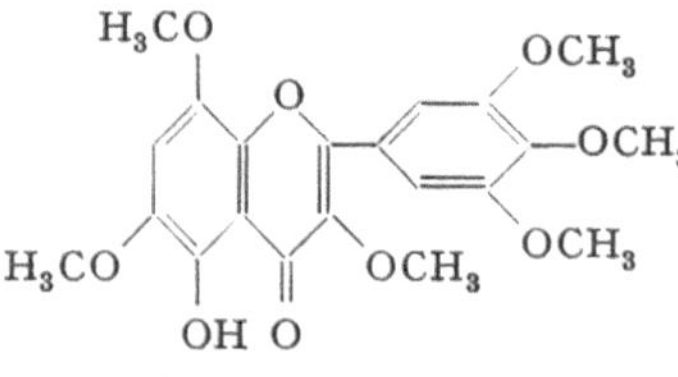          (LVIII.) Gardenin.

The position of attachment of a sugar residue in a flavone glycoside is usually determined by methylation, hydrolysis, and identification of the partially methylated flavone. A recent example in which color reactions and spectral evidence have been used in addition is azalein (*216, 110*). The presence of two or more adjacent phenolic hydroxyl groups was indicated by a dark brown color with chloropentamminocobaltic chloride. Hydrolysis of azalein yielded a tetrahydroxy-methoxyflavone (azaleatin) and rhamnose; methylation and hydrolysis gave quercetin 5,7,3',4'-tetramethylether. The rhamnose residue was therefore attached to the 3-hydroxyl. Azaleatin was different from the known quercetin

7-methyl ether (rhamnetin). Fluorescence in acetic anhydride showed
the absence of a free 5-hydroxyl group. The absorption spectrum of

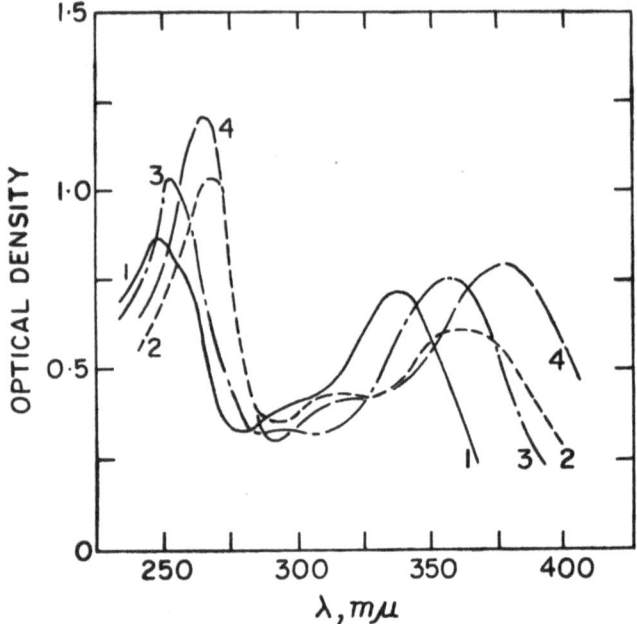

Fig. 4. Azalein, *(1)* ethanol, *(2)* ethanolic sodium acetate, *(3)* ethanolic boric acid-sodium acetate,
and *(4)* 0.002 *M* sodium ethylate.

azaleatin was very similar to that of fisetin (3,7,3′,4′-tetrahydroxyflavone).
Azaleatin therefore was quercetin 5-methyl ether and azalein its
3-rhamnoside (LIX).

**(LIX.)** R = H, Azaleatin.
          R = rhamnosyl, Azalein.

These conclusions were confirmed by the ultraviolet absorption
curves shown in *Figures 4* and *5 (110)*. The bathochromic shifts in the
azalein spectrum in ethanol produced by the addition of sodium acetate
and of boric acid-sodium acetate indicated the presence of 7-hydroxyl and
of *o*-dihydroxyl groups; the stability of azalein in sodium ethylate solution
(Fig. 4) and the instability of azaleatin (Fig. 5) proved that the rhamnoside
group was in the 3-position.

Other interesting examples of structure determination are erianthin (XXXII, p. 21) (*154*), the melicope pigments (*38*), and distemonanthin (XXX) (*122*).

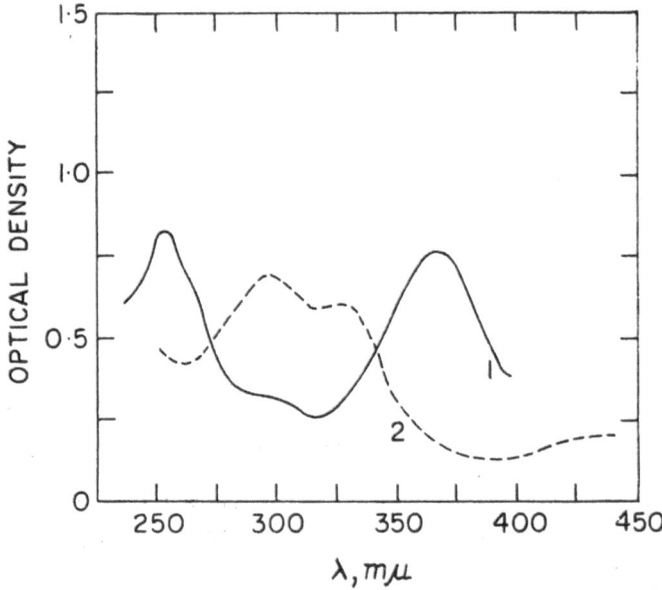

Fig. 5. Azaleatin, (*1*) ethanol, and (*2*) 0.002 *M* sodium ethylate.

*Degradation of Isoflavones.* Isoflavones are much more susceptible to alkaline hydrolysis than flavones unsubstituted in the 3-position. Mild hydrolysis, such as treatment with boiling 5% ethanolic KOH or baryta water for an hour, usually opens the γ-pyrone ring in an isoflavone; the products are formic acid and an *o*-hydroxydesoxybenzoin [*o*-hydroxyphenyl benzyl ketone; e. g., (LXI) from biochanin-A (LX)] (*31*). The identification of formic acid in the fission products constitutes

HO—⟨ ⟩—O / —OCH₃ → HO—⟨ ⟩—OH COCH₂—⟨ ⟩—OCH₃ + HCOOH
OH O | OH

(LX.) Biochanin-A.    (LXI.) Biochanetin-A.

HO—⟨ ⟩—OH + HOOC—CH₂—⟨ ⟩—OCH₃
OH

important evidence for an isoflavone structure. Drastic alkali fusion of an isoflavone, for instance by heating at 210–220° with caustic soda and a few drops of water, results in further degradation of the desoxybenzoin to a phenol and a phenylacetic acid.

Confirmation of the isoflavone character is obtained by recyclization of the desoxybenzoin (LXI) by ethyl formate or orthoformate, after protection of the 4,6-hydroxyl groups if necessary.

In connection with the action of alkali on isoflavones an observation made by MAHESH and SESHADRI (*144*) is the isomerization of 5-hydroxy-7,8-dimethoxy-isoflavone to the 6,7-ether by boiling with 2% alcoholic potash for 15 minutes.

## X. Synthesis of Flavones.

The first synthesis of a flavone (KOSTANECKI, 1898) was by the action of ethanolic potassium hydroxide on a 2′-acetoxychalcone dibromide; but the reaction was limited in scope because with some chalcones, derived for instance from phloracetophenone and *p*-alkoxybenzaldehyde, it took an alternative course leading to a 2-benzylidene-3-coumaranone (see Chart 1, p. 5). It has been shown recently (*58*) that benzylidenecoumaranones not derived from phloroglucinol undergo ring expansion to flavones on treatment with ethanolic potassium cyanide.

A second method, also due to KOSTANECKI (1898), was a reversal of the hydrolytic fission of flavones. Thus chrysin was obtained by condensing phloracetophenone trimethyl ether with ethyl benzoate and treating the resultant 1,3-diketone with hydriodic acid. MENTZER (*155*) has recently described a variation of this route in which phloroglucinol is condensed with ethyl benzoylacetate.

KOSTANECKI achieved the synthesis of several naturally occurring flavonols by a general method involving the conversion of a flavanone into the 3-isonitroso derivative and hydrolysis. Flavanones can be converted to flavonols by conversion to the 3-acetoxy compound with iodine and silver acetate and dehydrogenation with iodine and potassium acetate in glacial acetic acid (*73*).

TAHARA in 1892 observed that the prolonged action of boiling acetic anhydride and sodium acetate on resacetophenone led to 7-acetoxy-3-acetyl-2-methylchromone, which was then hydrolysed to 7-hydroxy-2-methylchromone. The possibilities of this reaction for the synthesis of flavones by using aromatic acid anhydrides were realised over thirty years later by ROBINSON (*4*). Quercetin (IV, p. 7) was thus synthesized by heating ω-methoxyphloracetophenone (LXII), prepared by a HOESCH reaction between phloroglucinol and methoxyacetonitrile, with veratric anhydride and sodium veratrate at 185°; after alkaline hydrolysis to remove O-benzoyl groups, the product was quercetin 3,3′,4′-trimethyl ether (LXIII), demethylated by the usual method to quercetin. Incidentally, the ether (LXIII) has proved to be a useful intermediate for further reactions in the flavone series.

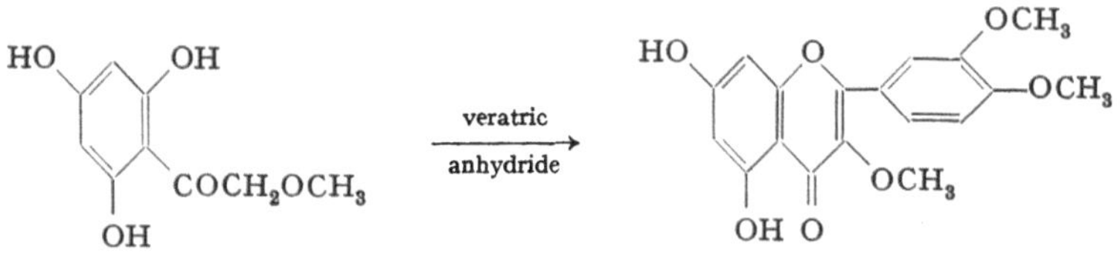

(LXII.) ω-Methoxyphloracetophenone.

(LXIII.) Quercetin 3,3′,4′-trimethyl ether.

Chrysin (XIII, p. 15) and other flavones unsubstituted in the 3-position can be prepared similarly; the flavone may be accompanied by the 3-aroyl derivative (LXIV); normally this C-aroyl group is removed during alkaline hydrolysis of the O-aroyl groups, but some 3-aroyl derivatives resist hydrolysis under conditions in which the $\gamma$-pyrone ring remains intact (23). In the course of studies on the mechanism of the ROBINSON reaction it was independently observed by BAKER (8) and MAHAL (142) that 2-aroyloxyacetophenone derivatives (LXV) undergo a transformation to the diketones (LXVI) by the action of potassium carbonate in boiling toluene (8) or sodamide in ether at 0° (142). WHEELER (170) has made an extensive study of this transformation and has shown that it is an intramolecular CLAISEN condensation effected by a wide variety of bases including KOH in pyridine and triphenylmethyl anion. SCHMID and BANHOLZER (194) have proved the intramolecular nature of the transformation of (LXV) to (LXVI) by using benzoyl chloride labelled with ¹⁴C in the carbonyl group.

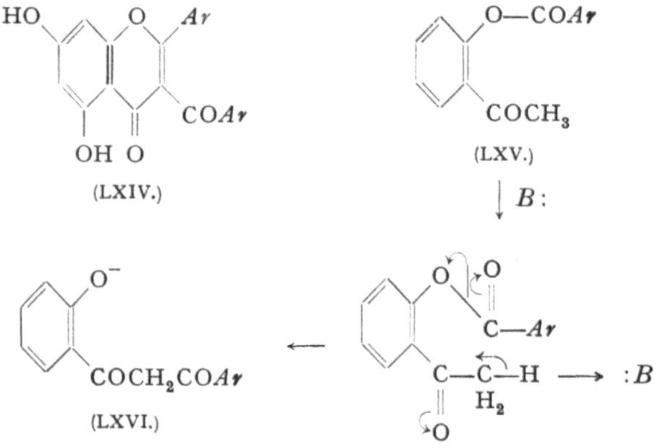

(LXIV.)

(LXV.)

(LXVI.)

Since the diketones (LXVI) readily yield the corresponding flavones by treatment with sulphuric acid, the procedure via the diketone is sometimes more convenient than the ROBINSON reaction; thus flavone may be prepared in an overall yield of 59–68% from *o*-hydroxyaceto-phenone (*222*). In favorable cases the diketone can be prepared from the phenolic ketone in one operation by heating the latter with the acid chloride and anhydrous potassium carbonate in acetone (*17*); longer treatment, especially when the ketone is ω-substituted, may lead directly to the flavone (*145*). *o*-Aroyloxyacetoarones (LXV) yield flavones when heated by themselves under reduced pressure or in glycerol at about 250° (*214*, *51*). The reaction involves a thermally induced BAKER-VENKATARAMAN transformation.

When the final step in the synthesis of a polyhydroxyflavone consists in treatment with hydriodic acid to effect demethylation, the rearrangement of a 5,8-dimethoxyflavone (LXVII) to the 5,6-isomer (LXVIII, scutellarein) (WESSELY-MOSER change) (*220*, *221*) has to be borne in mind; it has also been observed (*221*) that 2',5'-dimethoxyflavone undergoes a similar rearrangement and leads to 6,2'-dihydroxyflavone. The mechanism of these changes, which WHEELER (*221*) has discussed in detail, involves the opening of the pyrone ring and cyclization of the diketone (LXIX) in the alternate direction. Under normal conditions of demethylation with hydriodic acid, flavonols do not undergo the WESSELY-MOSER change, so that the original syntheses of gossypetin and other 5,8-dihydroxyflavonols remain valid; but on heating a 5,8-dimethoxyflavonol with hydriodic acid at 180° to 190° in a sealed tube rearrangement to the 5,6-isomer occurs (*221*). Demethylation of 5,8-dimethoxyflavones by aluminium chloride or bromide proceeds normally.

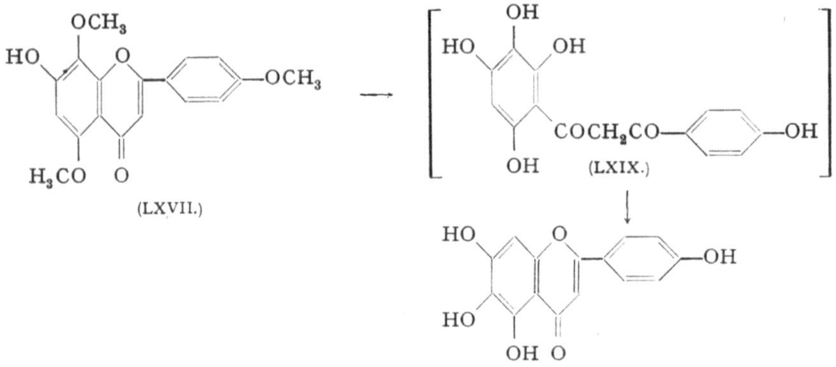

(LXVII.)

(LXIX.)

(LXVIII.) Scutellarein.

When 2,4-dihydroxy-3,6-dimethoxyacetophenone (LXX) is heated with anisic anhydride and sodium anisate, the ROBINSON reaction takes an unexpected

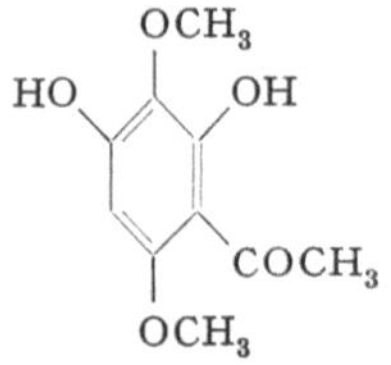

(LXX.)

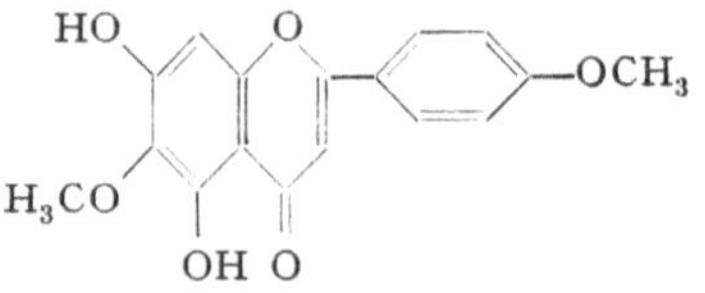

(LXXI.) Pectolinarigenin.

course (*220*). Partial demethylation occurs, followed by cyclization to the favored 5,6,7-orientation, and the product is pectolinarigenin (LXXI).

The device of protecting hydroxyl groups by benzoylation or benzylation and subsequently removing the protecting group by mild hydrolysis was adopted by ROBINSON for the synthesis of partially methylated polyhydroxyflavones such as isorhamnetin, kaempferide, diosmetin and chrysoeriol (*154*). Tricin (LXXII) was synthesized by carrying out the ROBINSON reaction on phloracetophenone and *O*-benzyl-syringic anhydride, followed by debenzylation with hydrochloric acid in acetic acid (*154*). Isopropylation and methoxymethylation have been recently suggested as more satisfactory (*207, 6*). Protection by means of dihydropyran appears to have been used for flavone synthesis only by SCHMID and BANHOLZER (*194*) in connection with their work on the mechanism of the BAKER-VENKATARAMAN transformation, and it should prove useful for the synthesis of naturally occurring flavones and isoflavones. Preferential demethylation of the 5-methoxyl group by aluminium chloride under suitable conditions (boiling benzene or nitrobenzene at about 100°), first demonstrated by BHARADWAJ in 1933 (*22*), is another approach to the synthesis of flavones containing both hydroxyl and methoxyl groups, which has been widely used. Wogonin (LXXIII) was thus obtained (*154*) from 7-hydroxy-5,8-dimethoxyflavone, and melisimplin (XXIX, p. 20) from melisimplexin (VIII, p. 9) (*38d*).

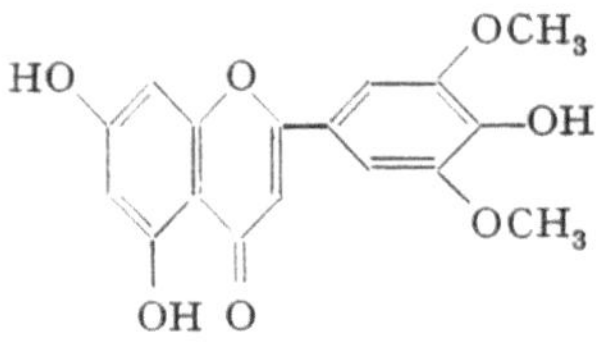

(LXXII.) Tricin.

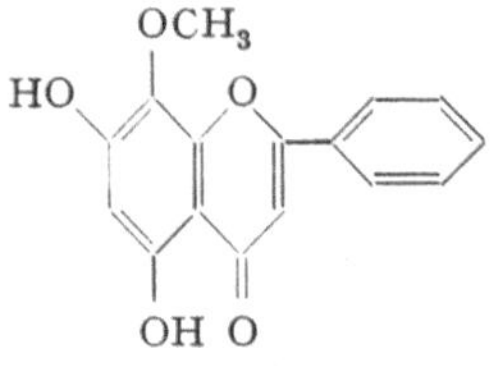

(LXXIII.) Wogonin.

The 3-methoxyl group in a flavone is also sensitive to the action of aluminium chloride in nitrobenzene at about 100° (*202*), and tambulin (LXXIV) has been prepared by this method from herbacetin pentamethyl ether (LXXV) (*14*).

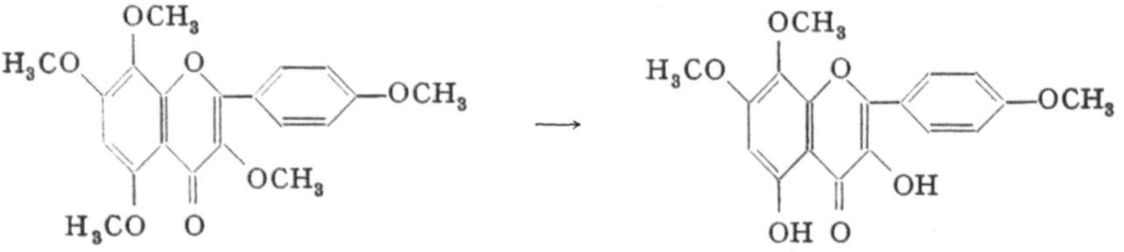

(LXXV.) Herbacetin pentamethyl ether

(LXXIV.) Tambulin.

In connection with the constitution of calycopterin (LI, p. 30) it was shown (*202*) that hydrogen bromide in acetic acid demethylates a 5-methoxyl group at room temperature and a 3-methoxyl at about 100°; such preferential demethylation of the 5-methoxyl in (LXXVI) has been used (*201*) for the synthesis of calycopterin. The ketone (LXXVII), which yielded (LXXVI) by the action of *p*-benzoyloxybenzoic anhydride and subsequent hydrolysis of the benzoyl group, was prepared by persulphate oxidation of (LXXVIII) followed by methylation of the new unchelated hydroxyl group (*161*).

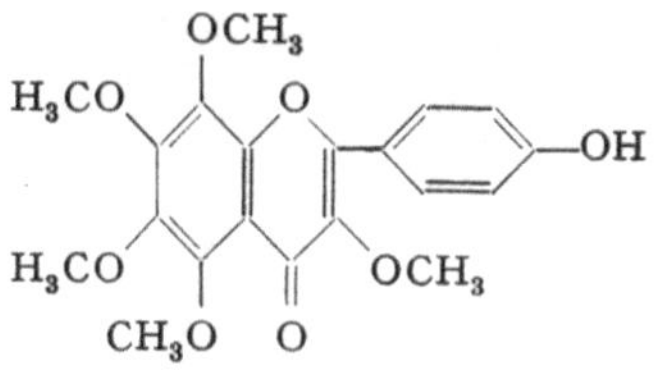

$\longrightarrow$ (LI.) Calycopterin (p. 30).

(LXXVI.)

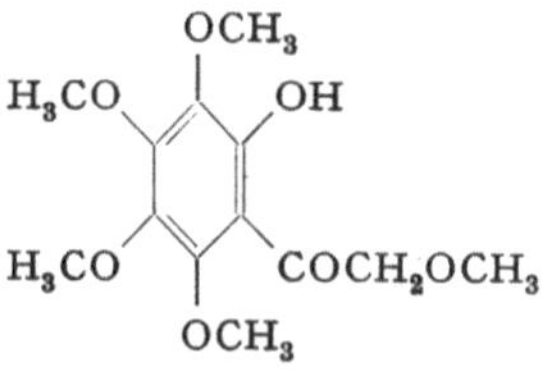

(LXXVII.)

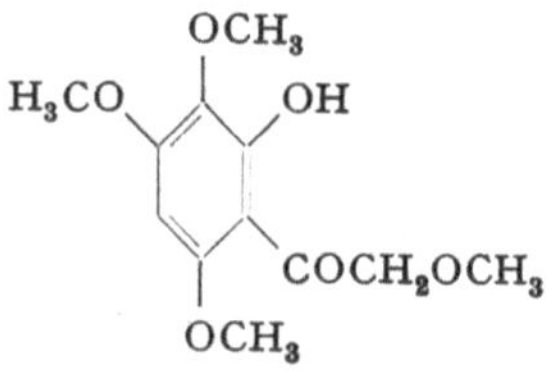

(LXXVIII.)

Hydroxyl groups in the 5- and 3',4'-positions resist methylation by diazomethane in the presence of monosodium orthoborate; quercetin thus yields the 3,7-dimethyl ether (*205*). Selective methylation of the fully acetylated compound is another method of preparing partial methyl ethers of polyhydroxyflavones, acetyl groups being replaced in the order 7 > 4' > 3 > 5 > 3'. Thus treatment of quercetin pentacetate with methyl iodide in acetone in the presence of potassium carbonate gives a high yield of rhamnetin tetracetate (*109*).

For the synthesis of flavones, as distinct from flavonols, perhaps the most convenient and generally applicable method is the oxidative cyclization of 2'-hydroxychalcones (*o*-hydroxyphenyl styryl ketones) with selenium dioxide (*141*) in boiling amyl alcohol. The likely mechanism of the reaction is cyclization of the chalcone to the flavanone, which then undergoes dehydrogenation to the flavone; flavanones are smoothly dehydrogenated to flavones by selenium dioxide. As stated by SIMPSON (*207*), this reaction has the advantages of being free from 3-acylation complications and of usually giving clean products in almost quantitative yield. Hydroxyl groups, other than the hydroxyl which is to provide the pyrone oxygen, are normally protected by methylation, benzylation or isopropylation (*141*, *207*), but with acetic anhydride in place of amyl alcohol as the solvent, flavanones containing free hydroxyl groups have been dehydrogenated, e. g., hesperetin (LXXIX) to diosmetin (LXXX), the corresponding flavonol (tamarixetin) being obtained as a by-product (*16*).

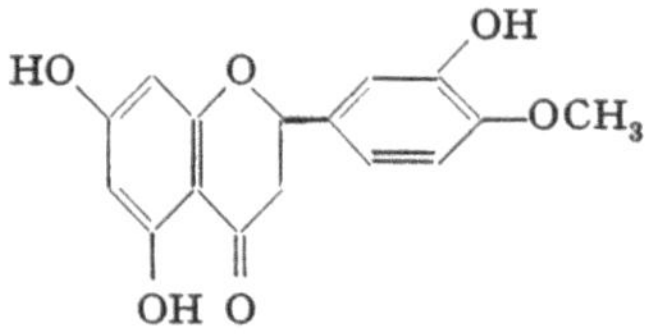

(LXXIX.) Hesperetin.

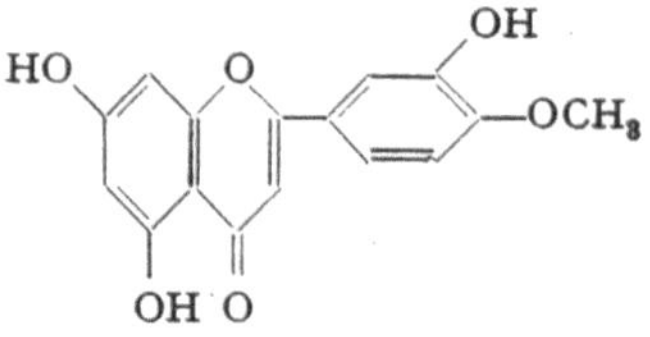

(LXXX.) Diosmetin.

An advance of the utmost importance in the synthesis of poly-hydroxyflavones is SESHADRI's application (1943) of the ELBS per-sulphate oxidation, and numerous naturally occurring and other flavones have been synthesized by ingenious modifications of the general procedure. In the first of a long series of papers SESHADRI (*198*) described the conversion of quercetin (IV, p. 7), via the 3,7,3',4'-tetramethyl ether (LXXXI) and gossypetin tetramethyl ether (LXXXII), to gossypetin (XXVII, p. 20). Primetin (XIV, p. 15) was similarly obtained by the persulphate oxidation of 5-hydroxyflavone (LXXXVI).

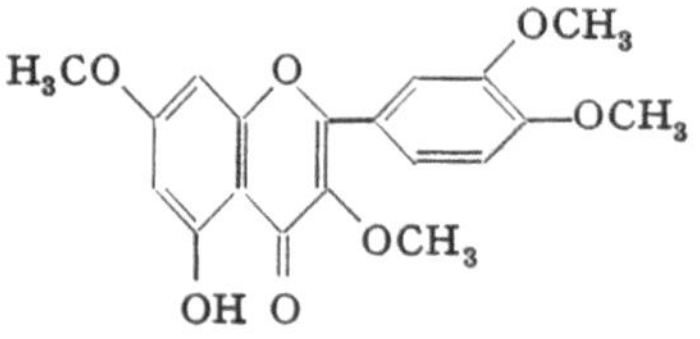

(LXXXI.)

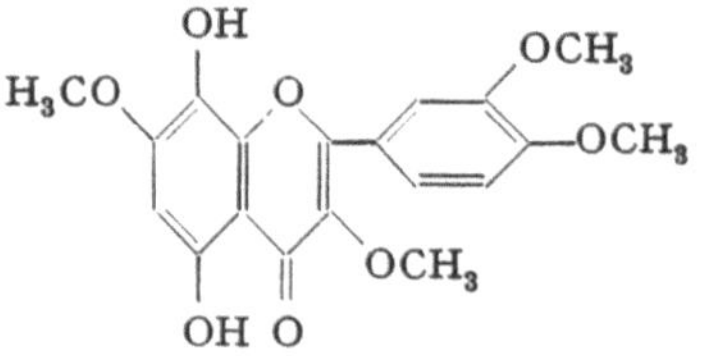

(LXXXII.)

A route to 5,6-dihydroxyflavone, which occurs in nature as the dimethyl ether, is to couple 6-hydroxyflavone with diazotized aniline, reduce the azo dye (LXXXIII) to the amine and replace the amino group by hydroxyl via the diazonium salt (*104*). When the amine (LXXXIV) is boiled with 32% hydrochloric acid for 24 hours, baicalein 7-methyl ether is obtained in good yield as the result of hydrolysis and a WESSELY-MOSER change (*104*).

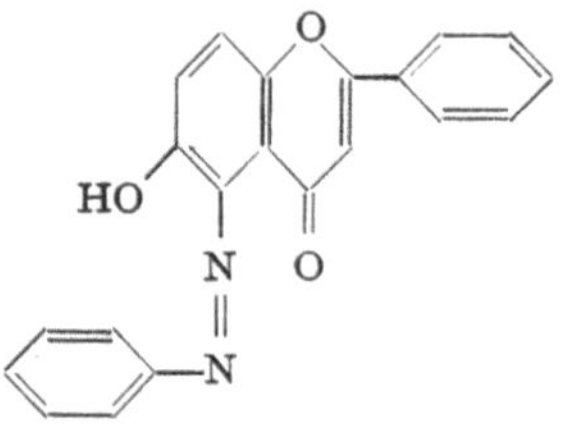

(LXXXIII.)

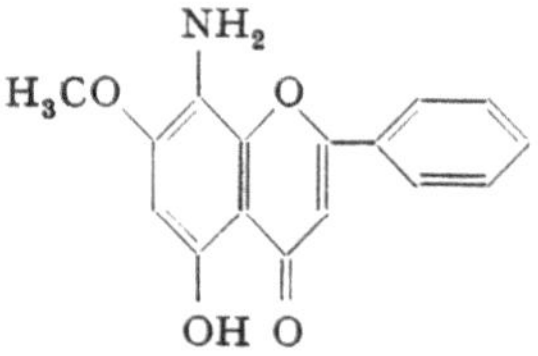

(LXXXIV.)

The reduction of a nuclear hydroxyl group in flavones by RANEY nickel hydrogenolysis of the tosylate was first described by RAMANATHAN and VENKATARAMAN (*178*), who showed that this is a convenient general method for the synthesis of 5-hydroxyflavones from 5,7-dihydroxyflavones. Chrysin (XIII, p. 15) was thus reduced to 5-hydroxyflavone (LXXXVI) by the action of RANEY nickel and hydrogen on the 7-tosylate (LXXXV; $R = H$); and galangin 3-methyl ether to 5-hydroxy-3-methoxyflavone (LXXXVI) via the tosylate (LXXXV; $R = OCH_3$).

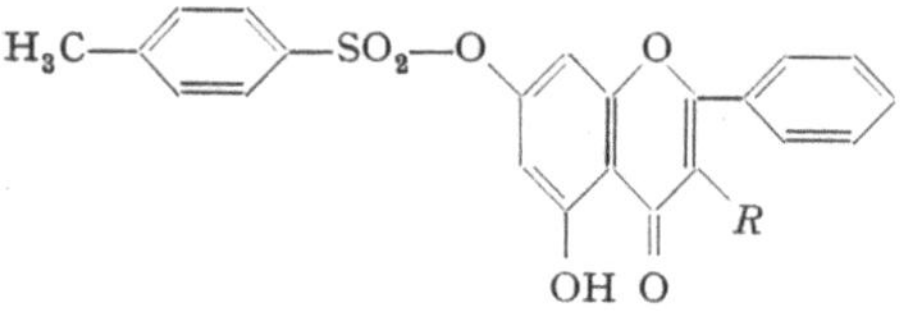

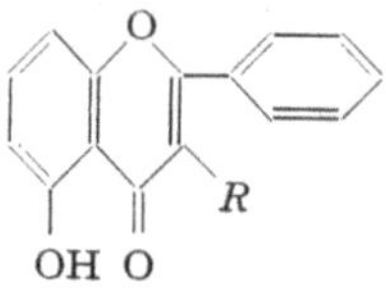

(LXXXV.)          (LXXXVI.) $R = H$, 5-Hydroxyflavone.
                  $R = OCH_3$, 5-Hydroxy-3-methoxyflavone.

SESHADRI's elegant synthesis of gardenin (LXXXVII) involves an application of this method in combination with persulphate oxidation and *ortho*-oxidation via the *ortho*-aldehyde *(Chart 2)*, starting from myricetin 3,3′,4′,5′-tetramethyl ether (LXXXVIII) obtained by the ROBINSON reaction (*2*).

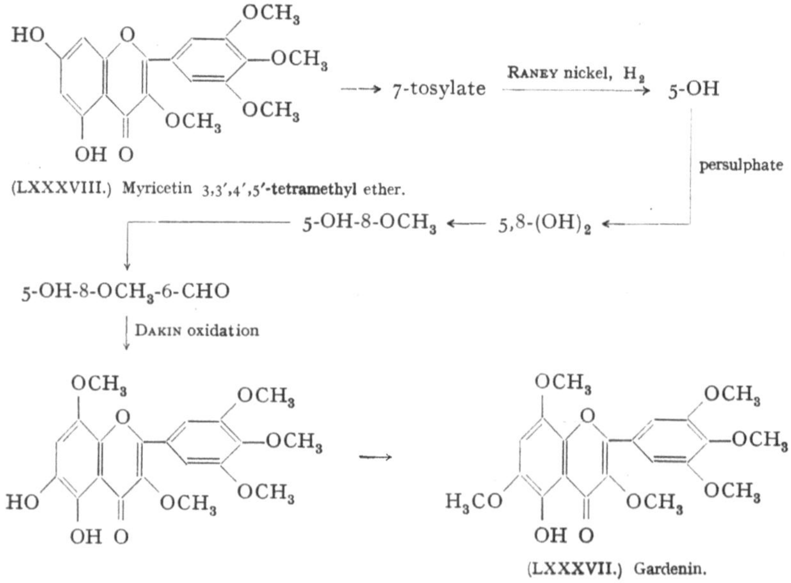

(LXXXVIII.) Myricetin 3,3′,4′,5′-tetramethyl ether.

5-OH-8-OCH$_3$-6-CHO

DAKIN oxidation

(LXXXVII.) Gardenin.

*Chart 2.* Synthesis of Gardenin (SESHADRI).

*The Algar-Flynn-Oyamada Reaction.* It was shown by ALGAR and FLYNN (*3*) and by OYAMADA, independently and about the same time (1934), that the action of hydrogen peroxide on *o*-hydroxyphenyl styryl ketones in hot alcoholic potash gave the corresponding flavonols. The flavanonol (e. g., LXXXIX) formed as an intermediate can be isolated in good yield (*148, 77*), and can then be oxidized to the flavonol by bismuth acetate (*77*). The ALGAR-FLYNN reaction takes an alternative course (*68*) if the chalcone carries a methoxyl group in the 6'-position, the product being the benzalcoumaranone (aurone); but if there is a free hydroxyl group in the 4-position as in (XC), the flavonol (XCI) is obtained (*5*). Selective demethylation of (XCI) yields rhamnazin (XCII).

The complexity of the ALGAR-FLYNN reaction, due partly to the reactivity of flavanonols towards alkali, has been discussed by WHEELER (*221*). The formation of flavonols by boiling 2'-hydroxychalcone dibromides with water (*136*) and by shaking *o*-hydroxyacetophenones with aromatic aldehydes in alcoholic sodium hydroxide has been observed (*137*).

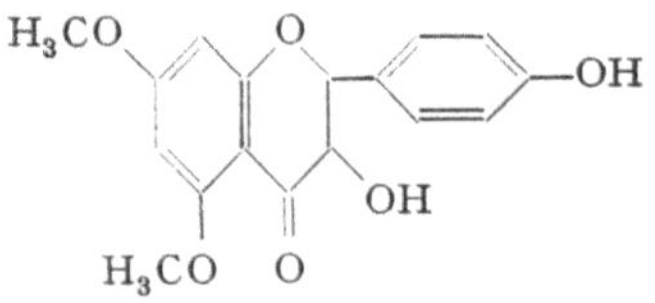

(LXXXIX.)

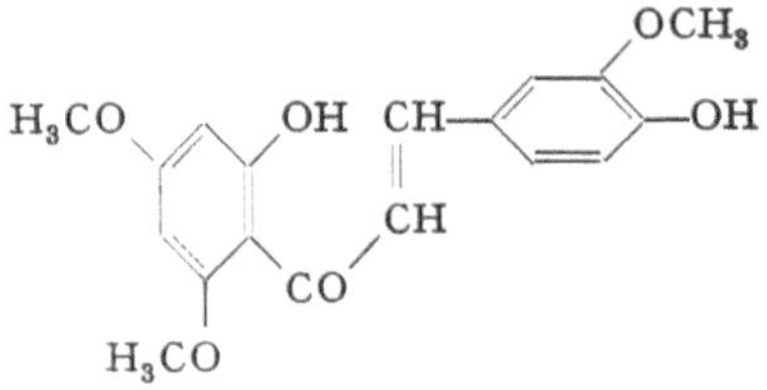

(XC.)

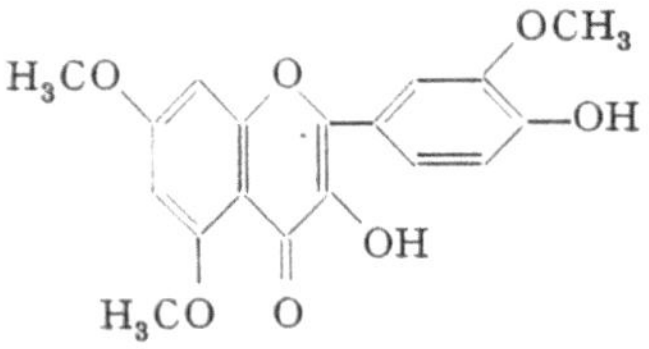

(XCI.)

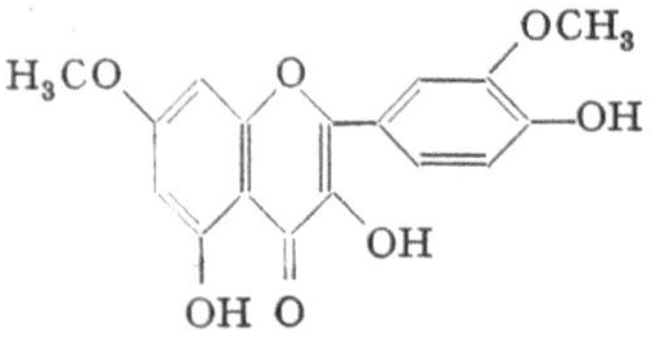

(XCII.) Rhamnazin.

Flavanones can be oxidized to flavonols by hydrogen peroxide and ferrous sulphate (FENTON's reagent) in dilute sulphuric acid (*143*).

## XI. Synthesis of Isoflavones.

The first synthesis of an isoflavone, 7-methoxyisoflavone (BAKER and ROBINSON, 1925), was effected by oxidation of the 2-styryl derivative, followed by decarboxylation (*12*). SPÄTH and LEDERER (cf. *154*) obtained pseudobaptigenin (XCIV) in a yield of about 8% by treating the desoxybenzoin (XCIII; *R* = H) with ethyl formate and sodium in a sealed tube at 100°; daidzein and formononetin were synthesized similarly.

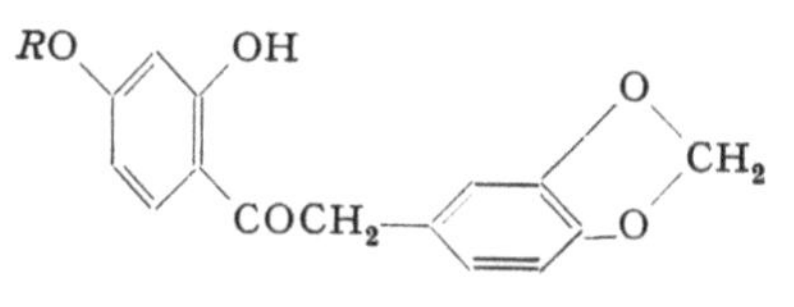

(XCIII.) 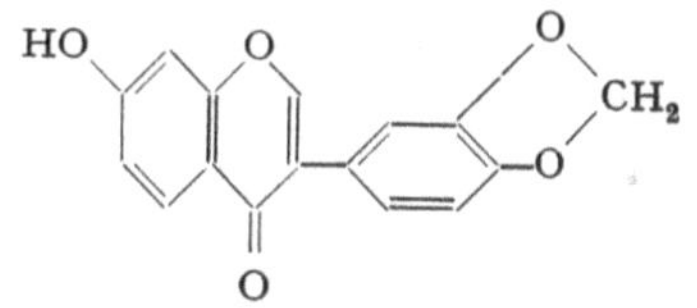            (XCIV.) Pseudobaptigenin.

The ethyl formate condensation proceeds smoothly at $0°$ (*107*) and some of the available evidence shows that the reaction proceeds directly to the isoflavone, but 2-hydroxyisoflavanones, which readily dehydrate by treatment with glacial acetic acid, have been isolated as intermediates in a few cases (*226, 218*). Numerous isoflavones, including the parent compound and several naturally occurring representatives, have been synthesized by this method.

It is desirable to protect all the hydroxyl groups in the desoxybenzoin, other than the hydroxyl *ortho* to the carbonyl group. Thus for the synthesis of pseudo-baptigenin the benzyl ether (XCIII; $R = C_6H_5CH_2$) was employed, debenzylation being effected by hydrochloric acid after the isoflavone cyclization. Santal (XCV) has been synthesized by ring closure of the desoxybenzoin (XCVI) with ethyl formate and sodium; treatment of the isoflavone thus obtained with aluminium bromide in nitrobenzene at $30°$ effected fission of the methylenedioxy group as well as demethylation in the 5-position (*103*).

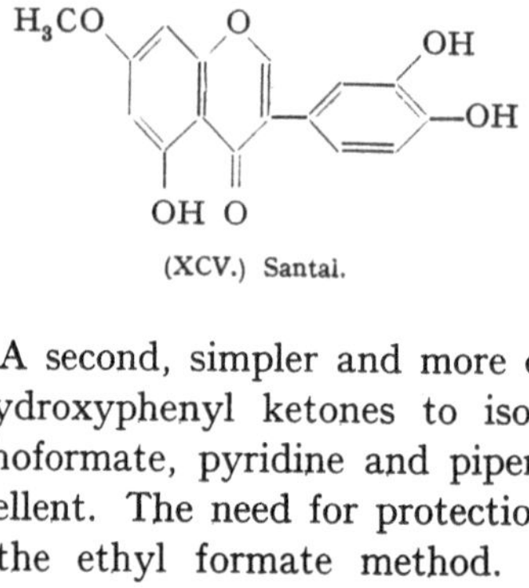

(XCV.) Santal. 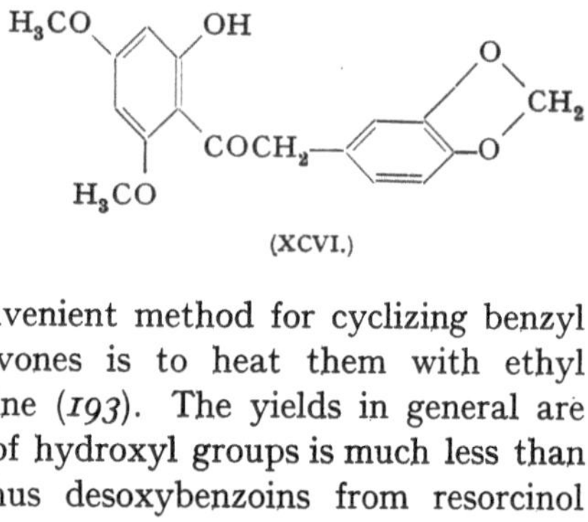            (XCVI.)

A second, simpler and more convenient method for cyclizing benzyl *o*-hydroxyphenyl ketones to isoflavones is to heat them with ethyl orthoformate, pyridine and piperidine (*193*). The yields in general are excellent. The need for protection of hydroxyl groups is much less than in the ethyl formate method. Thus desoxybenzoins from resorcinol (both $\beta$- and $\gamma$-substituted) and pyrogallol cyclize in good yields. Cyclization is particularly facile with *p*-nitrobenzyl ketones (*102*); 2,4-dihydroxyphenyl 4-nitrobenzyl ketone is converted to 7-hydroxy-4′-nitro-isoflavone in nearly quantitative yield.

The 7-methoxyl group in a polymethoxyisoflavone is the most resistant to demethylation; thus when 5,7,3′,4′-tetramethoxyisoflavone is heated with acetic anhydride and hydriodic acid at $140°$ for one hour, the product is a mixture of santal (XCV) and norsantal (orobol), separable by their difference in solubility in aqueous sodium carbonate (*164*). The relative resistance of the 7-methoxyl group has been used for the synthesis of muningin (XCVII). 5,6,7,4'-Tetramethoxyisoflavone, prepared by the ethyl orthoformate method, was demethylated to the 7-monomethyl ether (XCVIII), dibenzoylated in the 6- and 4'-positions, methylated in the 5-position, and the ester hydrolysed to muningin by shaking with methanolic potassium hydroxide (*113*).

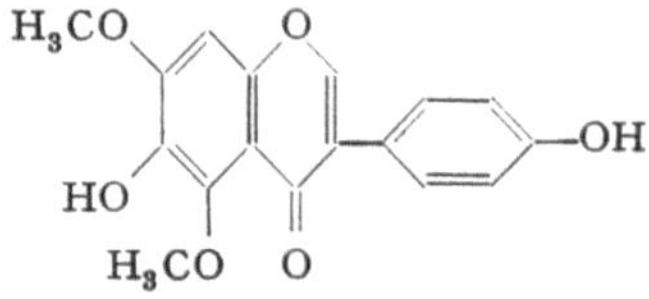

(XCVII.) Muningin. 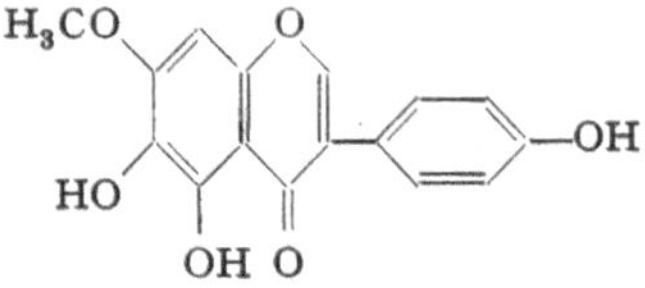   (XCVIII.)

In the conversion of a benzyl *o*-hydroxyphenyl ketone to an isoflavone the 2-carbon atom can be provided by a variety of reagents other than ethyl formate and orthoformate, such as formamide, formanilide (*75*), ethoxalyl chloride (*9*), and hydrogen chloride-zinc cyanide (*55*). The last two methods do not require the protection of hydroxyl groups. BAKER's ethoxalyl chloride method is illustrated by the synthesis of genistein (CI). The desoxybenzoin (IC) reacts with ethoxalyl chloride in pyridine at room temperature to form the 2-carbethoxyisoflavone (C), which is then hydrolysed and decarboxylated to genistein by heating the acid to about 10° above its melting point.

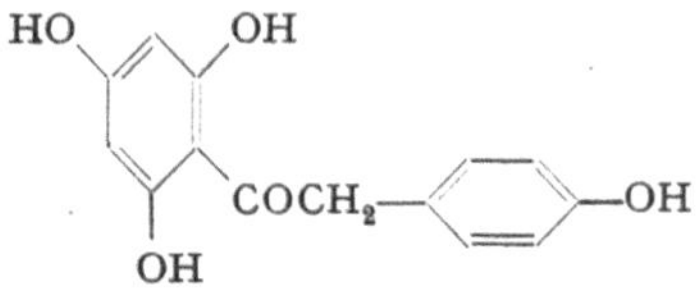

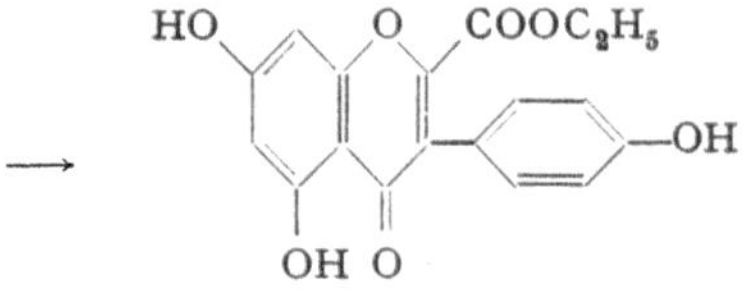

(IC.)   (C.)

2-Carboxylic acid.

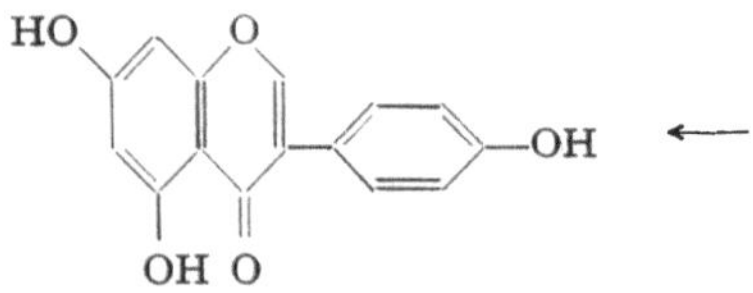

(CI.) Genistein.

The ethoxalyl chloride method is a useful supplement to the ethyl formate and ethyl orthoformate methods, but there is no justification for WHEELER's suggestion (*75*) that it is the method of choice. The "severe limitations" of the ethyl formate and ethyl orthoformate methods to which WARBURTON (*218*) refers are nonexistent. While the protection of hydroxyl groups in polyhydroxy-desoxy-benzoins is not necessary in the ethoxalyl chloride method, which is also true of the more recent hydrogen cyanide method, there is no particular difficulty in the protection of hydroxyls and the subsequent removal of the protecting groups. On the other hand, the ethyl formate and ethyl orthoformate reactions, especially the latter, have the great advantage of being facile one-step procedures usually giving much better yields than the ethoxalyl chloride method. There is no naturally occurring or other isoflavone which has been synthesized by the ethoxalyl chloride method which has not been or cannot be synthesized, usually more readily, by the older reactions. On the other hand, attempts to prepare tectorigenin (CII) by the action of ethoxalyl chloride on (CIII; $R = OH$) led to the isomeric 8-methoxyisoflavone (CIV) (*116, 10*); using ethyl orthoformate (CIII; $R = NO_2$) cyclized in the desired direction, yielding tectorigenin (CII) when the nitro group was replaced by hydroxyl via the amine and the diazonium salt (*111*).

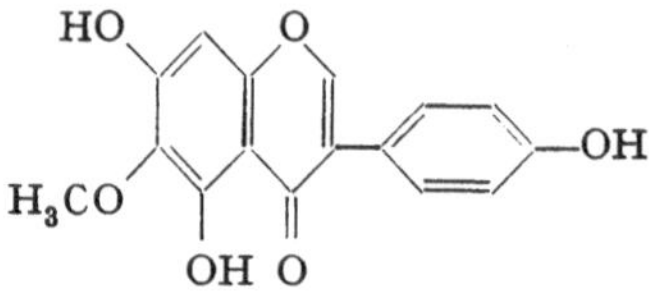

(CII.) Tectorigenin.

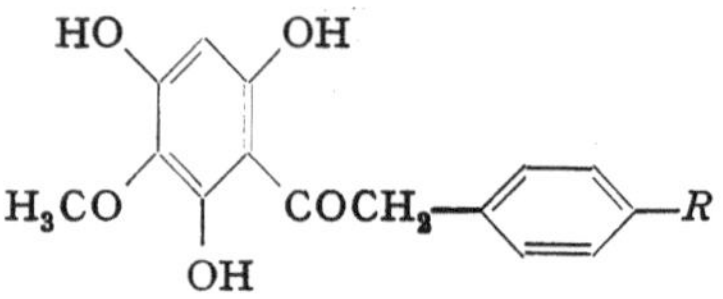

(CIII.)

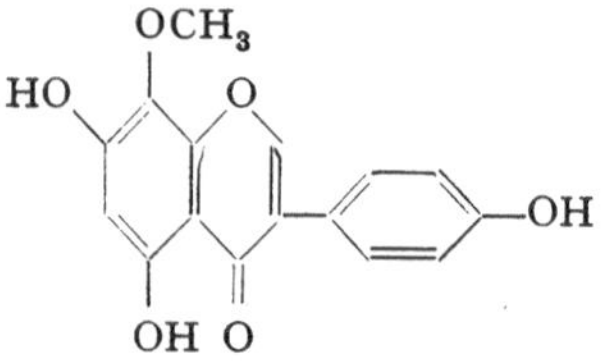

(CIV.)

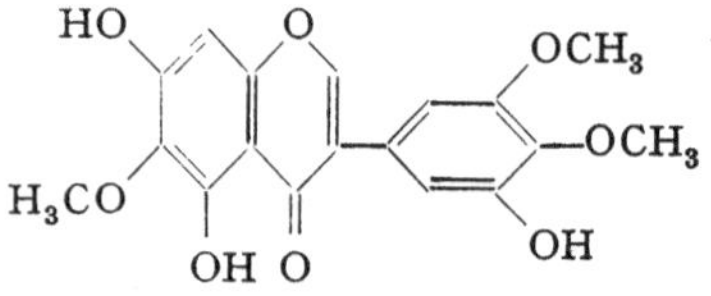

(CV.) Irigenin.

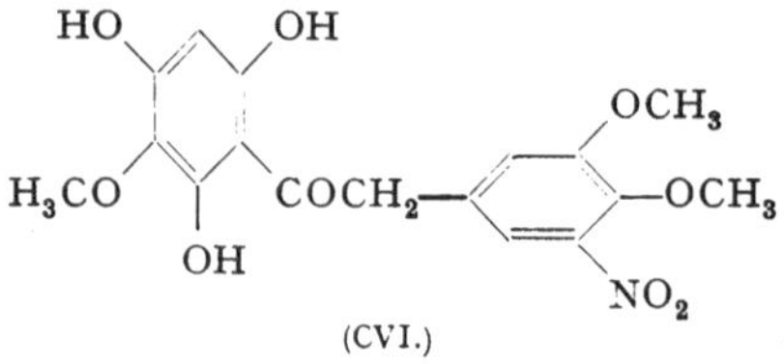

(CVI.)

The synthesis of irigenin (CV) by a similar procedure starting from the desoxybenzoin (CVI) is in progress.

## XII. Synthesis of Flavone and Isoflavone Glycosides.

Few flavone glycosides have been synthesized so far. HATTORI (1928) prepared 7- and 4'-β-D-glucosyloxyflavones by direct glucosylation of 7- and 4'-hydroxyflavone with tetracetyl α-D-glucosyl bromide (aceto-bromoglucose); ZEMPLÉN (*229, 228*) was able to synthesize toringin (chrysin 7-glucoside) and linarin (acacetin 7-rutinoside) similarly because of the relative inactivity of the 5-hydroxyl group.

Some of the naturally occurring flavanone glycosides (in the form of the fully acetylated compounds) have been dehydrogenated to the corresponding flavone glycosides, e. g., hesperidin (5,3'-dihydroxy-4'-methoxy-7-rutinosyloxyflavanone) to diosmin (diosmetin 7-rutinoside), and solipurposide (5-glucosyloxy-7,4'-dihydroxyflavanone) to apigenin

5-glucoside (occurring in the leaves of *Amorpha fruticosa*) by successive treatment with bromine and alkali (*233*). Rhoifolin, apigenin 7-rhamno-glucoside, isolated from the leaves of *Rhus succedanea*, has been prepared by bromination and dehydrobromination of naringin (*85*). The dehydrogenation step was improved by SESHADRI (*165*) who used iodine and sodium acetate in boiling ethanol, the reaction proceeding smoothly when the flavanone contained a 5-hydroxyl group.

REICHEL and MARCHAND (*186*) have shown that glycosides of 2'-hydroxy-chalcones can be oxidized to the corresponding flavonol glycosides by the ALGAR-FLYNN reaction, but their procedure does not appear to have been used so far for the synthesis of naturally occurring flavonol glycosides. The only synthesis of a naturally occurring flavonol glycoside recorded in the literature is that of isoquercitrin (quercetin 3-glucoside) by ICE and WENDER (*98*). They treated the monopotassium salt of quercetin (prepared from quercetin and alcoholic potassium acetate at pH 8) with acetobromoglucose in presence of liquid ammonia.

Four of the ten known isoflavone glycosides have been synthesized. Genistin, the 7-glucoside of genistein (CI), is obtained as the sole product by treatment of genistein in acetone and aqueous potassium hydroxide with acetobromoglucose under specified conditions, although genistein has hydroxyl groups in the 7- and 4'-positions in addition to the chelated and less reactive 5-hydroxyl (*230*). Ononin, formononetin 7-glucoside, is obtained similarly from formononetin (CXIV, p. 48), the reaction being free from any complication because there is only one hydroxyl group in formononetin (*232*). The constitution of sophoricoside as the 4'-glucoside of genistein, proved earlier by a degradation method, has been confirmed by synthesis from genistein by protecting the 7-hydroxyl with a *p*-nitrobenzyl group, condensing the 4'-hydroxyl with acetobromoglucose, deacetylating, and finally removing the benzyl group by hydrogenation (*27*).

Partial methylation of sophoricoside yields prunitrin, the constitution of which is therefore proved to be prunetin 4'-glucoside (*231*).

## XIII. Biogenesis of Flavones and Isoflavones.

The biogenesis of flavones and other flavonoids, which is the subject of a very recent and comprehensive review (*28*; cf. *25*), is part of the wider problem of aromatic biosynthesis. Biogenetic schemes based on the structural relations of natural products (*189*) and on laboratory analogies have served their purpose; they have frequently led to the structure of a plant constituent by short cuts and to new methods of synthesis (*199*). During the last few years experiments on plants using [14]C-labelled precursors have provided direct evidence of biosynthetic pathways for lignin, quercetin and tricin. Flavonoids, which include the leucoanthocyanins, and lignin are biogenetically related. Shikimic acid (CVII), cinnamic acid, ferulic acid (CVIII), phenylalanine, tyrosine, and *p*-hydroxyphenylpyruvic acid (CIX) have all been shown to be capable of functioning as lignin precursors (*28*). The transformation of shikimic acid to the phenylpropane derivatives proceeds through prephenic acid (CX).

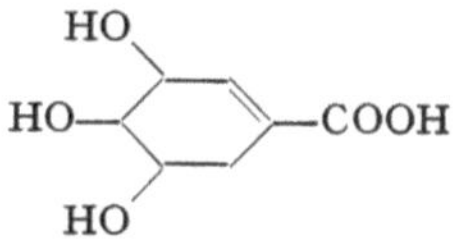

(CVII.) Shikimic acid.

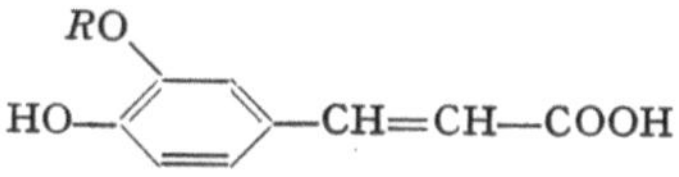

(CVIII.) $R = CH_3$, Ferulic acid.
$R = H$, Caffeic acid.

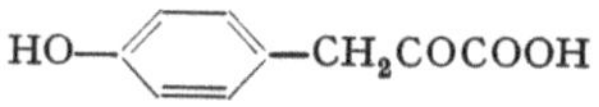

(CIX.) *p*-Hydroxyphenylpyruvic acid.

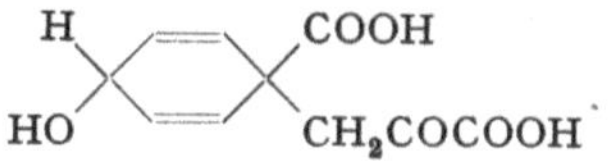

(CX.) Prephenic acid.

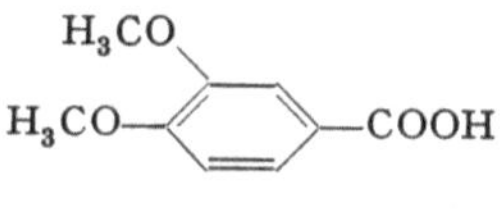

(CXI.) Veratric acid.

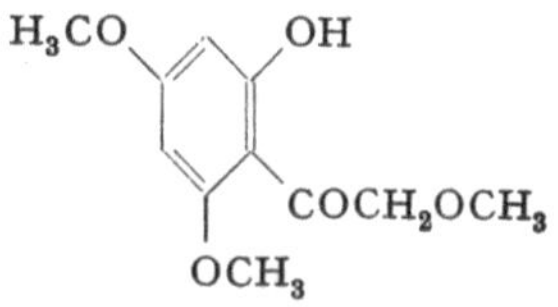

(CXII.) 2-Hydroxy-ω-4,6-trimethoxyacetophenone.

Three groups of workers (*187, 213, 71*) have reported the results of parallel experiments on the synthesis of tricin in wheat, quercetin (IV, p. 7) and caffeic acid (CVIII) in buckwheat (*Fagopyrum tartaricum; F. esculentum*), and of caffeic acid in tobacco (*Nicotiana tabacum*). Ferulic acid is incorporated into tricin (5,7,4'-trihydroxy-3',5'-dimethoxyflavone; LXXII) produced by wheat (*187*). The mechanism of quercetin biosynthesis was traced by alkaline hydrolysis of the pentamethyl ether to veratric acid (CXI) and 2-hydroxy-ω-4,6-trimethoxyacetophenone (CXII). Shikimic acid, phenylalanine, *p*-hydroxycinnamic acid, and cinnamic acid were very good precursors, followed in decreasing order of effectiveness by caffeic acid, sinapic acid, *m*-methoxycinnamic acid and ferulic acid. Quercetin formed from β- or ring-labelled cinnamic acid gave labelled veratric acid, and quercetin from α- or carboxyl-labelled cinnamic acid gave labelled (CXII). These results prove that the $C_6$—$C_3$ chain of the precursors is incorporated in quercetin for the *B*-ring and the 2, 3, and 4-carbon atoms, and that the ring hydroxylation pattern of a precursor determines its effectiveness (*213*).

Uniformly labelled sucrose or $^{14}CO_2$ (assimilated photosynthetically) yielded quercetin equally labelled in the *A*- and *B*-rings; sodium acetate labelled on either carbon gave quercetin labelled almost entirely in the *A*-ring, supporting the Birch hypothesis (*25, 203*).

Isotope competition experiments excluded the possibility of *meso*-inositol or phloroglucinol being precursors of the $A$-ring (*213*); this suggestion, originally a speculation based on the fact that phloroglucinol represents the invariable hydroxylation pattern of the $A$-ring in the numerous anthocyanins and the most frequent in the flavones, and for which experimental support was claimed by BIRCH, DONOVAN and MOEWUS, has now been withdrawn by BIRCH (*25*).

Adequate evidence is therefore available to show (a) that the $B$-ring with the attached $C_3$ group in quercetin, tricin and presumably other flavonoids is built from shikimic acid (and hence from carbohydrates) through several possible intermediates, and (b) that the $A$-ring is built from acetate units. More extensive biosynthetic experiments employing tracer technique may solve further problems of flavone biogenesis, such as the steps by which the $C_6$—$C_3$ unit is converted to the ultimate $C_{15}$ compound, the mechanism of nuclear hydroxylation, methoxylation, and removal of hydroxyl groups, and the relationship of the flavones to flavonoids at other oxidation levels. From the distribution of hydroxyflavones and their methyl ethers it would appear that direct methoxylation and methylation of hydroxyl groups (perhaps by means of methionine) are both possible; but no biogenetic proof for the direct introduction of methoxyl groups is available.

Mainly on the basis of the occurrence of a flavanone, flavone and isoflavone with the same pattern of methoxyl and hydroxyl groups (sakuranetin, genkwanin and prunetin) in the bark of *Prunus puddum*, GEISSMAN and HINREINER (*69*) have suggested that isoflavones may be formed in plants from a $C_6(B)$—C—C—C— fragment by a WAGNER-MEERWEIN rearrangement. When flavanones are oxidized by lead tetracetate, the acetate of the expected flavanonol is accompanied by the related flavone and isoflavone; ROBERTSON therefore concludes that the formation of both the flavone and isoflavone is more consistent with a radical mechanism than a WAGNER-MEERWEIN rearrangement (*40*). However, there is no biogenetic evidence for the transformation of a flavanone to a flavone or isoflavone; and irrespective of the precise mechanism of the reactions involved, GEISSMAN and HINREINER's view that the $C_6(B)$—C—C$\diagdown^C_C$ unit of an isoflavone may be derived from a $C_6(B)$—C—C—C precursor is unexceptionable.

KING (*124*) has isolated angolensin (CXIII) from the heartwood of *Pterocarpus angolensis*, which also contains the isoflavones muningin (XCVII, p. 43) and prunetin. Angolensin has the $C_{15}$ skeleton of isoflavones and may be regarded as a tetrahydro-derivative of formononetin (CXIV).

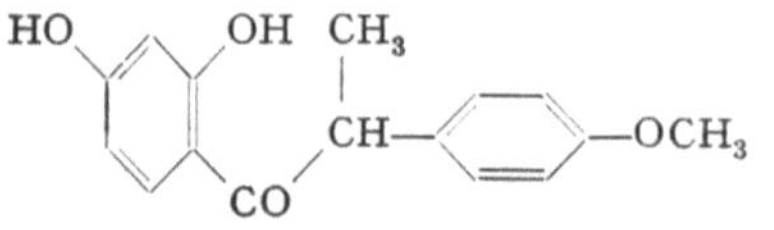

(CXIII.) Angolensin. 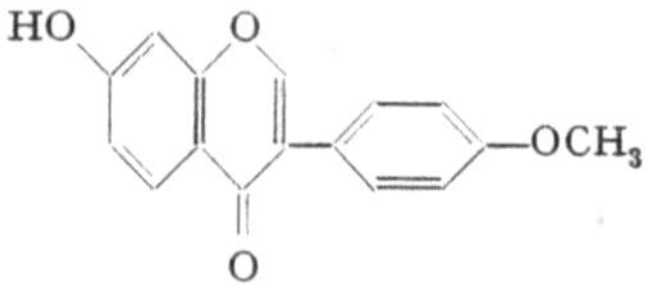       (CXIV.) Formononetin.

Starting from shikimic acid, several routes to isoflavones can be visualised. $p$-Hydroxyphenylpyruvic acid may be converted to $p$-hydroxyphenylacetic acid; a desoxybenzoin may then be formed by combination with acetate units, and cyclization to an isoflavone may take place by the intervention of formate. Alternatively, $p$-hydroxyphenylacetic acid may first formylate, or a 1-phenylpropane product from shikimic acid rearrange, to a hydratropic acid derivative, which then condenses with acetate units and finally cyclizes to an isoflavone.

## XIV. Physiological Properties of Flavones and Isoflavones.

*Flavones.* Because of their wide distribution, flavones have been encountered in many plants reputed to have medicinal properties, but no conclusive evidence has yet been obtained of any specific therapeutic action for which the flavone constituent is responsible.

WILLAMAN (*223*) has reviewed the biological effects of the flavonoids, and has listed thirty-three different manifestations of activity. Using the term "bioflavonoid" to cover those flavonoids which have biological activity, the chemistry, biochemistry, biogenesis, the role of the flavonoids in coumarin anticoagulant therapy, use of ascorbic acid and bioflavonoids in rheumatic fever, the clinical aspects of bioflavonoids and ascorbic acid, and perspectives for the bioflavonoids have been discussed (*152*).

The only flavone which is in clinical use is quercetin, both as the aglycone and the 3-rutinoside (rutin). Quercetin and rutin are employed (frequently in combination with ascorbic acid) because of their "vitamin P" activity in the treatment of conditions in which there is capillary bleeding due to increased capillary fragility (degenerative vascular disease, diabetes, allergic manifestations), although their actual efficacy like that of hesperidin is still controversial. Their use in the treatment of the common cold has been mentioned. Quercetin has been stated to be superior to rutin in the treatment of initial spontaneous capillary fault associated with hypertension, effective at lower dosage, and capable of correcting capillary fault in cases not responding to rutin therapy (*76*). Some pharmaceutical preparations of quercetin and rutin also contain aminophylline and reserpine.

Evidence regarding the mechanism by which flavonoids decrease capillary permeability has been conflicting. The in vitro protection of epinephrine by flavonoids was demonstrated by WILSON and DEEDS (*225*); CLARK and GEISSMAN (*45*) found that flavonoids able to form metal chelates inhibited epinephrine oxidation; among seventy compounds examined, the most potent was 3,3',4'-trihydroxyflavone which was 16 times as active as rutin. The property of potentiating the action of epinephrine could not be correlated with protective action against capillary fragility. A direct vasoconstrictor action on capillaries remains a possibility.

When rutin or quercetin is given orally to rabbits, substantial amounts are absorbed, metabolized, and largely excreted in the urine; evidence for the excretion of at least four compounds was obtained, and 3,4-dihydroxyphenylacetic acid was isolated in about 25% yield (*159*). Since rutin and quercetin are usually administered by the oral route, this evidence of their absorption from the gastro-intestinal tract is important. The formation of 3,4-dihydroxyphenylacetic acid (instead of protocatechuic acid as in chemical degradation) probably involves the reduction of quercetin to the flavanonol (taxifolin) as an intermediate stage. In later work it was found that taxifolin, as well as quercetin and 3,4-dihydroxyphenyl-alanine, led to 3,4-dihydroxyphenylacetic acid, its 3-methyl ether, and *m*-hydroxy-phenylacetic acid in the urine, when fed to rats, rabbits or humans (*29*).

Flavonols containing two *o*- or *p*-hydroxyl groups in the 2-phenyl ring have anti-oxidant properties; free hydroxyls in the 5,7-positions have a pro-oxidant effect (*87, 134, 208*). Quercetin is about equal to propyl gallate in inhibiting the aerobic oxidation of ethyl linoleate, and rhamnetin (the 7-methyl ether) is more effective. Gossypetin is about thrice as active as propyl gallate. Several synthetic hydroxyflavones (e. g. 3,7,2',5'-tetrahydroxyflavone and 3,7,8,2',5'-pentahydroxyflavone) greatly exceed propyl gallate in activity.

The Chinese drug "Yuen-hua", which contains apigenin and genkwanin, is believed to be a diuretic and anthelminthic. KOIKE (*128*) observed that the diuretic action of flavones increased with an increasing number of hydroxyl groups. This was confirmed by FUKUDA (*62*) who found further that flavones and their glycosides were cardiac stimulants and vasoconstrictors. In view of the diuretic action of "digitoflavone" (luteolin) isolated from *Digitalis purpurea*, WAGNER and LUCK (*217*) have assumed that rutin and kaempferol 3-rhamnoglucoside present in *Nerium oleander* take part in its overall pharmacological action.

Calycopterin, first reported to be toxic to roundworms, was later found to be devoid of anthelminthic activity (*140*).

Among other effects of some flavones which have been mentioned (*223*) are prevention of anaphylactic shock; protection against x-rays and other radiation injury; cure of frostbite; hypotensive, hypertensive, bacteriostatic and bactericidal activity.

Most flavones inhibit smooth-muscle movements, and lucerne contains a flavone with this property which has been identified as tricin (*57*). The remarkably potent and highly specific gymnotermone effect of isorhamnetin on the alga *Chlamydomonas* reported by KUHN (*131*) and the suggestion of MOEWUS (*158*) that isorhamnetin and the anthocyanin peonin are sex-determining hormones in *C. eugametos* have not been confirmed by later work (*81*).

A valuable characteristic of the flavones is their lack of toxicity, which justifies further investigation of their pharmacology and possible therapeutic applications.

*Isoflavones.* BRADBURY and WHITE (*35*) have shown that the estrogenic properties of subterranean clover *(Trifolium subterraneum)*, which are of great importance for sheep breeding in the drier areas of Western Australia, are mainly due to genistein, although the estrogenic activity is only about $10^{-5}$ of that of estrone. Since soybean oil meal is used in livestock feeding, its isoflavone content and estrogenic activity have been studied [CHENG et al. (*152*)]; genistin, genistein 7-glucoside, was isolated in about 0.1% yield, and it was found that genistein has about 1/50000 the estrogenic potency of diethylstilbestrol. BRADBURY and WHITE (*36*) regard a 5-hydroxyl group as essential for activity; but CHENG (*152*) found that daidzein is somewhat more active as an estrogen than genistein (and biochanin which is approximately equivalent), although formononetin is much less active. A 2-alkyl substituent greatly reduces the activity, probably as a result of distortion of the coplanarity of the 3-phenyl ring with the chromone ring, which is indicated by models and by the hypsochromic shift in the ultraviolet absorption of an isoflavone by the introduction of a 2-alkyl group (*36*). The structural requirements for estrogenic activity in isoflav-3-ens are more analogous to the stilbestrol series, a 2- or 4-alkyl substituent greatly increasing the activity (*36*). 4-Ethyl-7,4'-dimethoxy-2-methylisoflav-3-en, which has the carbon skeleton of stilbestrol dimethyl ether, has estrogenic activity of the same order (*133*).

# XV. Tables.

Table 1. Visible Color and $\lambda_{max}$ of Methanolic Solutions of some Flavones after Reduction with Mg and HCl ($\lambda_{max}$ of the unreduced solutions is given in parentheses).

| Flavone | Position of OH groups | Color | $\lambda_{max}$ in m$\mu$ |
|---|---|---|---|
| Luteolin......... | 5,7,3′,4′ | Orange | 495 (350) |
| Fisetin ........... | 3,7,3′,4′ | Reddish orange | 509 (360) |
| Kaempferol....... | 3,5,7,4′ | Reddish orange | 513 (367) |
| Morin........... | 3,5,7,2′,4′ | Red | 513 (357) |
| Quercetin........ | 3,5,7,3′,4′ | Red | 526 (372) |
| Myricetin........ | 3,5,7,3′,4′,5′ | Red-Violet | 542 (380) |

Table 2. Trihydroxyflavones and their Methyl Ethers.

| Name | Substitution | Occurrence | References Isolation, structure | Synthesis |
|---|---|---|---|---|
| Baicalein | 5,6,7-(OH)$_3$ | Root of *Scutellaria baicalensis* as the 7-glucuronic acid, root bark of *Oroxylum indicum* as the 6-glucoside | (*154*) | (*154*) |
| Oroxylin-A | 5,7-(OH)$_2$-6-OCH$_3$ | Root bark of *Oroxylum indicum* | (*154*) | (*162*) |
| Apigenin | 5,7,4′-(OH)$_3$ | Parsley *(Carum petroselinum)*, *Dahlia variabilis* and other plants, free and as glycosides and glucuronide | (*174*) | (*174*) |
| Acacetin | 5,7-(OH)$_2$-4′-OCH$_3$ | *Robinia pseudacacia, Linaria vulgaris*, free and as glycosides | (*174*) | (*154*) |
| Genkwanin | 5,4′-(OH)$_2$-7-OCH$_3$ | *Daphne genkwa* blossoms, bark of *Prunus* species, free and as the 5-glucoside | (*154*) | (*154*) |
| | 5-OH-7,4′-(OCH$_3$)$_2$ | Birch buds | (*154*) | (*174*) |
| Galangin | 3,5,7-(OH)$_3$ | Rhizome of *Alpinia officinalis* | (*174*) | (*174*) |
| | 5,7-(OH)$_2$-3-OCH$_3$ | Rhizome of *Alpinia officinalis* | (*174*) | (*154*) |
| Izalpinin | 3,5-(OH)$_2$-7-OCH$_3$ | Seeds of *Alpinia japonica, A. chinensis*; heartwood of *Pinus griffithii* | (*154*) | (*154*) |
| Wogonin | 5,7-(OH)$_2$-8-OCH$_3$ | Root of *Scutellaria baicalensis* | (*154*) | (*154*) |

4*

Table 3. Tetrahydroxyflavones and their Methyl Ethers.

| Name | Substitution | Occurrence | References Isolation, structure | References Synthesis |
|---|---|---|---|---|
| Scutellarein | 5,6,7,4'-(OH)$_4$ | Leaves and flowers of *Scutellaria altissima* as the glucuronide, and in other plants | (174) | (154) |
| Pectolinarigenin | 5,7-(OH)$_2$-6,4'-(OCH$_3$)$_2$ | Flowers of *Linaria vulgaris* as the 7-rutinoside and 7-dirutinoside | (115) | (220) |
| Luteolin | 5,7,3',4'-(OH)$_4$ | *Reseda luteola* (weld) and other plants, free and as glycosides | (174) | (174) |
| Chrysoeriol | 5,7,4'-(OH)$_3$-3'-OCH$_3$ | Leaves of *Eriodictyon glutinosum* and in celery seeds, free and as the 7-apioglucoside | (154) | (154) |
| Diosmetin | 5,7,3'-(OH)$_3$-4'-OCH$_3$ | *Hyssopus officinalis*, *Dahlia variabilis* and other plants as the 7-rutinoside | (154) | (154) |
| Artocarpetin | 5,2',4'-(OH)$_3$-7-OCH$_3$ | Heartwood of *Artocarpus integrifolia* | (47) | (47) |
| Kaempferol | 3,5,7,4'-(OH)$_4$ | *Delphinium consolida* flowers and many other plants, free and as glycosides | (174) | (174) |
| Kaempferide | 3,5,7-(OH)$_3$-4'-OCH$_3$ | Rhizome of *Alpinia officinarum* | (174) | (154) |
| Rhamnocitrin | 3,5,4'-(OH)$_3$-7-OCH$_3$ | Fruit of *Rhamnus cathartica* | (174) | (183) |
| Datiscetin | 3,5,7,2'-(OH)$_4$ | Leaves and root of *Datisca cannabina* and in *Paeonia albiflora* as the rutinoside | (174) | (154) |
| Ptaeroxylol | 3,5,7-(OH)$_3$-2'-OCH$_3$ | Bark of *Ptaeroxylon obliquum* as the glucoside | (176) | |
| Fisetin | 3,7,3',4'-(OH)$_4$ | Wood of *Rhus cotinus* as a glucoside-tannic acid, and in other plants | (174) | (174) |
| Desmethoxykanugin | 3,7-(OCH$_3$)$_2$-3',4'-CH$_2$O$_2$ | Bark of *Pongamia glabra* | (157) | (157) |
| Pratoletin | 3,5,8(or 6),4'-(OH)$_4$ | Flowers of *Trifolium pratense* | (84) | |

Table 4. Quercetin and its Derivatives.

| Name | Substitution | Occurrence | Isolation, structure | Synthesis |
|---|---|---|---|---|
| Quercetin | 3,5,7,3',4'-(OH)₅ | Numerous plants, free and as glycosides | (174) | (174) |
| Quercitrin | 3-Rhamnoside | Quercus tinctoria and other plants | (115) | |
| | ?-Rhamnoside | Leaves of Vaccinium myrtillus | (99) | (98) |
| Isoquercitrin | 3-Glucoside | Cotton flowers, leaves of Rhododendron flavum, and in other plants | (115) | |
| Quercimeritrin | 7-Glucoside | Cotton flowers, Prunus emarginata, etc. | (115) | |
| Spiraeosid | 4'-Glucoside | Flowers of Spiraea ulmaria L. | (115) | |
| Incarnatrin | ?-Glucoside | Trifolium incarnatum | (115) | |
| Hyperin (Hyperoside) | 3-Galactoside | Hypericum perforatum and other plants | (115) | |
| Rutin | 3-Rutinoside | Buckwheat (Fagopyrum esculentum), Sophora japonica and many other plants | (115) | |
| Meratin | 3-Diglucoside | Meratia praecox | (86) | |
| Avicularin | 3-Arabinoside | Polygonum aviculare | (115) | |
| Renoutrin | 3-Xyloside | Leaves of Polygonum reynoutria | (163) | |
| | Triglucoside | Tea leaves | (115) | |
| | 3-Rhamnodiglucoside | Tea leaves | (115) | |
| Quercituron | ?-Glucuronide | Leaves of Phaseolus vulgaris | (115) | |
| Azaleatin | 5-Methyl ether | Flowers of Rhododendron mucronatum G. Don as the 3-rhamnoside | (216) | (130) |
| Rhamnetin | 7-Methyl ether | Persian berries (Rhamnus species), free and as the 3-trirhamnoside | (174) | (130) |
| Isorhamnetin | 3'-Methyl ether | Asbarg (flowers of Delphinium zalil), senna leaves, and other plants, free, as the 3-potassium sulphate (persicarin), 3,4'-diglucoside, 3-rutinoside and 3-glucorhamnogalactoside | (174) | (154) |
| Tamarixetin | 4'-Methyl ether | Leaves of Tamarix troupii | (78) | (78) |
| Rhamnazin | 7,3'-Dimethyl ether | Rhamnus infectorius, Polygonum hydropiper, free and as the 3-potassium sulphate | (174, 89) | (130) |
| Ombuin | 7,4'-Dimethyl ether | Leaves of Phytolacca dioica as the 3-rutinoside | (150) | (49) |
| Ayanin | 3,7,4'-Trimethyl ether | Heartwood of Distemonanthus Benthamianus | (120) | (120) |
| Pinoquercetin | 6-Methyl | Bark of Pinus ponderosa | (132) | (145) |

Table 5. Pentahydroxyflavones other than Quercetin and their Methyl or Methylene Ethers.

| Name | Substitution | Occurrence | References Isolation, structure | Synthesis |
|---|---|---|---|---|
| Tricin | $5,7,4'$-$(OH)_3$-$3',5'$-$(OCH_3)_2$ | Khapli wheat (*Triticum dicoccum*) | (154) | (154) |
| Tangeretin (Ponkanetin) | $5,6,7,8,4'$-$(OCH_3)_5$ | Rind of *Citrus nobilis* var. *deliciosa* | (154, 74) | (195) |
| Morin | $3,5,7,2',4'$-$(OH)_5$ | Wood of *Artocarpus integrifolia, Chlorophora tinctoria, Maclura braziliensis, M. pomifera* | (174) | (154) |
| Robinetin | $3,7,3',4',5'$-$(OH)_5$ | Wood of *Robinia pseudacacia, Gleditschia monosperma* | (154) | (154) |
| Kanugin | $3,7,3'$-$(OCH_3)_3$-$4',5'$-$CH_2O_2$ | Seeds, flowers and root bark of *Pongamia glabra* | (177) | (182) |
| Penduletin | $5,4'$-$(OH)_2$-$3,6,7$-$(OCH_3)_3$ | *Brickelia pendula* | (59) | (60) |
| Herbacetin | $3,5,7,8,4'$-$(OH)_5$ | Cotton flowers (*Gossypium indicum, G. herbaceum*) as the 7-glucoside | (154) | (154) |
| Tambuletin | $3,5,7,4'$-$(OH)_4$-$8$-$OCH_3$ | Seeds of *Xanthoxylum acanthopodium* | (13) | (13) |
| Tambulin | $3,5$-$(OH)_2$-$7,8,4'$-$(OCH_3)_3$ | Seeds of *Xanthoxylum acanthopodium* | (154) | (14) |
| Flindulatin | $5$-$OH$-$3,7,8,4'$-$(OCH_3)_4$ | *Flindersia maculosa* | (39) | (39) |
| Auranetin | $3,6,7,8,4'$-$(OCH_3)_5$ | Rind of *Citrus aurantium* | (160) | (18) |
| Melanoxetin | $3,7,8,3',4'$-$(OH)_5$ | Heartwood of *Acacia melanoxylon* | (117) | |

Table 6. Hexahydroxyflavones and their Derivatives.

| Name | Substitution | Occurrence | Isolation, structure | Synthesis |
|---|---|---|---|---|
| Nobiletin | 5,6,7,8,3',4'-(OCH₃)₆ | Rind of *Citrus nobilis* | (154) | (95) |
| Acrammerin | 5,7,3',4,5'-(OH)₅-8-OCH₃ (?) | Pods of *Gleditschia triacanthos* | (66) | (66) |
| Myricetin | 3,5,7,3',4',5'-(OH)₆ | *Myrica rubra* and other plants, free and as the 3'-glucoside, 3-rhamnoside, and galactoside | (174) | (154) |
| Gossypetin | 3,5,7,8,3',4'-(OH)₆ | Flowers of *Gossypium* and *Hibiscus* species, free and as the 7- and 8-glucoside | (174) | (154) |
| Limocitrin | 3,5,7,4'-(OH)₄-8,3'-(OCH₃)₂ | *Citrus* rind | (96) | |
| Ternatin | 5,4'-(OH)₂-3,7,8,3'-(OCH₃)₄ | Bark of *Melicope ternata* | (38 a) | (38 c) |
| Wharangin | 5,3',4'-(OH)₃-3-(OCH₃)-7,8-CH₂O₂ | Bark of *Melicope ternata* | (38 a) | |
| Meliternin | 3,5,7,8-(OCH₃)₄-3',4'-CH₂O₂ | Bark of *Melicope ternata* | (38 a) | (38 b) |
| Quercetagetin | 3,5,6,7,3',4'-(OH)₆ | Flowers of *Tagetes patula, T. erecta*, free and as the 7-glucoside | (174) | (154). |
| Patuletin | 3,5,7,3',4'-(OH)₅-6-OCH₃ | Flowers of *Tagetes patula* | (191) | |
| Oxyayanin-B | 5,6,3'-(OH)₃-3,7,4'-(OCH₃)₃ | Heartwood of *Distemonanthus Benthamianus* | (121) | (72) |
| Chrysosplenetin | 3,5,4'-(OH)₃-6,7,3'-(OCH₃)₃ | *Chrysosplenium japonicum*, as the 3- or 5-glucoside | (163) | |
| Artemetin (Artemisetin) | 5-(OH)-3,6,7,3',4'-(OCH₃)₅ | *Artemisia arborescens; A. absinthium* | (43) | (154) |
| Melisimplin | 5-OH-3,6,7-(OCH₃)₃-3',4'-CH₂O₂ | Bark of *Melicope simplex* | (38 d) | (38 e) |
| Melisimplexin | 3,5,6,7-(OCH₃)₄-3',4'-CH₂O₂ | Bark of *Melicope simplex* | (38 d) | (38 e) |
| Meliternatin | 3,5-(OCH₃)₂-6,7-3',4'-(CH₂O₂)₂ | Bark of *Melicope ternata* | (38 a, 38 f) | |
| Calycopterin | 5,4'-(OH)₂-3,6,7,8-(OCH₃)₄ | Leaves of *Calycopteris floribunda, Digitalis Thapsi* | (154, 202) | (201) |
| Oxyayanin-A | 5,2',5'-(OH)₃-3,7,4'-(OCH₃)₃ | Heartwood of *Distemonanthus Benthamianus* | (121) | |
| Pinomyricetin | 3,5,7,3',4',5'-(OH)₆-6-CH₃ | Bark of *Pinus ponderosa* | (132) | (145) |

References

Table 7. Natural Isoflavones.

| Name | Substitution | Occurrence | Isolation, structure | Synthesis |
|---|---|---|---|---|
| | | | References | |
| Daidzein | $7,4'\text{-(OH)}_2$ | Soya bean (*Soja hispida*) as the 7-glucoside, daidzin | (154) | (154) |
| Formononetin | $7\text{-OH-}4'\text{-OCH}_3$ | *Ononis spinosa* as the glucoside, ononin; subterranean and red clover (*Trifolium subterraneum*; *T. pratense*); *Cicer arietinum* | (154, 115) | (154) |
| Pseudobaptigenin | $7\text{-OH-}3',4'\text{-CH}_2\text{O}_2$ | *Baptisia tinctoria* as the 7-rhamnoglucoside, pseudobaptisin | (154) | (154) |
| Maximin | $7\text{-OCH}_2\text{CH=C(CH}_3)_2\text{-}3',4'\text{-CH}_2\text{O}_2$ | Root of *Tephrosia maxima* | (181) | |
| Genistein | $5,7,4'\text{-(OH)}_3$ | *Genista tinctoria* and soya bean as genistin, the 7-glucoside; *Sophora japonica* as sophoricoside, the 4'-glucoside, and sophorabioside, the 4'-rhamnoglucoside; *Trifolium* species | (154, 115) | (154, 9) |
| Prunetin | $5,4'\text{-(OH)}_2\text{-}7\text{-OCH}_3$ | *Prunus* bark as the 4'-glucoside, prunitrin; heartwood of *Pterocarpus angolensis* | (154, 124) | (102) |
| Biochanin-A (Olmelin?) | $5,7\text{-(OH)}_2\text{-}4'\text{-OCH}_3$ | Germinated grain of *Cicer arietinum*; heartwood of *Ferreirea spectabilis*; clover | (31, 115) | (9) |
| Orobol (Norsantal) | $5,7,3',4'\text{-(OH)}_4$ | *Orobus tuberosus* as the glucoside, oroboside; *Lathyrus macrorrhizus* | (154, 188) | (164) |
| Santal | $5,3',4'\text{-(OH)}_3\text{-}7\text{-OCH}_3$ | Santalwood (*Pterocarpus santalinus*); Barwood (*Baphia nitida*) | (188) | (103, 164) |
| Tectorigenin | $5,7,4'\text{-(OH)}_3\text{-}6\text{-OCH}_3$ | Rhizome of *Iris tectorum* Max. and *Belamcanda chinensis* L. as the 7-glucoside, tectorigin | (154, 115) | (111) |
| Muningin | $6,4'\text{-(OH)}_2\text{-}5,7\text{-(OCH}_3)_2$ | Heartwood of *Pterocarpus angolensis* | (123) | |
| Tlatlancuayin | $5,2'\text{-(OCH}_3)_2\text{-}6,7\text{-CH}_2\text{O}_2$ | *Iresine celosioides* L. | (46) | (113) |
| Irigenin | $5,7,3'\text{-(OH)}_3\text{-}6,4',5'\text{-(OCH}_3)_3$ | Root of *Iris florentina, germanica, pallida*; root of *Belamcanda chinensis*, as iridin, the 7-glucoside | (154, 115) | |

Table 8. Ultraviolet Absorption Spectra of some Naturally Occurring Flavones in Ethanol.

| Name | $\lambda_{max}$, m$\mu$ | log $\varepsilon$ | $\lambda_{max}$, m$\mu$ | log $\varepsilon$ |
|---|---|---|---|---|
| Flavone | 297.5 | 4.20 | 250 | 4.07 |
| Chrysin | 318 | 4.08 | 270 | 4.45 |
| Tectochrysin | 330 | 3.88 | 270 | 4.40 |
| Baicalein | 324 | 4.18 | 276 | 4.42 |
| Apigenin | 340 | 4.31 | 265 | 4.25 |
| Galangin | 360 | 4.07 | 267.5 | 4.23 |
| Scutellarein | 339 | 4.26 | 286 | 4.22 |
| Luteolin | 355 | 4.28 | 258 | 4.22 |
| Artocarpetin | 355 | 4.42 | 286 | 4.01 |
|  |  |  | 264 | 4.28 |
|  |  |  | 254 | 4.26 |
| Kaempferol | 370 | 4.28 | 310 | — |
|  |  |  | 267.5 | 4.12 |
| Datiscetin | 360 | 3.90 | 262.5 | 4.14 |
| Fisetin | 370 | 4.43 | 315 | 4.22 |
|  |  |  | 252.5 | 4.33 |
| Quercetin | 375 | 4.32 | 258 | 4.32 |
| Quercitrin | 352 | 4.24 | 260 | 4.35 |
| Isoquercitrin | 360 | 4.32 | 310 | 4.01 |
|  |  |  | 258 | 4.41 |
| Quercimeritrin | 374 | 4.39 | 257 | 4.42 |
| Hyperin | 362.5 | 4.31 | 312 | 3.97 |
|  |  |  | 258 | 4.38 |
| Rutin | 361 | 4.28 | 310 | 3.96 |
|  |  |  | 258 | 4.35 |
| Rhamnazin | 375 | 3.27 | 255 | 4.37 |
| Morin | 380 | 4.15 | 263 | 4.22 |
| Kanugin | 341 | 4.32 | 320* | 4.25 |
|  |  |  | 237* | 4.32 |
| Flindulatin | 360 | 3.95 | 325 | 4.0 |
|  |  |  | 270 | 4.1 |
| Myricetin | 377 | 4.40 | 308 | 3.92 |
|  |  |  | 258 | 4.32 |
| Ternatin | 368 | 4.28 | 273 | 4.29 |
|  |  |  | 258 | 4.33 |
| Wharangin | 377 | 4.27 | 273 | 4.27 |
|  |  |  | 261 | 4.32 |
| Meliternin | 351 | 4.29 | 273 | 4.27 |
|  |  |  | 253 | 4.36 |
| Quercetagetin | 361 | 4.34 | 272* | 4.15 |
|  |  |  | 259 | 4.34 |
| Melisimplexin | 336 | 4.26 | 235 | 4.27 |
| Meliternatin | 336 | 4.41 | 269* | 4.11 |
|  |  |  | 248 | 4.25 |
| Pongapin | 332.5 | 4.34 | 251.5 | 4.34 |
| Karanjin | 303 | 4.23 | 261 | 4.40 |

* Inflection.

Table 9. Ultraviolet Absorption Spectra of some Naturally Occurring Isoflavones in Ethanol.

| Name | $\lambda_{max},$ m$\mu$ | log $\varepsilon$ | $\lambda_{max},$ m$\mu$ | log $\varepsilon$ |
|---|---|---|---|---|
| Formononetin .......... | 300 | 4.05 | 250 | 4.44 |
| Genistein .............. | 325* | — | 263 | 4.63 |
| Biochanin-A ........... | 325* | — | 263 | — |
| Prunetin .............. | 325* | — | 263 | — |
| Orobol................ | — | — | 287 | — |
| Santal ................ | — | — | 263 | — |
| Tlatlancuayin (in chloro-form) ............... | 320 | 3.86 | 245 | 4.34 |
|  |  |  | 278* | 4.12 |

\* Inflection.

Table 10. Hydroxyl and Carbonyl Frequencies of some Naturally Occurring Flavonoids.

| Compound | Hydroxyl, cm.$^{-1}$ | Carbonyl, cm.$^{-1}$ |
|---|---|---|
| *Flavones.* | | |
| Flavone.................................. | — | 1660 |
| Apigenin 7-rhamnoglucoside ............... | 3330 | 1658 |
| Apiin.................................... | 3240 | 1660 |
| Luteolin ................................ | 3220 | 1655 |
| Luteolin 7-glucoside ...................... | 3160 | 1658 |
| Pectolinarin ............................. | 3320 | 1658 |
| *Flavonols.* | | |
| Chrysosplenetin........................... | 3320 | 1658 |
| Chrysosplenin ........................... | 3300 | 1658 |
| Dactylin (Isorhamnetin 3,4'-diglucoside)...... | 3220 | 1655 |
| Isorhamnetin............................. | 3160 | 1655 |
| Quercetin ............................... | 3340 | 1655 |
| Quercitrin .............................. | 3280 | 1655 |
| Reynoutrin.............................. | 3140 | 1652 |
| Robinin................................. | 3240 | 1655 |
| Rutin .................................. | 3300 | 1655 |
| *Flavanones.* | | |
| Flavanone ............................... | — | 1680 |
| Hesperidin .............................. | 3340 | 1639 |
| Naringin................................. | 3360 | 1639 |

## References.

1. AHLUWALIA, V. K., N. R. KRISHNASWAMI, S. K. MUKERJEE, V. V. S. MURTI, T. R. SESHADRI and C. VENKATARAMANI: The Factors that Affect Alkali Colour Reactions of Flavonols: a Study of Flavonols of Uncommon Types. Proc. Indian Acad. Sci. 47 A, 230 (1958). See this paper for earlier references.
2. AHLUWALIA, V. K., S. K. MUKERJEE and T. R. SESHADRI: Nuclear Oxidation of Flavones and Related Compounds. Synthesis of Gardenin. J. Chem. Soc. (London) 1954, 3988.
3. ALGAR, J. and J. P. FLYNN: A New Method for the Synthesis of Flavonols. Proc. Roy. Irish Acad. 42 B, 1 (1934).
4. ALLAN, J. and R. ROBINSON: An Accessible Derivative of Chromonol. J. Chem. Soc. (London) 125, 2192 (1924); 1926, 2334.
5. ANAND, N., R. N. IYER and K. VENKATARAMAN: Synthetical Experiments in the Chromone Group. XXIII. A New Synthesis of Rhamnazin and a Synthesis of 3,4'-Dihydroxy-7-methoxyflavone. Proc. Indian Acad. Sci. 29 A, 203 (1949).
6. ARCOLEO, A., A. BELLINO and P. VENTURELLA: Synthesis of Hydroxychalcone and Hydroxyflavone Derivatives. I. Ann. chim. (Rome) 47, 66 (1957).
7. ARONOFF, S.: Some Structural Interpretations of Flavone Spectra. J. Organ. Chem. (USA) 5, 561 (1940).
8. BAKER, W.: Molecular Rearrangement of some o-Acyloxyacetophenones and the Mechanism of the Production of 3-Acylchromones. J. Chem. Soc. (London) 1933, 1381.
9. BAKER, W., J. CHADDERTON, J. B. HARBORNE and W. D. OLLIS: A New Synthesis of Isoflavones. I. J. Chem. Soc. (London) 1953, 1852.
10. BAKER, W., I. DUNSTAN, J. B. HARBORNE, W. D. OLLIS and R. WINTER: Orientation in the Isoflavone Series. Chem. and Ind. 1953, 277.
11. BAKER, W., R. HEMMING and W. D. OLLIS: The Structures of Buddleoflavonoloside (Linarin) and Buddleoflavonol (Acacetin). J. Chem. Soc. (London) 1951, 691.
12. BAKER, W. and W. D. OLLIS: Developments in the Synthesis of Isoflavones. Sci. Proc. Roy. Dublin Soc. 27, No. 6, 119 (1956).
13. BALAKRISHNA, K. J. and T. R. SESHADRI: Colouring Matter of Tambul Seeds. I, II, III, IV. Proc. Indian Acad. Sci. 25 A, 449 (1947); 26 A, 72, 214, 234 (1947).
14. — — Synthesis of 7,8,4'-O-Trimethyl-herbacetin Considered to be Identical with Tambulin. Proc. Indian Acad. Sci. 26 A, 296 (1947).
15. BALASUBRAMANIAN, S. K., S. NEELAKANTAN and T. R. SESHADRI: Synthetic Experiments in the Benzopyrone Series. LII. Constitution of Acrammerin. J. Sci. Industr. Res. (India) 14 B, 6 (1955).
16. BANNERJEE, N. R. and T. R. SESHADRI: Mechanism of Selenium Dioxide Dehydrogenation of Flavanones. Current Sci. (India) 25, 143 (1956).
17. BAPAT, D. S. and K. VENKATARAMAN: Potential Antitubercular Compounds. III. 7-Aminoflavone. Proc. Indian Acad. Sci. 42 A, 336 (1955).
18. BARGELLINI, G. und A. OLIVERIO: Flavon-Derivate des 1,2,3,4-Tetraoxybenzols. Ber. dtsch. chem. Ges. 75, 2083 (1942).
19. BATE-SMITH, E. C.: The Commoner Phenolic Constituents of Plants and their Systematic Distribution. Sci. Proc. Roy. Dublin Soc. 27, No. 6, 165 (1956).
20. BATE-SMITH, E. C., T. SWAIN and G. S. POPE: The Isolation of 7-Hydroxy-4'-methoxyisoflavone (Formononetin) from Red Clover (Trifolium pratense L.) and a Note on the Identity of Pratol. Chem. and Ind. 1953, 1127.

21. Bate-Smith, E. C. and R. G. Westall: Chromatographic Behaviour and Chemical Structure. I. Some Naturally Occurring Phenolic Substances. Biochim. Biophys. Acta 4, 427 (1950).
22. Bharadwaj, G. K. and K. Venkataraman: The Action of Aluminium Chloride on Polymethoxyflavones. Current Sci. (India) 2, 50 (1933).
23. Bhullar, A. S. and K. Venkataraman: Synthetical Experiments in the Chromone Group. II. 1,4-α-Naphthapyrones. J. Chem. Soc. (London) 1931, 1165.
24. Bielig, H. J.: Die Farbstoffe der Sanddornbeere. Ber. dtsch. chem. Ges. 77, 748 (1944).
25. Birch, A. J.: Biosynthetic Relations of Some Natural Phenolic and Enolic Compounds. Fortschr. Chem. organ. Naturstoffe 14, 186 (1957).
26. Blasdale, W. C.: The Composition of the Solid Secretion Produced by Primula denticulata. J. Amer. Chem. Soc. 67, 491 (1945).
27. Bognár, R. and V. Szabó: Synthesis of "Sophoricoside", one of the Glycosides of Sophora japonica L. Chem. and Ind. 1954, 518.
28. Bogorad, L.: The Biogenesis of Flavonoids. Annu. Rev. Plant Physiol. 9, 417 (1958).
29. Booth, A. N. and F. DeEds: The Toxicity and Metabolism of Dihydroquercetin. J. Amer. Pharmaceut. Assoc., Sci. Ed. 47, 183 (1958).
30. Bose, J. L.: A Note on the Possible Identity of Biochanin A and Pratensol. J. Sci. Industr. Res. (India) 15 B, 324 (1956).
31. Bose, J. L. and S. Siddiqui: Studies in the Constituents of Chana (Cicer arietinum): IV. The Identity of Biochanin B and Formononetin. J. Sci. Industr. Res. (India) 10 B, 291 (1951) and earlier papers.
32. Bose, P. K.: On a New Method for the Detection of some Polyhydroxyphenols. J. Indian Chem. Soc., Sir P. C. Ray Commemoration Vol., 1933, 65.
33. — Natural Flavones. V. Further Observations on the Constitution of Gardenin and Tambulin. J. Indian Chem. Soc. 22, 233 (1945).
34. Bose, P. K. and R. Nath: Natural Flavones. I. The Constitution of Gardenin. J. Indian Chem. Soc. 15, 139 (1939).
35. Bradbury, R. B. and D. E. White: The Chemistry of Subterranean Clover. I. Isolation of Formononetin and Genistein. J. Chem. Soc. (London) 1951, 3447.
36. — — The Chemistry of Subterranean Clover. II. Synthesis and Reduction of Isoflavones related to Genistein and Formononetin. J. Chem. Soc. (London) 1953, 871.
37. Briggs, L. H. and R. C. Cambie: The Extractives of Vitex lucens. I. Tetrahedron 3, 269 (1958).
38. Briggs, L. H. and R. H. Locker: Chemistry of New Zealand Melicope Species. I–VII. J. Chem. Soc. (London) 1949, (a) 2157, (b) 2162; 1950, (c) 864, (d) 2376, (e) 2379; 1951, (f) 3131, (g) 3136.
39. Brown, R. F. C., P. T. Gilham, G. K. Hughes and E. Ritchie: The Chemical Constituents of Australian Flindersia Species. V. The Constituents of Flindersia maculosa Lindl. Austral. J. Chem. 7, 181 (1954).
40. Carill, G. W. K., F. M. Dean, A. McGookin, (Miss) B. M. Marshall and A. Robertson: The Oxidation of Chromanones and Flavanones with Lead Tetra-acetate. J. Chem. Soc. (London) 1954, 4573.
41. Carruthers, (Mrs.) W. R., R. H. Farmer and R. A. Laidlaw: Dihydromorin from East African Mulberry (Morus lactea Mildbr.). J. Chem. Soc. (London) 1957, 4440.
42. Casteel, H. W. and S. H. Wender: Identification of Flavonoid Compounds by Filter Paper Chromatography. Analyt. Chemistry 25, 508 (1953).

43. ČEKAN, Z. and V. HEROUT: Isolation of 5-Hydroxy-3,6,7,3',4'-pentamethoxy-flavone from *Artemisia absinthium*. Chem. Listy **49**, 1053 (1955).

44. CHEN, F. C. and C. H. LIN: Ultraviolet Absorption of Flavone Series: Spectra of Halogenoflavones. J. Formosan Sci. **6**, 81 (1952).

45. CLARK, W. G. and T. A. GEISSMAN: Potentiation of Effects of Adrenaline by Flavonoid ("Vitamin P"-like) Compounds. Relation of Structure to Activity. J. Pharmacol. exp. Therapeut. **95**, 363 (1949).

46. CRABBÉ, P., P. R. LEEMING and C. DJERASSI: Naturally Occurring Oxygen Heterocyclics. III. The Structure of the Isoflavone Tlatlancuayin. J. Amer. Chem. Soc. **80**, 5258 (1958).

47. DAVE, K. G., S. A. TELANG and K. VENKATARAMAN: in press.

48. DAVE, K. G. and K. VENKATARAMAN: The Colouring Matters of the Wood of *Artocarpus integrifolia*: I. Artocarpin. J. Sci. Industr. Res. (India) **15 B**, 183 (1956).

49. DEULOFEU, V. e N. SCHOPFLOCHER: I glucosidi flavonici dell'Ombu (*Phytolacca dioica* LINN.). III. La sintesi dell'ombuina e di altri derivati flavonici della serie isovainiglica. Gazz. chim. ital. **83**, 449 (1953).

50. DOPORTO, M. L., K. M. GALLAGHER, J. E. GOWAN, A. C. HUGHES, (Mrs.) E. M. PHILBIN, T. SWAIN and T. S. WHEELER: Rearrangement in the Demethylation of 2'-Methoxyflavones. II. Further Experiments and the Determination of the Composition of Lotoflavin. J. Chem. Soc. (London) **1955**, 4249.

51. DUNNE, A. T. M., J. E. GOWAN, J. KEANE, B. M. O'KELLY, D. O'SULLIVAN, M. M. ROCHE, P. M. RYAN and T. S. WHEELER: Thermal Cyclization of *o*-Aroyl-oxyacetoarones. A New Synthesis of Flavones. J. Chem. Soc. (London) **1950**, 1252.

52. DUTTA, N. L.: Chemical Investigation of *Mundulea suberosa* BENTH. II. Constitution of Munetone, the Principal Crystalline Product of the Root Bark. J. Indian Chem. Soc. (in press).

53. ERDTMAN, H.: Flavonoid Heartwood Constituents of Conifers. Sci. Proc. Roy. Dublin Soc. **27**, No. 6, 129 (1956).

54. EVANS, W. H., A. McGOOKIN, L. JURD, A. ROBERTSON and W. R. N. WILLIAM-SON: Vitexin. I. J. Chem. Soc. (London) **1957**, 3510.

55. FARKAS, L.: Eine neue Isoflavonsynthese. Chem. Ber. **90**, 2940 (1957).

56. FEINSTEIN, L. and M. JACOBSON: Insecticides Occurring in Higher Plants. Fortschr. Chem. organ. Naturstoffe **10**, 423 (1953).

57. FERGUSON, W. S., DE B. ASHWORTH and R. A. TERRY: Identity of a Muscle-inhibiting Flavone in Lucerne. Nature (London) **166**, 116 (1950).

58. FITZGERALD, D. M., J. F. O'SULLIVAN, (Mrs.) E. M. PHILBIN and T. S. WHEELER: Ring Expansion of 2-Benzylidenecoumaran-3-ones. A Synthesis of Flavones. J. Chem. Soc. (London) **1955**, 860.

59. FLORES, S. E. and J. HERRÁN: The Structure of Pendulin and Penduletin: A New Flavonol Glucoside Isolated from *Brickelia pendula*. Tetrahedron **2**, 308 (1958).

60. FLORES, S. E., J. HERRÁN and H. MENCHACA: The Synthesis of Penduletin. Tetrahedron **4**, 132 (1958).

61. FORD, E. B.: Studies in the Chemistry of Pigments in the Lepidoptera with Reference to their Bearing on Systematics. 3. The Red Pigments of the Papilionidae. Proc. Roy. Ent. Soc. (London) **19 A**, 92 (1944).

62. FUKUDA, T.: Über die pharmakologische Wirkung der Flavonverbindungen. Arch. exp. Pathol. Pharmakol. **164**, 585 (1932).

63. FUNAOKA, K. and M. TANAKA: Flavonoids of *Zelkowa serrata* Wood. Mokuzai Gakkaishi **3**, 144 (1957).

64. Gage, T. B., C. D. Douglass and S. H. Wender: Identification of Flavonoid Compounds by Filter Paper Chromatography. Analyt. Chemistry 23, 1582 (1951).

65. Gakhokidze, A. M.: Synthesis of Olmelin. J. Appl. Chem. (USSR) 23, 559 (1950).

66. Gakhokidze, A. M. and N. D. Kutidze: Pigments of Gleditschia triacanthos. J. Appl. Chem. (USSR) 20, 899, 904 (1947).

67. Geissman, T. A.: Anthocyanins, Chalcones, Aurones, Flavones and Related Water-Soluble Plant Pigments. In: K. Paech and M. V. Tracey, Modern Methods of Plant Analysis, Vol. III, p. 450. Berlin: Springer-Verlag. 1955.

68. Geissman, T. A. and T. K. Fukushima: Flavanones and Related Compounds. V. The Oxidation of 2'-Hydroxychalcones with Alkaline Hydrogen Peroxide. J. Amer. Chem. Soc. 70, 1686 (1948).

69. Geissman, T. A. and E. Hinreiner: Theories of the Biogenesis of Flavonoid Compounds. I and II. Botan. Rev. 18, 77, 165 (1952).

70. Geissman, T. A. and H. Lischner: Flavanones and Related Compounds. VII. The Formation of 4,6,3',4'-Tetrahydroxy-2-benzylcoumaranone-3 by the Sodium Hydrosulphite Reduction of Quercetin. J. Amer. Chem. Soc. 74, 3001 (1952).

71. Geissman, T. A. and T. Swain: Biosynthesis of Flavonoid Compounds in Higher Plants. Chem. and Ind. 1957, 984.

72. Goel, R. N., A. C. Jain and T. R. Seshadri: Synthesis of Oxyayanin B. J. Chem. Soc. (London) 1956, 1369.

73. Goel, R. N. and T. R. Seshadri: New Synthesis of Tamarixetin, Alpinone and Izalpinin. Proc. Indian Acad. Sci. 47 A, 191 (1958).

74. Goldsworthy, L. J. and R. Robinson: A Correction respecting the Structure of Tangeretin. Chem. and Ind. 1957, 47.

75. Gowan, J. E., M. F. Lynch, N. S. O'Connor, (Mrs.) E. M. Philbin and T. S. Wheeler: The Synthesis of Isoflavones. J. Chem. Soc. (London) 1958, 2495.

76. Griffith, J. Q., C. F. Krewson and J. Naghski: Rutin and Related Flavonoids. Chemistry. Pharmacology. Clinical Applications. Easton, Pa.: Mack Publ. Co. 1955.

77. Guider, (Miss) J. M., T. H. Simpson and D. B. Thomas: Anthoxanthins. II. Derivatives of Katuranin and Kaempferol. J. Chem. Soc. (London) 1955, 170.

78. Gupta, S. R. and T. R. Seshadri: Survey of Anthoxanthins. VI. Colouring Matter of Tamarix troupii. Constitution of the Aglycone and its Synthesis. J. Chem. Soc. (London) 1954, 3063.

79. Harborne, J. B.: Use of Alkali and Aluminium Chloride in the Spectral Study of Phenolic Plant Pigments. Chem. and Ind. 1954, 1142.

80. Harper, S. H.: The Active Principles of Leguminous Fish-poison Plants. V. Derris malaccensis and Tephrosia toxicaria. J. Chem. Soc. (London) 1940, 1178.

81. Hartshorne, J. N.: Pigmentation and Sexuality in Plants. Nature (London) 182, 1382 (1958).

82. Hasegawa, M. and T. Shirato: Two New Flavonoid Glycosides from the Leaves of Phellodendron amurense Ruprecht. J. Amer. Chem. Soc. 75, 5507 (1953).

83. Hashimoto, Y., I. Mori and M. Kimura: Paper Electromigration of Flavonoids and Sugars using a High Constant-voltage Current. Nature (London) 170, 975 (1952).

*84.* HATTORI, S., M. HASEGAWA and M. SHIMOKORIYAMA: The Structure of the Glycoside, Trifolin, from the Blossoms of *Trifolium pratense* and Remarks concerning a Yellow Substance accompanying it. Acta Phytochim. (Japan) **13**, 99 (1943).

*85.* HATTORI, S. and H. MATSUDA: Rhoifolin, a New Flavone Glycoside, Isolated from the Leaves of *Rhus succedanea*. Arch. Biochem. Biophys. **37**, 85 (1952).

*86.* HAYASHI, K. and K. OUCHI: Yellow Colouring Matter in the Flower of *Meratia praeco*. Acta Phytochim. (Japan) **15**, 11 (1949).

*87.* HEIMANN, W. und F. REIFF: Zusammenhänge zwischen chemischer Konstitution und antioxygener Wirkung bei Flavonolen. Fette u. Seifen **55**, 451 (1953).

*88.* HÖRHAMMER, L. (private communication). — HÖRHAMMER, L., H. WAGNER und H. GÖTZ: Über das Vorkommen von Flavonen in einheimischen Umbelliferen. Arch. Pharmaz. **291/63**, 44 (1958); see also earlier papers and papers under publication.

*89.* HÖRHAMMER, L. und R. HÄNSEL: Isolierung eines Rhamnazinesters aus *Polygonum hydropiper*. Arch. Pharmaz. **286/58**, 153 (1953).

*90.* — — Zur Analytik der Flavone. II. Über das Komplexbildungsvermögen einiger Oxyflavone und die Konstitution des in *Polygonum hydropiper* L. vorkommenden Rhamnazinesters. Arch. Pharmaz. **286/58**, 425 (1953).

*91.* HÖRHAMMER, L. und K. H. MÜLLER: Zur Analytik der Flavone. VI. Das optische Verhalten der Reduktionsprodukte einiger Polyoxy-flavone. Arch. Pharmaz. **287/59**, 448 (1954).

*92.* HÖRHAMMER, L. und H. WAGNER: Die Gegenstromverteilung in der modernen Pflanzenanalyse. Pharmaz. Ztg. **102**, 278, 537 (1957).

*93.* — — Isolierung, Reinigung und Reinheitsprüfung von Flavonglykosiden durch Gegenstromverteilung. Arch. Pharmaz. **290/62**, 224 (1957).

*94.* HÖRHAMMER, L., H. WAGNER und W. LEEB: Über die Polyamidchromatographie von Flavonen, Flavonolen und deren Glykoside (im Druck).

*95.* HORII, Z.: Synthesis of Nobiletin. J. pharmac. Soc. Japan **60**, 614 (1940).

*96.* HOROWITZ, R. M.: Flavonoids of Citrus. II. Isolation of a New Flavonol from Lemons. J. Amer. Chem. Soc. **79**, 6561 (1957).

*97.* ICE, C. H. and S. H. WENDER: Adsorption Chromatography of Flavonoid Compounds. Analyt. Chemistry **24**, 1616 (1952).

*98.* — — The Synthesis of Isoquercitrin. J. Amer. Chem. Soc. **74**, 4606 (1952).

*99.* — — Quercetin and its Glycosides in Leaves of *Vaccinium myrtillus*. J. Amer. Chem. Soc. **75**, 50 (1953).

*100.* INGLETT, G. E.: Infrared Spectra of some Naturally Occurring Flavonoids. J. Organ. Chem. (USA) **23**, 93 (1958).

*101.* IRIARTE, J., F. A. KINCL, G. ROSENKRANZ and F. SONDHEIMER: The Constituents of *Casimiroa edulis* LLAVE et LEX. II. The Bark. J. Chem. Soc. (London) **1956**, 4170.

*102.* IYER, R. N., K. H. SHAH and K. VENKATARAMAN: Synthetical Experiments in the Chromone Group. XXIV. A Synthesis of Prunetin. Proc. Indian Acad. Sci. **33 A**, 116 (1951).

*103.* — — — Synthetical Experiments in the Chromone Group. XXV. A Synthesis of Santal. Proc. Indian Acad. Sci. **33 A**, 228 (1951).

*104.* IYER, R. N. and K. VENKATARAMAN: Synthetical Experiments in the Chromone Group. XXVII. Coupling of 5-Hydroxyflavone, 5-Hydroxy-6-methoxyflavone and Tectochrysin with Diazotized Aniline. New Synthesis ot 5:6-Dihydroxyflavone, Baicalein, and 5:6:8-Trihydroxyflavone. Proc. Indian Acad. Sci. **37 A**, 629 (1953).

*105.* Jatkar, S. K. K. and B. N. Mattoo: Absorption and Fluorescence Spectra of Flavones and Flavonols. J. Indian Chem. Soc. **33**, 623, 641 (1956).

*106.* Joshi, C. G. and A. B. Kulkarni: Anthoxanthins. V. A Convenient Method for the Synthesis of Flavan-3,4-diols: Synthesis of Flavan-3,4-diols Related to Melacacidin. J. Sci. Industr. Res. (India) **16** B, 307 (1957), and subsequent communications. See these papers for other references to the synthesis of flavandiols.

*107.* Joshi, P. C. and K. Venkataraman: Synthetical Experiments in the Chromone Group. XI. Synthesis of Isoflavone. J. Chem. Soc. (London) **1934**, 513.

*108.* Jurd, L.: A Spectrophotometric Method for the Detection of *o*-Dihydroxyl Groups in Flavonoid Compounds. Arch. Biochem. Biophys. **63**, 376 (1956).

*109.* — Plant Polyphenols. V. Selective Alkylation of the 7-Hydroxyl Group in Polyhydroxyflavones. J. Amer. Chem. Soc. **80**, 5531 (1958).

*110.* Jurd, L. and R. M. Horowitz: Spectral Studies on Flavonols—the Structure of Azalein. J. Organ. Chem. (USA) **22**, 1618 (1957).

*111.* Kagal, S. A., S. S. Karmarkar and K. Venkataraman: Synthetical Experiments in the Chromone Group. XXXII. A Synthesis of Tectorigenin. Proc. Indian Acad. Sci. **44** A, 36 (1956).

*112.* Kariyone, T. and K. Sawada: Complete Publication in Memory of Professor T. Kariyone, p. 16. 1956.

*113.* Karmarkar, S. S., K. H. Shah and K. Venkataraman: Synthetical Experiments in the Chromone Group. XXXI. A Synthesis of Muningin. Proc. Indian Acad. Sci. **41** A, 192 (1955).

*114.* Karrer, P. und G. Schwab: Über ein natürliches Vorkommen des 5-Oxy-flavons. Helv. Chim. Acta **24**, 297 (1941).

*115.* Karrer, W.: Konstitution und Vorkommen der organischen Pflanzenstoffe (exclusive Alkaloide). Basel: Birkhäuser. 1958.

*116.* Kawase, Y., Y. Fujino, Y. Ichioka and K. Fukui: The Synthesis of 4',5,7-Trihydroxy-8-methoxyisoflavone. Bull. Chem. Soc. Japan **30**, 689 (1957).

*117.* King, F. E. and W. Bottomley: The Chemistry of Extractives from Hardwoods. XVII. The Occurrence of a Flavan-3,4-diol (Melacacidin) in *Acacia melanoxylon*. J. Chem. Soc. (London) **1954**, 1399.

*118.* King, F. E., M. F. Grundon and K. G. Neill: The Chemistry of Extractives from Hardwoods. IX. Constituents of the Heartwood of *Ferreirea spectabilis*. J. Chem. Soc. (London) **1952**, 4580.

*119.* King, F. E., T. J. King and L. C. Manning: An Investigation of the Gibbs Reaction and its Bearing on the Constitution of Jacareubin. J. Chem. Soc. (London) **1957**, 563.

*120.* King, F. E., T. J. King and K. Sellars: The Chemistry of Extractives from Hardwoods. V. The Isolation of 3,7,4'-Trimethylquercetin (Ayanin) from the Heartwood of *Distemonanthus Benthamianus*. J. Chem. Soc. (London) **1952**, 92.

*121.* King, F. E., T. J. King and P. J. Stokes: The Chemistry of Extractives from Hardwoods. XIX. The Structures of Further New Flavones Occurring in Ayan *(Distemonanthus Benthamianus)*. J. Chem. Soc. (London) **1954**, 4587.

*122.* — — — The Chemistry of Extractives from Hardwoods. XX. Distemonanthin, a New Type of Flavone Pigment from *Distemonanthus Benthamianus*. J. Chem. Soc. (London) **1954**, 4594.

*123.* King, F. E., T. J. King and A. J. Warwick: The Chemistry of Extractives from Hardwoods. VI. Constituents of Muninga, the Heartwood of *Pterocarpus angolensis*. A.: 6,4'-Dihydroxy-5,7-dimethoxyisoflavone (Muningin). J. Chem. Soc. (London) **1952**, 96.

*124.* KING, F. E., T. J. KING and A. J. WARWICK: The Chemistry of Extractives from Hardwoods. VII. Constituents of Muninga, the Heartwood of *Pterocarpus angolensis.* B: 2,4-Dihydroxyphenyl 1-*p*-Methoxyphenylethyl Ketone (Angolensin). J. Chem. Soc. (London) 1952, 1920

*125.* KING, F. E. and K. G. NEILL: The Chemistry of Extractives from Hardwoods. X. The Constitution of Ferreirin and of Homoferreirin. J. Chem. Soc. (London) 1952, 4752.

*126.* KING, H. G. C. and T. WHITE: The Conversion of Flavonols into Anthocyanidins. J. Chem. Soc. (London) 1957, 3901.

*127.* KLEIN, G. und O. WERNER: Ein Beitrag zur Physiologie und Verbreitung der Flavone. Z. physiol. Chem. (Hoppe-Seyler) 143, 9 (1925).

*128.* KOIKE, H.: Pharmacological Investigation on the Flavone Compounds, particularly on their Diuretic Action. Folia Pharmacol. Japon. 12, 89 (1931).

*129.* KUBOTA, T. and I. ARAI: Studies on Flavanonols. II. Chemical Constitution of Fukugetin. J. Chem. Soc. Japan 76, 1069 (1955).

*130.* KUHN, R. und I. LÖW: Synthese von Rhamnetin und Rhamnazin. Ber. dtsch. chem. Ges. 77, 211 (1944).

*131.* KUHN, R., F. MOEWUS und I. LÖW: Über die pflanzenphysiologische Spezifität von Quercetinderivaten. Ber. dtsch. chem. Ges. 77, 219 (1944).

*132.* KURTH, E. F., V. RAMANATHAN and K. VENKATARAMAN: The Colouring Matters of *Ponderosa* Pine Bark: I. J. Sci. Industr. Res. (India) 15 B, 139 (1956).

*133.* LAWSON, W.: Oestrogenic Activity of some Derivatives of Isoflaven and Isoflavanol. J. Chem. Soc. (London) 1954, 4448.

*134.* LEA C. H. and P. A. T. SWOBODA: On the Antioxidant Activity of the Flavonols, Gossypetin and Quercetagetin. Chem. and Ind. 1956, 1426.

*135.* LIMAYE, D. B.: Syntheses in the Furo-coumarin Group. II. Rasayanam (Poona) 1936, 1.

*136.* LIMAYE, S. D.: Rasoda Reaction for the Synthesis of Flavonols from Chalkone Dibromides. Rasayanam (Poona) 2, 1 (1950).

*137.* LIMAYE, S. D. and D. B. LIMAYE: Ranjorwa Reaction for the Synthesis of Flavonols from *o*-Hydroxyacetophenones and Aldehydes. Rasayanam (Poona) 2, 41 (1952).

*138.* LINDSTEDT, G.: Constituents of Pine Heartwood. IX. Heartwood of *Pinus montana.* Acta Chem. Scand. 3, 755 (1949), and subsequent communications.

*139.* LINDSTEDT, G. and A. MISIORNY: Constituents of Pine Heartwood. XXV. Investigation of 48 *Pinus* Species by Paper Partition Chromatography. Acta Chem. Scand. 5, 121 (1951).

*140.* MAHAL, H. S.: Antiseptics and Anthelminthics. III. Pharmacology of Certain Flavones with Special Reference to their Anthelminthic Action. Proc. Indian Acad. Sci. 5 B, 186 (1937).

*141.* MAHAL, H. S., H. S. RAI and K. VENKATARAMAN: Synthetical Experiments in the Chromone Group. XVI. Chalkones and Flavanones and their Oxidation to Flavones by Means of Selenium Dioxide. J. Chem. Soc. (London) 1935, 866.

*142.* MAHAL, H. S. and K. VENKATARAMAN: A Synthesis of Flavones at Room Temperature. Current Sci. (India) 2, 214 (1933).

*143.* MAHESH, V. B. and T. R. SESHADRI: Hydroxylation of Flavanones in the 3-Position. J. Chem. Soc. (London) 1955, 2503.

*144.* — — Isomerization of Isoflavones in Alkaline Medium. J. Sci. Industr. Res. (India) 14 B, 671 (1955).

*145.* MANI, R., V. RAMANATHAN and K. VENKATARAMAN: The Colouring Matters of *Ponderosa* Pine Bark. II. A Synthesis of Pinoquercetin and Pinomyricetin. J. Sci. Industr. Res. (India) 15 B, 490 (1956).

*146.* Manjunath, B. L., A. Seetharamiah und S. Siddappa: Konstitution von Karanjin aus den Wurzeln von *Pongamia glabra* Vent. Ber. dtsch. chem. Ges. **72**, 93 (1939).

*147.* Mansfield, G. H., T. Swain and C. G. Nordström: Identification of Flavones by the Ultraviolet Absorption Spectra of their Ions. Nature (London) **172**, 23 (1953).

*148.* Marathey, M. G.: Synthesis of Dihydroflavonols and Flavonols. I and II. J. Univ. Poona No. 4, 73 (1953).

*149.* Marini-Bettolo, G. B. and A. Ballio: Su alcune reazioni cromatiche dei composti flavonici. Gazz. chim. ital. **76**, 410 (1946).

*150.* Marini-Bettolo, G. B., V. Deulofeu e E Hug: I glucosidi flavonici dell'Ombu (*Phytolacca dioica* L.). Isolamento della quercetina e di una nuova dimetil-quercetina (Ombuina). Gazz. chim. ital. **80**, 63 (1950).

*151.* Marsh, C. A.: Glucuronide Metabolism in Plants. 2. The Isolation of Flavone Glucosiduronic Acids from Plants. Biochemic. J. **59**, 58 (1955).

*152.* Martin, G. J. (Consulting Ed.): Conference on Bioflavonoids and the Capillary. Ann. New York Acad. Sci. **61**, Art. 3, 637 (1955).

*153.* Matsuura, S.: The Structure of Cryptostrobin and Strobopinin, the Flavanones from the Heartwood of *Pinus strobus*. Pharmac. Bull. (Japan) **5**, 195 (1957).

*154.* Mayer, F. and A. H. Cook: The Chemistry of Natural Colouring Matters. New York: Reinhold. 1943.

*155.* Mentzer, C. et D. Pillon: Nouvelle synthèse de la chrysine et d'autres colorants naturels oxyflavoniques. C. R. hebd. Séances Acad. Sci. **234**, 444 (1952).

*156.* Mirza, R. and R. Robinson: Conversion of Flavonols into Anthocyanidins. Nature (London) **166**, 997 (1950).

*157.* Mittal, O. P. and T. R. Seshadri: Demethoxykanugin: A New Crystalline Compound from *Pongamia glabra*. J. Chem. Soc. (London) **1956**, 2176.

*158.* Moewus, F.: Die Sexualstoffe von *Chlamydomonas eugametos*. Erg. Enzym-forsch. **12**, 173 (1951).

*159.* Murray, C. W., A. N. Booth, F. DeEds and F. T. Jones: Absorption and Metabolism of Rutin and Quercetin in the Rabbit. J. Amer. Pharmaceut. Assoc., Sci. Ed. **43**, 361 (1954).

*160.* Murti, V. V. S., S. Rangaswami and T. R. Seshadri: Chemical Investigation of Indian Fruits. V. Constitution of Auranetin. Proc. Indian Acad. Sci. **28 A**, 19 (1948).

*161.* Murti, V. V. S., L. R. Row and T. R. Seshadri: 5,6,7,8-Hydroxyflavonols. II. A Total Synthesis. Proc. Indian Acad. Sci. **23 A**, 233 (1946).

*162.* Murti, V. V. S. and T. R. Seshadri: Nuclear Oxidation in Flavones and Related Compounds. XVII. A Synthesis of Oroxylin-A. Proc. Indian Acad. Sci. **29 A**, 1 (1949).

*163.* Nakaoki, T. and N. Morita: Medicinal Resources. III. A New Glycoside of *Chrysosplenium japonicum*. J. Pharmac. Soc. Japan **76**, 320 (1956).

*164.* Narasimhachari, N. and T. R. Seshadri: Synthetic Experiments in the Benzopyrone Series. XIV; XXVII. Proc. Indian Acad. Sci. **32 A**, 342 (1950); **37 A**, 531 (1953).

*165.* — — A New Synthesis of Flavones. Proc. Indian Acad. Sci. **30 A**, 151 (1949).

*166.* Narayanaswami, S., S. Rangaswami and T. R. Seshadri: Chemistry of Pongamol. II. J. Chem. Soc. (London) **1954**, 1871.

*167.* NEELAKANTAM, K., L. R. Row and V. VENKATESWARLU: Fluorescence Reactions with Boric Acid and *o*-Hydroxycarbonyl Compounds and their Applicability in Analytical Chemistry. Proc. Indian Acad. Sci. 18 A, 364 (1943).

*168.* NEILL, K. G.: The Synthesis of Homoferreirin. J. Chem. Soc. (London) 1953, 3454.

*169.* NEU, R.: Separation of Chalcones, Flavanones and Flavonols by Chromatography with Polyamide. Nature (London) 182, 660 (1958).

*170.* NOWLAN, N. V., P. A. SLAVIN and T. S. WHEELER: The Effect of Bond Structure on the Transformation of *o*-Aroyloxyacetoarones into *o*-Hydroxydiaroylmethanes. Baker-Venkataraman Transformation. J. Chem. Soc. (London) 1950, 340, and subsequent communications.

*171.* PANKAJAMANI, K. S. and T. R. SESHADRI: Survey of Anthoxanthins. I. Proc. Indian Acad. Sci. 36 A, 157 (1952), and subsequent communications.

*172.* PARIS, R.: Chromatographie sur papier de quelques dérivés flavoniques. Bull. soc. chim. biol. (Paris) 34, 767 (1952).

*173.* PAVANARAM, S. K. and L. R. Row: New Flavones from *Pongamia pinnata* (L.) MERR. II. The Synthesis of Compounds C and D. Austral. J. Chem. 9, 132 (1956).

*174.* PERKIN, A. G. and A. E. EVEREST: The Natural Organic Colouring Matters. London: Longmans. 1918.

*175.* PEW, J. C.: A Flavonone from Douglas-Fir Heartwood. J. Amer. Chem. Soc. 70, 3031 (1948).

*176.* PRISTA, L. N.: Chemical-pharmacognostic Investigation of *Ptaeroxylon obliquum*. Annais fac. farm. Porto (Lisbon) 11, 81 (1951).

*177.* RAJAGOPALAN, S., S. RANGASWAMI, K. V. RAO and T. R. SESHADRI: Constitution of Kanugin. II. Proc. Indian Acad. Sci. 23 A, 60 (1946).

*178.* RAMANATHAN, V. and K. VENKATARAMAN: Synthetical Experiments in the Chromone Group. XXIX. A Method for the Reduction of 5,7-Dihydroxy-flavones to 5-Hydroxyflavones. Proc. Indian Acad. Sci. 38 A, 40 (1953).

*179.* RAMANUJAM, S. and T. R. SESHADRI: Components of the Bark of *Prunus puddum*. III. Synthesis of Padmakastein and its Derivatives. Proc. Indian Acad. Sci. 48 A, 175 (1957).

*180.* RANGASWAMI, S. and B. V. R. SASTRY: Constitution of Lanceolatin C and Lanceolatin B. Current Sci. (India) 24, 13 (1955).

*181.* — — Constitution of Maxima Substance B. Current Sci. (India) 24, 337 (1955).

*182.* RAO, K. V. and T. R. SESHADRI: Synthesis of Kanugin and Related Compounds. Proc. Indian Acad. Sci. 23 A, 147 (1946).

*183.* — — Synthesis of the Monomethyl Ethers of Kaempferol, and the Constitution of Rhamnocitrin. J. Chem. Soc. (London) 1947, 122.

*184.* RAO, P. R., P. S. RAO and T. R. SESHADRI: Synthesis of Hibiscetin. Proc. Indian Acad. Sci. 19 A, 88 (1944).

*185.* RAO, P. R. and T. R. SESHADRI: Isolation of Hibiscitrin from the Flowers of *Hibiscus sabdariffa*: Constitution of Hibiscetin. Proc. Indian Acad. Sci. 15 A, 148 (1942).

*186.* REICHEL, L. und J. MARCHAND: Synthesen einiger neuer Chalkon-Flavanon-Flavonol-glucoside. Chemie und Biochemie der Pflanzenstoffe. XI. Ber. dtsch. chem. Ges. 76, 1132 (1943).

*187.* REZNIK, H. und R. URBAN: Über den Metabolismus [14]C-markierter Ferula-säure im Pflanzenversuch. Naturwiss. 44, 13 (1957).

*188.* Robertson, A., C. W. Suckling and W. B. Whalley: The Chemistry of the "Insoluble Red" Woods. III. The Structure of Santal and a Note on Orobol. J. Chem. Soc. (London) **1949**, 1571.

*189.* Robinson, R.: The Structural Relations of Natural Products. Oxford: Clarendon Press. 1955.

*190.* Roux, D. G. and S. R. Evelyn: The Correlation between Structure and Paper Chromatographic Behaviour of some Flavonoid Compounds and Tannins. J. Chromatogr. **1**, 537 (1958).

*191.* Row, L. R. and T. R. Seshadri: Constitution of Patuletin. III. A Study and Synthesis of O-Pentaethylpatuletin. Proc. Indian Acad. Sci. **23** A, 140 (1946).

*192.* Sarin, P. S., J. M. Sehgal and T. R. Seshadri: Synthetic Experiments in the Benzopyrone Series. LXIII. Synthesis of α-Methyl-furano Isoflavones. J. Sci. Industr. Res. (India) **16** B, 61 (1957). See this paper for earlier references.

*193.* Sathe, V. R. and K. Venkataraman: A New Reaction for the Synthesis of Chromones and Isoflavones. Current Sci. (India) **18**, 373 (1949).

*194.* Schmid, H. und K. Banholzer: Nachweis der intramolekularen Natur der Baker-Venkataraman-Umlagerung. Helv. Chim. Acta **37**, 1706 (1954).

*195.* Sehgal, J. M., T. R. Seshadri and K. L. Vadehra: Synthetic Experiments in the Benzopyrone Series. LVII. Synthesis of 5,6,7,8,4'-Pentamethoxy Flavanone and Flavone: The Constitution of Ponkanetin. Proc. Indian Acad. Sci. **42** A, 252 (1955).

*196.* Seikel, M. K. and T. A. Geissman: The Flavonoid Constituents of Barley *(Hordeum vulgare)*. Arch. Biochem. Biophys. **71**, 17 (1957).

*197.* Seshadri, T. R.: Biochemistry of Natural Pigments (Exclusive of Haeme Pigments and Carotenoids). Annu. Rev. Biochem. **20**, 487 (1951).

*198.* — Nuclear Oxidation in the Flavones and Related Compounds. XIII. A Discussion of the Results. Proc. Indian Acad. Sci. **28** A, 1 (1948). See this paper for earlier references.

*199.* — Syntheses Following Possible Paths of Biogenesis. Lectures, XIVth Intern. Congr. Pure and Appl. Chem., Zürich, 1955.

*200.* Seshadri, T. R. and V. Venkateswarlu: Synthesis and Study of 5,6,7,8-Hydroxyflavonols. Proc. Indian Acad. Sci. **23** A, 192 (1946).

*201.* — — 5,6,7,8-Hydroxyflavonols. IV. A Synthesis of Calycopterin. Proc. Indian Acad. Sci. **24** A, 349 (1946).

*202.* Shah, R. C., V. V. Virkar and K. Venkataraman: The Constitution of Calycopterin, the Yellow Colouring Matter of the Leaves of *Calycopteris floribunda*. J. Indian Chem. Soc. **19**, 135 (1942).

*203.* Shibata, S. and M. Yamazaki: The Biogenesis of Rutin. Pharmac. Bull. (Tokyo) **5**, 501 (1957).

*204.* Shimizu, M.: A New Method for Distinguishing Flavonol-3-glycosides from Flavonols. J. pharmac. Soc. Japan **71**, 1329 (1951).

*205.* Shimizu, M. and G. Ohta: Partial Methylation of Rutin and Quercetin. J. pharmac. Soc. Japan **71**, 879 (1951), and subsequent communications.

*206.* Simmonds, N. W. and R. Stevens: Occurrence of the Methylene-dioxy Bridge in the Phenolic Components of Plants. Nature (London) **178**, 752 (1956).

*207.* Simpson, T. H.: Protecting Groups in the Synthesis of Flavones. Sci. Proc. Roy. Dublin Soc. **27**, No. 6, 75 (1956).

*208.* Simpson, T. H. and N. Uri: Hydroxyflavones as Inhibitors of the Aerobic Oxidation of Unsaturated Fatty Acids. Chem. and Ind. **1956**, 956.

*209.* Skarzynski, B.: Spektrographische Untersuchungen von Flavonfarbstoffen. Biochem. Z. **301**, 150 (1939).

*210.* STAMM, O. A., H. SCHMID und J. BÜCHI: Die Konstitution des Jamaicins. Helv. Chim. Acta **41**, 2006 (1958).

*211.* SWAIN, T.: Spectral Studies on Phenolic Compounds. Chem. and Ind. **1954**, 1480.

*212.* TAUBÖCK, K.: Über Reaktionsprodukte von Flavonolen mit Borsäure und organischen Säuren und ihre Bedeutung für die Festlegung des Bors in Pflanzenorganen. Naturwiss. **30**, 439 (1942).

*213.* UNDERHILL, E. W., J. E. WATKIN and A. C. NEISH: Biosynthesis of Quercetin in Buckwheat. I, II. Canad. J. Biochem. and Physiol. **35**, 219, 229 (1957).

*214.* VENKATARAMAN, K.: Synthetical Experiments in the Flavone and Isoflavone Groups. Proc. Nat. Inst. Sci. India **5**, 253 (1939).

*215.* WACHTMEISTER, C. A.: Paper Chromatography on Borate-impregnated Paper. Acta Chem. Scand. **5**, 976 (1951).

*216.* WADA, E.: On a Flavonol Glycoside Isolated from Flowers of a White Azalea (*Rhododendron mucronatum* G. DON.). J. Amer. Chem. Soc. **78**, 4725 (1956).

*217.* WAGNER, H. und R. LUCK: Zur Kenntnis der Inhaltsstoffe von *Nerium oleander.* Naturwiss. **42**, 607 (1955).

*218.* WARBURTON, W. K.: The Isoflavones. Quart. Rev. Chem. Soc. (London) **8**, 67 (1954).

*219.* WENDER, S. H. and O. NORMAN: Isolation of Flavonoid Compounds. U. S. Patent 2681907.

*220.* WESSELY, F. und G. H. MOSER: Synthese und Konstitution des Skutellareins. Monatsh. Chem. **56**, 97 (1930).

*221.* WHEELER, T. S.: Unsolved Problems in Flavonoid Chemistry. Record Chem. Progr. (Kresge-Hooker Sci. Libr.) **18**, 133 (1957).

*222.* — Flavone. Org. Syntheses **32**, 72 (1952).

*223.* WILLAMAN, J. J.: Some Biological Effects of the Flavonoids. J. Amer. Pharmaceut. Assoc., Sci. Ed. **44**, 404 (1955).

*224.* WILSON, C. W.: A Study of the Boric Acid Colour Reaction of Flavone Derivatives. J. Amer. Chem. Soc. **61**, 2303 (1939).

*225.* WILSON, R. H. and F. DEEDS: The *in vitro* Protection of Epinephrine by Flavonoids. J. Pharmacol. exp. Therapeut. **95**, 399 (1949).

*226.* WOLFROM, M. L., W. D. HARRIS, G. F. JOHNSON, J. E. MAHAN, S. M. MOFFETT and B. WILDI: Osage Orange Pigments. XI. Complete Structures of Osajin and Pomiferin. J. Amer. Chem. Soc. **68**, 406 (1946).

*227.* WOLFROM, M. L. and B. S. WILDI: Osage Orange Pigments. XII. Synthesis of Dihydro-iso-osajin and Dihydro-isopomiferin. J. Amer. Chem. Soc. **73**, 235 (1951).

*228.* ZEMPLÉN, G. und R. BOGNÁR: Konstitution und Synthese des Linarins und Pektolinarins aus ihren Aglykonen und aus Rutinose. Ber. dtsch. chem. Ges. **74**, 1818 (1941).

*229.* ZEMPLÉN, G., R. BOGNÁR und J. MECHNER: Synthese des Glucosids Toringin. Ber. dtsch. chem. Ges. **77**, 99 (1944).

*230.* ZEMPLÉN, G. und L. FARKAS: Synthese des Genistins. Ber. dtsch. chem. Ges. **76**, 1110 (1943).

*231.* — — Notiz über die Synthese des Prunitrins. Ber. dtsch. chem. Ges. **90**, 836 (1957).

*232.* ZEMPLÉN, G., L. FARKAS und A. BIEN: Synthese des Ononins. Ber. dtsch. chem. Ges. **77**, 452 (1944).

*233.* ZEMPLÉN, G. und L. MESTER: Synthese des Apigenin-glykosids-(5) der *Amorpha fruticosa* L. Ber. dtsch. chem. Ges. **76**, 776 (1943).

*(Received, February 24, 1959.)*

# Fortschritte der Chemie
# der Vitamine D und ihrer Abkömmlinge.

Von **H. H. Inhoffen** und **K. Irmscher**, Braunschweig.

## Inhaltsübersicht.

# I. Konstitution der Vitamine D und ihrer Abkömmlinge.

## 1. Vitamine D.

Durch Arbeiten, die vor allem von WINDAUS und seiner Schule durchgeführt wurden, ist die Konstitution der Vitamine D seit etwa 25 Jahren bekannt. Die verschiedenen Vitamine D unterscheiden sich in den Seitenketten ($R$). Allen gemeinsam sind das *trans*-verknüpfte Hydrindan-*C,D*-Ringsystem, der konjugierte Trienchromophor, der die beiden Brücken-C-Atome 6 und 7 (Steroidnomenklatur) und die exocyclische Methylengruppe am Ring *A* erfaßt, sowie die räumliche Lage der Hydroxylgruppe.

Unter Berücksichtigung der Drehbarkeit um die 6,7-Einfachbindung lassen sich zwei Konstellationen, (I) und (II), der Vitamine D formulieren.

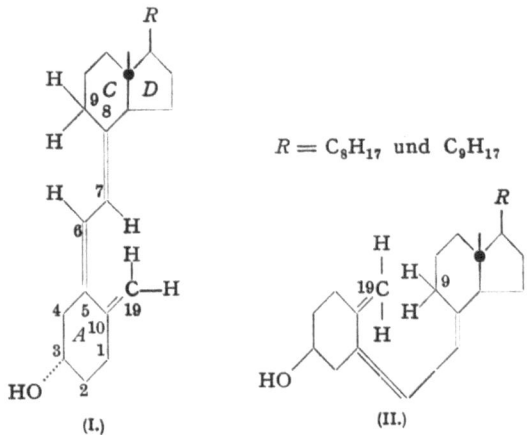

$R = C_8H_{17}$ und $C_9H_{17}$

Konstellationen der Vitamine D.

Nachdem auf Grund zwei-dimensionaler, röntgenographischer Messungen am Vitamin D₂-4-jod-3-nitrobenzoat für den kristallinen Zustand das Vorliegen der 6,7-*seco-trans*-Konstellation (I) bewiesen werden konnte, gewannen CROWFOOT HODGKIN und Mitarbeiter (*10*) durch die Anwendung drei-dimensionaler Elektronendichtebestimmungen am gleichen Vitamin D₂-Ester einen tieferen Einblick in die Feinstruktur des Vitamins D₂.

Danach liegt in kristallinem Zustand das *s-trans*-Diensystem 5,6—7,8 ziemlich in einer Ebene; die Bindung 5,10 ist verdreht. Die C-Atome 19, 10, 1 und 5 liegen fast in einer Ebene, die C-Atome 6, 5, 4 und 10 in einer zweiten. Beide Ebenen sind unter einem Winkel von 54° zueinander geneigt. Der Abstand der H-Atome an $C_{(7)}$ und $C_{(19)}$ beträgt 2,2 Å; das ist etwa der Betrag, der sich aus den van der Waalsschen Radien für sich berührende Wasserstoffatome errechnet.

Damit ist freilich nicht bewiesen, daß auch in Lösung die gleiche Konstellation (I) vorliegen muß. Vielmehr könnte durch den Fortfall der Gitterkräfte das Molekül durch Drehung um die Bindung 6, 7 der gegenseitigen sterischen Hinderung der H-Atome an $C_{(7)}$ und $C_{(19)}$ ausweichen. Diese Möglichkeit ist um so mehr zu diskutieren, als wir durch Kenntnis der Konstitution des Praecalciferols (vgl. S. 76) die Konstellation (II) nicht, wie früher angenommen, diesem, sondern dem Vitamin D selbst zuzuordnen haben. Also könnte die Konstellation (II), etwa in Form eines Gleichgewichtes der Konstellationen (I) und (II), am Zustand des Vitamins D in Lösung beteiligt sein. Da jedoch gerade bei der Konstellation (II) starke sterische Hinderungen, und zwar zwischen den H-Atomen an $C_{(9)}$ und $C_{(19)}$, auftreten, die eine freiwillige coplanare Anordnung der drei konjugierten Doppelbindungen weitgehend ausschließen, spricht nichts dagegen, die *s-trans*-Konstellation (I) auch in Lösung zumindest als bevorzugt anzunehmen. Havinga und Mitarb. (74) stützten diese Annahme durch halbquantitative Berechnungen. Es ist jedoch zu bedenken, daß beim Übergang in Praecalciferol (vgl. SS. 76 und 96) die 6,7-*s-cis*-Konstellation (II) durchlaufen werden dürfte. . Diese Gleichgewichtsreaktion läuft schon langsam bei Zimmertemperatur ab. Möglicherweise liegt die wirkliche Raumform der Vitamine D in Lösung zwischen den beiden durch Formeln darstellbaren extremen Konstellationen (I) und (II) (*18*).

Infolge der *cis*oiden Struktur der 19-6-Diengruppierung einerseits und der sterischen Behinderung der Coplanarität andererseits, weisen die Vitamine D eine sehr geringe Chromophorlänge auf, die sich in einer relativ kurzwellig liegenden Lichtabsorption schwacher Intensität äußert: $\lambda_{max} = 265\ m\mu$ mit $\varepsilon = 18\,300$. Die Bedeutung der beiden genannten Faktoren für die Absorption kann durch Vergleich mit den inzwischen bekanntgewordenen 5,6-*trans*-Vitaminen D (s. unten) erhärtet werden, bei denen die sterische Hinderung geringer ist und deren Absorption gegenüber der der Vitamine D um 7—8 $m\mu$ bathochrom verschoben ist. Auch die Intensität der Absorption der 5,6-*trans*-Vitamine D liegt höher: $\varepsilon = 24\,600$.

### 2. 5,6-*trans*-Vitamine D.

5,6-*trans*-Vitamine D sind auf zwei verschiedenen Wegen erhalten worden. Zuerst konnte das 5,6-*trans*-Vitamin D$_2$ als 3-Epimerengemisch partialsynthetisch erhalten werden, dem sich Formel (III) zuordnen ließ (*31, 32*). Inzwischen wurden auch die freien, sterisch einheitlichen 5,6-*trans*-Vitamine $D_2$ und $D_3$ synthetisiert (vgl. S. 108) (*29*).

HAVINGA und Mitarb. (*73*) erhielten 5,6-*trans*-Vitamin $D_2$ durch Isomerisierung von Vitamin $D_2$, (I) → (III), mit Jod in unpolaren Lösungsmitteln bei Gegenwart von Pyridin unter dem Einfluß von diffusem Tageslicht und isolierten die Substanz in Form eines kristallinen Esters. Nach einem abgewandelten Verfahren konnten die 5,6-*trans*-Vitamine $D_2$ und $D_3$ selbst kristallin erhalten werden (*36, 37*).

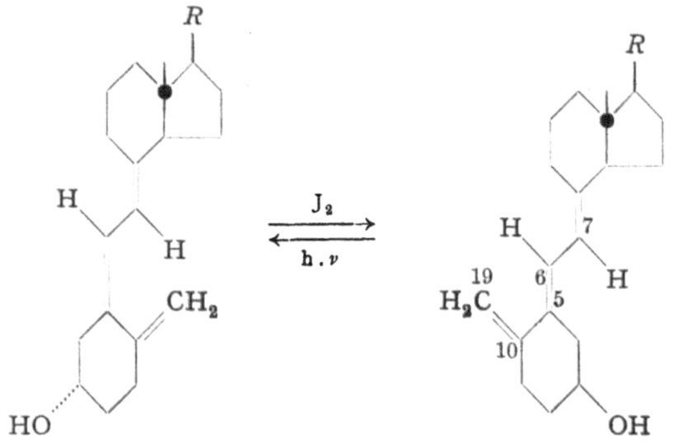

(I.)  (III.) 5,6-*trans*-Vitamin $D_2$.

Die 5,6-*trans*-Vitamine D (III) besitzen wie die Vitamine D einen Trien-chromophor mit drei exocyclischen Doppelbindungen. Das IR-Spektrum zeigt durch eine Bande bei 885 cm$^{-1}$ die Anwesenheit einer exocyclischen Methylengruppe an. Sie haben gleichfalls ein *cis*oides 19,10—5,6-Diensystem. An der 6,7-Einfachbindung dürfte *s-trans*-Konstellation vorliegen, wodurch sich das Triensystem einer coplanaren Lage annähert. Von den Vitaminen D unterscheiden sich die 5,6-*trans*-Vitamine D lediglich durch *cis-trans*-Isomerie an der 5,6-Doppelbindung. Die UV-Absorption bei 272—273 m$\mu$ mit $\varepsilon = 24600$ deutet im Vergleich

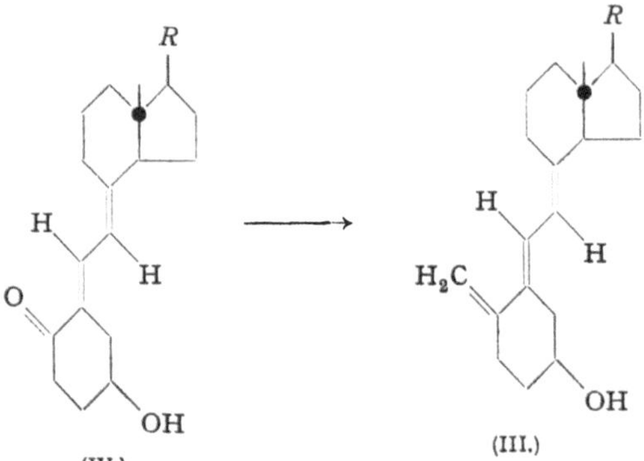

(IV.)  (III.)

zu den Vitaminen D auf eine sterische Entspannung und eine Verlängerung des Trien-chromophors hin.

Folgende Reaktionen bieten das Material für die Ableitung der Konstitution (III):

a) Die bereits besprochene Darstellbarkeit der 5,6-*trans*-Vitamine D aus den Vitaminen D sowie ihre Rückverwandlung in die Vitamine D durch Photo-isomerisierung (vgl. S. 109) (*37*).

b) Die Darstellbarkeit aus den entsprechenden Dienonen (IV) (vgl. S. 117) (*29, 31, 32*).

c) Die Aufhebung der 5,6-Stereoisomerie zu den Vitaminen D durch Alkoholat-Oxydation zu den Vitamin D-iso-Ketonen (V), die in gleicher Weise auch aus den Vitaminen D entstehen (*32, 36*).

(III.) ⟶     ⟵ (I.)

5,6-*trans*-
Vitamin D$_2$.

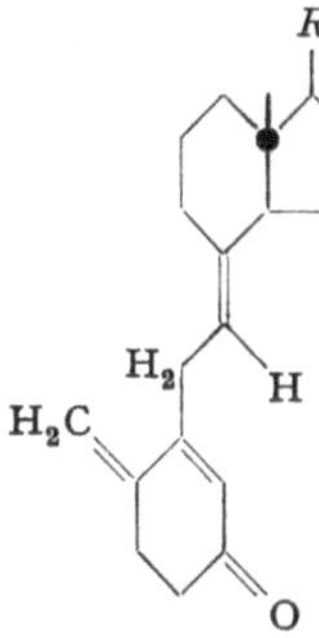

(V.)

d) Die Isomerisierung des 5,6-*trans*-Vitamins D$_2$ zum iso-Vitamin D$_2$ (VI) (*36*), bei der lediglich die exocyclische 19,10-Doppelbindung in den Ring *A* umklappt (vgl. S. 98).

(III.) $\xrightarrow{BF_3}$

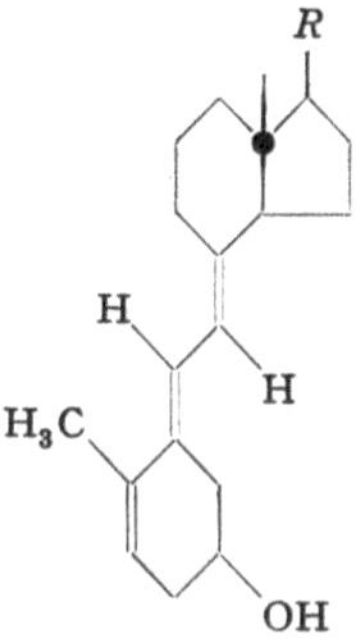

(VI.) iso-Vitamin D$_2$.

e) Der Verlauf der Diensynthese des 5,6-*trans*-Vitamins D$_2$ mit Maleinsäureanhydrid (*73*), die im Vergleich zu der analogen Reaktion der Vitamine D rasch verläuft, weil am Diensystem 19,10—5,6 kein *cis*-Substituent vorhanden ist.

### 3. iso-Vitamine D.

Der Begriff der iso-Vitamine D war ursprünglich für solche 9,10-*seco*-Steroide definiert worden, von deren konjugiertem Trien-system zwei Doppelbindungen zwischen den C-Atomen 5,6 und 7,8 liegen (*22*). In diesem Sinne wären auch die 5,6-*trans*-Vitamine D als iso-Vitamine D zu bezeichnen. Im engeren Sinne bezeichnet man jetzt als iso-Vitamine D 9,10-*seco*-Steroide mit dem Trien-system 1,10 — 5,6 — 7,8, gemäß der Formel (VI).

Das iso-Vitamin $D_2$, das als 3-Epimerengemisch partialsynthetisch durch Wasserabspaltung aus dem Dien-diol (VII) bereitet worden war (*22*), konnte als 3-Epimerengemisch inzwischen kristallin erhalten werden (*23*).

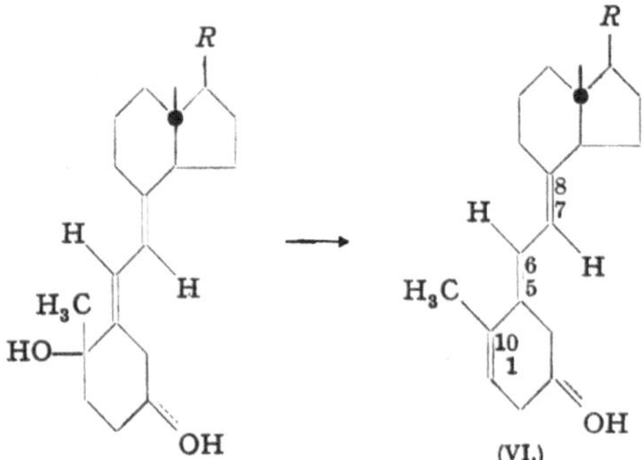

(VII.) → (VI.)

Für dieses Produkt hat sich später erwiesen, daß es nicht, wie zuerst vermutet, ein *cis*-verknüpftes *C,D*-System besitzt, sondern daß es die natürliche *trans*-Verknüpfung des Hydrindan-Gerüstes aufweist. Dieser wichtige Befund ergab sich, als im Laufe der Synthese des 5,6-*trans*-Vitamins $D_2$ bewiesen werden konnte, daß das sowohl für dieses als auch für das iso-Vitamin $D_2$-3-Epimerengemisch als Ausgangsmaterial dienende $C_{27}$-Keton (IV) am $C_{(14)}$-Wasserstoff die natürliche $\alpha$-Stellung aufweist (*32*).

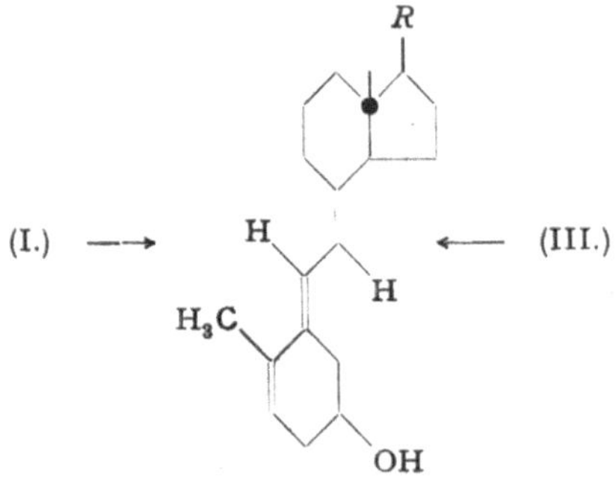

(I.) ⟶      ⟵ (III.)

(VI.) iso-Vitamin $D_2$.

Die am C-Atom 3 einheitlich $\beta$-substituierten iso-Vitamine D konnte Baron (3) durch eine gleichartige Wasserabspaltung aus den auf einem anderen Wege (vgl. S. 99) gewonnenen Dien-diolen (VII) gewinnen.

In kristallinem Zustand wurde das 3-$\beta$-iso-Vitamin $D_2$ durch zwei neue Isomerisierungsreaktionen erhalten, indem man einerseits Vitamin $D_2$ und andererseits 5,6-*trans*-Vitamin $D_2$ mit Bortrifluorid-ätherat behandelte (36).

Diese Isomerisierungsreaktionen stützen ihrerseits die Konstitution (VI) für die iso-Vitamine D, die schon früher durch die genannte Synthesereaktion einerseits sowie durch den Ozonabbau andererseits bewiesen worden war (22). Die aus der UV-Absorption bei 287 m$\mu$ mit $\varepsilon = 44100$ abgeleitete, sterisch freie all-*trans*-Feinstruktur des konjugierten Triensystems fand eine weitere Stütze in dem experimentellen Befund, daß das iso-Vitamin $D_2$ keine Diensynthese mit Maleinsäureanhydrid eingeht (23). Damit erscheint das Vorliegen einer 6,7-*s-cis*-Konstellation unwahrscheinlich, durch die allein eine Dienreaktion an $C_{(5)}$—$C_{(8)}$ ermöglicht würde. Eine Dienreaktion an $C_{(1)}$—$C_{(6)}$ . ist auf Grund der starren *trans*oiden Anordnung dieses Diensystems nicht möglich, da für den Eintritt der Diensynthese die Voraussetzung gilt, daß das Dien aus einer quasi-cyclischen Konstellation heraus reagieren kann.

### 4. Praecalciferole.

Nachdem erstmalig die Existenz der Praecalciferole durch Velluz und Mitarb. (72) erwiesen worden war und die experimentelle Handhabung der thermischen Gleichgewichte zwischen den Vitaminen D und den Praecalciferolen durch die französischen Forscher eine bequeme Bereitung größerer Mengen der letzteren gestattete, konzentrierte sich die Arbeit auf die Ermittlung der Konstitution der Praecalciferole.

Die Annahmen, daß den Praecalciferolen die Formeln der 6,7-*s-cis*-Isomeren der Vitamine D (II, S. 71) oder der 5,6-*trans*-Vitamine D (III) zukommen könnten, waren zunächst auf Grund der leichten Einstellung des Praecalciferol-Vitamin D-Gleichgewichts diskutiert worden. Abgesehen davon, daß die Formel (III, S. 73) durch die inzwischen erlangte Kenntnis der 5,6-*trans*-Vitamine D (vgl. S. 72) entfiel, konnten Velluz und Mitarb. (65) durch Abbau-Reaktionen am Praecalciferol$_2$ sowie durch IR-spektrale Untersuchungen nachweisen, daß sich die Verknüpfung der Ringe A und C beim Praecalciferol$_2$ von der des Vitamins $D_2$ mehr als nur durch räumliche Isomerie unterscheiden müsse. Die exocyclische $C_{(19)}$-Methylengruppe war nämlich weder im IR-Spektrum noch durch Ozonabbau zu finden. Ferner ergab weder die Chromsäure-Oxydation den $C_{21}$-Abbau-aldehyd wie Vitamin $D_2$, noch die Bleitetraacetat-Oxydation das aus dem Vitamin zu erhaltende 5,6-Dihydroxy-Vitamin $D_2$. Durch diese Ergebnisse konnten die Formeln (II) und (III) ausgeschieden werden. Die Formel der iso-Vitamine D (VI) war auf Grund der schon damals vorliegenden Kenntnis dieser Substanzen (vgl. S. 75) besonders wegen der Lage des UV-Absorptionsmaximums sowie der thermischen Stabilität der iso-Vitamine D (23) ebenfalls nicht in Betracht zu ziehen, zumal sich die Energiedifferenzen von endo-exo-Isomeren nicht mit der Leichtigkeit der Umwandlung der Praecalciferole in die Vitamine D vereinbaren ließen. Da die Anzahl der Doppelbindungen des Praecalciferols$_2$ damals noch nicht genau festlag, diskutierten VELLUZ und Mitarb. (65) die Entstehung der Praecalciferole im thermischen Gleichgewicht durch innere Diensynthese der Vitamine D und stellten vorübergehend die vorläufige Formel (VIII) zur Diskussion.

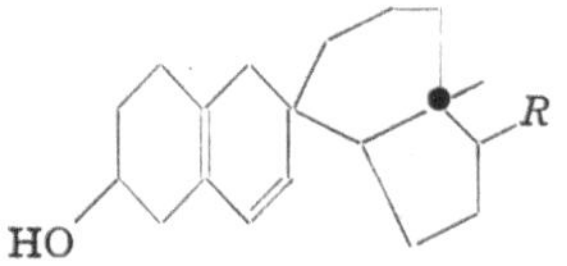

(VIII.) 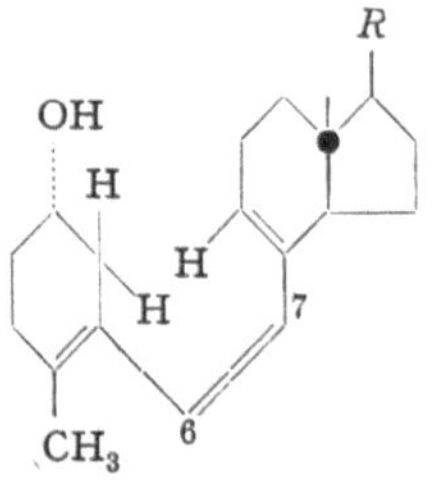        (IX a.) Praecalciferol.

Die Zuteilung der für die Praecalciferole heute allgemein angenommenen Formel (IX a) wurde erst möglich, nachdem auf Grund von Überlegungen von ALDER und Mitarb. (2) sowie von eigenen Modelluntersuchungen (34) den Tachysterinen statt der bis dahin angenommenen Formel (IX a) die 6,7-*trans*-Formel (X) zugeteilt werden konnte (23) (vgl. S. 79) und die Formel (IX a) damit für die Praecalciferole frei wurde.

VELLUZ und Mitarb. (66, 69) kamen zu der Formel (IX a) für die Praecalciferole, indem sie aus Praecalciferol ein Mono-epoxyd mit einer Dien-Konfiguration ($\lambda_{max}$ 244 m$\mu$) erhielten und damit die Existenz eines konjugierten Trien-systems bei den Praecalciferolen bewiesen. Durch Vergleich der UV-Absorption mit der der Modelltriene leitete VELLUZ auf der Basis seiner früheren Ergebnisse die Formel (IX a) ab. Die Isomerisierung des Praecalciferols (IX a) zum Tachysterin (X) konnte VELLUZ verwirklichen, indem er Praecalciferol mit UV-Licht bestrahlte (67, 69).

HAVINGA und Mitarb. (46) konnten gleichfalls den Praecalciferolen die Formel (IX a) zusprechen, als sie durch Isomerisierung von Praecalciferol$_2$ mit Jod in Petroläther, unter dem Einfluß von diffusem Tageslicht, das 6,7-*trans*-isomere Tachysterin$_2$ erhielten. In konzentrierteren Ätherlösungen entsteht bei der Jod-Isomerisierung (46) wie auch bei der Isomerisierung der benzolischen Lösung mit Bortrifluorid-ätherat (*36*) das weitgehend isomerisierte iso-Tachysterin₂ (XI) (vgl. S. 81). Schließlich ist auch das Fehlen der *trans*-Bande bei 957 cm$^{-1}$ im IR-Spektrum der Praecalciferole, die bei den Tachysterinen auftritt, eine weitere Stütze der Struktur (IXa) (*74*).

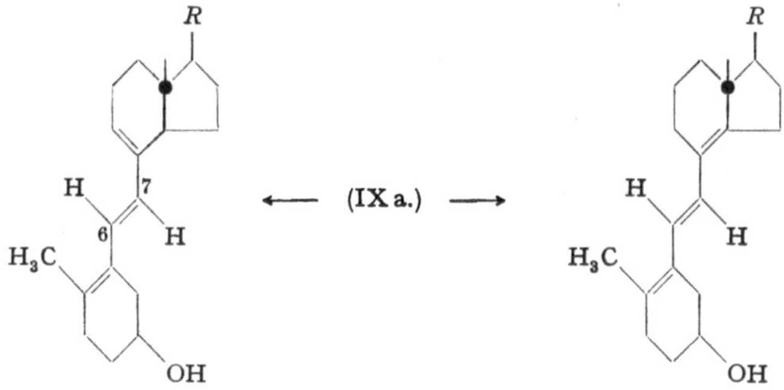

←— (IX a.) —→

(X.) Tachysterin.                    (XI.) iso-Tachysterin₂.

Die Diskussion der Feinstruktur der Praecalciferole (*18, 36, 71, 74*) geht um das Vorliegen der 5,6-s-*trans*-Form (IXa) bzw. der 5,6-s-*cis*-Form (IXb). Das wenig ausgeprägte UV-Absorptionsmaximum der Praecalciferole, mit 260—263 mµ und ε = 9000, das noch unterhalb der entsprechenden Werte der Vitamine D liegt und somit den kürzestwellig absorbierenden Trien-Chromophor im Bereich der Vitamin D-Chemie darstellt, weist auf eine starke Verkürzung der Chromophorlänge und sterische Behinderung der Coplanarität hin. Nun ist sowohl die Konstellation (IXa) sterisch behindert, und zwar durch die H-Atome an $C_{(9)}$ und $C_{(4)}$, als auch die Konstellation (IXb), und zwar durch das H-Atom an $C_{(9)}$ einerseits und die Methylgruppe an $C_{(10)}$ andererseits. Wenn auch die Behinderung bei (IXa) geringer ist als bei (IXb) und daher erstere Konstellation als bevorzugt anzusehen ist, so dürfte doch der reale räumliche Zustand zwischen den beiden durch Formeln darstellbaren Extremen liegen. Beim Übergang der Praecalciferole in die Vitamine D, der schon bei Zimmertemperatur stattfindet, dürfte die 5,6-s-*cis*-Form (IXb) durchlaufen werden (vgl. S. 96).

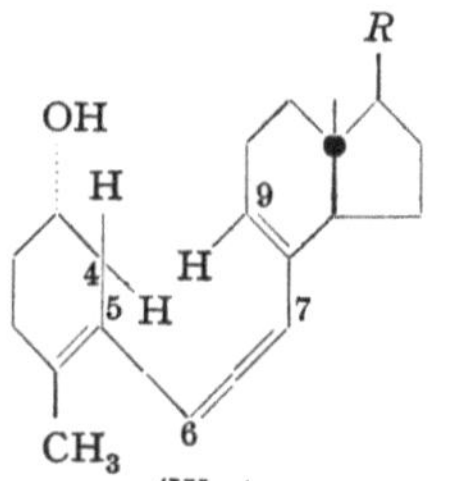

(IX a.) 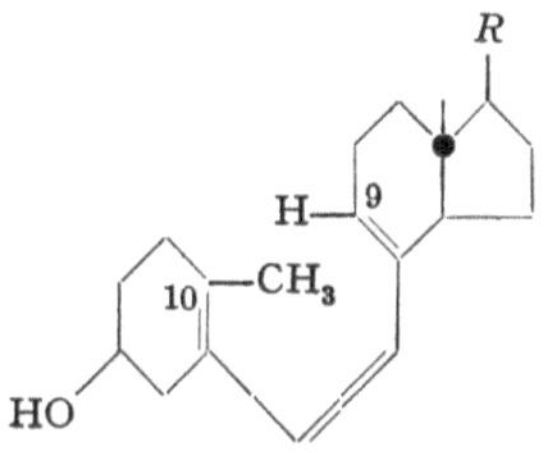                    (IX b.)

## 5. Tachysterine.

Die Konstitution der Tachysterine (X), die schon vor 25 Jahren durch Arbeiten der WINDAUSschen Schule aufgeklärt worden war, wurde durch folgende Arbeiten der letzten Jahre bestätigt: Die Lage der 8,9-Doppelbindung ließ sich durch die Isomerisierung des Tachysterins$_2$ (X) zum iso-Tachysterin$_2$ (XI) sichern (23).

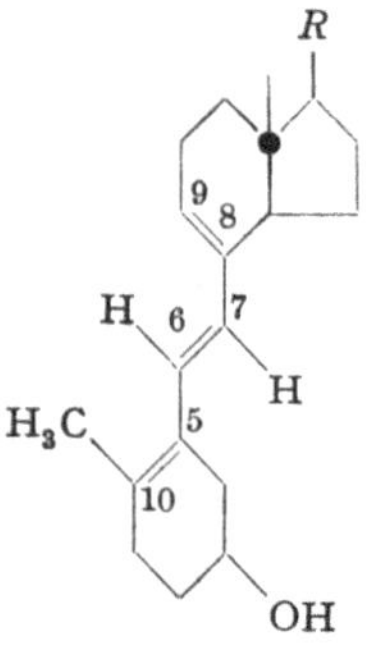

(X.) Tachysterin$_2$. 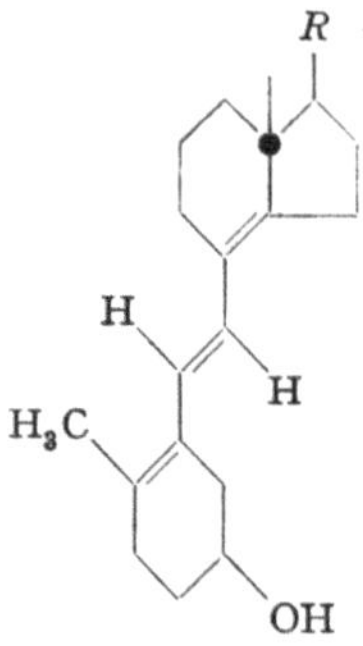          (XI.) iso-Tachysterin$_2$.

Die Lage der 10,5-Doppelbindung erhielt einen neuen Beweis durch die Kenntnis des u-Tachysterins$_2$ (XII) (25) (vgl. S. 79), das eine 4,5-Doppelbindung aufweist. Obwohl die Lage der ersteren Doppelbindung durch die Bildung von Dihydro-Vitamin D$_2$-I (XIII) bei der Einwirkung von Natrium und Propanol auf Tachysterin$_2$ gesichert schien (77), forderten BRAUDE und Mitarb. (7) die Formel (XII) für Tachysterin. Dabei gingen sie einerseits davon aus, daß das Tachysterin der Formel (X) als Derivat des 1-Methyl-2-vinyl-cyclohexens kein Maleinsäureanhydrid addieren dürfe, andererseits davon, daß die UV-Absorption der Tachysterine im Vergleich zu der des Modells (XIV) unverständlich bathochrom verschoben sei.

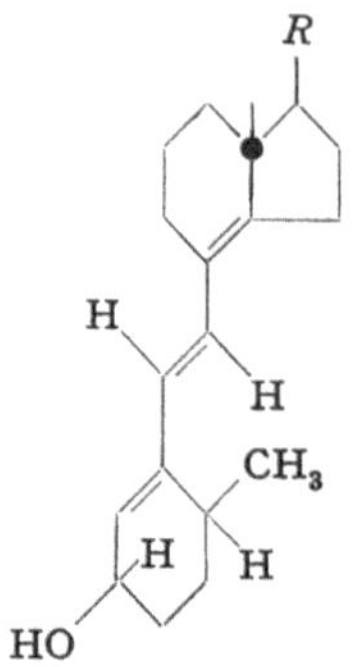

(XII.) u-Tachysterin$_2$. 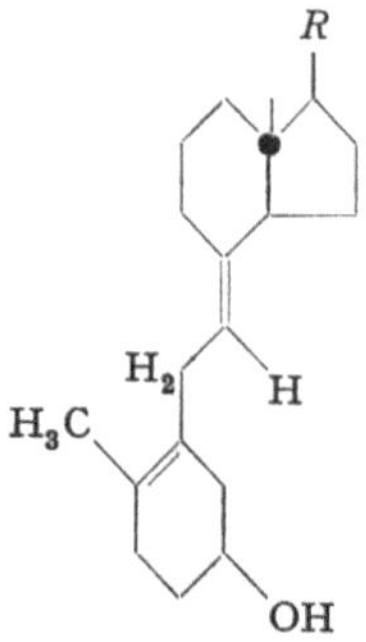     (XIII.) Dihydro-Vitamin D$_2$-I. 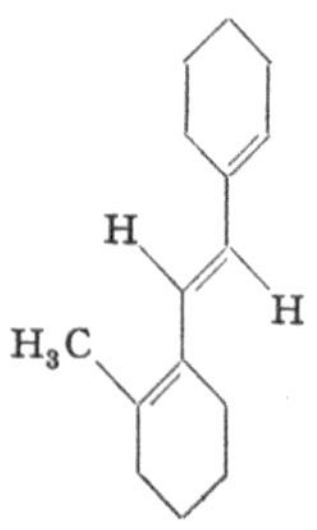     (XIV.)

Nun reagiert das Tachysterin, wie sich aus seiner Feinstruktur ergibt, sehr wahrscheinlich nicht mit der Diengruppierung 10 — 7, wie Braude fordert, sondern am Diensystem 6 — 9. Das geht auch daraus hervor, daß das Modell (XIV) im Gegensatz zu 1-Methyl-2-vinyl-cyclohexen sehr leicht mit Maleinsäureanhydrid reagiert (27). Die Darstellung des Triens (XIV) auf einem neuen Wege in sehr reiner Form ließ auch die scheinbaren Widersprüche in der UV-Absorption verschwinden (27). Bei einem solchen Vergleich der UV-Absorption der 9,10-seco-Steroide mit ihren Modellen muß man einen bathochromen Effekt von 5—6 m$\mu$ berücksichtigen, der auf den Einfluß des substituierten Ringes D mit der angulären Methylgruppe zurückzuführen ist (23, 28). Da schließlich auch noch das in seiner Struktur auf Grund der Darstellung eindeutige u-Tachysterin$_2$ in seinen Eigenschaften vom Tachysterin$_2$ verschieden ist, bestätigt sich die Lage des Trien-systems der Tachysterine gemäß Formel (X) als 10,5 — 6,7 — 8,9-Trien.

Die Frage der 6,7-cis-trans-Isomerie wurde ebenfalls geklärt. Während früher eine cis-Doppelbindung angenommen worden war, konnte man durch die Synthese einer Reihe von Modell-Kohlenwasserstoffen beweisen, daß eine trans-Doppelbindung zwischen den beiden Brücken-Kohlenstoffatomen vorliegt (23, 34), wie durch Vergleich der UV-Absorption hervorging. Im IR-Spektrum ist diese Doppelbindung durch eine Bande bei 955 cm$^{-1}$ gekennzeichnet (23), die auch beim iso-Tachysterin sowie dem Modell-Kohlenwasserstoff (XIV) auftritt (27).

Die Feinstruktur der Tachysterine ergibt sich daraus, daß die sterischen Hinderungen bei der 5,6-s-cis-Konstellation (zwischen der Methylgruppe an $C_{(10)}$ und dem Wasserstoff an $C_{(7)}$ und bei der 7,8-s-trans-Konstellation (zwischen den H-Atomen an $C_{(6)}$ und $C_{(15)}$) größer sind als bei den entgegengesetzten Konstellationen. Wahrscheinlich liegt daher die 5,6-s-trans-7,8-s-cis-Konstellation (X) im Isomerengleichgewicht bevorzugt vor. Damit ist das Trien-system der Tachysterine aus einer all-trans-Anordnung herausgedreht, was sich in der niedrigen Extinktion seines UV-Maximums sowie dessen relativ kurzwelliger Lage meßbar äußert ($\lambda_{max} = 281$ m$\mu$, $\varepsilon = 24600$).

Durch die 7,8-s-cis-Konstellation entsteht in 6-9-Stellung ein cisoides Diensystem mit trans-ständigen Substituenten, wie es für den Eintritt einer raschen Diensynthese Bedingung ist. So findet das stark ausgeprägte Reaktionsvermögen des Tachysterins bei der Umsetzung mit Philodienen eine plausible Erklärung. Die sterische Behinderung der 5,6-s-cis-Konstellation dagegen schließt eine 10-7-Diensynthese praktisch aus. Andererseits ist gerade in der Leichtigkeit der 6-9-Dienreaktion ein unabhängiger Hinweis auf die trans-Konfiguration der 6,7-Doppelbindung zu sehen (2).

*Literaturverzeichnis: SS. 118—123.*

## 6. iso-Tachysterine.

Das iso-Tachysterin$_2$ war seinerzeit durch Isomerisierung von Vitamin D$_2$ mit Bortrifluorid-ätherat erhalten und durch Abbaureaktionen sowie Erörterung der UV-Absorption in seiner Konstitution aufgeklärt worden, die zu der Formel (XI) geführt hatte (22). Die Messung des IR-Spektrums (23) bestätigte die abgeleitete Konstitution (XI); es tritt nämlich darin eine ausgeprägte Bande bei 955 cm$^{-1}$ auf, die der disubstituierten, konjugierten 6,7-*trans*-Doppelbindung zuzuordnen ist. Auch das iso-Tachysterin-keton$_2$ (XV), in kristalliner Form durch Alkoholat-oxydation aus iso-Tachysterin darstellbar (22, 23), zeigt in seinem IR-Spektrum diese Bande.

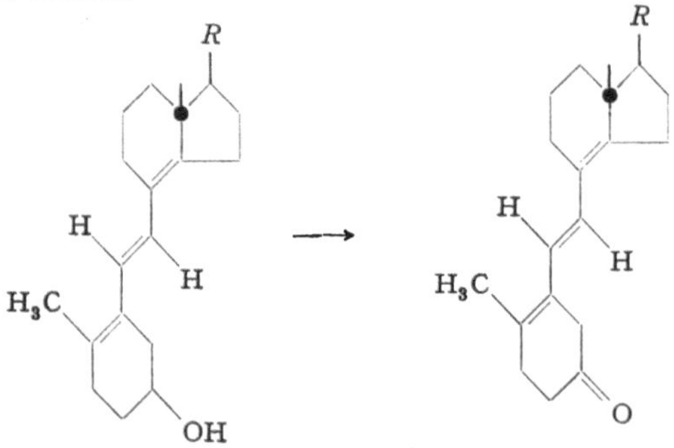

(XI.) iso-Tachysterin$_2$.

(XV.) iso-Tachysterin-keton$_2$.

Die iso-Tachysterine haben sich für eine große Anzahl von Trienen der Vitamin D-Reihe als der Endpunkt einer chemischen Isomerisierung erwiesen. So entstand iso-Tachysterin$_2$ bei der BF$_3$-ätherat-Behandlung von Vitamin D$_2$ (22), 5,6-*trans*-Vitamin D$_2$ (36), iso-Vitamin D$_2$ (22), Praecalciferol$_2$ (36), Tachysterin$_2$ (23). Diese Isomerisierungsreaktionen sind sämtlich von einer bathochromen Verschiebung der UV-Absorption begleitet, die auf eine sterische Entspannung unter Ausbildung eines all-*trans*-Chromophors in Verbindung mit einer Zunahme des Substitutions-

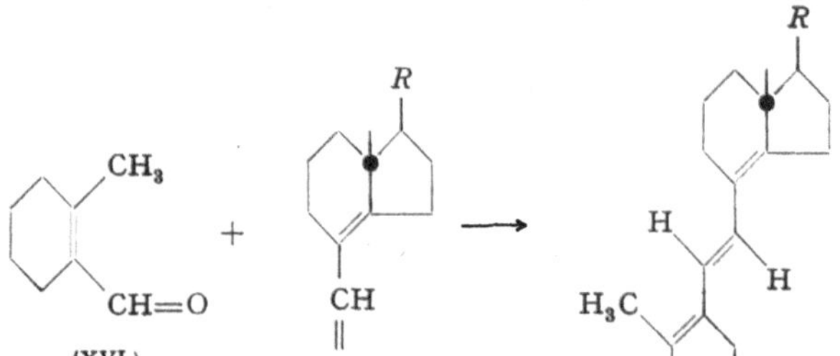

(XVI.)
Methyl-cyclohexen-aldehyd.

(XVII.)

(XVIII.)

grades des chromophoren Systems zurückzuführen ist, und bestätigen ihrerseits die Konstitution (XI). Ein weiterer Hinweis ergibt sich aus dem experimentellen Befund (23), daß das iso-Tachysterin nicht mit Maleinsäureanhydrid reagiert.

Auf synthetischem Wege wurde die Lage der Doppelbindungen 10,5 und 6,7 dadurch bestätigt, daß das Trien-system mit der UV-Absorption des iso-Tachysterins durch Wittig-Reaktion aus dem Methyl-cyclohexen-aldehyd (XVI) entsteht (18, 44).

### 7. u-Tachysterin.

Das u-Tachysterin$_2$ (XII) erhält man (25), wenn das Keton (XIX), das bei der sauren Spaltung des Semicarbazons des iso-Tachysterin-ketons$_2$ (XV) entsteht (23), mit Lithiumaluminiumhydrid reduziert wird.

Die Lage des Trien-systems des u-Tachysterins ergibt sich aus der Entstehung sowie aus der UV-Absorption des Ketons (XIX) bei 324 m$\mu$, bzw. seines Semicarbazons bei 334 m$\mu$.

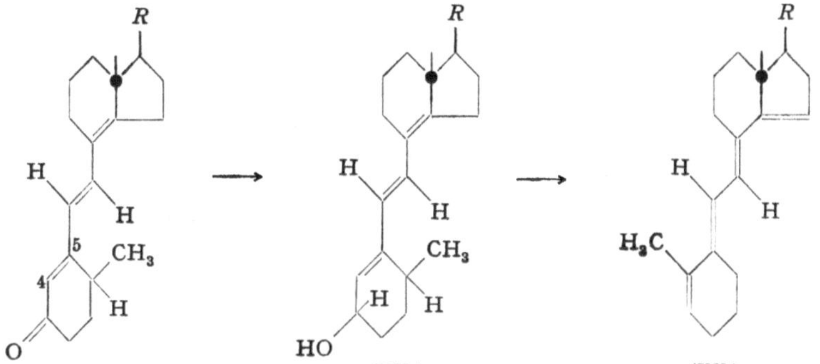

(XIX.)          (XII.)          (XX.)

Weiterhin ist für das u-Tachysterin kennzeichnend, daß es sehr leicht, schon bei der Destillation und unter milden Veresterungsbedingungen, Wasser abspaltet, so daß sich keine Derivate erhalten ließen. Dieses Verhalten stimmt mit der Allylstellung der OH-Gruppe überein. Dem entstehenden Kohlenwasserstoff $C_{28}H_{42}$ ist auf Grund seiner UV-Absorption bei 305 m$\mu$ wohl die Formel (XX) zuzuordnen. Die UV-Absorption des u-Tachysterins bei 285 m$\mu$ entspricht einem all-*trans*-konfiguriertem System. Die hypsochrome Verschiebung im Vergleich zum iso-Tachysterin ist darauf zurückzuführen, daß das u-Tachysterin am Chromophor einen Substituenten weniger trägt. Die bathochrome Verschiebung der Absorption gegenüber dem Tachysterin mit gleicher Substituentenzahl gibt einen Hinweis auf dessen 7,8-s-*cis*-Konstellation, während das u-Tachysterin sterisch freier ist.

## 8. Pyrotachysterin(e).

Während den Pyrotachysterinen früher die Formel (X) zugeordnet wurde, die, wie wir jetzt wissen, den Tachysterinen zukommt, hat eine nochmalige Darstellung eines reineren Pyrotachysterin$_2$-Präparates (*23*) verbesserte Daten für die UV-Absorption und die optische Drehung ergeben, die zu dem Schluß führten, daß es sich beim Pyrotachysterin$_2$ um iso-Vitamin D$_2$ (VI) handelt. Sowohl die Lage der UV-Absorptionsmaxima als auch die positive Richtung der optischen Drehung stimmen überein; in der molaren Extinktion sowie im Absolutwert der Drehung liegen die Daten des noch immer unreinen Pyrotachysterins niedriger. Beide Substanzen sind thermisch stabil.

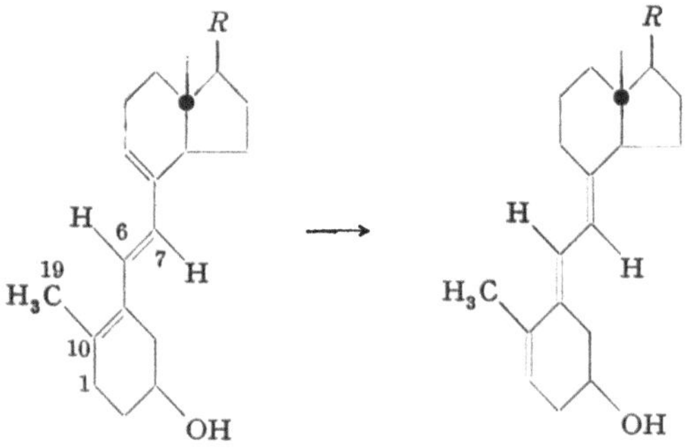

(X.) Tachysterin.          (VI.) iso-Vitamin D$_2$.

Während offenbar die 6,7-*cis*-Doppelbindung die Verschiebung des Triensystems über einen cyclischen Mechanismus in die C$_{(19)}$-Stellung ermöglicht, muß die analoge Verschiebung beim 6,7-*trans*-konfigurierten Tachysterin, das die Ausbildung eines cyclischen Mechanismus verbietet, unter größerem Widerstand in die C$_{(1)}$-Stellung ausweichen.

## 9. Suprasterine.

Jedes Vitamin D gibt nach WINDAUS bei längerer UV-Bestrahlung zwei Suprasterine, die als Suprasterin-I und Suprasterin-II bezeichnet werden. Während über die Suprasterine-I noch nicht genügend experimentelles Material vorliegt, um ihnen eine eindeutige Konstitution zuordnen zu können, ist die Konstitution des Suprasterins$_2$-II in jüngster Zeit durch Untersuchungen von DAUBEN und Mitarb. (*11*) aufgeklärt worden. Das Vorhandensein der normalen Ergosterin-Seitenkette konnte durch Ozonabbau zum Methyl-isopropyl-acetaldehyd bereits früher gesichert werden, ein Befund (*13*), der sich durch IR-Spektrum, Kernresonanzspektrum und partielle Hydrierung weiter bestätigen ließ (*11*). Die Anwesenheit einer acylierbaren Hydroxylgruppe war schon von Windaus und Mitarb. (83) nachgewiesen worden. Diese Autoren zeigten auch, daß es sich um ein Ergosterin-Isomeres handelt und daß Suprasterin$_2$-II kein konjugiertes Dien-system enthält. Weitere Arbeiten der Windausschen Schule hatten ergeben, daß das Ringsystem sowohl von dem des Vitamins D$_2$ als auch von dem des Ergosterins verschieden sein müsse. So konnte bei der Dehydrierung mit Selen oder Platin kein Diels-Kohlenwasserstoff C$_{18}$H$_{16}$ erhalten werden (50), und die Perhydrierung ergab ein Suprastanol-II, das sowohl vom Hexa-hydro-Vitamin D$_2$ als auch vom Ergostanol verschieden war (1, 83). Es mußte also eine Veränderung des Vitamin D-Gerüstes, wahrscheinlich eine Cyclisierung, eingetreten sein, bei der sich keine neue 9,10-Bindung ausgebildet hatte.

Eine C-Methyl-Bestimmung nach Kuhn-Roth ergab vier C-Methyl-gruppen, wodurch wahrscheinlich wurde, daß das C$_{(19)}$ nicht als Methyl-gruppe vorlag (11). Die quantitative Mikrohydrierung (11, 50) ergab die Aufnahme von 3 Molen Wasserstoff, so daß also mindestens vier Ringe vorliegen mußten. Bei milder Hydrierung über Pd/CaCO$_3$ wird nur die Seitenketten-doppelbindung hydriert, wie man durch das Ver-schwinden der IR-Bande bei 970 cm$^{-1}$ erkennt. Das Kernresonanz-spektrum des Dihydro-Derivates zeigt keine Vinyl-Wasserstoffatome mehr. Mit Osmiumtetroxyd erhielt Dauben (11) aus dem Suprasterin-II ein Triol, das nach Hydrierung der Seitenketten-doppelbindung keine Absorption im UV besaß. Folglich enthält das Suprasterin-II außer der *trans*-disubstituierten eine tetrasubstituierte Doppelbindung sowie fünf Ringe, von denen einer durch Hydrierung zu öffnen ist.

Das schon von der Windausschen Schule (1) erhaltene Tetrahydro-suprasterin-II, das bei der Hydrierung mit Platin in der Kälte entsteht, besitzt im UV eine hohe Endabsorption, im Kernresonanzspektrum aber keine Vinyl-Wasserstoffatome. Es enthält also noch die tetrasubstituierte Doppelbindung, die sich auch mit Osmiumtetroxyd erfassen läßt, während die Seitenkette abgesättigt sowie ein Ring aufhydriert sind. Die relativ leichte Hydrierbarkeit deutet auf einen Cyclopropanring hin.

Die UV-Absorption des Suprasterins-II sowie des 22-Dihydro-supra-sterins-II bei 210 m$\mu$ zeigt, daß die tetrasubstituierte Doppelbindung und der Cyclopropanring in Konjugation stehen. Ein aus dem Dihydro-Derivat durch Chromsäure-Oxydation erhältliches Keton läßt sich zu einem Keton mit $\lambda_{max} = 268$ m$\mu$ isomerisieren. Aus diesem Ergebnis folgt die $\beta,\gamma$-Stellung der tetrasubstituierten Doppelbindung zur OH-Gruppe sowie die Verknüpfung des Cyclopropanringes mit dem $\beta$-ständigen C-Atom. Dieser steht nämlich auch noch nach der Isomerisierung in Konjugation, wie besonders der Vergleich mit dem aus Tetrahydro-suprasterin-II erhältlichen Keton zeigt, das nach Isomerisierung bei 242 m$\mu$ absorbiert. Letzteres ergibt bei der Ozonisierung eine Ketosäure, deren Ester durch seine IR-Bande bei $1704\,\mathrm{cm^{-1}}$ ein Sechsringketon anzeigt. Das Keton hat also eine konjugierte trisubstituierte Doppelbindung, die exocyclisch zu einem sechs- oder mehrgliedrigen Ring steht, der ursprünglich das konjugierte Doppelbindung-Cyclopropan-System enthielt. DAUBEN teilt auf Grund dieser Befunde dem Suprasterin$_2$-II die Formel (XXI) zu und deutet seine Entstehung aus Vitamin D$_2$ durch folgende Umlagerung:

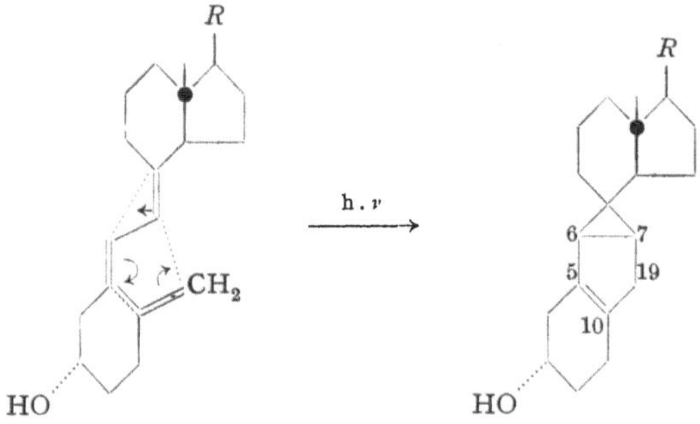

(I.) Vitamin D$_2$.    (XXI.) Suprasterin$_2$-II.

Das würde bedeuten, daß das Suprasterin$_2$-II bei der Ergosterin-Bestrahlung aus dem Vitamin D$_2$ selbst, nicht aber direkt aus dem Praecalciferol entsteht, wodurch sich eine Temperaturabhängigkeit der entstehenden Suprasterin-II-Menge ergeben müßte. Bemerkenswert angesichts der Formel (XXI) bleibt der WINDAUSsche Befund (83), daß bei der Benzopersäuretitration 3 Atome Sauerstoff verbraucht werden. Das würde bedeuten, daß hierbei der sonst als relativ oxydationsunempfindlich bekannte Dreiring aufgespalten werden müßte. Zu beachten ist die erste fruchtbare Verwendung von Kernresonanzspektren im Bereich der Vitamin D-Chemie.

## 10. Dihydro-Verbindungen.

Von den Vitaminen D und ihren Isomeren leitet sich eine Reihe von Dihydro-Verbindungen ab, deren Entstehung nicht nur davon abhängt, welches 9,10-seco-Steroid man zur Hydrierung einsetzt, sondern auch davon, welches Reduktionsmittel verwendet wird.

Am eingehendsten ist die Hydrierung der Vitamine D untersucht worden. Bei der Hydrierung mit Natrium in Äthanol oder Propanol entstehen, wie schon WINDAUS und Mitarb. (84, 87) zeigten, die sogenannten Dihydro-Vitamine D-I und D-II. Verwendet man dagegen einen höheren Alkohol, wie n-Butanol, so bildet sich daneben auch Dihydro-tachysterin (48), das sich durch seine antitetanische Wirkung als ein wertvolles pharmazeutisches Präparat bewährt hat, sowie ein weiteres Produkt, das Dihydro-Vitamin D-IV (*62*, *78*). Schubert und Mitarb. (*62*) isolierten noch zwei weitere Substanzen, nämlich das von ihnen so benannte Dihydro-Vitamin D-III und das Dihydro-tachysterin$_2$ 66, ein Befund, der von anderen Arbeitsgruppen bislang noch nicht bestätigt werden konnte. Mit Natrium und Glycerin in Xylol wurden als Reduktionsprodukte bisher nur Dihydro-Vitamin D-I und Dihydro-tachysterin isoliert (*62*, *76*). Ohne Dihydro-Vitamine D entsteht nach Untersuchungen der holländischen Autoren (*79*) das Dihydro-tachysterin bei der Reduktion mit Lithium in flüssigem Ammoniak.

Französische Autoren (*3*, *53*) wollen unter diesen Bedingungen noch ein weiteres Produkt gefaßt haben, das sie als 9,10-*seco*-Cholesterin bezeichnen. Bei der Hydrierung mit Raney-Nickel als Katalysator erhielt Schubert (*59*, *61*) Dihydro-Vitamin D-II, außerdem isolierten die holländischen Forscher Dihydro-Vitamin D-IV (*81*). Die gleichen Autoren (*81*) untersuchten auch die Reduktion mit Natrium und N-Methylanilin, wobei sie neben Dihydro-Vitamin D-I ein neues Dihydro-Vitamin D-V isolierten, sowie die Hydrierung mit Palladium auf Calciumcarbonat als Katalysator, wobei sie unter anderem ein neues Dihydro-Vitamin D-VI fanden. Auch diese beiden Stoffe sowie das 9,10-*seco*-Cholesterin sind bislang von anderen Arbeitsgruppen nicht bestätigt worden. Billet (*6*) erhielt bei der Reduktion mit Natrium und N-Methylanilin ein Produkt, das dem Dihydro-Vitamin D-V in seinen Eigenschaften ähnelt, aber etwas andere spektrale Daten aufweist.

Die Reduktion der 5,6-*trans*-Vitamine D mit Natrium und 2-Methylbutanol-2 in Xylol ergibt Dihydro-tachysterin und Dihydro-Vitamin D-II. Bei der Hydrierung mit Raney-Nickel als Katalysator entsteht Dihydro-tachysterin (*47*). Praecalciferol liefert nach Velluz und Mitarb. (*65*) bei der Hydrierung über Raney-Nickel ein unkonjugiertes, nicht kristallin isoliertes Dihydro-Derivat, das inzwischen als kristallines Allophanat isoliert werden konnte (*63*), während die holländischen Autoren (*79*) bei der Reduktion mit Lithium in flüssigem Ammoniak Dihydro-tachysterin fanden. Tachysterin ergibt mit Natrium und Propanol Dihydro-Vitamin D-I und Dihydro-tachysterin, wie mehrfach bestätigt wurde (*77*, *79*). Die holländischen Autoren (*79*) konnten auch Dihydro-Vitamin D-II nachweisen. Mit Lithium in flüssigem Ammoniak dagegen entsteht das Dihydro-tachysterin ohne Dihydro-Vitamine D (*75*, *79*). Daß aus dem Suprasterin-II ein Dihydro-Derivat mit hydrierter Seitenkette zu erhalten ist, war schon erwähnt worden (S. 83).

Die Konstitution der Dihydro-Vitamine D-I konnte durch Abbau- und Umwandlungsreaktionen von v. Reichel und Deppe (*58*) sowie von Windaus und Mitarb. (*86*) aufgeklärt werden und führte zu der Formel (XIII).

Dem Dihydro-Vitamin D-II ist auf Grund seiner UV-Absorption bei 251 m$\mu$ ein 5,7-Dien-system zuzuordnen. Unter der Annahme, daß es aus dem Vitamin D durch Hydrierung der 10,19-Methylengruppe ohne

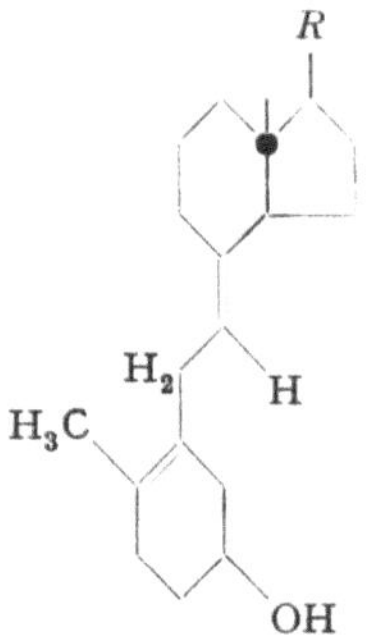

(XIII.) Dihydro-Vitamin D-I. 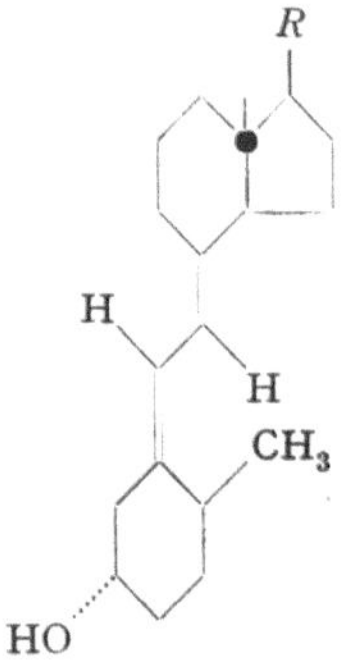     (XXII.) Dihydro-Vitamin D-II.

weitere Veränderung entsteht, müßte ihm die Formel (XXII) zukommen. Für die Annahme eines Primärproduktes mit einer 5,6-*cis*-Doppelbindung spricht die Empfindlichkeit des Dihydro-Vitamins D-II gegenüber isomerisierenden Bedingungen (*61*) in Verbindung mit der Tatsache, daß es bei der Reduktion in alkalischem Medium nur in geringer Menge auftritt. Die Stellung der $C_{(19)}$-Methylgruppe läßt sich aus dem IR-spektrographischen Befund ableiten, daß die OH-Gruppe des Dihydro-Vitamins D-II im Gegensatz zu allen anderen Dihydro-Derivaten eine axiale Stellung einnimmt (*61*). In der 6,7-*s-trans*-Form (I, S. 71) nimmt die OH-Gruppe des Vitamins D die äquatoriale $\alpha$-Stellung ein. Erfolgt nun bei der katalytischen Hydrierung die Anlagerung des Wasserstoffs von der $\alpha$-Seite her, so entsteht eine $\beta$-ständige, gleichfalls äquatoriale Methylgruppe. Nun läßt sich aus Modellbetrachtungen ersehen, daß äquatoriale $C_{(19)}$-Methylgruppen und der Wasserstoff an $C_{(7)}$ sich sterisch behindern. Dieser Hinderung kann durch eine konstellationelle Transformation des Ringes *A* ausgewichen werden, wodurch sowohl die Methylgruppe als auch die OH-Gruppe axial werden. Bei der Anlagerung des Wasserstoffs von der $\beta$-Seite her entsteht eine $\alpha$-ständige, axiale Methylgruppe, die keine sterische Hinderung auf den $C_{(7)}$-Wasserstoff ausübt.

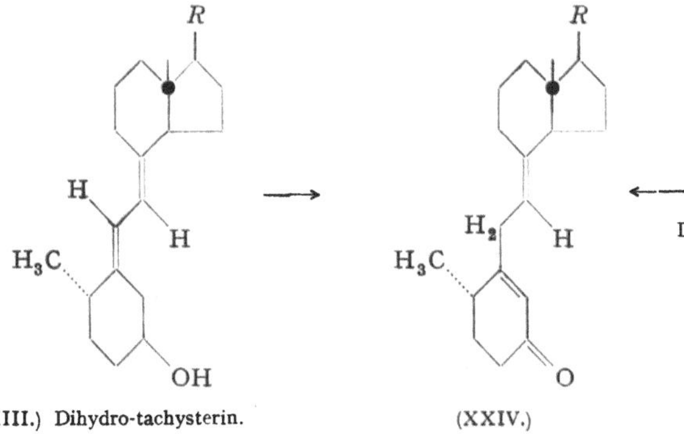

(XXIII.) Dihydro-tachysterin.     (XXIV.)

$\longleftarrow$ (XXII.)
Dihydro-Vitamin-D-II.

Da diese Hinderung nach einer konstellationellen Transformation auftreten würde, unterbleibt diese, so daß die OH-Gruppe ihre äquatoriale Lage beibehält. Das Auftreten einer axialen OH-Gruppe im Dihydro-Vitamin D-II ist also ein Beweis für die *trans*-Stellung von Hydroxyl und Methyl gemäß Formel (XXII).

Das Dihydro-Vitamin D-II läßt sich nach Oppenauer zu einem $\alpha,\beta$-ungesättigten Keton (XXIV) oxydieren. Das gleiche Keton entsteht, wie Westerhof und Keverling Buisman (*81*) fanden, durch Oppenauer-Oxydation aus Dihydro-tachysterin, dessen Lage der Doppelbindungen schon durch v. Werder (*77*) aufgeklärt worden war. Da die Äthylen-Isomerie der 5,6-Doppelbindung durch diese Reaktion aufgehoben wird (vgl. Konstitutionsaufklärung der 5,6-*trans*-Vitamine D; S. 72), kann sich das Dihydro-tachysterin nur durch 5,6-*cis-trans*-Isomerie vom Dihydro-Vitamin D-II unterscheiden. Folglich kommt ihm die Formel (XXIII) zu.

Das Dihydro-Vitamin D-IV dagegen, das ebenfalls im UV bei 251 m$\mu$ absorbiert, ergibt bei der Oppenauer-Oxydation ein anderes $\alpha,\beta$-ungesättigtes Keton (*81*), dem offenbar die Formel (XXVI) zuzusprechen ist.

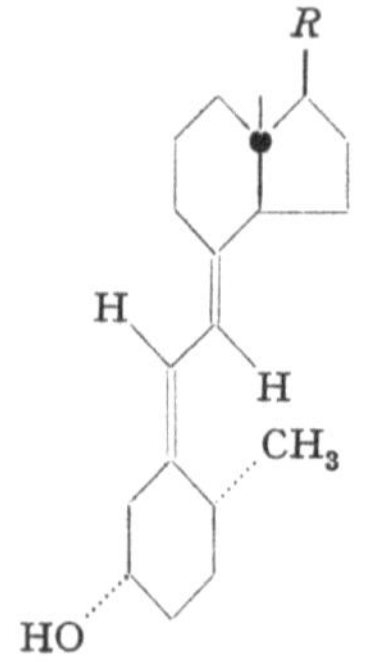

(XXV.) Dihydro-Vitamin D-IV. 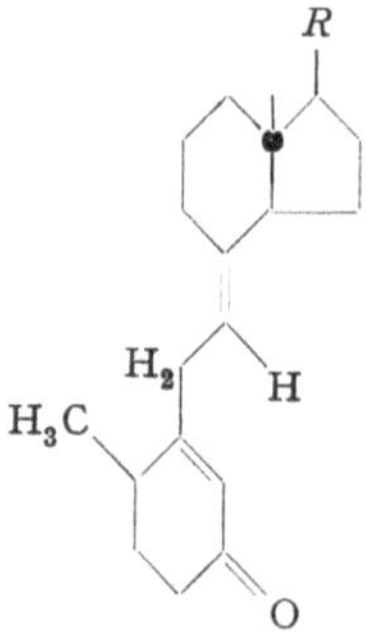    (XXVI.) 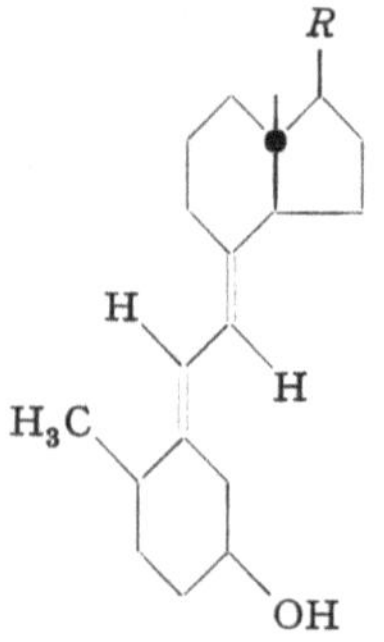    (XXVII.)

Für das Dihydro-Vitamin D-IV sind daher die Formeln (XXV) und (XXVII) zu diskutieren. Aus dem Auftreten des Dihydro-Vitamins D-IV neben D-II bei der Hydrierung mit Raney-Nickel (*81*) darf man wohl folgern, daß die Wasserstoffanlagerung an Vitamin D zum Teil auch von der $\beta$-Seite her erfolgt, so daß dem Dihydro-Vitamin D-IV die Formel (XXV) zuzuordnen wäre. Von den möglichen vier 5,6-7,8-Dienen bliebe dann noch eine Formel frei. Dieser stehen die beiden Schubertschen Substanzen (*62*) „Dihydro-Vitamin D-III" und „Dihydro-tachysterin$_2$ 66" mit UV-Absorption bei 251 m$\mu$ gegenüber. Sollte sich die Existenz beider Stoffe bestätigen, müßte man die Möglichkeit einer Epimerie an $C_{(3)}$ einbeziehen.

Dem 9,10-*seco*-Cholesterin, den Dihydro-Vitaminen D-V und D-VI sowie dem Billetschen Dien kommen auf Grund ihrer kurzwelligen UV-Absorption andere konjugierte Diensysteme zu, denen eine 6,7-Doppelbindung gemeinsam ist. Die *trans*-Konfiguration dieser Doppelbindung ist nur beim Dihydro-Vitamin D-V und bei BILLETS Dien einigermaßen sicher durch das IR-Spektrum (*6, 81*) nachgewiesen, die Lage der anderen Doppelbindung noch unsicher (vgl. S. 99).

## 11. $C_{(3)}$-Ketone.

WINDAUS und Mitarb. (*82*) erhielten bei der OPPENAUER-Oxydation des Vitamins $D_2$ ein Keton, für das sie die Formel (V, s. unten) als eine der möglichen Konstitutionen in Erwägung zogen. TRIPPETT (*64*) gelang es, dieses Keton in kristalliner Form darzustellen und die Konstitution (V) eines $\alpha,\beta; \gamma,\delta$-ungesättigten Ketons durch folgende Argumente und Reaktionen zu beweisen:

a) Durch die Lage des UV-Absorptionsmaximums des Semicarbazons bei 293 m$\mu$, b) durch die IR-Bande bei 1678 cm$^{-1}$ und c) durch die MEERWEIN-PONNDORF-Reduktion des Ketons (V) zum Alkohol (XXVIII), der eine Dien-absorption bei 235 m$\mu$ aufweist.

(I.) $\longrightarrow$
(S. 71)

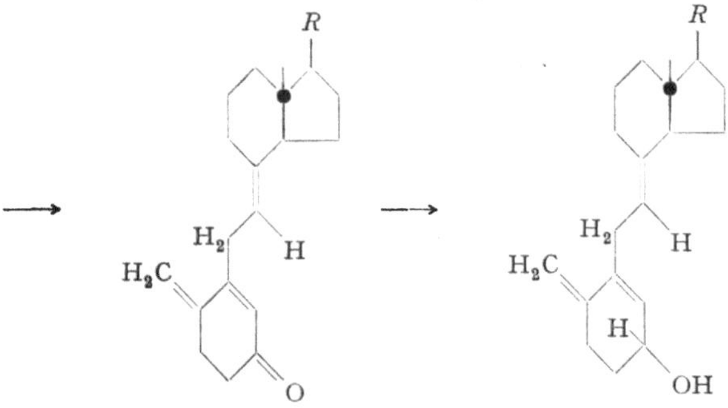

(V.)          (XXVIII.)

Bei der Spaltung des Semicarbazons von (V) mit Oxalsäure in Eisessig erhielten WINDAUS und Mitarb. (*82*) ein umgelagertes Keton un-

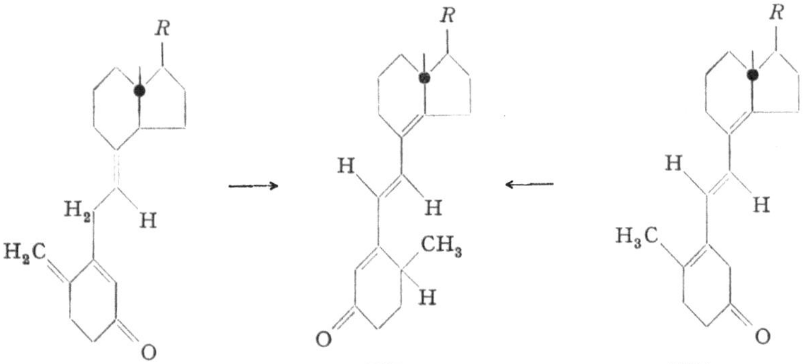

(V.)          (XIX.)          (XV.)

bekannter Konstitution. Es zeigte sich, daß dieses u-Keton in UV-Absorption, Spektrum des Semicarbazons sowie Schmelzpunkt und Misch-schmelzpunkt des Semicarbazons identisch ist (*44*) mit dem u-Keton (XIX), das bei der sauren Semicarbazonspaltung des iso-Tachysterin-ketons$_2$ (XV) entsteht (*23*). Dieses u-Keton (XIX) scheint demnach als Endpunkt der sauren Isomerisierung in der Reihe der 9,10-*seco*-Steroid-trienone-(3) die entsprechende Stellung einzunehmen, wie das iso-Tachysterin in der Trien-Reihe (vgl. S. 98).

## II. Neue Abbauprodukte der Vitamine D.

### 1. Die stereoisomeren 5,6-Dihydroxy-Vitamine D$_3$ (*29*.)

Bei der Oxydation der Vitamine D mit Bleitetraacetat und nachfolgenden Verseifung entstehen, wie Windaus und Mitarb. (*85*) zeigen

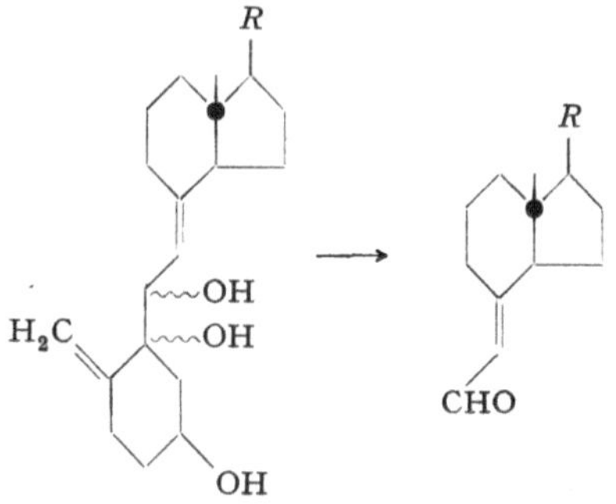

(XXIX.) 5,6-Dihydroxy-Vitamine D.   (XXX.) C$_{20}$-Abbau-aldehyd.

konnten, 5,6-Dihydroxy-Vitamine D (XXIX). Aus Vitamin D$_3$ erhielten diese Autoren ein Triol vom Schmp. 156°. Wie sich neuerdings zeigen ließ (*29*), entsteht daneben noch ein anderes Triol, das sich von dem ersteren nur durch die sterische Anordnung der OH-Gruppen in 5- und 6-Stellung unterscheidet. Für die Bildung dieser Isomeren gibt es zwei Möglichkeiten. Entweder erfolgt der Angriff des Bleitetraacetats sowohl von der α- als auch von der β-Seite her, oder es erfolgt vor dem Angriff des Oxydationsmittels eine teilweise Isomerisierung zum 5,6-*trans*-Vitamin D$_3$. Beide Triole bilden bei der Glykolspaltung mit Bleitetraacetat den schon von Brockmann und Mitarb. (*8*) beschriebenen C$_{20}$-Abbau-aldehyd (XXX), der durch diese Reaktionsfolge in präparativem Maße zugänglich ist.

### 2. Die *trans*-C,D-Abbau-alkohole (*39, 40*).

Durch Ozonabbau des Vitamins D$_3$ und des Dihydro-Vitamins D$_2$-I mit nachfolgender Lithiumalanat-Reduktion sind die im C,D-System *trans*-verknüpften Abbau-alkohole (XXXI) erhältlich (*39, 40*).

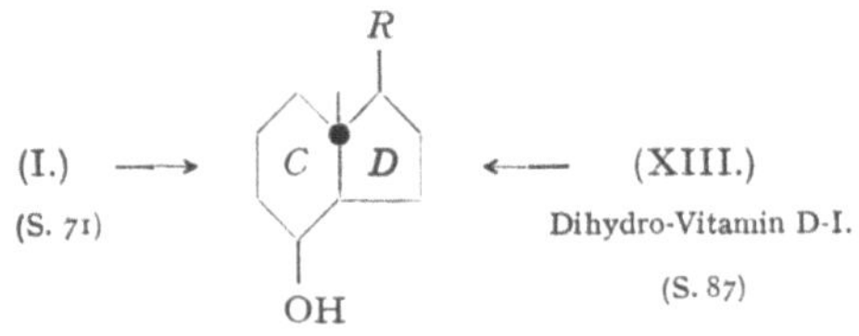

(I.) $\longrightarrow$    $\longleftarrow$ (XIII.)
(S. 71)    Dihydro-Vitamin D-I.
   (S. 87)

OH

(XXXI.) $R = C_8H_{17}$ und $C_9H_{17}$.

Im Falle der Vitamin $D_2$-Verbindung gelang dabei glatt der partielle Abbau an der 7,8-Doppelbindung unter Erhaltung der 22,23-Doppelbindung. Die Bedeutung dieses Abbauproduktes der $D_2$-Reihe liegt in der Möglichkeit, eine Verkürzung der Seitenkette durchzuführen, wobei weitere für die Synthese wichtige Relais-Verbindungen erhalten wurden (vgl. SS. 92 und 93). Ferner konnte die Isomerisierbarkeit der entsprechenden Ketone (XXXII) zu den $C,D$-cis-verknüpften Isomeren (XXXIII) studiert werden (40), ein Verhalten, das schon von DIMROTH und Mitarb. (12) beobachtet worden war.

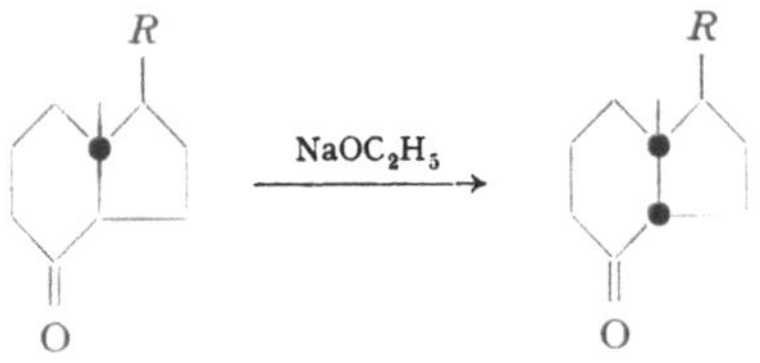

$NaOC_2H_5$

O    O

(XXXII.)    (XXXIII.)

Durch Umgehung der isomerisierbaren Ketonstufe bot der neue Reaktionsweg die Gewähr für die Erhaltung der trans-Verknüpfung in den Abbau-alkoholen (XXXI). Es zeigte sich (40), daß weder bei der Darstellung der Ketone (XXXII) aus (XXXI) durch Oxydation mit Chrom-(VI)-Oxyd/Pyridin noch bei der Durchführung von WITTIG-Reaktionen mit (XXXII) als Carbonylkomponente eine Isomerisierung zu (XXXIII) eintritt.

### 3. Die Seitenketten-alkohole (39, 45).

Der vollständige Ozonabbau des Dihydro-Vitamins $D_2$-I mit anschließender Lithiumalanat-Reduktion der Ozonide ergab das Diol (XXXIV) (39). Auch das optisch aktive, linksdrehende 2,3-Dimethyl-butanol-(1) (XXXV) konnte gefaßt und als Dinitrobenzoat charakterisiert werden (45).

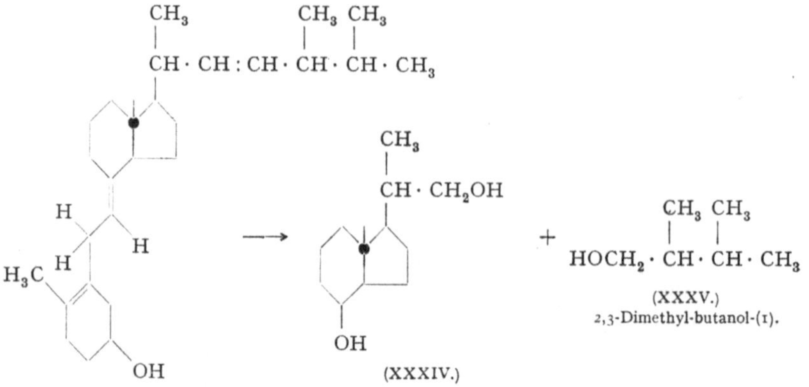

$$CH_3 \quad CH_3 \; CH_3$$
$$CH \cdot CH : CH \cdot CH \cdot CH \cdot CH_3$$

(XIII.) Dihydro-Vitamin D-I.

$$CH_3$$
$$CH \cdot CH_2OH$$

(XXXIV.)

$$+$$

$$CH_3 \; CH_3$$
$$HOCH_2 \cdot CH \cdot CH \cdot CH_3$$

(XXXV.)
2,3-Dimethyl-butanol-(1).

### 4. Seitenketten-acetoxy-aldehyde (39).

Durch Ozonisierung des Acetats des $C_{19}$-Abbau-alkohols (**XXXVI**) gelangt man zum Aldehyd (**XXXVII**), wenn das Ozonid mit desaktiviertem RANEY-Nickel gespalten wird. Bromierung zu (**XXXVIII**) und anschließende HBr-Abspaltung führt zum $\alpha,\beta$-ungesättigten Aldehyd (**XXXIX**) (39).

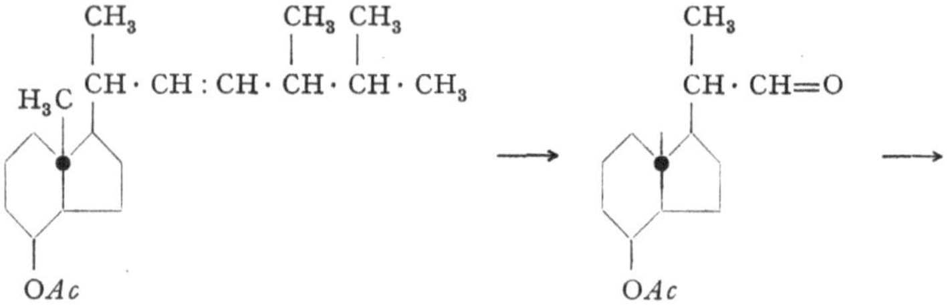

$$CH_3 \quad CH_3 \; CH_3$$
$$H_3C \quad CH \cdot CH : CH \cdot CH \cdot CH \cdot CH_3$$

OAc

(XXXVI.)

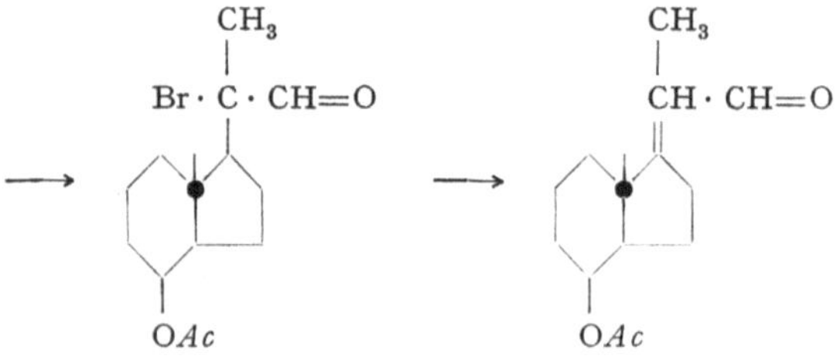

$$CH_3$$
$$CH \cdot CH{=}O$$

OAc

(XXXVII.)

$$CH_3$$
$$Br \cdot C \cdot CH{=}O$$

OAc

(XXXVIII.)

$$CH_3$$
$$CH \cdot CH{=}O$$

OAc

(XXXIX.)

### 5. Das 8-Methyl-*trans*-hydrindanol-(4)-on-(1) und die 8-Methyl-hydrindandione (39).

Die Darstellung des 8-Methyl-*trans*-hydrindanol-(4)-ons-(1) (**XL**) gelingt auf zwei Wegen (39). Einmal durch Ozonabbau des ungesättigten

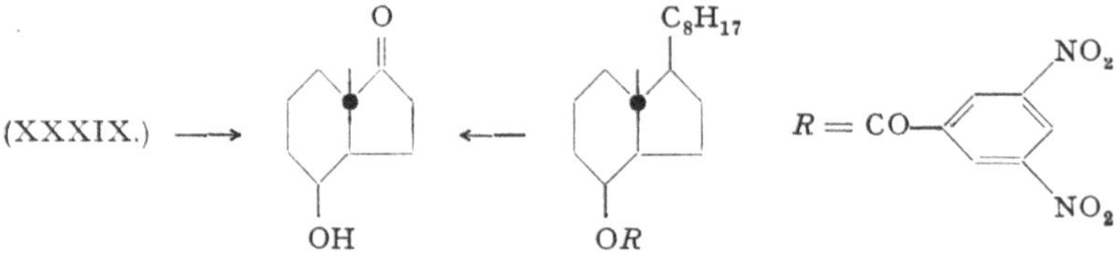

(XXXIX.) $\longrightarrow$     $\longleftarrow$     $R = CO-$

(XL.) 8-Methyl-*trans*-
hydrindanol-(4)-on-(1).

(XLI.)

Acetoxy-aldehyds (XXXIX) mit nachfolgender Verseifung, und zum anderen durch Chromsäure-Oxydation des Dinitrobenzoats des $C_{18}$-Abbaualkohols (XLI) mit nachfolgender Verseifung.

Durch Pyridin-chromat-Oxydation entstand aus diesem Olon das 8-Methyl-*trans*-hydrindandion-(1,4) (XLII), das alkalisch zum *cis*-verknüpften Dion (XLIII) isomerisiert werden konnte.

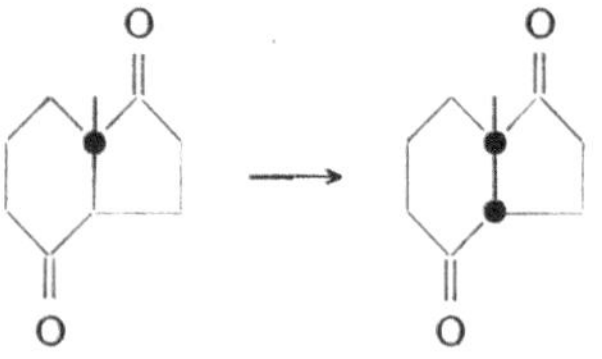

(XLII.)        (XLIII.)

### III. Photochemische Isomerisierung. '

Die photochemische Umwandlung der Provitamine D (Ergosterin, 7-Dehydro-cholesterin usw.) hatte bis zur Entdeckung des Praecalciferols (S. 76) durch das klassische Bestrahlungsschema gedeutet werden können, das aus den Arbeiten der WINDAUSschen Schule hervorgegangen war:

Provitamin D → Lumisterin → Tachysterin → Vitamin D → Suprasterine.

Nun hatten die Entdecker des Praecalciferols, VELLUZ und Mitarb. (72), gefunden, daß sich das Vitamin D durch eine nicht-photochemische Umwandlung aus dem Praecalciferol bildet. VELLUZ und Mitarb. (67—69) beobachteten ferner, daß bei der Bestrahlung von Praecalciferol Tachysterin, Ergosterin und Lumisterin entstehen. Die Bildung von Tachysterin und Ergosterin wurde inzwischen auch von HAVINGA (16) und der Philips-Arbeitsgruppe (57) bestätigt. Die wichtigste Folgerung aus diesem Ergebnis ist die, daß die Reaktionen, die während der Bestrahlung ablaufen, reversibel sind. Damit war fraglich geworden, ob das Lumisterin wirklich direkt aus Ergosterin durch eine Epimerisierung an $C_{(9)}$ und $C_{(10)}$ entsteht, oder ob es sekundär durch Re-cyclisierung aus einem 9,10-*seco*-Produkt gebildet wird. Ferner schien die Bildung von Tachysterin aus Praecalciferol zu erweisen, daß das Praecalciferol vor dem Tachysterin in der Bestrahlungsfolge zu stehen hat. Auf Grund dieser Befunde und Überlegungen stellte Velluz ein Bestrahlungsschema auf *(Formelübersicht 1)*. Hierin fungiert das Praecalciferol als zentrale Schlüsselsubstanz zwischen Provitamin D (XLIV), Lumisterin (XLV), Tachysterin (X) und Vitamin D (II).

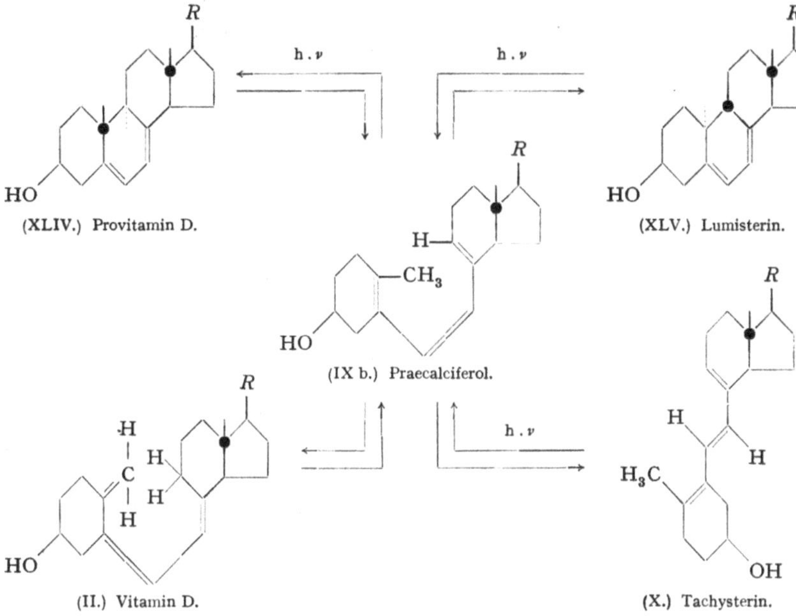

*Formelübersicht 1.* Bestrahlungsschema nach Velluz.

Die Reversibilität der beteiligten photochemischen Reaktionen ergibt sich auch aus dem neueren Befund von Rappoldt (55), daß sich bei längerer Bestrahlung von Ergosterin, Lumisterin oder Tachysterin ein stationärer Zustand einstellt, in dem Ergosterin, Tachysterin und Praecalciferol im gleichen, nur von der Wellenlänge des zur Bestrahlung verwendeten Lichtes abhängigen Verhältnis zu finden sind (57, 75). Das hat zur Folge, daß die Auffindung eines Stoffes im Bestrahlungsprodukt keineswegs ein Beweis für eine direkte Bildung dieses Stoffes aus dem Ausgangsmaterial ist. Es bedarf daher der Anwendung neuer Methoden, um herauszufinden, welche Übergänge zwischen den Substanzen des Bestrahlungsschemas tatsächlich stattfinden. Zwei solcher Methoden sind von Havinga und Mitarb. entwickelt und angewendet worden. Eine Methode ist die Analyse der Bestrahlungsgemische in Abhängigkeit von der Zeit, wodurch die primär entstehenden Produkte erkannt werden können. Die zweite Methode beruht auf folgendem Prinzip: Bestrahlt man ein Gemisch von $^{14}$C-haltigem Provitamin *(14)* mit inaktivem, vermeintlichem Zwischenprodukt A, so kann das Produkt B nur dann direkt aus dem Provitamin entstanden sein, wenn seine spezifische Aktivität höher als die von A ist *(15, 16)*.

Mit Hilfe dieser Methoden haben die holländischen Autoren die Reaktionen des klassischen und des VELLUZschen Bestrahlungsschemas überprüft. Für den neuesten, aber wahrscheinlich noch nicht letzten Stand der Forschung *(55, 56)* ergibt sich aus diesen Untersuchungen folgendes Bestrahlungsschema, wenn man für jeden der vier beteiligten Stoffe einen Anregungszustand formuliert *(Formelübersicht 2)*.

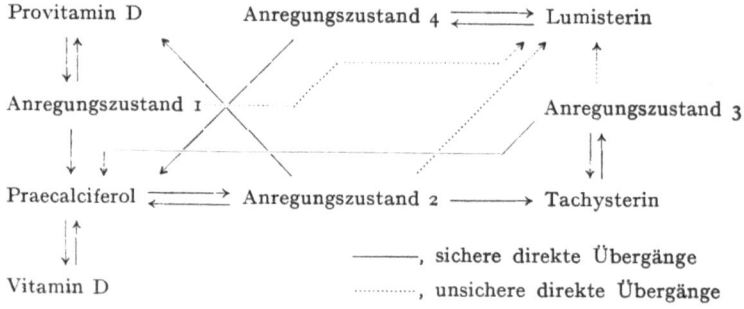

*Formelübersicht 2.* Bestrahlungsschema nach RAPPOLDT.

Von den Reaktionen der klassischen Bestrahlungsreihe ist der Übergang Lumisterin → Tachysterin nicht mehr enthalten, da das bei der Bestrahlung von Lumisterin gebildete Tachysterin nicht aus diesem direkt, sondern aus Praecalciferol sekundär entsteht. Die direkte Bildung von Praecalciferol aus Provitamin D und aus Lumisterin wurde von den holländischen Autoren nachgewiesen. Die von VELLUZ gefundenen Umwandlungen des Praecalciferols in Tachysterin und Provitamin erwiesen sich als direkte Übergänge, ebenso der Schritt vom Tachysterin zum Praecalciferol. Das VELLUZsche Bestrahlungsschema hat jedoch nur dann volle Gültigkeit, wenn das bei der Bestrahlung von Praecalciferol gebildete Lumisterin nur aus diesem direkt entsteht. Das ist jedoch noch nicht bewiesen.

Eine endgültige Klärung dieses Problems sowie der Fragen, ob die Schritte Provitamin → Lumisterin und Tachysterin → Lumisterin direkte Übergänge darstellen, und ob nicht auch in gewissem Umfang (unter bestimmten Bestrahlungsbedingungen) ein direkter Übergang vom Provitamin zum Vitamin möglich ist, steht noch aus.

Über die Natur der Anregungszustände existieren bislang nur Spekulationen. Physikalische Messungen, die einen tieferen Einblick in die Bestrahlungsvorgänge gestatten würden, liegen noch nicht vor. HAVINGA und Mitarb. (74) haben Phosphoreszenzprüfungen am Ergosterin in Angriff genommen.

Über die Rolle der übrigen, bei der Bestrahlung der Provitamine D auftretenden Isomeren ist noch wenig bekannt. Lediglich die Kenntnis des Suprasterins-II konnte durch seine Konstitutionsaufklärung (S. 83) in letzter Zeit gefördert werden. Die Untersuchung der schon von der WINDAUSschen Schule gefundenen Toxisterine wurde von der *Philips*-Arbeitsgruppe (80) wieder aufgenommen und diese Stoffgruppe durch Isolierung einiger neuer Stoffe bereichert.

Zur Photo-isomerisierung der 5,6-*trans*-Vitamine D zu den Vitaminen D durch Bestrahlung in Glas mit UV-Licht vgl. S. 118. Über die Bestrahlung von iso-Vitamin D (23) und iso-Tachysterin (22) liegen bislang nur orientierende Versuche vor.

## IV. Thermische Isomerisierung.

Die Vitamine D stehen mit den entsprechenden Praecalciferolen in einem thermischen Gleichgewicht. Während Temperaturerhöhung die Einstellung des Gleichgewichtes beschleunigt, läßt sich dieses andererseits bei tiefer Temperatur einfrieren, wodurch die Isolierung beider Komponenten möglich wurde. Nachdem nun die Konstitution der Praecalciferole festgelegt worden war (S. 76), begann man, sich Vorstellungen über den Mechanismus der thermischen Isomerisierung zu machen. Obwohl experimentelles Material bislang nicht vorliegt, nehmen sowohl VELLUZ als auch HAVINGA einen intramolekularen Mechanismus

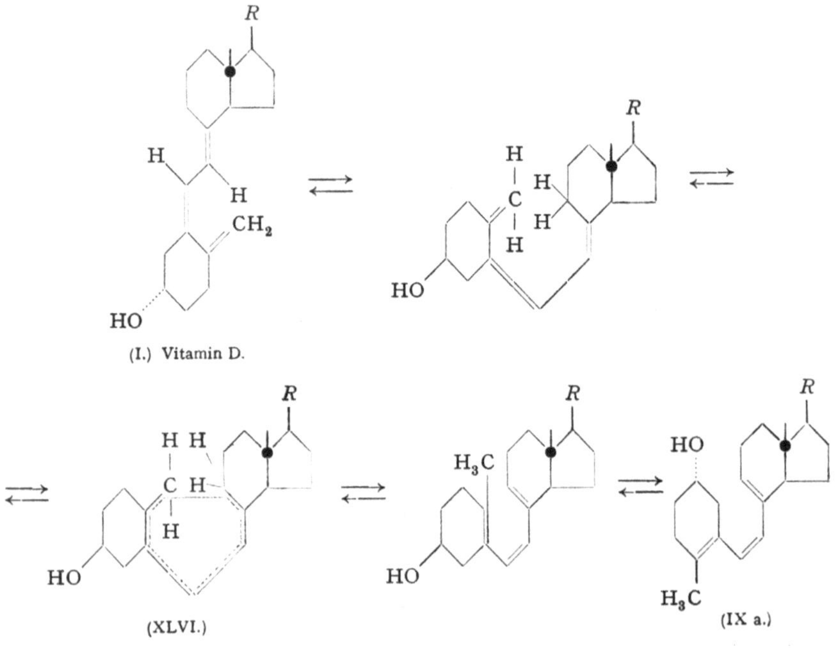

(I.) Vitamin D.

(XLVI.)

(IX a.)

*Formelübersicht 3.* Mechanismus der thermischen Isomerisierung von Vitamin D.

*Literaturverzeichnis: SS. 118—123.*

an. Nach Velluz (71) werden die s-cis-Formen (II) und (IX b) durchlaufen; Havinga (74) formuliert einen cyclischen Übergangszustand
(XLVI) *(Formelübersicht 3)*.

Ob ein solcher 8-gliedriger Übergangszustand wirklich existiert, bleibt
zu beweisen. Die Synthese der Vitamine D (S. 108) gestattet jetzt
durch die Möglichkeit der Deuterierung des C-Atoms 19, diese Verhältnisse
zu überprüfen. Für einen cyclischen Übergangszustand spricht die
Tatsache, daß das Tachysterin, dem auf Grund seiner 6,7-*trans*-Struktur
die Möglichkeit fehlt, sein Trien-system in eine solche Anordnung zu
bringen, sich nicht thermisch zum Vitamin D isomerisieren läßt. Infolgedessen enthält das Pyrotachysterin (VI, S. 75) auch nicht die exocyclische 10,19-Doppelbindung (23), deren Entstehung nur durch den
cyclischen Mechanismus plausibel wird, sondern die 1,10-Endodoppel-
bindung, während die Verschiebung des Trien-systems der Isomerisierung
des Praecalciferols sonst analog ist. Allerdings erfordert die Isomerisierung
des Tachysterins (X, S. 94) wesentlich höhere Temperaturen.

Die thermische Isomerisierung der Vitamine D zu den von Windaus
gefunden und untersuchten Pyro- und iso-Pyrovitaminen D, (XLVII)
und (XLVIII), ist nach Havinga (74) vielleicht als innere Diensynthese
des Praecalciferols aufzufassen.

(I.) ⟶
Vitamin D.

(IX b.) Praecalciferol.

(XLVII.) Pyrovitamin D.
9 α-Lumisterin.

(XLVIII.) iso-Pyrovitamin D.
9 β-Ergosterin.

(Nach E. R. H. Jones.)

Sollte sich diese Auffassung bestätigen*, so ist der zur photochemischen
Re-cyclisierung des Praecalciferols (S. 94) entgegengesetzte sterische
Verlauf der Reaktion an $C_{(9)}$ bemerkenswert.

Die *trans*-Vitamine D gehen bei höheren Temperaturen, langsam schon bei 65°,
in ein kristallines, hochschmelzendes Produkt über (17), dessen Konstitution noch
nicht geklärt werden konnte. Analyse, Schwerlöslichkeit, Schmelzpunkt und
Molekulargewichtsbestimmung weisen auf ein dimeres Produkt hin. Im UV zeigen
sich Maxima bei 245,5, 254 und 264 mμ.

Iso-Vitamin D (VI, S. 75) ist thermisch stabil (23). Das erscheint
verständlich, wenn es, wie angenommen, mit dem Pyrotachysterin

---

* Der sterische Verlauf intramolekularer Diensynthesen ist bislang wenig
bekannt. Sollte sich diese Cyclisierung als eine solche erweisen, so scheint ein zu
intermolekularen Diensynthesen entgegengesetzter sterischer Verlauf vorzuliegen.

identisch, d. h. Endprodukt einer thermischen Isomerisierung ist. Ebenfalls erweist sich das iso-Tachysterin als thermisch stabil (*22*). Das u-Tachysterin (S. 82) dagegen geht schon bei der Destillation im Hochvakuum in einen Kohlenwasserstoff $C_{28}H_{42}$ über (S. 82). Durch die Allylstellung der 3-OH-Gruppe ist bei diesem Stoff die Wasserabspaltung gegenüber einer thermischen Isomerisierung bevorzugt.

## V. Chemische Isomerisierung.

Die Einführung von zwei neuen Isomerisierungsmitteln in die Vitamin D-Chemie hat sich als sehr fruchtbar erwiesen, nicht nur weil dadurch die Vielzahl der Vitamin D-Isomeren noch erweitert und in einen genetischen Zusammenhang gebracht werden konnte, sondern auch weil erst durch ihre Anwendung die meisten Vitamin D-Isomeren auf relativ einfache Weise präparativ zugänglich geworden sind. Hinzu kommt, daß sich die beiden Agenzien in idealer Weise ergänzen. Es handelt sich um das Bortrifluorid-ätherat, dessen isomerisierende Wirkung durch Variation der Konzentration sowie durch Zusatz von Diäthyläther so verändert werden kann, daß sowohl totale als auch partielle Isomerisierungen durchführbar werden, und das von Havinga (*46*) eingeführte Jod/Pyridin, ein für *cis-trans*-Isomerisierungen vorzüglich geeignetes Mittel.

Schon relativ früh erwies sich das iso-Tachysterin (XI, S. 81) als Endpunkt der chemischen Isomerisierung mit Bortrifluorid-ätherat (*22*), wie durch die vollständige Isomerisierung von Vitamin $D_2$ (I) und partialsynthetisch gewonnenem iso-Vitamin $D_2$ (VI, S. 75) sowie wenig später auch (*23*) von Pyrotachysterin (S. 83) und Tachysterin$_2$ (X, S. 94) erhärtet werden konnte. Im Rahmen der systematischen Untersuchung der chemischen Isomerisierung (*36*) gelang auch die Totalisomerisierung von 5,6-*trans*-Vitamin $D_2$ (III, S. 73), reinem kristallinem iso-Vitamin $D_2$ (VI) sowie Praecalciferol$_2$ (IX a, S. 96) (*Formelübersicht 4*). Es ist wichtig zu beachten, daß Endisomerisierungen zum iso-Tachysterin (XI) auch von Säuren, wie z. B. Phosphorsäure (*22*), verursacht werden. Das hat zur Folge, daß beim Arbeiten mit den Vitaminen D und ihren zum Teil noch empfindlicheren Isomeren Protonen peinlich ausgeschlossen werden müssen. In manchen Fällen ist sogar ein Arbeiten in besonderen Räumen zweckmäßig, in denen die Luft nicht durch saure Gase verunreinigt ist.

(I.) Vitamin $D_2$ ⎯⎯⎯⎯⎯↘          ↙⎯ iso-Vitamin $D_2$ (VI.)

(III.) 5,6-*trans*-Vitamin $D_2$ → iso-Tachysterin ⟵⎯ Pyrotachysterin (VI.)

(IX a.) Praecalciferol ⎯⎯⎯↗ (XI.)          ↘⎯ Tachysterin (X.)

*Formelübersicht 4*. Endisomerisierungen mit Bortrifluorid-ätherat.

*Literaturverzeichnis: SS. 118—123.*

Wesentliche Einblicke in die Isomerisierungs-Verhältnisse wurden schließlich durch die partiellen Isomerisierungen erreicht. So konnte mit Bortrifluorid-ätherat Vitamin $D_2$ (I) sowie 5,6-*trans*-Vitamin $D_2$ (III) in iso-Vitamin $D_2$ (VI) übergeführt werden (*36*). Das iso-Vitamin $D_2$ wurde aus diesen Versuchen als kristalline Substanz isoliert. Dagegen ergab sich, daß 5,6-*trans*-Vitamin $D_2$ durch partielle Isomerisierung von Vitamin $D_2$ mit Bortrifluorid nicht zugänglich ist. Dieser Übergang läßt sich jedoch nach HAVINGA und Mitarb. (*73*) durch Jod/Pyridin-Isomerisierung realisieren, die nach einem anderen, wahrscheinlich radikalischen Mechanismus verläuft, während für die Bortrifluorid-Umlagerung ein ionischer Mechanismus ablaufen dürfte. Nach diesen Ergebnissen konnte eine Isomerisierungs-Reihe des Vitamins $D_2$ (*36*) aufgestellt werden, die vom Vitamin $D_2$ (I) über das 5,6-*trans*-Vitamin $D_2$ (III) und iso-Vitamin $D_2$ (VI) zum iso-Tachysterin (XI) verläuft und innerhalb derer sich jeder Schritt präparativ verwirklichen läßt.

Eine zweite Isomerisierungs-Reihe (*36*) läßt sich für das Prae-calciferol$_2$ (IX a) ableiten, aus dem sich nach HAVINGA und Mitarb. (*46*) durch Jod/Pyridin-Umlagerung Tachysterin$_2$ (X) darstellen läßt.

Die Kombination der chemischen Isomerisierungen mit Bortrifluorid-ätherat und Jod/Pyridin mit einigen thermischen Übergängen ergibt ein genetisches Schema der 9,10-*seco*-Steroide, das ein Gegenstück zu dem nachstehenden Bestrahlungsschema darstellt *(Formelübersicht 5)*.

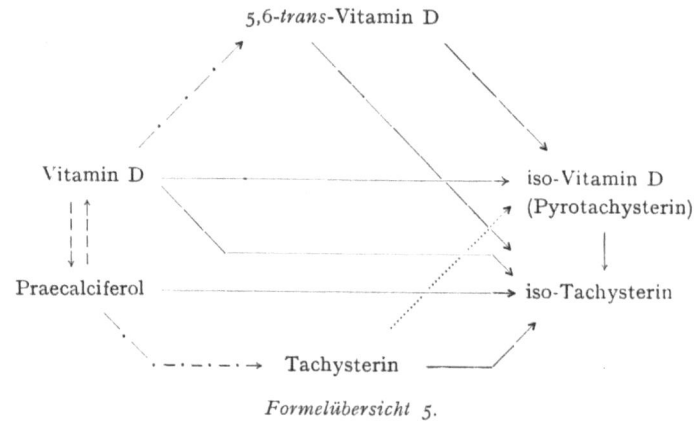

*Formelübersicht 5.*

————, Chemische Isomerisierung mit Bortrifluorid-ätherat und _._._., mit Jod/Pyridin; ........., thermische Übergänge.

## VI. Ergebnisse von BARON, LE BOULCH und RAOUL.

Die französischen Autoren befassen sich seit einigen Jahren mit der Aufgabe, die 1926 von C. N. BILLS beschriebene antirachitische Aktivierung,

die das Cholesterin unter bestimmten Bedingungen bei der Behandlung
mit aktiviertem Floridin, einer sauren Bleicherde, erfahren soll, auf-
zuklären. Dabei gelang es der Arbeitsgruppe, eine Reihe von Substanzen
zu isolieren, die nach ihrer Deutung zum Teil 9,10-*seco*-Steroide sind.
Das würde bedeuten, daß eine Ringöffnung auf chemischem Wege
zwischen den Kohlenstoffatomen 9 und 10 eines Steroids verwirklicht
worden wäre (*21*).

Diese Arbeiten gewinnen nunmehr an Interesse, zumal sie nach
Angabe der Autoren zu einer neuen Art antirachitisch wirksamer
Substanzen geführt haben, die auch in der Natur vorkommen sollen
und von den bisher bekannten Vitaminen D verschieden sind. Aus diesen
Gründen sollen hier die wesentlichen Punkte der französischen Arbeiten
gebracht werden, die bislang von anderen Arbeitsgruppen kaum beachtet
und infolgedessen auch nicht reproduziert worden sind.

Die Autoren aktivierten das Floridin, indem sie es eine Stunde auf
280° erhitzten, und absorbierten daran Cholesterin (XLIX) aus siedendem
Tetrachlorkohlenstoff. Wurde kurzzeitig gekocht, bildete sich außer
einer Reihe von Nebenprodukten die Verbindung (L), nach einstündiger
Versuchsdauer die Verbindung (LI), wobei bei (L) über die *cis-trans*-
Isomerie an der 6,7-Doppelbindung keine Klarheit besteht.

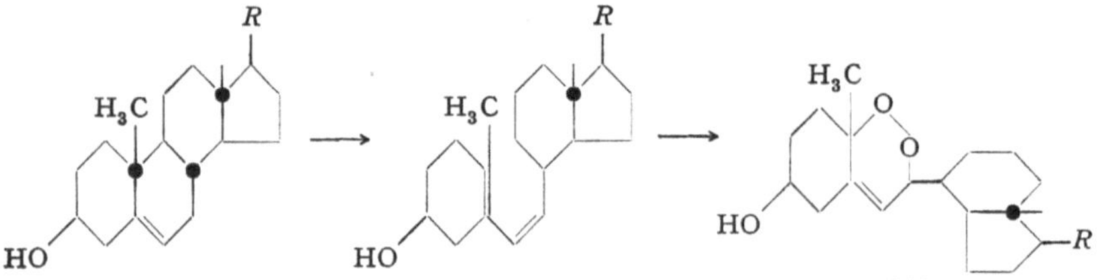

(XLIX.) Cholesterin.        (L.) *seco*-Cholesterin.        (LI.)

Bei der Behandlung mit methanolischer Kalilauge isomerisiert (LI)
zu (LII), einem $\alpha,\beta$-ungesättigten Keton, das nach der Lage seines UV-

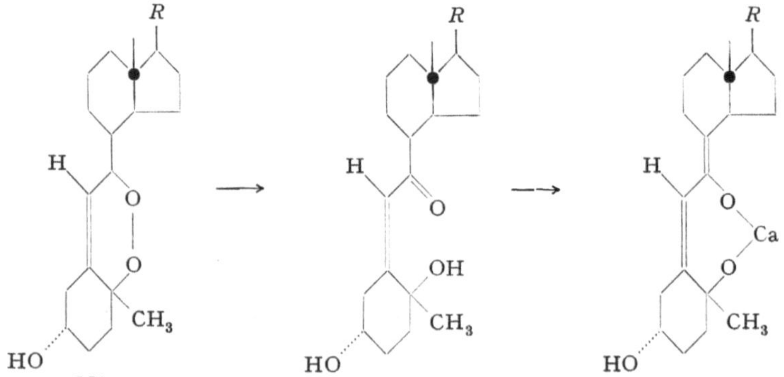

(LI.)                    (LII.)                    (LIII.)

*Literaturverzeichnis: SS. 118—123.*

Absorptionsmaximums als „Keton 250" benannt worden ist. Dieser Stoff soll in tierischen und pflanzlichen Ölen vorkommen und 10% der antirachitischen Wirksamkeit des Vitamins $D_3$ aufweisen (3).

Bei der Filtration einer Lösung des „Ketons 250" in Tetrachlorkohlenstoff über eine Calciumhydroxydsäule entsteht das Calciumenolat (LIII), dessen antirachitische Wirksamkeit der des Vitamins $D_3$ gleich sein soll (54). Das Primärprodukt (L) geht bei der Floridinbehandlung in (LI) über, ist also offenbar ein Zwischenprodukt bei dessen Entstehung (4).

Für die Konstitution der einzelnen Produkte geben die Autoren folgende Befunde an:

Auf den Stoff (L), seco-Cholesterin genannt, paßt die Analyse $C_{27}H_{46}O$. Es zeigt UV-Absorption bei $\lambda_{max} = 235$ m$\mu$ ($\varepsilon = 12000$) und besitzt eine acetylierbare OH-Gruppe (3). Die Hydrierung von (L) und seines Acetats ergibt das Vorhandensein von zwei Doppelbindungen. Das Acetat liefert ein Addukt mit Maleinsäureanhydrid (4). Luftsauerstoff oxydiert die Substanz zu (LI). Die Analyse der Substanz (LI) paßt auf $C_{27}H_{46}O_3$; sie zeigt kein UV-Absorptionsmaximum. Umsetzung mit Kaliumjodid und Titration des Jods zeigen zwei aktive Sauerstoffatome an (3). (LII) ist mit (LI) isomer; es absorbiert bei 250 m$\mu$ und gibt ein Semicarbazon, das bei 290 m$\mu$ absorbiert. In der Kälte liefert es ein Mono-, in der Wärme ein Diacetyl-Derivat. Letzteres wird als Enolacetat gedeutet. Die dritte Sauerstoff-funktion ist offenbar eine tertiäre OH-Gruppe. Beim Ozonabbau entsteht kein Abbau-keton (XXXII, S. 91), dagegen ist ein anderes Semicarbazon mit etwa 21% N faßbar, das wohl aus dem Ring $A$ entstanden ist (3). (LIII) ist gekennzeichnet durch seine Analyse sowie eine IR-Bande bei 1260 cm$^{-1}$; die UV-Absorption liegt bei 265 m$\mu$ (3).

Ebenso erstaunlich wie die oben genannten Reaktionen sind die folgenden: Bei der Behandlung des Peroxyds (LI) mit Natriumsulfit sollen nach Angabe der französischen Autoren iso-Tachysterin sowie Vitamin $D_3$ entstehen, was als Beweis für die behauptete 9,10-seco-Struktur der genannten Stoffreihe angeführt wird (3, 4).

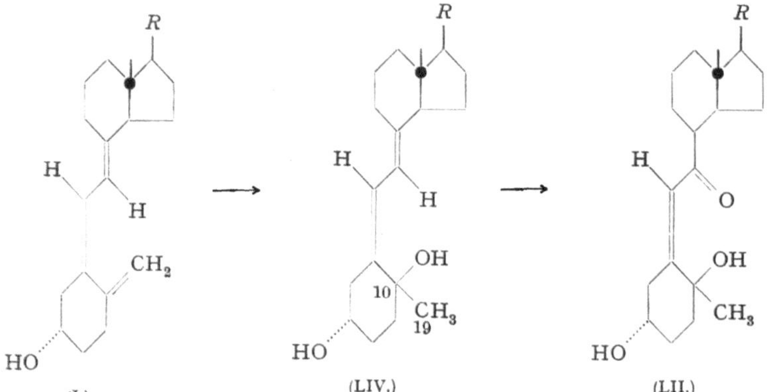

(I.)     (LIV.)     (LII.)

Es werden noch zwei andere Wege zur Darstellung des „Ketons 250" angegeben. Nach dem einen wird Vitamin D an Floridin in 10,19 hydratisiert (5) und das entstehende Diol (LIV) entweder durch weitere Floridin-Behandlung (3) oder durch Kochen mit methanolischer Kalilauge an der Luft (53) zum Keton (LII) oxydiert.

Nach dem anderen Weg wird Vitamin D mit Lithium in flüssigem Ammoniak reduziert, wobei, wie schon von der *Philips*-Arbeitsgruppe gefunden wurde (S. 86), Dihydro-tachysterin (XXIII) entsteht (53). Unter anderen Bedingungen, deren Abweichung von der üblichen Reduktionsweise mit Li/NH₃ nicht aus den experimentellen Beschreibungen ersichtlich ist, soll *seco*-Cholesterin (L) entstehen (3, 4). Beide Diene geben an Floridin das Peroxyd (LI).

Alle diese Experimente bedürfen der Überprüfung nicht nur wegen der unorthodoxen Reaktionsfolgen, sondern weil die Arbeiten der genannten Autoren sowohl mit offensichtlichen Fehlern als auch mit sich widersprechenden Angaben durchsetzt sind. So wird z. B. für das 2,4-Dinitrophenylhydrazon des Ketons (LII) $\lambda_{max}$ 365 m$\mu$ ($\varepsilon$ = 8000) angegeben (3). Eine derart kurzwellige Lage der Absorption mit derart niedriger Extinktion kann dem Derivat des Ketons (LII) jedoch nicht zukommen. Als Beispiel für widerspruchsvolle Angaben sei angeführt, daß für die Absorption des Diacetats (Enolacetats), das sich vom „Keton 250" (LII) ableitet, an verschiedenen Stellen für $\lambda_{max}$ 245 m$\mu$ (4), 255 m$\mu$ bzw. 270 m$\mu$ (3) angegeben wird.

Von Raoul und Mitarb. (53) sind noch andere Methoden zur Erlangung antirachitischer Stoffe aus Cholesterin beschrieben worden; doch sind diese Ergebnisse noch nicht klar genug, um hier referiert werden zu können.

# VII. Verschiedene Verbindungen.

## 1. Ätio-Analoga der Praecalciferole und der Vitamine D.

Durch UV-Bestrahlung von Ätio-ergosterin (XLIV) bzw. 17$\beta$-Hydroxyätio-ergosterin (XLIV) erhielten Velluz und Mitarb. (70, 71) Ätio-

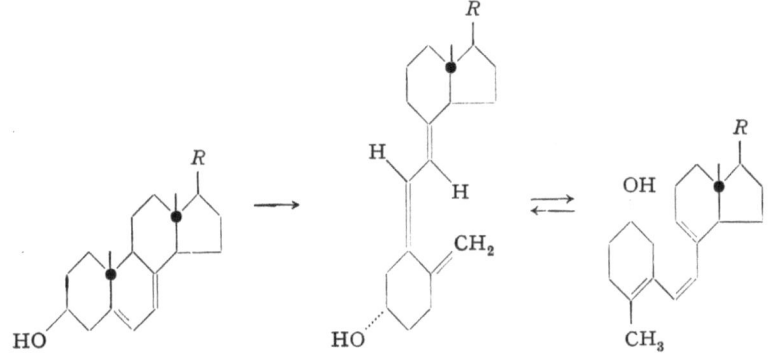

(XLIV.)  R = H, Ätio-ergosterin.
          R = OH, 17$\beta$-Hydroxy-
                   ätio-ergosterin.

(I.)  R = H, Ätio-Vitamin D.
       R = OH, 17$\beta$-Hydroxy-
                ätio-Vitamin D.

(IX a.)  R = H, Ätio-Praecalciferol.
          R = OH, 17$\beta$-Hydroxy-ätio-
                   praecalciferol.

Vitamin D (I) bzw. 17β-Hydroxy-ätio-Vitamin D (I), die als kristalline Ester isoliert werden. Durch thermische Isomerisierung der Ester wurden die entsprechenden Praecalciferole (IXa) gewonnen. Dabei gelang im Falle des 17β-Hydroxy-ätio-praecalciferols (IXa) erstmalig die Darstellung eines kristallinen Praecalciferols ($\lambda_{max}$ 263 mμ; $\varepsilon = 10000$).

### 2. Vitamin D-homologe Verbindung.

Milas und Mitarb. (*49*) synthetisierten eine Vitamin D-homologe Verbindung, indem sie Cholestanyliden-acetaldehyd (LV) mit *p*-Methoxy-cyclohexanon kondensierten, das *cis*- (LVI) und *trans*-Dienon (LVII) chromatographisch trennten und ersteres nach Wittig zum Trien (LVIII) umsetzten. Dieses Trien soll antirachitische Wirksamkeit aufweisen.

(LV.) Cholestanyliden-acetaldehyd. (LVI.) (LVII.) (LVIII.)

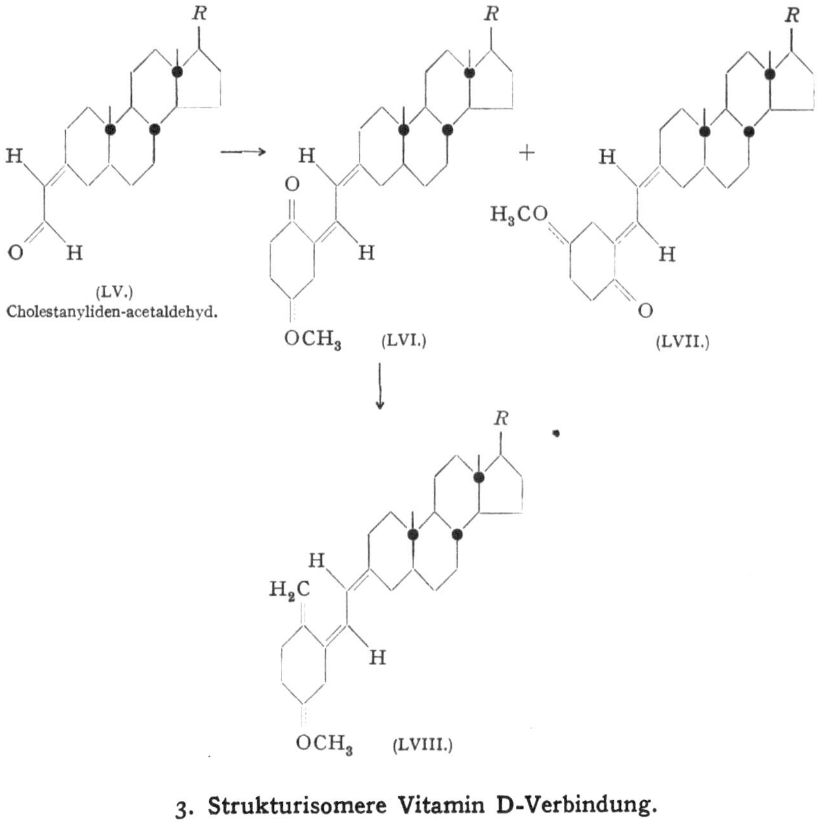

### 3. Strukturisomere Vitamin D-Verbindung.

Bei dem Versuch einer Synthese des Trien-systems des *trans*-Vitamins $D_2$ wurde eine Allylumlagerung bei der Umsetzung des Bromids (LIX) zum Phosphoniumbromid (LX) beobachtet (*24*). Das aus diesem erhältliche Ylen (LXI) gab bei der WITTIG-Reaktion mit dem Vitamin $D_2$-$C_{21}$-Abbau-aldehyd (XXX) die strukturisomere Vitamin D-Verbindung (LXII).

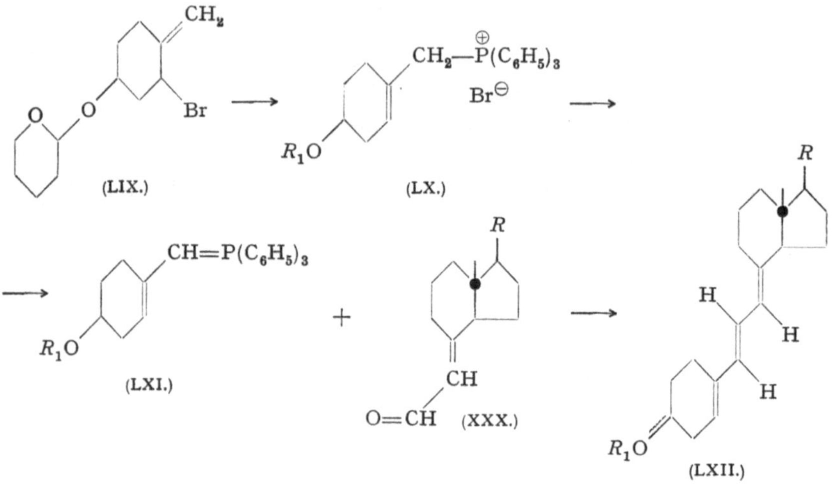

## 4. 9,10-*seco*-D-Homo-Steroid-Verbindung.

Eine D-homo-Verbindung (LXV) ließ sich auf folgendem Wege synthetisieren (*30*): An Dekalindion-1,5 lagerte man partiell Acetylen an, methylierte angulär und ketalisierte die verbliebene Ketogruppe. Das dabei erhaltene Produkt (LXIII) wurde über die lithiumorganische Verbindung mit Methylcyclohexanon zu (LXIV) umgesetzt. Wasserabspaltung und Ketalspaltung ergab das In-dien-keton (LXV), dessen Doppelbindungen gemäß der UV-Absorption bei 266 m$\mu$ keine einheitliche Lage aufwiesen.

*Literaturverzeichnis: SS. 118—123.*

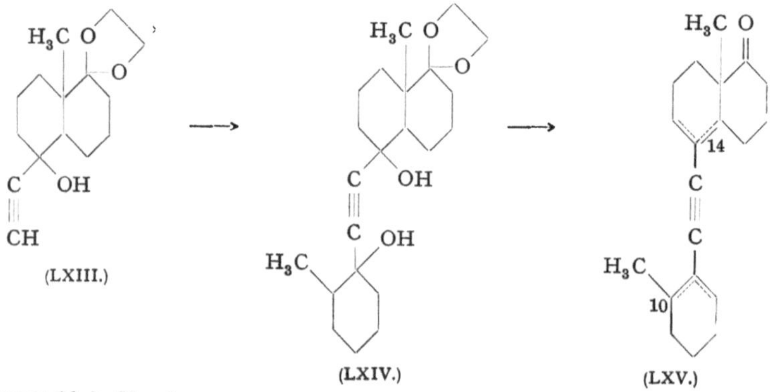

## 5. Vitamin D$_m$.

Aus den Bestrahlungsprodukten von Sterinen aus der Muschel *Modiolus demissus* DILLWYN isolierte PETERING (*52*) ein neues Vitamin D, das sogenannte Vitamin D$_m$, in Form eines kristallinen 3,5-Dinitrobenzoats. Die Substanz war zwar dem Vitamin D$_3$-Ester sehr ähnlich, wies aber doch deutliche Unterschiede auf. Die biologische Aktivität beträgt zwei Drittel der des Vitamins D$_3$. Über die Konstitution der Seitenkette ist noch wenig bekannt. Die C,H-Analysenwerte des Esters sowie die Molgewichtsbestimmung des Provitamins weisen auf eine C$_{29}$-Verbindung hin.

## VIII. Partialsynthesen.

### 1. 3-Desoxy-14 $\beta$-praecalciferol$_2$.

Methyl-äthinyl-cyclohexen (LXVI) wurde mit dem *cis*-C$_{19}$-Abbauketon (XXXIII, $R = C_9H_{17}$) kondensiert (*43*) *(Formelübersicht 6)*. Bei der Wasserabspaltung aus dem entstandenen En-in-ol (LXVII) erhielt man je nach dem verwendeten Agens, entweder Aluminiumoxyd oder

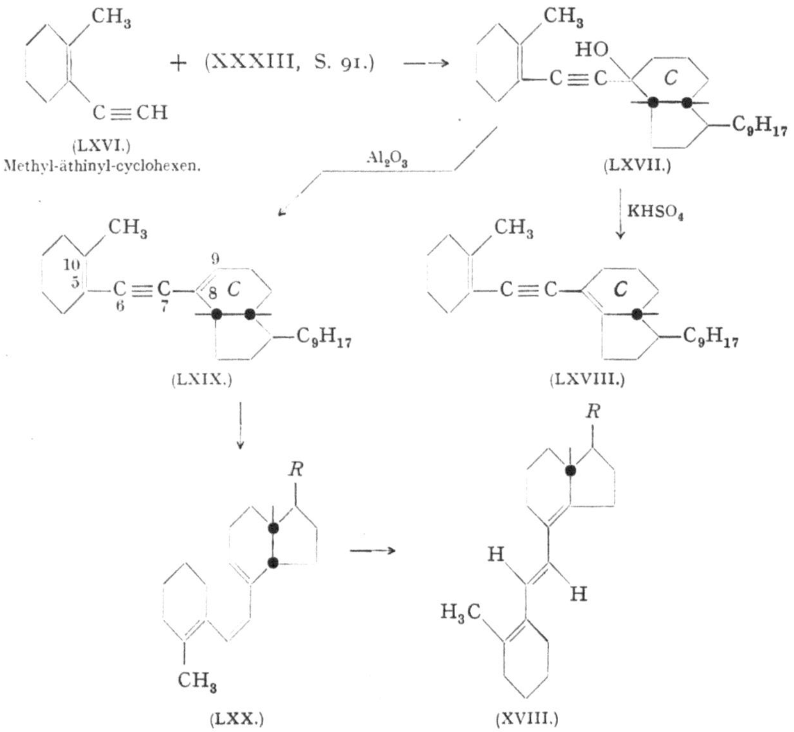

*Formelübersicht 6.*

Kaliumhydrogensulfat, In-diene mit verschieden liegenden UV-Absorptionsmaxima, nämlich 266 mμ (LXIX) bzw. 271 mμ (LXVIII). Dieser Befund kann so gedeutet werden, daß sich die beiden In-diene durch die Lage der Doppelbindung im Ring C unterscheiden. Der dadurch gegebene Unterschied der Substituentenzahl am chromophoren System würde die Verschiebung der Absorptionsmaxima gegeneinander erklären. Das kürzerwellig absorbierende In-dien (LXIX) wurde partiell zu (LXX) hydriert, dessen UV-Absorptionsmaximum mit dem des Praecalciferols übereinstimmte. Das Produkt ließ sich mit Jod zur all-*trans*-Verbindung (XVIII) mit iso-Tachysterin-Spektrum isomerisieren.

Indessen müssen diese Reaktionen noch weiter ausgebaut und gesichert werden.

### 2. iso-Vitamin $D_2$-methyläther-$C_{(3)}$-Epimerengemisch und iso-Tachysterin-methyläther-$C_{(3)}$-Epimerengemisch.

Nach der Partialsynthese des iso-Vitamin $D_2$-$C_{(3)}$-Epimerengemisches (VI, S. 75) (22) wurde auch eine unabhängige Synthese des

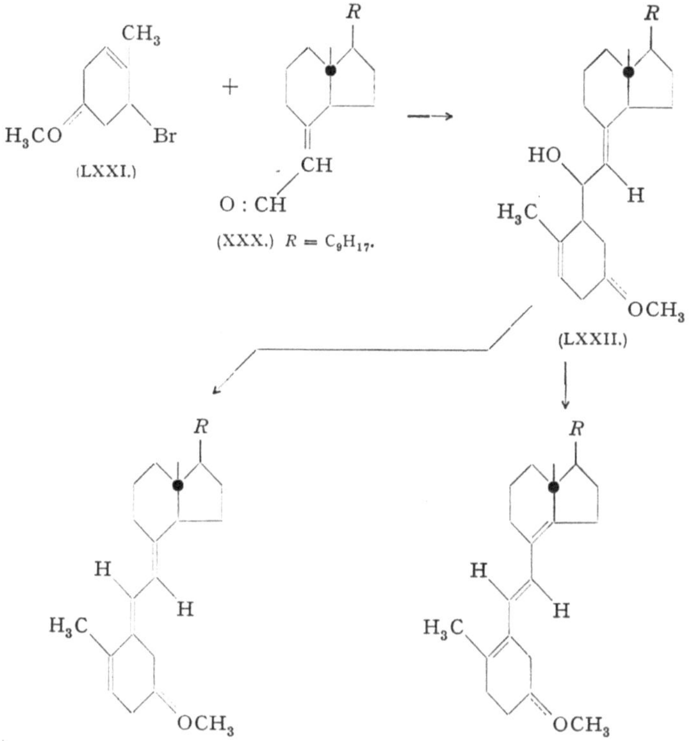

(LXXI.)    (XXX.) $R = C_9H_{17}$.    (LXXII.)

(LXXIII.) iso-Vitamin $D_2$-methyläther.          (LXXIV.) iso-Tachysterin-methyläther.

*Formelübersicht 7.*

entsprechenden Methyläthers erarbeitet (*43*) *(Formelübersicht 7)*. Dabei wurde das Bromid (LXXI) als GRIGNARD-Verbindung mit dem $C_{21}$-Abbau-aldehyd des Vitamins $D_2$ (XXX) umgesetzt, wobei das Carbinol (LXXII) entstand, dessen Pyrolyse den iso-Vitamin $D_2$-äther (LXXIII) ergab. Bei der Wasserabspaltung mit *p*-Toluolsulfonsäure dagegen entstand unter gleichzeitiger Isomerisierung der schon bekannte iso-Tachysterin-äther (LXXIV) (*42*).

### 3. 3-Desoxy-iso-tachysterine.

Verschiedene Abbau-ketone (XXXII) und (XXXIII) setzte man nach WITTIG zu den Methylen-Verbindungen (LXXV) um, die nach Bromierung mit N-Bromsuccinimid und Umsetzung mit Triphenyl-phosphin Phosphoniumbromide ergaben. Deren Ylene der Formel (XVII) bildeten mit Methylcyclohexen-aldehyd (XVI, S. 81) 3-Desoxy-iso-tachysterine (XVIII) (*18, 44*). Entgegen der Annahme, daß die Bromie-rung in 9-Stellung erfolgen und sich bei der Allylumlagerung eine 8,9-Doppelbindung ergeben würde, bildete sich an einer Stelle nach der Bromierung eine 8,14-Doppelbindung aus.

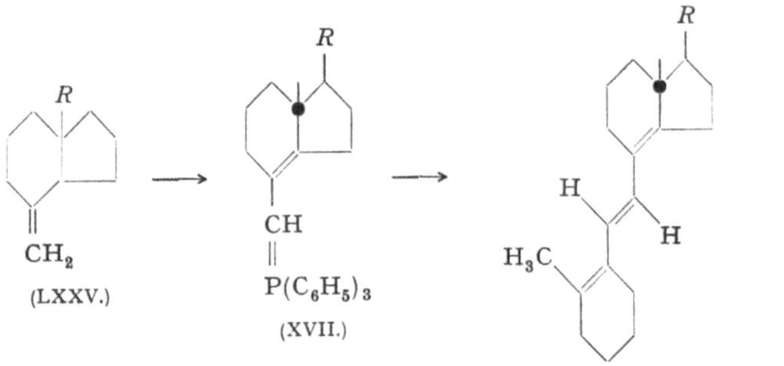

(LXXV.)

$$CH$$
$$\parallel$$
$$P(C_6H_5)_3$$

(XVII.)

(XVIII.) 3-Desoxy-iso-tachysterine.

### 4. 6,7-*cis*-iso-Tachysterin$_2$-methyläther-$C_{(3)}$-Epimerengemisch.

Das bisher unbekannte Trien-system des *cis*-iso-Tachysterins wurde partialsynthetisch nach *Formelübersicht 8* gewonnen (*43*): Das Acetylen-carbinol (LXXVI) wurde in Form seiner lithiumorganischen Verbindung mit dem *C,D-cis*-Vitamin$_2$-Abbauketon (XXXIII, $R = C_9H_{17}$) kondensiert. Die Dehydratisierung des entstandenen Acetylen-dicarbinols (LXXVII) ergab das In-dien (LXXVIII), dessen Partialhydrierung ein Produkt mit der UV-Absorption bei *272* m$\mu$ (LXXIX) lieferte, das allerdings durch Doppelbindungsisomerie und teilweise Methanolabspaltung in Form eines Gemisches vorlag.

(LXXVI.) (LXXVII.) (LXXVIII.) (LXXIX.)

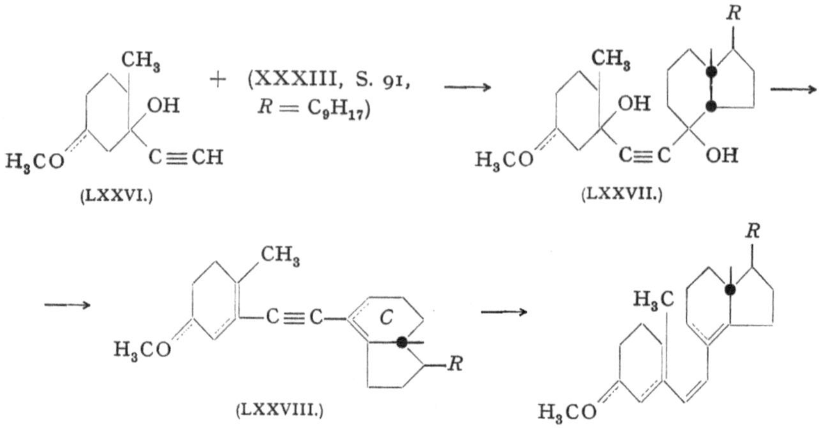

*Formelübersicht 8.*

## IX. Totalsynthese des Vitamins $D_3$.

Die nachstehend beschriebene Synthese des Vitamins $D_3$ (LXXX) (*19*) basiert auf folgender Dreiteilung des Gesamtproblems:

1. Aufbau der Hydrindan-Verbindung, des (+)-8-Methyl-*trans*-hydrindanol-(4)-ons-(1) (XL) (*33, 41*). Dieses wichtige Kernstück war bereits vorher durch Abbau der Vitamine $D_2$ und $D_3$ bekanntgeworden (*39*), so daß der Anschluß an die richtige sterische Situation und an die natürliche optische Aktivität schon in einem frühzeitigen Stadium gefunden werden konnte. Außerdem war hiermit eine erste Relais-Verbindung gegeben, womit das Abbau-Material für den rückläufigen Aufbau mit verwendbar wurde.

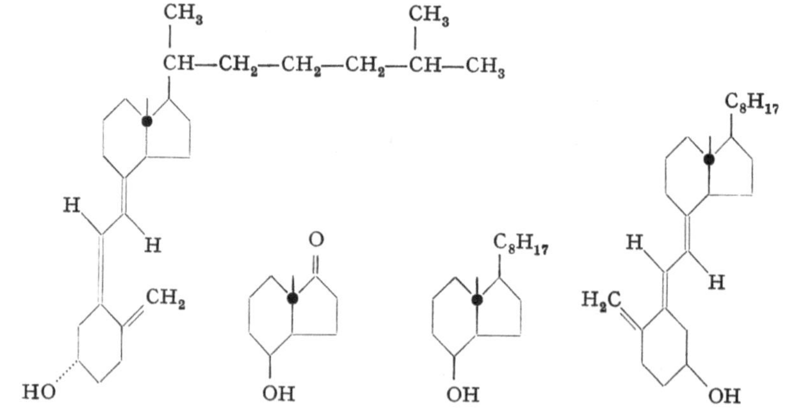

(LXXX.) Vitamin $D_3$. (XL.) (+)-8-Methyl-*trans*-hydrindanol-(4)-on-(1). (XXXI.) $C_{18}$-Abbau-alkohol. (III.) 5,6-*trans*-Vitamin $D_3$.

*Literaturverzeichnis: SS. 118—123.*

2. Anbau der Seitenkette an die Carbonyl-Funktion von (XL) unter Bildung des sogenannten $C_{18}$-Abbau-alkohols (XXXI) (*26*). Auch dieser Abbau-alkohol war, wie der Name besagt, bereits durch Ozonabbau des Vitamins $D_3$ erhalten worden und stellte die zweite wichtige Relais-Substanz dar (*40*).

3. Angliederung der restlichen neun Kohlenstoffatome unter gleichzeitiger Erzeugung des Triensystems, wobei nach der $C_{(3)}$-Epimerentrennung zunächst das 5,6-*trans*-Vitamin $D_3$ (III) (*29, 31, 32, 38*) erhalten wurde. Als letztes war schließlich die Photo-isomerisierung des 5,6-*trans*-Triens (III) zum 5,6-*cis*-Vitamin $D_3$ durchzuführen (*37*). Die hierzu insgesamt notwendige Reaktionsfolge steht nicht nur formal in Anbetracht der Empfindlichkeit des Trienchromophors am Ende der Synthese. Sämtliche Reaktionen des 3. Teiles konnten, da es sich wieder um Relais-Verbindungen handelte, durch partielle Abbau- und Aufbau-Kreisläufe bereits vor Erreichung des letzten Anschlußpunktes bewältigt werden.

### 1. Das Hydrindansystem (*33, 41*).

Ausgangsmaterial für das *trans*-Hydrindanolon (XL) war die NENITZESCU-Säure (LXXXI) (*51*), die uns in dankenswerter Weise von der Fa. E. Merck, Darmstadt, in größerer Menge zur Verfügung gestellt worden war. Das durch Anlagerung von Cyankali erhaltene Nitril (LXXXII) wurde ohne Isolierung des bei der stark salzsauren Aufarbeitung ent-

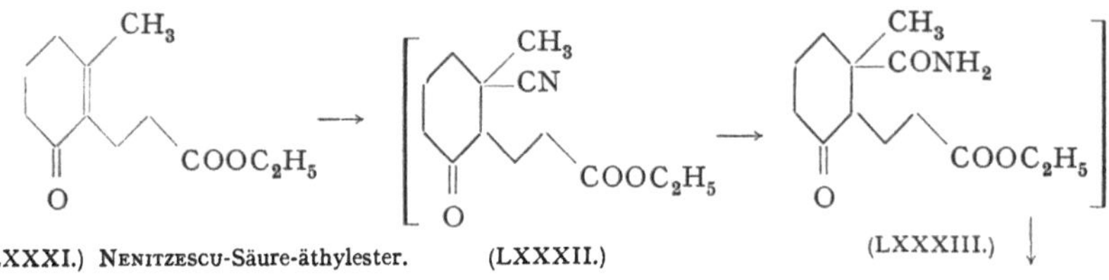

(LXXXI.) NENITZESCU-Säure-äthylester.  (LXXXII.)  (LXXXIII.)

stehenden Säureamids (LXXXIII) durch thermische Wasserabspaltung direkt in das vorzüglich kristallisierende, sogenannte cyclische Imid (LXXXIV) übergeführt. Da dieses keine Carbonylreaktion zeigt, wird die Lactam-lacton-Form bevorzugt.

Die saure Verseifung von (LXXXIV) führte unter Öffnung der beiden Ringe zu einem Gemisch der Racemate von *trans*- und *cis*-Dicarbonsäure, die im Gleichgewicht miteinander vorliegen. Die gewünschte Verschiebung des Gleichgewichts konnte dadurch erreicht werden, daß das ölige Säuregemisch in feuchtem Äther unter Einleiten von Chlorwasserstoff gekocht wurde. Nach dem Animpfen mit

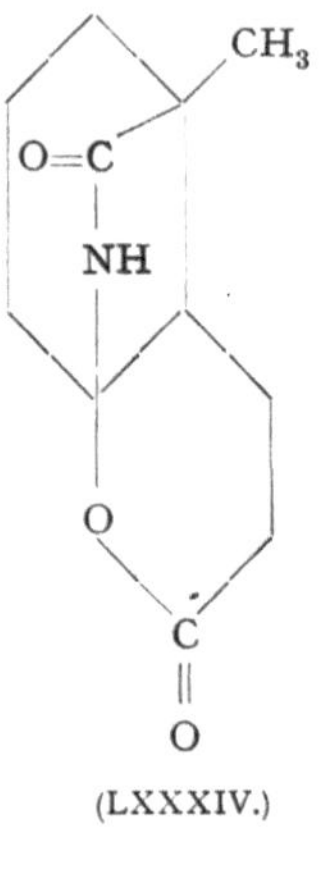

(LXXXIV.)

der *trans*-Säure (LXXXV) („Prinz-Säure") begann sie rasch zu kristalli-
sieren und wurde so dem Gleichgewicht entzogen. Die *cis*-Säure konnte
bisher nicht in reiner Form isoliert werden. Zur Racemat-trennung wurde
nur die Hälfte der berechneten Menge an Brucin verwandt und das kristalline Brucinsalz der unerwünschten (+)-Säure leicht und nahezu quantitativ
abgetrennt. Die Mutterlauge lieferte nach Filtration über eine kurze Kiesel-
gelsäure und nach Kristallisation die gesuchte (—)-Säure (LXXXVI)
mit der Drehung $[\alpha]_D^{20} = -16°$.

(LXXXV.)       ⟶ Racemat-trennung ⟶       (LXXXVI.) *(trans)*

Da der Dieckmann-Ringschluß mit dem Ester der Keto-säure zum
energetisch begünstigten 8-Methyl-*cis*-hydrindandion-(1,4) führt, mußte
die Ketogruppe zunächst reduziert werden, was sowohl mit Natrium-
borhydrid in Methanol als auch mit Platin in Eisessig gelang. Mit $NaBH_4$
entsteht neben dem Hydroxy-diester (LXXXVII) auch ein Lacton-ester,
der mit Platin in saurer Lösung praktisch ausschließlich erhalten wird.
Der Lacton-ester kann jedoch gleichfalls über das Dikaliumsalz mittels
Methyljodids in den gewünschten Hydroxy-diester (LXXXVII) übergeführt werden. Die Dieckmann-Reaktion liefert nunmehr den Hydroxy-
keto-ester (LXXXVIII), der über das Kalium-enolat abgetrennt wird
und durch Kochen mit Eisessig/Trifluoressigsäure unter Umesterung
und Decarboxylierung sowie durch anschließende alkalische Verseifung
in das erwartete, rechtsdrehende Hydrindanolon (XL) übergeht, das
in Schmelzpunkt und Misch-schmelzpunkt, im IR-Spektrum und in
der Drehung (+ 110°) mit dem Vitamin D-Abbauprodukt vollkommen
übereinstimmt (vgl. S. 92).

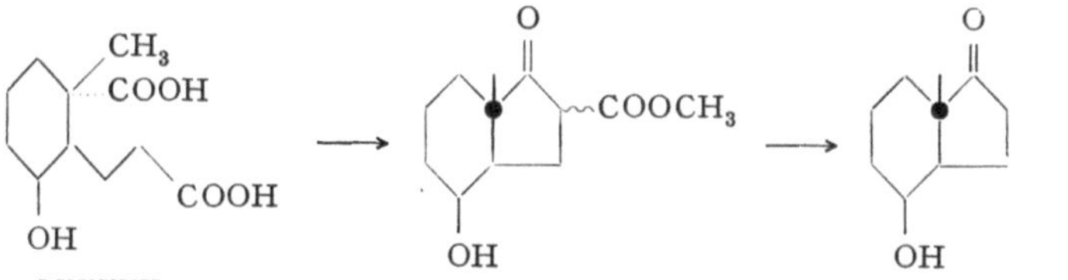

(LXXXVII.)       (LXXXVIII.)    (XL.) 8-Methyl-*trans*-hydrindanol-(4)-on-(1).

## 2. Die Seitenkette (26).

Für den Anbau der Seitenkette an die Carbonylfunktion von Formel
(XL) stand zunächst in Analogie zur Androstan-Reihe das Butenandt-
sche Cyanhydrin-Verfahren zur Diskussion, das eine dem Pregnenolon

analoge CO—CH$_3$-Seitenkette ergeben würde. Hieran hätte sich eine GRIGNARD-Reaktion mit nachfolgender Wasserabspaltung am C$_{(20)}$ und Hydrierung der gebildeten Doppelbindung anzuschließen (*88, 9*). Diese Hydrierung liefert sowohl auf Grund der Uneinheitlichkeit der Anhydro-Produkte als auch wegen ihres uneinheitlichen Verlaufs schlechte Ausbeuten an den gewünschten Stoffen und muß daher als der kritische Punkt angesehen werden, so daß einige Vorbetrachtungen am Platze erscheinen.

Vorweg ist die sterische Situation am gesättigten C-Atom 20 (Steroid-Nomenklatur) exakt zu definieren. Nach L. FIESER geschieht dies in der Weise, daß man den längsten Substituenten dieses C-Atoms (hier CH$_2$—C$_5$H$_{11}$) ganz nach hinten dreht. Bei der Projektion in die Ebene kommt bei natürlicher Anordnung das Methyl links und der Wasserstoff rechts zu stehen. Für unsere Betrachtungen ist es übersichtlicher, den Wasserstoff zum Bezugspunkt zu wählen und ihn oben in $\beta$-Stellung erscheinen zu lassen, was ebenfalls durch entsprechende Drehung der 17,20-Einfachbindung zu erreichen ist. Das Methyl kommt dann bei natürlicher Anordnung (a) wiederum links, aber der andere Substituent jetzt rechts zu stehen, während es beim Isomeren (b) umgekehrt ist.

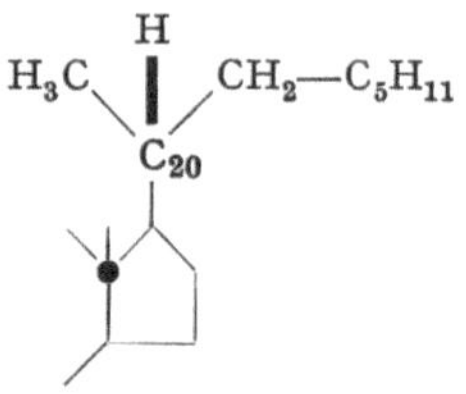

(a) 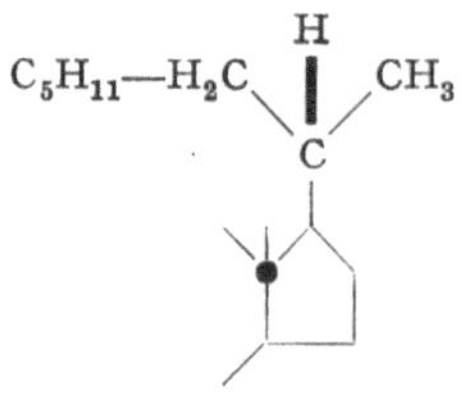      (b) 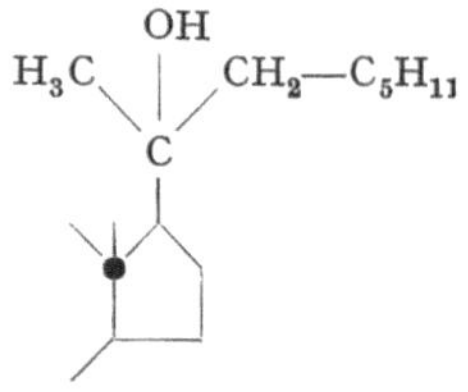      (c)

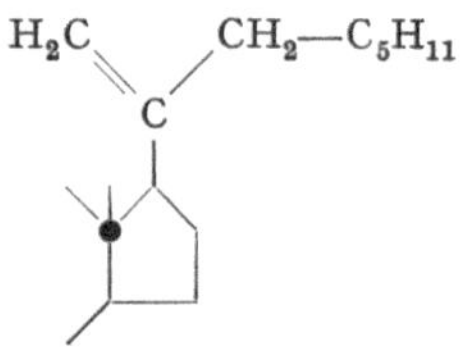

(d) 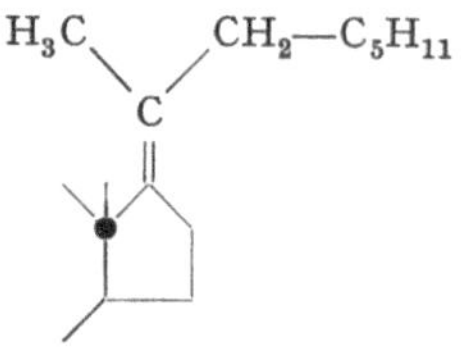      (e) 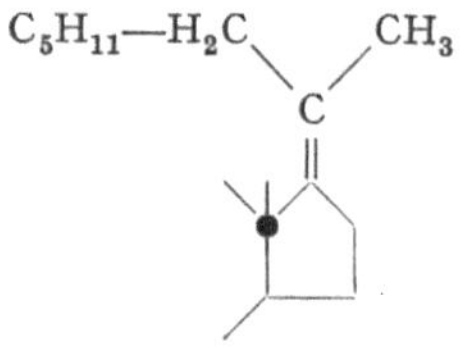      (f)

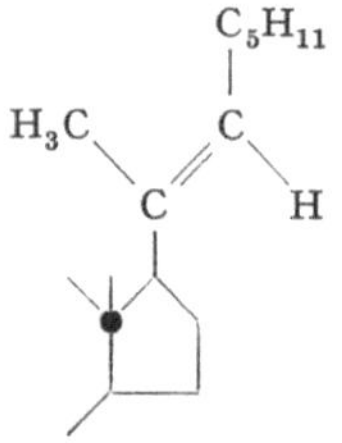

(g) 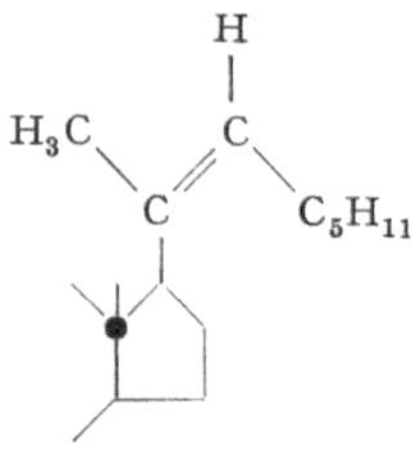      (h)

Die Wasserabspaltung aus dem tertiären Alkohol (c) kann theoretisch zu fünf isomeren Produkten führen, (d) bis (h), von denen (e) und (f) sowie (g) und (h) *cis-trans*-Isomere darstellen.

Modellbetrachtungen lassen unter Berücksichtigung der Erfahrungen aus der Steroid-Chemie erkennen, daß eine katalytische Hydrierung der 20,21- und der 20,22-Doppelbindungen (d, g, h) unter Bildung des asymmetrischen C-Atoms 20 die gewünschte Verbindung (a) neben dem unerwünschten Isomeren (b) wohl ergeben kann, da infolge der größeren Entfernung dieser Doppelbindungen vom angulären $C_{(18)}$-Methyl und dementsprechend geringerer räumlicher Behinderung keine ausschließlich stereospezifische Hydrierung zu erwarten ist. Beim 17,20-*trans*-Olefin (e) kann jedoch am $C_{(20)}$ bei katalytischer Hydrierung nur die verkehrte sterische Anordnung entstehen, falls der Wasserstoff an der Rückseite des Moleküls ($\alpha$-Stellung) addiert wird, was erfahrungsgemäß der Fall ist. Dagegen sollte die analoge Hydrierung des *cis*-Ens (f) das richtige Isomere ergeben.

Die Loslösung vom klassischen Reaktionsschema wurde noch durch einen zweiten Umstand begründet. Der durch stufenweisen Abbau des Dihydro-Vitamins $D_2$-I als Zwischenprodukt gewonnene Seitenketten-aldehyd (XXXVII) war bereits vorweg durch eine WITTIG-Reaktion (LXXXIX) mit anschließender Absättigung der disekundären Doppelbindung in den $D_3$-Abbau-alkohol (XXXI) zurückverwandelt worden. Andererseits hatte sich der gesättigte Aldehyd (XXXVII) in den ungesättigten (XXXIX) überführen lassen. Mit dem Aldehyd (XXXVII) hatte man daher eine neue Relais-Verbindung zur Hand (*39*).

(XXXVII.) (LXXXIX.)

(XXXVII.)

(XXXIX.)

*Literaturverzeichnis: SS. 118—123.*

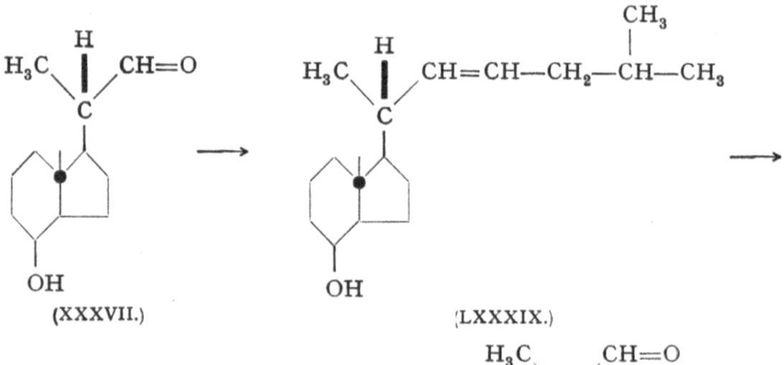

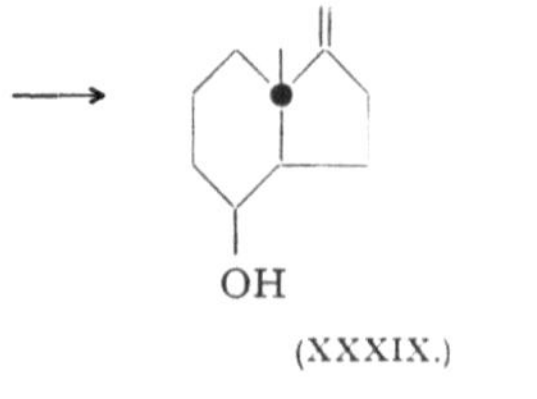

Experimentelles Ziel mußte es daher sein, den einleitenden Aufbau der Seitenkette auf das obige Dreikohlenstoffsystem auszurichten und ungesättigte Aldehyde mit eindeutiger Lage und Substitution der Doppelbindung, d. h. Olefine des Typs (e) bzw. (f), anzustreben.

Hierzu wurde folgender Weg eingeschlagen: Das Hydrindanolon (XL) wurde mit der GRIGNARD-Verbindung aus Crotylbromid umgesetzt, und zwar unter dem Gesichtspunkt, daß nach YOUNG und ROBERTS (*89, 20*) die Addition von Allyl-GRIGNARD-Reagenzien an behinderte Ketone als „anomale" Reaktion über eine cyclische Zwischenform verläuft, die im Fall substituierter Allyl-Verbindungen (durch intramolekulare Umlagerung) zur bevorzugten Bildung desjenigen Reaktionsproduktes führt, das zum weiteren Aufbau der Vitamin D-Seitenkette besonders geeignet ist. Hierbei war ferner anzunehmen, daß sich am C-Atom 17 die $\beta$-Hydroxy-$\alpha$-alkyl-Konfiguration (XC) ausbilden würde.

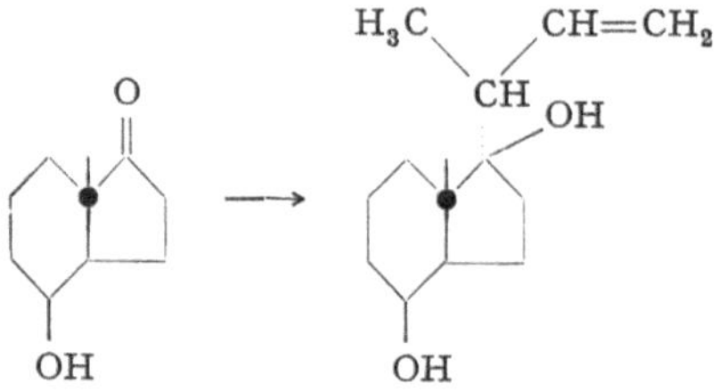

(XL.)　　　　　(XC.)

Damit waren die drei ersten C-Atome mit der Methyl-Verzweigung angegliedert, und die endständige Methylengruppe ermöglichte dazu eine Eliminierung des überzähligen Kohlenstoffatoms unter Bildung der gewünschten Aldehydgruppe.

Nach Erfahrungen in der Androstan-Reihe wurde das Gesamtprodukt, das im wesentlichen nur aus dem Isocrotyl-carbinol (XC) bestand, direkt nach CRIEGEE mit Osmiumsäure hydroxyliert und das nach Zerlegung des Osmiumesters erhaltene Glycol (XCI) der Bleitetraacetat-Spaltung unterworfen (XCII). Hierbei erhielt man den $\beta$-Hydroxyaldehyd (XCII) in guter Ausbeute.

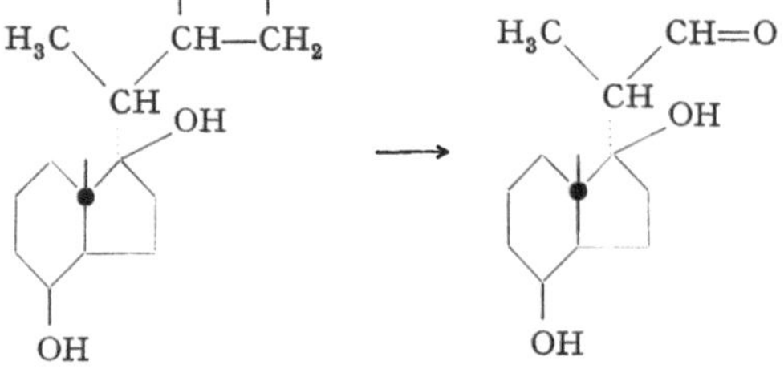

(XCII.)

Mit der Isocrotyl-GRIGNARD-Reaktion ist die Bildung zweier neuer Asymmetriezentren verbunden, die aber durch die angeschlossene Wasserabspaltung mittels methanolischer Phosphorsäure sogleich wieder ausgeschaltet werden konnten. Andererseits war in Anbetracht der racemischen Situation am tertiären C-Atom der Seitenkette sowie einer *trans*-Eliminierung bei der ionogenen Wasserabspaltung von vornherein mit der Bildung zweier *cis-trans*-isomerer ungesättigter Aldehyde zu rechnen. Die chromatographische Auftrennung der Reaktionsprodukte an Hand der UV-Absorption gestattete in der Tat leicht die eindeutige Isolierung zweier ungesättigter Aldehyde (XXXIX und XCIII), die bei 247—248 bzw. 244 m$\mu$ absorbierten. Die hieraus gebildeten 2,4-Dinitrophenylhydrazone mit den Schmelzpunkten 206,5—207,5° und 187—189° kristallisierten vorzüglich. Elementaranalysen wiesen sie als Isomere aus. Das höher schmelzende Hydrazon des längerwellig bei 247—248 m$\mu$ absorbierenden Aldehyds (XXXIX) erwies sich als identisch mit dem entsprechenden Derivat des aus Dihydro-Vitamin D$_2$-I durch stufenweisen Abbau erhaltenen ungesättigten Aldehyds. Da für diesen die *trans*-Stellung von Aldehyd-Gruppe und Sechsring an der Doppelbindung anzunehmen ist, gilt das gleiche für die synthetische Verbindung. Der kürzerwellig bei 244 m$\mu$ absorbierende ungesättigte Aldehyd ist bisher unbekannt; ihm ist *cis*-Struktur an der Doppelbindung zuzuerteilen (XCIII).

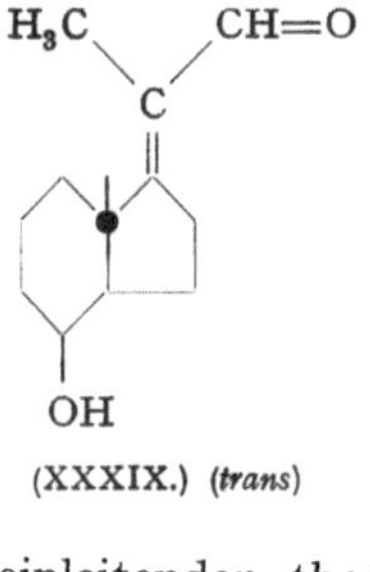

(XXXIX.) *(trans)* 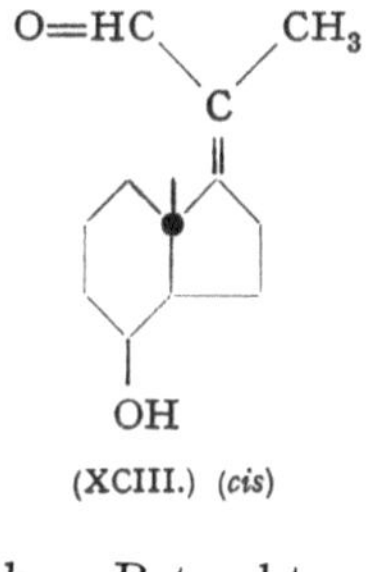        (XCIII.) *(cis)*

Wie aus den einleitenden theoretischen Betrachtungen hervorgeht, ist nur der ungesättigte *cis*-Aldehyd für eine stereospezifisch-katalytische Wasserstoffanlagerung an die Doppelbindung geeignet, während für den *trans*-Aldehyd eine andere Methode gewählt werden muß.

Der *cis*-Aldehyd (XCIII) wurde zunächst mit dem Ylen aus Isoamylbromid umgesetzt und das erwartete Dien (XCIV) in guter Ausbeute erhalten. Der Vorzug dieser Verbindung liegt darin, daß mit dem Verschwinden der Aldehydgruppe die Isomerisierbarkeit aufgehoben ist. Das 17,20-*cis*-22-Dien (XCIV) (Steroid-Nomenklatur) lieferte nach katalytischer Hydrierung, Verseifung und Veresterung mit 3,5-Dinitrobenzoylchlorid einen Ester, der mit dem entsprechenden Derivat des C$_{18}$-Abbau-alkohols (XXXI) aus Vitamin D$_3$ vollkommen identisch war.

(XCIII.) ⟶

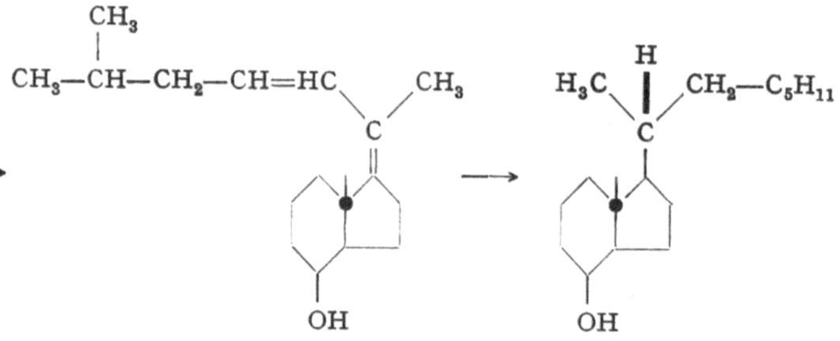

$$CH_3$$
$$CH_3-CH-CH_2-CH=HC$$ $$CH_3$$
$$C$$

XCIV.)

OH

⟶

$$H_3C$$ $$H$$ $$CH_2-C_5H_{11}$$
$$C$$

(XXXI.)

OH

Schließlich konnte auch der ungesättigte *trans*-Aldehyd (XXXIX) mittels einer speziellen Reaktion in den gesättigten n-Aldehyd und weiterhin in den $C_{18}$-Abbau-alkohol (XXXI) übergeführt werden (*28a*). Zur Absättigung der Doppelbindung wurde für diesen Fall eine Behandlung mit Lithium in flüssigem Ammoniak gewählt. Als Zwischenprodukt war das Lithium-enolat der 17-Lithium-Verbindung (XXXVIIa) (Steroid-Nomenklatur) eines gesättigten Aldehyds zu erwarten, wobei am C-Atom 17 die α-Stellung des Lithiums angenommen werden konnte.

Das Verhältnis, in welchem sich die beiden an $C_{(20)}$ epimeren gesättigten Aldehyde bei der Re-aldehydisierung des Enols bilden, ist von ihrem Stabilitätenverhältnis abhängig. Darüber hinaus wäre nach ZIMMER-MAN (*90*) bei schonender Hydrolyse eine weitere Anreicherung des instabileren Epimeren denkbar.

(XXXIX.) ⟶

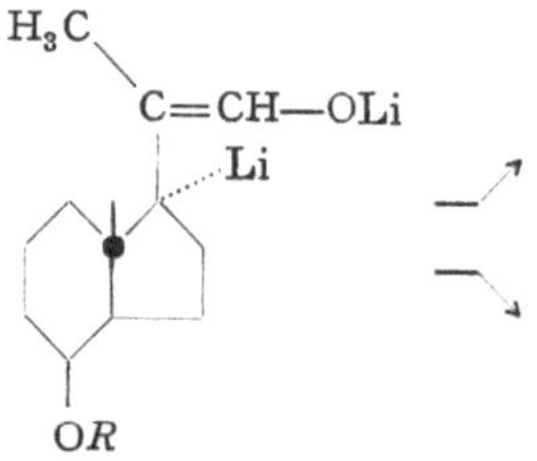

$$H_3C$$
$$C=CH-OLi$$ (XXXVII.)
$$Li$$ n (normal)

OR

(XXXVIIa.)

(XXXVII b.)
(iso)

Um zunächst Aufschluß über die Stabilität des gesättigten n-Aldehyds (XXXVII) und damit zugleich über die Gleichgewichtslage von n- und iso-Aldehyd zu erlangen, wurde der n-Aldehyd einer sauren Behandlung unterworfen. Nach Dinitrophenylhydrazon-Fällung und anschließender chromatographischer Auftrennung konnte das Vorliegen eines Gemisches von 38% n- (XXXVII) und 62% iso-Aldehyd (XXXVIIb) festgestellt werden. Das gleiche Verhältnis ließ sich auch von der iso-Seite einstellen. Die Gleichgewichtseinstellung verläuft in einer Reaktion I. Ordnung.

8*

Wenn man den ungesättigten *trans*-Aldehyd (XXXIX), wie oben angegeben, mit Lithium in flüssigem Ammoniak behandelte, so wurden die beiden an $C_{(20)}$ isomeren Aldehyde in der gleichen Mischung erhalten. Eine Bevorzugung des instabileren n-Aldehyds nach Zimmerman liegt also praktisch nicht vor. Das Gleichgewichtsgemisch 38/62 der beiden gesättigten Aldehyde war daher als neues Relais anzusprechen, womit die material- und zeitverbrauchende Bereitung größerer Mengen des ungesättigten Aldehyds eingespart werden konnte. Schließlich konnte noch festgestellt werden, daß auch die ungesättigten Aldehyde durch saure Behandlung ineinander übergeführt werden können, so daß somit die Umwandelbarkeit aller vier Aldehyde ineinander vorliegt:

Für die Beendigung dieses zweiten Aufbaus der $D_3$-Seitenkette wurde nun das Gleichgewichtsgemisch in der üblichen Weise mit dem Ylen aus iso-Amylbromid umgesetzt, das erhaltene Produkt an der 22,23-Doppelbindung katalytisch hydriert, und nach Veresterung mit 3,5-Dinitrobenzoylchlorid das Estergemisch fraktioniert kristallisiert. Der Umstand, daß der natürliche Ester schwerer löslich war und höher schmolz, ließ seine Trennung vom isomeren Ester und seine Identifizierung ohne Schwierigkeiten durchführen: (XXXVII) → (XXXIX) → (XXXI).

### 3. Trienchromophor, Epimeren-trennung, *trans* → *cis*-Isomerisierung.

Nach der Synthese des $C_{18}$-Abbau-alkohols (XXXI) mußten die weiteren Reaktionen am C-Atom 8 ansetzen. Die Pyridin/Chromat-

Oxydation des sekundären Hydroxyls nach SARETT lieferte zunächst das *trans*-$C_{18}$-Abbau-keton (XXXII) in hoher Reinheit und guter Ausbeute (*40*). Anschließend erbrachte wiederum eine WITTIG-Reaktion, hier mit dem Ylen aus Allylbromid, den gewünschten Fortschritt (XCV). Das überzählige Kohlenstoffatom des so gewonnenen Butadien-Derivats ließ sich durch partielle Ozonisierung eliminieren, wobei so gearbeitet

$C_8H_{17}$ $\qquad$ $C_8H_{17}$ $\qquad$ $C_8H_{17}$ $\qquad$ $C_8H_{17}$

8

O

(XXXII.) $\qquad$ H $\qquad$ HOCH$_2$ H $\qquad$ O=CH H

CH$_2$ $\qquad\qquad$ (XCVI.) $\qquad$ (XXX.) $C_{20}$-Abbau-aldehyd

(XCV.) $\qquad\qquad\qquad\qquad\qquad\qquad$ aus Vitamin D$_3$

wurde, daß die reduktive Spaltung des Ozonids direkt zum $\alpha,\beta$-ungesättigten primären Alkohol (XCVI) führte. Dessen unmittelbare Oxydation mit Braunstein nach MORTON ergab einen $\alpha,\beta$-ungesättigten $C_{20}$-Aldehyd (XXX), der mit dem $C_{20}$-Abbau-aldehyd aus Vitamin D$_3$ identisch war (*8, 38*).

Die Aldolkondensation von (XXX) mit *p*-Oxy-cyclohexanon führte in einer Stufe zur Angliederung des Ringes *A* mit der sekundären Hydroxylgruppe unter gleichzeitiger Ausbildung der zweiten exocyclischen 5,6-Doppelbindung (IV und IV a) (*22, 29*). Die gleichfalls am richtigen Platz stehende Carbonylfunktion gestattete zum drittenmal, und zwar hier entscheidend, die Anwendung eines WITTIG-Reagens, des Triphenyl-phosphin-methylens, wodurch die charakteristische Methylengruppe erzeugt wurde. Hierdurch entstand praktisch am Schluß der Gesamtreaktionsfolge das empfindliche Trien-system unter außerordentlich milden Reaktionsbedingungen (*29, 31, 32*).

Vor dieser WITTIG-Reaktion war jedoch noch eine Epimeren-Trennung durchzuführen, da durch die Aldolkondensation am Hydroxyl-Kohlenstoffatom ein neues Asymmetriezentrum entstanden war. Diese ließ sich in einfacher Weise durch Chromatographie durchführen. Die beiden einheitlichen, an $C_{(3)}$ epimeren Hydroxy-dienone (IV und IV a) konnten nunmehr getrennt in die WITTIG-Reaktionen eingesetzt werden und ergaben die beiden erwarteten $C_{(3)}$-epimeren *trans*-Vitamine D$_3$ (III und III a). Von diesen erwies sich (III) als identisch mit dem bereits bekannten, nach HAVINGA durch Jod-Isomerisierung dargestellten Präparat (*29*).

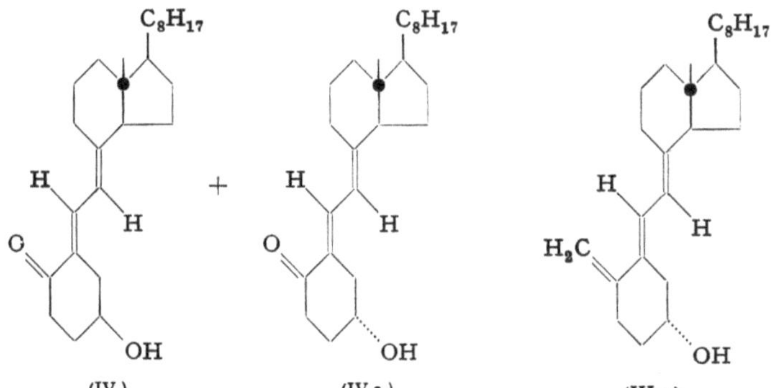

(IV.)            (IV a.)            (III a.)

Nachdem sich bei der Aldolkondensation infolge der sterischen Verhältnisse die weniger behinderte *trans*-Konfiguration an der 5,6-Doppelbindung ausgebildet hatte, blieb als letzter Schritt noch eine *trans* → *cis*-Isomerisierung. Diese wurde mit durch Glas gefiltertem

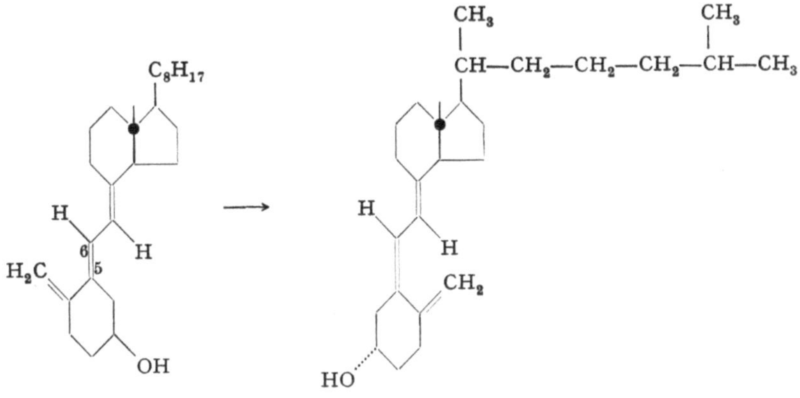

(III.)            (LXXX.) Vitamin D$_3$.

UV-Licht bewerkstelligt. Durch das Glas wird das kurzwellige Ultraviolett weggefiltert, gerade so, daß die längerwellige Absorption des maximal bei 272—273 m$\mu$ absorbierenden *trans*-Triens erfaßt wird und das gebildete *cis*-Vitamin (I) mit der Absorption bei 265 m$\mu$ sich gerade der weiteren Einwirkung der Strahlung entzieht (*35, 37*).

Das so gewonnene Vitamin D$_3$ gab den Phenylazobenzoesäureester, der in allen seinen Eigenschaften mit dem bekannten Präparat identisch war.

### Literaturverzeichnis.

*1.* Ahrens, G., E. Fernholz und W. Stoll: Über Hexahydroderivate von Bestrahlungsprodukten des Ergosterins. Liebigs Ann. Chem. **500**, 109 (1933).

*2.* Alder, K. und M. Schumacher: Anwendungen der Dien-Synthese für die Erforschung von Naturstoffen. Fortschr. Chem. organ. Naturstoffe **10**, 1 (1953).

3. BARON, C.: Comparaisons entre la structure chimique et l'activité biologique de composés voisins des vitamines D. Ann. Chim. [13] 1, 897 (1956).

4. BARON, C. et N. LE BOULCH: Description et propriétés des intermédiaires entre le cholestérol et la „cétone 250" antirachitique. Bull. soc. chim. France 1958, 300.

5. BARON, C., N. LE BOULCH et Y. RAOUL: Diols d'hydratation des vitamines $D_2$ et $D_3$. Bull. soc. chim. France 1955, 948.

6. BILLET, D.: Sur une dihydrovitamine $D_3$. C. R. hebd. Séances Acad. Sci. 244, 1794 (1957).

7. BRAUDE, E. A. and O. H. WHEELER: Studies in the Vitamin D Field. Part I. Synthesis of Trienes containing the Tachysterol Chromophore. J. Chem. Soc. (London) 1955, 320.

8. BROCKMANN, H. und A. BUSSE: Die Konstitution des antirachitischen Vitamins der Thunfischleber. Z. physiol. Chem. (Hoppe-Seyler) 256, 252 (1938).

9. CARDWELL, H. M. E., J. W. CORNFORTH, S. R. DUFF, H. HOLTERMANN and R. ROBINSON: Experiments on the Synthesis of Substances related to the Sterols. Part LI. Completion of the Syntheses of Androgenic Hormones and of the Cholesterol Group of Sterols. J. Chem. Soc. (London) 1953, 361.

10. CROWFOOT HODGKIN, D., M. S. WEBSTER and J. D. DUNITZ: Structure of Calciferol. Chem. and Ind. 1957, 1148.

11. DAUBEN, W. G., I. BELL, T. W. HUTTON, G. F. LAWS, A. RHEINER, Jr. and H. URSCHELER: Structure of Suprasterol-II. J. Amer. Chem. Soc. 80, 4116 (1958).

12. DIMROTH, K. und H. JONSSON: Die sterische Verknüpfung der Ringe C und D bei den Steroiden. Ber. dtsch. chem. Ges. 74, 520 (1941).

13. GUITERAS, A., Z. NAKAMIYA und H. H. INHOFFEN: Über die Einwirkung einiger Oxydationsmittel auf Derivate des Ergosterins. Liebigs Ann. Chem. 494, 116 (1932).

14. HAVINGA, E. and J. P. L. BOTS: Studies on Vitamin D. I. The Synthesis of Vitamin $D_3$-3-$C^{14}$. Rec. trav. chim. Pays-Bas 73, 393 (1954).

15. HAVINGA, E., A. L. KOEVOET and A. VERLOOP: Studies on Vitamin D and Related Compounds. IV. The Pattern of the Photochemical Conversion of the Provitamins D. Rec. trav. chim. Pays-Bas 74, 1230 (1955).

16. HAVINGA, E., A. VERLOOP and A. L. KOEVOET: Studies on Vitamin D and Related Compounds. V. Corroboration of the Scheme Proposed for the Photochemical Conversion of the Provitamins D, in Particular with Regard to the Place of the Previtamins D. Rec. trav. chim. Pays-Bas 75, 371 (1956).

17. HIRSCHFELD, H.: Partialsynthese der Vitamine $D_2$ und $D_3$. Dissert., Techn. Hochschule Braunschweig, 1958.

18. INHOFFEN, H. H.: Synthetische Versuche in der Vitamin D-Reihe. Festschrift Arthur Stoll, S. 419. Basel: Birkhäuser. 1957.

19. — Totalsynthese des Vitamins $D_3$. Angew. Chem. 70, 576 (1958). — Akad. der Wiss. und Literatur, Sitz Mainz, Jahrbuch 1958, S. 225.

20. INHOFFEN, H. H., F. BOHLMANN und E. REINEFELD: Über Grignard-Reaktionen mit „Crotylbromid". Chem. Ber. 82, 313 (1949).

21. INHOFFEN, H. H. und K. BRÜCKNER: Probleme und neuere Ergebnisse in der Vitamin D-Chemie. Fortschr. Chem. organ. Naturstoffe 11, 83 (1954), spez. S. 118.

22. INHOFFEN, H. H., K. BRÜCKNER und R. GRÜNDEL: Studien in der Vitamin D-Reihe: Umlagerung des Vitamins $D_3$ zu einem iso-Tachysterin und Partialsynthese eines iso-Vitamins $D_2$. Chem. Ber. 87, 1 (1954).

23. INHOFFEN, H. H., K. BRÜCKNER, R. GRÜNDEL und G. QUINKERT: Studien in der Vitamin D-Reihe. V: Weitere Untersuchungen an 9,10-seco-Ergostatetraenen, zugleich ein Beitrag zur Konstitution der Vitamin D-Isomeren. Chem. Ber. 87, 1407 (1954).

*24.* INHOFFEN, H. H., K. BRÜCKNER und H.-J. HESS: Studien in der Vitamin D-Reihe. XIII: Bildung einer strukturisomeren Vitamin D-Verbindung. Chem. Ber. **88,** 1850 (1955).

*25.* INHOFFEN, H. H., K. BRÜCKNER, K. IRMSCHER und G. QUINKERT: Studien in der Vitamin D-Reihe. XII: Die Konstitution des Tachysterins, zugleich Darstellung des u-Tachysterins, eines neuen 9,10-*seco*-Steroid-triens. Chem. Ber. **88,** 1424 (1955).

*26.* INHOFFEN, H. H., H. BURKHARDT und G. QUINKERT: Studien in der Vitamin D-Reihe. XXIX: Aufbau der Seitenkette des Vitamins $D_3$. Zugleich Beendigung der Totalsynthese des Vitamins $D_3$. Chem. Ber. **92,** im Druck (1959).

*27.* INHOFFEN, H. H. und K. IRMSCHER: Studien in der Vitamin D-Reihe. XV: Neue Aufbaumethode für 9,10-*seco*-Steroide mit dem Triensystem des Tachysterins. Chem. Ber. **89,** 1833 (1956).

*28.* — — Zur UV-Absorption Vitamin D-analoger Chromophore. Naturwiss. **45,** 86 (1958).

*28 a.* INHOFFEN, H. H., K. IRMSCHER, G. FRIEDRICH, D. KAMPE und O. BERGES: Studien in der Vitamin D-Reihe. XXX: Zweite Beendigung der Totalsynthese des Vitamins $D_3$. Chem. Ber. **92,** im Druck (1959).

*29.* INHOFFEN, H. H., K. IRMSCHER, H. HIRSCHFELD, U. STACHE und A. KREUTZER: Studien in der Vitamin D-Reihe. XXVI: Partialsynthese der Vitamine $D_2$ und $D_3$. Chem. Ber. **91,** 2309 (1958).

*30.* INHOFFEN, H. H. und J. KATH: Studien in der Vitamin D-Reihe. VII: Darstellung neuer Dekalinsysteme und deren Verwendung zu Modellsynthesen von 9,10-*seco*-D-Homo-Steroid-Verbindungen. Chem. Ber. **87,** 1589 (1954).

*31.* INHOFFEN, H. H., J. KATH und K. BRÜCKNER: Partialsynthese einer „trans"-Vitamin $D_2$-Verbindung mit Hilfe der Reaktion von Wittig. Angew. Chem. **67,** 276 (1955).

*32.* INHOFFEN, H. H., J. KATH, W. STICHERLING und K. BRÜCKNER: Studien in der Vitamin D-Reihe. XVIII: Partialsynthese des 5,6-*trans*-Vitamins $D_2$. Liebigs Ann. Chem. **603,** 25 (1957).

*33.* INHOFFEN, H. H. und E. PRINZ: Studien in der Vitamin D-Reihe. IV: Synthese der 3-Methyl-3-carboxy-cyclohexanon-(1)-propionsäure-(2β) sowie ihre Überführung in das 8-Methyl-hydrindan-dion-(1,4). Chem. Ber. **87,** 684 (1954).

*34.* INHOFFEN, H. H. und G. QUINKERT: Studien in der Vitamin D-Reihe. VI: Modellsynthesen zur Frage der *cis-trans*-Isomerie beim Tachysterin. Chem. Ber. **87,** 1418 (1954).

*35.* INHOFFEN, H. H., G. QUINKERT und H.-J. HESS: Partialsynthese des Vitamins $D_2$. Naturwiss. **44,** 11 (1957).

*36.* INHOFFEN, H. H., G. QUINKERT, H.-J. HESS und H.-M. ERDMANN: Studien in der Vitamin D-Reihe. XVI: Die chemische Isomerisierungsreihe des Vitamins $D_2$. Chem. Ber. **89,** 2273 (1956).

*37.* INHOFFEN, H. H., G. QUINKERT, H.-J. HESS und H. HIRSCHFELD: Studien in der Vitamin D-Reihe. XXIV: Photo-Isomerisierung der *trans*-Vitamine $D_2$ und $D_3$ zu den Vitaminen $D_2$ und $D_3$. Chem. Ber. **90,** 2544 (1957).

*38.* INHOFFEN, H. H., G. QUINKERT und S. SCHÜTZ: Studien in der Vitamin D-Reihe. XXIII: Aufbau des *trans*-Hydrindan-Ketons aus Vitamin $D_3$ zum vinylogen Aldehyd. Chem. Ber. **90,** 1283 (1957).

*39.* INHOFFEN, H. H., G. QUINKERT, S. SCHÜTZ, G. FRIEDRICH und E. TOBER: Studien in der Vitamin D-Reihe. XXV: Abbau der Vitamine $D_2$ und $D_3$ zum 8-Methyl-*trans*-hydrindanol-(4)-on-(1). Chem. Ber. **91,** 781 (1958).

*40.* INHOFFEN, H. H., G. QUINKERT, S. SCHÜTZ, D. KAMPE und G. F. DOMAGK: Studien in der Vitamin D-Reihe. XXI: Hydrindan-Verbindungen aus Vitamin $D_3$. Chem. Ber. **90,** 664 (1957).

*41.* INHOFFEN, H. H., S. SCHÜTZ, P. ROSSBERG, O. BERGES, K.-H. NORDSIEK, H. PLENIO und E. HÖROLDT: Studien in der Vitamin D-Reihe. XXVII: Synthese des (+)- und (—)-8-Methyl-*trans*-hydrindanol-(4)-ons-(1). Chem. Ber. 91, 2626 (1958).

*42.* INHOFFEN, H. H. und K. WEISSERMEL: Studien in der Vitamin D-Reihe. II: Partialsynthese des *iso*-Tachysterinmethyläthers. Chem. Ber. 87, 187 (1954).

*43.* INHOFFEN, H. H., K. WEISSERMEL, G. QUINKERT und K. IRMSCHER: Studien in der Vitamin D-Reihe. X: Partialsynthesen von Tachysterin-Analoga. Chem. Ber. 88, 1321 (1955).

*44.* IRMSCHER, K.: Über Konstitutionsbeweise und Partialsynthesen in der Tachysterin-Reihe. Dissert., Techn. Hochschule Braunschweig, 1957.

*45.* KAMPE, D.: Dissert., Techn. Hochschule Braunschweig, 1959.

*46.* KOEVOET, A. L., A. VERLOOP and E. HAVINGA: Studies on Vitamin D and Related Compounds. II. Preliminary Communication on the Interconversion and the Possible cis/trans Isomerism of Previtamin D and Tachysterol. Rec. trav. chim. Pays-Bas 74, 788 (1955).

*47.* KOEVOET, A. L., A. VERLOOP und J. A. KEVERLING BUISMAN: Verfahren zur Herstellung eines Isomerisierungsproduktes von Vitamin D oder von einem Ester desselben (sogenanntes *trans*-Vitamin D) bzw. von Dihydrotachysterin oder dessen Estern. Deutsches Patentamt, Auslegeschrift 1026748 (eingereicht von der Firma N. V. Philips' Gloeilampenfabrieken, Eindhoven, Ndl.).

*48.* LINSERT, O.: Verfahren zur Darstellung eines physiologisch hochwirksamen Produktes. D. R. Pat. 730017 (eingereicht von der I. G. Farbenindustrie A. G.) 1942.

*49.* MILAS, N. A. and C. PRIESING: Studies in the Synthesis of the Antirachitic Vitamins. IV. The Synthesis of a Biologically Active Vitamin D Homolog. J. Amer. Chem. Soc. 79, 3610 (1957).

*50.* MÜLLER, M.: Zur Kenntnis des Vitamins D und seiner thermischen und photochemischen Umwandlungsprodukte. Z. physiol. Chem. (Hoppe-Seyler) 233, 223 (1935).

*51.* NENITZESCU, C. D. und E. CIORANESCU: Durch Aluminiumchlorid katalysierte Reaktionen. XXIII: Versuche zur Synthese von Verbindungen mit dem Steringerüst. Ber. dtsch. chem. Ges. 75, 1765 (1942).

*52.* PETERING, H. G.: Isolation of Vitamins Dm and Vitamin $D_3$ from the Irradiation Products Obtained from the Sterols of the Mussel, *Modiolus Demissus*, DILLWYN. J. Organ. Chem. (USA) 22, 808 (1957).

*53.* RAOUL, Y.: La cétone 250, nouveau type de vitamine antirachitique. Rev. internat. vitaminologie 28, 306 (1958).

*54.* RAOUL, Y., N. LE BOULCH, C. BARON, R. BAZIER et A. GUERILLOT-VINET: Formes énolique et calcique de la ,,cétone 250'' d'activité antirachitique supérieure à celle du calciférol-3. C. R. hebd. Séances Acad. Sci. 242, 3004 (1956).

*55.* RAPPOLDT, M. P.: Photochemische reacties van provitaminen D en verwante verbindingen. Dissert., Univ. Leiden, 1958.

*56.* RAPPOLDT, M. P., J. A. KEVERLING BUISMAN and E. HAVINGA: Studies on Vitamin D and Related Compounds. VIII. The Photoisomerisations of Provitamin D and its Irradiation Products. Short Communication. Rec. trav. chim. Pays-Bas 77, 327 (1958).

*57.* RAPPOLDT, M. P., P. WESTERHOF, K. H. HANEWALD and J. A. KEVERLING BUISMAN: Investigations on Sterols. X. The Conversion of Ergosterol and Pre-ergocalciferol by U. V. Light of 254 mμ. Rec. trav. chim. Pays-Bas 77, 241 (1958).

58. Reichel, S. v. und M. Deppe: Über die Konstitution des Dihydro-Vitamins $D_2$. Z. physiol. Chem. (Hoppe-Seyler) **239**, 143 (1936).
59. Schubert, K.: Kristallisiertes Dihydro-Vitamin $D_2$-II. Naturwiss. **41**, 231 (1954).
60. — Zur Konfiguration des Dihydrovitamins $D_2$-II. Biochem. Z. **326**, 132 (1954).
61. — Über die partielle Hydrierung des Vitamins $D_2$. Ein Verfahren zur Herstellung von kristallisiertem Dihydrovitamin $D_2$-II. Biochem. Z. **327**, 507 (1956).
62. Schubert, K. und K. Wehrberger: Zum Auftreten von Isomeren des Dihydrovitamins $D_2$-II bzw. Dihydrotachysterins. Biochem. Z. **328**, 199 (1956).
63. Schütz, S.: Untersuchungen am Präcalciferol. Dipl. Arbeit, Techn. Hochschule Braunschweig, 1955.
64. Trippett, S.: The Oppenauer Oxydation of Ergocalciferol. J. Chem. Soc. (London) **1955**, 370.
65. Velluz, L. et G. Amiard: Recherches sur le précalciférol. Bull. soc. chim. France **1955**, 205.
66. Velluz, L., G. Amiard et B. Goffinet: Époxydes dérivés du calciférol et du précalciférol. C. R. hebd. Séances Acad. Sci. **240**, 2076 (1955).
67. — — — Transformation photochimique du précalciférol en tachystérol. C. R. hebd. Séances Acad. Sci. **240**, 2156 (1955).
68. — — — Régression photochimique du précalciférol vers le lumistérol et l'ergostérol. C. R. hebd. Séances Acad. Sci. **240**, 2326 (1955).
69. — — — Le précalciférol. Structure et photochimie. Son rôle dans la genèse du calciférol et des photoisomères de l'ergostérol. Bull. soc. chim. France **1955**, 1341.
70. — — — Analogues étio du précalciférol. C. R. hebd. Séances Acad. Sci. **244**, 1794 (1957).
71. — — — Analogues étio du précalciférol. Bull. soc. chim. France **1957**, 882.
72. Velluz, L., G. Amiard et A. Petit: Le précalciférol. Ses relations d'équilibre avec le calciférol. Bull. soc. chim. France **1949**, 501.
73. Verloop, A., A. L. Koevoet and E. Havinga: Studies on Vitamin D and Related Compounds. III. Short Communication on the *cis-trans* Isomerisation of Calciferol and the Properties of *"trans"*-Vitamin $D_2$. Rec. trav. chim. Pays-Bas **74**, 1125 (1955).
74. — — — Studies on Vitamin D and Related Compounds. VII. Some Remarks on the Stereochemistry of the Trienoic Systems of Vitamin D and Related Compounds. Rec. trav. chim. Pays-Bas **76**, 689 (1957).
75. Vliervoet, J. L. J. van de, P. Westerhof, J. A. Keverling Buisman and E. Havinga: Studies on Vitamin D and Related Compounds. VI. The Synthesis and Properties of Dihydrotachysterol₃. Rec. trav. chim. Pays-Bas **75**, 1179 (1956).
76. Wander, A., A. G.: Hydrierungsprodukt des Vitamins $D_2$. Schweiz. Pat. 246835 (1948).
77. Werder, F. v.: Über Dihydrotachysterin. Z. physiol. Chem. (Hoppe-Seyler) **260**, 119 (1939).
78. — Zur Kenntnis der Dihydrovitamine $D_2$. Liebigs Ann. Chem. **603**, 15 (1957).
79. Westerhof, P. and J. A. Keverling Buisman: Investigations on Sterols. VI. The Preparation of Dihydrotachysterol₂. Rec. trav. chim. Pays-Bas **75**, 453 (1956).
80. — — Investigations on Sterols. VIII. Some Hitherto Unknown Irradiation Products of Ergosterol. Rec. trav. chim. Pays-Bas **75**, 1243 (1956).

*81.* WESTERHOF P. and J. A. KEVERLING BUISMAN: Investigations on Sterols. IX. Dihydroderivates of Ergocalciferol. Rec. trav. chim. Pays-Bas **76**, 679 (1957).

*82.* WINDAUS, A. und K. BUCHHOLZ: Über ein Keton des Vitamins $D_2$. Z. physiol. Chem. (Hoppe-Seyler) **256**, 273 (1938).

*83.* WINDAUS, A., J. GAEDE, J. KÖSER und G. STEIN: Über einige kristallisierte Bestrahlungsprodukte aus Ergosterin und Dehydro-ergosterin. Liebigs Ann. Chem. **483**, 17 (1930).

*84.* WINDAUS, A., O. LINSERT, A. LÜTTRINGHAUS und G. WEIDLICH: Über das krystallisierte Vitamin $D_2$. Liebigs Ann. Chem. **492**, 226 (1932).

*85.* WINDAUS, A. und U. RIEMANN: Über die Einwirkung von Bleitetraacetat auf einige Sterinderivate. Z. physiol. Chem. (Hoppe-Seyler) **274**, 206 (1942).

*86.* WINDAUS, A. und C. ROOSEN-RUNGE: Zur Konstitution des Dihydro-Vitamins $D_2$ und $D_3$. Z. physiol. Chem. (Hoppe-Seyler) **260**, 181 (1939).

*87.* WINDAUS, A., F. v. WERDER und A. LÜTTRINGHAUS: Über das Tachysterin. Liebigs Ann. Chem. **499**, 188 (1932).

*88.* WOODWARD, R. B., F. SONDHEIMER, D. TAUB, K. HEUSLER and W. M. McLAMORE: The Total Synthesis of Steroids. J. Amer. Chem. Soc. **74**, 4223 (1952).

*89.* YOUNG, W. G. and J. D. ROBERTS: Allylic Rearrangements. XXI. Further Studies Related to the Nature of the Butenyl Grignard Reagent. J. Amer. Chem. Soc. **68**, 1472 (1946).

*90.* ZIMMERMAN, H. E.: The Stereochemistry of the Ketonization Reaction of Enols. II. J. Amer. Chem. Soc. **78**, 1168 (1956).

*(Eingelaufen am 6. Februar 1959.)*

# Neuere Ergebnisse der Chemie pflanzlicher Bitterstoffe.

Von **F. Korte, H. Barkemeyer** und **I. Korte**, Bonn.

Mit 1 Abbildung.

### Inhaltsübersicht.

# I. Einleitung.

Als Bitterstoffe bezeichnet man bitter schmeckende, N- und S-freie, O-haltige Substanzen, die noch nicht in andere große Verbindungsklassen, wie Alkaloide, Farbstoffe, Sterine usw., eingegliedert sind. Bis heute ist es nicht gelungen, eine Beziehung zwischen dem bitteren Geschmack und der chemischen Konstitution zu entdecken (75). Wenn auch viele Bitterstoffe Keton- oder Lactongruppierungen enthalten, so ist doch das Vorhandensein derartiger Gruppen zur Erregung der auf „bitter" ansprechenden Geschmackssinnes-zellen keine notwendige Voraussetzung.

Das Vorkommen der Bitterstoffe ist nicht auf bestimmte Pflanzenfamilien beschränkt, wenn sie auch in einigen besonders zahlreich vertreten sind, wie z. B. in den Gentianaceen oder den Compositen. Entstehung oder eventuelle Aufgabe der Bitterstoffe in den Pflanzen sind völlig unbekannt. Lediglich bei den Gentianaceen sind bisher Versuche zur Lokalisierung des Bildungsortes durch I. KORTE (91) unternommen worden. Hier konnte sowohl durch Pfropfungsversuche wie auch durch Wurzelkulturen gezeigt werden, daß weder Sproß noch Wurzel allein für die Bitterstoffbildung verantwortlich ist. Im allgemeinen sind die Bitterstoffe in allen Organen der Pflanzen zu finden (137); bei den stets bitteren Gentianaceen sind die Samen bitterstoff-frei. Im Verlauf der Vegetationsperiode kann es zu Schwankungen im Bitterstoffgehalt kommen, hierüber liegen aber ebenfalls nur wenige Untersuchungen vor (101). Es hat ferner in manchen Fällen den Anschein, als stiege mit zunehmendem Alter einer Pflanze auch die Bitterkeit an (75).

Die Existenz der Bitterstoffe ist schon seit langer Zeit bekannt, bereits im 15. Jahrhundert werden sie in deutschen Kräuterbüchern beschrieben (110). Wenn man sich schon im Mittelalter mit den „Amara" beschäftigte, so vor allem deshalb, weil man ihnen therapeutischen Wert beimaß. Für die moderne Medizin haben sie jedoch erheblich an Bedeutung verloren.

Aus mehreren Gründen sind die pflanzlichen Bitterstoffe Gegenstand wissenschaftlicher Forschung. Zunächst ist es notwendig, die Verbindungen rein darzustellen, um ihre chemischen und pharmakologischen Eigenschaften kennenzulernen; ferner handelt es sich um chemisch interessante Substanzen, die oft gegen Luftsauerstoff und in der Regel gegen pH-Änderungen empfindlich sind (Lactonöffnung im alkalischen Bereich, Verlust der Bitterkeit nach Alkalibehandlung). Hinzu kommt eine häufig beobachtete, große Umlagerungsbereitschaft, die durch den meist komplizierten Bau der Moleküle bedingt ist. Für die Isolierung aus pflanzlichem Material ist der bittere Geschmack ein ebenso einfacher wie empfindlicher Test, wenn auch die Reindarstellung in vielen Fällen auf Schwierigkeiten stößt.

Bitterstoffe sind mehrfach bei systematischen Untersuchungen zur chemischen Klassifizierung von Pflanzen als charakteristische Inhaltsstoffe angesehen worden (*85, 54, 57*). Nicht zuletzt haben pflanzliche Bitterstoffe auch wirtschaftliche Bedeutung.

Im einleitenden Teil dieses Übersichtsartikels wird über therapeutische Verwendung und wirtschaftliche Bedeutung der Bitterstoffe berichtet, anschließend soll an Hand einiger Beispiele die Problematik der „Bitterstoffchemie" dargelegt werden. Die dabei diskutierten Verbindungen sind nach Pflanzenfamilien geordnet.

Eine vollständige Übersicht der bis 1953 beschriebenen pflanzlichen Bitterstoffe gibt Korte (*75*). Bei vielen der dort zusammengestellten Substanzen ist jedoch der Bitterstoffcharakter zweifelhaft, da die Auswahl zum Teil auf Grund sehr alter Literaturangaben getroffen wurde. Für eine Zuordnung in die Klasse der Bitterstoffe sollte ein „Mindestbitterwert" gefordert werden. Da entsprechende Bitterwertsbestimmungen aber erst seit kurzem durchgeführt werden, kann dieses Kriterium hier noch nicht berücksichtigt werden.

### Der bittere Geschmack und die Bestimmung des Bitterwertes.

Da bei der chemischen Bearbeitung der Bitterstoffe der Geschmackstest eine wichtige Rolle spielt, sei hier kurz die Frage erörtert, welche Vorgänge sich bei der Erregung des menschlichen Geschmacksorgans, der Zunge, abspielen. Trotz zahlreicher experimenteller Arbeiten weiß man hierüber nur sehr wenig.

Von den zirka 9000 Geschmackssinnes-zellen, die der Mensch besitzt, befindet sich der größte Teil auf der Zunge. Sie sind aber besonders bei Erwachsenen nicht gleichmäßig über die Zungenoberfläche verteilt, sondern in bestimmten Regionen dicht angeordnet. Die vier Hauptgeschmackseindrücke: salzig, sauer, süß und bitter, werden deshalb an verschiedenen Stellen der Zunge sehr unterschiedlich empfunden. *Abb. 1* zeigt schematisiert diejenigen Stellen, an denen die Rezeptoren angehäuft sind. Der bittere Geschmack wird weitaus am stärksten empfunden. *Tabelle 1* gibt die Grenzkonzentrationen an, bei denen der betreffende Geschmack gerade noch wahrnehmbar ist.

Tabelle 1. Grenzkonzentrationen
    des Geschmacksempfindens.

| Geschmack | Grenzkonzentration (%) | |
|---|---|---|
| sauer | 0,045 | HCl |
| salzig | 0,055 | NaCl |
| süß | 0,7 | Rohrzucker |
| | 0,001 | Saccharin |
| bitter | 0,0001 | Brucin |
| | 0,00005 | Chinin |
| | 0,000007 | Amarogentin |

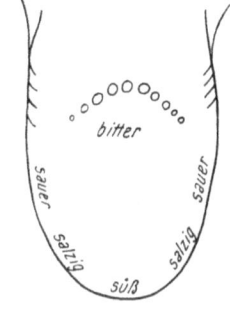

Abb. 1. Die geschmacks-empfindlichen Zentren der Zunge.

Die Zahlen stellen nur Relativ-werte dar, da die Sinneseindrücke stark variieren. So wird der Phenylthioharnstoff von etwa 70% der Menschen als bitter, von 30%

als geschmacksfrei empfunden (*75*). Das Geschmacksvermögen ist ferner abhängig von Witterungseinflüssen und atmosphärischem Luftdruck; mit abnehmendem Luftdruck sinkt auch die Reizschwelle (*75, 108*).

Wie heterogen in chemischer Hinsicht bitter schmeckende Substanzen sind, zeigt die folgende Gegenüberstellung: Caesiumjodid, Magnesiumsulfat, Pikrinsäure, Phenylharnstoff, Glucose-triacetat, Strychnin.

Man gewinnt hier den Eindruck, daß die chemische Konstitution, soweit sie sich in den heute üblichen Strukturformeln ausdrückt, nicht den primären Faktor für die Auslösung des Geschmacksreizes darstellt. Die physikalisch-chemischen Eigenschaften einer Substanz scheinen hierfür viel entscheidender zu sein. Dahingehend sind vermutlich auch von ALLEN und WEINBERG (*5*) durchgeführte Versuche zu deuten, bei denen es gelang, die vier Geschmacksempfindungen durch elektrische Reize hervorzurufen.

*Bitterwertsbestimmung.* Um quantitative Aussagen über die Bitterkeit einer Substanz zu machen, bedient man sich mit Vorteil der Methode von WASICKY (*155*), bei welcher die Grenzkonzentration des noch wahrnehmbaren bitteren Geschmacks ermittelt wird. Ein individueller Fehler wird dadurch ausgeschaltet, daß man die Probe mit einer Standard-Brucinlösung, die auf 1 : 4 800 000 verdünnt ist, vergleicht. Eine modifizierte Form des Verfahrens erlaubt die Bitterwertsbestimmung von Drogen, die als geschmacks-beeinträchtigende Begleitstoffe Tannin oder ätherische Öle enthalten. Nach Entfernung dieser Substanzen durch Behandlung mit entfettetem Caseinpulver ist eine Bestimmung des Bitterwertes möglich (*68*).

### Therapeutische Verwendung der Bitterstoffe.

Die Kräuterbücher des Mittelalters schreiben den bitterstoffhaltigen Pflanzen eine Reihe heilkräftiger Wirkungen zu, so wird z. B. ihre Anwendung als Fiebermittel oder Digestiva empfohlen. Die moderne Medizin kann die fiebersenkende Eigenschaft der Amara nicht bestätigen (*110*). Wenn trotzdem auch heute noch bitterstoffhaltige Pflanzenextrakte in den Pharmakopöen der verschiedenen Kulturstaaten offizinell sind, so deshalb, weil sich die Vorliebe für bittere Arzneien beim Volke erhalten hat („was nicht bitter schmeckt, ist keine Medizin"). In unserer Zeit werden die Amara überwiegend als appetitanregende Mittel verwendet. Experimente nach PAVLOFF an Hunden zeigen, daß Rohauszüge von bitteren Drogen die Magensaftsekretion steigern (Enzian, Condurangorinde, Colombowurzel) (*61*) und die Acidität des Magensaftes erhöhen (*3*). Bitterstoffe sind Hauptbestandteile in antidiabetischen Tees.

Nach JUNKMANN (*65*) steigern die Bitterstoffe aus *Achillae* und *Menyanthes* die Erregbarkeit des N. sympathicus. Sie erinnern in ihrer Wirkung an die des Uzarons (*75*), führen aber nicht wie dieses zu einer direkten Erregung, sondern nur zu einer Erregbarkeits-steigerung des sympathischen Nervensystems. Im gleichen Sinne läßt sich auch der Einfluß auf den Uterus deuten. Dies erklärt die Anwendung einiger

Bitterstoffdrogen bei Menstruationsstörungen oder deren Benutzung als Abortiva.

Die hier gemachten Angaben basieren auf Erfahrungen, die zum großen Teil durch Einsatz unreiner Substanzen oder Rohextrakte gewonnen sind. Bevor man ein sicheres Urteil über therapeutische Verwendbarkeit pflanzlicher Bitterstoffe wird abgeben können, sind entsprechende pharmakologische und klinische Versuche mit Reinsubstanzen durchzuführen. Als ein Anfang in dieser Hinsicht ist der Befund zu werten, daß die Bitterstoffe des Hopfens, Humulon und Lupulon, das Wachstum von *Mycobacterium tuberculosis* hemmen (*30*).

### Wirtschaftliche Bedeutung der Bitterstoffe.

Bitterstoffe werden in großem Maße zur Herstellung bitterer Getränke, wie ,,Bitters'', Aperitifs, Bier, benötigt. In einigen Ländern wurde früher sogar wegen Hopfenmangels dem Bier zur Erzeugung der erwünschten Bitterkeit das sehr giftige Picrotoxin in kleinsten Dosen hinzugefügt (*44*). Außerdem finden Bitterstoffe auch als Geschmacks-korrigentien Verwendung. Sehr häufig aber ist Bitterstoffbildung Ursache wirtschaftlicher Schäden, da durch sie die Verwendung vieler Früchte und Pflanzen als Nahrungs- und Futtermittel stark beeinträchtigt wird. Dies sei an einigen Beispielen erläutert.

Bei den Orangen ist geschmacklich die Navel- der Valencia-Sorte überlegen, jedoch ist ein großer Nachteil, daß Navel-Preßsaft kurze Zeit nach dem Ausquetschen bitter und damit für Handelszwecke unbrauchbar wird. Aus Samen von Navel-Orangen konnten in reiner Form zwei Bitterstoffe, Limonin und Nomilin, isoliert werden (*45, 139*). Die nicht bitteren Früchte enthalten eine Vorstufe des Limonins in der Albedoschicht. Es handelt sich dabei um eine wasserlösliche Lactoncarbonsäure oder um eine Dicarbonsäure (*92*). Beim Ausquetschen der Frucht wird die Albedoschicht zerrissen und die Limonin-Vorstufe gelangt in den Fruchtsaft. Hier erfolgt im sauren Milieu Lactonisierung zum Bitterstoff (*100*). Marsh (*103*) wies nach, daß Ausmaß und Geschwindigkeit der Bitterstoffbildung von der Art des Wurzelstocks abhängen, auf dem die betreffende Pflanze wächst.

Auch im Hafer kann es zu einer unerwünschten Bildung von Bitterstoffen kommen. Auf Grund experimenteller Ergebnisse vertritt Rothe (*129*) die Ansicht, daß sich die chemischen Reaktionen, die zum Bitterstoff führen, an den Doppelbindungen der ungesättigten Fettsäuren abspielen. Analytisch ließ sich bisher nur Polymerisation der Lipoidkomponente nachweisen, jedoch sind als Vorläufer der bitteren Substanzen Peroxyde anzunehmen. Die Beteiligung von Lipoxydasen an dieser Reaktion ist wahrscheinlich. Die eigentliche Bitterstoffbildung, d. h. die Polymerisation der Peroxyde, soll durch den ,,fett-antioxygenen Komplex'' ermöglicht werden.

Aus bitter gewordenen Karotten konnten Dodson und Mitarbeiter (*40*) durch Extrahieren mit Aceton einen kristallinen Bitterstoff gewinnen. Chicoree könnte in weit größerem Maße, als es zur Zeit der Fall ist, als Gemüsepflanze Verwendung finden, falls es gelänge, die sporadisch auftretende Bitterstoffbildung zu verhindern. Über die Natur des Bitterstoffes oder die Ursachen seiner Entstehung ist wenig bekannt.

Kuhmilch kann manchmal derartig bitter sein, daß sie für den Genuß unbrauchbar wird. Die Ursache dafür dürfte in einem in bestimmten Unkräutern enthaltenen Bitterstoff zu suchen sein, dessen Isolierung Herzer (*58 b*) gelang.

Es hat nicht an Versuchen gefehlt, die Bildung dieser störenden Substanzen, z. B. durch Züchtung, zu verhindern.

Als erfolgreiches Experiment dieser Art ist die Züchtung der Süßlupine bekannt; allerdings wird die Bitterkeit der gewöhnlichen Lupine nicht durch einen Bitterstoff im chemischen Sinne, sondern durch ein Alkaloid verursacht. Über Züchtungsversuche berichtet ALMEIDA (6) in einer Arbeit, die außerdem eine Liste mit 87 Gräsern, Gemüsen und anderen Nutzpflanzen und Hinweise auf deren Gehalt an schädlichen oder toxischen Begleitstoffen enthält.

Abschließend sei noch über einen Fall unerwünschter Bitterstoffbildung beim Lagern von Rotwein berichtet. Die Synthese dieser Bitterstoffe erfolgt nach RENTSCHLER und TANNER (119) in zwei voneinander unabhängigen Schritten: (a) Aus Glycerin entsteht auf enzymatischem Wege Acrolein; (b) Acrolein reagiert mit Polyphenolen unter Bitterstoffbildung. Die Polyphenole entstammen den Gerb- und Farbstoffen des Weines. Es gelang den Autoren, durch Umsatz von Acrolein mit *D,L*-Epicatechin (dem Grundkörper des Birngerbstoffes), Quercetin, Phloroglucin und anderen Polyphenolen, synthetische Bitterstoffe zu erzeugen. Ersatz des Acroleins durch andere ungesättigte Aldehyde, z. B. Crotonaldehyd, führte ebenfalls zur Bildung von Bitterstoff.

Diese wenigen Beispiele zur wirtschaftlichen Bedeutung der Bitterstoffe mögen zeigen, daß die chemische Untersuchung dieser interessanten Stoffklasse nicht nur von theoretischem Interesse ist. Das Inhibieren der Bitterstoffbildung in Fällen wie den oben angeführten, sowie die Beseitigung störender Bitterstoffe wird mit der Kenntnis der Konstitution der betreffenden Substanz sicher erleichtert.

## II. Bitterstoffe der Gentianaceen.

Die Familie der Gentianaceen besteht nach HEGI (56) aus zwei Unterfamilien, den Gentianoideae und den Menyanthoideae. WETTSTEIN (159) dagegen faßt die Menyanthaceen als selbständige Familie neben den Gentianaceen auf. Bei den Menyanthaceen ist als Bitterstoff das Meliatin aus *Menyanthes trifoliata* beschrieben worden. Konstitutionsvorschläge sind von BIRCH und SMITH (19) und von MERZ und Mitarbeitern (105, 106) gemacht worden. Nach ROSENTHALER (128) ist Meliatin mit dem aus *Strychnos nux vomica* isolierten Loganin (S. 133) identisch.

Als Bitterstoffe der Gentianaceen werden in der Literatur Gentiamarin (146, 115), Gentiin (146), Erythramarin (93), Erytaurin (66), Chiratin (48), Swertiamarin (96), Gentiopikrin (76) und Amarogentin (79) angegeben. Als einheitliche Individuen können nur Gentiopikrin und Amarogentin angesehen werden. Amarogentin ist der Stoff mit dem bisher höchsten bekannten Bitterwert; noch in einer Verdünnung von 1 : 58 Millionen ist der bittere Geschmack deutlich wahrnehmbar (88).

Die Beschreibung so vieler Bitterstoffe bei den Gentianaceen ist sicherlich dem Umstand zuzuschreiben, daß die Substanzen sehr schwer zur Kristallisation zu bringen sind. Es besteht die hohe Wahrscheinlichkeit, daß das extrem bittere Amarogentin adsorptiv in kleinsten Mengen als Verunreinigung festgehalten werden kann und dadurch eine nicht bittere Substanz als „Bitterstoff" erscheint.

Die Rechnung ergibt, daß bei einer nicht bitteren Substanz eine Verunreinigung durch 0,02% Amarogentin ausreicht, um noch in einer Verdünnung von 1 : 12 000 bitteren Geschmack hervorzurufen; dieser Wert entspricht der Bitterkeit des Gentiopikrins. Der analytische Nachweis einer so kleinen Amarogentinmenge wäre schwierig.

## 1. Gentiopikrin.

Das Gentiopikrin wurde 1862 aus der frischen Enzianwurzel *(Gentiana lutea)* isoliert. Die Isolierungsmethode wurde später von Bourquelot und Hérissey (20) und von Tanret (146) verbessert; außerdem stellten sie fest, daß Gentiopikrin ein Glykosid mit Lactoncharakter ist. Korte und Schiffer (83) gelang kürzlich die Isolierung des Gentiopikrins aus nicht stabilisierter Trockendroge.

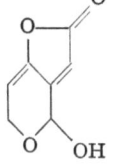

Patulin.

Chemisch verhält sich Gentiopikrin dem Schimmelpilz-Produkt Patulin (163, 164) sehr ähnlich (76). Wie dieses neigt es zu unübersichtlichen Umlagerungen, die zur Zerstörung des Moleküls führen; schon in neutraler wäßriger Lösung, besonders aber im sauren oder alkalischen Gebiet, erleidet es Zersetzung. Aus diesem Grunde liegen bis heute nur Vorschläge zur Konstitution des Gentiopikrins vor, die von Korte (76), Sakurai (132) sowie Canonica und Pelizzoni stammen (26). Die von Sakurai vorgeschlagene Formel (III) entspricht nicht den optischen Eigenschaften des Gentiopikrins und denen der Tetrahydro-Derivate.

(I.) Gentiopikrin (Korte). *(Gluc = D-Glucose.)*

(II a.)

(III.) Gentiopikrin (Sakurai).

(II b.) Gentiopikrin (Canonica).

*Reaktionen des Gentiopikrins.*

*Hydrierung.* Gentiopikrin enthält drei C,C-Doppelbindungen; auf Grund UV-spektroskopischer Beobachtungen stehen zwei Doppelbindungen miteinander und zur Carbonylgruppe in Konjugation, die dritte steht isoliert (76). Zwei Doppelbindungen sind gleichwertig und werden durch katalytisch erregten Wasserstoff innerhalb weniger Minuten hydriert, dagegen werden zur Absättigung der dritten Doppelbindung mehrere Stunden benötigt (76). Bei der Tetrahydrierung des Tetraacetyl-gentiopikrins entstehen nach Asahina (9) und Korte (76) zwei Isomere, die mit α und β bezeichnet werden und sich durch Schmelzpunkt und

*Literaturverzeichnis: SS. 175—182.*

Drehwert voneinander unterscheiden. Führt man die Tetrahydrierung am Gentiopikrin selbst durch und acetyliert nachträglich, so entsteht nur das α-Isomere (9). Die Annahme ASAHINAS, daß es sich um optische Isomere handle, ist nach KORTE unzutreffend, da die UV- und IR-Spektren beider Verbindungen ebenfalls verschieden sind. Wie CANONICA und PELIZZONI (26) feststellten, liefern α- und β-Derivat dasselbe Hexahydroprodukt bei der Hydrierung mit Rhodium in Methanol. Die Entstehung der beiden isomeren Tetrahydroverbindungen läßt sich durch Annahme einer 1,2- bzw. einer 1,4-Addition erklären (26):

(II a, b.) (als Tetraacetat) $\xrightarrow{+ 2 H_2}$

α- (II c.) Tetrahydro-gentiopikrin. β-

*Ozonolyse.* Bei der Ozonspaltung des Tetraacetyl-gentiopikrins wird zirka 1 Mol Formaldehyd gebildet (76), wodurch eine endständige Methylengruppe bewiesen ist.

*Enzymspaltung.* Bei der Spaltung des Gentiopikrins mit Emulsin entsteht neben 1 Mol Glucose ein uneinheitliches Öl, das als „Mesogentiogenin-Fraktion" bezeichnet wird (9). Da auch unter schonendsten Bedingungen erhaltene Präparate optisch inaktiv sind, muß während der Enzymspaltung Racemisierung eingetreten sein. Nach KORTE (76 a) läßt sich dieser Befund dadurch deuten, daß der Lacton-ring geöffnet wird und damit Racemisierung am Halbacetal-Kohlenstoffatom erfolgen kann (Analogie zum Patulin und den Zuckern). „Mesogentiogenin" gibt eine tiefblaue $FeCl_3$-Reaktion (9); der Glucose-Rest muß deshalb mit dem Genin über ein enolisches Hydroxyl verknüpft sein. Aus der Mesogentiogenin-Fraktion konnte ASAHINA (8, 9) nach längerem Stehen zwei kristalline, ebenfalls optisch inaktive Substanzen isolieren, die durch Isomerisierung bzw. Dimerisierung entstanden sind, das gelbe Eugentiogenin und das dimere Gentiogenin. Die Bildung der beiden Verbindungen läßt sich aus der von KORTE vorgeschlagenen Konstitutions-formel erklären (76 a).

Während es bisher nicht gelungen ist, durch Enzymspaltung des Gentiopikrins zum primären Genin zu gelangen, führt die Enzymolyse des Tetrahydro-gentio-pikrins mit 10% Ausbeute zum optisch aktiven, kristallinen Tetrahydro-proto-gentiogenin, das sich zum optisch aktiven Hexahydro-protogentiogenin hydrieren läßt. Dasselbe Hexahydroderivat entsteht auch durch direkte Hexahydrierung des Gentiopikrins in Wasser (8). Nach CANONICA und PELIZZONI (26) soll bei der Bildung des Hexahydro-protogentiogenins eine Veränderung des Grundgerüstes erfolgen, wie sich u. a. durch Vergleich der molaren Drehwerte des Tetraacetyl-tetrahydro-gentiopikrins und des Hexahydro-protogentiogenins ergibt.

Hydrierung bewirkt offensichtlich eine Stabilisierung des Aglykons. Zu diesem Schluß führen auch Untersuchungen über die Stabilität des Lactonringes.

*Der Lactonring.* Lactontitration führt beim Gentiopikrin zu dem einer Lacton-gruppe entsprechenden Wert. Bei den optisch inaktiven Verbindungen („Meso-gentiogenin-Fraktion", Eugentiogenin, dimeres Gentiogenin) läßt sich der Lacton-ring bereits durch Behandlung mit $KHCO_3$ in Alkohol öffnen; in den optisch aktiven Verbindungen erweist sich das Lacton als merklich stabiler (76 a). Daraus ergibt sich, daß die bei der Bildung des Mesogentiogenins beobachtete Racemisierung über die Öffnung des Lactonringes erfolgt (76 a).

*Reaktion mit Triphenyl-tetrazoliumchlorid („TTC") (76).* Gentiopikrin, seine Hydrierungsprodukte und alle nach enzymatischer Hydrolyse erhaltenen Derivate liefern mit TTC einen roten Formazan-Farbstoff. Die Reaktion eignet sich für quantitative Bestimmungen; nach Erhitzen der reinen Substanzen mit TTC in Alkali erhält man stets den einer CHO-Gruppe entsprechenden Farbwert (Bezugssubstanz: Gentiopikrin). Formel (I, S. 130) vermag diese Reaktion durch Überführung des nach Lactonöffnung entstandenen Cyclohalbacetals in den entsprechenden Aldehyd zu erklären. Die gleiche Reaktion wird bei Patulin, Allopatulin und $\gamma$-Lactonen von Cyclohalbacetalen beobachtet.

Nach neuesten IR-spektroskopischen Befunden *entspricht keiner der angeführten Strukturvorschläge dem Gentiopikrin.* Gentiopikrin und Hexahydro-protogentiogenin zeigen im IR Carbonylfrequenzen von 1727 bzw. 1730 cm$^{-1}$ und entsprechen damit den Carbonsäureestern oder $\gamma$-Lactonen. Die für ein $\gamma$-Lacton [entsprechend Formel (I)] abnorm tiefe Frequenzlage war bei Aufstellung der Formel mit der Halbacetal-lacton-Struktur gedeutet worden (76). Inzwischen gelang Korte und Mitarbeitern (83, 86, 89) die Synthese des der Formel (I) entsprechenden Grundgerüstes (IV) und analoger bicyclischer Lactone. Die CO-Frequenzen liegen bei den synthetischen Produkten zwischen 1780 und 1795 cm$^{-1}$ und entsprechen damit dem für $\gamma$-Lactone typischen Wert. Patulin, das ein doppelt ungesättigtes $\gamma$-Lactonsystem enthält, absorbiert im gleichen Bereich (1770 und 1785 cm$^{-1}$). Damit ist eine $\gamma$-Lactonformulierung im Gentiopikrin ausgeschlossen.

(IV.)

Aus den bisher vorliegenden experimentellen Befunden muß man nach Meinung der Verfasser im Gentiopikrin auf folgende Strukturelemente schließen: Doppelt ungesättigtes, mindestens sechsgliedriges Lacton. — Isolierte, endständige Methylengruppe. — Enol-Glucosid. — Durch Lactonisierung stabilisiertes Cyclohalbacetal.

Ferner müssen sich die folgenden Reaktionen und Beobachtungen aus der Konstitutionsformel des Gentiopikrins ableiten lassen: (a) Das Auftreten strukturisomerer $\alpha$- und $\beta$-Tetraacetyl-tetrahydroderivate und deren Überführbarkeit in dasselbe Hexahydroprodukt. (b) Die Racemisierung des Gentiopikrins bei enzymatischer Hydrolyse. Bildung von Eugentiogenin und dimerem Gentiogenin. (c) Die Stabilisierung des Lactonringes durch Hydrierung. (d) Der anormale Drehwert des Hexahydro-protogentiogenins. (e) Bildung des Hexahydro-gentiogenin-methyläthers (132, 76b) aus Hexahydrogentiogenin mit alkoholischem HCl bei Zimmertemperatur. (Die Reaktion ist nach Sakurai (132) in wäßriger Salzsäure rückläufig.)

## 2. Amarogentin.

Mit Hilfe der TTC-Reaktion (s. oben) wurden die in der älteren Literatur beschriebenen Bitterstoff-Fraktionen Gentiamarin, Erytaurin und Swertiamarin (77) auf ihren Gehalt an Gentiopikrin untersucht. Dabei zeigte es sich, daß in der von Tanret beschriebenen Gentiamarin-Fraktion mit Sicherheit Gentiopikrin anwesend war (77). Desgleichen konnte für das Erytaurin der Nachweis erbracht werden, daß es ein mit einer nicht bitteren Substanz verunreinigtes Gentiopikrin darstellt (77). Die Namen Gentiamarin und Erytaurin sind aus der Literatur zu streichen.

Als später die TTC-Reaktion quantitativ durchgeführt wurde, stellte es sich heraus, daß in den Gentianaceen neben dem Gentiopikrin noch ein weiterer Bitterstoff enthalten sein muß, da die hohen Bitterwerte durch das Vorhandensein des Gentiopikrins allein nicht erklärbar sind. Die Isolierung eines neuen Bitterstoffes, des Amarogentins, gelang zuerst bei *Swertia chirata*, und zwar sowohl aus der frischen als auch aus der getrockneten Pflanze (79). Der Bitterstoff befindet sich in der Gentiamarin-Fraktion TANRETS. Wie erwähnt, ist das Amarogentin die bis heute bitterste Substanz.

Aus *Ophelia chirata* (identisch mit *Swertia chirata*) hatten FLÜCKIGER und HÖHN (48) eine bittere Substanz, das Chiratin, isoliert. Dieser Stoff konnte später von MAJUMDAR und GUHA (102) in zwei Komponenten zerlegt werden, die beide stark bitter, aber noch uneinheitlich waren. Da es sich bei den isolierten Substanzen um Amarogentin (79) handelte, ist der Name Chiratin ebenfalls zu streichen.

In einer kürzlich erschienenen Arbeit beschreiben KUBOTA und TOMITA (96) die Isolierung des bereits 1927 von KARIYONE und MATSUSHIMA (66) aus *Swertia japonica* erhaltenen Swertiamarins, das nach Emulsinspaltung in 1% Ausbeute Erythrocentaurin liefert und dessen Konstitution der eines Dihydrocumarin-aldehyds entspricht (97). Ob in der genannten Pflanze auch Amarogentin enthalten ist, steht noch offen.

*Isolierung des Amarogentins.* Nach KORTE (79) erhält man aus 10 kg handels-üblichem Chirettakraut (Herba chirettae, *Swertia chirata*) 0,5 g reines Amarogentin. Die Trennung vom Gentiopikrin gelingt durch Chromatographie an Zellulosepulver mit wassergesättigtem Amylalkohol. Es gelang bisher nicht, das Amarogentin kristallin zu erhalten; seine Einheitlichkeit ergibt sich aber daraus, daß die nach verschiedensten Aufarbeitungsverfahren gewonnenen Reinstfraktionen nach C,H-Bestimmung, dem Schmelzpunkt sowie den spektroskopischen Daten identisch sind. Dem Amarogentin kommt die Formel $C_{20}H_{24}O_{10}$ zu (88).

*Acetylierung* (88). In Pyridin mit Acetanhydrid wird ein nicht bitter schmecken-des Tetraacetat erhalten. Da letzteres durch Spaltung mit $KHCO_3$ in Methanol wieder in Amarogentin zurückverwandelt wird, treten bei der Acetylierung keine Veränderungen am Grundgerüst auf.

*Lactontitration* (88). Das Molekül enthält keine freie Carboxylgruppe; bei der Lactontitration werden zwei Mole NaOH verbraucht. Amarogentin schmeckt im alkalischen Bereich nicht mehr ·bitter.

*Spaltung* (88). Säure- und Enzymspaltung verlaufen uneinheitlich wie bei Gentiopikrin. Es konnte lediglich 1 Mol Glucose papierchromatographisch nach-gewiesen werden.

*Spektren* (79). In Methanol gelöst, zeigt Amarogentin im UV drei Maxima: 228 m$\mu$ (log $\varepsilon$ = 4,35), 266 m$\mu$ (log $\varepsilon$ = 3,97) und 304 m$\mu$ (log $\varepsilon$ = 3,67). Im IR (in KBr gepreßt) deuten die Absorptionen bei 1700 cm$^{-1}$ auf eine mit einer Doppel-bindung in Konjugation stehende Carbonylgruppe hin. Aromatische Bauelemente sind nicht vorhanden.

## 3. Loganin.

Der Bitterstoff Loganin wurde erstmalig von DUNSTAN und SHORT (41) aus dem Fruchtmus von *Strychnos nux vomica* (Loganiaceae) isoliert.

Nach Rosenthaler (*128*) ist Loganin identisch mit dem von Bridel aus *Menyanthes trifoliata* isolierten Meliatin (*22*). (Der letzte Name ist zu streichen.) Mit verd. $H_2SO_4$ wird der Bitterstoff in 1 Mol Glucose und das Aglykon „Loganetin" gespalten (*41*).

*Konstitutionsvorschläge.* In der neueren Literatur finden sich zwei voneinander sehr abweichende Vorschläge zur Konstitution des Loganins. Offenbar sind die einerseits von Birch und Smith (*19*) und andererseits von Merz und Krebs (*105*) bzw. Merz und Lehmann (*106*) untersuchten „Loganine" nicht identisch, da sowohl die UV- und IR-Spektren als auch die Schmelzpunkte der Pentaacetylderivate verschieden sind. Das von Birch und Smith untersuchte Loganin wurde aus *Strychnos lucida*, das von Merz untersuchte aus *S. nux vomica* isoliert. Papier-chromatographische Untersuchungen Jaminets (*64*) an Extrakten verschiedener Strychnosarten bestärken in dem Verdacht, daß die beiden Forschungsgruppen nicht dieselbe Substanz in Händen hatten. Nach Birch und Smith (*19*) entspricht dem von ihnen untersuchten Loganin die Formel (V).

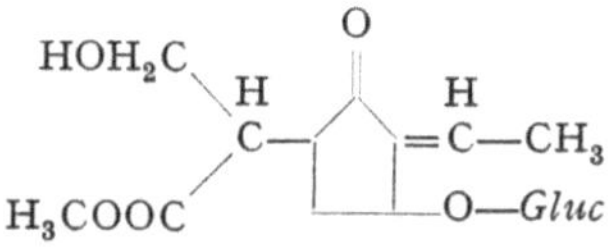

(V.) Loganin (Birch und Smith).

Aus der Substanz läßt sich ein Pentaacetat darstellen; bei der Ozonolyse wird Acetaldehyd gebildet. Die Carbonylgruppe kann durch Bildung eines 2,4-Dinitrophenylhydrazons sowie die IR-Absorption bei $1708 \text{ cm}^{-1}$ nachgewiesen werden. Die konjugierte Doppelbindung gibt sich durch die UV-Absorption bei 242 m$\mu$ und eine IR-Bande bei $1648 \text{ cm}^{-1}$ zu erkennen. Außerdem konnten Birch und Smith eine Carbomethoxygruppe nachweisen.

Das Loganin aus *Strychnos nux vomica* ist nach Merz und Mit-arbeitern ein Lacton. Für das Vorhandensein einer Carbomethoxy-gruppe finden sie keinen Anhaltspunkt, desgleichen kann auch die CO-Gruppe nicht nachgewiesen werden. Der Glucose-Rest ist über eine primäre OH-Gruppe mit dem Genin (Loganetin) verknüpft, wie sich daraus ergibt, daß bei der Tritylierung von Glykosid und Aglykon jeweils eine Monotritylverbindung erhalten wurde. Im Falle einer Verknüpfung über das sekundäre Hydroxyl muß bei Tritylierung des Glykosids eine Ditritylverbindung entstehen. Merz und Mitarbeiter (*105, 106*) fassen ihre Ergebnisse in der Formel (VI) für Loganin aus *S. nux vomica* zusammen:

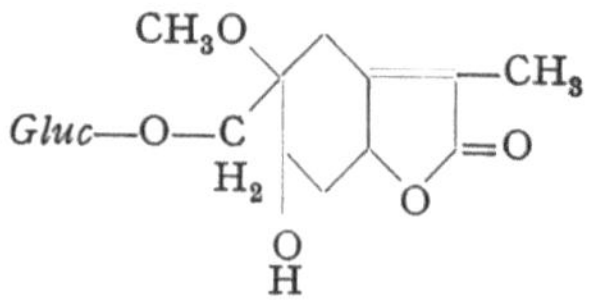

(VI.) Loganin (MERZ und Mitarb.).

Auffallend ist die für ein γ-Lacton sehr niedrige Wellenzahl der IR-Absorption (1709 cm⁻¹). Nachdem KORTE und Mitarbeiter *(83, 86, 89)* durch Synthese und IR-Messung von Vergleichssubstanzen zeigen konnten, daß die γ-Lactonformulierung im Gentiopikrin zugunsten einer δ-Lactongruppe zu ändern ist (S. 130), wird man auch beim Loganin mit einem α,β-ungesättigten δ-Lacton rechnen müssen. Für diese Annahme spricht ferner die bei bicyclischen α,β-ungesättigten δ-Lactonen gemessene CO-Bande bei 1709 cm⁻¹ *(84)*.

## III. Bitterstoffe der Asclepiadaceen.

Aus der Familie der Asclepiadaceen sind bis jetzt zwei Bitterstoffe bekannt, das Kondurangin aus *Marsdenia condurango* und das Vincetoxin (Asclepiadin, Cynanchin) aus *Cynanchum vincetoxicum*. Bereits 1885 wurden Kondurangin von VULPIUS *(154)* und Vincetoxin von TANRET isoliert, jedoch waren beide Substanzen, die sich als Glykoside erwiesen, noch nicht rein. Einen kurzen geschichtlichen Überblick zur Chemie der beiden Stoffe gibt KORTE *(78)*. Auf Grund ihres Vorkommens in der gleichen Pflanzenfamilie wurde erwartet, daß Kondurangin und Vincetoxin den von WINKLER und REICHSTEIN *(161)* aus *Dregea volubilis* isolierten Samenglykosiden (den Drevosiden *A, B, C, D*) ähnlich sind, was sich im Verlauf der Untersuchungen bestätigte. Über diese Beziehung der Drevoside zum Kondurangin berichten F. und I. KORTE *(85)*.

### 1. Kondurangin.

Die Gewinnung des Kondurangins in reiner Form ist bis heute noch nicht gelungen. Als Spaltprodukte konnten KUBLER *(94)* Zimtsäure und KERN und MOMSEN *(70)* sowie KERN und HASELBECK *(69)* ein Disaccharid isolieren, das sie Kondurangobiose nannten.

KORTE *(78)* konnte zeigen, daß Kondurangin ein Zimtsäureester ist; ein durch Gegenstromverteilung gereinigtes Konduranginpräparat enthält zirka 1 Mol Zimtsäure *(84a)*. Die Esterbindung ist derart instabil, daß der Zimtsäuregehalt einer Konduranginprobe bereits beim Umfällen aus Petroläther abnimmt.

Kondurangin zeigt abnorme Löslichkeitseigenschaften: Während es in kaltem Wasser klar löslich ist, trübt sich die wäßrige Lösung beim Erwärmen, um bei 40° zu einer Gallerte zu erstarren. Bei Abkühlung wird die Lösung wieder klar.

Durch Alkalispaltung in Methanol entsteht das Descinnamyl-kondurangin *(90)*. Durch schonende, saure Methanolyse läßt sich hieraus ein rohes „Descinnamyl-kondurangogenin" gewinnen, dessen Zerlegung in zwei kristalline Genine durch Gegenstromverteilung KORTE und RIPPHAHN gelang (87). Die IR-Spektren der beiden Genine („I" und „II") zeigen gewisse Ähnlichkeit mit denen der Drevogenine A und B (161), ohne jedoch mit ihnen identisch zu sein. Keines der Genine enthält eine Carbonylgruppe, dagegen sprechen beim Genin „II" die Absorptionen bei 1675 cm$^{-1}$ und bei 207 m$\mu$ für eine C=C-Doppelbindung, die jedoch mit Platin in Eisessig nicht hydrierbar ist. Auch Genin „I" ist nicht hydrierbar. Die Selen-Dehydrierung des Genins „I" lieferte zwei kristalline Kohlenwasserstoffe (A und B), deren Trennung auf chromatographischem Wege möglich war (87). Auf Grund der Analyse, Molgewichtsbestimmung, Schmelzpunkt, UV- und IR-Spektren ergab sich die Identität des Kohlenwasserstoffs A mit einem von WINKLER und REICHSTEIN (161) durch Selen-Dehydrierung des Desisovaleryltetrahydrodrevogenins A erhaltenen Kohlenwasserstoff. Derselbe war schon früher durch Dehydrierung des Jervins und des Veratramins (63) entstanden und als ein dialkyliertes 1,2-Benzfluoren angesprochen worden.

Kohlenwasserstoff B liefert dieselben Analysendaten wie A, jedoch sind Molgewicht, Schmelzpunkt, UV- und IR-Spektren verschieden. Das UV-Spektrum von B entspricht dem eines alkylierten 2,3-Benzfluorens.

Bei der Selen-Dehydrierung des Genins „II" konnte nur ein öliger Kohlenwasserstoff gefaßt werden, dessen UV-Spektrum auffällige Ähnlichkeit mit einem gleichfalls öligen Kohlenwasserstoff zeigt, den WINKLER und REICHSTEIN (neben dem oben erwähnten kristallinen Kohlenwasserstoff) bei der Dehydrierung des Desisovaleryl-tetrahydrodrevogenins A erhalten hatten. Auf Grund des UV-Spektrums könnte es sich dabei um ein Fluorenderivat handeln.

KORTE und WEITKAMP (90) hatten früher die Vermutung ausgesprochen, daß Kondurangin ein einfaches Perhydro-fluorenonderivat sei. Von KORTE und BEHNER (82) daraufhin an Modellsubstanzen durchgeführte Dehydrierungsversuche zeigten, daß bei der Dehydrierung sowohl von Perhydro-fluorenen als auch von Perhydro-fluorenonen nur Fluorene entstehen. Eine Vergrößerung des Kohlenstoffskeletts durch pyrolytische Reaktion, wie sie für eine Benzfluorenbildung gefordert werden müßte, konnte durch diese Versuche ausgeschlossen werden. Daraus ergibt sich, daß Kondurangin kein einfaches Perhydro-fluorenon-Derivat ist.

Nach ORCHIN und Mitarbeitern (111, 112) können bei der Dehydrierung des (2-Methoxyphenyl)-naphthalins 1,2- und 2,3-Benzfluoren entstehen (Chromoxyd-Aluminiumoxyd-Katalysator, 400°). Damit ist die Möglichkeit gegeben, daß es sich beim Genin „I" um ein (2-Methoxyphenyl)-naphthalinderivat handelt. Andererseits könnte auch das Fluorengerüst vorgebildet sein und durch Ringschluß einer Seitenkette ein 1,2- oder 2,3-Benzfluoren entstehen (VII—VIII):

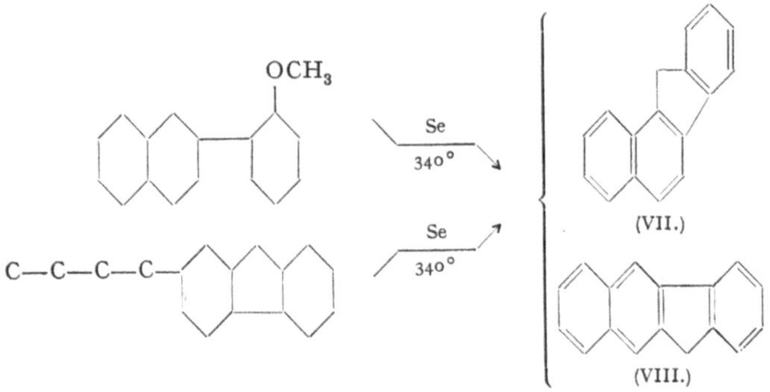

$OCH_3$

C—C—C—C—

Se 340°

Se 340°

(VII.)

(VIII.)

· Als Zuckerkomponenten des Kondurangins konnte KORTE (*78*) *D*-Cymarose, *D*-Glucose und *D*-Thevetose nachweisen. Die Kondurango-biose ist ein Biosid aus *D*-Glucose und *D*-Thevetose.

## 2. Vincetoxin.

Auch das Vincetoxin konnte bislang nicht kristallin erhalten werden. Wie bei Kondurangin wird beim Erwärmen einer wäßrigen Lösung des Vincetoxins eine Abnahme der Löslichkeit beobachtet. FENUELLE und HARNACK bezeichnen einen Bitterstoff aus *Cynanchum vincetoxicum* als Asclepiadin. GRAM und TANRET erkannten die Glykosidnatur der Substanz und bezeichneten sie als Vincetoxin. KUBLER (*94*) stellte nach Hydrolyse einen vergärbaren Zucker fest und fand für das Vincetoxin ein Molgewicht von 915.

Nach KORTE (*78*) läßt sich die Isolierung des Vincetoxins analog der des Kondurangins durchführen. Durch saure Methanolyse des Vincetoxins konnten KORTE und RIPPHAHN (*87*) neben dem amorphen Vincetoxogenin die vier Zuckerkomponenten des Vincetoxins isolieren. Es handelt sich um *D*-Glucose, *D*-Thevetose, *L*-Oleandrose und einen noch unbekannten Desoxyzucker.

Die nähere Untersuchung des Vincetoxogenins zeigte, daß es sich papierchromatographisch in zwei Genine aufspalten läßt. Selen-dehydrierung des Geningemisches führte zur Isolierung zweier öliger Kohlenwasserstoffe, deren UV-Spektren denen des kristallinen Kohlen-wasserstoffes *B* aus Descinnamyl-kondurangogenin „I" und dem des öligen Kohlenwasserstoffs aus Descinnamyl-kondurangogenin „II" ähnlich sind (S. 136).

## IV. Bitterstoffe der Compositen.

Von den zahlreichen, aus Compositen isolierten Bitterstoffen sollen nur die folgenden näher beschrieben werden, da die zu ihrer Konstitutionsaufklärung durchgeführten Arbeiten neueren Datums sind: Absinthin, Anabsinthin, Cnicin, Lactucin, Lactucopikrin, Tenulin, Helenalin und Alantolacton (Helenin). Keine dieser Substanzen ist ein Glykosid. Damit ist die in der älteren Literatur häufig vertretene Ansicht, typische pflanzliche Bitterstoffe besäßen Glykosid-charakter, hinfällig geworden.

### 1. Absinthin und Anabsinthin.

Das Anabsinthin wurde zuerst von ADRIAN und TRILLAT (2) aus Wermut erhalten. Der von ihnen isolierte Stoff schmeckte stark bitter und schmolz bei 258—260°. Im Rahmen ihrer Arbeiten über die Inhaltsstoffe des Wermuts untersuchten ŠORM, NOVOTNY und HEROUT (141, 142) auch das Anabsinthin. Nach ihren Ergebnissen liegt Anabsinthin nicht als solches in der Pflanze vor, sondern es entsteht im Laufe der Aufarbeitung aus einer anderen, ebenfalls sehr bitteren Substanz, dem Absinthin, $C_{15}H_{20}O_3 \cdot {}^1/_2 H_2O$, das von den tschechischen Forschern rein dargestellt werden konnte.

Absinthin und Anabsinthin lassen sich noch in einer Verdünnung von 1 : 10 Millionen (Grenzkonzentration) durch ihren Geschmack nachweisen. Auf Grund des Bitterwertes beträgt nach KORTE (80) der Gehalt einer im Raum Bonn gewachsenen Frischpflanze an Absinthin in den Blättern 0,3%, in den Blüten 0,16% und in den Stengeln 0,006%. Die Wurzeln enthalten keinen Bitterstoff. Absinthin schmilzt nach Umkristallisation aus Methanol oder Äthanol bei 182—183°. Nach dem Umkristallisieren aus Benzol schmilzt es in solvatisierter Form bei 108°. Die Darstellung einer Analysenprobe ist ungewöhnlich schwierig, da das Lösungs-mittel sehr fest gehalten wird.

Abgesehen von der durch den Mehrgehalt an Wasser bedingten Differenz liefern Absinthin und Anabsinthin gleiche Analysenwerte, es sind also Isomere. Beim Versuch, Absinthin aus höher siedenden Lösungs-mitteln umzukristallisieren, ging es in Anabsinthin über (141).

Absinthin und Anabsinthin gehören zu einem neuen Typ sesqui-terpenischer Lactone. Ihre IR-Spektren sind einander sehr ähnlich, obwohl die Verbindungen sich in anderen physikalischen Eigenschaften und in ihrem chemischen Verhalten deutlich voneinander unterscheiden. So liefert nur das Absinthin die sog. „Chamazulen-Reaktion", d. h. nach alkalischer Hydrolyse und anschließender Destillation aus saurem

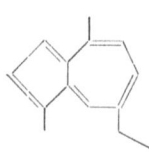
(IX.) Chamazulen.

Milieu in Gegenwart von Luftsauerstoff wird Cham-azulen (IX) gebildet (141, 149), dessen Konstitution von MEISELS und WEIZMANN (104) geklärt wurde. Positive Chamazulen-Reaktion zeigt auch das dem Absinthin isomere Artabsin (X) (58), ferner das Matricin (XI) (29).

*Literaturverzeichnis: SS. 175—182.*

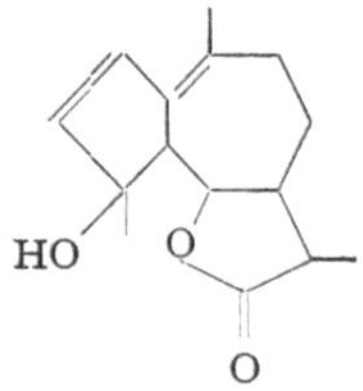

(X.) Artabsin.

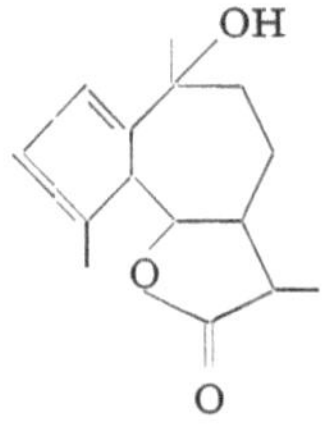

oder

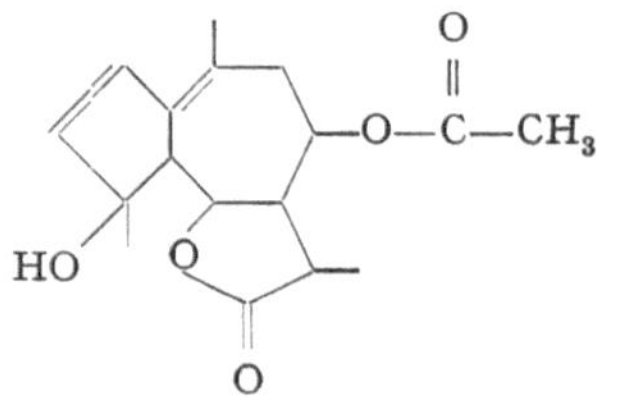

(XI.) Matricin.

Anabsinthin verhält sich im Chamazulen-Test negativ. In diesem Zusammenhang ist die Beobachtung von Interesse, daß nur das Absinthin eine freie OH-Gruppe enthält (IR-Bande bei 3410 cm⁻¹).
Beide Verbindungen weisen die für $\gamma$-Lactone typische Absorption bei 1774 cm⁻¹ auf. Nach einem Vorschlag von HEROUT und ŠORM (58a) sollen Lactone, deren Grundgerüst sich auf das Guaiazulen (XII) (117) zurückführen läßt, als „Guaianolide" bezeichnet werden.

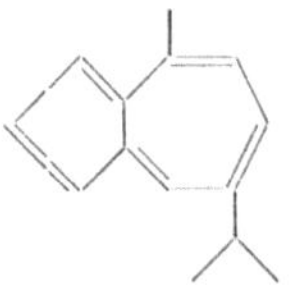

(XII.) Guaiazulen.

Außer Absinthin und Anabsinthin sind auch die Bitterstoffe Helenalin, Tenulin und Isotenulin zu dieser Gruppe zu zählen (SS. 144—147). In der Familie der Compositen sind Guaianolide häufig vertreten, ein weiterer Hinweis, daß zwischen Pflanzenverwandtschaft und chemischen Inhaltsstoffen eine enge Beziehung besteht.

## 2. Cnicin.

Die Isolierung der im Benediktinerkraut *(Cnicus benedictus)* enthaltenen Bitterstoffe war, trotz zahlreicher Versuche, bis vor kurzem nicht gelungen. Erst jetzt konnten KORTE und BECHMANN (81) durch Anwendung chromatographischer Verfahren den Bitterstoff „Cnicin" aus handelsüblichen Drogen in 0,27% Ausbeute kristallin gewinnen.

Analyse und Molgewichtsbestimmung liefern für die Formel $C_{25}H_{32}O_9$ passende Werte. Nach KUHN-ROTH läßt sich eine C—CH₃-Gruppe bestimmen. Nach ZEREWITINOFF erhält man den Wert für vier aktive H-Atome. Die katalytische Hydrierung führt mit Raney-Nickel zum Tetrahydro-, mit PtO₂ zum Hexahydroderivat. Das mit Acetanhydrid in Pyridin gewonnene Acetat läßt sich durch KHCO₃ in Methanol nicht wieder in Cnicin zurückverwandeln. Glucose läßt sich weder nach enzymatischer Spaltung noch nach Hydrolyse mit HCl nachweisen. Bei der Lactontitration verbraucht Cnicin 2 Äquivalente NaOH.

Neben Cnicin konnte bei der Aufarbeitung handelsüblicher Drogen verschiedener Herkunft in einigen Fällen noch ein weiterer Bitterstoff papierchromatographisch nachgewiesen werden, der als „Benedictin" bezeichnet wird.

Laut einer Privatmitteilung haben M. SUCHÝ, V. HEROUT und F. ŠORM* ebenfalls Cnicin kristallin erhalten. Die Substanz entspricht in den chemischen und physikalischen Eigenschaften den von KORTE

---

* Chem. and Ind. (im Druck).

und Bechmann erhaltenen. Auf Grund der Molekulargewichtsbestimmung und der katalytischen Hydrierung gelangen die tschechischen Autoren zu der Summenformel $C_{20}H_{26}O_7$. Nach den genannten Autoren handelt es sich beim Cnicin um ein Hydroxyguaienolid, wie sich u. a. aus folgenden Beobachtungen ergibt: Nach Reduktion des Cnicins mit $LiAlH_4$ entstehen bei der Selen-Dehydrierung Chamazulen (IX) und Artem-

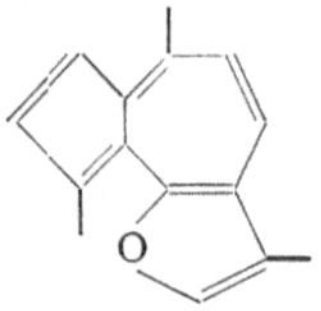

(XII a). Artemazulen.

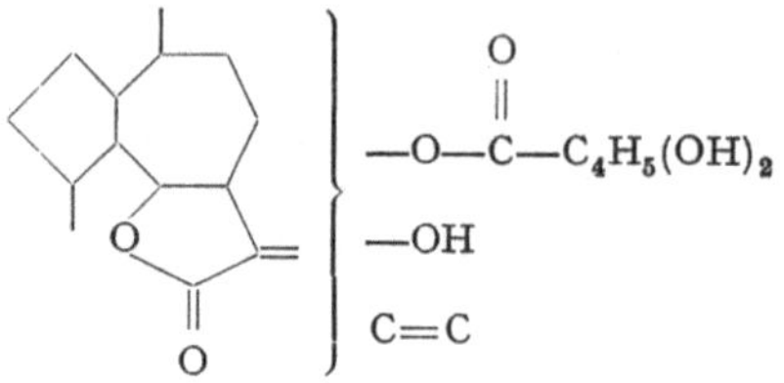

$$-O-\overset{O}{\overset{\|}{C}}-C_4H_5(OH)_2$$

$-OH$

$C=C$

(XII b.)

azulen (XIIa). Bei der katalytischen Hydrierung werden 3,5 $H_2$ aufgenommen. Neben noch nicht identifizierten Hydrogenolyse-Produkten entsteht dabei ein kristallines Hexahydroderivat (Mol.-Gew. 407). Alkalische Hydrolyse des Cnicins führt zu einem $C_{15}$-Lakton (Hydroxyguaienolid) und einem $C_5$-Hydroxylakton.

Auf Grund dieser Ergebnisse wird von Suchý, Herout und Šorm für das von ihnen isolierte Cnicin die Formel eines Hydroxyguaienolids (XIIb) vorgeschlagen, das mit einer $\alpha,\beta$-ungesättigten Dihydroxypentensäure verestert ist. (Die Stellung der Doppelbindung in der Säurekomponente wird aus dem IR-Spektrum gefolgert.)

Es ist bisher nicht bekannt, ob das Hydroxyguaienolid in den Pflanzen mit verschiedenen Säuren verestert vorkommt. Der Befund, daß es sich beim Cnicin um ein Azulenderivat handelt, bedeutet einen weiteren Hinweis auf die enge Beziehung zwischen Pflanzenfamilien und chemischen Inhaltsstoffen (85).

### 3. Lactucin und Lactucopikrin.

Eine bis 1956 reichende Literaturübersicht ist im Zitat (15) enthalten.

Lactucin und Lactucopikrin sind die Hauptbestandteile des früher als „Lactucarium germanicum" offizinell gewesenen Milchsaftes aus Gift-Lattich (Lactuca virosa). Lactucopikrin ist identisch mit Neolactucin und Intybin (aus Cichorium-Arten) (133). Nach Zellner (165) sind Lactucin und Lactucopikrin auch in anderen milchsaft-führenden Pflanzen vorhanden, die sämtlich der Untergruppe der Cichoriaceen angehören. Es ist möglich, daß Lactucin erst während der Lagerung der Droge aus dem primären Bitterstoff Lactucopikrin gebildet wird (16). Über die Zusammenhänge zwischen den beiden Substanzen s. unten.

Mit der Konstitutionsaufklärung von Lactucin (XIII) haben sich Späth, Lorenz, Kuhn (143) sowie Wessely, Lorenz, Kuhn (157) beschäftigt. Die endgültige Strukturaufklärung gelang BARTON und NARAYANAN (*15*).

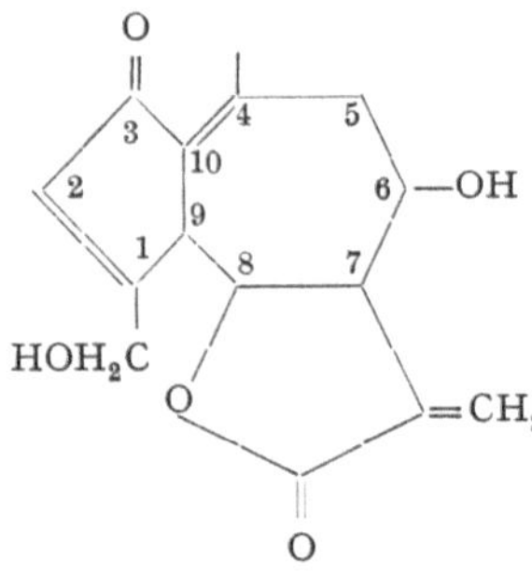

(XIII.) Lactucin.

*Dehydrierung.* Nach vollständiger Hydrierung (Pt/Eisessig) des Lactucins entsteht bei der Dehydrierung mit Pd/Kohle Chamazulen (*104*).

*Funktionelle Gruppen.* Lactucin enthält eine $\gamma$-Lactongruppierung (Lactontitration des Hexahydrolactucins; IR-Absorption bei 1755 cm$^{-1}$). Das Molekül enthält ferner zwei leicht acylierbare OH-Gruppen und eine $\alpha,\beta$-ungesättigte Ketongruppierung ($\lambda_{max} = 257$ m$\mu$). Nach KUHN-ROTH läßt sich eine C—CH$_3$-Gruppe bestimmen; bei der Ozonolyse wird Formaldehyd gebildet. Bei der katalytischen Hydrierung werden rasch 3 Mole H$_2$ aufgenommen. Das entstandene Hexahydrolactucin reagiert mit Carbonyl-Reagenzien und bildet bei Permanganat-Oxydation Methylbernsteinsäure. Trotz der drei C=C-Doppelbindungen verbraucht Lactucin bei 0° keine Persäuren.

Die von BARTON und NARAYANAN angegebene Lactucin-Formel (XIII) vermag die in der *Formelübersicht 1* verzeichneten Beobachtungen zu erklären.

Die katalytische Hydrierung (Pd/CaCO$_3$ in CH$_3$OH) kann jeweils nach Aufnahme von 1, 2 oder 3 H$_2$ abgebrochen werden. Es entstehen dabei die Dihydroverbindung (XIV), die beiden isomeren Tetrahydroderivate (XV), (XVI) und das Hexahydrolactucin (XVII).

Aus der UV-Absorption von (XIV) folgt, daß im chromophoren System gegenüber (XIII) keine Änderung eingetreten ist. Nach KUHN-ROTH lassen sich bei (XIV) zwei C—CH$_3$-Gruppen bestimmen; bei der Ozonolyse wird kein Formaldehyd mehr gebildet. Damit ist eine isolierte, endständige C=CH$_2$-Gruppe nachgewiesen.

Die isomeren Tetrahydroverbindungen (XV) und (XVI) zeigen unterschiedliche UV-Absorption. Der Wert von 227 m$\mu$ (XVI) entspricht dem eines $\alpha,\beta$-ungesättigten Ketons mit höchstens einem Substituenten an der Doppelbindung. (XV) ist an der Doppelbindung höher substituiert (252 m$\mu$). Bei Annahme einer CO-Gruppe im Fünfring bleibt nur die vorstehend gewählte Formulierung.

Behandelt man (XV) und (XVI) (das letztere in Form seines Diacetats) mit KOH in Alkohol, so entstehen (XXI) und (XXII). In beiden Fällen ist eine zur konjugierten Doppelbindung $\beta$-ständige Hydroxylgruppe eliminiert worden (UV-

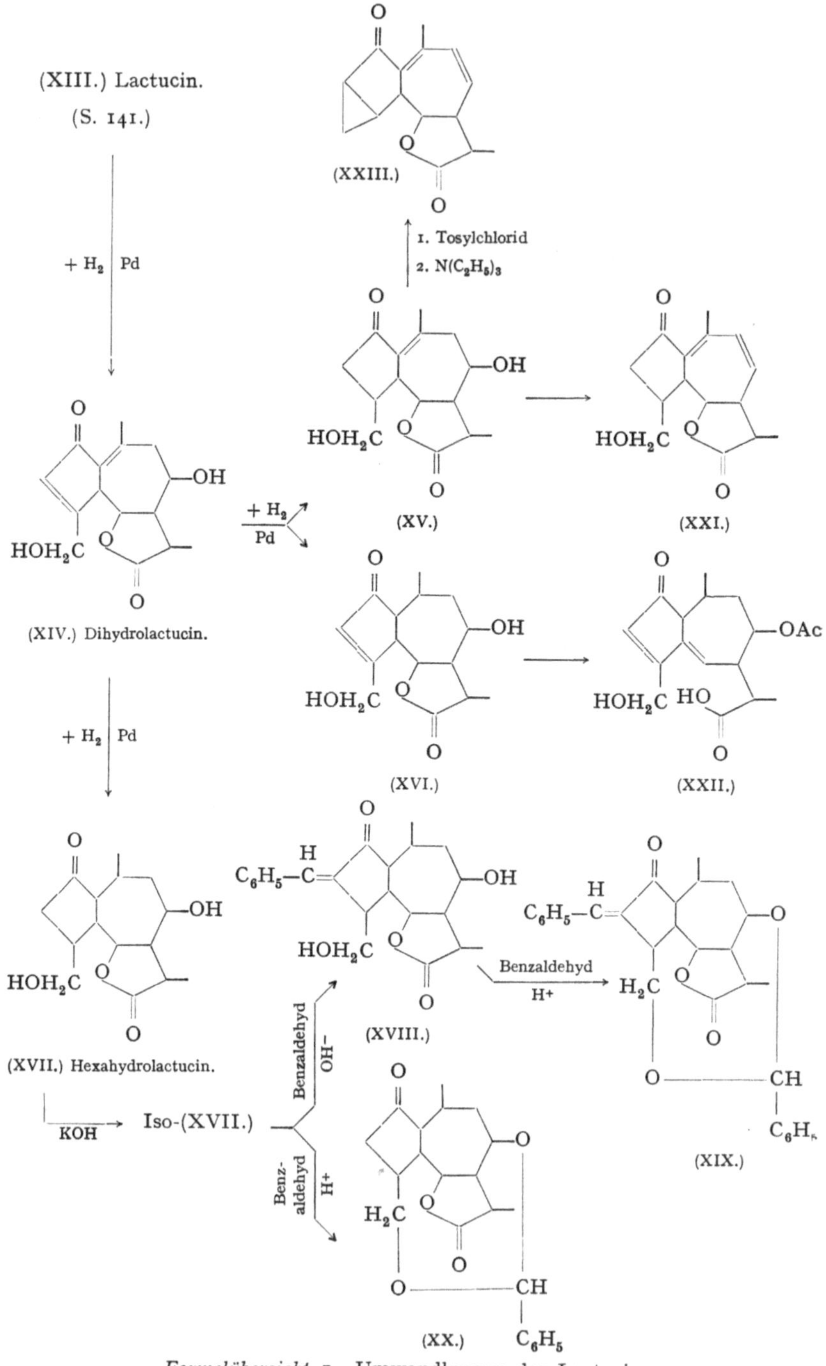

(XIII.) Lactucin.

(S. 141.)

(XXIII.)

1. Tosylchlorid
2. N(C₂H₅)₃

+ H₂ | Pd

(XV.)

(XXI.)

HOH₂C

(XIV.) Dihydrolactucin.

+ H₂
Pd

(XVI.)

(XXII.)

+ H₂ | Pd

(XVII.) Hexahydrolactucin.

KOH → Iso-(XVII.)

C₆H₅—C

(XVIII.)

Benzaldehyd
OH⁻

Benzaldehyd
H⁺

Benzaldehyd
H⁺

C₆H₅—C

(XIX.)

C₆Hₖ

(XX.)    C₆H₅

*Formelübersicht 1.* Umwandlungen des Lactucins.

*Literaturverzeichnis: SS. 175—182.*

Absorption), damit ist die Stellung des Lacton-Hydroxyls festgelegt. Durch den Einfluß des Alkalis wird der am primären Hydroxyl stehende Acetylrest (in XVI) abgespalten.

Für die Existenz eines primären Hydroxyls spricht der bei der KUHN-ROTH-Oxydation der hydrierten Verbindungen erhaltene Wert von zwei C—CH$_3$-Gruppen. Ein sekundäres Hydroxyl am C$_{(2)}$ erscheint ausgeschlossen, da in diesem Fall mit der Carbonyl-Gruppe ein Endiol-System entstehen müßte, das im UV-Spektrum erkennbar wäre. Einen direkten Beweis für den Sitz der HOCH$_2$-Gruppe am C$_{(1)}$ liefert die Reaktion (XV) → (XXIII). Durch Einwirkung von abs. Triäthylamin auf das Ditosylat von (XV) entsteht (XXIII) mit einem dem Fünfring angegliederten Cyclopropansystem. (Die Absorptionsbande des Cyclopentanons verlagert sich durch den Cyclopropanring von 1728 nach 1718 cm$^{-1}$.) Die Bildung des Cyclo-propan-Derivates wäre mit einer C$_{(2)}$-ständigen HOCH$_2$-Gruppe ausgeschlossen, da die Abspaltung der $p$-Toluolsulfosäure aktivierten Wasserstoff benötigt, der nur an C$_{(2)}$, nicht aber an C$_{(1)}$ vorhanden ist. Die Gruppierung (XXIV) konnte dadurch ausgeschlossen werden, daß die Jodoform-Probe nach Ozonolyse von (XVI) negativ ver-lief. Unter gleichen Bedingungen entstanden aus (XV) 0,4 Mole CHJ$_3$.

(XXIV.)

Die Reaktionen des Hexahydrolactucins (XVII) mit Benzaldehyd erlauben, Angaben über den stereochemischen Bau dieser Verbindung zu machen. (XVII) reagiert mit Benzaldehyd nur nach Vorbehandlung mit KOH in Methanol (unter sehr milden Bedingungen); wahrscheinlich erfolgt dabei Epimerisierung an C$_{(10)}$. Die Bildung des O—O-Benzyliden-derivates (XX) kann nur erfolgen, wenn eine räumliche Anordnung entsprechend (XXV) gegeben ist.

(XXV.) (Oder Spiegelbild.)

Zur Darstellung von (XX) wird in 6 $N$-HCl gearbeitet. Läßt man Benzaldehyd im schwach alkalischen Gebiet mit „Iso-XVII" reagieren, so erfolgt lediglich Reaktion an C$_{(2)}$ unter Bildung von (XVIII). Wenn anschließend (in Gegenwart von über-schüssigem Benzaldehyd) angesäuert wird, entsteht die der Verbindung (XX) homologe Substanz (XIX).

Angenommen, die katalytische Hydrierung der C$_{(4)}$—C$_{(10)}$-Doppel-bindung erfolge über $cis$-Addition der H-Atome und die Epimerisierung bewirke Inversion an C$_{(10)}$, so müssen die H-Atome an C$_{(4)}$ und C$_{(10)}$ in $trans$-Stellung zueinander stehen (XXV).

Lactucin ist biogenetisch interessant, da es das erste Sesquiterpen-lacton ist, für das eine CO-Gruppe an C$_{(3)}$ bewiesen werden konnte (*15*).

Das *Lactucopikrin* wurde bereits 1862 von LUDWIG und KROMAYER (*99*) beschrieben. ZINKE und HOLZER (*166*) konnten den Zusammenhang zwischen Lactucin und Lactucopikrin aufklären: Lactucopikrin ist der p-Hydroxyphenylessigsäure-monoester des Lactucins.

Dies Ergebnis basiert auf folgenden Befunden: Lactucopikrin wird pyrogen oder mit verd. Alkalien unter Bildung von p-Hydroxy-phenylessigsäure gespalten. Hexahydro-lactucopikrin liefert bei der Hydrolyse ebenfalls p-Hydroxy-phenylessigsäure. Daneben entsteht unter Öffnung des Lactonrings die bekannte Hexahydro-lactucinsäure (*157*).

## 4. Tenulin.

Tenulin wurde 1939 von Clark (*32*) aus *Helenium tenuifolium, H. elegans, H. badium* und *H. montanum* isoliert. Die Substanz schmeckt bitter, steigert den Niesreiz und ist ein Fischgift. Außer Clark beschäftigten sich Ungnade und Hendley (*151*), Ungnade, Hendley und Dunkel (*152*) sowie Barton und de Mayo (*14*) mit der Konstitutionsaufklärung. Tenulin (XXVI) wird durch milde Alkalibehandlung zu Isotenulin (XXVII) isomerisiert. Bei der katalytischen Hydrierung gehen beide in die Dihydroverbindungen über. Beide Verbindungen bilden Phenylhydrazone. Pyrolyse von (XXVI) führt über Anhydrotenulin (XXXII, S. 146) zu Pyrotenulin (XXXIII). (XXVII) spaltet bei Umsatz mit $H_2SO_4$ Essigsäure ab, das entstandene Desacetyl-isotenulin kann durch Acetylierung wieder in (XXVII) überführt werden. Oxydation von (XXVI) oder (XXVII) mit $H_2O_2$ im alkalischen Milieu liefert Tenulinsäure (XXXIV, S. 146), die nach Acetylierung in die Acetyltenulinsäure übergeht; letztere Verbindung wird auch direkt bei der Permanganat-Oxydation von (XXVI) oder (XXVII) gebildet (*32*). Auf Grund der UV-Spektren sind (XXVI) und (XXVII) α,β-ungesättigte Ketone. Lactontitration und IR-Absorption (1772 cm$^{-1}$) zeigen, daß beide Verbindungen einen γ-Lactonring enthalten (*151*, *152*).

Unter Berücksichtigung dieser und eigener experimenteller Ergebnisse stellten Barton und de Mayo (*14*) die Formeln (XXVI) und (XXVII) auf.

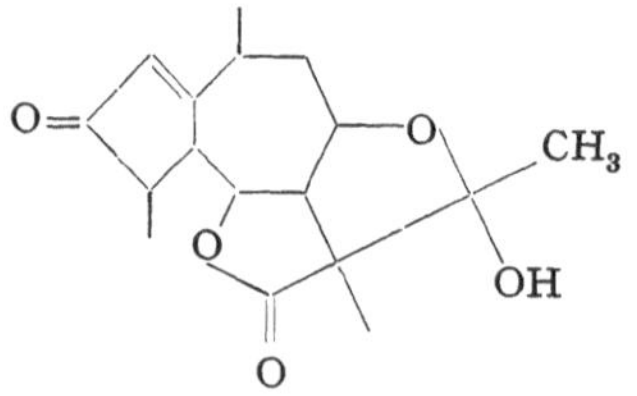

(XXVI.) Tenulin. 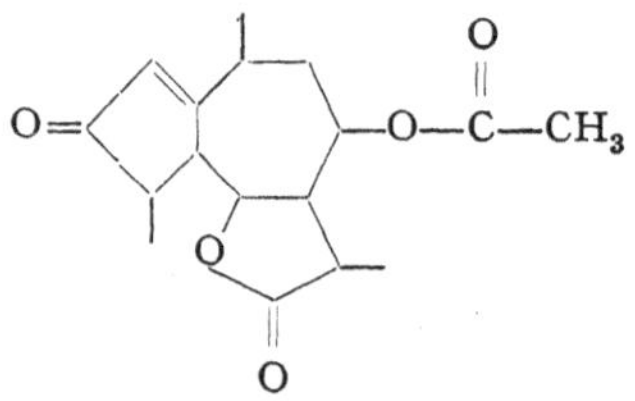                    (XXVII.) Isotenulin.

α,β-*Ungesättigte Ketongruppe.* Die Absorptionen bei 1708 und 1595 cm$^{-1}$ sprechen für ein Cyclopentenon-derivat. Die UV-Absorption

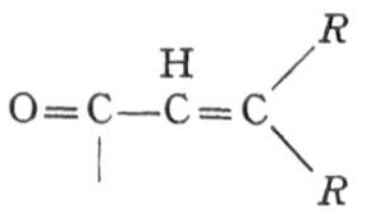

*Literaturverzeichnis: SS. 175—182.*

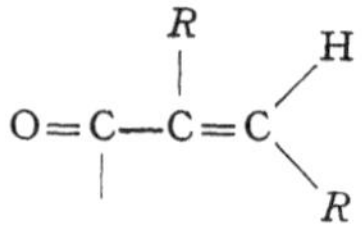

bei 226 m$\mu$ deutet auf ein $\alpha,\beta$-ungesättigtes Keton hin mit nur zwei Substituenten an der Doppelbindung. Die Entscheidung zwischen den beiden möglichen Formeln $A$ und $B$ ist durch das Ergebnis der Ozonolyse möglich. Da hierbei Ameisensäure auftritt, kann $B$ ausgeschlossen werden.

*Grundgerüst.* Nach Reduktion des Isotenulins mit KBH$_4$ entsteht bei der Dehydrierung mit Pd/Kohle Chamazulen (S. 138). Nach Reduktion des Dihydroisotenulins mit LiAlH$_4$ führt Dehydrierung zum Linderazulen (XXVII a) (*145*), in welchem mit Ausnahme des Acetylrestes sämtliche C-Atome erhalten geblieben sind.

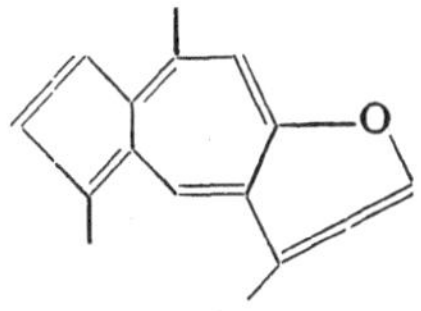

(XXVII a.) Linderazulen.

*Stellung der Ketogruppe.* Entsprechend der Gruppierung $A$ kann die Ketogruppe die Positionen $C$ oder $D$ einnehmen. Die Richtigkeit von $C$ wird durch *Formelübersicht 2* bewiesen (*14*).

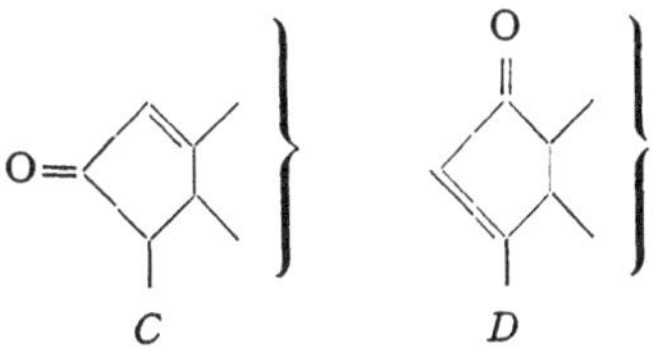

$C$ $D$

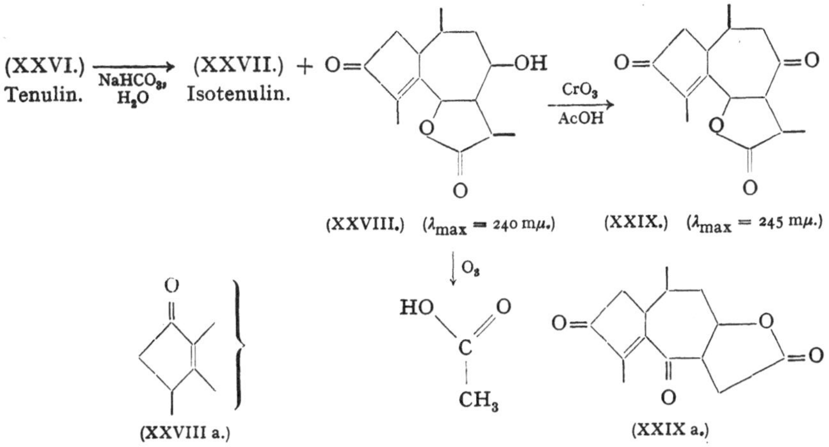

(XXVI.) $\xrightarrow[\text{H}_2\text{O}]{\text{NaHCO}_3,}$ (XXVII.) Isotenulin.
Tenulin.

(XXVIII.) ($\lambda_{max} = 240$ m$\mu$.) $\xrightarrow[\text{AcOH}]{\text{CrO}_3}$ (XXIX.) ($\lambda_{max} = 245$ m$\mu$.)

(XXVIII a.)

(XXIX a.)

*Formelübersicht 2.* Umwandlungen des Tenulins.

Aus (XXVI) wird außer dem isomeren (XXVII) bei der Behandlung mit NaHCO$_3$ in Wasser auch die als Desacetyl-neotenulin bezeichnete Verbindung (XXVIII) gebildet, die im UV bei 240 m$\mu$ absorbiert. (XXVIII) ist ein $\alpha,\beta$-ungesättigtes Keton, das an der Doppelbindung drei Substituenten trägt. Bei Ozonolyse bildet es Essigsäure.

Wäre die Formulierung $D$ zutreffend, so müßte bei der Isomerisierung (XXVIIIa) entstehen, das nach der Ozonspaltung keine Essigsäure liefern dürfte.

*Lactonring.* Der $\gamma$-Lactonring kann über die an $C_{(6)}$ oder $C_{(8)}$ stehenden O-Atome geschlossen sein. Bei der Oxydation mit $CrO_3$ in Eisessig bei 20° entsteht aus (XXVIII) das Desacetyl-dehydrotenulin (XXIX) ($\lambda_{max} = 245$ m$\mu$). Bei Lactonbildung über das an $C_{(6)}$ stehende O-Atom würde die Oxydation an $C_{(8)}$ unter Bildung des En-1,4-dions (XXIXa) erfolgen, dessen UV-Spektrum dem des Cholest-4-en-3,6-dions ($\lambda_{max} = 254$ m$\mu$) entsprechen müßte.

*Verhältnis zum Isotenulin.* Im IR-Spektrum unterscheiden sich (XXVI) und (XXVII) dadurch, daß nur bei dem vorigen eine OH-Bande (3450 cm$^{-1}$) sichtbar ist. Dagegen weist (XXVII) die für die Acetylgruppe charakteristischen Banden bei 1748 und 1238 cm$^{-1}$ auf. Damit steht in Übereinstimmung, daß nur aus (XXVII) bei saurer Hydrolyse 1 Mol Essigsäure abgespalten wird.

Barton und de Mayo (*14*) formulieren Tenulin (XXVI) als Halbketal. Der für die Addition an die Carbonylgruppe erforderliche aktivierte Wasserstoff befindet sich in $\alpha$-Stellung zum Lactoncarbonyl. Beweisend für die Formulierung als Halbketal sind die folgenden Reaktionen:

Dihydro-(XXVI).

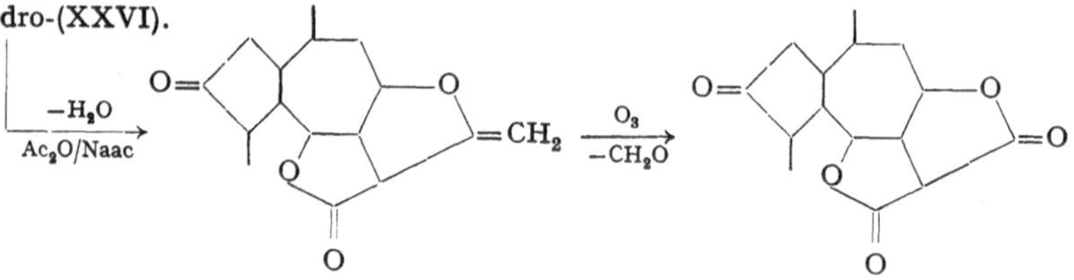

(XXX.) Anhydro-dihydrotenulin.

(XXXI.)

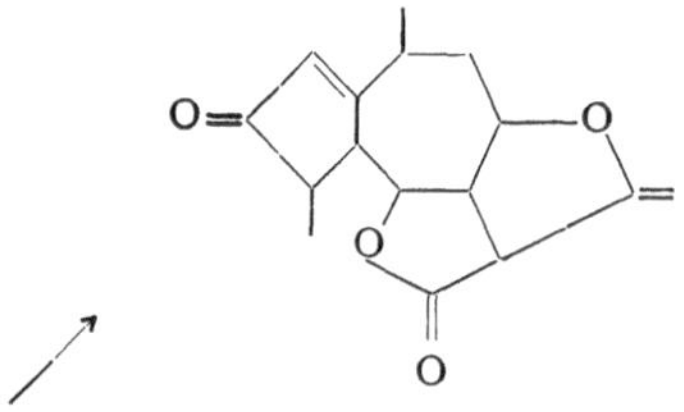

(XXXII.) Anhydrotenulin.

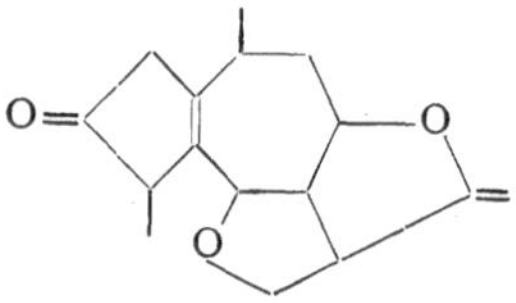

(XXXIII.) Pyrotenulin.

(XXVI.) Tenulin.

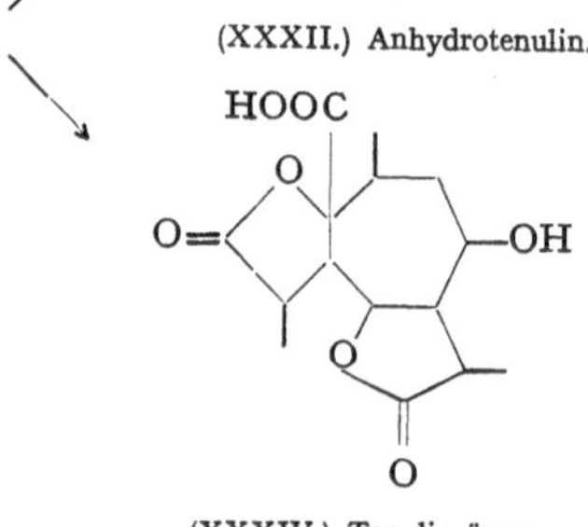

(XXXIV.) Tenulinsäure.

Pyrotenulin (XXXIII) verbraucht bei der Titration nur 1 Äquivalent Alkali. Genau so verhalten sich Anhydrotenulin (XXXII) und Anhydrodihydrotenulin (XXX), das man durch Erhitzen des Dihydrotenulins mit Na-acetat in Acetanhydrid erhält. Unterwirft man (XXX) der Ozonolyse, so entsteht neben Formaldehyd das Bis-$\gamma$-lacton (XXXI). Die letztere Formel steht in Übereinstimmung mit dem IR-Spektrum der Substanz: Absorptionen bei 1765 und 1800 cm$^{-1}$ für $\gamma$-Lactone mit gemeinsamem C-Atom und bei 1740 cm$^{-1}$ für den Cyclopentanonring.

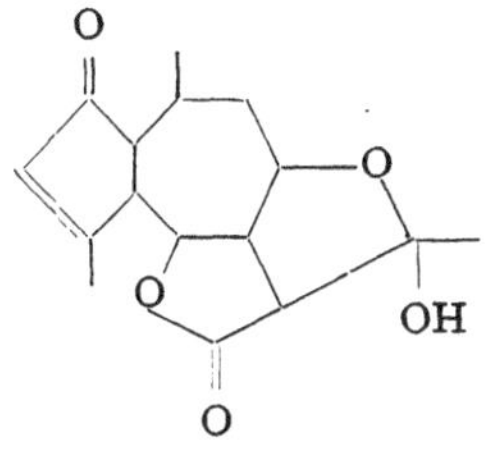

Basierend auf der Formel (XXVI, S. 144) für Tenulin ergeben sich für Anhydrotenulin, Pyrotenulin und Tenulinsäure die Formelbilder (XXXII), (XXXIII) und (XXXIV).

Die von BRAUN, HERZ und RABINDRAN (21) für Tenulin vorgeschlagene Struktur $E$ mit einer Carbonylfunktion an $C_{(3)}$ anstatt an $C_{(2)}$ ist unzutreffend [Privatmitt. von W. HERZ an BARTON und DE MAYO (15)].

### 5. Helenalin.

Der Bitterstoff Helenalin wurde erstmalig 1910 von REEB als „acide hélénique" aus *Helenium autumnale* isoliert. Nach CLARK (32) wird Helenalin auch in *H. macrocephalum* und *H. quadridentalum* gefunden. Das von ADAMS und HERZ (1) untersuchte Helenalin stammte aus *H. microcephalum*. Viele *Helenium*-Arten gelten als unliebsame Unkräuter, da sie auf das Weidevieh toxisch wirken. Wirksames Prinzip ist das Helenalin, das von LAMSON [vgl. (1)] auf seine physiologische Wirkung geprüft wurde: Es wirkt stark auf die Schleimhäute, ruft unangenehme Magen- und Darmentzündungen hervor und lähmt das Herz und andere Muskeln. Helenalin ist ein Fischgift und Insekticid und wurde als Wurmmittel vorgeschlagen.

Nach ADAMS und HERZ (1) enthält das Helenalin die folgenden vier Strukturelemente:

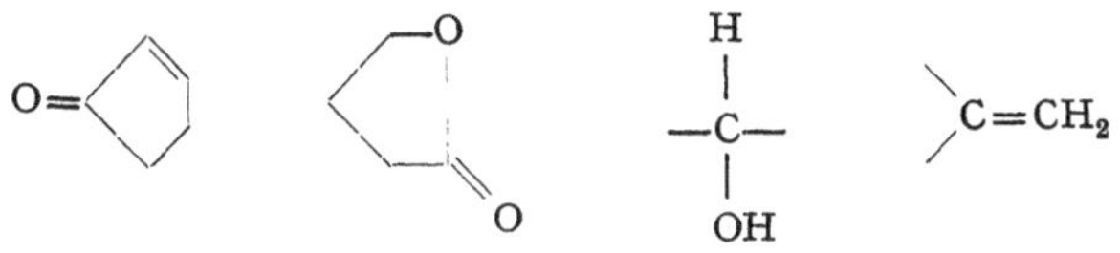

Auf Grund weiterer Versuche gelangten BÜCHI und ROSENTHAL (23) für Helenalin zu zwei möglichen Strukturformeln, deren eine sich durch quantitative Auswertung der bei der magnetischen Kernresonanz erhaltenen Spektren als zutreffend erwies.

Grundgerüst des Helenalins ist das Guaiazulen (XII, S. 139). Dieser Kohlenwasserstoff entsteht nach Reduktion des Tetrahydrohelenalins mit LiAlH$_4$, Acetylierung des dabei gebildeten Tetrols und Dehydrierung mit Pd. Den Beweis, daß das Helenalin eine der Formel (XXXV) entsprechende Struktur besitzt, führen Büchi und Rosenthal auf dem in *Formelübersicht 3* verzeichneten Wege.

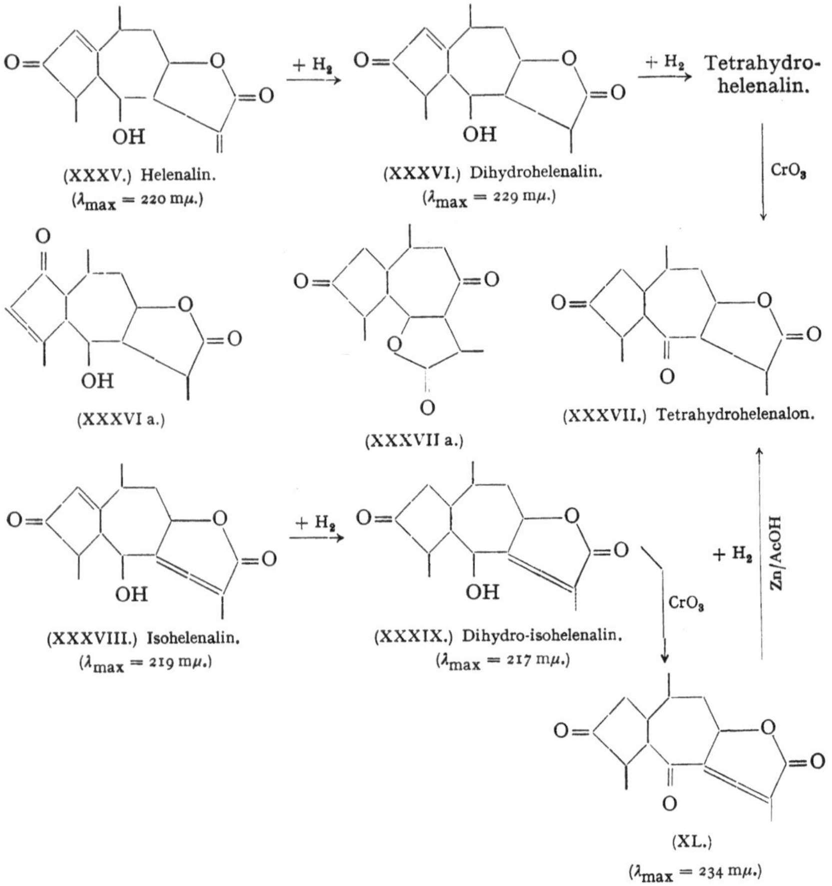

*Formelübersicht 3.* Struktur des Helenalins.

Subtrahiert man die UV-Absorptionskurve des Dihydrohelenalins (XXXVI) von der des Helenalins (XXXV), so entsteht die Kurve des ungesättigten Chromophors, das die endständige Methylengruppe konjugiert zum Lactoncarbonyl enthält ($\lambda_{max}$ = 210 m$\mu$, $\varepsilon$ = 10000).

Aus *H. microcephalum* konnten Büchi und Rosenthal (*23*) außerdem eine Substanz isolieren, die sie als Isohelenalin bezeichnen (XXXVIII). Im Gegensatz zum Helenalin, das bei der katalytischen Hydrierung 2 $H_2$ aufnimmt, reagiert (XXXVIII) nur mit 1 Mol $H_2$ unter Bildung des Dihydro-isohelenalins (XXXIX); letzteres kann, nach Chromsäure-

*Literaturverzeichnis: SS. 175—182.*

oxydation zu (XL), durch Behandeln mit Zn/Essigsäure zum Tetrahydrohelenalon (XXXVII) reduziert werden. (XXXVII) wurde bereits
von ADAMS und HERZ *(1)* dargestellt. Der von diesen Autoren nachge-
wiesene Cyclopentenon-teil könnte außer gemäß (XXXVI) auch entsprechend (XXXVIa) mit dem übrigen Teil des Moleküls verknüpft
sein. Quantitative Auswertung des magnetischen Kernresonanzspektrums
entscheidet zugunsten der Formel (XXXVI). Nach derselben Methode
ist eine Entscheidung zwischen den beiden für Tetrahydrohelenalon
denkbaren Formeln (XXXVII) und (XXXVIIa) möglich, demnach
ist (XXXVII) zutreffend.

Die Richtigkeit von (XXXV) für Helenalin ist damit bewiesen.

### 6. Alantolactone.

Im ätherischen Öl der Alantwurzel *(Inula helenium)* sind drei Bitterstoffe vorhanden, die als Alantolacton (XLI), Isoalantolacton (XLII)
und Dihydro-isoalantolacton (XLIII) bezeichnet werden. Das früher
gehandelte „Helenin" ist ein Gemisch dieser drei Lactone. Die
antihelminthische Wirksamkeit der Alantwurzel ist auf die Lactone zurückzuführen.

Mit Na-Amalgam entstehen aus (XLI) und (XLII) die Dihydro-
Verbindungen, wobei das aus (XLII) gebildete Derivat mit dem im
natürlichen Helenin (auch als „Alantcampher" bezeichnet) vorkommenden (XLIII) identisch ist. (XLI), (XLII) und (XLIII) liefern bei katalytischer Hydrierung dasselbe Tetrahydroprodukt *(55)*. Die von
TSUDA u. a. *(150)* für Alantolacton vorgeschlagene Formel (XLI) sei
den folgenden Ausführungen vorangestellt.

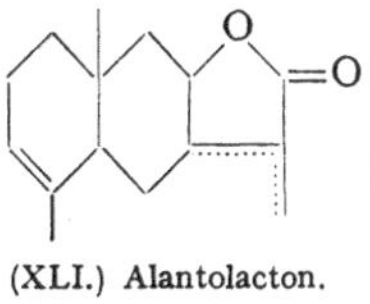

(Die Lage der zum Lactoncarbonyl α-ständigen
Doppelbindung konnte noch nicht geklärt werden.)

(XLI.) Alantolacton.

Zur Bestimmung des Grundgerüstes führten RUZICKA und VAN
MELSEN *(130)* an (XLI) und (XLII) Selen-Dehydrierungen durch. In
beiden Fällen entstand in guten Ausbeuten 1-Methyl-7-äthyl-naphtha-
lin (XLIV). Nach Reduktion des Lactoncarbonyls im Tetrahydroprodukt
mit Na in Alkohol zum Glykol, Veresterung mit HBr und anschließender

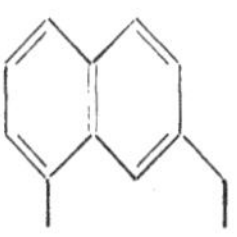

(XLIV.)

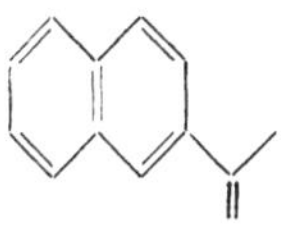

(XLV.)

HBr-Abspaltung in Chinolin konnten dieselben Autoren bei der Dehydrierung das 1-Methyl-7-isopropenyl-naphthalin (XLV) erhalten.

Bei Reduktion mit Na-Amalgam wird die zum Lactoncarbonyl α-ständige Doppelbindung hydriert. Da die aus (XLI) und (XLII) gebildeten Dihydroverbindungen voneinander verschieden, die Tetrahydroverbindungen dagegen identisch sind, muß die Verschiedenheit von (XLI) und (XLII) durch die Lage der zweiten Doppelbindung bedingt sein. Diese Frage konnte Hansen (55) mit Hilfe der Ozonspaltung klären.

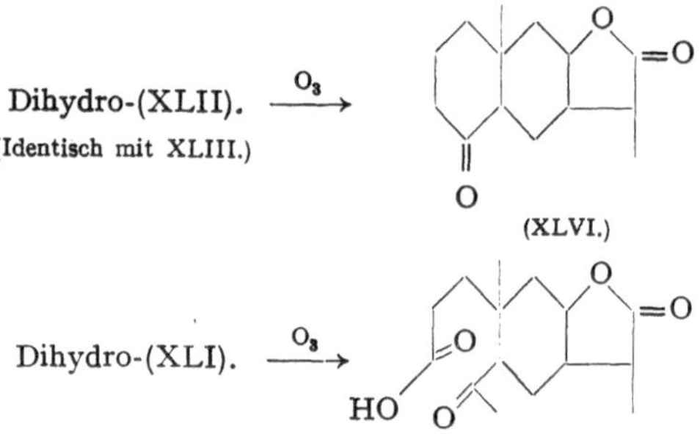

Dihydro-(XLII). $\xrightarrow{O_3}$
(Identisch mit XLIII.)

(XLVI.)

Dihydro-(XLI). $\xrightarrow{O_3}$

(XLVII.)

Aus der Bildung des Ketolactons (XLVI) und der Ketolactoncarbonsäure (XLVII) ergibt sich die Lage der Doppelbindungen in Alantolacton und Isoalantolacton. Die Festlegung des Lactonhydroxyls gelang erst in jüngster Zeit der japanischen Forschergruppe (150) auf folgendem Wege:

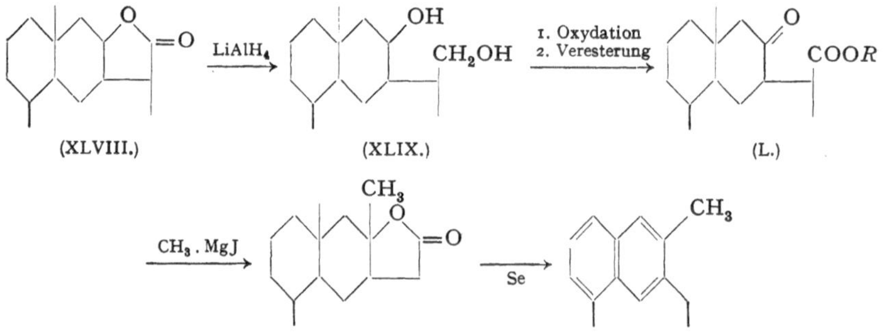

(XLVIII.)    $\xrightarrow{\text{LiAlH}_4}$    (XLIX.)    $\xrightarrow[\text{2. Veresterung}]{\text{1. Oxydation}}$    (L.)

$\xrightarrow{\text{CH}_3 \cdot \text{MgJ}}$    (LI.)    $\xrightarrow{\text{Se}}$    (LII.)

Im Dehydrierungsprodukt ist das ursprünglich die OH-Gruppe tragende C-Atom durch die CH₃-Gruppe markiert.

Abschließend sei noch auf einen Befund von Ruzicka, Pieth, Reichstein und Ehmann (131) hingewiesen: Das aus Tetrahydro-alantolacton dargestellte Glykol (LIII) erleidet bei der Reaktion mit HBr und Chinolin eine Umlagerung, so daß als Hauptprodukt der anschließend durchgeführten Selen-Dehydrierung das Naphthalin-Derivat (LV) erscheint. Es hat also eine Art Retropinakolin-

Umlagerung zwischen der am quartären Kohlenstoffatom stehenden Methylgruppe und dem Hydroxyl stattgefunden. Ein analoger Vorgang ist auch beim Santonin beobachtet worden (*33*).

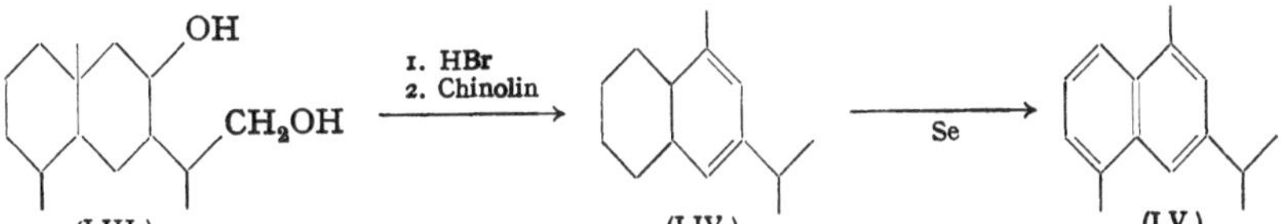

(LIII.)     1. HBr   2. Chinolin     (LIV.)     Se     (LV.)

## V. Bitterstoffe der Menispermaceen.

Von den in der Familie der Menispermaceen beschriebenen Bitterstoffen sei hier über die Strukturaufklärungen des Columbins und des Picrotoxinins berichtet, da die entsprechenden Arbeiten neueren Datums sind. Die Bitterstoffe Chasmanthin und Palmarin sind chemisch dem Columbin sehr eng verwandt. Eine Abhandlung, die sich mit den Zusammenhängen zwischen diesen drei Substanzen beschäftigt, ist bereits angekündigt (*13*).

### 1. Columbin.

Literaturübersicht: (*13*).

Die Wurzel von *Jatrorrhiza palmata* (Colombowurzel) enthält drei neutrale Bitterstoffe: Columbin $C_{20}H_{22}O_6$, Chasmanthin $C_{20}H_{22}O_7$ und Palmarin $C_{20}H_{22}O_7$. Columbin wurde bereits 1830 von WITTSTOCK beschrieben. Besonders FEIST (*47*) und WESSELY (*158*) und Mitarb. haben

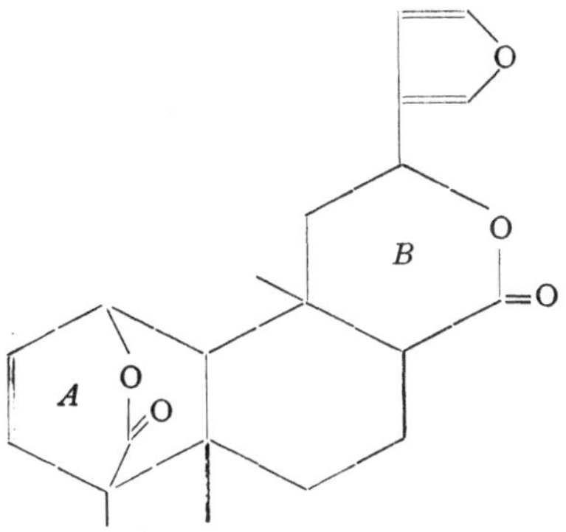

(LVI.) Columbin.
(LVII.) Isocolumbin (vgl. S. 154).

sich mit der Konstitutionsaufklärung der Substanz beschäftigt. Die endgültige Formel für Columbin (LVI) konnten BARTON und ELAD (*13*) aufstellen.

*Allgemeine Reaktionen.* Durch milde Alkalibehandlung geht Columbin in Isocolumbin (LVII) über. Gegen Alkalien verhalten sich beide als

Dilactone. Während die Reaktion mit dem ersten Mol Alkali reversibel ist, werden nach Einwirkung eines zweiten Mols keine definierten Produkte gebildet. Katalytische Hydrierung überführt (LVI) und (LVII) in Octahydroderivate; in beiden Fällen tritt Hydrogenolyse zu Monocarbon-säuren ein, die nach dem Ergebnis der Titration scheinbar keine Lacton-gruppe mehr enthalten. Bei der Acetylierung mit Na-acetat/Acet-anhydrid liefern (LVI) und (LVII) dasselbe Monoacetat, das sich von Isocolumbin (LVII) ableitet. Entsprechend verläuft die Methylierung mit Dimethylsulfat (*156*): der Monomethyläther läßt sich von (LVII) ableiten. Auf Grund dieser Reaktionen muß der Hydroxylgruppe des Columbins eine gewisse Acidität zugesprochen werden, jedoch ist sie nicht phenolisch oder enolisch. — Beim Erhitzen über den Schmelzpunkt spalten Columbin und Isocolumbin je 1 Mol $CO_2$ ab unter Bildung der Decarboxy-verbin-dungen. Da diese weder acetyliert noch methyliert werden können, liegt die OH-Gruppe nicht mehr als solche vor. Die Decarboxylierung gelingt indessen sowohl bei den Methyläthern als auch bei den Acetaten. In den Decarboxy-verbindungen ist noch eine Lactongruppe vorhanden.

*Grundgerüst.* Bei der Zinkstaubdestillation des Columbins konnten Feist und Mitarbeiter neben o-Kresol das 1,2,5-Trimethyl-naphthalin isolieren [vgl. (*13*)]. Bei der Alkalispaltung entstanden aus Columbin 2,4-Dimethylbenzoesäure und 2-Methylterephthalsäure; Oxydation mit $MnO_2/H_2SO_4$ führte zur Bildung von Benzol-1,2,3-tricarbonsäure und

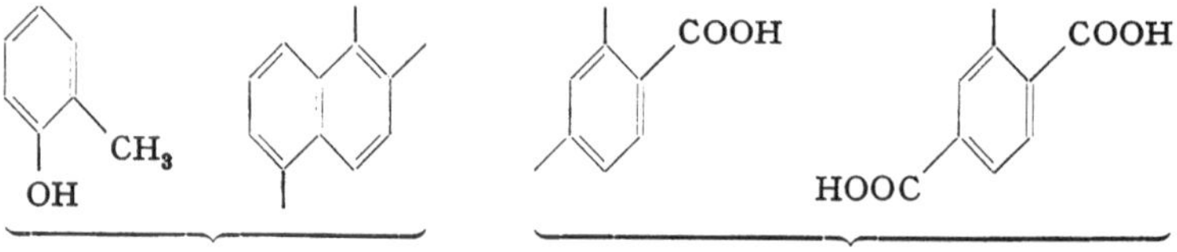

Zinkstaubdestillation                    KOH-Spaltung

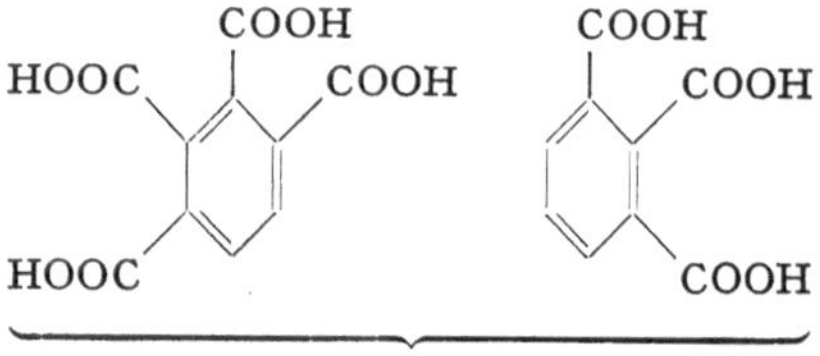

$MnO_2/H_2SO_4$

Benzol-1,2,3,4-tetracarbonsäure. Das Auftreten von Trimethylnaphthalin deutet auf ein substituiertes bicyclisches System hin.

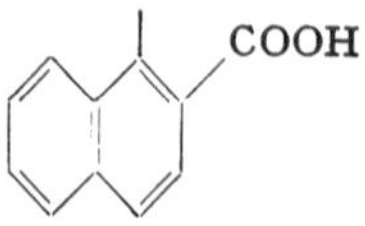

Durch Reduktion der Decarboxy-octahydro-colum-binsäure (LXIII, S. 154) nach Wolff-Kishner (durch Beseitigung von OH- und CO-Gruppen lassen sich Umlagerungen bei der Selen-Dehydrierung weitgehend vermeiden) und anschließende Selen-Dehydrierung

(LVIII.)

konnten Barton und Elad (*13*) 1-Methyl-naphthalin-2-carbonsäure (LVIII) fassen. Hierdurch ist die Stellung des einen Lactonringes festgelegt.

*C—CH₃-Gruppen.* Deren Bestimmung führte bei verschiedenen Derivaten des Columbins zu Werten, die nur wenig höher lagen als die für eine Gruppe berechneten. Quantitative IR-Messungen (gegen Stearinsäure als Bezugssubstanz) bewiesen aber eindeutig die Existenz von zwei C—CH₃-Gruppen.

*Lactonringe.* Durch selektive Hydrierung über Pd in Essigester gelangt man zu Dihydro-columbin und -isocolumbin, bei denen die Decarboxylierung nicht mehr möglich ist. Die Decarboxylierung des [in (LVI, S. 151) mit *A* bezeichneten] Lactons läßt sich durch Annahme einer β,γ-ständigen Doppelbindung deuten. Wie sich aus den Spektren ergibt, liegt in den Decarboxy-Verbindungen ein β,γ-ungesättigtes Keton vor, dies erlaubt den Schluß auf eine in Columbin und Isocolumbin vorhandene tertiäre OH-Gruppe. Für die Decarboxylierung läßt sich somit das folgende Schema aufstellen:

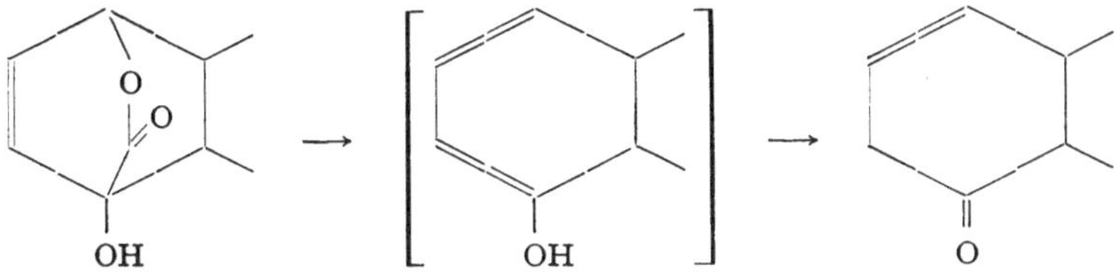

Die Stellung der Doppelbindung ergibt sich daraus, daß nach Überführung in das entsprechende Glykol mit OsO₄ bei der anschließend durchgeführten Spaltung 2 Mole Bleitetraacetat verbraucht werden. Die in *A* gewählte Formulierung (S. 151) entspricht diesem Ergebnis. Lactonring *A* erfährt durch Hydrierung eine Stabilisierung, wie die Titration erkennen läßt, die Öffnung von *A* gelingt nur unter verschärften Bedingungen. Lactonring *B* erleidet leicht Hydrogenolyse. Dies wird erklärt durch die Allylstellung des Ringsauerstoffs zur Doppelbindung des Furan-kerns. (Vinylstellung erscheint wegen der Reversibilität der Öffnung von *B* ausgeschlossen.)

*Furan-kern.* Bei der Ozonspaltung des Dihydrocolumbins entsteht außer Ameisensäure eine C₁₇-Carbonsäure und daneben in kleineren Mengen eine C₁₈-α-Ketocarbonsäure (isoliert als Methylester).

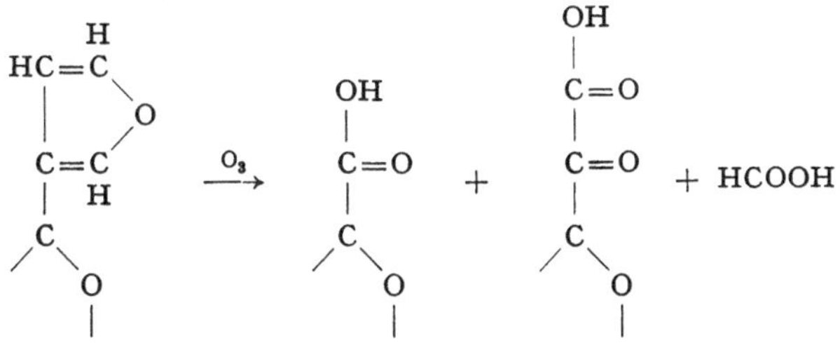

Einen direkten Beweis für die Anwesenheit eines Furan-kerns im Columbin konnten KUBOTA und MATSUURA (95) erbringen. Der Bitterstoff wird mit $LiAlH_4$ zur Pentahydroxyverbindung reduziert, an die

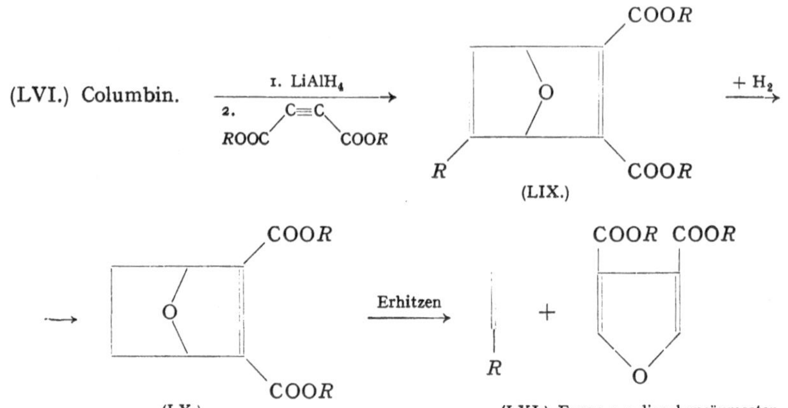

(LVI.) Columbin. $\xrightarrow[\text{2.}]{\text{1. LiAlH}_4}$ ... $\xrightarrow{+ H_2}$

(LIX.)

(LX.)    Erhitzen    (LXI.) Furan-3,4-dicarbonsäureester.

sich Acetylen-dicarbonsäureester anlagern läßt. Partielle Hydrierung und thermische Spaltung des Adduktes nach ALDER und RICKERT (4) führt zum Furan-3,4-dicarbonsäureester (LXI).

*Die Isomerie zwischen Columbin und Isocolumbin* wird durch eine Epimerisierung (13) des zur Carbonylgruppe des Lactonringes B α-ständigen C-Atoms verursacht. Die Überführung des Columbins in Isocolumbin gelingt durch kurzes Erhitzen in wäßrig-alkoholischem KOH. Die UV-Spektren beider Verbindungen sind identisch; der $[\alpha]_D$-Wert wird verändert; im IR wird Verschiebung einer Lactonfrequenz von 1727 nach 1748 cm$^{-1}$ beobachtet.

*Derivate.* Mit der für Columbin vorgeschlagenen Struktur (LVI, S. 151) läßt sich die Bildung der Derivate (LXII)—(LXV) erklären.

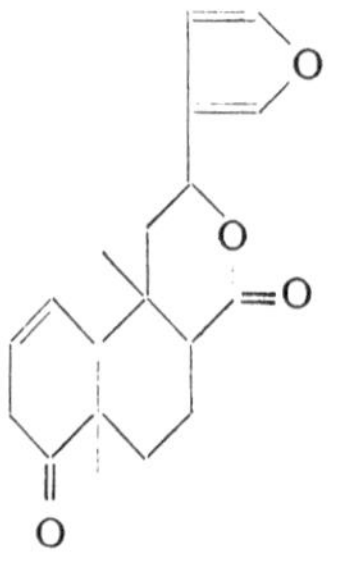

(LXII.) Decarboxy-columbin. 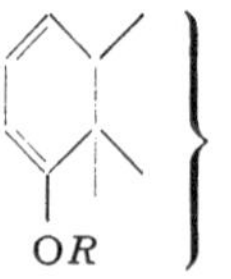    (LXII a.) $R = CH_3$. Decarboxy-iso-columbin-O-methyläther.    (LXIII.) Decarboxy-octahydro-columbinsäure.
    $R = Ac$. Decarboxy-isocolumbinacetat.

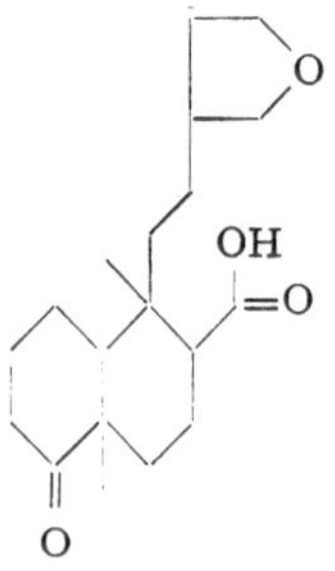

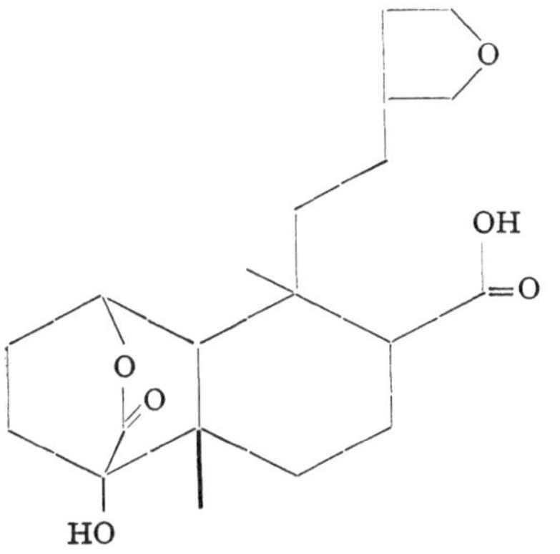

(LXIV.) Octahydro-columbinsäure.

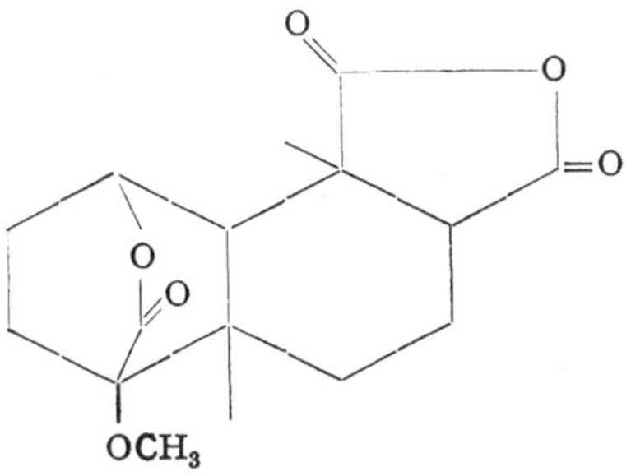

(LXV.) (Entstanden durch MnO₄⁻-Oxydation des 2-Methyl-isocolumbins.)  (IR: 1850, 1777 cm⁻¹.)

Die von CAVA und SOBOCZENSKI (28) als Arbeitshypothese vorgeschlagene Columbin-Formel ist durch diese Arbeiten widerlegt.

## 2. Picrotoxinin (Picrotoxin).

Der Bitterstoff Picrotoxin befindet sich in den sogen. „Kokkelskörnern", den Früchten von *Anamirta cocculus*, *Menispermum cocculus* oder *Cocculus indicus*. Die erste Isolierung erfolgte durch BOULLEY [vgl. (144)]. In der Folge beschäftigten sich zahlreiche Forscher mit der Substanz, bis es CONROY (38) 1957 gelang, für das Picrotoxinin, die physiologisch aktive Komponente des „Picrotoxins", eine allen experimentellen Ergebnissen gerecht werdende Strukturformel aufzustellen. Eine Übersicht über die umfangreiche Literatur bis 1949 geben SUTTER und SCHLITTLER (144).

1880 konnten BARTH und KRETSCHY (11) nach mühevollen Versuchen das bis dahin als einheitlichen Körper betrachtete Picrotoxin in zwei Komponenten, Picrotin und Picrotoxinin, zerlegen. Während Picrotin physiologisch unwirksam ist, ist das Picrotoxinin hochaktiv. Ein elegantes Verfahren zur Abtrennung des Picrotoxinins vom Picrotin stammt von MEYER und BRUGER (107). Durch Bromieren in wäßriger Lösung läßt sich das Picrotoxinin in Form seines Monobromproduktes quantitativ vom Picrotin trennen. Nach HORRMANN (59, 59a) kann Picrotoxinin aus der Bromverbindung durch Behandlung mit Zinkstaub in wäßrig-alkoholischer Lösung regeneriert werden.

Den Streit, ob es sich bei Picrotoxin um ein Gemenge (schwankender Zusammensetzung) von isomorphen Kristallen oder um eine im Molverhältnis 1 : 1 kristallisierende, leicht spaltbare Molekülverbindung handle, konnte SIELISCH (140) im Sinne der letzteren Auffassung entscheiden. Ihm gelang die „Synthese" des Picrotoxins aus den Komponenten; als Kriterium diente der optische Drehwert, der mit wechselnder Zusammensetzung starken Schwankungen unterliegt. Kryoskopische

Molgewichtsbestimmung lieferte ihm den auf die Summenformel $C_{30}H_{34}O_{13}$ passenden Wert 602. Von früheren Autoren ebullioskopisch durchgeführte Molgewichtsbestimmungen liegen zwischen 200 und 280, da unter diesen Bedingungen bereits Spaltung in die Komponenten erfolgt.

Formal unterscheidet sich das Picrotin vom Picrotoxinin durch den Mehrgehalt von 1 Mol $H_2O$. Die chemische Verwandtschaft beider Körper ist noch ungeklärt, auch ist es noch nicht gelungen, den einen Stoff in den anderen umzuwandeln. Wichtig ist, daß Picrotoxinin eine Doppelbindung enthält, während Picrotin gesättigt ist. Auf diesem Umstand beruht das von Meyer und Bruger (107) entwickelte Trennverfahren.

*Physiologische Wirkung.* Ursprünglich dienten die das Picrotoxin enthaltenden Kokkelskörner zum Fischfang (in Deutschland verboten). Picrotoxin ist ein Fisch- und Krampfgift. In kleinen Dosen führt es zu einer Erregung der parasympathischen Zentren (Herzverlangsamung u. a.) und zur Erregung der spinalen Schweißzentren; dadurch tritt Absinken der Körpertemperatur ein. Picrotoxin wird als Weckmittel und als Therapeuticum bei schwersten Schlafmittelvergiftungen verwendet (44).

## Konstitution des Picrotoxinins.

Die Formel des Picrotoxinins (LXVI), $C_{15}H_{16}O_6$, stellte Conroy (38) 1957 auf, nachdem er die Struktur mehrerer Abbauprodukte geklärt hatte. Die Aufklärung des Kohlenstoffgerüstes gelang Conroy (37) mit der Totalsynthese des Picrotoxadiens (LXXIII), eines Kohlenwasser-

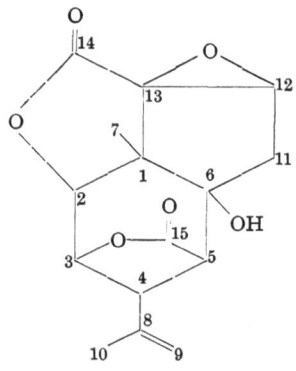

(LXVI.) Picrotoxinin.

stoffs, den er durch Abbau des Picrotoxinins erhalten hatte (36). Mit der heutigen Kenntnis vom Bau der Zwischenprodukte läßt sich der Abbau nach *Formelübersicht 4* darstellen.

(LXVI) geht bei der Bromierung in Brompicrotoxinin (LXVII) über (107). Da bei der Reaktion ein neues Asymmetriezentrum (an $C_{(8)}$) entsteht, tritt (LXVII) in zwei stereomeren Formen auf, die mit α und β

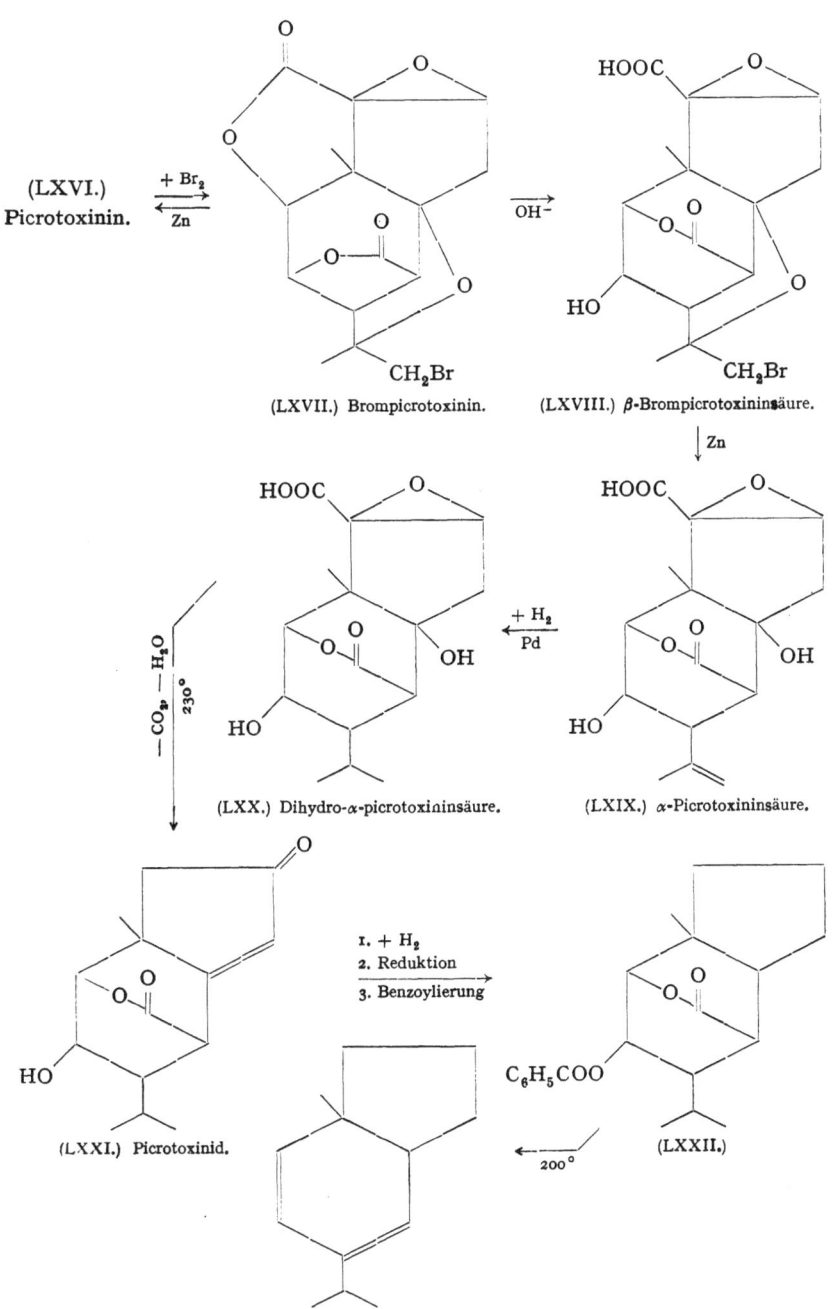

(LXVI.) Picrotoxinin.

(LXVII.) Brompicrotoxinin.

(LXVIII.) β-Brompicrotoxininsäure.

(LXX.) Dihydro-α-picrotoxininsäure.

(LXIX.) α-Picrotoxininsäure.

(LXXI.) Picrotoxinid.

(LXXII.)

(LXXIII.) Picrotoxadien.

*Formelübersicht 4.* Abbau des Picrotoxinins.

bezeichnet werden. Die Halogenierung verläuft unter Einbeziehung der tertiären OH-Gruppe und Bildung eines cyclischen Äthers (*126*).

Aus (LXVII) entsteht bei der Behandlung mit Alkali ein Gemisch der an $C_{(8)}$ stereomeren Brompicrotoxininsäuren (LXVIII) (Umlactonisierung, SS. 157, 144), die nach Horrmann (*59, 59a*) mit Zink in die sogen. α-Picrotoxininsäure (LXIX) überführbar sind. Horrmann beobachtete, daß die Dihydro-α-picrotoxininsäure (LXX) beim Schmelzen (230°) Zersetzung erleidet. Unter $CO_2$-Verlust entsteht daraus Picrotoxinid (LXXI) (*36*), ein α,β-ungesättigtes Keton, das noch eine Lactongruppe enthält. Nach katalytischer Hydrierung ist aus (LXXI) eine Dibenzyliden-Verbindung darstellbar; damit ist die Carbonylfunktion an $C_{(12)}$ festgelegt. Dihydropicrotoxinid reagiert erst nach Alkali-Vorbehandlung mit 1 Mol $HJO_4$, damit ist gezeigt, daß nach Öffnung des Lactons ein 1,2-Glykol entsteht.

Für den weiteren Abbau wird die Carbonylgruppe im Dihydro-picrotoxinid über das Äthylenmercaptal mit Raney-Nickel zur Methylen-gruppe reduziert und die nach Benzoylierung der freien OH-Gruppe entstandene Verbindung (LXXII) einer vorsichtigen Pyrolyse bei 200° unterworfen. Neben $CO_2$ und Benzoesäure entsteht dabei der flüssige Kohlenwasserstoff Picrotoxadien (LXXIII, S. 157), aus dem sich mit Maleinsäureanhydrid in guter Ausbeute ein kristallines Addukt gewinnen läßt (*36*).

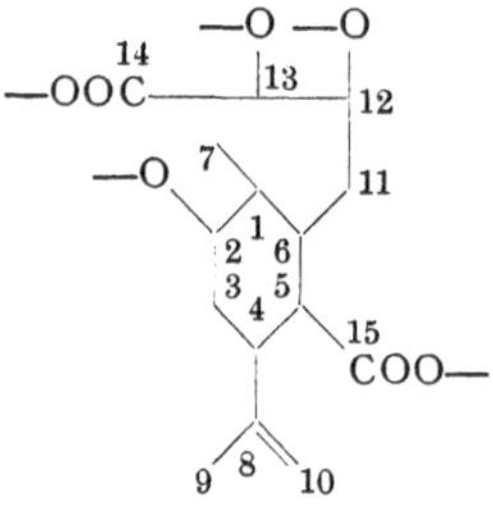

(LXXIV.)

*Sauerstoff-Funktionen.* Robertson und Mitarbeiter (*123, 125*) schlugen für Picrotoxinin (LXVI) die Partialformel (LXXIV) vor, zu der sie auf Grund folgender Befunde gelangten. Die Stellung der Doppelbindung ergibt sich durch den Verlauf der Ozonolyse (*59, 59a*). Neben Formaldehyd entsteht ein Keton, das Picrotoxinon. Aus letzterem entsteht bei der Behandlung mit rotem Phosphor und HJ u. a. Hydroxy-nor-picrotinsäure (LXXV).

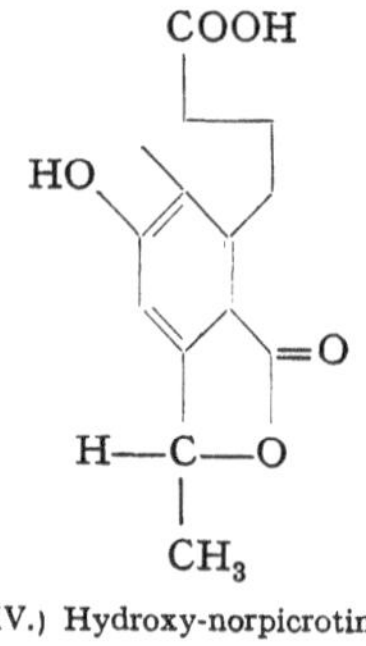

(LXXV.) Hydroxy-norpicrotinsäure. 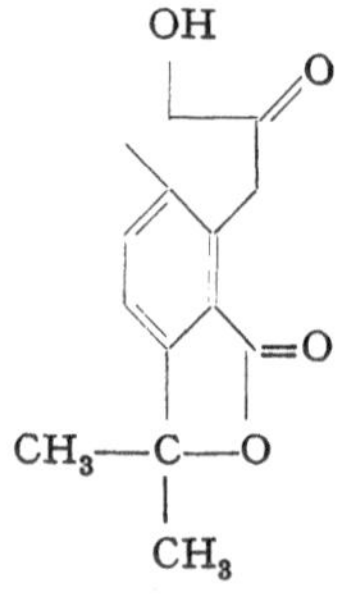    (LXXXVI.) Picrotonol.

*Literaturverzeichnis: SS. 175—182.*

Die anderen Sauerstoff-Funktionen ergaben sich aus der von TETT-WEILER und DRISHAUS (*148*) ermittelten Struktur des Picrotonols (LXXVI), das ANGELICO (*7*) durch Erhitzen der Picrotinsäure mit 40%iger $H_2SO_4$ erhalten hatte. SUTTER und SCHLITTLER (*144*) gelang die Spaltung des Dihydro-α-picrotoxinins durch Alkaliabbau; unter Erhaltung aller Kohlenstoffatome entstanden die Substanzen *a* und *b*, deren Konstitution

bewiesen werden konnte. Die Auffindung dieser Spaltprodukte veranlaßte CONROY (*38*), in die Partialformel (LXXIV) an $C_{(3)}$ und $C_{(6)}$ weitere Sauerstoff-Funktionen einzuführen. Unter Berücksichtigung der für Picrotoxadien ermittelten Struktur (LXXIII, S. 157) ergab sich damit die erweiterte Partialformel (LXXVII).

*Zuordnung der Sauerstoff-Funktionen. Lactone.* Bereits HORRMANN (*59, 59a*) konnte im Picrotoxinin zwei Lactongruppen nachweisen. Die IR-Banden, bei 1796 und 1776 cm⁻¹, entsprechen der Absorption von γ-Lactonen.

*Die tert. OH-Gruppe.* Es handelt sich um ein tert. Hydroxyl, da es unter normalen Bedingungen nicht acylierbar ist und auch nicht zum Keton oxydiert werden kann. Bei der Bestimmung des aktiven Wasserstoffs nach ZEREWITINOFF erhält man für Picrotoxinin den 1,5 OH-Gruppen entsprechenden Wert (*59, 59a*). Da die OH-Gruppe in den bromierten Derivaten des Picrotoxinins nicht mehr nachweisbar ist, muß sie in die Bromierungsreaktion einbezogen sein (*126*). Aus sterischen Gründen ist für den Sitz der OH-Gruppe das tert. $C_{(6)}$ anzunehmen (*38*) (Absorption im IR bei 3450 cm⁻¹).

*Äther-Gruppe.* Das sechste Sauerstoffatom könnte nach ROBERTSON (*124*) als Äther vorliegen. Gegen die von CONROY vorgeschlagene Epoxyd-Formulierung sind Bedenken erhoben worden (*17*). Der Äthylenoxydring erwies sich gegen Mineralsäuren als recht beständig, so bleiben z. B. konz. HCl in der Kälte und heißes verdünntes, mit $MgCl_2$ gesättigtes HCl ohne Einwirkung. Des weiteren ist unter den zahlreichen Derivaten des Picrotoxinins nirgends ein Paar zu finden, das die einfache Beziehung zwischen dem Epoxyd und einem daraus entstandenen vicinalen Glycol wiedergeben würde.

Es zeigt sich hier sehr deutlich der Unterschied im reaktiven Verhalten bestimmter Gruppierungen (hier des Äthylenoxyds), in einfachen Verbindungen und in derart komplex gebauten „Käfig-Strukturen", wie sie bei Picrotoxinin und seinen Derivaten vorliegen. Wie die folgende Betrachtung zeigen wird, sind tatsächlich sterische Gründe für das ungewöhnliche Verhalten des Epoxyds verantwortlich.

Den exakten chemischen Nachweis der Äthylenoxydgruppe führte CONROY (*38*) bei der β-Brompicrotoxininsäure (LXVIII, S. 160). Unter Berücksichtigung der

Partialformel (LXXVII, S. 159) lassen sich die Sauerstoff-Funktionen in (LXVIII) wie folgt zuordnen: Auf Grund der IR-Absorption enthält die Verbindung denselben Lactonring wie (LXXI). Der Brom-Äther-Ring wird mit dem tert. OH an $C_{(6)}$ (nicht an $C_{(13)}$) gebildet, da die Alkalispaltung nach Sutter und Schlittler (S. 159) nicht an den bromierten Verbindungen durchführbar ist; diese Reaktion setzt vielmehr ein freies OH an $C_{(6)}$ voraus. Die sekundäre OH-Gruppe in (LXVIII) ist acetylierbar und kann zum Keton oxydiert werden. $C_{(14)}$ liegt als freie Carboxylgruppe vor, da er bei der Bildung von (LXXI, S. 157) als $CO_2$ abgespalten wird. Aus (LXVIII) entsteht nach Reduktion mit $NaBH_4$ in guter Ausbeute die um zwei H-Atome reichere Dihydro-$\beta$-brompicrotoxininsäure (LXXVIII), aus der ein

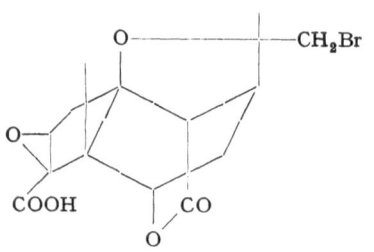

(LXVIII, S. 157.)

(LXXVIII.) Dihydro-$\beta$-brompicrotoxininsäure.          (LXXIX.)

Diacetat darstellbar ist. Der Umsatz von (LXXVIII) mit $PbO_2$ in Eisessig führt unter $CO_2$-Verlust zum Keton (LXXIX), das keine $\alpha$-ständige Methylengruppe enthält, da man daraus keine Benzylidenverbindung gewinnen kann. [Mit Bleitetraacetat reagiert (LXXVIII) nicht.]

Die sterische Betrachtung macht die Öffnung des Äthylenoxydrings in der Reaktion (LXVIII) → (LXXVIII) verständlich.

Eine Raumformel läßt sich für die $\beta$-Brompicrotoxininsäure (LXVIII) auf Grund folgender Überlegungen aufstellen: Die OH-Gruppe an $C_{(3)}$ steht in *trans*-Stellung zur Isopropylgruppe, da Reaktionen wie die Eliminierung der Benzoesäure (LXXII → → LXXIII, S. 157) immer nur zwischen *cis*-ständigen Substituenten erfolgen. Die Lactonbrücke muß zum Brom - Äther - Ring *trans*-angeordnet sein, da eine andere Formulierung sterisch unmöglich ist. Der Cyclopentanring ist aus Spannungsgründen

(LXVIII.) $\beta$-Brompicrotoxininsäure.

mit dem starren Oxabicyclo-(2,2,2)-octanon bevorzugt *cis*-verknüpft. Eine Bindung zwischen OH an $C_{(2)}$ und der $C_{(14)}$-Carboxylgruppe zum $\gamma$-Lacton kann nur bei der hier diskutierten Konformation erfolgen. Raumformel (LXVIII) erklärt auch die Stabilität des Epoxyds gegen

Hydrolyse: Ein nucleophiles Agens muß sich von der Seite der geringsten Substitution her nähern, bei (LXVIII) also von unten. Durch die nahe Lactonbrücke findet hier aber besonders starke Abschirmung statt. Macht man indessen durch geeignete Substitution die Lactongruppe selbst zum Agens, so kann sie, eben wegen ihrer räumlichen Nähe, den Oxydring öffnen.

Die *Konstitution des Picrotoxinins* (LXVI, S. 156) folgt aus der für (LXVIII) ermittelten Struktur (*38*).

. Da (LXVI) aus (LXVII) wieder regenerierbar ist, hat in der Reaktion (LXVI) → → (LXVII) noch keine irreversible Umlagerung stattgefunden. Bei dem von (LXVII) nach (LXVIII) führenden Schritt (S.157, Einwirkung von 1 Äquivalent Alkali) erfolgt neben der Öffnung des einen Lactonringes außerdem eine Umlactonisierung, wie sich aus den IR-Spektren ergibt. Von den in (LXVII) vorhandenen beiden $\gamma$-Lactonen ist in (LXVIII) keines mehr vorhanden. Raumformel (LXVIII) läßt für das neu entstandene $\delta$-Lacton nur die Verknüpfung zwischen $C_{(15)}$-Carboxyl und $C_{(2)}$-Hydroxyl zu.

Für Brompicrotoxinin mit seinen beiden $\gamma$-Lactonringen ergibt sich damit die Raumformel (LXVII), bei der der Cyclohexanring in der Form eines gespannten Halbbootes vorliegt.

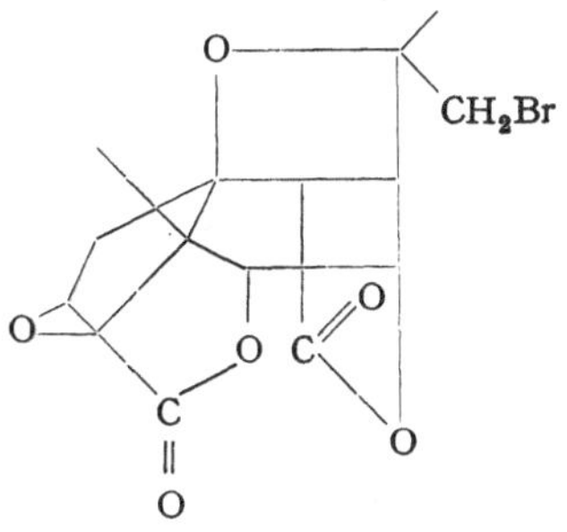

(LXVII.) Brompicrotoxinin. 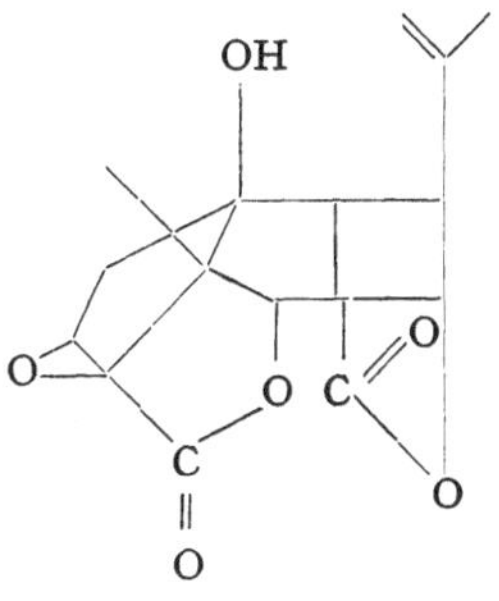          (LXVI.) Picrotoxinin.

Als treibende Kraft für die Reaktion (LXVII) → (LXVIII) ist neben der Ausbildung des $C_{(14)}$-Carboxylat-ions und dem damit verbundenen Gewinn an Resonanzenergie der Übergang aus der spannungsreichen Halbbootform in die energieärmere Bootform anzusehen. Durch Öffnung des Brom-Äther-Ringes in (LXVII) erhält man die Raumformel des Picrotoxinins (LXVI).

# VI. Bitterstoffe der Coriariaceen.
## Tutin und Coriamyrtin.

Tutin ist in den Samen von *Coriaria thymifolia*, *C. ruscifolia* und *C. angustissima* enthalten (*43*). Coriamyrtin wurde (neben Tutin) aus den Blättern von *C. japonica* (*67*) isoliert. Nach KINOSHITA (*72, 73*) ist

Tutin mit dem Coriarin aus *C. japonica* identisch, jedoch wird dies von Slater (*140a*) bezweifelt. Tutin $C_{15}H_{18}O_6$ und Coriamyrtin $C_{15}H_{18}O_5$ sind chemisch wie physiologisch dem Picrotoxinin $C_{15}H_{16}O_5$ außerordentlich ähnlich. Auf Grund eines Abbaues zum Coriaria-dilacton (*Formelübersicht 5*), dessen Struktur durch Synthese bewiesen wurde, schlagen Kariyone und Okuda (*67, 109*) in Anlehnung an die von Conroy für Picrotoxinin vorgeschlagene Konstitution (S. 161) für Coriamyrtin die Struktur (LXXX) vor. Der letzte Reaktionsschritt, die Oxydation der Coriarinsäure mit Permanganat in saurer Lösung, ist allgemeiner Natur. Wie durch Versuche an Modellsubstanzen bewiesen werden konnte (*67*), gehen tertiäre Carbonsäuren mit α-ständigem Benzolkern dabei unter $CO_2$-Verlust in den korrespondierenden tert. Alkohol über.

(LXXX.) Coriamyrtin. (LXXXIII.) Coriaria-dilacton.

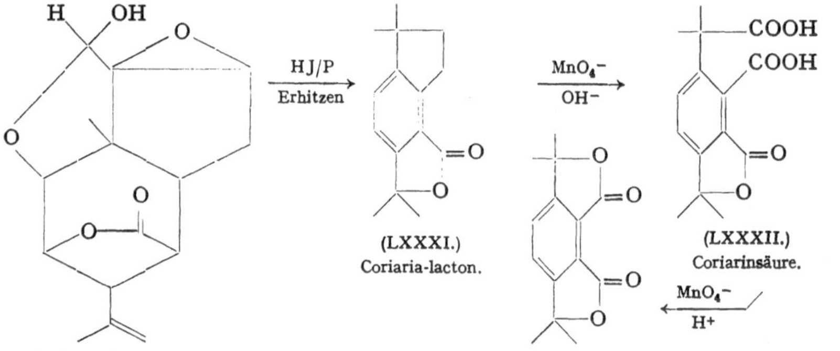

*Formelübersicht 5.* Abbau des Coriamyrtins.

## VII. Bitterstoffe der Urticaceen.

### Hopfenbitterstoffe: Humulon, Lupulon, Humulinon.

Im Hopfen (*Humulus lupulus*) sind zahlreiche Bitterstoffe vorhanden, die sich auf die drei Grundkörper Humulon (LXXXIV), Lupulon (LXXXV) und Humulinon (LXXXVI) zurückführen lassen. Von jedem dieser Grundkörper leiten sich Derivate ab, die jeweils durch die Vorsilben

(LXXXIV.) Humulon.

(LXXXV.) Lupulon.

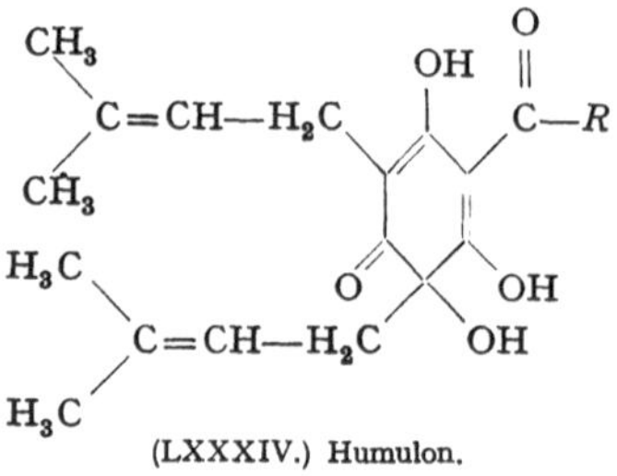

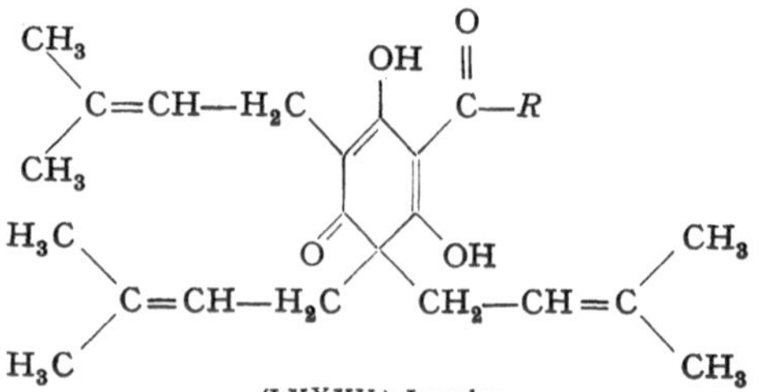

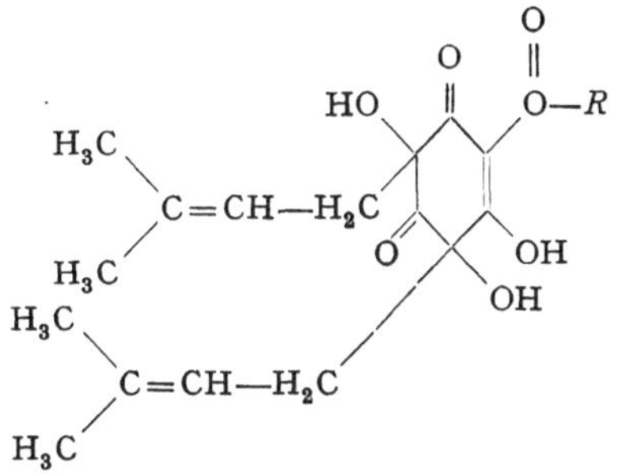

$$H_3C$$
$$C=CH-H_2C$$
$$H_3C$$

(Struktur LXXXVI)

(LXXXVI.) Humulinon.

Co- oder Ad- bezeichnet werden und die sich durch den Acylrest unterscheiden:

| Humulon | Lupulon | Humulinon | $R = -CH_2-CH\begin{smallmatrix}CH_3\\CH_3\end{smallmatrix}$ |
| Cohumulon | Colupulon | Cohumulinon | $R = -CH\begin{smallmatrix}CH_3\\CH_3\end{smallmatrix}$ |
| Adhumulon | Adlupulon | Adhumulinon | $R = -\underset{\underset{CH_3}{\mid}}{CH}-CH_2-CH_3$ |

Die Hopfenbitterstoffe sind u. a. deshalb von Interesse, weil ihnen beträchtliche antibiotische Wirksamkeit zukommt (*30, 127*).

Humulon und Lupulon sind die Hauptbestandteile der Bitterstoff-Fraktion und zu je 4—10% im Hopfen enthalten. Ihre Konstitutions-

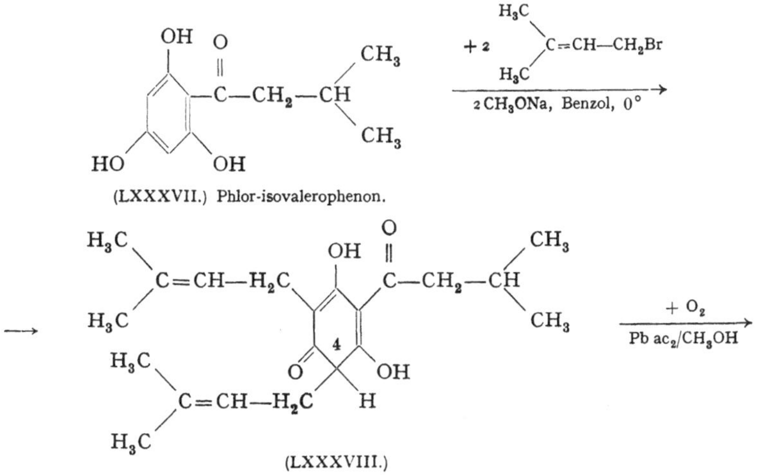

(LXXXVII.) Phlor-isovalerophenon.

(LXXXVIII.)

→ (LXXXIV.) *D,L*-Humulon (als Pb-Salz).

*Formelübersicht 6.* Synthese des Humulons.

11*

aufklärung gelang 1923—1926 Wieland und Mitarbeitern (*160*). Jedoch erfuhren die von ihnen aufgestellten Strukturformeln 1949 durch Govaert und Verzele (*53*) eine Korrektur im Sinne der oben angeführten Formelbilder, da die von Wieland angegebene Lage der Doppelbindung in den Isoamylen-resten zu den Ergebnissen der später durchgeführten Ozonolyse in Widerspruch stand. Den endgültigen Beweis für die Richtigkeit der Formeln führte Riedl (*120*) mit der Synthese des *D,L*-Humulons und des Lupulons (*Formelübersicht 6*, S. 163).

Die Einführung der Isoprenreste erfolgte nach dem von Claisen (*31*) entwickelten Verfahren, bei dem ungesättigte Alkylhalogenide mit den Alkalisalzen von Phenolen im inerten Lösungsmittel umgesetzt werden. Je lockerer das Halogen gebunden ist, desto besser verläuft die Reaktion; bei dem hier verwendeten „Prenylbromid" liegen die Verhältnisse besonders günstig. Komplikationen können bei derartigen Synthesen durch Chroman-Ringschluß, besonders im sauren Gebiet, erfolgen:

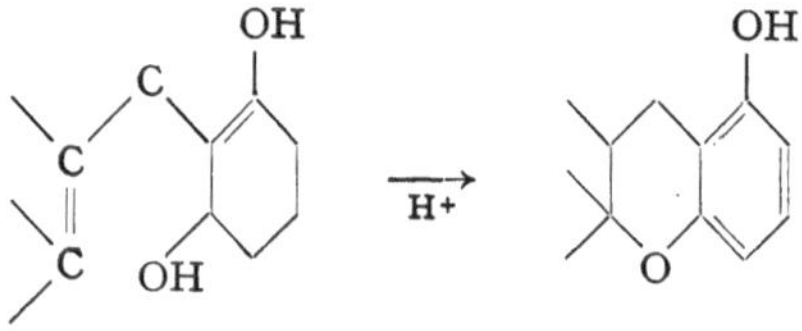

Die Oxydation in der aktivierten 4-Stellung wird nach Wöllmer (*162*) mit Luftsauerstoff und $Pb^{++}$ als Katalysator durchgeführt.

Bei der Synthese des Humulochinons und Humulohydrochinons, zweier bereits von Wieland und Mitarbeitern bei der Konstitutionsaufklärung des Humulons erhaltener Substanzen, führten Riedl und Leucht (*120 a*) die Sauerstoff-Funktion an $C_{(4)}$ durch Umsatz des entsprechenden Phloroglucinderivates mit der äquimolaren Menge Diazoaminobenzol ein.

Bei der katalytischen Hydrierung mit Pd in Methanol erleiden Humulon und Lupulon Hydrogenolyse und spalten Isopentan ab (*121*). Nach Wieland (*160*) läßt sich dieser Befund als das Bestreben eines Chinols, in den aromatischen Zustand überzugehen, deuten. Nach Riedl und Nickl (*121*) wird die Hydrogenolyse bei Humulon und Lupulon außerdem durch eine Prädissoziation der C—C-Bindung zwischen dem Ringkohlenstoffatom und dem C-Atom der $\beta,\gamma$-ungesättigten Seitenkette gefördert. Bei Verwendung von $PtO_2$ als Hydrierungskatalysator tritt keine Hydrogenolyse ein.

Humulinon (LXXXVI) wurde erst 1950 von Cook und Harris (*39*) isoliert. Der Konstitutionsbeweis auf chemischem Wege gelang Howard und Slater (*60*). Nach diesen Autoren entsteht durch katalytische Hydrierung und Clemmensen-Reduktion des Cohumulinons dasselbe Phenol, das auch durch Clemmensen-Reduktion des Tetrahydro-

cohumulons gebildet wird. Mit diesem Ergebnis ist die von ALDER-
WEIRELDT und VERZELE [vgl. (60)] für Humulinon vorgeschlagene Fünf-
ring-Formel widerlegt.

Die obigen Hopfenbitterstoffe haben sich im Bier bislang nicht nachweisen lassen. Die Träger der Bittere des Bieres, die sogen. Weichharze
oder „Isokörper", enthalten einen Fünfring. Nach CARSON (27) besitzt
das Isohumulon die Struktur (LXXXIX). Es konnte bisher nur als Öl
erhalten werden. Die Substanz schmeckt intensiv bitter und ist anti-
biotisch unwirksam.

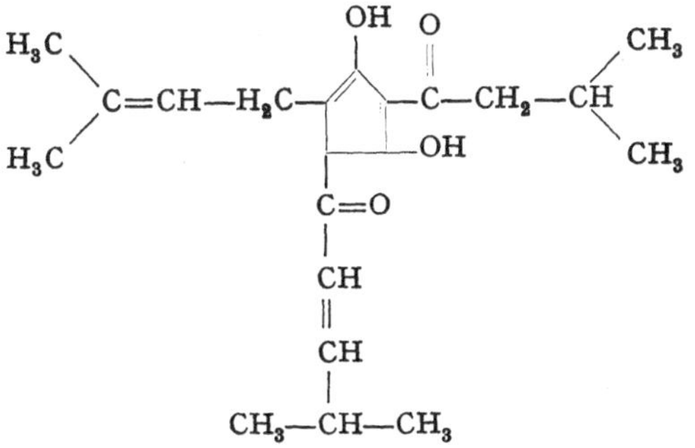

(LXXXIX.) Isohumulon.

## VIII. Bitterstoffe der Labiaten.

### Marrubiin.

Marrubiin ist der Bitterstoff aus *Marrubium vulgare*. Das Kraut
der Pflanze *(Herba marrubii)* ist ein seit altersher geschätztes Heilmittel,
das äußerlich zur Förderung der Wundheilung, innerlich bei Magen-
und Darmkatarrhen, bei Leber- und Gallenleiden
sowie als Emmenagogum benutzt wird. Als wirksames Prinzip ist der Bitterstoff zu betrachten.
Auf die Drüsen der Atemwege wirkt Marrubiin
sekretionsfördernd ohne ein Brechmittel zu sein, es
wird daher als Expectorans verwendet. Daneben
steigert Marrubiin die Leberfunktion, größere Dosen
erzeugen Herzrhythmus-störungen. Nach COCKER
u. a. (35) besitzt das Marrubiin, $C_{20}H_{28}O_4$, die Struktur (XC).

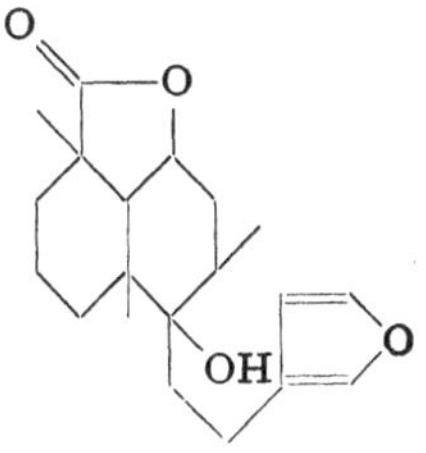

(XC.) Marrubiin.

Die Ergebnisse früherer Autoren (vgl. *34)* lassen sich wie folgt zusammen-
fassen: Marrubiin ist ein zweifach ungesättigtes Lacton, das mit Alkali
in die korrespondierende Carbonsäure überführbar ist (Marrubinsäure).
Die Hydrierung führt zu Tetrahydroderivaten. Das Molekül enthält eine
nicht acylierbare Hydroxylgruppe, die durch dehydratisierende Reagenzien
leicht als $H_2O$ eliminierbar ist. Das vierte Sauerstoffatom ist inert,

liegt also wahrscheinlich als Äther-O vor. Bei der Dehydrierung entsteht aus Marrubiin das 1,2,5-Trimethylnaphthalin.

*Konstitutionsbeweis.* Im UV zeigt Marrubiin drei Absorptionsmaxima (208, 212 und 216 m$\mu$). In Übereinstimmung mit IR-Befunden deutet dies auf ein substituiertes Furansystem hin. Im IR-Spektrum erkennt man ferner den $\gamma$-Lactonring (1765 cm$^{-1}$) und die OH-Gruppe (3470 cm$^{-1}$). Einen Überblick über die Strukturaufklärung des Marrubiins vermittelt die *Formelübersicht 7.*

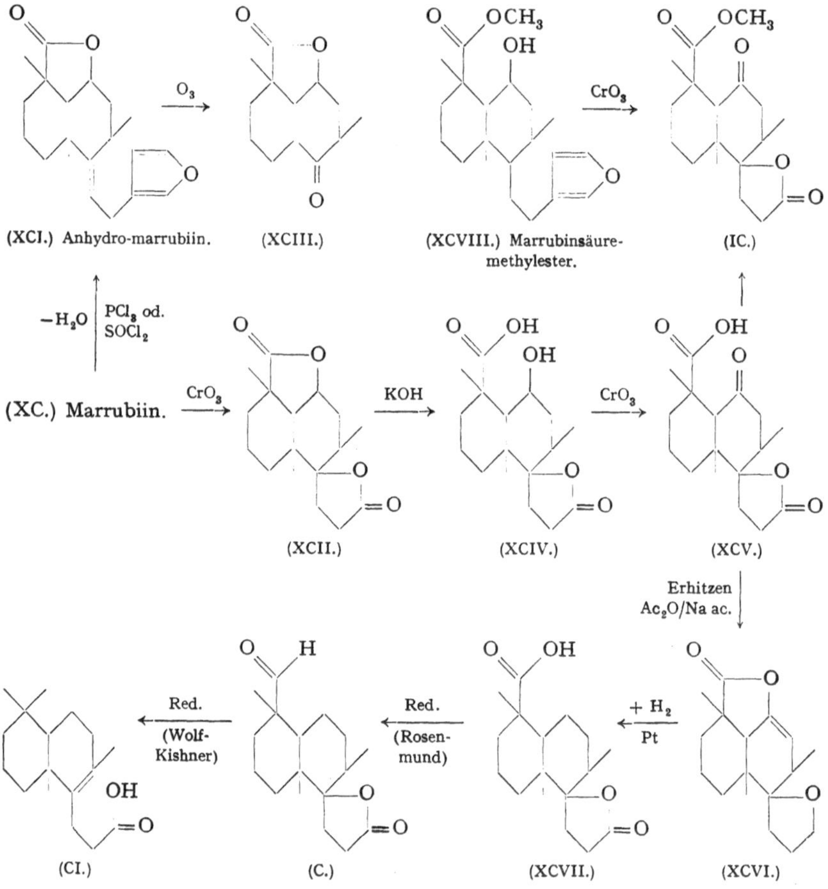

O$_3$ →

CrO$_3$ →

(XCI.) Anhydro-marrubiin.    (XCIII.)    (XCVIII.) Marrubinsäure-methylester.    (IC.)

−H$_2$O | PCl$_3$ od. SOCl$_2$

(XC.) Marrubiin.    CrO$_3$ →    KOH →    CrO$_3$ →

(XCII.)    (XCIV.)    (XCV.)

Erhitzen Ac$_2$O/Na ac.

Red. (Wolf-Kishner) ←    Red. (Rosenmund) ←    + H$_2$ Pt ←

(CI.)    (C.)    (XCVII.)    (XCVI.)

*Formelübersicht 7.* Strukturaufklärung des Marrubiins.

Das aus Marrubiin (XC) durch Chromsäure-Oxydation gebildete Dilacton (XCII) entsteht über die aus dem Furanring gebildete Carbonsäure, die sofort mit dem tert. OH zum Lacton cyclisiert. Dadurch ist die $\gamma$-Stellung der OH-Gruppe zum Furan-kern bewiesen.

*Literaturverzeichnis: SS. 175—182.*

Unterwirft man Anhydro-marrubiin (XCI) der Ozonolyse, so erweist sich die Carbonylgruppe in (XCIII) als sterisch stark behindert. Sie kann nicht nach CLEMMENSEN reduziert werden und bildet kein Semicarbazon, nur ein Oxim läßt sich darstellen. Da die aus (XCIII) durch Verseifung entstehende Keto-hydroxy-säure nicht decarboxylierbar ist, kann es sich nicht um eine $\beta$-Ketosäure handeln; damit kommt von den drei Substitutionsmöglichkeiten für die Carboxylgruppe ($C_{(1)}$, $C_{(5)}$ oder $C_{(7)}$) nur die an $C_{(1)}$ in Frage. Daß das Carboxyl der Marrubinsäure (vgl. XCVIII) an einem tert. C-Atom haftet, ergibt sich daraus, daß mit konz. $H_2SO_4$ Kohlenmonoxyd abgespalten wird und die Säure unter normalen Bedingungen nicht zu verestern ist.

Im Dilacton (XCII) wird der ursprünglich vorhandene Lactonring durch Hydrolyse leichter geöffnet als der über Chromsäure gebildete. Dies ergibt sich daraus, daß (XCIV) mit $CrO_3$ oxydierbar ist. Die Lacton-ketocarbonsäure (XCV) liefert nach Umsatz mit $CH_3J$ denselben Methylester (IC), der auch durch Chrom-säure-Oxydation des Marrubinsäure-methylesters (XCVIII) erhalten wird.

Aus (XCV) kann das Enollacton (XCVI) durch Kochen mit Acetanhydrid/Na-acetat dargestellt werden, das durch katalytisch erregten Wasserstoff unter Hydrogenolyse in (XCVII) übergeführt wird. [Durch die Bildung von (XCVI)

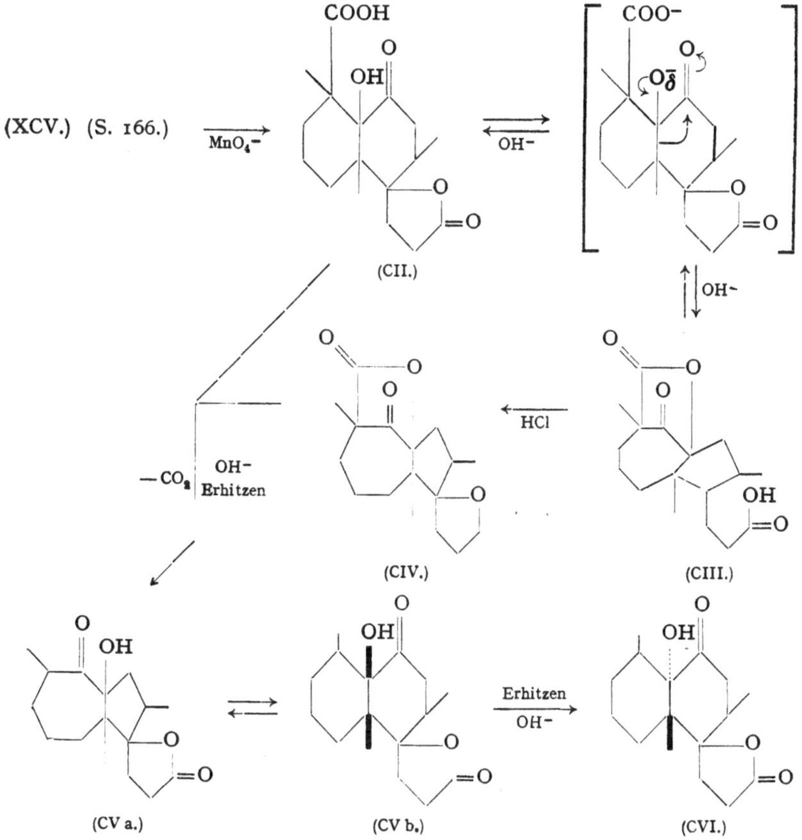

*Formelübersicht 8. Ring-Ring Tautomerie.*

ist aus sterischen Gründen die Möglichkeit ausgeschlossen, daß der Lacton-Sauerstoff an $C_{(3)}$ sitzt, da in diesem Fall die Bildung eines Enollactons sterisch unmöglich ist.] Die Konstitution des Marrubiins entsprechend der Formel (XC) ist
damit bis auf die Substitutionsstelle im Furanring gesichert. Eine weitere Stütze
für die Formel ergibt sich daraus, daß es Burn und Rigby (25) gelang, das
Hydrogenolyse-Produkt (XCVII) über den durch Rosenmund-Reaktion erhaltenen
Aldehyd (C) in eine Substanz bekannter Struktur (CI) zu überführen. [Das letztere
entsteht aus (C) als Hauptprodukt der nach Huang-Minlon modifizierten Wolff-
Kishner-Reduktion.] (CI) wurde von Ruzicka und Lardon (129a) als Abbauprodukt des Ambreins erhalten.

Im Zusammenhang mit den Arbeiten zur Konstitutionsaufklärung des Marrubiins
sei noch über einen interessanten Fall von Ring-Ring-Tautomerie berichtet (Formelübersicht 8). Wenn man (XCV) nach Ghigi (50) mit Permanganat oxydiert, so
entsteht (vermutlich über die Enolform) die um ein Sauerstoffatom reichere Verbindung (CII), die sich im Gleichgewicht mit dem Ring-Ring-Tautomeren (CIII)
befindet. Sowohl die Bildung des Dilactons (CIV) als auch die durch Erhitzen
mit wäßrigem Alkali erfolgende Decarboxylierung sind nur über die Formel (CIII)
verständlich.

Burn, Moody und Rigby (24) konnten nach der durch Erhitzen mit wäßrigem
Alkali eingetretenen Decarboxylierung des Dilactons (CIV), neben dem schon von
Ghigi (50) erhaltenen (CVI), die Verbindung (CV) isolieren, die bei längerem Erhitzen im alkalischen Medium in (CVI) übergeht. Für (CV) werden die beiden
tautomeren Formeln (CVa) und (CVb) angegeben.

Über einen Fall ähnlicher Tautomerie berichteten kürzlich Shoppee u. a. (138).

# IX. Bitterstoffe der Apocynaceen.

## Plumierid.

Bereits im Jahre 1870 gelang Peckolt die Isolierung eines Bitterstoffes aus der Rinde von *Plumiera lancifolia*, den er „Agoniadin" nannte.
1899 gewann Franchimont (49) aus der Rinde von *P. acutifolia* ebenfalls
einen Bitterstoff, für den er den Namen „Plumierid" vorschlug und
dessen Identität mit dem Agoniadin er feststellte. Auf seinen Vorschlag
hin wurde der Name Agoniadin gestrichen. Auf Franchimont gehen die
ersten Untersuchungen des Plumierids zurück. Es handelt sich um ein
$D$-Glucosid der Formel $C_{21}H_{26}O_{12}$, das eine esterartig gebundene $CH_3O$-
Gruppe trägt. Die endgültige Strukturaufklärung
des Glucosids (CVII) gelang Schmid und Mitarbeitern
(134, 135) (Formelübersicht 9, S. 171).

*Allgemeine Reaktionen.* Die *Hydrolyse* des Plumierids
mit verd. Mineralsäure oder mit Schneckenferment führt zur
Abspaltung von $D$-Glucose. Der Aglykonteil wird unter
Bildung humin-artiger Substanzen zerstört.

*Acetylierung.* Mit Acetanhydrid in Pyridin entsteht ein
Pentaacetat. Da unter milden Bedingungen gearbeitet wird,
muß das Aglykon eine prim. oder sek. OH-Gruppe enthalten.

*Hydrierung.* Bei der energischen Hydrierung des
Pentaacetats werden 4 $H_2$ aufgenommen, dabei tritt
Hydrogenolyse des im Aglykon befindlichen O-Atoms

(CVII.) Plumierid.

ein. Plumierid enthält demnach drei Doppelbindungen, die als $\alpha$, $\beta$ und $\gamma$ bezeichnet werden. Die $\alpha$-Doppelbindung steht isoliert und läßt sich bereits unter milden Bedingungen (Pd/CH$_3$OH) hydrieren. Die isolierte Stellung ergibt sich daraus, daß die UV-Absorption des Dihydroderivates (CVIII) der des Plumierids entspricht. Die Hydrierung der $\beta$-Doppelbindung gelingt mit PtO$_2$ in Eisessig. Unter diesen Bedingungen tritt Hydrogenolyse des Aglykonsauerstoffs ein, und man erhält das Tetrahydro-desoxyplumierid (CIX), $\lambda_{max} = 236$ m$\mu$. Dieses Maximum entspricht dem von 5,6-Dihydropyran-3-carbonsäuren und ist durch die „$\gamma$''-Doppelbindung bedingt. Die $\gamma$-Doppelbindung ist schwer hydrierbar. In Methanol oder Eisessig gelöstes Brom wird unter Bildung der sehr unbeständigen Brommethoxy- oder Bromacetoxy-körper an diese Doppelbindung angelagert. Das aus (CIX) gebildete Additionsprodukt zeigt im UV keine Absorption, das aus (CVIII) entstandene absorbiert bei 210—218 m$\mu$, dies entspricht der Absorption eines $\alpha$,$\beta$-ungesättigten Lactons.

COOCH$_3$     $+ Br_2$     Br COOCH$_3$

in CH$_3$OH oder AcOH     —OR

O     O

R = CH$_3$ oder Acetyl.

*Hydrolyse der hydrierten Verbindungen.* Im Gegensatz zur Hydrolyse der ungesättigten Verbindungen (CVII)—(CIX) läßt sich nach der Hydrolyse des Hexahydro-desoxyplumierids (CX) ein kristallines Aglykon gewinnen. Die Abbau-Reaktionen wurden deshalb an der perhydrierten Verbindung durchgeführt.

Die *Hydrogenolyse* des Aglykon-Sauerstoffs erfolgt bei der Hydrierung von (CVIII). Demnach ist das O-Atom als zur $\beta$-Doppelbindung allylständig anzusehen.

*Lactontitration, Sauerstoff-Funktionen.* Bei der Lactontitration werden 2 Äquivalente Alkali verbraucht. Barytspaltung führt zu Methanol-Abspaltung (aus der Carbonestergruppe). Damit ist die Funktion von sechs der sieben im Aglykon vorhandenen O-Atome geklärt: vier sind in Lacton- und Estergruppe enthalten, eines in der Glucosidbindung und eines in Allylstellung zur $\beta$-Doppelbindung.

Bei der *Ozonspaltung* des Pentaacetyl-plumierids entsteht L-(+)-Milchsäure.

*Kuhn-Roth-Oxydation.* Aus (CVII) und (CVIII) wird Essigsäure gebildet; aus den höher hydrierten Verbindungen entsteht daneben noch Propionsäure. Dies beweist, daß durch die Hydrogenolyse eine Äthylgruppe entstanden ist.

*Abbau-Reaktionen* (Formelübersicht 9, S. 171). Die Spaltung des Hexahydro-desoxyplumierids (CX) gelingt in Ausbeuten bis 80% durch Kochen mit 0,1-$N$ abs. methanolischer Salzsäure. Es treten drei Spaltprodukte (CXI a, b, c) auf, die durch Chromatographie trennbar sind. Die kristallinen Substanzen a und b enthalten je zwei, das Öl c drei CH$_3$O-Gruppen. In a und b sind je eine, in c zwei Carbomethoxygruppen vorhanden. In allen drei Fällen handelt es sich um Methyläther von Halbacetalen. Durch Behandlung mit methanolischer Salzsäure gelingt die Überführung der Substanzen a und b ineinander, daneben entsteht immer das ölige c. Substanzen a und b unterscheiden sich nur durch die räumliche Stellung

O     O

C     ROH/HCl     C

RO   H     H   OR

(CXI a.)     (CXI b.)

des $CH_3O$-Restes am Acetalkohlenstoff. Die potentielle Aldehydgruppe läßt sich durch Reaktion mit 2,4-Dinitrophenylhydrazin in saurer Lösung nachweisen. Substanzen a und b liefern dabei dasselbe Phenylhydrazon. In Übereinstimmung mit den spektroskopischen Daten wird daraus geschlossen, daß es sich bei a und b um die Aglykon-methyläther des Hexahydro-desoxyplumierids (CX) handelt.

Substanz c kann außer aus a und b auch aus (CX) direkt erhalten werden, wenn bei der Spaltung die HCl-Konzentration erhöht wird. Substanz c enthält eine C=C-Doppelbindung und zwei Carbomethoxygruppen. Es handelt sich um ein nicht trennbares Gleichgewichtsgemisch der am Acetalkohlenstoff epimeren Methyläther der Verbindung (CXIc), die durch Isomerisierung des $\gamma$-Lactons zur Carbonsäure (unter Ausbildung der Doppelbindung) und Veresterung entstanden ist:

(CXI a, b.)                (CXI c.)

Nach Verseifung entsteht aus c dasselbe 2,4-Dinitrophenylhydrazon wie aus a und b. Bei der Reaktion mit 2,4-Dinitrophenylhydrazin wird also (unter saurer Katalyse) der Lactonring zur Carbonsäure isomerisiert. Dies Verhalten ist erklärbar, wenn man annimmt, daß der Lacton-Sauerstoff an einem zum Acetalkohlenstoff $\beta$-ständigen C-Atom steht. (Analog verläuft der protonen-katalysierte Übergang von $\beta$-Hydroxy-carbonylverbindungen in die ungesättigten Körper.)

Mit dem Nachweis der Acetalgruppierung ist auch die Funktion des siebenten, im Aglykon enthaltenen O-Atoms geklärt.

Das Aglykon des Hexahydro-desoxyplumierids (CXI d) läßt sich durch kurzes Erhitzen der Aglykon-methyläther a oder b oder von (CX) mit verdünnter, wäßriger Mineralsäure erhalten. Es entsteht ein Gleichgewichtsgemisch der am Acetalkohlenstoff epimeren Cyclohalbacetale.

Substanz d läßt sich mit methanolischer Salzsäure in die Methyläther a und b zurückverwandeln. Außerdem gelang durch Umsatz von d mit $\alpha$-$D$-Acetobromglucose die Synthese des Tetraacetyl-hexahydro-desoxyplumierids. Dadurch ist bewiesen, daß (mit Ausnahme des Acetalkohlenstoffs) durch die säure-katalysierte Solvolyse keine Umlagerungen im Molekül stattfinden.

Oxydation von d mit $CrO_3$ in Pyridin überführt das Halbacetal in den entsprechenden Dilactonester (CXII). Im Gegensatz zum $\gamma$-Lacton läßt sich der $\delta$-Lactonring schon kurz oberhalb pH 7 öffnen. Beweisend für die Konstitution von (CXII) ist die durch 2 Äquivalent $CH_3ONa$ erfolgende Isomerisierung zur doppelt ungesättigten Tricarbonsäure (CXVII), die in ihren optischen und chemischen Eigenschaften einer Substanz dieser Struktur entspricht (katalytische Hydrierung, Titration). Bei Ozonolyse entsteht neben Formaldehyd das Spiro-dilacton (CXIII) (identisch mit synthetischem Racemat).

Oxydation von d mit $CrO_3$ in $H_2SO_4$ liefert Propionsäure, Bernsteinsäure, (+)-Äthylbernsteinsäure und (CXIII). Auch läßt sich ein weiteres Oxydationsprodukt, das Anhydridlacton (CXIV), fassen. Seine Konstitution ergibt sich aus folgenden Abbau-Reaktionen: Nach Überführung in den Lactondimethylester (CXV) gelingt die Isomerisierung zu einer ungesättigten Diestercarbonsäure (CXVI) durch Kochen mit 1 Äquivalent $CH_3ONa$ in $CH_3OH$. Die Lage der Doppelbindung folgt aus dem Ergebnis der Chromsäure-Oxydation, die zur Bildung von Bernsteinsäure,

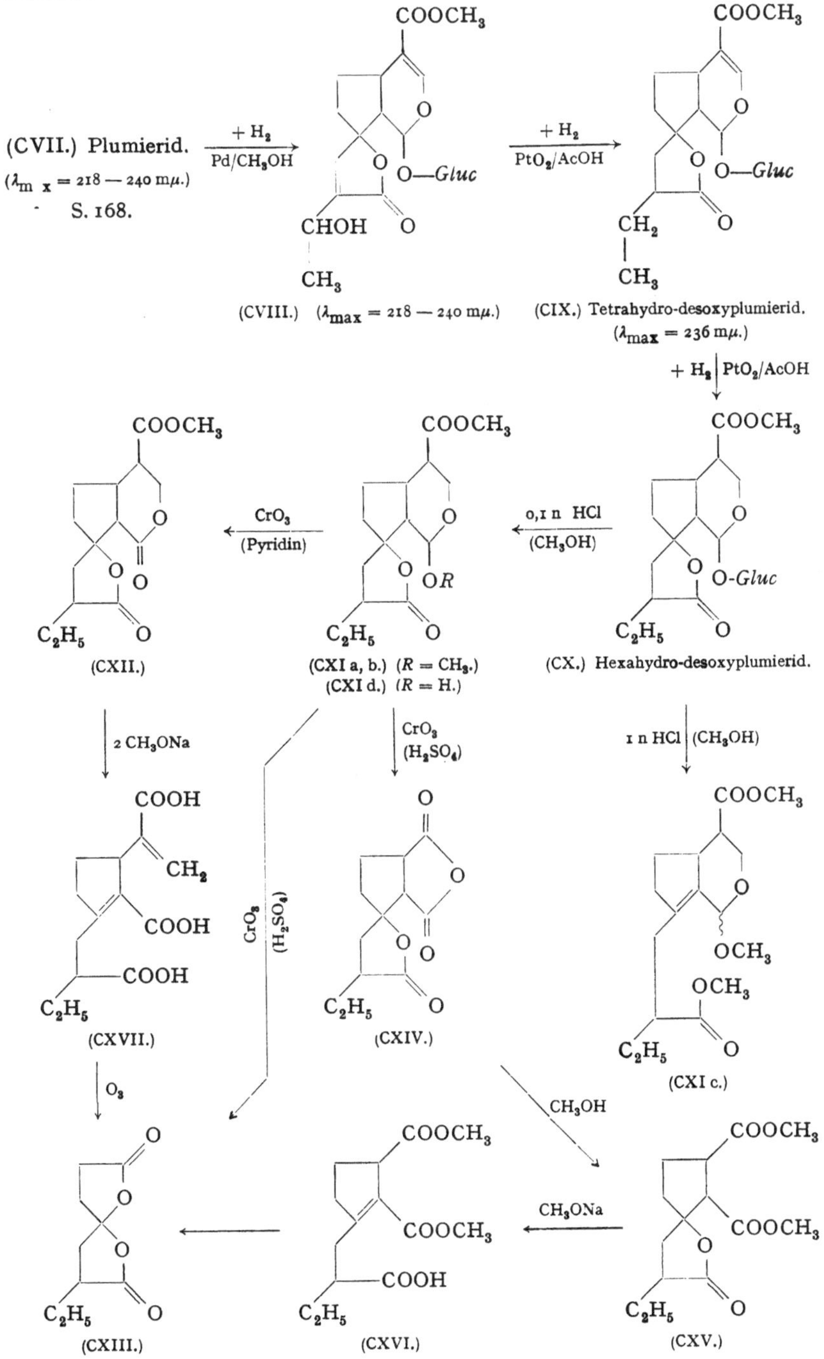

Formelübersicht 9. Abbau und Strukturaufklärung des Plumierids.

(+)-Äthylbernsteinsäure und des Spiro-dilactons (CXIII) führt. Die C=C-Doppelbindung befindet sich also in Konjugation zu einer der beiden Carbomethoxygruppen; dem entspricht auch die UV-Absorption von (CXVI) mit $\lambda_{max} = 230$ m$\mu$. Die Doppelbindung ist unter dem Einfluß des basischen Katalysators durch $\beta$-Eliminierung der Acyloxygruppe des $\gamma$-Lactonrings entstanden.

Nach der Festlegung des C-Grundgerüstes bleibt für die $\alpha$-Doppelbindung im Plumierid nur die Anordnung zwischen $C_{(6)}$ und $C_{(7)}$ übrig. Als Ergebnis weiterer Untersuchungen wird für Plumierid die Stereoformel (CVII) wahrscheinlich gemacht.

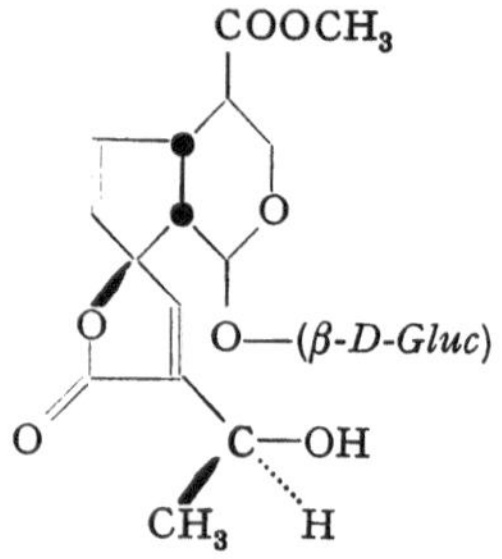

(CVII.) Plumierid (Raumformel).

# X. Tabellen.

## Pflanzliche Bitterstoffe.

| Name | Pflanzenfamilie | Literatur |
|---|---|---|
| Absinthin | Compositen | S. 138 |
| Acorin | Araceen | (113) |
| Adansonin | Berberidaceen | (113) |
| Aesculin | Hippocastanaceen | (75) |
| Ailantin | Simarubaceen | (75) |
| Alantopikrin | Compositen | (51) |
| Alantolactone | Compositen | S. 149 |
| Aloin | Liliaceen | (113) |
| Alstonialactone | Apocynaceen | (75) |
| Ammoresinol | Umbelliferen | (75) |
| Anabsinthin | Compositen | S. 138 |
| Anamyrtin | Menispermaceen | (75) |
| Andrographolid | Acanthaceen | (74) |
| Anemonin | Ranunculaceen | (75) |
| Angelicin | Umbelliferen | (75) |
| Angosturin | Rutaceen | (75) |
| Arnicin | Compositen | (113) |
| Artemisin | Compositen | (75) |
| Athamantin | Umbelliferen | (75) |
| Aurantiamarin | Rutaceen | (113) |
| Aurapten | Rutaceen | (75) |
| Ayapin | Compositen | (75) |
| Barbaloin | Liliaceen | (75) |
| Benedictin | Compositen | S. 139 |
| Bergamottin | Rutaceen | (75) |
| Bergapten | Rutaceen | (75) |
| Bergenin | Saxifragaceen | (75) |
| Bonducin | Leguminosen | (75) |
| Byak-Angelicin | Umbelliferen | (75) |
| Cailcedrin | Meliaceen | (75) |
| Cascarillin | Euphorbiaceen | (113) |
| Castelamarin | Simarubaceen | (75) |

| Name | Pflanzenfamilie | Literatur |
|---|---|---|
| Cedrelin | Meliaceen | (75) |
| Ceroxylin | Palmaceen | (113) |
| Cetrarin | (Kryptogamen) | (75) |
| Chasmanthin | Menispermaceen | S. 151 |
| Chellol-Glucosid | Umbelliferen | (75) |
| Chinovin | Rubiaceen | (75) |
| Cichoriin | Compositen | (75) |
| Citrolimonin | Rutaceen | (75) |
| Clerodin | Verbenaceen | (75) |
| Cnicin | Compositen | S. 139 |
| Columbin | Menispermaceen | S. 151 |
| Condurangin | Asclepiadaceen | S. 135 |
| Coniferin | (Gymnospermen) | (75) |
| Corchoritin | Berberidaceen | (75) |
| Corchsularin | Berberidaceen | (71) |
| Coriamyrtin | Coriariaceen | S. 161 |
| Cucumin | Cucurbitaceen | (75) |
| Cucurbitacine | Cucurbitaceen | (57, 46, 98) |
| Cuscutalin | Cuscutaceen | (75) |
| Darutin | Compositen | (118) |
| Elaterin | Cucurbitaceen | (57, 46, 98) |
| Erythramarin, Erytaurin | Gentianaceen | S. 129 |
| Esenbeckin (Chinovin) | Rutaceen | (75) |
| Fraxidin (+ Isofraxidin) | Oleaceen | (75) |
| Fraxin | Oleaceen | (75) |
| Fraxinol | Oleaceen | (75) |
| Geigerin | Compositen | (75) |
| Gentiin | Gentianaceen | S. 129 |
| Gentiopikrin | Gentianaceen | S. 129 |
| Helenalin | Compositen | S. 147 |
| Humulon | Urticaceen | S. 162 |
| Hyananchin | Euphorbiaceen | (75) |

| Name | Pflanzenfamilie | Literatur |
| --- | --- | --- |
| Ibamarin | Cruciferen | (136) |
| Imperatorin | Umbelliferen | (75) |
| Intybin | Compositen | (75) |
| Juniperin | (Gymnospermen) | (113) |
| Kawain | Piperaceen | (75) |
| Kirondrin | Simarubaceen | (75) |
| Kondurangin | Asclepiadaceen | S. 135 |
| Lactucin | Compositen | S. 140 |
| Lactucopikrin | Compositen | S. 140 |
| Lichesterinsäure | Kryptogamen | (75) |
| Limettin | Rutaceen | (75) |
| Limonin | Rutaceen | (78) |
| Loganin | Loganiaceen | S. 133 |
| Loroglossin | Palmaceen | (75) |
| Lupulon | Urticaceen | S. 162 |
| Luvangetin | Umbelliferen | (75) |
| Mansonin | Berberidaceen | (75) |
| Marrubiin | Labiaten | S. 165 |
| Meliatin | Gentianaceen | S. 129 |
| Mesuol | Hypericaceen | (42) |
| Murrayosid | Rutaceen | (75) |
| Nodakenin | Umbelliferen | (75) |
| Nomilin | Rutaceen | (75) |
| Ombelliprenin | Umbelliferen | (75) |
| Oreosolon | Umbelliferen | (75) |
| Osthenol | Umbelliferen | (75) |
| Osthol | Umbelliferen | (75) |
| Palmarin | Menispermaceen | S. 151 |
| Peucedanin | Umbelliferen | (75) |
| Peucenin | Umbelliferen | (75) |
| Phyllantin | Euphorbiaceen | (113) |
| Picein | Salicaceen | (113) |
| Picrocrocin | Iridaceen | (113) |
| Picroretin | Menispermaceen | (75) |
| Picrotoxin | Menispermaceen | S. 155 |

| Name | Pflanzenfamilie | Literatur |
| --- | --- | --- |
| Pimpinellin | Umbelliferen | (75) |
| Pinipikrin | (Gymnospermen) | (113) |
| Plumierid | Apocynaceen | S. 168 |
| Podophyllotoxin | Berberidaceen | (55a) |
| Picropodophyllin | Berberidaceen |  |
| Protoanemonin | Ranunculaceen | (75) |
| Psoralen | Leguminosen | (75) |
| Pyrethrosin | Compositen | (75) |
| Quassin | Simarubaceen | (122) |
| Salicoside | Saliaceen | (113) |
| Samaderin | Simarubaceen | (75) |
| Sandaricin | (Gymnospermen) | (113) |
| Santonin | Compositen | (12, 33, 35a) |
| Seselin | Rutaceen | (75) |
| Simarubin, Simarubein | Simarubaceen | (75) |
| Skimmin | Rutaceen | (75) |
| Swertiamarin | Gentianaceen | S. 129 |
| Syringin | Oleaceen | (75) |
| Tanacetin | Compositen | (113) |
| Taraxacin | Compositen | (113) |
| Temisin | Compositen | (75) |
| Tenulin | Compositen | S. 144 |
| Tiliadin, Tilicin | Berberidaceen | (113) |
| Toddalolacton | Rutaceen | (75) |
| Tutin = Oxycoriamyrtin | Coriariaceen | S. 161 |
| Umbelliferon | Umbelliferen | (75) |
| Valdivin | Simarubaceen | (75) |
| Veratramarin | (Monokotyledonen) | (113) |
| Verbenalin | Verbenaceen | (75) |
| Vernonin | Compositen | (113) |
| Vincetoxin | Asclepiadaceen | S. 137 |
| Visnagin | Umbelliferen | (75) |
| Xanthotoxin | Rutaceen | (75) |
| Xanthyletin | Rutaceen | (75) |

*Noch unbenannte Bitterstoffe.*

| Bitterstoff aus | Literatur |
|---|---|
| *Avena sativa* (Hafer) .................. | (*129*) |
| *Brucea sumatrana* ..................... | (*153*) |
| *Citrus natsudaidai* ................... | (*62*) |
| *Daucus carota* (Mohrrübe) ............ | (*40, 52*) |
| *Glycine histida* (Sojabohne) .......... | (*147*) |
| *Maesa emirnensis* .................... | (*114*) |
| Saké (Reiswein) ...................... | (*10*) |
| *Tinospora cordifolia* ................. | (*18*) |
| *Trichilia heudelotii* ................... | (*116*) |

### Literaturverzeichnis.

*1.* ADAMS, R. and W. HERZ: Helenalin. I. Isolation and Properties. II. Helenalin Oxide. III. Reduction and Dehydrogenation. J. Amer. Chem. Soc. **71**, 2546, 2551, 2554 (1949).

*2.* ADRIAN, M. M. et A. TRILLAT: Sur l'anabsinthine. Bull. soc. chim. France [3] **21**, 234 (1899).

*3.* ALADASHVILI, V. A.: Effect of Some Bitter Substances on Secretion of Gastric Juice. Therapevt. Arkh. **24**, No. 5, 58 (1952) [Chem. Abstr. **47**, 1847 (1953)].

*4.* ALDER, K. und H. F. RICKERT: Zur Kenntnis der Dien-Synthese, II. Mitt.: Über den thermischen Zerfall der Additionsprodukte des Acetylen-dicarbonsäureesters. Ber. dtsch. chem. Ges. **70**, 1354 (1937).

*5.* ALLEN, F. and M. WEINBERG: The Gustatory Sensory Reflex. Quart. J. exp. Physiol. **15**, 385 (1925).

*6.* ALMEIDA, J. DE: The Problem of Toxic and Bitter Substances in the Breeding of Forage Plants. Herbage Abstr. **10**, 253 (1940) [Chem. Abstr. **36**, 2639 (1942)].

*7.* ANGELICO, F.: Ricerche sulla picrotossina. Gazz. chim. ital. **40**, I, 391 (1910).

*8.* ASAHINA, Y., J. ASANO, Y. TANASE und Y. UENO: Über das Gentiopikrin. 1. Mitt. Ber. dtsch. chem. Ges. **69**, 771 (1936).

*9.* ASAHINA, Y. und Y. SAKURAI: Über das Gentiopikrin. 2. Mitt. Ber. dtsch. chem. Ges. **72**, 1534 (1939).

*10.* ASO, K., T. NAKAYAMA and M. MAKI: Studies on the Bitter Components in Alcoholic Drinks. I, II. The Tyrosol Content in "Saké". J. Fermentation Technol. (Japan) **31**, 43 (1953); **32**, 52 (1954).

*11.* BARTH, L. und M. KRETSCHY: Untersuchungen über das Picrotoxin. Monatsh. Chem. **1**, 99 (1880).

*12.* BARTON, D. H. R.: The Stereochemistry of Santonin, $\beta$-Santonin, and Artemisin. J. Organ. Chem. (USA) **15**, 466 (1950).

*13.* BARTON, D. H. R. and D. ELAD: Colombo Root Bitter Principles. Part I. The Functional Groups of Columbin. Part II. The Constitution of Columbin. J. Chem. Soc. (London) **1956**, 2085, 2090.

*14.* BARTON, D. H. R. and P. DE MAYO: Sesquiterpenoids. Part VII. The Constitution of Tenulin, a Novel Sesquiterpenoid Lactone. J. Chem. Soc. (London) **1956**, 142.

*15.* BARTON, D. H. R. and C. R. NARAYANAN: Sesquiterpenoids. Part X. The Constitution of Lactucin. J. Chem. Soc. (London) **1958**, 963.

*16.* BAUER, K. H. und K. BRUNNER: Die Bitterstoffe des Milchsaftes von *Lactuca virosa*. III. Mitt. über Lactucarium. Ber. dtsch. chem. Ges. **70**, 261 (1937).

17. Benstead, J. C., H. V. Brewerton, J. R. Fletcher, M. Martin-Smith, S. N. Slater and A. T. Wilson: Picrotoxin and Tutin. Part IV. The Reducing Properties and Functional Groups. J. Chem. Soc. (London) 1952, 1042.
18. Bhide, B. V., N. L. Phanilkar and K. Paranjpe: Chemical Investigation of Tinospora cordifolia (Miers). J. Univ. Bombay 10, 89 (1941).
19. Birch, A. J. and E. Smith: Loganin. I. Some Observations on the Structure. Austral. J. Chem. 9, 234 (1956) [Chem. Zbl. 1957, 13046].
20. Bourquelot, H. et L. Hérissey: Sur la préparation de la gentiopicrine, glucoside de la racine fraîche de gentiane. C. R. hebd. Séances Acad. Sci. 131, 113 (1900).
21. Braun, B. H., W. Herz and K. Rabindran: The Structure of Tenulin. J. Amer. Chem. Soc. 78, 4423 (1956).
22. Bridel, M.: La méliatine, nouveau glucoside, hydrolysable par l'émulsine, retiré du Trèfle d'eau (Menyanthes trifoliata L.). C. R. hebd. Séances Acad. Sci. 152, 1694 (1911).
23. Büchi, G. and D. Rosenthal: Terpenes. VI. The Structures of Helenalin and Isohelenalin. J. Amer. Chem. Soc. 78, 3860 (1956).
24. Burn, D., D. P. Moody and W. Rigby: The Structure of Marrubiin. Chem. and Ind. 1956, 928.
25. Burn, D. and W. Rigby: The Structure of Marrubiin. Chem. and Ind. 1955, 386.
26. Canonica, L. e F. Pelizzoni: Ricerche sulla struttura del genziopicroside. Gazz. chim. ital. 87, 1251 (1957).
27. Carson, J. F.: The Alkaline Isomerization of Humulone. J. Amer. Chem. Soc. 74, 4615 (1952).
28. Cava, M. P. and E. J. Soboczenski: Bitter Principles of Plants. I. Columbin: Preliminary Structural Studies. J. Amer. Chem. Soc. 78, 5317 (1956).
29. Čekan, Z., V. Herout and F. Šorm: Structure of Matricin. Chem. and Ind. 1956, 1234.
30. Chin, Y.-C., N.-C. Chang and H. H. Anderson: Factors Influencing the Antibiotic Activity of Lupulon. J. Clin. Investigation 28, 909 (1949).
31. Claisen, L., F. Kremers, F. Roth und E. Tietze: Über C-Alkylierung (Kernalkylierung) von Phenolen. Liebigs Ann. Chem. 442, 210 (1925).
32. Clark, E. P.: The Constituents of Certain Species of Helenium. II. Tenulin. III. The Ester Nature of Tenulin. IV. Concerning the Compound Melting at 233–234° Obtained from Helenium tenuifolium. J. Amer. Chem. Soc. 61, 1836 (1939); 62, 597, 2154 (1940).
33. Clemo, G. R., R. D. Haworth and E. Walton: The Constitution of Santonin. Part I. The Synthesis of dl-Santonous Acid. Part II. The Synthesis of Racemic desmotropoSantonin. J. Chem. Soc. (London) 1929, 2368; 1930, 1110.
34. Cocker, W., B. E. Cross, S. R. Duff, J. T. Edward and T. F. Holley: The Constitution of Marrubiin. Part I. J. Chem. Soc. (London) 1953, 2540.
35. Cocker, W., J. T. Edward, T. F. Holley and D. M. S. Wheeler: Ring-Ring Tautomerism in Some Degradation Reactions of Marrubiin. Chem. and Ind. 1955, 1484.
35a. Cocker, W. and T. B. H. McMurry: The Absolute Configuration of $\psi$-Santonin. Proc. Chem. Soc. (London) 1958, 147.
36. Conroy, H.: Picrotoxin. I. The Skeleton of Picrotoxinin. The Degradation to Picrotoxadiene. J. Amer. Chem. Soc. 74, 491 (1952).
37. — Picrotoxin. II. The Skeleton of Picrotoxinin. The Total Synthesis of dl-Picrotoxadiene. J. Amer. Chem. Soc. 74, 3046 (1952).
38. — Picrotoxin. III. α-Picrotoxininic and Bromopicrotoxininic Acids; Apopicrotoxininic Dilactone. J. Amer. Chem. Soc. 79, 1726 (1957).

39. COOK, A. H. and G. HARRIS: The Chemistry of Hop Constituents. Part I. Humulinone, a New Constituent of Hops. J. Chem. Soc. (London) 1950, 1873.
40. DODSON, A., H. N. FUKUI, C. D. BALL, R. L. CAROLUS and H. M. SELL: Occurrence of a Bitter Principle in Carrots. Science (Washington) 124, 984 (1956).
41. DUNSTAN, W. R. und F. W. SHORT: Ein neues Glykosid aus *Strychnos nux vomica*. Pharm. J. Trans. 14, 1025 [Ber. dtsch. chem. Ges. 17, Referate 359 (1884)].
42. DUTT, P., N. C. DEB and P. K. BOSE: A Preliminary Note on Mesuol, the Bitter Principle of *Mesua ferrea*. J. Indian Chem. Soc. 17, 277 (1940) [Chem. Abstr. 34, 6297 (1940)].
43. EASTERFIELD, T. H. and B. C. ASTON: Tutu. Part I. Tutin and Coriamyrtin. J. Chem. Soc. (London) 79, 120 (1901).
44. EICHHOLTZ, F.: Lehrbuch der Pharmakologie, 9. Aufl., S. 341. Berlin: Springer-Verlag. 1957.
45. EMERSON, O. H.: Bitter Principles of Citrus. II. Relation of Nomilin and Obacunone. III. Some Reactions of Limonin. J. Amer. Chem. Soc. 73, 2621 (1951); 74, 688 (1952).
46. ENSLIN, P. R. and D. E. A. RIVETT: The Cucurbitacins, an Interesting New Group of Natural Products. South African Ind. Chemist 11, 75 (1957).
47. FEIST, K. und W. VÖLKSEN: Über die Bitterstoffe der Colombowurzel. VI. Liebigs Ann. Chem. 534, 41 (1938).
48. FLÜCKIGER, F. und H. HÖHN: Untersuchung der *Ophelia Chirata*. Arch. Pharmaz. 189, 213, 229 (1869).
49. FRANCHIMONT, A. P. N.: La Plumiéride et son identité avec l'Agoniadine. Rec. trav. chim. Pays-Bas 19, 350 (1900).
50. GHIGI, E. e A. DRUSIANI: Sulla costituzione molecolare della marrubina. Gazz. chim. ital. 86, 682 (1956).
51. GIZYCKI, F. v.: Alantopikrin, ein Bitterstoff aus den Blättern des Alant. Arch. Pharmaz. 287, 57 (1954).
52. GIZYCKI, F. v. und H. HERRMANNS: Untersuchung des Krautes von *Daucus carota* L. Arch. Pharmaz. 284, 8 (1951).
53. GOVAERT, F. and M. VERZELE: The Constitution of the β Hops Acid. Bull. soc. chim. Belges 58, 432 (1949).
54. HÄNSEL, R.: Pflanzenchemie und Pflanzenverwandtschaft. Arch. Pharmaz. 289, 619 (1956).
55. HANSEN, K. FR. W.: Über Bitterstoffe aus der Alantwurzel (vorl. Mitt.). — Über die Bitterstoffe der Alantwurzel (II. Mitt. über Bitterstoffe). — Zur Konstitution des Isoalantolactons (III. Mitt. über Bitterstoffe). Ber. dtsch. chem. Ges. 64, 67, 943 1904 (1931). — Über die Bitterstoffe der Alantwurzel (IV. Mitt. über Bitterstoffe). J. prakt. Chem. 136, 176 (1933).
55a. HARTWELL, J. L. and A. W. SCHRECKER: The Chemistry of Podophyllum. Fortschr. Chem. organ. Naturstoffe 15, 83 (1958).
56. HEGI, E.: Flora von Mitteleuropa, Bd. 5. München: J. F. Lehmann. 1953.
57. HEGNAUER, R.: Chemotaxonomische Übersichten. 4. Cucurbitaceae. Pharmaceut. Acta Helv. 32, 334 (1957).
58. HEROUT, V., L. DOLEJŠ and F. ŠORM: The Structure of Artabsin, the Prochamazulenogen from *Artemisia absinthium* L. Chem. and Ind. 1956, 1236.
58a. HEROUT, V. and F. ŠORM: On Terpenes. LXI. Contribution to the Constitution of Pro-chamazulenogen, the Natural Precursor of Chamazulene in *Artemisia absinthium*. Collect. Czechoslov. Chem. Communs. 19, 792 (1954).
58b. HERZER, F. H.: Bitterweed Studies. Assoc. Southern Agr. Workers, Proc. Annu. Convention 43, 112 (1942) [Chem. Abstr. 37, 1789 (1943)].

59. Horrmann, P.: Über die Zusammensetzung des Picrotoxinins. — Über Derivate des α- und β-Brompicrotoxinins. Ber. dtsch. chem. Ges. **45**, 2090 (1912); **46**, 2793 (1913). — Beiträge zur Kenntnis des Picrotoxins. Liebigs Ann. Chem. **411**, 273 (1916).

59a. Horrmann, P. und H. Wächter: Über die Aufspaltung des Pikrotoxins mit methylalkoholischer Kalilauge und über die Pikrotoxinsäure. Ber. dtsch. chem. Ges. **49**, 1554 (1916).

60. Howard, G. A. and C. A. Slater: The Chemistry of Hop Constituents. Part XII. The Structure of Humulinone. J. Chem. Soc. (London) **1958**, 1460.

61. Ikuta, M.: Influence of Bitter Drugs on Secretion of Gastric Juice. Osaka Igakkai Zassi **39**, 2072 (1940) [Chem. Abstr. **38**, 6384 (1944)].

62. Inagaki, C.: Isolation of a New Bitter Substance from Citrus Natsudaidai (So-Called Natsumikan). Nat. Sci. Rep. Ochanomizu Univ. (Tokyo) **2**, 133 (1951).

63. Jacobs, W. A. and Y. Sato: The Veratrine Alkaloids. XXX. A Further Study of the Structure of Veratramine and Jervine. J. Biol. Chem. **181**, 55 (1949).

64. Jaminet, F.: A New Strychnos Heteroside: the Logasonic Acid. Lejeunia **15**, 23 (1951) [Chem. Abstr. **46**, 10549 (1952)].

65. Junkmann, K.: Über die Wirkung der sog. „Bitterstoffe". Arch. exp. Pathol. Pharmakol. **143**, 368 (1929).

66. Kariyone, T. und Y. Matsushima: Über die Bestandteile von *Swertia japonica* Makino. J. pharmac. Soc. Japan **1927**, Nr. 540, 25 [Chem. Zbl. **1927**, I, 2660].

67. Kariyone, T. and T. Okuda: Studies on the Compounds of Japanese Toxic Plants. I. On the Toxic Components of *Coriaria Japonica*. Bull. Inst. Chem. Res., Kyoto Univ. **31**, 387 (1953).

68. Kedvessy, G. and B. Fürstner: Determination of the Value of Tinctures Containing Bitter Principles in the Presence of Ethereal Oils and Tannic Acids. Ber. ungar. pharm. Ges. **19**, 492 (1943) [Chem. Abstr. **41**, 4891 (1947)].

69. Kern, W. und W. Haselbeck: Über die Inhaltsstoffe der Kondurangorinde, VI. Mitt. Die Isolierung eines neuen Disaccharids aus dem Glykosid Kondurangin. Arch. Pharmaz. **283**, 105 (1950).

70. Kern, W. und H. Momsen: Über die Inhaltsstoffe der Kondurangorinde. I. Mitt. Arch. Pharmaz. **276**, 463 (1938).

71. Khalique, M. A. and M. Ahmed: Corchsularin, a New Bitter from Jute Seeds. I. Its Isolation and Constitution of Corchsularose. J. Organ. Chem. (USA) **19**, 1523 (1954). — II. 2-Deoxy-*D*-ribose from Corchsularose. J. Indian Chem. Soc. **32**, 510 (1955) [Chem. Abstr. **50**, 10007 (1956)].

72. Kinoshita, K.: Constituents of *Coriaria Japonica* Gray. II. Coriarine, a Poisonous Constituent. J. Chem. Soc. Japan **51**, 99 (1930) [Chem. Abstr. **26**, 731 (1932)].

73. — Constituents of *Coriaria Japonica* A. Gray. III. Catalytic Reduction of Tutin. J. Chem. Soc. Japan **52**, 171 (1931) [Chem. Abstr. **26**, 5100 (1932)].

74. Kondo, H. and A. Ono: Structure of Andrographolide. III, IV. Annu. Rept. Itsuu Lab. **4**, 78 (1953); **5**, 89 (1954) [Chem. Abstr. **49**, 962, 15918 (1955)].

75. Korte, F.: Die pflanzlichen Bitterstoffe. Arch. Pharmaz. **286**, 257, 295 (1953).

76. — Über neue glykosidische Pflanzeninhaltsstoffe, II. Mitt. Zur Konstitution des Gentiopikrins. Chem. Ber. **87**, 512 (1954).

76a. — Über neue glykosidische Pflanzeninhaltsstoffe, IV. Mitt. Die Konstitution des Gentiopikrins und seiner enzymatischen Spaltprodukte. Chem. Ber. **87**, 769 (1954).

*76b.* KORTE, F.: Über neue glykosidische Pflanzeninhaltsstoffe, V. Mitt. Zur Konstitution des Gentiopikrins und des Hexahydrogentiogenin-methyläthers. Chem. Ber. **87**, 780 (1954).

*77.* — Über glykosidische Pflanzeninhaltsstoffe, VI. Mitt. Die Bitterstoffe der Gentianaceen. Chem. Ber. **87**, 1357 (1954).

*78.* — Zur chemischen Klassifizierung von Pflanzen, XI. Mitt. Zur Konstitution des Kondurangins und Vincetoxins. Chem. Ber. **88**, 1527 (1955).

*79.* — Charakteristische Pflanzeninhaltsstoffe, IX. Mitt. Amarogentin, ein neuer Bitterstoff aus Gentianaceen. Chem. Ber. **88**, 704 (1955).

*80.* — Neuere Ergebnisse über pflanzliche Bitterstoffe. Angew. Chem. **69**, 207 (1957).

*81.* KORTE, F. und G. BECHMANN: XVII. Mitt. zur chemischen Klassifizierung von Pflanzen. Über die Bitterstoffe aus *Cnicus benedictus*. Naturwiss. **45**, 390 (1958).

*82.* KORTE, F. und O. BEHNER: Zur chemischen Klassifizierung von Pflanzen, XVIII. Mitt. Zur Synthese und Dehydrierung von Perhydrofluoren-Verbindungen. Liebigs Ann. Chem. (im Druck).

*83.* KORTE, F., K. H. BÜCHEL und L. SCHIFFER: Zur chemischen Klassifizierung von Pflanzen, XVI. Mitt. Zur Kenntnis des Gentiopikrins. Chem. Ber. **91**, 759 (1958).

*84.* KORTE, F., J. FALBE und A. ZSCHOCKE: Synthese des *d,l*-Iridomyrmecins und verwandter bicyclischer Lactone. Tetrahedron (im Druck).

*84a.* KORTE, F. und A. GROEBEL (unveröffentlicht).

*85.* KORTE, F. und I. KORTE: Charakteristische Pflanzeninhaltsstoffe, VIII. Mitt. Über die Beziehung zwischen morphologischer Systematik und chemischen Inhaltsstoffen bei den Asclepiadaceen. Z. Naturforsch. **10 b**, 223 (1955).

*86.* KORTE, F. und H. MACHLEIDT: Zur chemischen Klassifizierung von Pflanzen, XV. Mitt. Zur Konstitution des Gentiopikrins. Chem. Ber. **90**, 2276 (1957).

*87.* KORTE, F. und J. RIPPHAHN: Zur chemischen Klassifizierung von Pflanzen. XIX. Mitt. Zur Konstitution des Kondurangins und Vincetoxins. Liebigs Ann. Chem. (im Druck).

*88.* KORTE, F. und H. G. SCHICKE: Zur chemischen Klassifizierung von Pflanzen, XII. Mitt. Zur Kenntnis des Amarogentins. Chem. Ber. **89**, 2404 (1956).

*89.* KORTE, F. und K. TRAUTNER: (unveröffentlicht).

*90.* KORTE, F. und H. WEITKAMP: Zur chemischen Klassifizierung von Pflanzen, XIII. Mitt. Zur Konstitution des Kondurangins. Chem. Ber. **89**, 2669 (1956).

*91.* KORTE, I.: Über die Entstehung der Bitterstoffe in Gentianaceen. Dissertation, Bonn, 1957.

*92.* KOTIDI, E. P.: The Chemical Nature of Substances that Produce Bitter Taste in the Food Products from Tangerines. Doklady Akad. Nauk S. S. S. R. **73**, 763 (1950) [Chem. Abstr. **45**, 781 (1951)].

*93.* KROEBER, L.: Studienergebnisse einer Reihe von Fluidextrakten aus heimischen Arzneipflanzen. Extractum Erythraeae centaurii fluidum. Pharmaz. Zentralhalle **69**, 807 (1928).

*94.* KUBLER, K.: Beiträge zur Chemie der Kondurangorinde. Arch. Pharmaz. **246**, 620 (1908).

*95.* KUBOTA, T. and T. MATSUURA: Confirmation of the Presence of a Furan Ring in Columbin. Proc. Chem. Soc. (London) **1957**, 262.

*96.* KUBOTA, T. and Y. TOMITA: Swertiamarin. Chem. and Ind. **1958**, 229.

*97.* — — Structure of Erythrocentaurin. Chem. and Ind. **1958**, 230.

*98.* LAVIE, D. and Y. SHVO: A Degradation Product of Elatericin A. Proc. Chem. Soc. (London) **1958**, 220.

*99.* LUDWIG, H. und A. KROMAYER: Gewinnung des Lactucins. Arch. Pharmaz. **161**, 1 (1862).

*100.* McColloch, R. J.: Preliminary Studies on Debittering Navel Orange Products. California Citrograph, May 1950.

*101.* Madaus, G. und H. Schindler: Untersuchungen über die Gehaltsschwankungen einiger Arzneipflanzen im Verlaufe der Vegetationsperiode. Arch. Pharmaz. **276**, 280 (1938) [Chem. Abstr. **32**, 6803 (1938)].

*102.* Majumdar, D. N. und P. C. Guha: Indische Heilpflanzen. II. *Swertia chirata*. J. Indian Inst. Sci. A **16**, 34 (1933) [Chem. Zbl. **1934**, I, 236].

*103.* Marsh, G. L.: Bitterness in Navel Orange Juice. Food Technol. **7**, 145 (1953).

*104.* Meisels, A. and A. Weizmann: The Structure of Chamazulene. J. Amer. Chem. Soc. **75**, 3865 (1953).

*105.* Merz, K. W. und K. G. Krebs: Zur Kenntnis des Loganins. 1. Mitt. Arch. Pharmaz. **275**, 217 (1937).

*106.* Merz, K. W. und H. Lehmann: Zur Kenntnis des Loganins. 2. Mitt. Arch. Pharmaz. **290**, 543 (1957).

*107.* Meyer, R. J. und P. Bruger: Zur Kenntnis des Picrotoxins. Ber. dtsch. chem. Ges. **31**, 2958 (1898).

*108.* Moncrieff, R. W.: The Chemical Senses. London: L. Hill, Ltd. 1951.

*109.* Okuda, T.: Studies on the Components of *Coriaria Japonica*. XIII. Pharm. Bull. (Japan) **2**, 185 (1954) [Chem. Abstr. **50**, 830 (1956)].

*110.* Olbricht, G. und L. Lendle: Über die Verwendung von bitterstoffhaltigen Heilpflanzen als Fiebermittel in der Medizin der Kräuterbücher des 15. bis 17. Jahrhunderts und in der Volksmedizin. Die Pharmazie **5**, 241 (1950).

*111.* Orchin, M. and R. A. Friedel: Structure of the Benzfluorenes and Benzfluorenones. J. Amer. Chem. Soc. **71**, 3002 (1949).

*112.* Orchin, M. and L. Reggel: Aromatic Cyclodehydrogenation. V. A Synthesis of Fluoranthrene. — VII. Rearrangements in the Phenylnaphthalene Series. J. Amer. Chem. Soc. **69**, 505 (1947); **70**, 1245 (1948).

*113.* Paris, R.: Les principes amers des végétaux. Ann. pharm. franç. **4**, 207 (1946).

*114.* Paris, R. and C. Rabenoro: Two Myrsinaceae from Madagascar Used as Vermifuges. Ann. pharm. franç. **8**, 380 (1950) [Chem. Abstr. **44**, 10265 (1950)].

*115.* Pesonen, S. and E. Ramstad: Studies on Gentiopicrin. J. Amer. Pharmaceut. Assoc., Sci. Edit. **45**, 522 (1956) [Chem. Abstr. **50**, 15521 (1956)].

*116.* Planche, O.: *Trichilia heudelotii*. Ann. pharm. franç. **7**, 460 (1949) [Chem. Abstr. **44**, 801 (1950)].

*117.* Plattner, Pl. A., A. Fürst, L. Marti und H. Schmid: Zur Kenntnis der Sesquiterpene und Azulene. 87. Mitt. Synthese des Guaj-azulens. I. Helv. Chim. Acta **32**, 2137 (1949).

*118.* Pudles, J., A. Diara et E. Lederer: Sur l'isolement et la constitution chimique du darutoside, principe amer de *Siegesbeckia orientalis*. C. R. hebd. Séances Acad. Sci. **244**, 472 (1957).

*119.* Rentschler, H. und H. Tanner: Das Bitterwerden der Rotweine. Mitt. Lebensmittelunters. Hyg. (Bern) **42**, 463 (1951).

*120.* Riedl, W.: Konstitution und Synthese der Hopfenbitterstoffe *d,l*-Humulon und Lupulon sowie einiger Analoga. V. Mitt. Chem. Ber. **85**, 692 (1952).

*120 a.* Riedl, W. und E. Leucht: Synthese des Humulo-chinons und -hydrochinons. Chem. Ber. **91**, 2784 (1958).

*121.* Riedl, W. und J. Nickl: Die C,C-Hydrogenolyse der Hopfenbitterstoffe. Chem. Ber. **89**, 1838 (1956).

*122.* Robertson, A., R. J. S. Beer, D. B. G. Jaquiss and W. E. Savige: Quassin and *neo*Quassin. Part II. J. Chem. Soc. (London) **1954**, 3672.

*123.* ROBERTSON, A. and J. C. HARLAND: Picrotoxin. Part III. J. Chem. Soc. (London) **1939**, 937.

*124.* ROBERTSON, A. and D. MERCER: Picrotoxin. Part II. Picrotone and Picrotonol. J. Chem. Soc. (London) **1936**, 288.

*125.* ROBERTSON, A., D. MERCER and R. S. CAHN: Picrotoxin. Part I. The Constitution of Picrotic Acid and the C-Skeleton of Picrotoxinin and Picrotin. J. Chem. Soc. (London) **1935**, 997.

*126.* ROBERTSON, A., R. W. H. O'DONNELL and J. C. HARLAND: Picrotoxin. Part IV. J. Chem. Soc. (London) **1939**, 1261.

*127.* ROHNE, K. und H. RISCHE: Ergebnisse experimenteller Untersuchungen über bacteriostatische Stoffe aus Hopfen. Brauwissensch. **1952**, 221.

*128.* ROSENTHALER, L.: Über Loganin. Schweiz. Apoth.-Ztg. **61**, 398 (1923) [Chem. Zbl. **1923**, III, 932].

*129.* ROTHE, M.: Über das Bitterwerden von Cerealien. II. Zusammenhang zwischen Fett-Autoxydation und Bitterstoff-Bildung. Fette, Seifen, Anstrichmittel **56**, 667 (1954).

*129a.* RUZICKA, L. und F. LARDON: Zur Kenntnis der Triterpene. 105. Mitt. Über das Ambreïn, einen Bestandteil des grauen Ambra. Helv. Chim. Acta **29**, 912 (1946).

*130.* RUZICKA, L. und J. A. VAN MELSEN: Höhere Terpenverbindungen. XLV. Zur Kenntnis des Alantolactons und des Iso-alantolactons. Helv. Chim. Acta **14**, 397 (1931).

*131.* RUZICKA, L., P. PIETH, T. REICHSTEIN und L. EHMANN: Polyterpene und Polyterpenoide. LXXX. Zur Kenntnis der Alantolactone. Synthese des 1,4-Dimethyl-6-isopropyl- und des 1,5-Dimethyl-7-isopropyl-naphthalins. Helv. Chim. Acta **16**, 268 (1933).

*132.* SAKURAI, Y. and K. YOSHINA: Constitution of Gentiopicrin. J. Pharmac. Soc. Japan **71**, 55 (1951).

*133.* SCHENCK, G. und H. GRAF: Zur Kenntnis des Lactucariums. II. Mitt. Arch. Pharmaz. **275**, 36 (1937).

*134.* SCHMID, H., H. BICKEL und TH. M. MEIJER: Zur Kenntnis des Plumierids, 1. Mitt. Helv. Chim. Acta **35**, 415 (1952).

*135.* SCHMID, H. und O. HALPERN: Zur Kenntnis des Plumierids, 2. Mitt. Helv. Chim. Acta **41**, 1109 (1958).

*136.* SCHULTZ, O. E. und R. GMELIN: Das Senfölglukosid „Glukoiberin" und der Bitterstoff „Ibamarin" von *Iberis amara* L. (Schleifenblume). Arch. Pharmaz. **287**, 404 (1954).

*137.* SHIELDS, L. M.: Distribution of the Bitter Principle in the Shoot of *Helenium tenuifolium*. Bot. Gaz. **113**, 471 (1952).

*138.* SHOPPEE, C. W., D. N. JONES, J. R. LEWIS and G. H. R. SUMMERS: Steroids and Walden Inversion. Part XXVI. 4β-Methoxy-cholest-5-ene, 6β-Methoxy-cholest-4-ene, and Related Compounds. J. Chem. Soc. (London) **1955**, 2876.

*139.* SIDDAPPA, G. S., J. S. PRUTHI and P. T. TAKARKHEDE: Bitter Principles in Citrus Fruits. Indian J. Hort. **10**, 133 (1953) [Chem. Abstr. **48**, 8484 (1954)].

*140.* SIELISCH, J.: Über das Picrotoxin. Liebigs Ann. Chem. **391**, 1 (1912).

*140a.* SLATER, S. N.: Tutin. J. Chem. Soc. (London) **1943**, 50.

*141.* ŠORM, F., L. NOVOTNY and V. HEROUT: The Bitter Principle of *Artemisia absinthium* L. Chem. and Ind. **1955**, 569.

*142.* — — — Plant Substances. V. Isolation of Further Crystalline Compounds from Wormwood. Chem. Listy **50**, 591 (1956).

*143.* SPÄTH, E., R. LORENZ und H. KUHN: Über das Lactucin. 1. Mitt. Monatsh. Chem. **82**, 114 (1951).

*144.* Sutter, M. und E. Schlittler: Pikrotoxin. 3. Mitt. Sodaspaltung von α-Dihydro-pikrotoxinin. — 4. Mitt. Bariumhydroxydspaltungen von Pikrotoxinin und α-Dihydro-pikrotoxinin. Helv. Chim. Acta 32, 1855, 1860 (1949).

*145.* Takeda, K. and W. Nagata: Components of the Root of *Lindera strychinifolia*. V. Azulenes Isolated from Linderene by Zinc-dust Distillation. Pharm. Bull. (Japan) 1, 164 (1953) [Chem. Abstr. 48, 7716 (1954)].

*146.* Tanret, G.: Sur la gentiopicrine. Bull. soc. chim. France [3] 33, 1059 (1905). — Sur la gentiamarine. Bull. soc. chim. France [3] 33, 1071 (1905).

*147.* Teeter, H. M., L. E. Gast, E. W. Bell, W. J. Schneider and J. C. Cowan: Investigations on the Bitter and Beany Components of Soybeans. J. Amer. Oil Chem. Soc. 32, 390 (1955).

*148.* Tettweiler, K. und I. Drishaus: Über die Konstitution der aromatischen Umwandlungsprodukte des Picrotoxins. Liebigs Ann. Chem. 520, 163 (1935).

*149.* Thieme, H.: Azulene Precursors of Chamomile, Milfoil, and Absinthium *(Artemisia absinthium)*. Planta Med. 6, 70 (1958).

*150.* Tsuda, K., K. Tanabe, I. Iwai and K. Funakoshi: On the Structure of Alantolactone. J. Amer. Chem. Soc. 79, 1009 (1957). — The Structure of Alantolactone. J. Amer. Chem. Soc. 79, 5721 (1957).

*151.* Ungnade, H. E. and E. C. Hendley: The Bitter Principle of *Helenium tenuifolium*. J. Amer. Chem. Soc. 70, 3921 (1948).

*152.* Ungnade, H. E., E. C. Hendley and W. Dunkel: Tenulin. II. Anhydrotenulin and Pyrotenulin. J. Amer. Chem. Soc. 72, 3818 (1950).

*153.* Uno, H.: Components of the Seeds of *Brucea Sumatrana*. J. pharmac. Soc. Japan 63, 579 (1943) [Chem. Abstr. 45, 1731 (1951)].

*154.* Vulpius, G.: Über Condurango-Glycosid. Arch. Pharmaz. 223, 299 (1885).

*155.* Wasicky, R., E. Barbieri and H. Weber: Method of Determining the Active Principle in Drugs and Preparations by their Bitterness. Anais fac. farm. odontol. Univ. São Paulo 3, 113 (1942–43) [Chem. Abstr. 39, 5396 (1945)].

*156.* Wessely, F. und K. Jentzsch: Zur Kenntnis der Bitterstoffe der Colombowurzel. V. Über die Methylierung des Columbins. Monatsh. Chem. 70, 30 (1937).

*157.* Wessely, F., R. Lorenz und H. Kuhn: Über das Lactucin. (II. Mitt.) Monatsh. Chem. 82, 322 (1951).

*158.* Wessely, F. und K. Schönol: Zur Kenntnis des Chasmanthins. Monatsh. Chem. 71, 10 (1938).

*159.* Wettstein, K.: Handbuch der systematischen Botanik. Leipzig u. Wien: Franz Deuticke. 1935.

*160.* Wieland, H. und E. Martz: Über die chemische Natur der Hopfenharzsäuren (III.). Ber. dtsch. chem. Ges. 59, 2352 (1926).

*161.* Winkler, R. E. und T. Reichstein: Die Glykoside der Samen von *Dregea volubilis* (L.) Benth. ex Hook. Glykoside und Aglykone, 131. Mitt. Helv. Chim. Acta 37, 721 (1954).

*162.* Wöllmer, W.: Über die Bitterstoffe des Hopfens. Ber. dtsch. chem. Ges. 58, 672 (1925).

*163.* Woodward, R. B. and G. Singh: The Synthesis of Patulin. J. Amer. Chem. Soc. 72, 1428 (1950).

*164.* — — *Allo*patulin. J. Amer. Chem. Soc. 72, 5351 (1950).

*165.* Zellner, J.: Beiträge zur vergleichenden Pflanzenchemie. XVI. Zur Chemie milchsaftführender Pflanzen. Monatsh. Chem. 47, 681 (1926).

*166.* Zinke, A. und K. Holzer: Über die Bitterstoffe der Zichorie (*Cichorium intybus* L.) I. Kurze Mitt.: Lactucin und Lactucopikrin. Monatsh. Chem. 84, 212 (1953).

*(Eingelaufen am 15. Dezember 1958.)*

# Alkaloide aus Calebassencurare und südamerikanischen Strychnosarten.

## Von K. BERNAUER, Zürich.

Mit 6 Abbildungen.

## Inhaltsübersicht.

# I. Einleitung.

Der Ausdruck Curare, von europäischen Zungen vermutlich aus einem indianischen Wort geformt (65), stellt ursprünglich einen Sammelbegriff für die Pfeilgifte der südamerikanischen Indianer dar. Im naturwissenschaftlich-medizinischen Sinne gebraucht, bezeichnet er jedoch heute nur noch Pfeilgifte, die auf Grund ihres Gehaltes an bestimmten Alkaloiden muskellähmend ("curarisierend") wirken.

Im Anschluß an den Pharmakologen BOEHM pflegte man drei Arten solcher Gifte zu unterscheiden. BOEHM fand nämlich einen Zusammenhang zwischen der chemischen Natur der Gifte und der Art der Gefäße, in welchen sie in den Handel kamen. Er prägte die Begriffe Calebassen-, Topf- und Tubo-curare (32, 33).

Calebassencurare war in ausgehöhlte Flaschenkürbisse (Calebassen), Topfcurare in unglasierte Tontöpfchen, Tubocurare in Bambusröhren abgefüllt. Der Wert dieser Einteilung wird schon seit längerem in Zweifel gezogen (z. Beisp. 42, 58, 95). Tatsächlich hat es keinen Sinn, von "Topfcurare" zu sprechen, denn ein genauer chemischer Vergleich zwischen Curareproben, die aus der gleichen Gegend stammten, aber teils in Calebassen, teils in Tontöpfchen verpackt waren, ergab keinen Unterschied (53). Andererseits wies von KING untersuchtes "Topfcurare" große Ähnlichkeit mit Tubocurare auf; die daraus isolierten Alkaloide Protocuridin und Neoprotocuridin sind aufs engste mit Tubocurarin verwandt (54).

Bambusröhren und Calebassen als Verpackung haben jedoch auch heute noch einen gewissen diagnostischen Wert. Bis jetzt sind nämlich in Tubocurare Calebassen-Alkaloide noch nicht nachgewiesen worden; umgekehrt hat man nur sehr selten in Calebassencurare Tubocurare-

Alkaloide gefunden (*101*). Bezüglich der Hauptinhaltsstoffe stimmten alle bis heute näher untersuchten Proben von Calebassencurare weitgehend überein.

Demnach bestehen keine Bedenken, die ohnehin fest eingebürgerten Begriffe Tubocurare und Calebassencurare weiterhin zu verwenden. Sie lassen sich an Hand der Ergebnisse der chemischen und botanischen Forschung noch schärfer umreißen: Tubocurare wird aus Chondrodendron-Arten (Familie Menispermaceae) bereitet. Seine Alkaloide können als dimere Abkömmlinge des Coclaurins (I) aufgefaßt werden. *d*-Tubocurarin (II), das zuerst aus Tubocurare, dann aus *Ch. tomentosum* isoliert worden ist, findet seit Jahren in der Anästhesie Verwendung.

Eine Zusammenfassung ,,Die Alkaloide der Menispermaceae-Pflanzen'' von TOMITA ist in Band IX dieser Reihe erschienen (*90 a*).

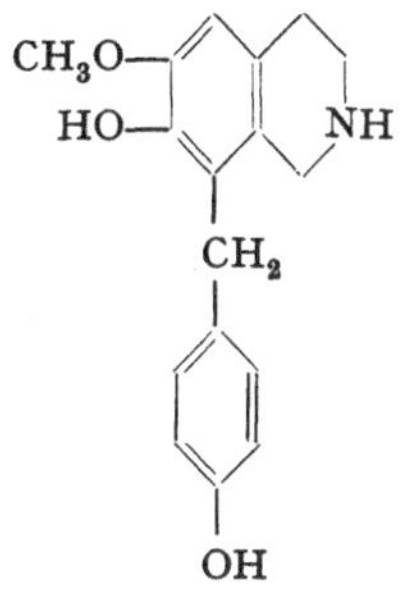

(I.) Coclaurin.

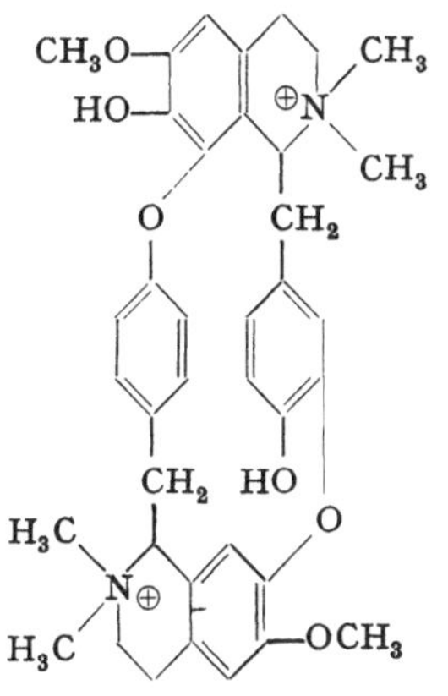

(II.) Tubocurarin.

Zwischen den stark wirksamen Tubo- und Calebassencurare-Alkaloiden besteht eine konstitutionelle Gemeinsamkeit: das Vorhandensein zweier quartärer Ammoniumgruppierungen pro Molekül. Alle bis jetzt ganz oder teilweise aufgeklärten Calebassen-Alkaloide leiten sich vom Tryptamin ab und fügen sich vorzüglich in die für die Klasse der Indolalkaloide entwickelten Biogenese-vorstellungen ein.

Der pflanzliche Ursprung des Calebassencurare ist noch nicht in den Einzelheiten abgeklärt. Es kann jedoch als sicher gelten, daß die Indianer bei der Herstellung von Calebassencurare hauptsächlich Rinden von Strychnosarten (Familie Loganiaceae) verwenden.

In einer Arbeit (*110*) aus neuester Zeit findet sich die Angabe, daß die Piaroa-Indianer (oberer Orinoko) als einziges wichtiges Pflanzenmaterial die Rinde von *Str. toxifera* ROB. SCHOMBURGK bei der Bereitung von Calebassencurare verwenden.

Das Calebassencurare kommt aus dem nördlichen Teil Südamerikas, hauptsächlich aus Nordbrasilien, Venezuela und Columbien. Besonders häufig genannt werden die Gebiete des oberen Orinoko und oberen Rio Negro.

Die Toxizitäten roher Curareproben schwanken in weiten Grenzen. Meist ist das rohe Calebassencurare wirksamer als das rohe Tubocurare. Charakteristische Toxizitätsunterschiede zwischen den reinen Hauptalkaloiden beider Gifte bestehen nicht, wie aus einem Vergleich der Letaldosen hervorgeht (*52, 94*):

$d$-Tubocurarin-dichlorid ...... 0,13 mg/kg Maus
C-Curarin-I-dichlorid......... 0,05  ,,  ,,
C-Calebassin-dichlorid........ 0,30  ,,  ,,

*Das „C" ist eine Abkürzung für „Calebassen-" (100).*

Einzelne Calebassen-Alkaloide sind allerdings extrem giftig; so z. B. die Alkaloide E und G, die mit 0,001—0,012 mg/kg Maus letal wirken (*52, 94*).

## II. Die Entwicklung der Chemie der Calebassen-Alkaloide.

Übersichtsreferate: (*106a, 49*).

Schon in der ersten Hälfte des vorigen Jahrhunderts versuchte man, die wirksamen Prinzipien des Calebassencurare zu isolieren. Jedoch hat erst Boehm eine stark curare-aktive (aber amorphe) Substanz aus diesem Material gewonnen (1897).

Die eigentliche Chemie des Calebassencurare nimmt ihren Anfang mit den Arbeiten H. Wielands und seiner Schüler (ab 1937). Der Münchner Arbeitskreis hat erstmals die Isolierung kristallisierter, aktiver Verbindungen beschrieben. Eine Feinauftrennung des Calebassencurare und damit die Isolierung vieler neuer Alkaloide gelang 1952—1953 mit Hilfe der Papierchromatographie (Kebrle, Schmid und Karrer; Th. Wieland und Merz). Mit Arbeiten zur Konstitutionsaufklärung hat schon H. Wieland begonnen. Sie werden seither von Th. Wieland, insbesondere aber durch den Kreis um Schmid und Karrer fortgesetzt.

Den ersten tieferen Einblick in die Struktur zweier Calebassen-Alkaloide erhielten Bickel, Schmid und Karrer (*31*) bei ihren Untersuchungen an C-Fluorocurin und C-Mavacurin (1955). Diese beiden $C_{20}$-Verbindungen sind jedoch nicht curare-aktiv. Bald darauf erkannten v. Philipsborn, Schmid und Karrer (*74*), daß dem C-Curarin-I und anderen toxischen Calebassen-Alkaloiden Formeln mit 40 Kohlenstoffatomen und zwei quartären Ammoniumgruppierungen zukommen.

Das Prinzip, nach welchem in den einfachsten toxischen $C_{40}$-Alkaloiden zwei $C_{20}$-,,Hälften" verknüpft sind, ist zuerst 1957 von Karrer ausgesprochen und 1958 am C-Dihydrotoxiferin bewiesen worden [Bernauer, Schmid und Karrer (*28*)]. Aus Arbeiten der Züricher Schule über

C-Fluorocurarin, Caracurin-VII und C-Dihydrotoxiferin hat sich ergeben, daß die hochwirksamen Calebassen-Alkaloide mit dem Strychnin verwandt sind. Seit kurzem können die meisten dieser Alkaloide partialsynthetisch aus „WIELAND-GUMLICH-Aldehyd", einem Abbauprodukt des Strychnins, erhalten werden.

Südamerikanische Strychnospflanzen sind von H. WIELAND, von KING, von SCHLITTLER, von SCHMID und KARPER sowie von einem Kreis um BOVET und MARINI-BETTOLO untersucht worden. Diejenige Strychnosart, der die mengenmäßig bedeutendsten Calebassen-Alkaloide entstammen, ist noch nicht mit Sicherheit bekannt.

## III. Die Auftrennung des Calebassencurare in Einzelalkaloide.

Das Calebassencurare ist ein Konglomerat von unbekannten Neutralstoffen, von wenigen tertiären und zahlreichen quartären Alkaloiden. Über die Säuren, die mit den Basen zu Salzen vereinigt sind — Calebassenextrakte reagieren neutral oder schwach sauer —, ist kaum etwas bekannt. [Vorkommen von Protocatechusäure, vgl. (99)].

Zu einem bedeutenden Erfolg bei der Auftrennung dieses Gemisches und zu der Isolierung der ersten kristallisierten Calebassen-Alkaloide gelangte H. WIELAND, indem er sich der Chromatographie der Alkaloid-Reineckate an Aluminiumoxyd bediente. Seine Schule entwickelte im Zuge ihrer Arbeiten (96, 99—102) einen Trennungsgang.

Wie für andere Gebiete bedeutete für die Chemie der Calebassen-Alkaloide die Einführung der Papierchromatographie einen großen Fortschritt. Von den in verschiedenen Arbeitskreisen für Calebassen-Alkaloide entwickelten Chromatographiegemischen (85, 86, 104, 110, 36) haben sich besonders die Gemische „C" und „D" der Züricher Schule bewährt.

Gemisch „C": wassergesättigtes Methyläthylketon mit 1 bis 3 Vol.-% Methanol.
Gemisch „D": Essigester-Pyridin-Wasser (7,5 : 2,3 : 1,65).

Das mit diesen Lösungsmitteln erhaltene zweidimensionale Chromatogramm einer durch Reineckat-Fällung vorgereinigten Calebassencurare-Probe vermittelt einen Begriff davon, wie komplex das Gemisch der Calebassen-Alkaloide ist *(Abb. 1)*.

Als Bezugssubstanz dient wegen seiner leichten Identifizierbarkeit und großen Stabilität meist C-Curarin-I-dichlorid, dessen relative Wanderungsstrecke = 1 gesetzt wird. Die relativen Wanderungsstrecken, bezogen auf C-Curarin-I-dichlorid (Tabelle 1, S. 236), werden als $R_c$-Werte bezeichnet (86). $R_c$ = Wanderungsstrecke des betreffenden Alkaloid-chlorids/Wanderungsstrecke des C-Curarin-I-dichlorids.

Zum Sichtbarmachen der Alkaloidflecke auf Papier führten SCHMID und KARRER eine Lösung von Cer(IV)-sulfat in 2-n Schwefelsäure als Sprühreagens ein (50, 85). Die meisten Calebassen-Alkaloide geben mit diesem Oxydationsmittel Farbreaktionen, die teilweise recht charakteristisch sind (siehe Tabelle 1, ferner

S. 206 ff.). Auch die Art, wie die mit dem Cer-Reagens erhaltenen Farben sich mit der Zeit verändern, kann wertvolle Hinweise liefern („Verblassung"). Nach Th. Wieland und Merz (*104*) eignet sich auch Zimtaldehyd als Reagens. Die damit besprühten Papiere werden in eine Salzsäureatmosphäre gehängt, worauf

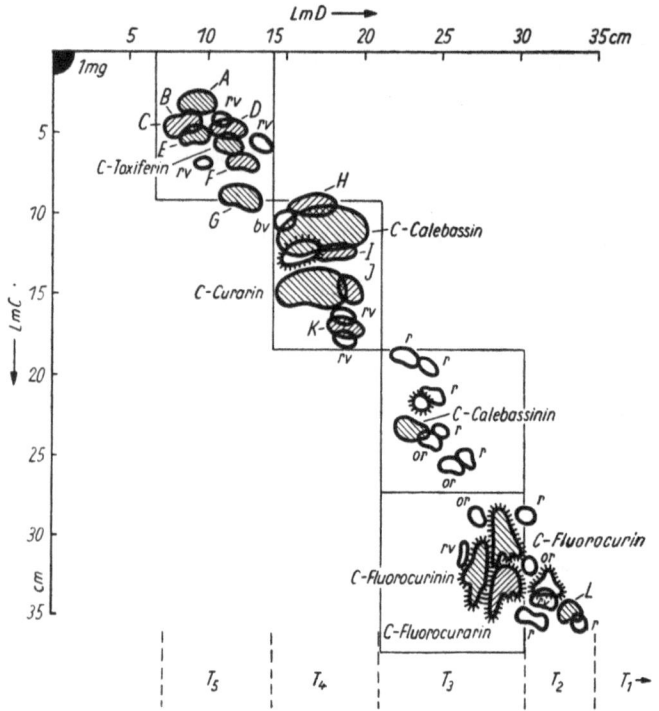

Abb. 1. Papierchromatogramm von 1 mg gereinigter Chloride aus einem Calebassencurare, nach Schmid, Keeble und Karrer. Die einzelnen Flecke wurden durch Ansprühen mit Cer(IV)-sulfat- bzw. Jodlösung kenntlich gemacht. ⌊⌊⌊⌊ = Fluoreszenz im UV. — *r* = rote, *bv* = blauviolette, *rv* = rotviolette, *or* = orange Cer(IV)-sulfat-Reaktion. Die schraffierten Flecke repräsentieren isolierte Alkaloide. [Aus: Helv. Chim. Acta *35*, 1866 (1952).]

die Alkaloide als farbige Flecke erscheinen. Wo die beiden genannten Reagenzien versagen, ist man auf Kaliumjodo-platinat-lösung (*81*) oder alkoholische Jodlösung angewiesen.

Auch die Papierelektrophorese, manchmal kombiniert mit der Papierchromatographie, ist wiederholt mit Erfolg für analytische und mikropräparative Arbeiten an Calebassen- und südamerikanischen Strychnos-Alkaloiden angewandt worden (*92, 93, 64, 36*). Gelegentlich hat man sich auch der Elektrophorese an Papiersäulen und der Gegenstromverteilung bedient (*35*).

Die beste Methode für die Trennung der Calebassen-Alkaloide in präparativem Maßstab ist heute zweifellos die Verteilungschromatographie der über die Reineckate (*86, 48*), Pikrate (*110*), oder durch Filtration über Aluminiumoxyd (*104*) vorgereinigten Alkaloidchloride an Säulen von Papierpulver [TH. WIELAND und MERZ (*104*); KEBRLE, SCHMID und KARRER (*86*)].

In *Tabelle 1* (S. 236) sind die 42 Alkaloide zusammengestellt, die bis heute aus Calebassencurare isoliert worden sind. Einige sind dadurch ausgezeichnet, daß sie im Calebassencurare regelmäßig und unabhängig von dessen Provenienz angetroffen werden. Es sind dies das C-Curarin-I, C-Calebassin und C-Fluorocurarin. Die beiden erstgenannten Substanzen übertreffen mengenmäßig meist alle anderen Alkaloide.

## IV. Die Alkaloide südamerikanischer Strychnosarten.

### 1. Übersicht.

Wie erwähnt, nimmt man an, daß das Calebassencurare hauptsächlich aus Rinden von Strychnospflanzen bereitet wird. Diese Auffassung hat insofern eine Bestätigung erfahren, als sich herausgestellt hat, daß eine Reihe von Calebassen-Alkaloiden, darunter alle stark curare-aktiven, biogenetisch in engster Beziehung zum Strychnin steht, dem Hauptalkaloid der in Vorder- und Hinterindien und Nordaustralien heimischen . Species *Strychnos nux vomica*. In Südamerika ist die Gattung Strychnos (Familie Loganiaceae) durch mehr als 50 Arten vertreten, deren eindeutige Identifizierung allein an Hand der Morphologie offenbar nicht immer möglich ist (s. unten). Um herauszufinden, welche dieser Strychnosarten den Indianern bei der Curareherstellung dienen, müssen ihre Inhaltsstoffe ermittelt werden.

Bisher sind 22 Strychnosarten mit den im vorigen Kapitel besprochenen Techniken auf ihren Gehalt an Alkaloiden geprüft worden; Strychnin wurde in keiner gefunden. In 13 Arten konnte man Calebassen-Alkaloide nachweisen (*Tabelle 2*, S. 237). Jedoch nur sechs Calebassen-Alkaloide, nämlich C-Toxiferin-I, Toxiferin-II, C-Mavacurin, C-Fluorocurin, C-Alkaloid-Y und Caracurin-II, nicht aber die Hauptalkaloide C-Curarin-I und Calebassin, sind bis jetzt i n S u b s t a n z aus südamerikanischen Strychnospflanzen isoliert worden (*Tabelle 3*, S. 238).

Neben den wenigen Calebassen-Alkaloiden hat man aus Strychnospflanzen über 30 n e u e Indolalkaloide gewonnen (Tabelle 3). Darüber hinaus ist die Existenz zahlreicher weiterer Alkaloide papierchromatographisch nachgewiesen (*1, 34, 76, 64, 63, 47, 61, 62, 71, 35, 60*).

### 2. Die Alkaloide aus Strychnos toxifera.

*Strychnos toxifera* ist die am häufigsten untersuchte südamerikanische Strychnosart. H. WIELAND und seine Schüler (*100, 96*) isolierten aus einer von King zur Verfügung gestellten, aus Britisch-Guayana stammenden Rinde dieser Pflanze die Alkaloide Toxiferin-I und Toxiferin-II.

King hat gleichfalls eine *Str. toxifera*-Rinde aus Britisch-Guayana bearbeitet (*55a*) und daraus 11 Alkaloide, die Toxiferine-I bis X und das Toxiferin-XII isoliert*. Sein Toxiferin-I ist identisch mit C-Toxiferin-I (*82*), welches seinerseits höchstwahrscheinlich dem Toxiferin-I H. Wielands entspricht (*83*). Wie King vermutet, ist sein Toxiferin-II mit dem Toxiferin-II H. Wielands identisch. Seine anderen Toxiferine können nicht ohne weiteres mit den Calebassen-Alkaloiden verglichen werden. Battersby und Hodson (*15*) haben aus einer ebenfalls aus Britisch-Guayana bezogenen *Str. toxifera*-Rinde neben C-Toxiferin-I ein Alkaloid „A 8" isoliert, das sich als $N_{(b)}$-Metho-Wieland-Gumlich-Aldehyd (LII, S. 216) herausgestellt hat und auch bei der Hydrolyse von C-Toxiferin-I entsteht.

Bei der (venezolanischen) *Str. toxifera*-Pflanze, deren Rinde von Asmis, Schmid und Karrer (*8*, *10*) untersucht worden ist, handelte es sich offenbar um eine Varietät, die die meisten ihrer Alkaloide am $N_{(b)}$-Atom** nicht methyliert. Denn der größte Teil der daraus isolierten Alkaloide, nämlich Nor-dihydrotoxiferin und die Caracurine-I bis IX, ist tertiär. Das gleichfalls aus dieser Rinde stammende Fedamazin ist eine Anhydroniumbase. Caracurin-V ist isomer mit Nor-C-toxiferin-I und kann leicht in dieses übergeführt werden (*7*, *20*). Caracurin-VI ist vermutlich Nor-C-alkaloid-H. Caracurin-VII hat sich als Wieland-Gumlich-Aldehyd (XLII, S. 217) herausgestellt.

Von den restlichen Caracurinen sind nur die UV-Spektren und provisorische Summenformeln bekannt.

### 3. Die Alkaloide aus Str. melinoniana Baillon.

Unter den zahlreichen Alkaloiden aus *Str. melinoniana* Baillon [Schlittler und Hohl (*81*); Bächli, Vamvacas, Schmid und Karrer (*11*)] sind nur zwei Calebassen-Alkaloide, nämlich C-Mavacurin und C-Fluorocurin, aufgefunden worden.

Die Melinonine-A, B, E und G sind teils aufgeklärt, teils sind dafür Konstitutionen in Vorschlag gebracht worden (S. 202 ff.). Melinonin-C ist identisch mit *l*-Narcotin, Melinonin-D mit Thebain. (Diese beiden

---

* Nach einer Privatmitteilung von Herrn Prof. A. R. Battersby an Herrn Prof. H. Schmid ist Toxiferin-XI (von King) mit C-Toxiferin-I identisch.

** Bei Alkaloiden, die sich vom Tryptamin ableiten, bezeichnet man den Indol-stickstoff als $N_{(a)}$, den anderen als $N_{(b)}$.

Alkaloide stammen vermutlich nicht aus *Str. melinoniana*, sondern aus einer beim Mahlen der Rinde eingeschleppten Verunreinigung.) Melinonin-F, $C_{13}H_{13}N_2{}^+$, ist mit $N_{(b)}$-Metho-harman (III) identisch, welches damit erstmals in der Natur gefunden worden ist.

Das quartäre Melinonin-H, $C_{20}H_{21-23}ON_2{}^+$, ist seinem UV-Spektrum nach interessanterweise ein Chinolin-Abkömmling. Es enthält 1 N—CH$_3$ und 1 C—CH$_3$. Mit $N_{(b)}$-Metho-cinchonin oder -cinchonidin ist es nicht identisch.

(III.)  Melinonin-F.

Die Melinonine-I und K sind Verbindungen mit Indolspektrum. Sie sind zersetzlich und bisher nur in kleinen Mengen gefaßt worden.

Melinonin-L, $C_{22}H_{26}O_4N_2$, enthält je eine OCH$_3$-, N—CH$_3$- und C—CH$_3$-Gruppe und zwei aktive Wasserstoffatome. Es liefert eine Monoacetylverbindung und eine Monomethoverbindung. Sein UV-Spektrum ist demjenigen von Melinonin-A sehr ähnlich. In alkalischer Lösung, ferner in Dioxan, Methylcellosolve und Chloroform zeigt es jedoch im Gegensatz zu diesem ein langwelliges Spektrum.

### 4. Die Alkaloide aus Str. amazonica Kruk., Str. macrophylla und Str. guianensis (Aubl.) Mart.

Das „Alkaloid γ" aus *Str. amazonica* besitzt ein Indolspektrum; seinem Chlorid wird die Formel $C_{34}H_{45}N_3O_2Cl_2$ zugeschrieben (*35*). Für Macrophyllin-A aus *Str. macrophylla* BARB. R. kommt die Formel $C_{20}H_{23}O_2N_2{}^{++}$ in Frage. Sein UV-Spektrum ist wenig charakteristisch (*47*). Guiacurarin-III (*62*) ist als Chlorid und Pikrat kristallisiert; es zeichnet sich durch sein langwelliges UV-Spektrum aus.

### 5. Diabolin und Desacetyl-diabolin aus Str. diaboli Sandw.

Diabolin, $C_{21}H_{26}O_3N_2$, ist eine $N_{(a)}$-Acetylverbindung. Es enthält eine Hydroxylgruppe und wahrscheinlich eine Äthergruppierung. Die Desacetyl-verbindung ist ein Indolin. Man erhielt sie einerseits durch Verseifen des Diabolins, andererseits aus einem Salzsäurehydrolysat der *Str. diaboli*-Rinde; und zwar in zwei polymorphen Formen. Desacetyl-diabolin zeigt ähnliche Farbreaktionen wie Vomicin. Durch Methyljodid wird es in eine quartäre Base umgewandelt. Sein $N_{(b)}$-Atom ist demnach trisubstituiert (*13*).

Anmerkung bei der Korrektur: Das Desacetyl-diabolin ist, wie BATTERSBY und HODSON (*15 a*) jetzt gefunden haben, mit WIELAND-GUMLICH-Aldehyd (XLII, S. 217) identisch. Diabolin ist $N_{(a)}$-Acetyl-WIELAND-GUMLICH-Aldehyd.

## V. Die Elektronenspektren der Alkaloide aus Calebassencurare und südamerikanischen Strychnosarten.

Unter den in den Kapiteln III und IV erwähnten zahlreichen Alkaloiden trifft man, was die Lichtabsorption im ultravioletten und sichtbaren Spektralbereich angeht, bezeichnenderweise nur einige wenige Typen an, denen allen das Indolskelett (a) gemeinsam ist. Es sind dies:

(a)

Indole (b), α-Methylenindoline (c), Indoline (d), Oxindole (e), N-Acylindoline (f), ψ-Indoxyle (g), β-Carboliniumderivate (h) und Verbindungen
mit Curarin-chromophor.

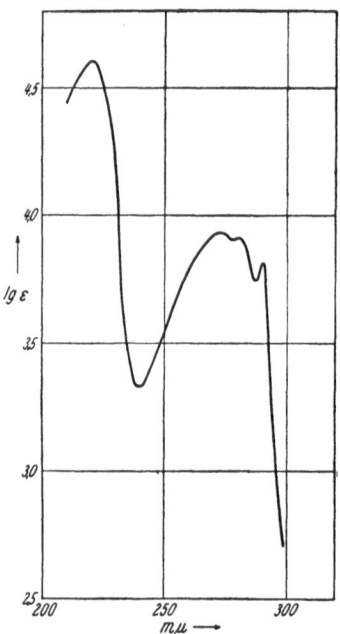

Für die Spektren der wichtigsten dieser Chromophore werden nachstehend Beispiele angeführt *(Abb. 2—6)*.

Zwischen Oxindolen (e) und N-Acyl-
indolinen (f) läßt sich oft an Hand des
UV-Spektrums allein nicht unterscheiden.

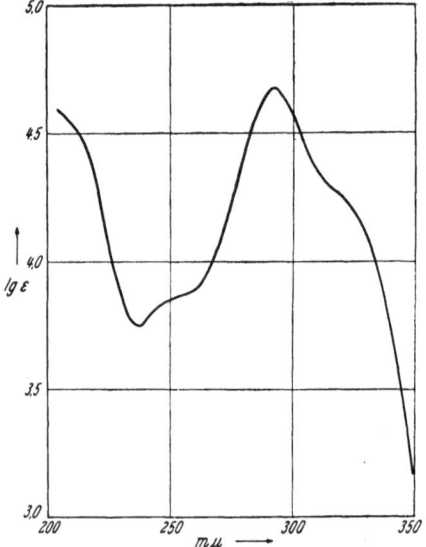

Abb. 2. UV-Spektrum eines Alkaloids mit Indolchromophor b [Melinonin-B-chlorid (XXII,
S. 203) in 95-proz. Äthanol].

Abb. 3. UV-Spektrum eines Alkaloids mit α-Methylen-
indolinchromophor c [C - Dihydrotoxiferin - dichlorid
(XXXIX, S. 212) in abs. Methanol].

Bei β-Carbolinium-derivaten (Chromophor h) kann mit Hilfe der
Spektroskopie festgestellt werden, ob das $N_{(a)}$-Atom substituiert ist oder

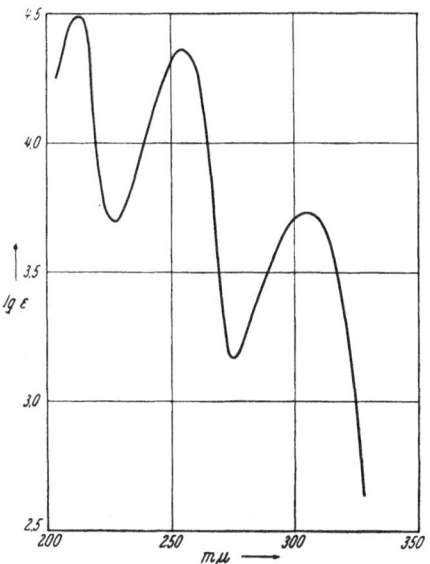

Abb. 4. UV-Spektrum eines Alkaloids mit Indolinchromophor d [C-Calebassin-dichlorid (LVIII, S. 221)
in 96-proz. Äthanol].

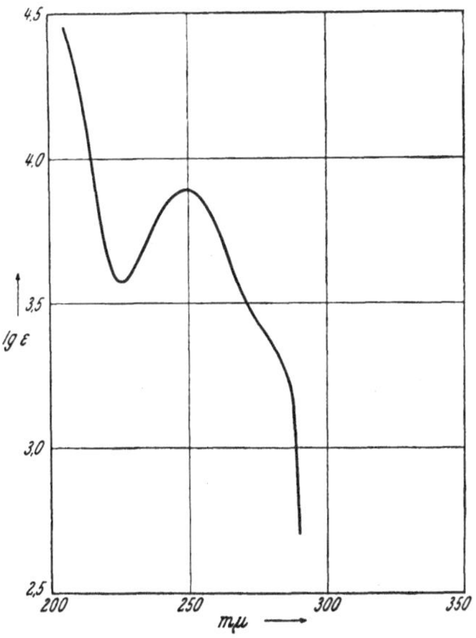

Abb. 5. UV-Spektrum eines Alkaloids mit
Oxindolchromophor e [C-Alkaloid-M-chlorid in
Wasser].

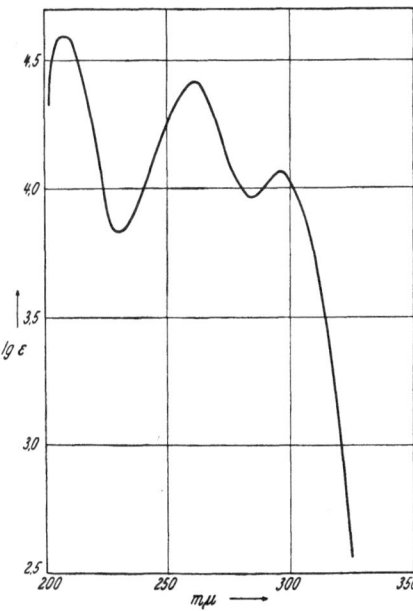

Abb. 6. UV-Spektrum eines Alkaloids mit Curarin-chromophor [C-Curarin-I-dichlorid (LIV, S. 219) in
95-proz. Äthanol].

nicht. Im letzteren Fall (IV) wird nämlich die Absorptionskurve nach längeren Wellen verschoben, sobald man die Lösung alkalisch macht. Diese Erscheinung ist auf Abspaltung eines Protons vom $N_{(a)}$-Atom und Bildung einer Anhydroniumbase (IV a) zurückzuführen.

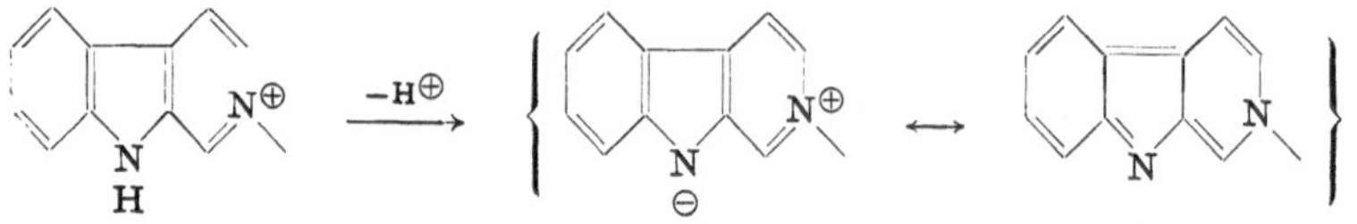

(IV.)                    (IV a.) Anhydroniumbase.

Das dem C-Curarin-I und seinen Verwandten zugrunde liegende chromophore System ist nicht bekannt (vgl. S. 218).

In *Tabelle 4*, S. 239, sind die Calebassen- und Strychnosalkaloide nach Spektren geordnet aufgeführt.

## VI. Konstitution von Calebassen- und südamerikanischen Strychnos-Alkaloiden.

### Vorbemerkung.

Die über 70 bisher aus Calebassencurare und südamerikanischen Strychnosarten isolierten Alkaloide, welche in den *Tabellen 1* und *3* (SS. 236 und 238) zusammengestellt sind, lassen sich, soweit über ihre Konstitution etwas bekannt ist, biogenetisch fast ausnahmslos auf ein System zurückführen, das als Bausteine Tryptamin, Formaldehyd und einen $C_8$-Aldehyd umfaßt:

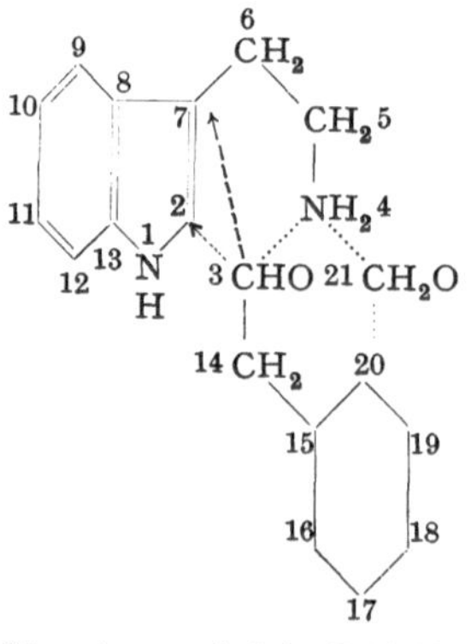

(Numerierung wie beim Yohimbin.)

Von diesem System leiten sich zwei Alkaloid-Grundtypen ab: Je nachdem ob das C-Atom 3 (Aldehyd-C) des $C_8$-Körpers eine Bindung mit dem C-Atom 2 oder 7 des Tryptamins eingegangen ist, liegt entweder ein Alkaloid vom Yohimbin- oder vom Strychnin-Typus vor.

# A. Alkaloide vom Yohimbin-Typus.

## 1. Übersicht.

Fünf von den insgesamt elf Calebassen- und Strychnos-Alkaloiden, die nach heutiger Kenntnis zum Yohimbin-Typus gehören, nämlich C-Mavacurin, Lochnerin, Lochneram, Melinonin-A und Melinonin-B, sind Körper mit dem durch sein charakteristisches UV-Spektrum (Abb. 2, S. 192) leicht nachweisbaren Indolchromophor. Melinonin-A ist vollständig aufgeklärt; für die übrigen Alkaloide sind Formelvorschläge gemacht worden. C-Alkaloid-Y, ein Indolin, und das $\psi$-Indoxyl C-Fluorocurin sind Folgeprodukte des C-Mavacurins und durch übersichtliche Reaktionen mit diesem verknüpft. Die Melinonine-E und G, für welche ebenfalls Konstitutionen in Vorschlag gebracht worden sind, stellen $\beta$-Carbolin-Derivate mit unsubstituierten $N_{(a)}$-Atomen dar.

## 2. C-Mavacurin, C-Alkaloid-Y und C-Fluorocurin.

C-Mavacurin, $C_{20}H_{25}ON_2^+$, wurde erstmals von TH. WIELAND und MERZ (*104*) aus der südamerikanischen Droge Mavacure isoliert. C-Fluoro-curin, $C_{20}H_{25}O_2N_2^+$, wurde von SCHMID und KARRER (*84*) im Calebassen-curare entdeckt. Die beiden eng verwandten, physiologisch nur schwach wirksamen Alkaloide sind auch in einigen südamerikanischen Strychnos-Arten gefunden worden. Durch Arbeiten der Züricher Schule [BICKEL, GIESBRECHT, KEBRLE, SCHMID und KARRER (*30*); BICKEL, SCHMID und KARRER (*31*)] wurde ihre Struktur weitgehend aufgeklärt. Fluoro-curin (V) ist seinem IR- und UV-Spektrum nach ein $\psi$-Indoxyl (Va)*. Bei der Reduktion mit $NaBH_4$ geht es in ein $\beta$-Hydroxy-indolin (VIa) über, das als Hydrofluorocurin (VI, S. 199) bezeichnet wird. Dieses erleidet unter Säureeinwirkung eine WAGNER-MEERWEIN-Umlagerung. Das hierbei gebildete Indol (VIIa) ist identisch mit Mavacurin.

WITKOP (*106*) hat auf Grund von Erfahrungen an einfacheren Ver-bindungen frühzeitig die Vermutung ausgesprochen, daß in der Pflanze umgekehrt aus Mavacurin (VII) das Fluorocurin (V) entsteht. Wie FRITZ und TH. WIELAND zeigen konnten (*40*), läßt sich diese Umwandlung in vitro durch Oxydation des Mavacurins (VII) mit Sauerstoff und Pt in Wasser oder verdünnter Essigsäure bewerkstelligen. Als isolierbares Zwischenprodukt tritt dabei C-Alkaloid-Y (VIII) auf, welches deshalb auch als Profluorocurin bezeichnet wird. Diesem ist die Teilformel (VIIIa) eines $\alpha,\beta$-Dihydroxy-indolins zuzusprechen.

---

* Müssen von ein und derselben Verbindung mehrere Partialformeln angegeben werden, so erhalten diese zur Erleichterung der Übersicht dieselbe römische Ziffer wie die betreffende Verbindung selbst und zusätzlich einen kleinen Buchstaben, z. B. (Va).

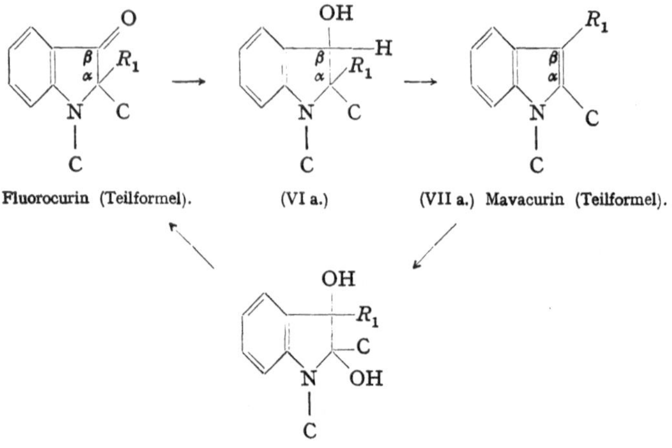

(V a.) Fluorocurin (Teilformel).    (VI a.)    (VII a.) Mavacurin (Teilformel).

(VIII a.) C-Alkaloid-Y (Teilformel).

C-Mavacurin (VII) und C-Alkaloid-Y (VIII, s. oben) haben gleiches Kohlenstoffgerüst. C-Fluorocurin (V) weicht nur insofern ab, als der in Mavacurin (VII) in $\beta$-Stellung befindliche Rest $R_1$ in der $\alpha$-Stellung steht. Unter Berücksichtigung dieser Tatsachen können Ergebnisse, die an einem der Alkaloide erhalten worden sind, auf die anderen übertragen werden.

Beim Erhitzen im Vakuum geht Fluorocurin-chlorid (vgl. V) unter Verlust von $CH_3Cl$ in das tert. Nor-fluorocurin (IX, S. 199) über; dieses kann mit $CH_3J$ zu Fluorocurin-jodid remethyliert werden. Am quartären $N_{(b)}$-Atom der drei Alkaloide haftet also eine Methylgruppe. O-Methylgruppen oder weitere N-Methylgruppen sind nicht vorhanden. Fluorocurin (V) und seine Norverbindung (IX) lassen sich in Monoacetyl-Derivate überführen, ohne daß dabei eine Änderung der UV-Spektren einträte. Dadurch ist die Anwesenheit einer Hydroxylgruppe nachgewiesen, die keine Beziehung zum chromophoren System hat.

Mit der Kuhn-Roth-Methode findet man in den Alkaloiden 1 C—$CH_3$. Ozonabbau des Fluorocurins (V) und Mavacurins (VII) liefert Acetaldehyd und eliminiert also die C—$CH_3$-haltige Gruppierung. Eine selektive Hydrierung, die in einheitlicher Reaktion allein diese erfaßt, war nur bei dem tertiären $\beta$-Hydroxy-indolin Nor-hydrofluorocurin (X) möglich, welches durch Reduktion von Nor-fluorocurin (IX) mit $LiAlH_4$ oder $NaBH_4$ entsteht. (X) geht bei der katalytischen Hydrierung unter Aufnahme von 1 $H_2$ in Tetrahydro-nor-fluorocurin (XI) über. Diese Verbindung gibt im Gegensatz zu Fluorocurin (V) und Nor-hydrofluorocurin (X) bei der „modifizierten Mikro-chromsäureoxydation" (31)*

---

* Dieses Verfahren besteht darin, daß man während der Chromsäureoxydation nicht wie bei der Kuhn-Roth-Methode zunächst unter Rückfluß kocht, sondern

neben Essigsäure auch Propionsäure. Damit ist erwiesen, daß das Fluorocurin (V) eine Äthylidengruppe enthält.

Katalytische Hydrierung des quartären Hydro-fluorocurins (VI) in Alkohol mit Platinkatalysator ergibt zur Hauptsache eine Verbindung $C_{20}H_{30}O_2N_2$, $\varepsilon_1$-Hexahydro-fluorocurin (XII). Diese tertiäre Base liefert mit Ozon keinen Acetaldehyd. Bei der Chromsäureoxydation konnte man Propionsäure und Methyläthylessigsäure nachweisen. $\varepsilon_1$-Hexahydro-fluorocurin (XII) ist demnach durch EMDE-Abbau und gleichzeitige Hydrierung der Äthylidendoppelbindung entstanden:

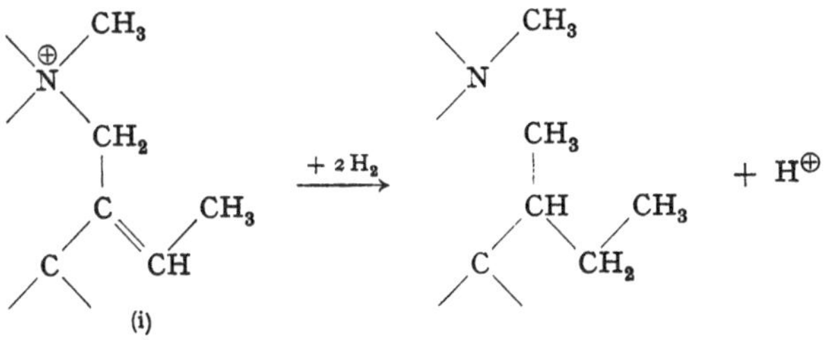

Über die Verknüpfung der somit für die drei Alkaloide abgeleiteten Anordnung (i) mit dem Indolsystem gibt eine Selendehydrierung des Nor-mavacurins (XIII) Auskunft. Bei dieser Reaktion erhielt man eine Base (XIV), die als Pikrat kristallisierte. Nach dem UV-Spektrum handelt es sich um ein $\beta$-Carbolinderivat. Da sich das UV-Spektrum des Chlormethylates von (XIV) unter dem Einfluß von OH-Ionen nicht ändert, kann weiter geschlossen werden, daß (XIV) ein $N_{(a)}$-substituiertes $\beta$-Carbolin ist (vgl. S. 194).

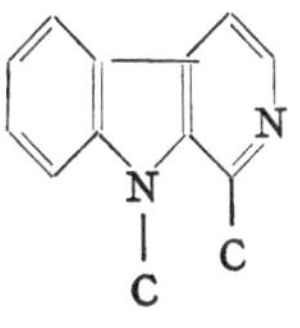

Ein $N_{(a)}$-substituiertes $\beta$-Carbolin-derivat wurde später auch als Nebenprodukt bei der oben schon erwähnten Oxydation von Mavacurin (VII) unter milden Bedingungen erhalten (40).

Aus den bis jetzt angeführten Tatsachen folgt für Mavacurin die Partialformel (VIIb, S. 198), die 18 von 20 C-Atomen sowie die zwei N-Atome des Alkaloids erfaßt. Nach Summenformel und Anzahl der Doppelbindungen (3 aromatische, 1 Indol- und 1 Äthylidendoppelbindung) muß Mavacurin fünf Ringe enthalten. In die Formel (VIIb) sind daher die zwei noch fehlenden Kohlenstoffatome so einzufügen, daß zwei weitere C-methylfreie Ringe entstehen. Die Zahl der dafür gegebenen Möglichkeiten läßt sich an Hand einer tertiären Base des Mavacurins (VII), die man bei Hydrierung mit Platinkatalysator in 0,1-n alkoholischer

gleich mit dem Abdestillieren beginnt. Dabei gehen höhere Fettsäuren, die primär bei der Oxydation entstehen, mit dem Wasser über und werden der Weiteroxydation entzogen. Ihr Nachweis neben Essigsäure erfolgt papierchromatographisch.

Kalilauge erhält, wesentlich einschränken. Diese Verbindung, $\varepsilon_2$-Dihydro-mavacurin (XV, S. 199), besitzt ein Indolspektrum, in saurer Lösung jedoch ein Indolinspektrum. Sie enthält keine zusätzliche C-Methylgruppe; bei ihrer Entstehung ist folglich eine Bindung zwischen dem $N_{(b)}$-Atom und einem tertiären oder quartären Kohlenstoffatom hydrogenolysiert worden:

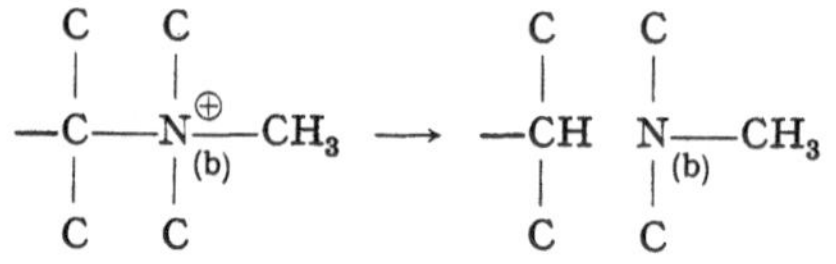

$$
\begin{array}{ccc}
\text{C} & \text{C} & \quad\quad \text{C} \quad\quad \text{C} \\
| & | & \quad\quad | \quad\quad | \\
-\text{C}-\overset{\oplus}{\text{N}}-\text{CH}_3 & \longrightarrow & -\text{CH} \quad \text{N}-\text{CH}_3 \\
| & |_{(b)} & \quad\quad | \quad\quad |_{(b)} \\
\text{C} & \text{C} & \quad\quad \text{C} \quad\quad \text{C}
\end{array}
$$

Dieser Emde-Abbau läßt vermuten, daß das $N_{(b)}$-Atom des Mavacurins über die in (XV) gelöste Bindung mit einer allylständigen Doppelbindung (Indol-Doppelbindung) verknüpft ist.

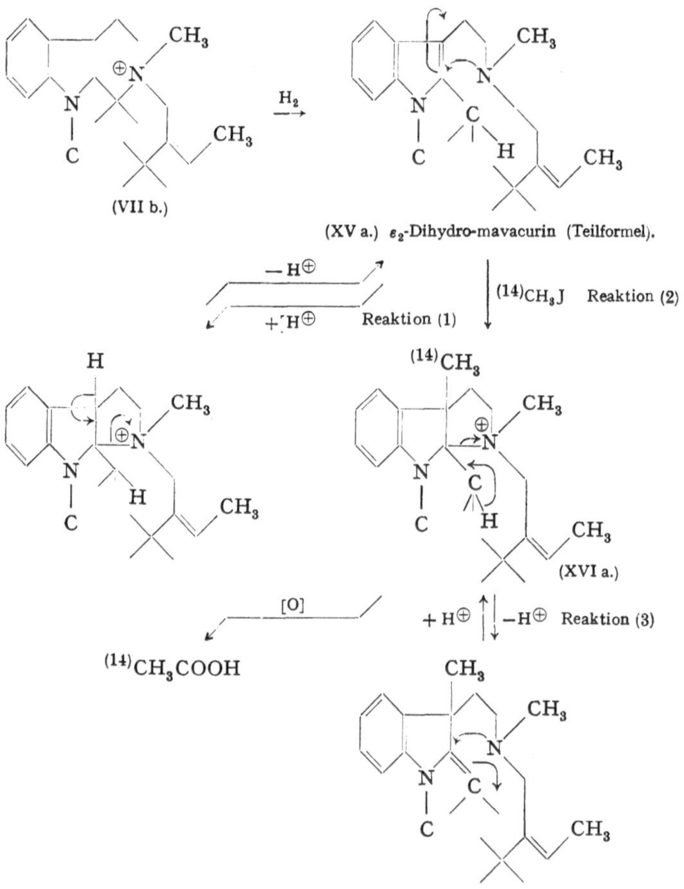

(VII b.)

(XV a.) $\varepsilon_2$-Dihydro-mavacurin (Teilformel).

$-H^{\oplus}$ / $+H^{\oplus}$  Reaktion (1)

$^{(14)}CH_3J$  Reaktion (2)

(XVI a.)

$+H^{\oplus}$ / $-H^{\oplus}$  Reaktion (3)

[O]

$^{(14)}CH_3COOH$

*Formelübersicht 1.*

*Literaturverzeichnis: SS. 241—247.*

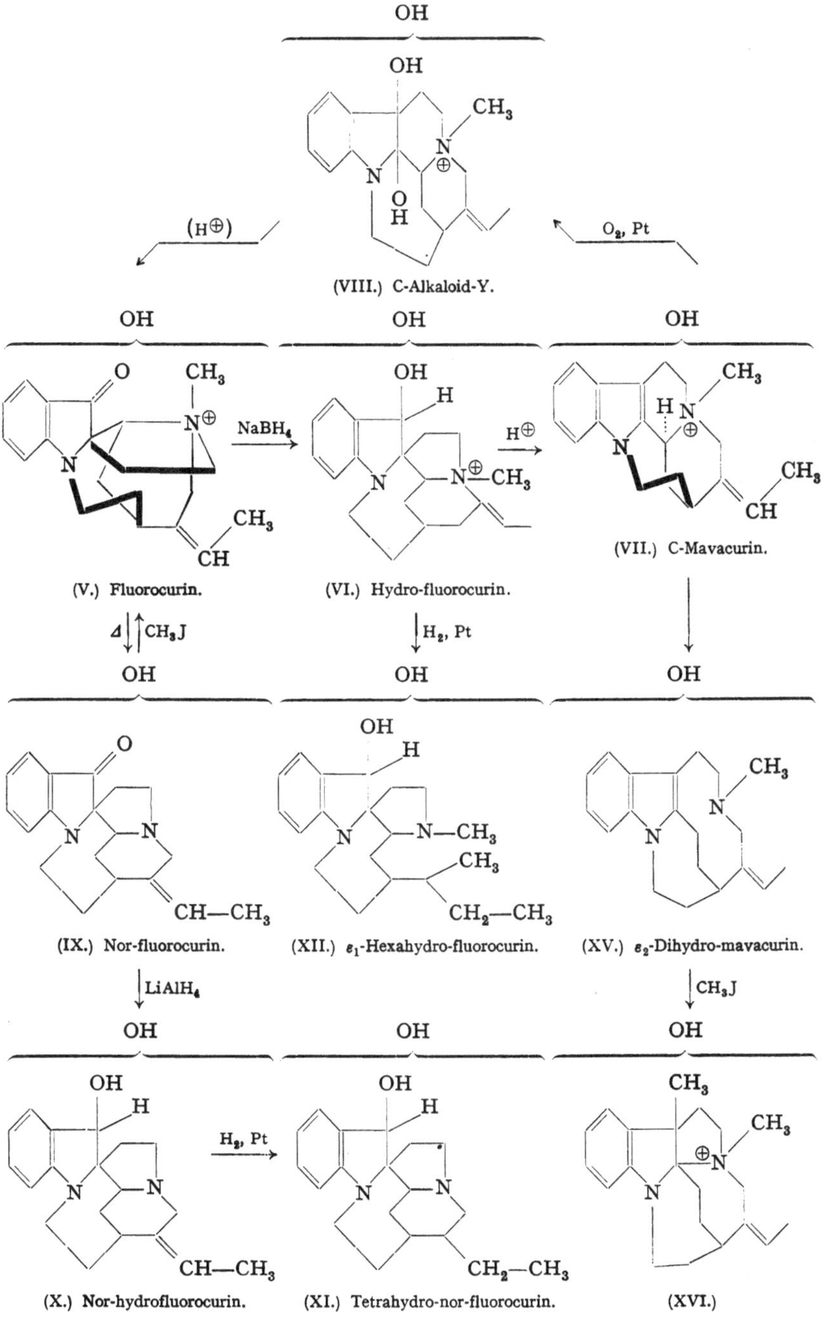

OH

(VIII.) C-Alkaloid-Y.

$(H \oplus)$ ⟋    ⟍ $O_2$, Pt

OH                    OH                    OH

(V.) Fluorocurin.    (VI.) Hydro-fluorocurin.    (VII.) C-Mavacurin.

NaBH₄    $H \oplus$

$\Delta \big\uparrow \big| CH_3 J$    $\big| H_2$, Pt

OH                    OH                    OH

(IX.) Nor-fluorocurin.    (XII.) $\varepsilon_1$-Hexahydro-fluorocurin.    (XV.) $\varepsilon_2$-Dihydro-mavacurin.

$\big|$ LiAlH₄    $\big|$ CH₃J

OH                    OH                    OH

(X.) Nor-hydrofluorocurin.    (XI.) Tetrahydro-nor-fluorocurin.    (XVI.)

$H_2$, Pt

*Formelübersicht 2.* C-Fluorocurin, C-Mavacurin, C-Alkaloid-Y und ihre Derivate.

Verbindung (XV) addiert Methyljodid; das Produkt enthält jedoch keine zusätzliche N-Methyl-, sondern eine neue C-Methylgruppe und ist ein Indolin (XVIa). Verwendet man für die Methylierung $^{14}CH_3J$, so erhält man markiertes (XVI), das bei der Chromsäureoxydation radioaktive Essigsäure liefert. Bildung, Protonisierung und (radioaktive) Methylierung von $\varepsilon_2$-Dihydro-mavacurin (XV) kann durch die in *Formelübersicht 1* (S. 198) angegebenen Partialformeln ausgedrückt werden.

Das Methylierungsprodukt (XVI) zeigt in alkoholischer Lauge das UV-Spektrum eines $\alpha$-Methylenindolins. Diese Erscheinung kann durch Neutralisation der Lösung rückgängig gemacht werden und findet durch die Annahme eines besonders glatt verlaufenden, reversiblen HOFMANN-Abbaus [Reaktion (3), S. 198] eine Erklärung.

Die Umsetzungen (1), (2) und (3) sind transannulare Reaktionen und setzen voraus, daß die beteiligten Atome in (XV) derart in einem 8-, 9- oder 10-gliedrigen Ring angeordnet sind, daß bei den ,,Überbrückungen" zwei spannungsarme kleinere Ringe entstehen.

Ergänzt man Partialformel (VIIb) an Hand der letzterwähnten Ergebnisse, so bleiben einige wenige Formulierungsmöglichkeiten. Aus biogenetischen Gründen bevorzugen BICKEL, SCHMID und KARRER für Mavacurin das Konstitutionsbild (VII), das auch der *Formelübersicht 2* (S. 199) zugrunde gelegt ist. (Die in dieser Übersicht angegebenen Stereoformeln folgen aus Modellbetrachtungen.)

Die Stellung der acetylierbaren Hydroxylgruppe im Mavacurin bzw. Fluorocurin ist nicht bekannt. Die zum $N_{(b)}$-Atom $\alpha$-ständigen C-Atome dürften jedoch als Haftstellen nicht in Frage kommen.

### 3. Lochnerin und Lochneram.

Lochnerin ist auch als C-Alkaloid-T (5) und (provisorisch) als Alkaloid C (46) bezeichnet worden.

Das tertiäre, eine phenolische Methoxylgruppe enthaltende Lochnerin (XVII), $C_{20}H_{24}O_2N_2$, ist bis jetzt erst einmal in Calebassencurare gefunden worden (5). Aus dem gleichen Material isolierte man später die entsprechende Methoverbindung Lochneram, $C_{21}H_{27}O_2N_2^+$ (4).

Lochnerin (XVII) ist zuerst aus einer brasilianischen Apocynacee, *Lochnera (Vinca) rosea* (L.) REICHB., var. *alba* HUBD., dann auch aus der in Madagaskar beheimateten Apocynacee *Lochnera (Vinca) rosea* (L.) REICHB. (*Catharanthus roseus* G. DON) var. *alba* isoliert worden (68, 46). Die entsprechende Verbindung mit phenolischer Hydroxyl- statt Methoxylgruppe (77) Sarpagin, $C_{19}H_{22}O_2N_2$, kommt in zahlreichen Rauwolfia-Arten vor (80).

Lochnerin und Sarpagin wurden etwa gleichzeitig in verschiedenen Laboratorien auf ihre Konstitution geprüft (77, 88, 5, 89).

Lochnerin ist seinem UV- und IR-Spektrum nach, ein 5-Methoxyindol. Es besitzt eine leicht acetylierbare Hydroxylgruppe. Das O-Acetat hat das gleiche UV-Spektrum wie Lochnerin. Neben diesem Monoacetat erhält man in geringerer Menge noch ein Diacetat, dessen UV- und IR-Spektrum $N_{(a)}$-Acetylierung anzeigen. Das $N_{(a)}$-Atom des Lochnerins ist demnach nicht substituiert. Lochnerin ist frei von N-Methyl. Da es beim Behandeln mit Methyljodid in Lochneram übergeht, muß sein $N_{(b)}$-Atom tertiär sein. Lochnerin enthält eine C—$CH_3$-Gruppe; mit Chromsäure liefert es nur Essigsäure. Ozonabbau gibt Acetaldehyd und Formaldehyd, Ozonabbau des quartären Lochnerams dagegen nur Acetaldehyd (4). Die Entstehung des Formaldehyds setzt also tertiäres $N_{(b)}$-Atom voraus, eine Erscheinung, die bis jetzt keine Erklärung gefunden hat.

Bei gelinder Hydrierung geht Lochnerin (XVII) in eine Dihydroverbindung über. Ozonolyse von Dihydro-lochnerin liefert weder Acetaldehyd noch Formaldehyd. Mit Chromsäure erhält man Propionsäure. Diese Ergebnisse beweisen, daß Lochnerin eine Äthylidengruppe enthält.

Die Hydroxylgruppe in (XVII) läßt sich tosylieren. Reduktion des Tosylats mit $LiAlH_4$ führt zu Desoxy-lochnerin (XVIII). Dieses besitzt im Gegensatz zu Lochnerin zwei C-Methylgruppen, liefert aber wie (XVII) beim Chromsäureabbau nur Essigsäure. Damit ist bewiesen, daß Lochnerin eine Hydroxy-methylgruppe enthält; diese haftet wahrscheinlich an einem tertiären oder quartären Kohlenstoffatom.

Wie aus der Perhydrierung hervorgeht, ist außer den vier Doppelbindungen des Indolsystems in Lochnerin nur noch die Doppelbindung der Äthylidengruppe vorhanden. An Hand der Summenformel errechnet sich dann, daß es fünf Ringe enthalten muß.

Diese Befunde veranlaßten ARNOLD, V. PHILIPSBORN, SCHMID und KARRER, für Lochnerin die Formel (XVII) in Vorschlag zu bringen (5). POISSON, LE MEN und JANOT (77) stellten die gleiche Formel zur Diskussion. Zur entsprechenden Formel (XIX) kamen STAUFFACHER, HOFMANN und SEEBECK (88) für Sarpagin, desgleichen TALAPATRA und CHATTERJEE (89), die sich auch mit dem räumlichen Bau dieses Alkaloids beschäftigt haben (90).

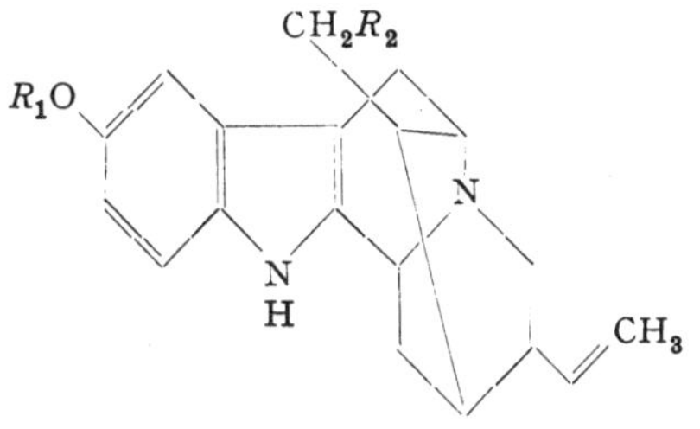

(XVII.) $R_1$ = $CH_3$, $R_2$ = OH. Lochnerin (?).
(XVIII.) $R_1$ = $CH_3$, $R_2$ = H.   Desoxy-lochnerin (?).
(XIX.) $R_1$ = H,  $R_2$ = OH. Sarpagin (?).

### 4. Melinonin-A.

Melinonin-A (XX), $C_{22}H_{27}O_3N_2{}^+$, das Hauptalkaloid aus *Str. melinoniana* BAILLON, ist von SCHLITTLER und HOHL (*81*) entdeckt und aufgeklärt worden. Es enthält 1 N—$CH_3$-, 1 O—$CH_3$- und 1 C—$CH_3$-Gruppe.

Durch Destillation im Hochvakuum wird Melinonin-A-chlorid unter Abspaltung von Methylchlorid zur Norverbindung tertiärisiert. Zur Ermittlung des Ringgerüstes unterwarf man Nor-melinonin-A der Zinkstaubdestillation, wobei man $\beta$-Methyl- und $\beta$-Äthylindol faßte. Bei der Selendehydrierung liefert Nor-melinonin-A das Alstyrin (XXI).

UV- und IR-Spektren wiesen auf die Anwesenheit der Gruppierung $CH_3O$—CO—CH=CH—C—O— in Melinonin-A hin; deshalb war anzunehmen, daß es in die Gruppe der Alkaloide Serpentin, Corynanthein und Alstonin gehört. Die UV-Spektren von Nor-melinonin-A und py-Tetrahydro-alstonin (*56*) decken sich praktisch. Ein Vergleich beider Substanzen ergab ihre Identität, d. h. Melinonin-A (XX) ist $N_{(b)}$-Metho-py-tetrahydro-alstonin.

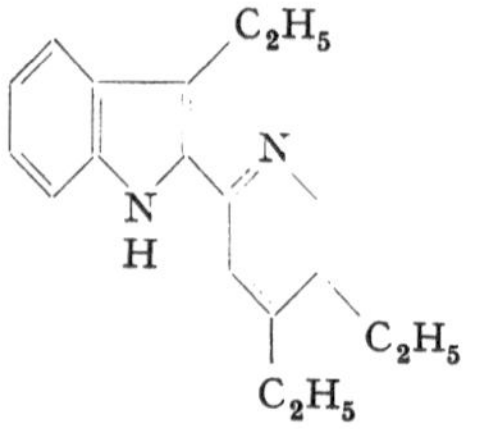

(XXI.) Alstyrin. 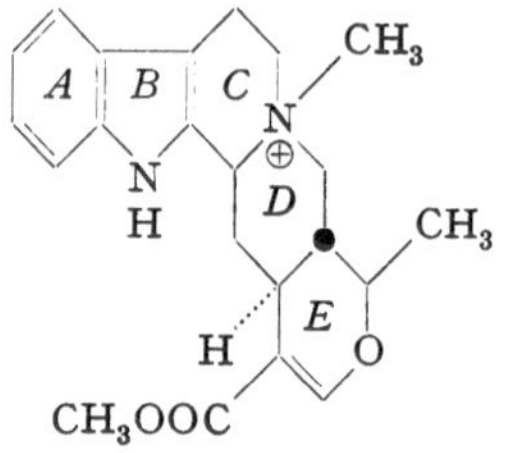  (XX.) Melinonin-A.

Die Lage der Doppelbindung im Ring $E$ der Verbindung (XX) ist exakt erst durch BADER (*12*) bewiesen worden, der Tetrahydro-alstonin mit einer Reihe von Modellsubstanzen UV- und IR-spektroskopisch verglich.

Zum sterischen Bau des Tetrahydro-alstonins s. (*95*).

### 5. Melinonin-B.

Melinonin-B (XXII, S. 203), $C_{20}H_{27}ON_2{}^+$, ist von SCHLITTLER und HOHL (*81*) neben Melinonin-A aus der Rinde von *Str. melinoniana* BAILLON isoliert worden. Eine eingehende Untersuchung des Alkaloids verdankt man VAMVACAS, v. PHILIPSBORN, SCHLITTLER, SCHMID und KARRER (*91*).

Das UV-Spektrum ist typisch dasjenige eines Indolalkaloids mit quartärem $N_{(b)}$-Atom. An letzterem haftet eine Methylgruppe, die sich thermisch abspalten läßt. Das u. a. entstehende Nor-melinonin-B wird durch Methyljodid in das Ausgangsalkaloid zurückverwandelt. Weiteres N-Methyl enthält Melinonin-B nicht; O-Methyl und C-Methyl fehlen. Dem IR-Spektrum nach besitzt Melinonin-B eine Hydroxyl und eine NH-Gruppe. Dementsprechend läßt es sich in ein N,O-Diacetat überführen mit dem für N-Acylindole charakteristischen UV-Spektrum und ist also am $N_{(a)}$-Atom nicht substituiert.

Einen weiteren Einblick in die Konstitution gewährten Dehydrierungsversuche. Mit Palladium-Norit erhielt man aus (XXII) Yobyrin (XXIII), welches alle Kohlenstoffatome des Nor-melinonins-B enthält. Mit Selen liefert (XXII) und ebenso sein Dihydroderivat (XXIV) ein 2,2'-Pyridylindol, für welches Formel (XXV) in Frage kommt. Die Dehydrierungsprodukte (XXIII) und (XXV) (s. unten) repräsentieren zusammen fünf Ringe. Melinonin-B kann indessen nur vier Ringe enthalten, da es bei Perhydrierung $5 H_2$ aufnimmt (4 Mole für das Indolsystem und 1 Mol für eine isolierte Doppelbindung). Ein Ring muß bei den Dehydrierungen entstanden sein.

Die isolierte Doppelbindung des Melinonins-B (XXII) liegt in einer Gruppierung (k) vor, wie sich aus folgenden Befunden ergibt.

(a) Melinonin-B (XXII) spaltet beim Ozonisieren Formaldehyd ab. Dihydromelinonin-B (XXIV), das bei milder Hydrierung aus (XXII) entsteht, gibt keinen Formaldehyd. (b) (XXIV) enthält im Gegensatz zu (XXII) eine C-Methylgruppe; beim modifizierten Mikro-chromsäureabbau liefert es neben Essigsäure Propionsäure (aber keine höhere Fettsäure). (c) Melinonin-B (XXII) zeigt im IR-Spektrum eine Vinylbande, die Dihydroverbindung (XXIV) nicht.

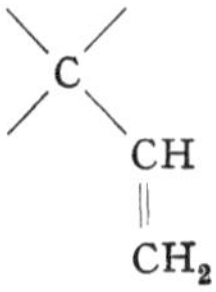

$$\begin{array}{c} \diagdown \quad \diagup \\ C \\ \diagup \quad \diagdown \\ CH \\ \| \\ CH_2 \\ (k) \end{array}$$

Dem Dehydrierungsprodukt Yobyrin (XXIII) fehlt eine der Vinylgruppe entsprechende Seitenkette. Die Autoren (91) nehmen deshalb an, daß der Ring D des Yobyrins der „überzählige", im Melionin-B (XXII) nicht vorhandene Ring ist, und erst bei der Dehydrierung aus der Vinyl-

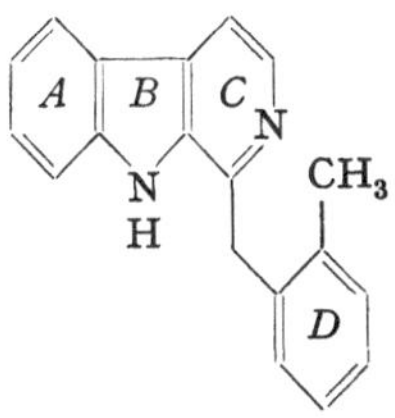

(XXIII.) Yobyrin.

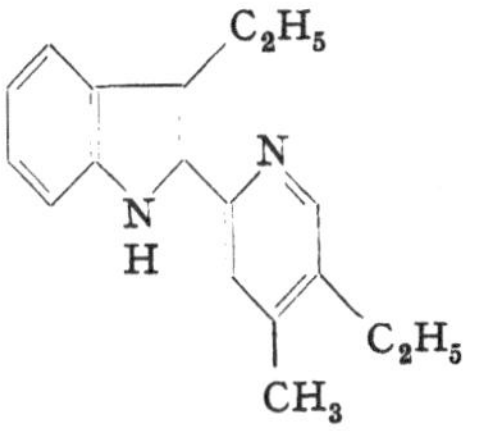

(XXV.)

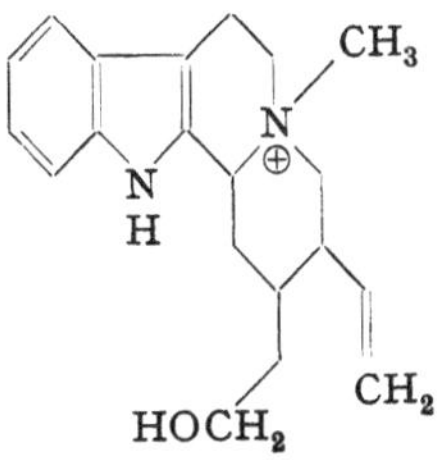

(XXII.) Melinonin-B (?).

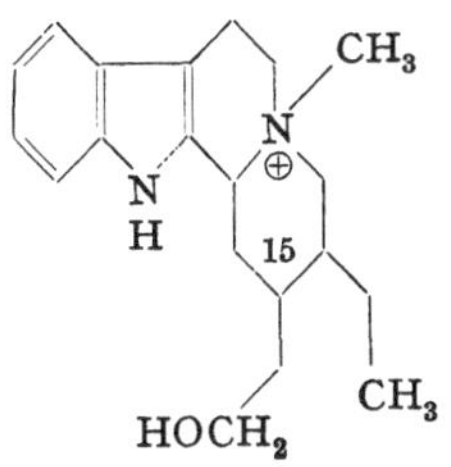

(XXIV.) Dihydro-melinonin-B (?).

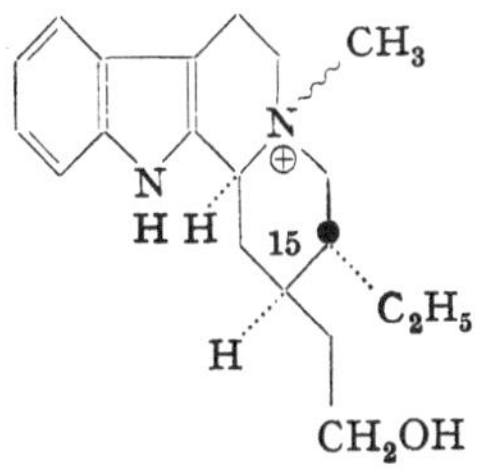

(XXVI.)

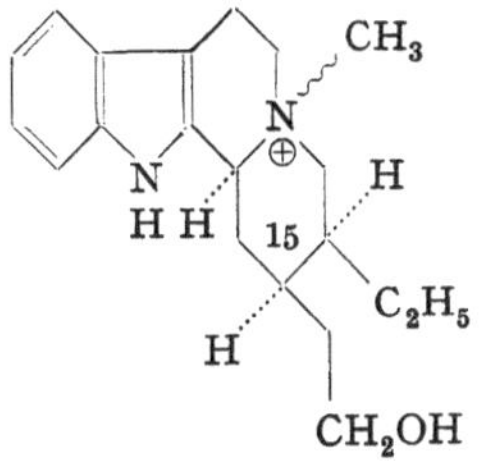

(XXVII.)

gruppe und einer zweiten zweigliedrigen Seitenkette gebildet wird. Da letztere frei von C-Methyl sein muß, kann es sich dabei nur um eine Hydroxyäthylgruppe handeln.

Nach dieser Hypothese verbleiben für (XXII) noch die auf S. 203 angegebene und eine alternative Formel, bei welcher Vinyl- und Hydroxyläthylseitenketten vertauscht sind. Die Autoren (*91*) halten die erstgenannte für wahrscheinlicher. Dem Dihydro-melinonin-B wäre dann die Konstitution (XXIV) zuzuerteilen. Von einer Verbindung dieser Formel sind 8 Racemate denkbar*. Die Verbindungen (XXVI) und (XXVII) wurden aus Corynanthein-Abkömmlingen dargestellt. Keine davon ist identisch mit (XXIV).

## 6. Melinonin-E.

Melinonin-E, $C_{20}H_{23-25}ON_2{}^+$, aus *Str. melinoniana* BAILLON ist das Salz einer Anhydroniumbase. Es ist frei von C-, N- und O-Methyl- und

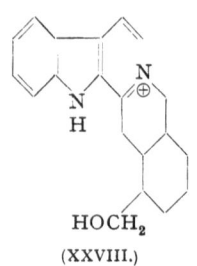

Vinylgruppen. Es liefert ein O-Acetylderivat; der Sauerstoff liegt also in einer Hydroxylgruppe vor. Auf Grund dieser Befunde ist die Konstitution (XXVIII) in Erwägung gezogen worden (*11*).

(XXVIII.)
Melinonin-E (?).

## 7. Melinonin-G.

Melinonin-G, $C_{17}H_{15}N_2{}^+$, aus *Str. melinoniana* BAILLON ist wie E das Salz einer Anhydroniumbase. Es enthält keine N—CH$_3$-Gruppe. Seinem UV- und IR-Spektrum nach muß es dem Sempervirin (XXIX) nahestehen. Bei katalytischer Hydrierung in wäßrig-alkalischem Milieu geht Melinonin-G in ein Produkt mit Indolspektrum über, welches mit Chromsäure Propionsäure liefert. Da Melinonin-G im IR-Spektrum keine Vinylbande zeigt, kommt Formel (XXX) in Betracht. Nach dieser wäre es mit dem kürzlich aufgeklärten und synthetisierten Flavopereirin aus *Geissospermum laeve* und *G. vellosii* identisch (*16, 45, 44, 78*).

(XXX.) Flavopereirin [Melinonin-G (?)].        (XXIX.) Sempervirin.

---

* Nach WENKERT (*95*) besitzen alle bisher daraufhin untersuchten Indol-Alkaloide das H-Atom an $C_{(15)}$ in α-Lage. Dies für (XXIV) vorausgesetzt, wäre nur noch mit 8 *Diastereomeren* zu rechnen.

*Literaturverzeichnis: SS. 241—247.*

## B. Alkaloide vom Strychnin-Typus.

### 1. Übersicht.

Bei den Alkaloiden vom Strychnin-Typus sind monomere und dimere Verbindungen zu unterscheiden, mit 19 bzw. 38 und 20 bzw. 40 C-Atomen, je nachdem, ob es sich um tertiäre oder quartäre Vertreter handelt.

v. PHILIPSBORN, SCHMID und KARRER (*74*) haben einen einfachen Test entwickelt, der es erlaubt, zwischen monomeren und dimeren *tertiären* Alkaloiden zu unterscheiden. (Quartäre Alkaloide müssen für den Test in tertiäre Derivate übergeführt werden.) Er hat zur Voraus-

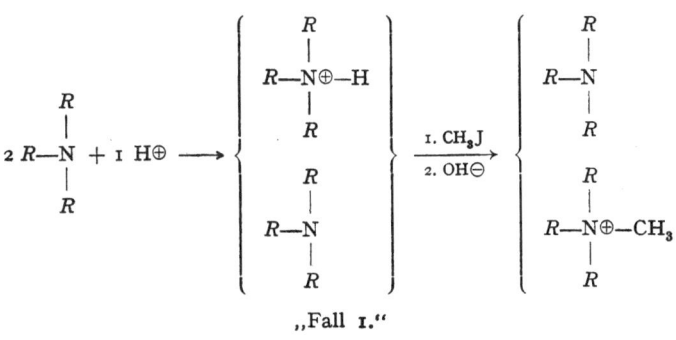

„Fall 1."

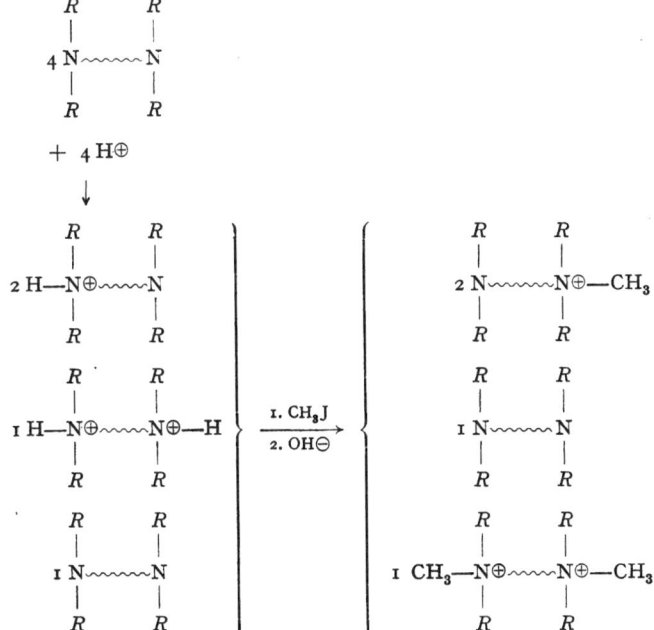

„Fall 2."

setzung, daß die monomeren Alkaloide nur e i n stark basisches N-Atom,
die dimeren zwei (etwa gleich) stark basische N-Atome enthalten.

Aus dem monomeren Alkaloid („Fall 1") erhält man beim Versetzen mit
$^1/_2$ Äquivalent Säure ein Gleichgewichtsgemisch 1 : 1 von freier Base und
protonisierter Base; durch Methyljodid-zugabe erreicht man eine „Fixierung"
in der Weise, daß die freie Base in die Methoverbindung übergeht, wogegen die
protonisierte Base unverändert bleibt (Schema S. 205). Aus dem dimeren Alkaloid
(„Fall 2") entsteht bei Halbneutralisation ein Gleichgewichtsgemisch von freier
Base, monoprotonisierter Base und diprotonisierter Base (zirka 1 : 2 : 1). Fixierung
mit Methyljodid gibt ein Gemisch von biquartärer Base, monoquartär-mono-
protonisierter Base und diprotonisierter Base (Schema S. 205). In praxi stellt man
durch Papierchromatogramme der in die Chloridform übergeführten Reaktionsprodukte fest, ob „Fall 1" oder „Fall 2" vorliegt: Mit dem basischen Lösungs-
mittelgemisch „D" (S. 187), welches aus Hydrochloriden die entsprechenden
Basen in Freiheit setzt, erhält man im „Fall 1" zwei Flecke (tertiäre Base schnell-
wandernd, quartäre Base langsamwandernd), im „Fall 2" drei Flecke (bitertiäre
Base, monotertiär-monoquartäre Base, biquartäre Base). Dieser Test wird in
der Folge als „Methode der partiellen Quartärisierung" erwähnt.

In die bis jetzt spärliche Reihe der monomeren Alkaloide vom
Strychnin-Typus gehören die Indoline Caracurin-VII, $C_{19}H_{22}O_2N_2$
[= Wieland-Gumlich-Aldehyd (XLII, S. 213)] und „Alkaloid A 8",
$C_{20}H_{25}O_2N_2^+$ [= $N_{(b)}$-Metho-Wieland-Gumlich-Aldehyd (LII, S. 216)]
als Verbindungen bekannter Konstitution; ferner das weitgehend auf-
geklärte C-Fluorocurarin, $C_{20}H_{23}ON_2^+$, das ein spezielles, erstmals bei
ihm angetroffenes chromophores System besitzt.

Die Reihe der zur Zeit als solche erkannten dimeren Alkaloide vom
Strychnin-Typus unterteilt man zweckmäßig in vier Gruppen (a—d)
von Verbindungen gleicher UV-Spektren und Farbreaktionen, die nachstehend aufgeführt sind.

*a) Die C-Dihydrotoxiferin-Gruppe.* UV-Spektren: Abb. 3
S. 192. Chromophor nebenstehend. $[\alpha]_D \approx -600°$ (Alkohol oder
Wasser). Charakteristische IR-Bande: 6,05 $\mu$ [$\nu$ (C=C) der Grup-
pierung >N—C=C<]. Farbreaktion mit Cer(IV)-sulfat: rot-
bis blauviolett → farblos; Farbreaktion mit $FeCl_3$—$H_2SO_4$: blau.
Vertreter:

| quartäre | | |
|---|---|---|
| C-Dihydrotoxiferin, | $C_{40}H_{46}N_4^{++}$ | |
| C-Alkaloid-H, | $C_{40}H_{45}(OH)N_4^{++}$ | (?) |
| C-Toxiferin-I, | $C_{40}H_{44}(OH)_2N_4^{++}$ | |
| C-Iso-dihydrotoxiferin-I, | ? | |
| Alkaloid 2, | ? | |

| tertiäre | | |
|---|---|---|
| Nor-dihydrotoxiferin, | $C_{38}H_{40}N_4$ | |
| Caracurin-VI, | $C_{38}H_{39}(OH)N_4$ | (?) |
| Nor-C-toxiferin-I, | $C_{38}H_{38}(OH)_2N_4$ | |
| C-Alkaloid-S, | ? | |

In enger Beziehung zu Nor-C-toxiferin-I steht das damit isomere Caracurin-V.

*Literaturverzeichnis: SS. 241—247.*

*b) Die C-Curarin-I-Gruppe.* UV-Spektren: Abb. 6, S. 193. Chromophor: unbekannt. $[\alpha]_D \approx +70°$. Charakteristische IR-Bande: 6,05 $\mu$ Doppelbindungsbande der Gruppierung $>C=C-N<$ oder $-\overset{|}{C}-\overset{|}{C}=N-$. Mit starken Mineralsäuren: rotviolette Halochromiefärbung. Farbreaktion mit Cer(IV)-sulfat: blau → blaßgrün. Vertreter:

quartäre $\begin{cases} \text{C-Curarin-I,} & C_{40}H_{44-46}ON_4^{++} \\ \text{C-Alkaloid-G,} & C_{40}H_{43-45}(OH)N_4^{++} \ (?) \\ \text{C-Alkaloid-E,} & C_{40}H_{42-44}(OH)_2N_4^{++} \\ \text{C-Guaianin,} & ? \end{cases}$

*c) Die C-Calebassin-Gruppe.* UV-Spektren: Abb. 4, S. 193; in alkalischer Lösung Rotverschiebung um 10—12 m$\mu$. Chromophor: nebenstehend. $[\alpha]_D \approx +70°$. Farbreaktion mit Cer(IV)-sulfat: blauviolett → orange → olivgrün. Beim Erhitzen mit konz. Salzsäure erhält man leuchtend gelbe Lösungen der Isoverbindungen (*19*). Vertreter:

quartäre $\begin{cases} \text{C-Calebassin,} & C_{40}H_{46}(OH)_2N_4^{++} \\ \text{C-Alkaloid-F,} & C_{40}H_{45}(OH)_3N_4^{++} \ (?) \\ \text{C-Alkaloid-A,} & C_{40}H_{44}(OH)_4N_4^{++} \\ \text{Toxiferin-II,} & ? \ \text{(wahrscheinlich identisch mit C-Alkaloid-A)} \end{cases}$

*d) Die B,C,D-Gruppe.* Chromophor (?): nebenstehend. Farbreaktion mit Cer(IV)-sulfat: rotviolett → farblos. Vertreter:

quartäre $\begin{cases} \text{C-Alkaloid-B} \\ \text{C-Alkaloid-C} \\ \text{C-Alkaloid-D} \end{cases}$ tertiäre $\begin{cases} \text{Caracurin-I} \\ \text{Caracurin-II} \\ \text{Caracurin-III} \end{cases}$

Aufgeklärt sind das C-Dihydrotoxiferin, das C-Toxiferin-I und das Caracurin-V. Ausführliche Untersuchungen liegen über C-Calebassin und C-Curarin-I vor. Über die Konstitution der sogen. B,C,D-Alkaloide ist kaum etwas bekannt.

Auf die zahlreichen und wichtigen experimentellen *Verknüpfungen* zwischen den Strychnin-Typus-Alkaloiden wird in Abschnitt 7 eingegangen (S. 226).

## 2. C-Fluorocurarin.

C-Fluorocurarin (XXXI, S. 208), $C_{20}H_{23}ON_2^+$, so genannt, weil es im ultravioletten Licht hellblau fluoresziert [= C-Curarin-III (*101*)], ist ein schwach gelb gefärbtes, nur wenig toxisches Alkaloid, das regelmäßig in Calebassencurare angetroffen wird. Daraus ist es erstmals von H. WIELAND, PISTOR und BÄHR (*101*) isoliert worden. Es entsteht auch als Abbauprodukt von (dimeren) Calebassen-Alkaloiden: aus C-Calebassin oder Desoxy-calebassin, wenn man darauf bei Gegenwart von Sauerstoff Ameisensäure-essigsäure-anhydrid einwirken läßt (*93, 29*); aus C-Curarin-I beim Behandeln mit konz. HCl oder HBr (*110, 39, 68a*). Wegen dieser Verknüpfung mit den Hauptalkaloiden des Calebassencurare ist das Fluorocurarin konstitutionell von besonderem Interesse.

v. PHILIPSBORN, MEYER, SCHMID und KARRER (73) leiteten eine Teilformel (XXXI) für Fluorocurarin ab (s. auch 37). Obige Summenformel wurde bestätigt, das Molekulargewicht nach der Methode der partiellen Quartärisierung (S. 206) gesichert. (XXXI) enthält eine N-Methylgruppe am quartären $N_{(b)}$-Atom. Eine C-Methylgruppe liegt in einer Äthylidenseitenkette vor, da beim Ozonisieren von (XXXI) Acetaldehyd abgespalten wird.

Das recht charakteristische UV-Spektrum des Fluorocurarins mit einem langwelligen Maximum bei 360 m$\mu$ erfährt durch Lauge eine Rot-

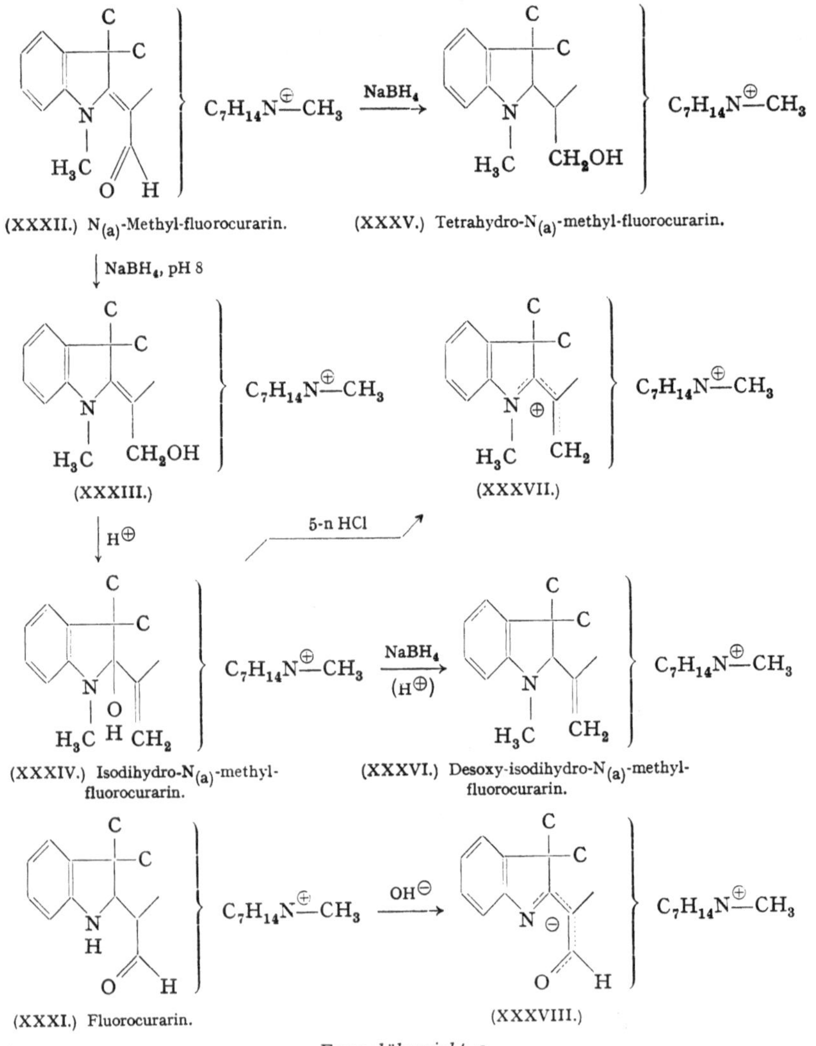

(XXXII.) $N_{(a)}$-Methyl-fluorocurarin.        (XXXV.) Tetrahydro-$N_{(a)}$-methyl-fluorocurarin.

(XXXIII.)        (XXXVII.)

(XXXIV.) Isodihydro-$N_{(a)}$-methyl-fluorocurarin.        (XXXVI.) Desoxy-isodihydro-$N_{(a)}$-methyl-fluorocurarin.

(XXXI.) Fluorocurarin.        (XXXVIII.)

*Formelübersicht 3.*

verschiebung. Das Alkaloid enthält demnach am Chromophor aciden Wasserstoff. Beim Behandeln mit Dimethylsulfat und Lauge geht es in ein Monomethylderivat (XXXII) über (*Formelübersicht 3*, S. 208). Diese Verbindung gibt die neu eingetretene Methylgruppe erst unter HERZIG-MEYER-Bedingungen wieder ab und ist folglich $N_{(a)}$-Methyl-fluorocurarin. Die UV-Kurve von (XXXII) gleicht derjenigen des Fluorocurarins, wird jedoch durch OH-Ionen nicht beeinflußt. Daraus ist zu schließen, daß die Rotverschiebung beim Fluorocurarin auf der Abspaltung eines Protons vom $N_{(a)}$-Atom beruht.

Die Schlüsselreaktion für das weitere Eindringen in die Konstitution des Fluorocurarins war die Reduktion von (XXXII) mit Natrium-borhydrid unter pH-Kontrolle. In sehr verdünnter, auf pH 8 gepufferter Lösung geht (XXXII) hierbei in eine Substanz (XXXIII) mit reinem $\alpha$-Methylen-indolin-Spektrum über, die jedoch nicht gefaßt werden kann. Schon in schwach saurer Lösung lagert sie sich um in eine Verbindung mit Indolinchromophor. Bei der Reduktion von (XXXII) *in präparativem Maßstab* entsteht wegen konzentrations-bedingter schlechterer pH-Kontrolle das $\alpha$-Methylenindolin (XXXIII) von vornherein in Mischung mit einem Indolin. Nach Säurebehandlung isoliert man aus einem solchen Ansatz drei Substanzen, jede mit Indolinspektrum: Isodihydro-$N_{(a)}$-methyl-fluorocurarin (XXXIV), $C_{21}H_{27}ON_2{}^+$; Tetra-hydro-$N_{(a)}$-methyl-fluorocurarin (XXXV), $C_{21}H_{29}ON_2{}^+$; und Desoxy-isodihydro-$N_{(a)}$-methyl-fluorocurarin (XXXVI), $C_{21}H_{27}N_2{}^+$.

Aus der Art ihrer Entstehung, aus ihren Reaktionen und spektro-skopischen Daten lassen sich für diese Verbindungen die Partialstrukturen ableiten, die in Formelübersicht 3 (S. 208) angegeben sind.

Verbindung (XXXIV) liefert bei Ozonolyse neben Acetaldehyd (aus der Äthylidenseitenkette) Formaldehyd, was auf die Anwesenheit einer Vinylgruppe deutet. Das IR-Spektrum bestätigt dies und beweist ferner, daß eine Hydroxylgruppe anwesend ist. Diese befindet sich in der $\alpha$-Stellung des Indolinsystems, wie sich aus folgenden Befunden ergibt:

(a) Das UV-Spektrum von (XXXIV) wird durch OH-Ionen nach längeren Wellen verschoben. (b) Diese Rotverschiebung bleibt aus, wenn (XXXIV) mit verd. methanolischer Salzsäure behandelt worden ist. (c) Eine Acetylierung der Hydroxylgruppe gelingt nicht. (d) (XXXIV) zeigt in 5-n Salzsäure ein Absorptions-spektrum, welches dem $\alpha$-Vinyl-indoleninium-ion (XXXVII) zugeschrieben werden kann, wodurch zugleich die Lage der Vinylgruppe bewiesen wird.

Die Rotverschiebung mit Lauge wird auch bei anderen $\alpha$-Hydroxy-indolinen beobachtet und kann als typisch für solche Verbindungen gelten (*40*). Das Ausbleiben der Rotverschiebung nach Einwirkung von methanolischer Salzsäure auf (XXXIV) beruht auf dem Austausch der (aciden) Hydroxylgruppe gegen Methoxyl. Ein solcher leichter Austausch ist charakteristisch für Aminohalbacetal-gruppierungen.

In den Punkten (a)—(d) besteht eine völlige Analogie zwischen (XXXIV) und C-Calebassin. (XXXIV) entsteht durch Allylumlagerung aus dem Indolenin (XXXIII) beim Ansäuern der Reduktionslösung. Tetrahydro-$N_{(a)}$-methyl-fluorocurarin (XXXV) dürfte aus (XXXIII) durch Borhydridreduktion der $>N—C=C<$-Doppelbindung entstehen. Sein UV-Spektrum wird durch Lauge nicht beeinflußt, geht hingegen schon in 1-n Salzsäure in ein Indoliniumspektrum über. Die Hydroxylgruppe in (XXXV) läßt sich leicht acetylieren und gibt sich auch im IR-Spektrum zu erkennen. Die (XXXV) entsprechende, am $N_{(a)}$-Atom nicht methylierte Verbindung entsteht, wenn Fluorocurarin (XXXI) in verdünnter Natronlauge mit Natriumborhydrid reduziert wird (75a).

Desoxy-isodihydro-$N_{(a)}$-methylfluorocurarin (XXXVI) wird aus (XXXIV) beim Ansäuern der Reduktionslösung, die noch überschüssiges $NaBH_4$ enthält, gebildet (Herausreduzieren der OH-Gruppe). Sein UV-Spektrum wird durch Lauge nicht beeinflußt. Für vollständige $N_{(a)}$-Protonisierung ist 7-n Säure erforderlich. Im IR zeigt (XXXVI) eine Vinylbande.

Die in die Partialformeln (S. 208) eingezeichnete Disubstitution der β-Stellung ist eine selbstverständliche Konsequenz der Strukturen (XXXIII) und (XXXIV) (die sonst in Indolstrukturen übergehen würden).

Die für Fluorocurarin (XXXI) angegebene Teilformel erklärt befriedigend das langwellige UV-Spektrum; die Rotverschiebung des Spektrums in alkalischer Lösung ist auf die Ausbildung der mesomeren, partiell anionischen Struktur (XXXVIII) zurückzuführen. Als α,β-ungesättigter β-Aminoaldehyd zeigt Fluorocurarin im Bereich zwischen 6,0 und etwa 6,25 μ intensive Doppelbindungsabsorption. In Alkohol-Wasser gibt es mit Eisen(III)-chlorid eine grüne Farbreaktion.

Die Fluorocurarin-Formel (XXXI) wird weiterhin dadurch bestätigt, daß das Alkaloid ein Oxim liefert und von alkalischem Wasserstoffperoxyd zu einer Verbindung mit Oxindolspektrum abgebaut wird.

Anmerkung bei der Korrektur: Durch Partialsynthese aus Wieland-Gumlich-Aldehyd (XLII, S. 213) (37a) und Überführung in

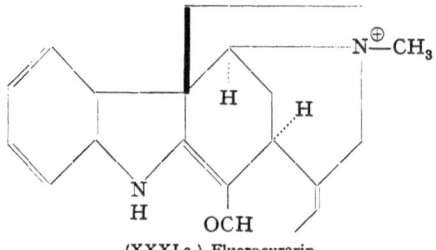

(XXXI a.) Fluorocurarin.

C-Dihydrotoxiferin (XXXIX, S. 212) (72a) wurde für C-Fluorocurarin die schon von den Autoren (73) vermutete Konstitution (XXXI a) bewiesen.

### 3. C-Dihydrotoxiferin.

Das C-Dihydrotoxiferin (XXXIX)* (*Formelübersicht 4*, S. 212) haben erstmals H. WIELAND, BÄHR und WITKOP (*96*) aus Calebassencurare isoliert, wo es zwar häufig aber nicht immer angetroffen wird**. Seine Nor-verbindung (XL) ist von ASMIS, WASER, SCHMID und KARRER (*10*) in der Rinde von *Str. toxifera* gefunden worden. Dihydrotoxiferin zeichnet sich durch hohe Linksdrehung aus. Es zählt zu den stark toxischen Calebassen-Alkaloiden. Die vorgeschlagene Summenformel (*96*) $C_{20}H_{23}N_2^+$ ist zu verdoppeln (*74*).

Entscheidend für die Aufklärung des Dihydrotoxiferins [BERNAUER, BERLAGE, v. PHILIPSBORN, SCHMID und KARRER (*28, 20*)] war die Beobachtung, daß das durch Thermolyse aus (XXXIX) erhältliche Nordihydrotoxiferin (XL) durch 1-n Schwefelsäure in der Wärme hydrolysiert wird. Dabei entsteht ein Indolin (XLI, S. 213), welches sich durch sein IR-Spektrum als Aldehyd ausweist. Auf Grund der für C-Fluorocurarin abgeleiteten Teilformel (XXXI, S. 208) war für (XLI) von vornherein die Partialstruktur (XLIa) in Betracht zu ziehen.

Hinsichtlich UV-Spektrum und Farbreaktionen gleicht der Aldehyd (XLI) dem Alkaloid Caracurin-VII, welches in der selben Strychnos-Rinde vorkommt, die auch Nor-dihydrotoxiferin führt (vgl. S. 190). Caracurin-VII zeigt jedoch im IR-Spektrum keine Aldehydbanden. BERNAUER, PAVANARAM, v. PHILIPSBORN, SCHMID und KARRER (*23*) vermuteten daher Maskierung der Aldehydgruppe in Caracurin-VII und stießen dadurch auf die Identität von Caracurin-VII und dem sogen. WIELAND-GUMLICH-Aldehyd (XLII), dessen Aldehydgruppe nach ANET und ROBINSON (*2*) einer Halbacetalgruppierung angehört.

Um die darnach naheliegende Annahme zu beweisen, daß der Aldehyd (XLI) 18-Desoxy-WIELAND-GUMLICH-Aldehyd ist, führten ihn die Autoren (*20*) durch Reduktion mit Natriumborhydrid in den entsprechenden primären Alkohol $C_{19}H_{24}ON_2$ (XLIII) über, der aus WIELAND-GUMLICH-Aldehyd (XLII) auf folgendem übersichtlichem Weg erhalten wurde: (XLII) geht bei $NaBH_4$-Reduktion in das Glykol (XLIV) über (*2*). Die allylständige Hydroxylgruppe an $C_{(18)}$ in (XLIV) läßt sich mit HBr-Eisessig selektiv durch Brom ersetzen. Reduktion des (nicht isolierten) Bromids (XLV) mit Zinkstaub in Eisessig ergibt das O-Acetat des Alkohols (XLIII). Nach Verseifung in methanolischer Salzsäure erhält man (XLIII) selbst.

---

* Ausführlich: C-Dihydrotoxiferin-I (*96*). Dieser Name gibt die Beziehung zwischen (XXXIX, S. 212) und C-Toxiferin-I (XLIX, S. 216) nicht richtig wieder. Synonymum: C-Alkaloid-K (*52*).

** C-Dihydrotoxiferin ist empfindlich gegenüber Säuren und Oxydationsmitteln. Vielleicht wird es oft bei der Curareherstellung zerstört.

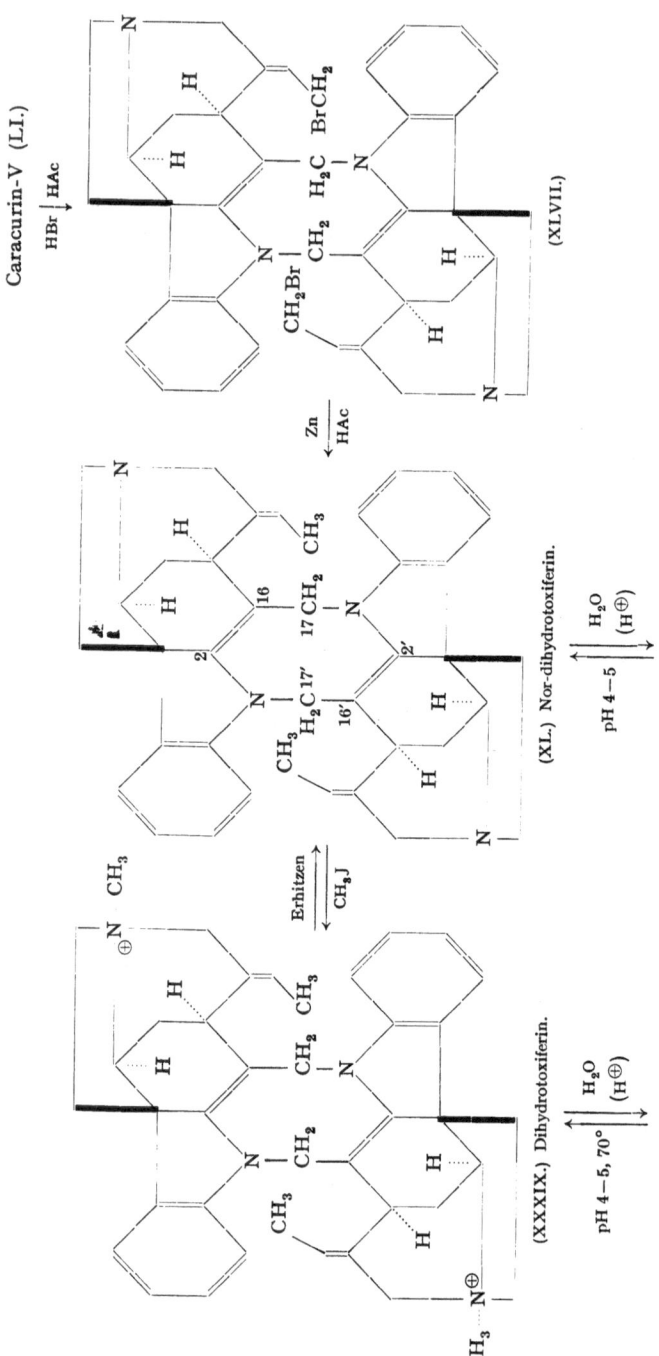

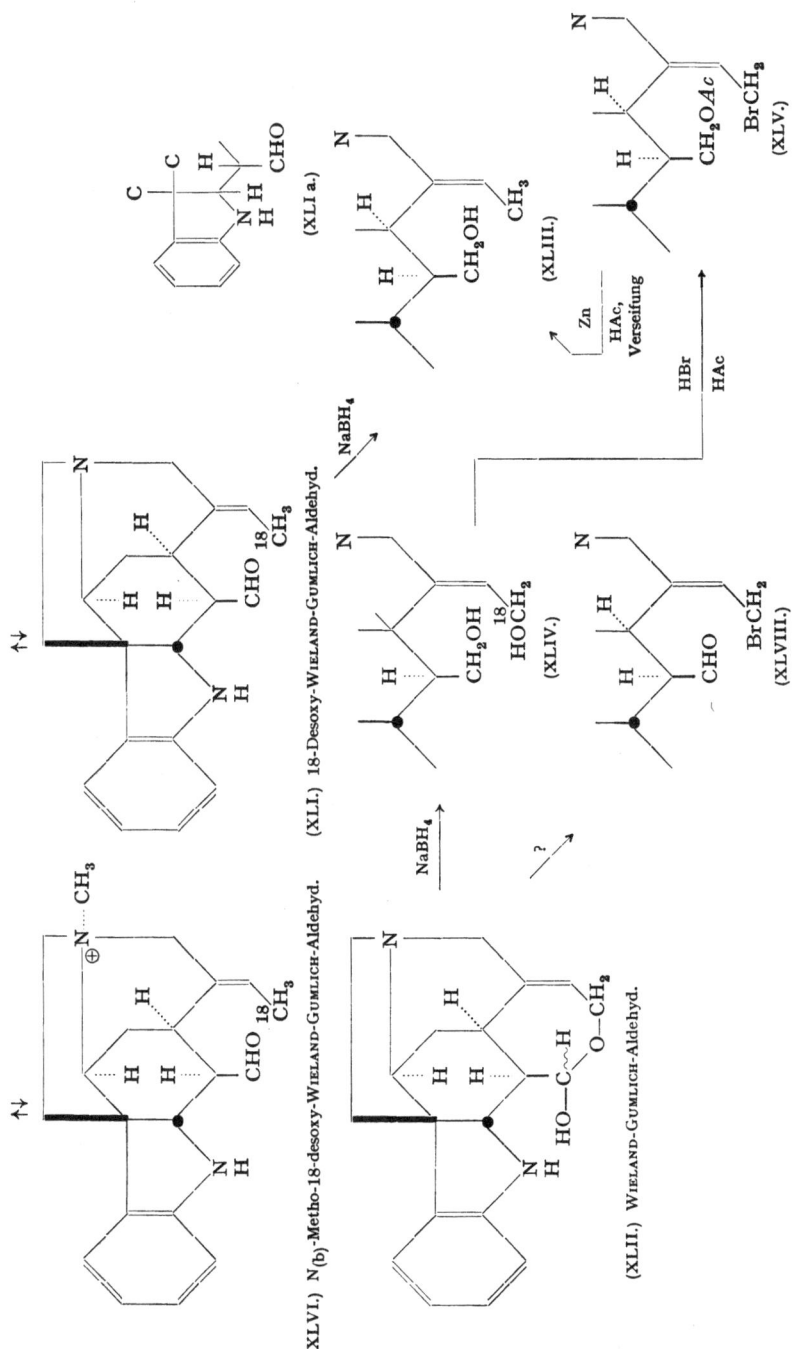

*Formelübersicht 4.* Aufklärung des C-Dihydrotoxiferins.

Säureeinwirkung auf Dihydrotoxiferin (XXXIX) liefert die dem Aldehyd (XLI) entsprechende $N_{(b)}$-Methoverbindung (XLVI). Die Rückumwandlung gelingt in guter Ausbeute beim Erwärmen von (XLVI) auf 70° im Acetatpuffer pH 4—5 unter Luftausschluß. Unterwirft man ein äquimolekulares Gemisch der Aldehyde (XLI) und (XLVI) diesen Reaktionsbedingungen, so erhält man nebeneinander Dihydrotoxiferin (XXXIX), Nor-dihydrotoxiferin (XL) und als Produkt „gemischter Kondensation" $N_{(b)}$-Monometho-nor-dihydrotoxiferin. Diese Reaktionen beweisen, daß (XXXIX) und (XL) bei der Säureeinwirkung in z w e i g l e i c h e Teile zerfallen.

Nor-dihydrotoxiferin (XL) enthält dem IR-Spektrum nach keine $N_{(a)}$-H-Gruppierung und ist O-frei. Als Summenformel ist zunächst $C_{38}H_{40-42}N_4$ in Betracht zu ziehen. Es ergibt sich somit, daß bei seiner Hydrolyse in zwei Moleküle (XLI), $C_{19}H_{22}ON_2$, pro $C_{19}$ 1 Mol Wasser angelagert, eine Bindung $C-N_{(a)}$ gespalten und eine Aldehydgruppe gebildet wird.

Daraus folgt für Nor-dihydrotoxiferin die Formel $C_{38}H_{40}N_4$ und die Anwesenheit von zwei Doppelbindungen, zusätzlich zu den beiden Äthyliden-doppelbindungen in 19—20- bzw. 19'—20'-Stellung, die beim Ozonisieren Acetaldehyd liefern (10). Diese zusätzlichen Doppelbindungen sind in zwei Vinylamingruppen enthalten und geben sich im IR-Spektrum durch eine intensive $\nu$ (C=C)-Bande bei 6,05 $\mu$ zu erkennen. Von den zwei verbleibenden Möglichkeiten für Nor-dihydrotoxiferin (XL) mit 2—16- und 2'—16'- bzw. 16—17- und 16'—17'-Stellung der beiden Doppelbindungen ist auf Grund des UV-Spektrums die erstgenannte [Struktur (XL)] vorzuziehen. Für Dihydrotoxiferin ergibt sich Konstitution (XXXIX).

Der Wieland-Gumlich-Aldehyd (XLII) enthält noch fünf von den sechs Asymmetriezentren des Strychnins, wenn man das Halbacetal-C-atom 17 außer Betracht läßt. Da er durch Kondensation mit Malonsäure in Strychnin zurückverwandelt werden kann (3), dürfte er den gleichen sterischen Bau wie dieses besitzen*. Es kann als sicher gelten, daß der Aldehyd (XLI) in sterischer Hinsicht mit dem Wieland-Gumlich-Aldehyd (XLII) übereinstimmt. Für Nor-dihydrotoxiferin (XL) und Dihydrotoxiferin (XXXIX) folgen daraus die angegebenen *Raum*formeln (S. 212).

Die Synthese des Nor-dihydrotoxiferins (XL) gelang ausgehend von dem aus Wieland-Gumlich-Aldehyd erhältlichen Caracurin-V (LI, S. 217), welches mit Bromwasserstoff-Eisessig in die (nicht isolierte) Bromverbindung (XLVII) umgewandelt wird. Reduktion von (XLVII)

---

* Zur Stereochemie des Strychnins siehe (59); abs. Konfiguration (70).

mit Zinkstaub in Eisessig liefert Nor-dihydrotoxiferin (XL) [Bernauer, Berlage, v. Philipsborn, Schmid und Karrer (21)].

Bei einem Versuch, aus (XLII) über die Bromverbindung (XLVIII) den 18-Desoxy-Wieland-Gumlich-aldehyd (XLI) und daraus Dihydro-toxiferin (XXXIX) zu synthetisieren, wurde eine Verbindung erhalten, die sich von dem Naturstoff wahrscheinlich durch Isomerie an den 19—20- und 19'—20'-Doppelbindungen unterscheidet (20).

## 4. C-Toxiferin-I und Caracurin-V.

C-Toxiferin-I (XLIX, S. 216), $C_{40}H_{46}O_2N_4^{++}$, ist von King (55a) aus Str. toxifera und von Schmid und Karrer (83) aus Calebassencurare isoliert worden. Sehr wahrscheinlich ist es identisch mit dem Toxi-ferin-I H. Wielands (96, 83). Es ist eines der am stärksten wirksamen Curare-Alkaloide. Seine Konstitution wurde von Bernauer, Berlage, v. Philipsborn, Schmid und Karrer (20) sowie von Battersby und Hodson (15) aufgeklärt.

Als Nor-C-toxiferin-I (L), $C_{38}H_{40}O_2N_4$, hat sich das Alkaloid Cara-curin-V a erwiesen (20, 15). Dieses entsteht in säure-katalysierter Reaktion aus dem isomeren Alkaloid Caracurin-V (LI, S. 217), welches aus Str. toxifera gewonnen worden ist (7, 20).

Bei längerer Säureeinwirkung zerfällt Caracurin-Va (L) in zwei Mol. Caracurin-VII, das, wie oben erwähnt, mit Wieland-Gumlich-Aldehyd (XLII) identifiziert worden ist. Dieser Vorgang entspricht vollständig der Hydrolyse des Nor-dihydrotoxiferins (XL) zu 18-Desoxy-Wieland-Gumlich-Aldehyd (XLI) (S. 213ff.). Damit ergeben sich für Caracurin-Va (Nor-C-toxiferin-I) (L) und für C-Toxiferin-I (XLIX) die angeführten Formeln (Formelübersicht 5, S. 216).

Caracurin-V (LI) ist im Gegensatz zu Caracurin-Va (L) ein Indolin. Da sein IR-Spektrum keine OH-Bande, hingegen starke Absorption in der Äther-region zeigt, ist für Caracurin-V eine Struktur mit verätherten Hydroxy-äthyliden-Resten anzunehmen. Von verschiedenen denkbaren Formeln ist Struktur (LI) am wahrscheinlichsten (20), da sie das gleiche stabile Siebenring-System wie der Wieland-Gumlich-Aldehyd (XLII) besitzt.

Caracurin-V (LI) entsteht in guter Ausbeute aus seinem Spalt-produkt (XLII) bei 15stündigem Erhitzen in Eisessig-Natriumacetat auf 80° unter Luftausschluß, neben Nor-C-toxiferin-I (L) (20). Aus Wieland-Gumlich-Aldehyd-chlormethylat (LII) erhält man ein Gemisch von C-Toxiferin-I (XLIX), von dessen Diacetylverbindung und Caracurin-V-dichlormethylat (LIII) (20, 17, 15). Letzteres läßt sich durch Säurekatalyse in C-Toxiferin-I überführen (20, 17).

In Anbetracht der Totalsynthese des Strychnins durch Woodward und Mitarbeiter (109) und der Tatsache, daß Wieland-Gumlich-Aldehyd

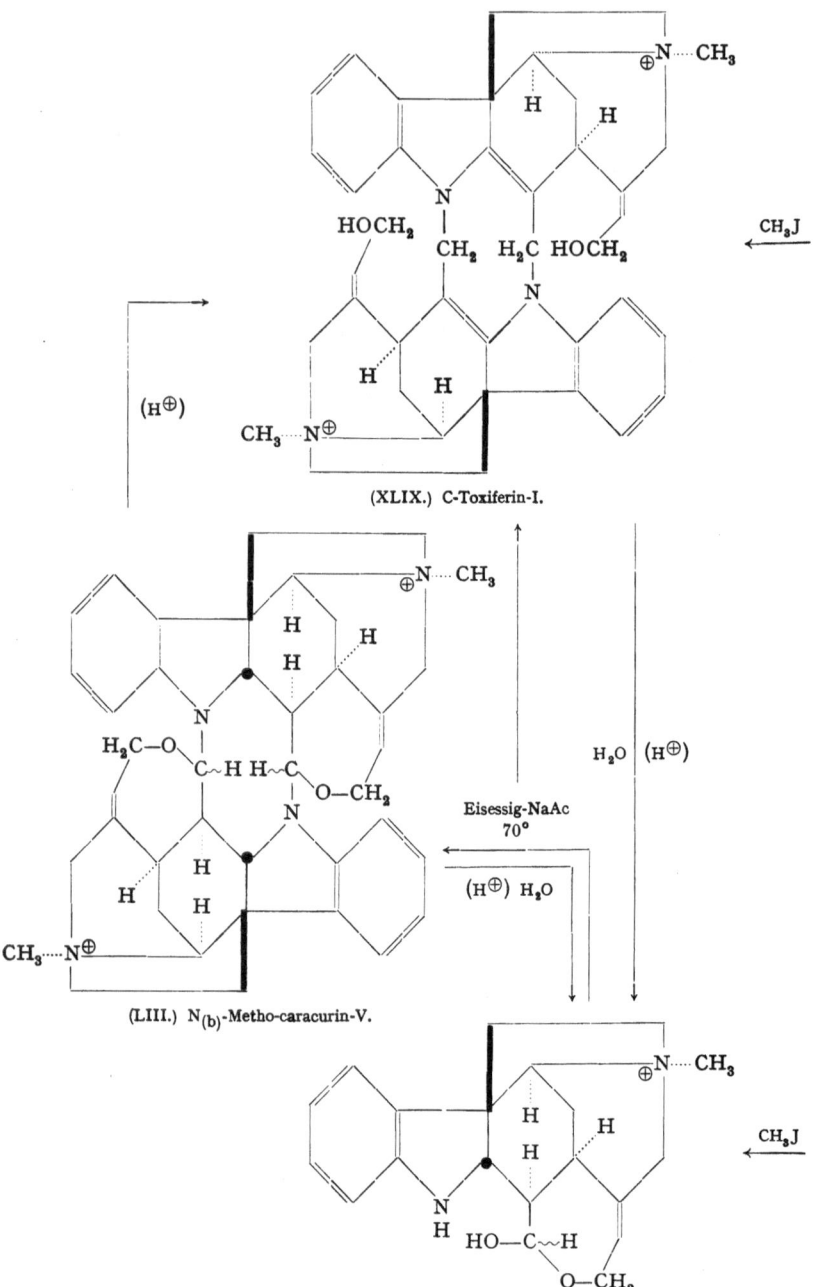

(XLIX.) C-Toxiferin-I.

(LIII.) N(b)-Metho-caracurin-V.

Eisessig-NaAc
70°

(LII.) N(b)-Metho-Wieland-Gumlich-Aldehyd (= Alkaloid A 8).

*Formelübersicht 5.* Aufklärung

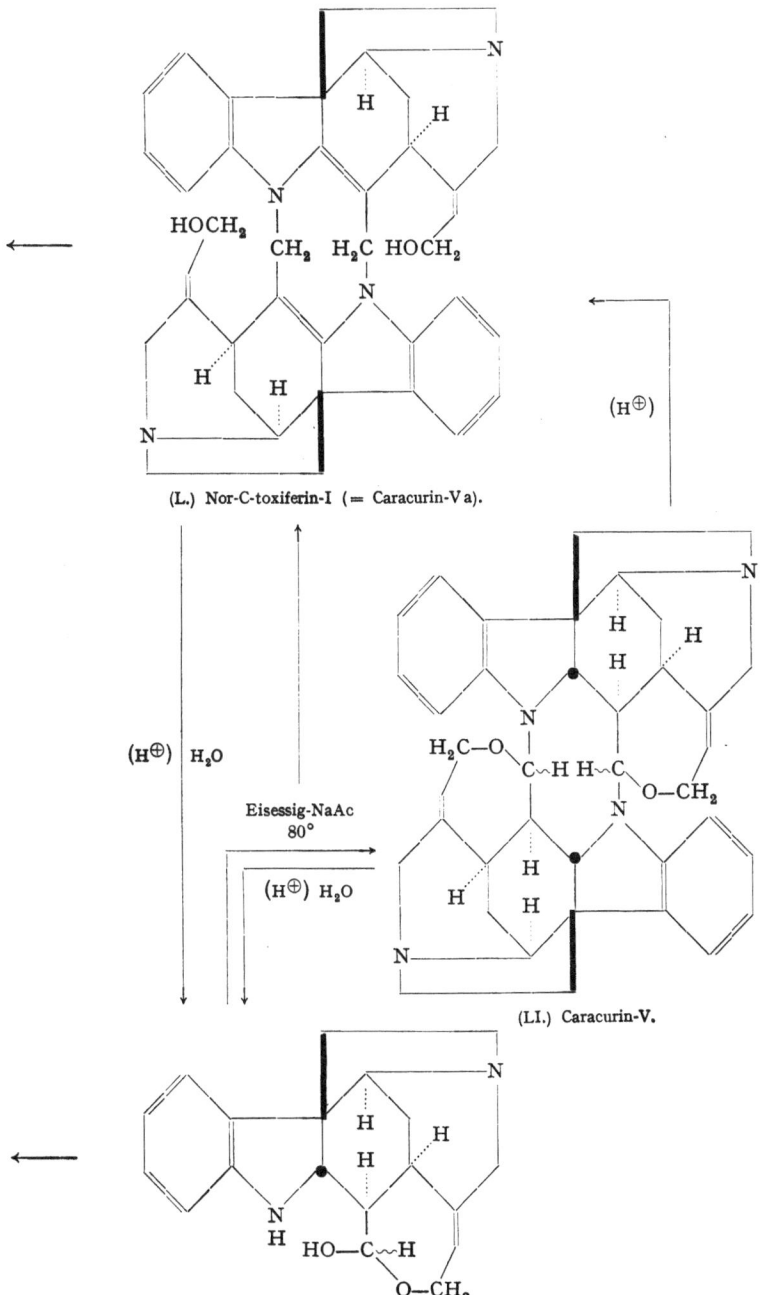

(L.) Nor-C-toxiferin-I (= Caracurin-V a).

(LI.) Caracurin-V.

(XLII.) Wieland-Gumlich-Aldehyd (= Caracurin-VII).

und Synthese von C-Toxiferin-I und Caracurin-V.

(XLII) aus Strychnin zugänglich ist (*97, 98, 2*), repräsentieren die geschilderten Synthesen des Nor-dihydro-toxiferins (XL) und damit des C-Dihydrotoxiferins (XXXIX, S. 212), des Caracurins-V (LI), des Nor-C-toxiferins-I (L), des C-Toxiferins-I (XLIX) und des Caracurin-V-dichlormethylats (LIII) formal Totalsynthesen; sie dürften aber auch praktische Bedeutung erlangen.

Die besprochenen Ergebnisse stellen einen Höhepunkt in der Erforschung der Calebassen-Alkaloide dar. Die Identifizierung des Caracurins-VII als WIELAND-GUMLICH-Aldehyd, der damit erstmals in der Natur gefunden worden ist, und die Strukturen (XXXIX), (XLIX) und (LI) bestätigen glänzend die Anschauungen WOODWARDS über die Strychnin-Biogenese. Das Prinzip, nach welchem in den Toxiferinen $C_{19}$- bzw. $C_{20}$-Alkaloide zu „dimeren" Molekülen verknüpft sind, ist neuartig. Seine Kenntnis ist auch für andere Calebassen-Alkaloide, die sich von der Dihydrotoxiferin-Gruppe ableiten (Curarin, Calebassin und ihre Analoga), von größter Wichtigkeit.

### 5. C-Curarin-I.

Das stark toxische C-Curarin-I (LIV) ist von H. WIELAND, KONZ und SONDERHOFF als erstes Alkaloid aus Calebassencurare kristallisiert erhalten und ursprünglich als „Toxiferin" bezeichnet worden (*99*). Man hat es seither in nahezu allen Calebassen gefunden und meist in bedeutend größerer Menge als andere Inhaltsstoffe.

Die Bemühungen um die Aufklärung der Curarin-Struktur dauern seit der Entdeckung des Alkaloids an (*100—102, 82, 75, 74, 72*). Indessen kann zur Zeit eine sichere Aussage nur über den Bau gewisser begrenzter Ausschnitte des Curarin-Moleküls gemacht werden.

Curarin läßt sich durch Photooxydation aus C-Dihydrotoxiferin (XXXIX, S. 212) gewinnen. Damit steht endgültig fest, daß es eng mit den Strychnos-Alkaloiden verwandt ist, wie schon SCHMID, EBNÖTHER und KARRER (*82*) vermutet haben, als sie bei der Zinkstaubdestillation des Nor-curarins (LV) die Abbauprodukte β-Äthylpyridin, β-Äthylindol, β-Methylindol, Carbazol und 1-Methylcarbazol faßten.

Wie die meisten Calebassen-Alkaloide, ist Curarin (LIV) eine quartäre Ammoniumbase. Es enthält 1 N—$CH_3$ pro $C_{20}$. Es wurde ihm zunächst die Formel $C_{20}H_{21}N_2^+$ zugeschrieben (*100, 101*). Diese blieb unwidersprochen, bis v. PHILIPSBORN, SCHMID und KARRER (*74*) bei der Untersuchung der erwähnten Norverbindung (LV) auf Widersprüche stießen. Mit der „Methode der partiellen Quartärisierung" (S. 206) wiesen die Autoren nach, daß Nor-curarin (LV) zwei stark basische N-Atome besitzt. Damit und aus weiteren analytischen Daten ergab sich für (LV), welches bei Hochvakuumdestillation aus Curarin-chlorid erhalten wird, die

Summenformel $C_{38}H_{38-40}ON_4$. Curarin (LIV) als die entsprechende Dimethoverbindung hat die Formel $C_{40}H_{44-46}ON_4^{++}$.

Curarin (LIV) enthält zwei C-Methylgruppen, die bei der Ozonolyse als Acetaldehyd eliminiert werden (103). Unter der Einwirkung von starkem Alkali wird es tertiärisiert (100), wobei die Base (LVI), $C_{40}H_{42-44}ON_4$, entsteht (72, 75, 101). Diese ist frei von C-Methyl und gibt mit Ozon keinen Acetaldehyd, statt dessen aber in 50%iger Ausbeute (bezogen auf vier Vinylgruppen) Formaldehyd. Im IR zeigt (LVI) Banden bei 10,12, 11,03 und 11,13 $\mu$, welche den Gruppierungen (l) und (m) zugeordnet werden können.

$$-CH=CH_2 \qquad >C=CH_2$$
$$\text{(l)} \qquad\qquad \text{(m)}$$

Bei katalytischer Hydrierung geht es in einen Octahydrokörper (LVII) über (101, 72, 75), der mit $O_3$ weder Formaldehyd noch Acetaldehyd gibt und nachweislich 3—4 $CH_3$-Gruppen enthält. Beim modifizierten Mikrochromsäureabbau liefert (LVII) Methyläthylessigsäure und Propionsäure neben Essigsäure. Diese Befunde sind nach v. PHILIPSBORN, SCHMID und KARRER (74) so zu deuten, daß Curarin (LIV) die Partialformel (LIVa) erhält. Der Übergang in (LVI), dessen Dienchromophore UV-spektroskopisch nachweisbar sind, ist ein vinyloger HOFMANNscher Abbau. Bei katalytischer Hydrierung der HOFMANN-Base (LVI) werden deren vier Vinyldoppelbindungen abgesättigt. Daß die Octahydrobase (Partialformel LVIIa) bei der Oxydation Methyläthylessigsäure geben muß, ist evident.

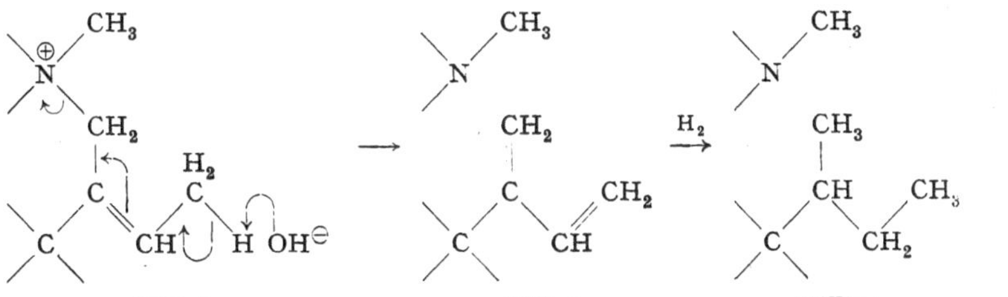

(LIVa.)        (LVIa.)        (LVIIa.)

Die Base (LVI) läßt sich durch Reduktion mit Natrium in Amylalkohol in ein Tetrahydroderivat überführen. Dieses entsteht unter den genannten Bedingungen auch aus Curarin selbst (72, 75) und stellt das Produkt teilweiser 1,2- und teilweiser 1,4-Addition von Wasserstoff an die beiden Diensysteme der primär gebildeten, bzw. in die Reaktion eingesetzten Base (LVI) dar.

Die Basen (LVI), (LVII) und die Tetrahydrobase, ferner das Norcurarin (LV) zeigen im IR keine OH-Banden. Die drei erstgenannten Verbindungen besitzen jedoch mittelstarke Banden bei 8,55—8,57 $\mu$, das Nor-curarin hat eine etwas schwächere Bande bei 8,35 $\mu$, weshalb anzunehmen ist, daß das Sauerstoffatom all dieser Substanzen in einer Äthergruppierung vorliegt.

Es ist ein Handikap der Curarin-Chemie, daß das UV-Spektrum des Alkaloids nicht ohne weiteres gedeutet werden kann. Stellt man in Rechnung, daß eine sehr intensive IR-Bande des Curarins bei 6,05 $\mu$ sehr wahrscheinlich entweder einer Gruppierung $>C=C-N<$ oder einer Gruppierung $>C=N-$ zugeschrieben werden muß (vergl. 57), so stehen drei Chromophore (n), (o) und (p) zur Diskussion, disubstituierte $\beta$-Stellung vorausgesetzt. Da keines der zugehörigen Spektren überzeugende Ähnlichkeit mit der Curarin-Kurve aufweist, muß geschlossen werden,

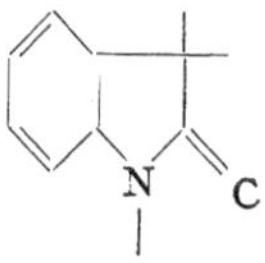

(n) 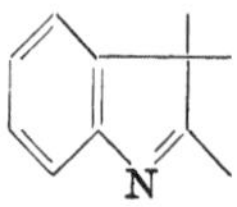　　　　(o)

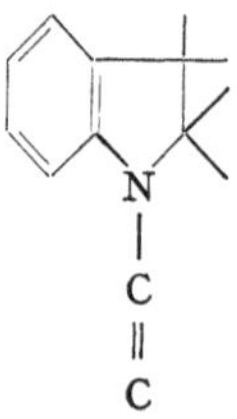

(p)

daß das Curarinspektrum entweder von zwei *verschiedenen*, in der Molekel vereinigten Chromophoren herrührt oder dadurch zustande kommt, daß das $\pi$-Elektronensystem eines der Chromophore (n), (o) oder (p) unter dem Einfluß einer benachbarten Gruppe ein abnormales Absorptionsverhalten zeigt.

Die HOFMANN-Base (LVI) und ihre Hydrierungsprodukte besitzen ähnliche UV-Spektren wie Curarin und Nor-curarin. Überraschenderweise erhält man beim Protonisieren oder Quartärisieren Verbindungen eines *neuen Spektraltyps* (ähnlich n). Es ließ sich indessen zeigen, daß beim Quartärisieren Methylierung an den $N_{(b)}$-Atomen, also nicht unmittelbar am Chromophor, eintritt, so daß die spektrale Änderung einen sekundären Effekt darstellt (72). Erwähnt sei in diesem Zusammenhang noch, daß Curarin und Nor-curarin mit wäßriger Mineralsäure Halochromie zeigen, die HOFMANN-Basen hingegen nicht.

Unter der Einwirkung von konz. Salzsäure (110, 39) oder HBr (68a) entsteht aus Curarin (LIV) in etwa 10% Ausbeute C-Fluorocurarin (XXXI, S. 208). Daneben wurden in je etwa 20% Ausbeute zwei weitere Substanzen, die sogen. Ultracurarine-A und B, gefaßt. Beide scheinen noch $C_{40}$-Verbindungen zu sein (38, 75a). Ihre Spektren weisen Ähnlichkeit mit demjenigen des C-Fluorocurarins auf. Im Ultracurarin-A dürfte neben einem Fluorocurarin-artigen Chromophor ein Indolchromophor vorliegen (75a).

### 6. C-Calebassin.

C-Calebassin (LVIII, S. 222) [C-Toxiferin-II (96), C-Strychnotoxin-I (104)] wird wie C-Curarin-I regelmäßig im Calebassencurare gefunden und ist nach diesem das mengenmäßig bedeutendste Calebassen-Alkaloid. Erstmals isoliert worden ist es von H. WIELAND, BÄHR und WITKOP (96).

*Literaturverzeichnis: SS. 241—247.*

Es ist sieben- bis achtmal weniger toxisch als C-Curarin-I. Die Aufklärung seiner Struktur ist noch nicht abgeschlossen. Nach den jetzt vorliegenden Ergebnissen kommt dem Calebassin die Formel $C_{40}H_{48}O_2N_4^{++}$ zu. Dem Folgenden ist diese Formel zugrunde gelegt ohne Rücksicht auf die in den Originalarbeiten verwendeten Formeln.

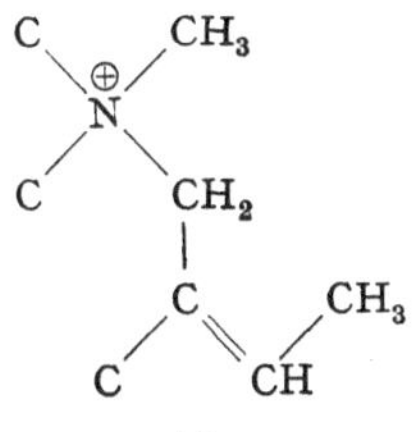

Wie C-Dihydrotoxiferin dürfte Calebassin (LVIII) aus zwei gleichen $C_{20}$-„Hälften" zusammengesetzt sein. Wie dieses auch das C-Curarin-I enthält es (zweimal) die Gruppierung (q).

Calebassin besitzt zwei C-Methyl- und zwei N-Methylgruppen. Bei der Ozonolyse liefert es Acetaldehyd (*103a*, *24*). Hydrierung mit Platinkatalysator in Wasser führt zu einem Tetrahydroderivat (LIX), welches mit Ozon keinen Acetaldehyd abspaltet (*83*, *103a*, *24*).

Den exakten Nachweis der Anordnung (q) brachten zwei parallellaufende Reaktionsfolgen in der Calebassin- und Tetrahydro-calebassin-Reihe [BERNAUER, SCHMID und KARRER (*24*)].

Beim Erwärmen mit starker Mineralsäure entsteht aus Calebassin (LVIII) das leuchtend gelbe Säureaddukt (LX), $C_{40}H_{49}O_2N_4^{+++}(A^-)_3$, einer Isocalebassin genannten, bis jetzt noch nicht in Substanz isolierten Verbindung $C_{40}H_{48}O_2N^{++}$. (LX), welches unten ausführlicher besprochen wird, spaltet wie Calebassin (LVIII) mit Ozon Acetaldehyd ab. Der hier interessierende Molekülbereich bleibt also beim Übergang (LVIII) → (LX) unverändert. Bei katalytischer Hydrierung in Natronlauge verbraucht (LX) rasch 2 $H_2$ und geht in die EMDE-Base (LXI) über, welche bei modifizierter Mikro-chromsäureoxydation nur Essigsäure liefert. Das zu (LX) analoge Iso-*tetrahydro*-calebassin-Säureaddukt (LXII) gibt mit Ozon *keinen* Acetaldehyd; aus seiner EMDE-Base (LXIII) entstehen bei der Oxydation die flüchtigen Säuren Methyläthylessigsäure, Propionsäure und Essigsäure.

Die Äthylgruppe der Methyläthylessigsäure geht offensichtlich auf zwei Äthylgruppen des Tetrahydro-calebassins (LIX) bzw. letzten Endes auf zwei Äthylidengruppen des Calebassins (LVIII) zurück; die Methylgruppe repräsentiert ein C-Methylpaar, das beim EMDE-Abbau von (LXII) neu entstanden sein muß. Damit können die beiden Reaktionsfolgen in Partialformeln ausgedrückt werden (*Formelübersicht 6*, S. *222*).

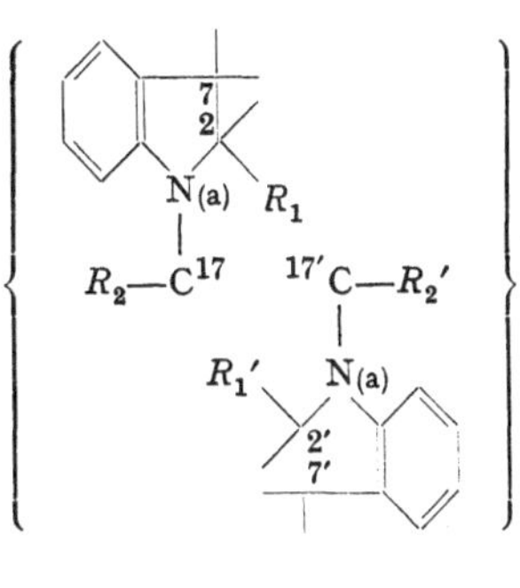

(LVIIIa.) Reste unspezifiert.
(LVIIIb.) $R_1 = R_1' = OH$
$R_2 = R_2' = H$
(LVIIIc.) $R_1 = R_1' = ?$
$R_2 = R_2' = OH$

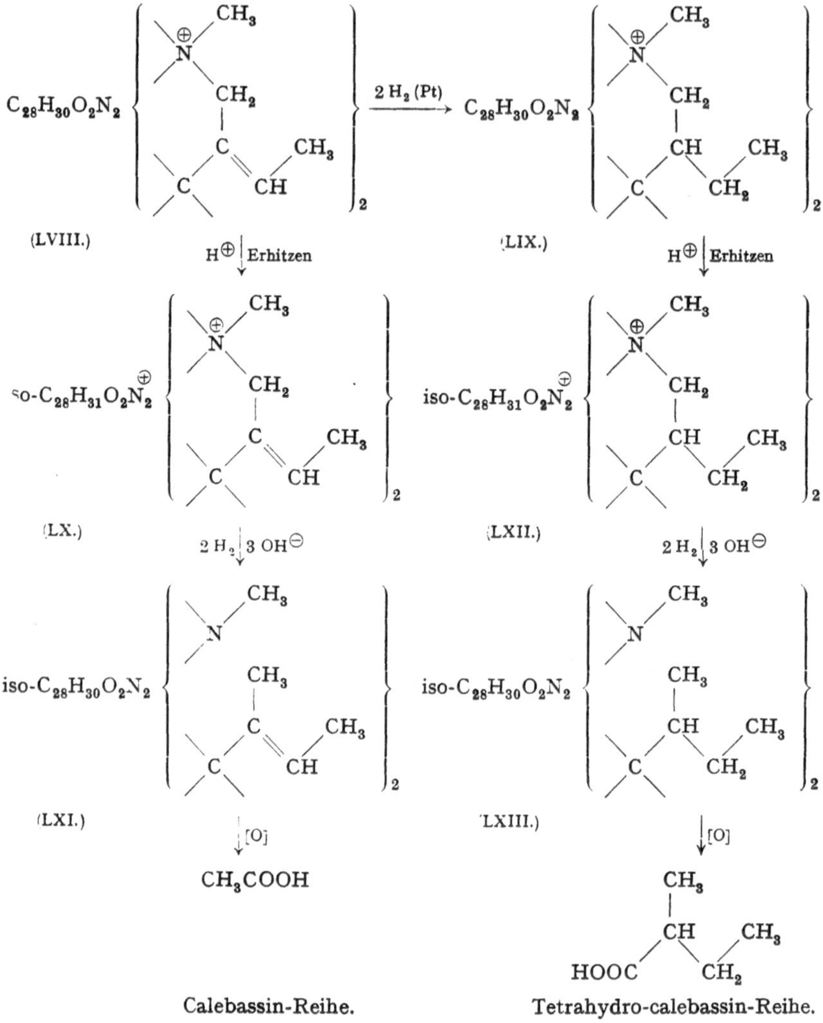

$C_{28}H_{30}O_2N_2$ (LVIII.)

$$\xrightarrow{\text{2 H}_2 \text{ (Pt)}}$$

$C_{28}H_{30}O_2N_2$ (LIX.)

$H^{\oplus}$ | Erhitzen

$H^{\oplus}$ | Erhitzen

iso-$C_{28}H_{31}O_2N_2^{\oplus}$ (LX.)

iso-$C_{28}H_{31}O_2N_2^{\oplus}$ (LXII.)

2 H$_2$ | 3 OH$^{\ominus}$

2 H$_2$ | 3 OH$^{\ominus}$

iso-$C_{28}H_{30}O_2N_2$ (LXI.)

iso-$C_{28}H_{30}O_2N_2$ (LXIII.)

↓ [O]

↓ [O]

$CH_3COOH$

Calebassin-Reihe.

Tetrahydro-calebassin-Reihe.

*Formelübersicht 6.*

Calebassin zeigt die UV-Absorption eines N-substituierten Anilins bzw. eines Indolins (Abb. 4, S. 193). Da es sich von Dihydrotoxiferin (XXXIX, S. 212) ableitet, darf geschlossen werden, daß es die Partialstruktur (LVIIIa) (mit den $N_{(a)}$-Atomen in Fünfringen) besitzt.

Das UV-Spektrum des Calebassins erleidet unter dem Einfluß von OH-Ionen eine Verschiebung nach längeren Wellen (*52*), was für aciden Wasserstoff am chromophoren System spricht. Nach Th. Wieland und Merz (*104, 105*) bleibt diese Rotverschiebung aus, wenn man Calebassin bei Gegenwart von Spuren Säure mit Methanol behandelt hat, eine Erscheinung, welche von BERNAUER, SCHMID und KARRER (27) aufgeklärt worden ist. Diese Autoren fanden, daß Calebassin mit Methanol in säure-katalysierter Reaktion äußerst leicht zu einem Dimethyläther (LXIV), $C_{40}H_{46}(OCH_3)_2N_4^{++}$, reagiert, der seinerseits schon in schwach saurem *wäßrigem* Milieu in Calebassin (LVIII) zurückverwandelt wird.

Calebassin (LVIII) enthält demnach an den Chromophoren zwei (schwach saure) Hydroxylgruppen, die den Charakter von Halbacetalhydroxylen besitzen. Als Haftstellen kommen die C-Atome 2 oder 17 bzw. 2′ oder 17′ in Frage [Partialformeln (LVIIIb) und (LVIIIc); Numerierung wie beim C-Dihydrotoxiferin (S. 212)].

Die 7- bzw. 7′-Stellung scheidet aus, da β-Hydroxy-indoline, z. B. Hydrofluorocurin (VI, S. 199) mit Lauge keine Rotverschiebung geben.

In beiden Fällen besäße Calebassin (LVIII) zwei Carbinol- (Aminohalbacetal-) Gruppierungen, wodurch nicht nur seine leichte Verätherbarkeit und die leichte Verseifbarkeit des Calebassin-dimethyläthers (LXIV) erklärt würden, sondern auch die erstmals von VOLZ und TH. WIELAND (92) beobachtete Reduzierbarkeit des Calebassins zu einer Desoxyverbindung.

Desoxy-calebassin (LXV), $C_{40}H_{48}N_4^{++}$, entsteht aus Calebassin mit Zinkstaub-Eisessig (92), Zinkstaub-konz. Salzsäure oder Ameisensäure bzw. Ameisensäure-essigsäure-anhydrid (29) und wird bei sensibilisierter Photooxydation wieder in dieses zurückverwandelt (22). Im Gegensatz zu Calebassin ist es gegen Säureeinwirkung sehr stabil. Seine Lösung in Essigsäure-ameisensäure-anhydrid ist äußerst oxydationsempfindlich. Aus dem Gemisch der Oxydationsprodukte läßt sich bemerkenswerterweise das $C_{20}$-Alkaloid C-Fluorocurarin (XXXI, S. 208) isolieren (29). Bei der Thermolyse seines Dichlorids geht Desoxy-calebassin (LXV) in die entsprechende Norverbindung $C_{38}H_{42}N_4$ über (29); durch Methyljodid-Methanol wird es $N_{(a)}$-quartärisiert (92). Nach BERNAUER, SCHMID und KARRER (26) entsteht dabei $N_{(a)}$-*Mono*metho-desoxy-calebassin

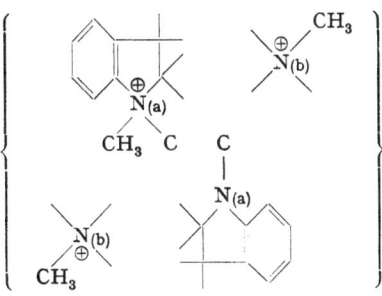

(LXVI.) $N_{(a)}$-Monometho-desoxy-calebassin.

(LXVI), $C_{40}H_{48}N_4 \cdot CH_3^{+++}$, eine Verbindung, die schlagend die lange angezweifelte *dimere* Formel des Calebassins beweist und ferner zeigt, daß dessen $N_{(a)}$-Atome substituiert sind.

Die für (LXVI) angegebene Partialformel wird durch das UV-Spektrum von dessen Trichlorid bestätigt, welches bei 294 m$\mu$ (Indolinmaximum) die halbe Extinktion des Desoxy-calebassins (LXV) zeigt, in Übereinstimmung mit der Tatsache, daß in (LXVI) nur noch ein Indolinchromophor vorhanden ist.

Der Umstand, daß Desoxy-calebassin (LXV) in eine $N_{(a)}$-*Mono*methoverbindung übergeht, scheint die Annahme, Calebassin sei aus zwei gleichen $C_{20}$-,,Hälften" zusammengesetzt, zu widerlegen. Der Widerspruch ist indessen nur ein scheinbarer. Dies ergibt sich, wenn man Lösungen der Verbindungen (LXV) und (LXVI) in 9-n Salzsäure UV-spektroskopisch vergleicht. (LXV) zeigt dabei ein reines Indoliniumspektrum, womit quantitative $N_{(a)}$-Protonisierung bewiesen ist. $N_{(a)}$-Monometho-desoxy-calebassin (LXVI) hingegen läßt unter diesen Bedingungen noch Indolin-Absorption erkennen und ist also nur unvollständig protonisiert. Dieser Befund kann einzig gedeutet werden, indem man annimmt, daß eine $CH_3$-Gruppe an einem der $N_{(a)}$-Atome das zweite $N_{(a)}$-Atom räumlich abschirmt und dadurch dessen Protonisierung erschwert. Ist schon die Protonisierung aus sterischen Gründen erschwert, so ist es ohne weiteres verständlich, daß die Methylierung ausbleibt (26).

Eine endgültige Entscheidung zwischen den Calebassin-Partialformeln (LVIIIb) und (LVIIIc) (S. 221) hat sich noch nicht treffen lassen. Nach beiden ist zu erwarten, daß Calebassin in stark saurer Lösung unter Ausbildung von Immoniumgruppierungen zwei Hydroxylionen abspalte. Das UV-Spektrum von Calebassin in 9-n Salzsäure steht damit nicht ohne weiteres in Einklang. Es ist bedeutend langwelliger als dasjenige eines Indoleniniumions (r) und spricht dafür, daß der

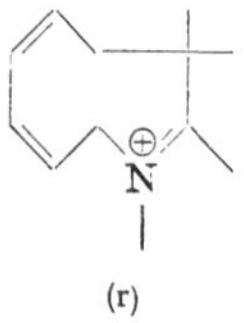

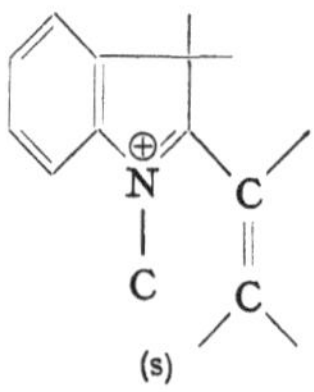

Chromophor (s) vorliegt. In guter Übereinstimmung mit dieser Annahme zeigt Isodihydro-$N_{(a)}$-methyl-fluorocurarin (XXXIV, S. 208) in saurer Lösung ein ähnliches Spektrum wie Calebassin.

Bernauer, Schmid und Karrer (27) haben für Calebassin die Partialformel (LVIIId) in Vorschlag gebracht. Diese bedarf vielleicht insofern einer Einschränkung, als die vom $C_{(16)}$ ausgehende Doppelbindung möglicherweise erst in stark saurer Lösung (auf Kosten eines Ringes) gebildet wird.

*Literaturverzeichnis: SS. 241—247.*

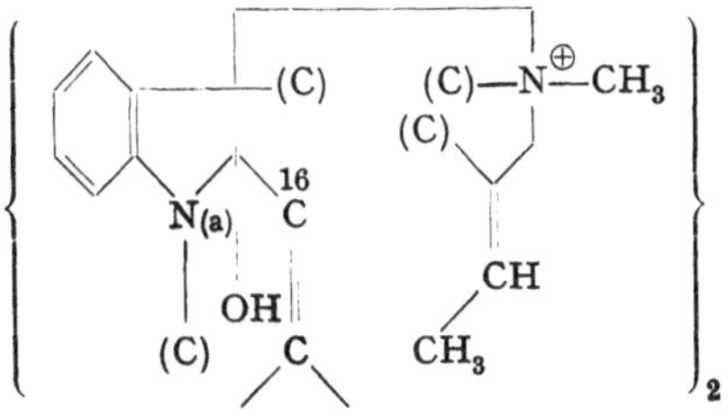

(LVIII d.) Calebassin (?).

**Beim** Erwärmen mit starker Mineralsäure geht Calebassin, wie erwähnt, in ein leuchtend gelb gefärbtes Kation $C_{40}H_{49}O_2N_4^{+++}$ (LX) über (*19, 24*) *(Formelübersicht 7)*. Dieses in Form des Trijodids (*24*) und Triperchlorats (*26*) faßbare Säureaddukt läßt sich mit 0,1-n Natronlauge als einbasische Säure titrieren (*29*), wodurch abermals die dimere Calebassinformel bewiesen wird.

Sein Spektrum zeichnet sich durch ein hohes Maximum bei 450 m$\mu$ aus (log $\varepsilon = 4,4$). Eine stark alkalische Lösung des dem Säureaddukt (LX) zugrunde liegenden, nicht in Substanz gefaßten Isocalebassins zeigt eine ganz ähnliche

Iso-tetrahydro-calebassin-EMDE-base (LXIII)

$\uparrow$ OH$^-$, H$_2$ (Pt)

Iso-tetrahydro-calebassin-Säureaddukt (LXII)
$(C_{40}H_{53}O_2N_4^{+++})$

$\uparrow$ H$^+$, Erhitzen

Tetrahydro-calebassin (LIX)          Nor-desoxy-calebassin
$(C_{40}H_{52}O_2N_4^{++})$          $(C_{38}H_{42}N_4)$

$\uparrow$ H$_2$ (Pt)          $\uparrow$

*Calebassin* (LVIII)   Zn, H$^+$   Desoxy-calebassin (LXV)
$(C_{40}H_{48}O_2N_4^{++})$  $\overrightarrow{\phantom{Photooxydation}}$ $\overleftarrow{\text{Photooxydation}}$   $(C_{40}H_{48}N_4^{++})$

$\downarrow$ CH$_3$J—CH$_3$OH

$\downarrow$ H$^+$, Erhitzen          N$_{(a)}$-Monometho-desoxy-calebassin (LXVI)
$(C_{40}H_{48}N_4 \cdot CH_3^{+++})$

Isocalebassin-Säureaddukt (LX)
$(C_{40}H_{49}O_2N_4^{+++})$

$\downarrow$ OH$^-$ \\\_\_\_\_ − 1 H$^+$

Zwitterion (LXVII)          Isocalebassin (instabil)
$(C_{40}H_{47}O_2N_4^{+-})$          $(C_{40}H_{48}O_2N_4^{++})$

$\downarrow$ (CH$_3$)$_2$SO$_4$          $\downarrow$ O$_2$

Methyl-isocalebassin (LXVIII)          Oxydationsprodukt (LXIX)
$(C_{40}H_{47}O_2N_4 \cdot CH_3^{++})$          $(C_{40}H_{48}O_4N_4^{++}?)$

*Formelübersicht 7.* C-Calebassin und Derivate.

Absorptionskurve, wogegen das langwellige Maximum einer *neutralen* (optisch instabilen) Isocalebassin-Lösung bei etwa 400 m$\mu$ liegt (*29*).

Aus den spektroskopischen Befunden ist zu schließen, daß es sich bei Isocalebassin um eine instabile, amphotere Verbindung handelt, die in saurer Lösung unter Aufnahme eines Protons das resonanz-stabilisierte (wahrscheinlich in bezug auf die Verteilung der $\pi$-Elektronen symmetrische) Kation (LX), in alkalischer Lösung unter *Abgabe* eines Protons eine partiell anionische, gleichfalls resonanz-stabilisierte Verbindung $C_{40}H_{47}O_2N_4^{++-}$ (LXVII) gibt. In guter Übereinstimmung mit dem spektralen Verhalten steht der Befund, daß Isocalebassin beim Methylieren mit Dimethylsulfat und Lauge eine Monomethylverbindung (LXVIII), $C_{41}H_{51}O_2N_4^{++}$, liefert, welche keine sauren Eigenschaften mehr besitzt (kein Unterschied zwischen den Spektren neutraler und alkalischer Lösungen), jedoch immer noch in der Lage ist, ein Proton zu addieren (*29*).

In schwach saurem bis neutralem Milieu (pH 4—7) ist Isocalebassin äußerst oxydabel und geht schon mit Luftsauerstoff rasch in eine rot gefärbte Verbindung (LXIX) über (*19, 29*).

*Formelübersicht 7* (S. 225) gibt einen Überblick über die wichtigsten Reaktionen des Calebassins und seiner Abkömmlinge.

## 7. Experimentelle Verknüpfung und Systematik der Alkaloide vom Strychnin-Typus.

Wie gezeigt worden ist, sind die Alkaloide C-Toxiferin-I (XLIX, S. 216) und C-Dihydrotoxiferin (XXXIX, S. 212) dimere Kondensationsprodukte des $N_{(b)}$-Metho-Wieland-Gumlich-Aldehyds (LII, S. 216) bzw. seiner 18-Desoxy-verbindung (XLI, S. 213) und mithin Verbindungen vom Strychnin-Typus. Es kann weiterhin als wahrscheinlich gelten, daß das C-Alkaloid-H, welches in die gleiche Spektralgruppe gehört, hälftig aus $N_{(b)}$-Metho-Wieland-Gumlich-Aldehyd- und 18-Desoxy-$N_{(b)}$-Metho-Wieland-Gumlich-Aldehyd-Resten zusammengesetzt ist, d. h. ein Produkt „gemischter Kondensation" darstellt. C-Alkaloid-H wäre demnach 18-Hydroxy-C-dihydrotoxiferin; C-Toxiferin-I ist 18,18'-Di-hydroxy-dihydrotoxiferin.

Daß die Alkaloide der C-Curarin-I-Gruppe, der C-Calebassin-Gruppe und der B,C,D-Gruppe zum Strychnin-Typus gehören, hat sich aus experimentellen Verknüpfungen mit den Alkaloiden der C-Dihydro-toxiferin-Gruppe ergeben. C-Curarin-I entsteht aus C-Dihydrotoxiferin, wenn man dieses in festem Zustand bei Gegenwart von Sauerstoff be-lichtet. Vermutlich gilt für diesen Vorgang folgende Gleichung:

$$C_{40}H_{46}N_4^{++} + O_2 \xrightarrow{h\nu} C_{40}H_{44}ON_4^{++} + H_2O.$$

In entsprechender Weise erhält man aus C-Toxiferin-I das C-Alkaloid-E, das als 18,18'-Dihydroxy-C-curarin-I aufgefaßt werden kann. Es ist anzunehmen, daß sich an C-Alkaloid-H analog C-Alkaloid-G als 18-Hydroxy-C-curarin-I anschließt [Bernauer, Berlage, Schmid und Karrer (*25, 22*)].

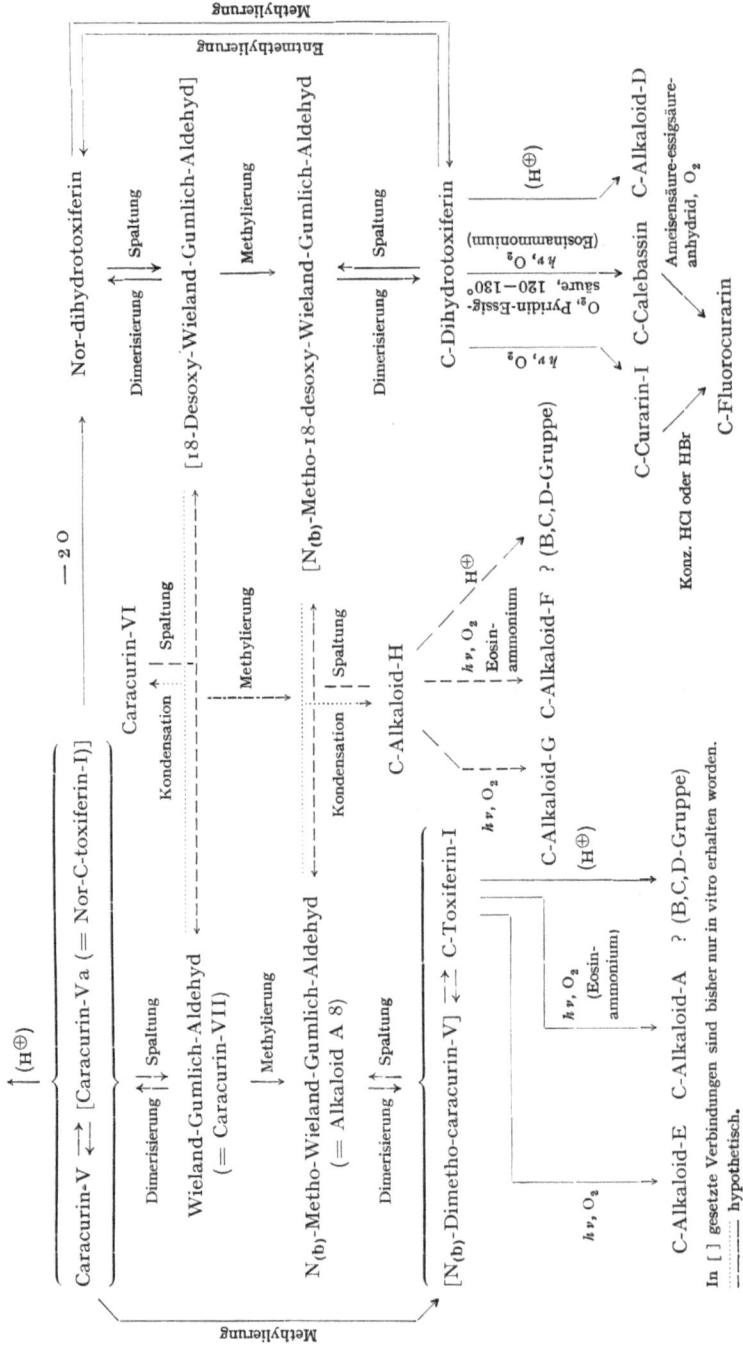

*Formelübersicht 8.* Verknüpfung der Alkaloide vom Strychnin-Typus.

In [ ] gesetzte Verbindungen sind bisher nur in vitro erhalten worden.

------ hypothetisch.

Erhitzt man C-Dihydrotoxiferin in Anwesenheit von Sauerstoff in Pyridin-Essigsäure, so geht es in C-Calebassin über [ASMIS, SCHMID und KARRER (9)]. Die gleiche Überführung kann man auch photochemisch bewerkstelligen, indem man C-Dihydrotoxiferin in Methanol-Wasser bei Gegenwart von Eosin-ammonium und Sauerstoff belichtet (25):

$$C_{40}H_{46}N_4^{++} + \frac{1}{2}O_2 + H_2O \xrightarrow{h\nu} C_{40}H_{48}O_2N_4^{++}.$$

C-Toxiferin-I liefert bei der Eosin-ammonium sensibilisierten Photooxydation das C-Alkaloid-A (demnach 18,18'-Dihydroxy-calebassin) (22). Als vom C-Alkaloid-H abgeleitetes Calebassin-Analogon ist das C-Alkaloid-F (demnach 18-Hydroxy-C-calebassin) zu betrachten. Sensibilisiert man die Photooxydation des C-Dihydrotoxiferins statt mit Eosin-ammonium mit Bengalrosa, so erhält man ,,Lumi-dihydrotoxiferin-I'', $C_{40}H_{48-50}ON_4^{++}$, das bisher nicht in der Natur gefunden worden ist (18).

Bei längerwährender Einwirkung von verdünnter Mineralsäure geht C-Dihydrotoxiferin u. a. in C-Alkaloid-D über (7). Aus Caracurin-Va (= Nor-C-toxiferin-I) (Formel L, S. 217) erhält man in entsprechender Weise das Caracurin-II (7). Über die Vorgänge, die diesen beiden Umwandlungen zugrunde liegen, ist nichts bekannt.

Bezieht man in die Betrachtung noch die früher schon erwähnten Überführungen von C-Curarin-I und von C-Calebassin in C-Fluorocurarin (XXXI, S. 218), von Caracurin-V (LI, S. 217) in Caracurin-Va (L) und von Caracurin-V-dichlormethylat (LIII) in C-Toxiferin-I (XLIX, S. 216), sowie die Übergänge der Norverbindungen in die entsprechenden $N_{(b)}$-Metho-verbindungen (und umgekehrt) mit ein, so läßt sich das in *Formelübersicht 8* dargestellte Bild von den Verknüpfungen zwischen den Alkaloiden des Strychnin-Typus entwerfen.

In Abschnitt B. 1, S. 205, ist die Zugehörigkeit der Alkaloide vom Strychnin-Typus zu bestimmten Spektralgruppen als Ordnungsprinzip benutzt worden. Faßt man die Alkaloide nun noch nach ihrer Abkunft vom gleichen ,,Stamm-Alkaloid'' der C-Dihydrotoxiferin-Gruppe zu ,,Familien'' zusammen (22), so erhält man das in *Tabelle 5*, S. 240, gegebene System der Alkaloide vom Strychnin-Typus. Zur Vervollständigung der Übersicht sind die monomeren Alkaloide mit aufgenommen.

## VII. Zur Biogenese der Alkaloide aus Calebassencurare und südamerikanischen Strychnosarten.

Nach den Vorstellungen von BARGER (14), HAHN (43), ROBINSON (79) und WOODWARD (107, 108) kommt das Gerüst der Indolalkaloide vom Yohimbin-Typus und vom Strychnin-Typus durch Kondensation der Bausteine Tryptamin, (Di)hydroxy-phenylacetaldehyd und Formaldehyd oder äquivalenter Verbindungen zustande (*Formelübersicht 9*).

WENKERT und BRINGI (95) haben neuerdings diese Hypothese modifiziert; und zwar u. a. deshalb, weil sich herausgestellt hat, daß alle

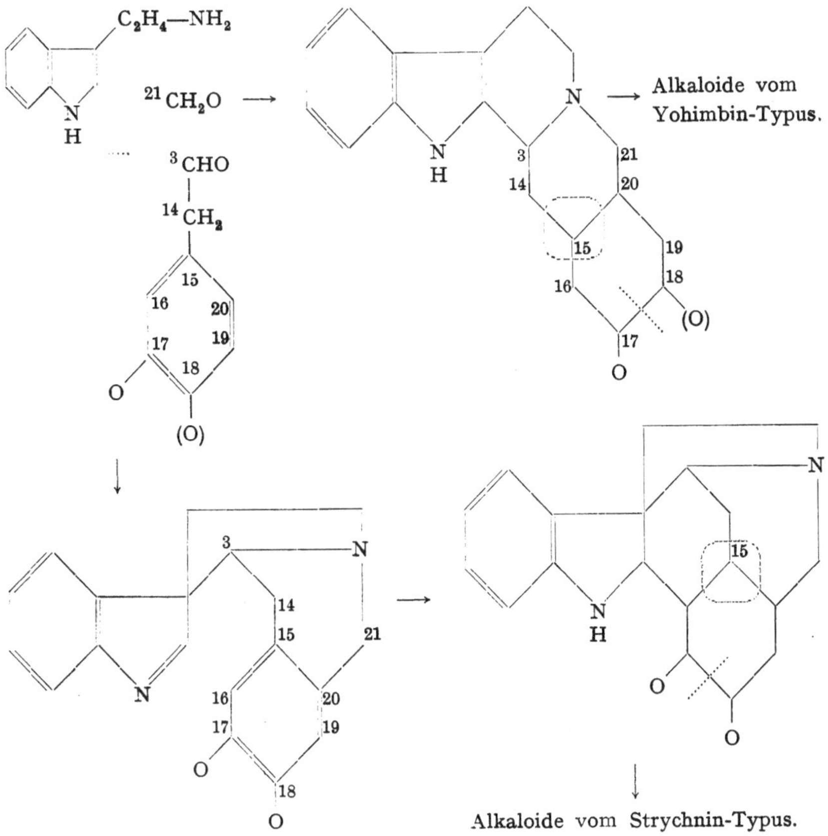

*Formelübersicht 9.*

Alkaloide, die man sich nach Formelübersicht 9 entstanden denken kann, soweit sie daraufhin untersucht sind, am $C_{(15)}$ die gleiche absolute Konfiguration besitzen, was nach den Autoren nur schwer verständlich wäre, stammte das $C_{(15)}$ aus einem aromatischen Sechsring. Sie nehmen daher an, daß die C-Atome 15 bis 20 von einem *hydro*aromatischen Sechsring herrühren: Eine aus Shikimisäure (LXX) und Brenztraubensäure entstandene hydratisierte Prephensäure (LXXI) reagiert mit einem Formaldehyd-Äquivalent* zu (LXXII, S. 232), von dem aus sich alle einschlägigen Alkaloide ableiten lassen.

*Formelübersicht 10,* S. 230 zeigt, wie sich die Alkaloide aus Calebassencurare und südamerikanischen Strychnosarten an (LXXII) anschließen. Es sei betont, daß die eingangs erwähnten Biogenesetheorien rein strukturell zu den gleichen Ergebnissen führen.

---

* In der Formelübersicht der Einfachheit halber Formal-tryptamin.

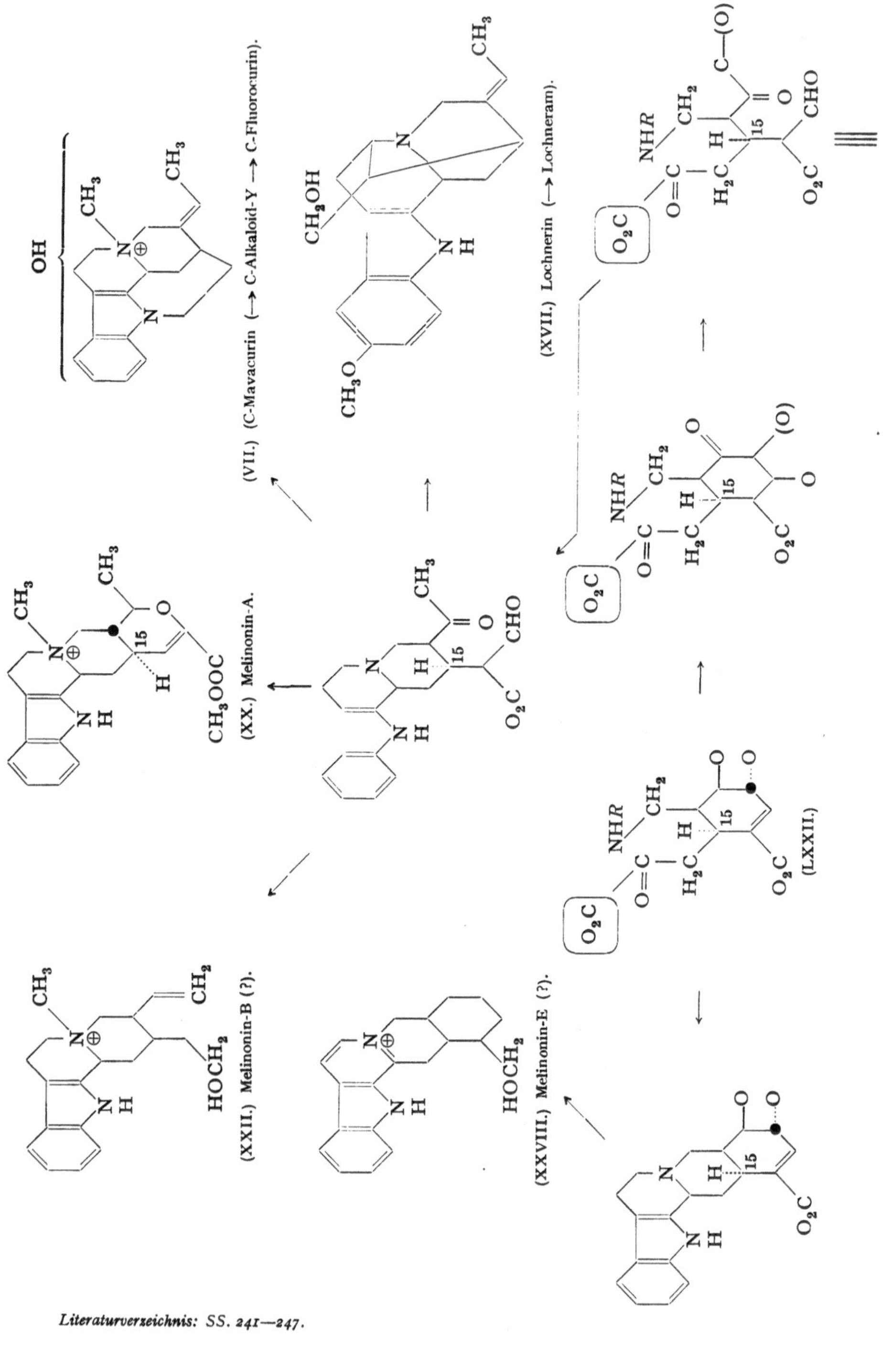

(VII.) (C-Mavacurin (→ C-Alkaloid-Y → C-Fluorocurin).

(XVII.) Lochnerin (→ Lochneram).

(XX.) Melinonin-A.

(XXII.) Melinonin-B (?).

(XXVIII.) Melinonin-E (?).

(LXXII.)

*Literaturverzeichnis:* SS. 241—247.

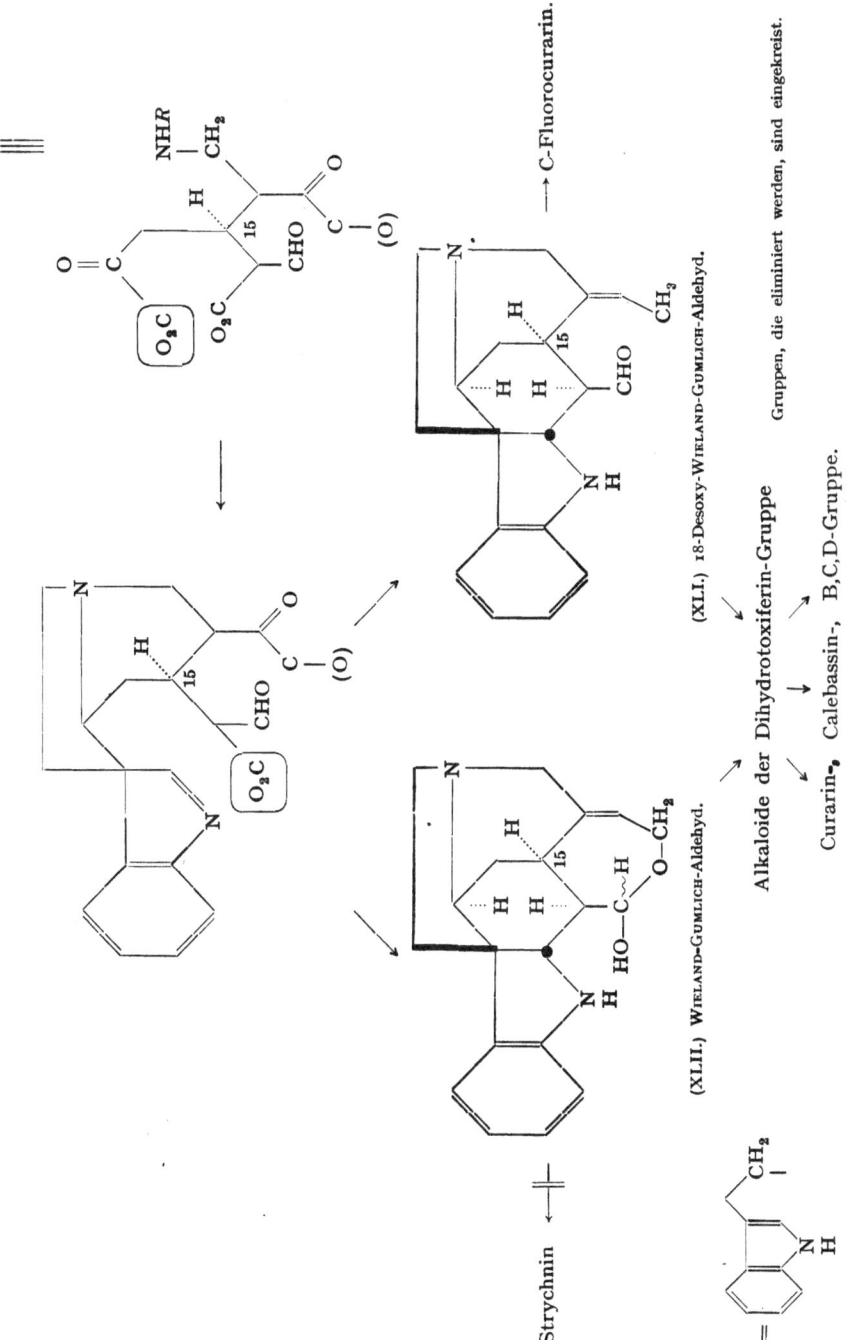

(XLI.) 18-Desoxy-Wieland-Gumlich-Aldehyd.

(XLII.) Wieland-Gumlich-Aldehyd.

Alkaloide der Dihydrotoxiferin-Gruppe

Curarin-, Calebassin-, B,C,D-Gruppe.

Gruppen, die eliminiert werden, sind eingekreist.

*Formelübersicht 10.* Biogenese-Schema.

Strychnin

$R =$

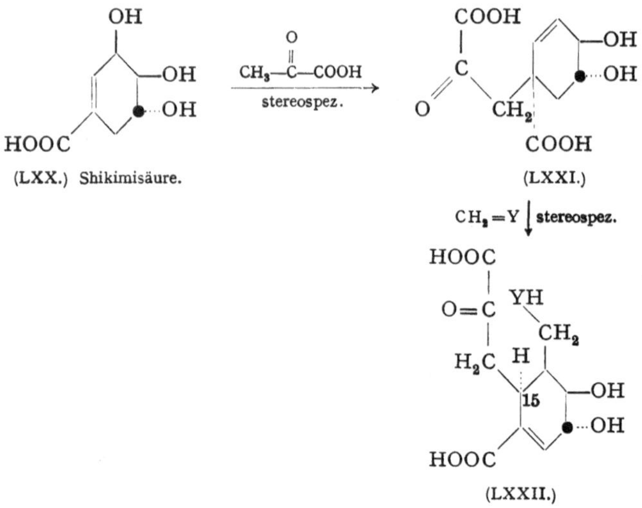

OH

‖ O
CH₃—C—COOH
──────────→
stereospez.

COOH

(LXX.) Shikimisäure.

(LXXI.)

$CH_3 = Y$ | stereospez.

(LXXII.)

# VIII. Zur Pharmakologie der Alkaloide aus Calebassencurare und südamerikanischen Strychnosarten*.

Übersichtsreferate über curare-aktive Stoffe: (33 a).

## 1. Allgemeines.

Curare-Alkaloide wirken muskellähmend. Die unmittelbare Todesursache bei letaler Curarevergiftung ist Atemlähmung. Die erwünschte Wirkung bei medizinischer Anwendung von Curare-Alkaloiden ist Stilllegung und Entspannung der Muskulatur, z. B. bei Operationen, oder die Aufhebung der Dauerkontraktion (Tetanus) bei der Behandlung von Starrkrampf-fällen.

Zum Verständnis dieser Erscheinungen muß kurz auf die Vorgänge im normalen und im curarisierten Muskel eingegangen werden: Zu jeder Faser eines quergestreiften Muskels gehört eine motorische Nervenfaser. Nervenfaser und Muskelfaser sind in der sogen. motorischen Endplatte miteinander verbunden. Erhält die Nervenfaser einen motorischen Impuls, so wird an ihrer Endigung (an der motorischen Endplatte) Acetylcholin gebildet. Dieses tritt an die cholinergischen Receptoren der Endplatte und leitet dadurch die Kontraktion der Muskelfaser ein. Ein in der Endplatte anwesendes Ferment Cholinesterase baut Acetylcholin rasch ab.

Tubocurarin und die curareaktiven Calebassen-Alkaloide haben wie Acetylcholin eine „Affinität" zu den cholinergischen Receptoren der

* Der Verfasser dankt Herrn Professor P. WASER für die kritische Durchsicht dieses Kapitels.

Endplatte, ohne indessen wie dieses eine Muskelkontraktion auslösen zu können. Sind sie in ausreichender Konzentration in der Endplatte anwesend, so besetzen sie die Receptoren und halten das Acetylcholin von seinem Wirkungsort fern. Im curarisierten Muskel sind Muskelfaser und motorische Nervenfaser und Acetylcholinbildung intakt. Blockiert ist lediglich die motorische Endplatte, genauer die cholinergischen Receptoren darin. Hemmt man die Cholinesterase, so daß eine Erhöhung der Acetylcholin-Konzentration eintritt, so kommt die Muskelkontraktion wieder zustande. Darauf beruht die Wirkung der Curareantagonisten, z. B. des Eserins.

## 2. Der Maustest.

Zur Testung auf Curarewirksamkeit werden die Alkaloide Versuchstieren injiziert. Für *quantitative* Zwecke verwendet man heute vornehmlich Mäuse.

Nach einem von WASER (*94*) ausgearbeiteten Verfahren wird die wäßrige Probelösung (bis zu 0,25 ml) in die Schwanzvene einer weißen Maus gespritzt, worauf man das Tier auf einer Glasplatte laufen läßt. Die einsetzende Curarisierung äußert sich zunächst im unsicheren Gang des Tieres; im weiter fortgeschrittenen Stadium sinkt der Kopf ab (,,*head-drop*''), später ist das Tier nicht mehr imstande, sich aus der *Seitenlage*, in die man es bringt, aufzurichten. Schließlich tritt bei ausreichender Dosis Tod durch Atemlähmung ein. Andernfalls erholt sich das Tier, wobei die geschilderten Stadien in umgekehrter Reihenfolge durchlaufen werden.

Die Minimaldosen, die nötig sind, um das Versuchstier in die genannten Stadien zu versetzen, werden gewöhnlich in mg oder $\gamma$/kg Maus ausgedrückt und als Head-drop-Dosis (HD), Seitenlagendosis (SL) und Dosis minimalis letalis (DML) bezeichnet. Als *therapeutische Breite* definiert man den Quotienten DML/HD. Als weitere charakteristische Größe kann noch die *Lähmungsdauer* festgestellt werden. WASER definiert diese als die Zeit, während welcher der Head-drop anhält, wenn man eine Dosis gegeben hat, die in der Mitte zwischen HD und DML liegt.

Eine Zusammenstellung der genannten Größen findet man in den Arbeiten (*52*) und (*94*).

## 3. Curare-Wirkung und Nebenwirkungen.

Die Wirkung der Curare-Alkaloide ist nicht auf die myoneuralen Synapsen (motorischen Endplatten) beschränkt, sondern kann auch andere Synapsen (die Nervenganglien) erfassen. Interessant ist hierbei ein Vergleich der Dosen, die für Paralyse einerseits und für Ausschaltung der Ganglien andererseits notwendig sind. WASER (*94*) hat an der Katze Versuche angestellt, um das Verhältnis der Ganglionblock-Dosis zur die Endplatten paralysierenden Dosis festzulegen.

Einige Alkaloide, vor allem das C-Toxiferin-I, zeichnen sich durch besondere Elektivität aus. Umgekehrt greift das C-Alkaloid-B bevorzugt an den Ganglien an. Es ist nach Waser eher als ein „Ganglionblocker" denn als eine Curaresubstanz anzusprechen.

Elektiv wirkende Curare-Alkaloide sind unterhalb der paralysierenden Dosis frei von Nebenwirkungen, oft auch weit darüber hinaus. Eine Blutdruck-senkende Wirkung z. B. tritt mit dem sehr spezifischen C-Toxiferin-I an der (künstlich beatmeten) Katze erst mit $150\,\gamma$/kg ein (paralytische Dosis = $5,4\,\gamma$/kg). Günstig liegen die Verhältnisse auch noch bei C-Alkaloid-H und dem C-Dihydrotoxiferin, ferner bei den Alkaloiden der Curarin-Gruppe. Die Alkaloide der Calebassin-Gruppe rufen, noch ehe vollständige Muskellähmung erreicht ist, eine starke Blutdrucksenkung hervor. Der „Ganglionblocker" C-Alkaloid-B zeichnet sich durch besonders intensive Blutdruckwirkung aus (94).

Eine Untersuchung der *zentralen* Wirkung reiner Calebassen- und südamerikanischer Strychnos-Alkaloide steht noch aus.

### 4. Resorption, Verteilung und Ausscheidung von Calebassen-Alkaloiden.

Subkutane und intramuskuläre Injektionen führen, wie zu erwarten, zu einer langsam einsetzenden und bei entsprechender Dosis lang anhaltenden Curarisierung. [Besonders sei auf den Befund hingewiesen, daß C-Curarin-I auch *enteral* resorbiert wird (94)].

Die Verteilung von mit [14]C markiertem C-Curarin-I (87) auf die Katze ist von Waser, H. Schmid und K. Schmid (94c) untersucht worden. Die größten Curarin-Konzentrationen wurden in der Leber und Niere gefunden. Die quergestreifte Muskulatur wies einen auffallend niedrigen Curarin-Gehalt auf.

Die Ausscheidung der Calebassen-Alkaloide erfolgt hauptsächlich über die Niere, zu einem geringeren Teil in der Galle. Nach Versuchen mit Radio-C-Curarin (94, 94c) tritt ein oxydativer Abbau der Calebassen-Alkaloide nicht ein.

### 5. Nachweis der Fixierung des C-Curarins-I an den motorischen Endplatten.

Waser und Lüthi (94b) haben Radio-C-Curarin-I Mäusen injiziert und von den Zwerchfellen Autoradiographien hergestellt. Die Autoradiographien wurden mit den Bildern verglichen, die man durch Cholinesterase- (d. h. Endplatten-) spezifische Anfärbung der jeweiligen Präparate erhalten hatte. Die gute Übereinstimmung zeigt, daß das Alkaloid an den Endplatten fixiert wird.

### 6. Konstitution und Curare-Wirksamkeit.

Von den Calebassen- und südamerikanischen Strychnos-Alkaloiden bekannter oder teilweise bekannter Konstitution besitzen nur die quartären, dimeren Verbindungen vom Strychnin-Typus starke Curare-Aktivität, d. h. die Alkaloide der Dihydrotoxiferin-, Curarin-, Calebassin-

und B,C,D-Gruppe. Monomere oder tertiäre Alkaloide sind praktisch unwirksam. Da die dimeren Alkaloide im Papierchromatogramm bedeutend langsamer wandern als die monomeren Alkaloide, ergibt sich die Regel, daß Alkaloide mit kleinem $R_C$-Wert (< 1,5) im allgemeinen aktiv, solche mit hohem $R_C$-Wert ($\gtrsim$ 2) wenig oder nicht aktiv sind (52, 94).

Innerhalb der einzelnen Alkaloid-Gruppen nimmt die Toxizität mit der Anzahl der Hydroxylgruppen (und damit mit abnehmenden $R_C$-Werten) zu.

Ein erstaunlicher und theoretisch vielleicht interessanter Toxizitätsunterschied besteht zwischen den isomeren und eng verwandten Alkaloiden C-Toxiferin-I (XLIX, S. 216) und Caracurin-V-dichlormethylat (LIII, S. 216) (17). Beachtung verdient auch die.Tatsache, daß *mono*quartäre dimere Alkaloide eine ganz beträchtliche Toxizität besitzen können, wie das Beispiel des Nor-C-Curarin-I-monochlormethylates lehrt (94a).

# IX. Tabellen.

Tabelle 1. Alkaloide aus Calebassencurare*.

| Alkaloid | $R_C$ (Gemisch „C") | Cer(IV)-Reaktion auf Papier sofort/nach 5—10 sec. | HD** | Isolierung | Struktur (Seite) |
|---|---|---|---|---|---|
| C-Alkaloid-A | 0,23 | blauviolett/orange | 7 | (52, 86) | 228 |
| ,,  B | 0,34 | rotviolett/farblos | 280 | (86, 52) | 207 |
| ,,  C | 0,34 | rotviolett/farblos | 240 | (86, 52) | 207 |
| ,,  D | 0,35 | rotviolett/farblos | 1 100 | (86, 52) | 207 |
| ,,  E | 0,36 | blau/hellgrün | 0,3 | (86, 52) | 226 |
| C-Toxiferin-I | 0,42 | rotviolett/farblos | 9 | (83) | 216 |
| C-Alkaloid-F | 0,49 | rotviolett/orange | 75 | (86, 52) | 228 |
| ,,  G | 0,65 | blau/hellgrün | 0,6 | (86, 52) | 226 |
| ,,  R | 0,68 | violett | | (66) | ← |
| ,,  H | 0,71 | rotviolett/farblos | 16 | (86, 52) | 226 |
| C-Calebassin | 0,80 | blauviolett/orange | 240 | (96, 50) | 228 |
| (= C-Toxiferin-II = | | | | (104) | |
| C-Strychnotoxin-I) | | | | | |
| C-Alkaloid-I | 0,89 | blauviolett/rot | 174 | (86, 52) | ← |
| C-Curarin-I | 1,00 | blau/hellgrün | 30 | (99, 50) | 218 |
| C-Alkaloid-J | 1,04 | blaß rotviolett/farblos | 290 | (86, 52) | ← |
| C-Guaianin | 1,12 | blau/blaßgrün | 35 | (41) | 207 |
| C-Dihydrotoxiferin | 1,22 | rotviolett/farblos | 30 | (96, 104, 86) | 212 |
| (= C-Alkaloid-K) | | | | (52) | |
| C-Alkaloid-M | 1,45 | keine | | (6, 4) | ← |
| ,,  Y | 1,59 | rotviolett | | (6) | 199 |
| (= Profluorocurin) | | | | | |
| C-Calebassinin | 1,68 | keine | 22 000 | (86, 52) | ← |
| C-Fluorocurin | 2,10 | orange | 4 400 | (84) | 199 |
| C-Pseudofluorocurin | 2,10 | orange | | (66) | ← |
| C-Fluorocurinin | 2,23 | blaß karmin | 2 750 | (86, 52) | ← |
| C-Fluorocurarin | 2,25 | blaß hellblau | 1 800 | (101, 86, 52) | 208 |
| (= C-Curarin-III) | | | | | |
| C-Alkaloid-L | 2,50 | rot | 1 900 | (86, 52) | ← |
| C-Mavacurin | 2,70 | karminrot | | (104, 6) | 199 |
| C-Alkaloid-O | 3,95 | | 7 000 | (41) | ← |
| ,,  P | | | 2 500 | (41) | ← |
| Lochneram | 3,1 | | | (4) | 200 |
| C-Alkaloid-X | | rot | | (84) | ← |
| C-Xanthocurin | | | | (41) | ← |
| C-Isodihydro-toxiferin | | | | (96) | 206 |
| Toxiferin-II | | | | (96) | 207 |
| ,,  IIa*** | | | | (96) | ← |
| ,,  IIb*** | | | | (96) | ← |
| C-Curarin-II | | | | (101) | ← |
| Alkaloid 1 | | | | (104) | ← |
| ,,  2 | | | | (104) | 206 |

*Literaturverzeichnis: SS. 241—247.*

*(Fortsetzung von Tabelle 1.)*

| Alkaloid | $R_C$ (Gemisch „C") | Cer(IV)-Reaktion auf Papier sofort/nach 5—10 sec. | HD** | Isolierung | Struktur (Seite) |
|---|---|---|---|---|---|
| C-Alkaloid-Q† | | | | *(66)* | ← |
| ,, S† | | | | *(66)* | 206 |
| Lochnerin* | | | | *(5)* | 200 |
| Caracurin-II† | | | | *(110)* | 207 |
| *l*-Curin | | | | *(101)* | ← |

\* Hat ein Alkaloid mehrere Bezeichnungen, so wird die charakteristischeste gewählt. — Die $R_C$-Werte beziehen sich auf Hydrochloride.

** „Head-drop"-Dosis in γ/kg Maus (S. 233).

*** Umwandlungsprodukte des Toxiferins-II.

† Tertiäres Alkaloid.

← verweist auf in der Spalte „Isolierung" angeführte Originalarbeiten.

Tabelle 2. Strychnosarten, die Calebassen-Alkaloide führen
[teilweise nach *(60)*] (+: papierchromatographisch nachgewiesen; i: isoliert).

| Alkaloid* | tri | div | $tox_1$ | $tox_2$ | tom | sol | Mi | Fro | rub | mel | am | gu | sub |
|---|---|---|---|---|---|---|---|---|---|---|---|---|---|
| C-Dihydrotoxiferin . | + | | | | | | | + | | | | | |
| C-Curarin-I | + | + | | | + | + | + | | | | | + | |
| C-Calebassin | + | + | | | | + | + | | + | | | | |
| C-Alkaloid-D | | | | | | + | + | | | | | | |
| C-Fluorocurarin | + | + | | | + | + | + | | + | | | | + |
| C-Alkaloid-H | + | | | | + | | | | | | | | |
| ,, G | | | | | + | | | | | | | | |
| ,, F | | | | | + | | | | | | | | |
| C-Toxiferin-I | | i | | | + | | | + | | | | | |
| Toxiferin-I** | | i | | | | | | | | | | | |
| C-Alkaloid-E | | | | | | + | + | + | | | | | |
| Toxiferin-II*** | | | i | | | | | | | | | | |
| C-Mavacurin | | + | i | | | | | + | + | + | i | + | + |
| C-Alkaloid-Y | | | i | | | | | | | | | | |
| C-Fluorocurin | | | i | | + | + | + | | | i | | | + |
| C-Alkaloid-C | | | | | | + | | | | | | | |
| Caracurin-II | | | | i | | | | | | | | | |
| C-Alkaloid-I | + | | | | | | | | + | + | | | |
| ,, J | | | | | | | | | | + | | | |
| C-Fluorocurinin | + | | | | + | | + | | + | + | | | |
| C-Calebassin | | | | | + | | | | | | | | |

tri = *Str. trinervis* (AUBL.) MART. *(1)*.

div = *Str. divaricans* DUCKE *(63)*.

$tox_1$ = *Str. toxifera* ROB. SCHOMB. *(100, 96, 55a, 15)*.

\* Bis zum C-Fluorocurin nach Familien geordnet *(22)* (vgl. S. 228).

** Wahrscheinlich identisch mit C-Toxiferin-I *(83)*.

*** Wahrscheinlich identisch mit C-Alkaloid-A *(22)*.

(Fortsetzung von Tabelle 2.)

tox₂ = *Str. toxifera* (?) *(8, 10)*.  rub = *Str. rubiginosa* DC. *(60)*.

tom = *Str. tomentosa* (Benth.) *(64, 76)*.  mel = *Str. melinoniana* Baill. *(11)*.

sol = *Str. solimoesana* Kruk. *(61)*.  am = *Str. amazonica* Kruk. *(35)*.

Mi = *Str. Mitscherlichii* *(53, 63)*.  gu = *Str. guianensis* (Aubl.) Mart. *(62a)*.

Fro = *Str. Froesii* Ducke *(76)*.  sub = *Str. subcordata* Spruce *(71)*.

Tabelle 3. Aus südamerikanischen Strychnosarten isolierte Alkaloide.

(Nur kristallisierte Alkaloide, die analysiert oder eindeutig identifiziert worden sind, werden angeführt. Kursiv gedruckte Alkaloide sind auch in Calebassencurare gefunden worden.)

| *Strychnos* | Alkaloid | Isolierung | Struktur (Seite) |
|---|---|---|---|
| *toxifera* bearbeitet von King | *Toxiferin-I* (C-Toxiferin-I) Toxiferine-III bis X (Toxiferin-XI = Toxiferin-I) Toxiferin-XII | *(55a)* | 216 190 190 |
| *toxifera* bearbeitet von H. Wieland | Toxiferin-I (C-Toxiferin-I ?) *Toxiferin-II*\* (C-Alkaloid-A ?) | *(96, 100)* | 216 207 |
| *toxifera* bearbeitet von Schmid und Karrer | *C-Mavacurin* *C-Fluorocurin* *C-Alkaloid-Y* Caracurin-I *Caracurin-II* Caracurin-III ,, IV ,, V ,, VI ,, VII (W.-G.- Aldehyd) Caracurin-VIII\*\*\* ,, IX\*\*\* Nor-dihydrotoxiferin Fedamazin | $R_C$\*\* | 199 199 199 207 207 207 |
| | | 0,7 0,8 0,8 1,0 1,4 1,6 | 217 190 |
| | | 2,1 1,4 1,1 | 217 ← ← |
| | | (8, 10) 1,9 | 212 ← |
| *toxifera* bearbeitet von Battersby und Hodson | *C-Toxiferin-I* ,,Alkaloid A 8'' (N_(b))-Metho- W.-G.-Aldehyd) | *(15)* | 216 216 |
| *melinoniana* Baillon | *C-Mavacurin* *C-Fluorocurin* Melinonin-A ,, B ,, C ,, D ,, E | $R_B$† | 199 199 202 203 190 190 204 |
| | | 1,46 1 (81, 11) 0,8—0,9 | |

*Literaturverzeichnis: SS. 241—247.*

*(Fortsetzung von Tabelle 3.)*

| Strychnos | Alkaloid | $R_B$† | Isolierung | Struktur (Seite) |
|---|---|---|---|---|
| | Melinonin-F | 0,74 | | 199 |
| | ,,       G | 1,1—1,2 | | 204 |
| | ,,       H | 0,54 | *(81, 11)* | 191 |
| | ,,       I | 0,74 | | 191 |
| | ,,       K | 0,56 | | 191 |
| | ,,       L | | | 191 |
| | ,,       M | | | ← |
| *amazonica* KRUK. | ,,Alkaloid γ'' | | *(35)* | 191 |
| *macrophylla* | Macrophyllin-A | | *(47)* | 191 |
| *guianensis* (AUBL.) MART. | Guiacurarin-III | | *(62)* | 191 |
| *diaboli* SANDW. | Diabolin | | *(13)* | 191 |

\* Gibt Umwandlungsprodukte Toxiferin-II a und Toxiferin-II b.

\*\* Definition auf S. 187. Mit Hydrochloriden und Gemisch ,,C'' bestimmt.

\*\*\* Als Chlormethylat isoliert; $R_C$-Wert des Chlormethylates.

† $R_B$ = Wanderungsstrecke des Alkaloid-chlorids/Wanderungsstrecke des Melinonin-B-chlorids.

← verweist auf in Spalte ,,Isolierung'' angeführte Originalarbeiten.

Tabelle 4. Calebassen- und Strychnos-Alkaloide nach UV-Spektren geordnet. (Die Spektren der kursiv gedruckten Alkaloide sind in den Abb. 2—6, SS. 192—193 wiedergegeben.)

*Indole* (Chromophor b): C-Mavacurin, *Melinonin-B*, ,,Alkaloid γ'', Melinonin-L, Melinonin-I, Melinonin-K, Melinonin-A, Lochnerin, Lochneram, Caracurin-VIII, Caracurin-IX, C-Alkaloid-J, C-Alkaloid-L, C-Alkaloid-Q.

*α-Methylen-indoline* (Chromophor c): *C-Dihydrotoxiferin*, C-Alkaloid-H, C-Toxiferin-I, C-Iso-dihydrotoxiferin, Alkaloid-2, C-Alkaloid-S, Nor-dihydrotoxiferin, Caracurin-VI. C-Fluorocurarin ist ein α-Methylenindolin mit einer Aldehydgruppe in Konjugation zu der Doppelbindung.

*Indoline* (Chromophor d): *C-Calebassin*, C-Alkaloid-F, C-Alkaloid-A, C-Alkaloid-B, C-Alkaloid-C, C-Alkaloid-D, Caracurin-II, Caracurin-III, Caracurin-I, C-Alkaloid-I, C-Alkaloid-Y, C-Alkaloid-X, C-Alkaloid-R, Alkaloid ,,A 8'', Caracurin-VII, Caracurin-V, Caracurin-IV, C-Curarin-II.

*Oxindole und/oder N-Acylindoline* (Chromophore e und f): *C-Alkaloid-M*, C-Calebassinin, Diabolin.

*ψ-Indoxyle* (Chromophor g): C-Fluorocurin, ψ-Fluorocurin, C-Fluorocurinin (?).

*β-Carbolinium-derivate* (Chromophor h): Melinonin-E, Melinonin-F, Melinonin-G, Melinonin-M, Fedamazin, C-Xanthocurin.

*Alkaloide mit C-Curarin-I-Spektrum*: *C-Curarin-I*, C-Alkaloid-G, C-Alkaloid-E, C-Guaianin.

Tabelle 5. Natürliche Alkaloide vom Strychnin-Typus.

Horizontalreihen = Familien. Vertikalreihen = Gruppen; in ( ) $R_C$-Werte, die von oben nach unten einen charakteristischen Gang zeigen.

*Dimere Alkaloide:* •

| Dihydrotoxiferin-Gruppe | Calebassin-Gruppe | Curarin-Gruppe | B,C,D-Gruppe |
|---|---|---|---|
| tertiäre | | | |
| Nor-dihydrotoxiferin | | | |
| Caracurin-V a (Nor-C-toxiferin-I) ⇌ Caracurin-V | | | Caracurin-II |
| Caracurin-VI C-Alkaloid-S | | | Caracurin-I Caracurin-III |
| quartäre | | | |
| C-Dihydrotoxiferin (1,22) | C-Calebassin (0,80) | C-Curarin-I (1,00) | C-Alkaloid-D (0,35) |
| C-Alkaloid-H (0,71) | C-Alkaloid-F (0,49) | C-Alkaloid-G (0,65) | |
| C-Toxiferin-I (0,42) | C-Alkaloid-A (0,23) | C-Alkaloid-E (0,36) | |
| C-Isodihydrotoxiferin Alkaloid-2 | | C-Guianin | C-Alkaloid-B (0,34) C-Alkaloid-C (0,34) |

*Monomere Alkaloide:*

tertiär

Caracurin-VII (WIELAND-GUMLICH-Aldehyd)

quartär

„Alkaloid A 8" ($N_{(b)}$-Metho-WIELAND-GUMLICH-Aldehyd)

C-Fluorocurarin

### Literaturverzeichnis.

*1.* ADANK, K., D. BOVET, A. DUCKE e G. B. MARINI-BETTOLO: Ricerche sugli alcaloidi curarizzanti di varie specie di Strychnos del Brasile. Gli alcaloidi della *Strychnos trinervis* (VELL.) MART. Nota I. Gazz. chim. ital. **83**, 966 (1953).

*2.* ANET, F. A. L. and R. ROBINSON: Alkaloids of Australian *Strychnos* Species. Part II. The Constitution of Strychnospermine and Spermostrychnine. J. Chem. Soc. (London) **1955**, 2253.

*3.* — — Conversion of the Wieland-Gumlich Aldehyde into Strychnine. Chem. and Ind. **1953**, 245.

*4.* ARNOLD, W., F. BERLAGE, K. BERNAUER, H. SCHMID und P. KARRER: Über Lochneram, ein neues Calebassenalkaloid, und über C-Alkaloid M. 35. Mitt. über Calebassen-Alkaloide. Helv. Chim. Acta **41**, 1505 (1958).

*5.* ARNOLD, W., W. v. PHILIPSBORN, H. SCHMID und P. KARRER: Über C-Alkaloid T [Sarpaginmethyläther, Lochnerin (?)]. 23. Mitt. über Calebassen-Alkaloide. Helv. Chim. Acta **40**, 705 (1957).

*6.* ASMIS, H., E. BÄCHLI, E. GIESBRECHT, J. KEBRLE, H. SCHMID und P. KARRER: Über weitere aus Calebassen isolierte quartäre Alkaloide. 11. Mitt. über Curare-Alkaloide aus Calebassen. Helv. Chim. Acta **37**, 1968 (1954).

*7.* ASMIS, H., E. BÄCHLI, H. SCHMID und P. KARRER: Umwandlung von Calebassen- und Strychnos-Alkaloiden unter der Einwirkung verdünnter Säuren. 14. Mitt. über Curare-Alkaloide aus Calebassen und verwandte Verbindungen. Helv. Chim. Acta **37**, 1993 (1954).

*8.* ASMIS, H., H. SCHMID und P. KARRER: Über Alkaloide aus einer Strychnostoxifera-Rinde aus Venezuela. 13. Mitt. über Curare-Alkaloide aus Calebassen und verwandte Verbindungen. Helv. Chim. Acta **37**, 1983 (1954).

*9.* — — — Umwandlung von C-Dihydro-toxiferin in C-Calebassin (= C-Toxiferin II). 18. Mitt. über Curare-Alkaloide aus Calebassen. Helv. Chim. Acta **39**, 440 (1956).

*10.* ASMIS, H., P. WASER, H. SCHMID und P. KARRER: Zur Kenntnis der Caracurine, des Nor-C-dihydrotoxiferins und C-Dihydrotoxiferins. 17. Mitt. über Curare-Alkaloide aus Calebassen. Helv. Chim. Acta **38**, 1661 (1955).

*11.* BÄCHLI, E., C. VAMVACAS, H. SCHMID und P. KARRER: Über die Alkaloide aus der Rinde von *Strychnos melinoniana* BAILLON. 25. Mitt. über Calebassen-Alkaloide. Helv. Chim. Acta **40**, 1167 (1957).

*12.* BADER, F. E.: Die Lage der Doppelbindung im Ring E des Alstonins. Helv. Chim. Acta **36**, 215 (1953).

*13.* BADER, F. E., E. SCHLITTLER und H. SCHWARZ: Zur Konstitution des Alkaloids Diabolin. Helv. Chim. Acta **36**, 1256 (1953).

*14.* BARGER, G. und C. SCHOLZ: Über Yohimbin. Helv. Chim. Acta **16**, 1343 (1933).

*15.* BATTERSBY, A. R. and H. F. HODSON: Studies on Toxiferine-I. Proc. Chem. Soc. (London) **1958**, 287.

*15 a.* — — The Structure of Diaboline. Proc. Chem. Soc. (London) **1959**, 126.

*16.* BEJAR, O., R. GOUTAREL, M.-M. JANOT et A. LE HIR: Constitution de la flavopéreirine, alcaloïde du *Geissospermum laeve* (VELLOZO) BAILLON (Apocynacées). C. R. hebd. Séances Acad. Sci. **244**, 2066 (1957).

*17.* BERLAGE, F., K. BERNAUER, W. v. PHILIPSBORN, P. WASER, H. SCHMID und P. KARRER: Notiz zur Synthese des C-Toxiferins-I aus Wieland-Gumlich Aldehyd. Toxizitätsvergleich bei synthetischen und natürlichen Calebassen-Alkaloiden. 38. Mitt. über Calebassen-Alkaloide. Helv. Chim. Acta **42**, 394 (1959).

*18.* BERLAGE, F., K. BERNAUER, H. SCHMID und P. KARRER: Lumi-dihydro-toxiferin-I, ein neues Bestrahlungsprodukt des C-Dihydro-toxiferins. 30. Mitt. über Calebassen-Alkaloide. Helv. Chim. Acta 41, 683 (1958).

*19.* BERNAUER, K., E. BÄCHLI, H. SCHMID und P. KARRER: Zur Kenntnis des C-Calebassins und der Alkaloide der Calebassin-Gruppe. 22. Mitt. über Calebassen-Alkaloide. Angew. Chem. 69, 59 (1957).

*20.* BERNAUER, K., F. BERLAGE, W. v. PHILIPSBORN, H. SCHMID und P. KARRER: Über die Konstitution der Calebassen-Alkaloide C-Dihydro-toxiferin und C-Toxiferin-I und des Alkaloids Caracurin-V aus *Strychnos toxifera*. Synthetische Versuche mit Wieland-Gumlich-Aldehyd als Ausgangsstoff. 36. Mitt. über Calebassen-Alkaloide. Helv. Chim. Acta 41, 2293 (1958).

*21.* — — — — — C-Dihydro-toxiferin aus Caracurin-V. 37. Mitt. über Calebassen-Alkaloide. Helv. Chim. Acta 42, 201 (1959).

*22.* BERNAUER, K., F. BERLAGE, H. SCHMID und P. KARRER: Umwandlung von C-Toxiferin-I in die C-Alkaloide E und A und von Desoxy-calebassin in C-Calebassin. Die Einordnung der Calebassen-Alkaloide in Familien. 31. Mitt. über Calebassen-Alkaloide. Helv. Chim. Acta 41, 1202 (1958).

*23.* BERNAUER, K., S. K. PAVANARAM, W. v. PHILIPSBORN, H. SCHMID und P. KARRER: Identifizierung von Caracurin-VII mit dem Wieland-Gumlich Aldehyd. 33. Mitt. über Calebassen-Alkaloide. Helv. Chim. Acta 41, 1405 (1958).

*24.* BERNAUER, K., H. SCHMID und P. KARRER: Zur Kenntnis des C-Calebassins und Isocalebassins. 24. Mitt. über Calebassen-Alkaloide. Helv. Chim. Acta 40, 731 (1957).

*25.* — — — Photochemie des C-Dihydro-toxiferins (C-Alkaloids-K). Präparative Umwandlung in C-Curarin-I und C-Calebassin. 27. Mitt. über Calebassen-Alkaloide. Helv. Chim. Acta 40, 1999 (1957).

*26.* — — — Ein triquartäres Calebassin-Derivat. Beitrag zur Festlegung der Bruttoformel des C-Calebassins. 28. Mitt. über Calebassen-Alkaloide. Helv. Chim. Acta 41, 26 (1958).

*27.* — — — Zur Kenntnis des C-Calebassins: Calebassin-dimethyläther. 29. Mitt. über Calebassen-Alkaloide. Helv. Chim. Acta 41, 673 (1958).

*28.* — — — Gegenseitige Umwandlung von C-Dihydrotoxiferin und Hemi-dihydrotoxiferin. Zur Konstitution der $C_{40}$-Calebassen-Alkaloide. 34. (vorl.) Mitt. über Calebassen-Alkaloide. Helv. Chim. Acta 41, 1408 (1958).

*29.* — — — Unveröffentlichte Versuche.

*30.* BICKEL, H., E. GIESBRECHT, J. KEBRLE, H. SCHMID und P. KARRER: Zur Kenntnis des Fluorocurins. 10. Mitt. über Curare-Alkaloide aus Calebassen. Helv. Chim. Acta 37, 553 (1954).

*31.* BICKEL, H., H. SCHMID und P. KARRER: Zur Kenntnis des Fluorocurins und Mavacurins. 15. Mitt. über Curare-Alkaloide aus Calebassen. Helv. Chim. Acta 38, 649 (1955).

*32.* BOEHM, R.: Das südamerikanische Pfeilgift Curare in chemischer und pharmakologischer Beziehung. I. Teil. Abhandl. kgl. sächs. Gesellsch. Wissensch. 22, 201 (1895).

*33.* — Das südamerikanische Pfeilgift Curare in chemischer und pharmakologischer Beziehung. II. Teil. Abhandl. kgl. sächs. Gesellsch. Wissensch. 24, 1 (1897).

*33a.* BOVET, D., F. BOVET-NITTI and G. B. MARINI-BETTOLO (Ed.): Curare and Curare-like Agents. Amsterdam-London-New York-Princeton: Elsevier Publ. Co. 1959.

*34.* BOVET, D., A. DUCKE, K. ADANK e G. B. MARINI-BETTOLO: Ricerche sugli alcaloidi curarizzanti di varie specie di Strychnos del Brasile. Nota II. Studi preliminari su sette nuove specie di Strychnos. Gazz. chim. ital. 84, 1141 (1954).

35. CASINOVI, G. C.: Ricerche sugli alcaloidi quaternari delle Strychnos del Brasile. Nota XI. Gli alcaloidi della ,,S. amazonica" KRUK. Applicazione dei metodi di distribuzione in controcorrente e di elettroforesi su colonna. Gazz. chim. ital. 87, 1174 (1957).

36. CASINOVI, G. C., M. LEDERER e G. B. MARINI-BETTOLO: Ricerche sugli alcaloidi curarizzanti di varie specie di Strychnos del Brasile. Nota VI. Separazione degli alcaloidi quaternari delle Strychnos per cromatografia su carta con solventi acquosi. Gazz. chim. ital. 86, 342 (1956).

37. FRITZ, H., E. BESCH und TH. WIELAND: Über die Alkaloide aus Calebassen-Curare. XV. Zur Kenntnis von C-Curarin-III (C-Fluorocurarin). Liebigs Ann. Chem. 617, 166 (1958).

37 a. — — — Synthese von C-Curarin-III (C-Fluorocurarin) aus Strychnin. Über die Alkaloide aus Calebassen-Curare. XVI. Angew. Chem. 71, 126 (1959).

38. FRITZ, H. und H. MEYER: Über die Alkaloide aus Calebassen-Curare. XIV. Ultracurarin-A und Ultracurarin-B, zwei neue Umwandlungsprodukte aus C-Curarin-I. Liebigs Ann. Chem. 617, 162 (1958).

39. FRITZ, H. und TH. WIELAND: Über Alkaloide aus Calebassen-Curare. XIII. Umwandlung von C-Curarin-I in C-Curarin-III. Liebigs Ann. Chem. 611, 277 (1958).

40. FRITZ, H., TH. WIELAND und E. BESCH: Über die Alkaloide aus Calebassen-Curare. XII. Umwandlung von C-Mavacurin in C-Fluorocurin über das C-Alkaloid-Y. Liebigs Ann. Chem. 611, 268 (1958).

41. GIESBRECHT, E., H. MEYER, E. BÄCHLI, H. SCHMID und P. KARRER: Über einige neue Calebassen-Alkaloide. 12. Mitt. über Curare-Alkaloide aus Calebassen. Helv. Chim. Acta 37, 1974 (1954).

42. GILL, R. C.: Curare: Misconceptions Regarding the Discovery and Development of the Present Form of the Drug. Anesthesiol. 7, 14 (1946).

43. HAHN, G. und H. WERNER: Synthese von Tetrahydro-harman-(4-Carbolin)-Systemen unter physiologischen Bedingungen. III. Mitt. Synthese des Yohimbin-Gerüstes. Liebigs Ann. Chem. 520, 123 (1935).

44. HUGHES, N. A. and H. RAPOPORT: Flavopereirine, an Alkaloid from *Geissospermum vellosii*. J. Amer. Chem. Soc. 80, 1604 (1958).

45. JANOT, M.-M., R. GOUTAREL, A. LE HIR et O. BEJAR: Flavopéreirine. I. Extraction et structure. Ann. pharm. franç. 16, 38 (1958).

46. JANOT, M.-M. et J. LE MEN: Sur les alcaloïdes cristallisés du *Lochnera* (Vinca) *rosea* (L.) REICHB. ou *Catharanthus roseus* G. DON. C. R. hebd. Séances Acad. Sci. 243, 1789 (1956).

47. JORIO, M. A., O. CORVILLON, H. MAGALHÃES ALVES e G. B. MARINI-BETTOLO: Ricerche sugli alcaloidi curarizzanti di varie specie di Strychnos del Brasile. Nota VII. Gli alcaloidi della *S. macrophylla* BARB. R. Gazz. chim. ital. 86, 923 (1956).

48. KAPFHAMMER, J. und C. BISCHOFF: Acetylcholin und Cholin aus tierischen Organen. Z. physiol. Chem. (Hoppe-Seyler) 191, 179 (1930).

49. KARRER, P. und H. SCHMID: Neuere Arbeiten über Curare, insbesondere Calebassen-Curare und Alkaloide aus Strychnos-Rinden. Angew. Chem. 67, 361 (1955).

50. — — Über Curare-Alkaloide aus Calebassen. 1. Mitt. Helv. Chim. Acta 29, 1853 (1946).

51. KEBRLE, J., H. SCHMID und P. KARRER: Untersuchung eines Extraktes aus Strychnos-toxifera-Rinde. Helv. Chim. Acta 36, 1384 (1953).

52. KEBRLE, J., H. SCHMID, P. WASER und P. KARRER: Über Curare-Alkaloide aus Calebassen. 8. Mitt. Helv. Chim. Acta 36, 102 (1953).

53. Kebrle, J., H. Schmid, P. Waser und P. Karrer: Über die Herkunft der Calebassen-Curare-Alkaloide. Untersuchung verschiedener Calebassen. 9. Mitt. über Calebassen-Alkaloide. Helv. Chim. Acta 36, 345(1953).
54. King, H.: Curare Alkaloids. Part III. Pot-curare. J. Chem. Soc. (London) 1937, 1472.
55. — Curare Alkaloids. Part IX. Examination of Some Strychnos Species from British Guiana: Characterisation of Diaboline, an Alkaloid from Strychnos diaboli Sandwith. J. Chem. Soc. (London) 1949, 955.
55a. — Curare Alkaloids. Part X. Some Alkaloids of Strychnos toxifera Rob. Schomb. J. Chem. Soc. (London) 1949, 3263.
56. Leonard, N. J. and R. C. Elderfield: Alstonia Alkaloids. I. Degradation of Alstonine to β-Carboline Bases and the Reduction of Tetrahydroalstonine with Sodium and Butyl Alcohol. J. Organ. Chem. (USA) 7, 556 (1942).
57. Leonard, N. J. and V. W. Gash: Unsaturated Amines. I. Determination of the Proximity of Nitrogen to a Double Bond by Infrared Absorption Spectra. J. Amer. Chem. Soc. 76, 2781 (1954).
58. Lewin, L.: Die Pfeilgifte. Leipzig: J. A. Barth. 1923.
59. Manske, R. H. F. and H. L. Holmes: The Alkaloids: Chemistry and Physiology, Vol. II, p. 536ff. New York: Academic Press. 1952.
60. Marini-Bettolo, G. B.: Contribution à l'étude des alcaloïdes des Strychnos du Brésil. Festschrift Arthur Stoll, S. 257. Basel: Birkhäuser. 1957.
61. Marini-Bettolo, G. B., P. de Berredo Carneiro e G. C. Casinovi: Ricerche sugli alcaloidi curarizzanti delle Strychnos del Brasile. Nota VIII. Gli alcaloidi della S. solimoesana Kruk. Gazz. chim. ital. 86, 1148 (1956).
62. Marini-Bettolo, G. B. e M. A. Jorio: Ricerche sugli alcaloidi curarizzanti di varie specie di Strychnos del Brasile. Nota IX. Gli alcaloidi della S. guianensis (Aubl.) Mart. Gazz. chim. ital. 86, 1305 (1956).
62a. — — Résumés trav. XIV Congr. intern. chim. pure et appl., Zürich, 1955, p. 152.
63. Marini-Bettolo, G. B., M. A. Jorio, A. Pimenta, A. Ducke e D. Bovet: Ricerche sugli alcaloidi curarizzanti di varie specie di Strychnos del Brasile. Nota V. Studi cromatografici su carta degli alcaloidi quaternari della S. guianensis (Aubl.) Mart., S. divaricans Ducke e S. Mitscherlichii Rich. Schomb. Gazz. chim. ital. 84, 1161 (1954).
64. Marini-Bettolo, G. B., M. Lederer, M. A. Jorio e A. Pimenta: Ricerche sugli alcaloidi curarizzanti di varie specie di Strychnos del Brasile. Nota IV. Metodi elettrocromatografici di separazione e riconoscimento degli alcaloidi. Gazz. chim. ital. 84, 1155 (1954).
65. McIntyre, A. R.: Curare, its History, Nature, and Clinical Use. Chicago, Ill.: Univ. of Chicago Press. 1947.
66. Meyer, H., H. Schmid und P. Karrer: Weitere neue Calebassenalkaloide (C-Alkaloide Q, R, S und Pseudo-fluorocurin). 20. Mitt. über Calebassen-Alkaloide. Helv. Chim. Acta 39, 1208 (1956).
67. Meyer, H., H. Schmid, P. Waser und P. Karrer: Calebassen-Alkaloide aus der Rinde einer südamerikanischen Pflanze. 21. Mitt. über Calebassen-Alkaloide. Helv. Chim. Acta 39, 1214 (1956).
68. Mors, W. B., P. Zaltzman, J. J. Beereboom, S. C. Pakrashi and C. Djerassi: Alkaloids of two Brazilian Apocynaceae: Rauwolfia grandiflora Mart. and Lochnera (vinca) rosea (L.) Reichb. var. alba (Sweet) Hubbd. Chem. and Ind. 1956, 173.
68a. Nagyvari, J., H. Schmid und P. Karrer: Unveröffentlicht.
69. Panouse, J. J.: Contribution à l'étude des reineckates de nicotine et de pyridine. Bull. soc. chim. France 1949, 594.

*70.* PEERDEMAN, A. F.: The Absolute Configuration of Natural Strychnine. Acta Crystallogr. **9**, 824 (1956).

*71.* PENNA, A., M. A. JORIO, S. CHIAVARELLI e G. B. MARINI-BETTOLO: Ricerche sugli alcaloidi curarizzanti di varie specie di Strychnos del Brasile. Nota X. Gli alcaloidi della „*S. subcordata*" SPRUCE. Gazz. chim. ital. **87**, 1163 (1957).

*72.* PHILIPSBORN, W. v.: Zur Konstitution des C-Curarins und über Reaktionen von Verbindungen mit Indolenin-Struktur. Dissert., Zürich, 1956.

*72a.* PHILIPSBORN, W. v., K. BERNAUER, H. SCHMID und P. KARRER: Die Struktur des Fluorocurarins. 39. Mitt. über Calebassenalkaloide. Helv. Chim. Acta **42** 461 (1959).

*73.* PHILIPSBORN, W. v., H. MEYER, H. SCHMID und P. KARRER: Zur Kenntnis des C-Fluorocurarins. 32. Mitt. über Calebassen-Alkaloide. Helv. Chim. Acta **41**, 1257 (1958).

*74.* PHILIPSBORN, W. v., H. SCHMID und P. KARRER: Über die Bruttoformeln der Curare-Alkaloide aus Calebassen und Strychnos-Arten. 19. Mitt. über Curare-Alkaloide aus Calebassen. Helv. Chim. Acta **39**, 913 (1956).

*75.* — — — Zur Kenntnis des C-Curarins. 16. Mitt. über Curare-Alkaloide aus Calebassen. Helv. Chim. Acta **38**, 1067 (1955).

*75a.* — — — Unveröffentlicht.

*76.* PIMENTA, A., M. A. JORIO, K. ADANK e G. B. MARINI-BETTOLO: Ricerche sugli alcaloidi di varie specie di Strychnos del Brasile. Nota III. Gli alcaloidi della *Strychnos tomentosa* (BENTH.), della *Strychnos Froesii* DUCKE e della *Strychnos rubiginosa* (D. C.). Gazz. chim. ital. **84**, 1147 (1954).

*77* POISSON, J., J. LE MEN et M.-M. JANOT: Sur la structure de la sarpagine et de la lochnérine. Bull. soc. chim. France **1957**, 610.

*78.* PRASAD, K. B. and G. A. SWAN: The Constitution of Yohimbine and Related Alkaloids. Part X. The Synthesis of Some 12 H-Indolo[2:3-α]pyridocolinium Salts, including Flavocoryline and Flavopereirine. J. Chem. Soc. (London) **1958**, 2024.

*79.* ROBINSON, R.: The Structural Relations of Natural Products. Oxford: Clarendon Press. 1955.

*80.* SCHLITTLER, E.: In: R. E. WOODSON, Jr., H. W. YOUNGKEN, E. SCHLITTLER and J. A. SCHNEIDER, Rauwolfia: Botany, Pharmacognosy, Chemistry and Pharmacology. Boston, Toronto: Little, Brown & Co. 1957.

*81.* SCHLITTLER, E. und J. HOHL: Über die Alkaloide aus *Strychnos melinoniana* BAILLON. Helv. Chim. Acta **35**, 29 (1951).

*82.* SCHMID, H., A. EBNÖTHER und P. KARRER: Über Curare-Alkaloide aus Calebassen. 5. Mitt. Helv. Chim. Acta **33**, 1486 (1950).

*83.* SCHMID, H. und P. KARRER: Über Curare-Alkaloide aus Calebassen. 2. Mitt. Helv. Chim. Acta **30**, 1162 (1947).

*84.* — — Über Curare-Alkaloide aus Calebassen. 3. Mitt. Helv. Chim. Acta **30**, 2081 (1947).

*85.* — — Über Curare-Alkaloide aus Calebassen. 4. Mitt. Helv. Chim. Acta **33**, 512 (1950).

*86.* SCHMID, H., J. KEBRLE und P. KARRER: Über Curare-Alkaloide aus Calebassen. 7. Mitt. Helv. Chim. Acta **35**, 1864 (1952).

*87.* SCHMID, H., K. SCHMID, P. WASER und A. EBNÖTHER: Über Curare-Alkaloide aus Calebassen. 6. Mitt. Radio-C-Curarin-I-chlorid. Helv. Chim. Acta **34**, 2042 (1951).

*88.* STAUFFACHER, D., A. HOFMANN und E. SEEBECK: Über Sarpagin. Helv. Chim. Acta **40**, 508 (1957).

*89.* Talapatra, S. K. and A. Chatterjee: Constitution of Sarpagine, the Minor Alkaloid of *Rauwolfia micrantha* Hook and *Rauwolfia beddomei* Hook. Science and Culture (India) **22**, 692 (1957).

*90.* — — The Stereoconfiguration of Sarpagine. Naturwiss. **45**, 58 (1958).

*90a.* Tomita, M.: Die Alkaloide der Menispermaceae-Pflanzen. Fortschr. Chem. organ. Naturstoffe **9**, 175 (1952).

*91.* Vamvacas, C., W. v. Philipsborn, E. Schlittler, H. Schmid und P. Karrer: Über die Konstitution des Melinonins-B. 26. Mitt. über Calebassen-Alkaloide. Helv. Chim. Acta **40**, 1793 (1957).

*92.* Volz, H. und Th. Wieland: Die Sauerstoff-Funktion im C-Toxiferin-II. Naturwiss. **44**, 376 (1957).

*93.* — — Umwandlung von C-Toxiferin-II in C-Curarin-III. Liebigs Ann. Chem. **604**, 1 (1957).

*94.* Waser, P. G.: Calebassen-Curare. Helv. Physiol. Acta **11**, Suppl. VIII (1953).

*94a.* — Curare and Cholinergic Receptors in the Motor End-Plate. In: D. Bovet, F. Bovet-Nitti and G. B. Marini-Bettolo, Curare and Curare-like Agents, p. 219. Amsterdam-London-New York-Princeton: Elsevier Publ. Co. 1959.

*94b.* Waser, P. G. und U. Lüthi: Autoradiographische Lokalisation von $^{14}$C-Calebassen-Curarin-I und $^{14}$C-Decamethonium in der motorischen Endplatte. Arch. int. Pharmacodyn. **112**, 272 (1957).

*94c.* Waser, P. G., H. Schmid und K. Schmid: Resorption, Verteilung und Ausscheidung von Radio-Calebassen-Curarin bei Katzen. Arch. int. Pharmacodyn. **96**, 386 (1954).

*95.* Wenkert, E. and N. V. Bringi: A Stereochemical Interpretation of the Biosynthesis of Indole Alkaloids. J. Amer. Chem. Soc. **81**, 1474 (1959).

*96.* Wieland, H., K. Bähr und B. Witkop: Über die Alkaloide aus Calebassen-Curare. IV. Liebigs Ann. Chem. **547**, 156 (1941).

*97.* Wieland, H. und W. Gumlich: Über einige Reaktionen der Strychnos-Alkaloide. XI. Liebigs Ann. Chem. **494**, 191 (1932).

*98.* Wieland, H. und K. Kaziro: Abbauversuche vom Isonitroso-strychnin aus. Über Strychnos-Alkaloide. XIII. Liebigs Ann. Chem. **506**, 60 (1932).

*99.* Wieland, H., W. Konz und R. Sonderhoff: Über das Curarin aus Calebassen-Curare. Liebigs Ann. Chem. **527**, 160 (1937).

*100.* Wieland, H. und H. J. Pistor: Über das Curarin aus Calebassen-Curare. II. Liebigs Ann. Chem. **536**, 68 (1938).

*101.* Wieland, H., H. J. Pistor und K. Bähr: Über die Alkaloide aus Calebassen-Curare. III. Liebigs Ann. Chem. **547**, 140 (1941).

*102.* Wieland, H., B. Witkop und K. Bähr: Über die Alkaloide aus Calebassen-Curare. V. Liebigs Ann. Chem. **558**, 144 (1947).

*103.* Wieland, Th. und H. Fritz: Ozonabbau von Curare-Alkaloiden. Naturwiss. **42**, 297 (1955).

*103a.* Wieland, Th., H. Fritz, K. Hasspacher und A. Bauer: Über die Alkaloide aus Calebassen-Curare. VIII. Zur Struktur von C-Toxiferin-II. Liebigs Ann. Chem. **588**, 1 (1954).

*104.* Wieland, Th. und H. Merz: Über die Alkaloide aus Calebassencurare. VI. Mitt. Chem. Ber. **85**, 731 (1952).

*105.* — — Über Alkaloide aus Calebassen-Curare. VII. Mitt. Liebigs Ann. Chem. **580**, 204 (1953).

*106.* Witkop, B.: Neuere Arbeiten über Indol-Alkaloide (Vortragsreferat). Angew. Chem. **65**, 466 (1953).

*106a.* — Neuere Arbeiten über Pfeilgifte. Angew. Chem. **55**, 85 (1942).

*107.* WOODWARD, R. B.: Biogenesis of the Strychnos Alkaloids. Nature (London) **162**, 155 (1948).

*108.* — Neuere Entwicklungen in der Chemie der Naturstoffe. Angew. Chem. **68**, 13 (1956).

*109.* WOODWARD, R. B., M. P. CAVA, W. D. OLLIS, A. HUNGER, H. U. DAENIKER and K. SCHENKER: The Total Synthesis of Strychnine. J. Amer. Chem. Soc. **76**, 4749 (1954).

*110.* ZÜRCHER, A., O. CEDER and V. BOEKELHEIDE: Calebash Curare of the Piaroa Indians. Conversion of C-Curarine-I to C-Curarine-III. J. Amer. Chem. Soc. **80**, 1500 (1958).

*(Eingelaufen am 4. Januar 1959.)*

# Occurrence and Metabolism of Simple Indoles in Plants.

By **Bruce B. Stowe**, Cambridge, Massachusetts.

### Contents.

The preparation of this article was initiated during the author's tenure of a LALOR Fellowship. It was also assisted by grants to Professor K. V. THIMANN by the National Science Foundation and the American Cancer Society. The author is indebted to Professor THIMANN for his interest, many suggestions, and criticism of the manuscript and acknowledges helpful suggestions by Drs. W. A. ANDREAE and J. B. GREENBERG.

## Introduction.

Indole compounds and their biochemistry have been attracting increasing attention in recent years. Application of new techniques has not only detected many more indoles in nature, but more importantly, the hormonal nature of some of these compounds has been widely established in both plants and animals. On these grounds alone, it is timely to consider the present status of higher plant indoles and their biochemistry.

Although botanists have been aware for twenty-five years of the potent effects of indoleacetic acid on higher plants, it has only been by the application of several modern methods that the presence of other indole hormones has been conclusively established. In addition, paper chromatography has demonstrated a bewildering array of apparently indolic spots whose identities and places in indole metabolism still await unraveling.

Likewise, although indole alkaloids and a few simple indoles have excited the attention of numerous pharmacologists and organic chemists for decades, it has been the recent demonstration of hydroxyindoles acting as neurohormones throughout the animal kingdom which has brought widespread interest to the simple indoles and their metabolism.

Thus, in this area of rapid advances there may be much that physiologists and biochemists of animals and plants may learn from each other. The occurrence of many indole compounds in both kingdoms indicates that comparison of metabolic pathways may be of mutual benefit, despite the fact that the plant hormones have little activity as animal hormones and vice versa. In particular, the need for students of the urinary indoles to know the indolic constituents occurring in diets of plant materials has already been emphasized (*1, 6, 260*).

Therefore, it is the intent of this Review to summarize the development of knowledge concerning simple indoles in higher plants. The more complex indole alkaloids have been the subject of many recent reviews both of their chemistry (*24, 126, 146, 193, 217, 242, 247, 280*) and metabolism (*55, 190, 203*). The animal indole hormones have also been subject to broad scrutiny (*175, 220*), and likewise will be treated here only insofar as it seems especially pertinent to the situation in higher plants.

The indoles of algae, fungi, and bacteria have been less satisfactorily monographed and necessarily will be discussed at greater length. Yet, inasmuch as present studies with these organisms have for the most part concerned themselves with tryptophan metabolism and with pigment formation, the analogies with higher plants are still few. Thus a complete survey of the lower plant investigations will not be attempted.

Possibly in part due to the hormonal affinities of many of these indole compounds, the quantities found in plants have often been small. The evidence cited for purposes of identification has thus in many cases been scanty or indirect, and not at all the secure evidence of crystals and derivatives demanded by the organic chemist. Similarly, only rudimentary information on indole metabolism in higher plants is often available. In fact, on only three of the many routes of indole metabolism have highly purified enzymes been prepared and their characteristics established to the satisfaction of the enzymologist. Therefore, this is a field which demands tolerance in assessing many types of evidence, but is nonetheless now reaching a stage where an overall picture of simple indoles can be put forward with reasonable confidence.

## I. Volatile Indoles.

### 1. Indole.

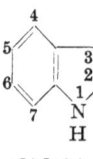

(I.) Indole.

Indole, due to its odor and ubiquitous distribution in fecal material, has been known to the organic chemist for nearly a hundred years. It was early recognized to be a product of the bacterial decomposition of proteins. When Dunstan (59) announced in 1889 the isolation of its 3-methyl derivative, skatole (V, p. 254), from the wood of a Javanese tree, this was the first recorded isolation of an indole from a higher plant. The discovery was soon followed by the identification of indole itself, hypaphorine (XIX, p. 260), and somewhat later, tryptophan (IX, p. 255), in vegetable matter. The number of simple indole derivatives has slowly increased until at present nineteen are well established as plant constituents, eight more can provisionally be identified, and there are a number of other probably indolic compounds which are as yet inadequately described.

The presence of indole itself in plant tissues was conclusively shown by Hesse (128). His extensive work on flower oil constituents was successful in demonstrating that oils from flowers of several species of *Jasminium* and *Citrus* contained 0.1–2.5% indole, and that it is a constituent of the perfume of these flowers. Inasmuch as Hesse worked with the oil obtained from several thousand kilograms of flowers, he had no difficulty in isolating crystalline indole via the picrate derivative from jasmine (*129*), and after vacuum distillation from the orange (*131*).

As often happens in perfume production, the composition of the oil collected by enfleurage (absorption of the volatile perfume from the air into neutral fat or another absorbent) was different from that obtained by direct extraction of jasmine. In this case, freshly plucked jasmine gave the least percentage of indole in the oil after extraction, a greater percentage was obtained in that from wilted flowers, and the highest proportion was obtained in the oil from enfleurage (*130*). The orange oil, however, remained roughly the same in indole concentration whatever the method of obtaining it. An earlier unpublished isolation from the same source was claimed (*246*). CERIGHELLI (*47*) found the indole produced by jasmine varied with the time of day. Several other isolations of indole from *Jasminium* and *Citrus* have been carried out (*179, 244, 301*).

ELZE (*62*) has isolated relatively large amounts of indole from ethereal extracts of *Robinia pseudoacacia* L. blossoms, and other definitive purifications have been made by KUMMERT from *Cheiranthus cheiri* L. (*165*), by VON SODEN from *Narcissus jonquilla* L. (*265*), and by LOUVEAU from *Cheimonanthus fragrans* LINDL (*180*).

An interesting sidelight to HESSE's work (*128*) is his discovery that indole can form a very stable crystalline bisulfite complex from which it can easily be regenerated. This procedure was repeated by ELZE in his isolation (*62*), and seems since to have been generally forgotten. This raises the intriguing possibility that indoles may generally be able to form bisulfite complexes, a situation which has an important bearing on the work with indoleacetaldehyde reported later in this review.

PLUGGE, who made steam distillations of the leaves of the Rubiaceous plant *Paederia foetida* L., is posthumously reported as concluding that indole was present in this tissue (*226*). This is the only case of indole being identified as such in non-floral material, and as no details are cited it is impossible to be sure that bacterial infection was not a factor in producing the indole. A closely related species, *Chione glabra*, has wood with a fecal odor, but investigators were unable to isolate indole from it (*60*).

More numerous reports of the presence of indole in plants are those which are based on various reagents which form colored compounds with indoles in the presence of acid.

The EHRLICH reagent, *p*-dimethylaminobenzaldehyde, which produces red to purple tints with indoles (*6, 80, 152, 276*), is generally considered the most specific, but is subject to various errors in quantitative estimations (*259, 268*). The SALKOWSKI reagent produces a larger range of colors (*6, 80, 152, 276*), but it, too, is subject to interference by other compounds (*224*), and some non-indolic materials give strong reactions (*36, 209, 277*). Nonetheless, both reagents are useful for rapid qualitative work and the application of the EHRLICH test in particular has made possible the screening of a large number of perfumes for indolic constituents.

VERSCHAFFELT was the first to suggest that indole in perfumes could be detected simply by enclosing a flower under a bell jar with a wad of cotton soaked in acid (*306*). Formation of a rose to violet color was considered indicative of indole or skatole, and he showed that jasmine flowers and commercial orange perfume would initiate the reaction; a large number of other flowers were ineffective, however. The perfume-producing parts of the flower were located in the corolla lobes and the tubule. Weehuizen improved this method by using the Ehrlich reagent in the same way (*315*). He detected indole in perfume from *Citrus decumana* Murr. (a racemic variety of *Caladium*) and *Murraya exotica* L., the last identification being further substantiated by proving that the reacting substance could be steam-distilled (*316*). Sack (*238*) has probably made the most extensive use of this technique, discovering indole in the odors of six out of eight *Citrus* species, and three out of four *Coffea* species. *Hevea brasiliensis* Muell. and *Randia formosa* (Jacq.) R. Schum. were also positive, while many other flowers investigated did not cause a reaction (*239*). Borzi (*37*) reported an indole reaction in *Visnea mocanera* L. but believed it to be skatole (see below). In addition to skatole, other indole compounds, as for instance the lower esters of indoleacetic acid, are appreciably volatile under these conditions (*340*). Hence, these reports require more definitive confirmation, such as could be provided by paper chromatographic examination of blossom extracts (*277*).

Baccarini has reported the presence of indole in the flowers of twenty plant species (*11*), as well as in vegetative tissue of *Tilia* (*12*). However, his report is based on boiling tissue with the Ehrlich reagent, a treatment which could easily result in a reaction with protein tryptophan or its degradation products. The lack of specificity of this method is indicated by his remark that material which had been preserved in alcohol for a year (a procedure which would certainly elute all simple indoles), gave the same reaction as fresh tissue.

An inconclusive report from a fungus has also been made (*181, 182*).

More recently, the possible presence of indole in plants has been shown by several workers in the dairy industry, who have attributed the "weedy" flavor in some samples of butter to indole. This is believed to result from the ingestion by cattle of two Crucifers, *Thlaspe arvense* and peppergrass, *Lepidium virginicum*. Ingle (*141*) was able to demonstrate that 70 µg of indole per gram dry weight of *Thlaspe* were present after a 12 hour aqueous extraction at 38°, and the yield was increased to 350 µg/g when proteolytic treatment analogous to bovine digestion was used. Hussong and Quam (*140*) showed that in *Lepidium* the indole resulted from bacterial action; this was surely also the case in Ingle's experiments. Herter's report (*127*) of indole in the wood of *Celtis reticulosa* Miq. is subject to similar objections. Although these results are unlikely to be due to free indole in the plant tissues they do indicate that relatively large amounts of a labile indole precursor must be present.

Conclusive evidence for the route of formation of indole in higher plants is lacking. Mann and Smithies (*191*) showed that pea amine oxidase can act on *o*-aminophenylethylamine (IIc), whereupon the *o*-aminophenylacetaldehyde presumably produced immediately cyclizes to indole. However, this substrate is not known to be present in higher plants.

In lower plants, Yanofsky (*333*) has isolated indole-3-glycerol phosphate (IV) from certain strains of *Escherichia coli* and has shown that indole and a triose phosphate can be formed from it; similar observations have been made by others (*103*). The indole-3-glycerol phosphate is derived from anthranilic acid (IIa). Yanofsky's suggestion that a ribulose anthranilate (similar to IIIa) is an intermediate has recently been strengthened by the discovery that some tryptophan requiring strains of *Aerobacter aerogenes* accumulate 1-deoxy-1-N-*o*-carboxyphenyl-ribulose (IIIb) (*89*). Evidence for a very similar compound

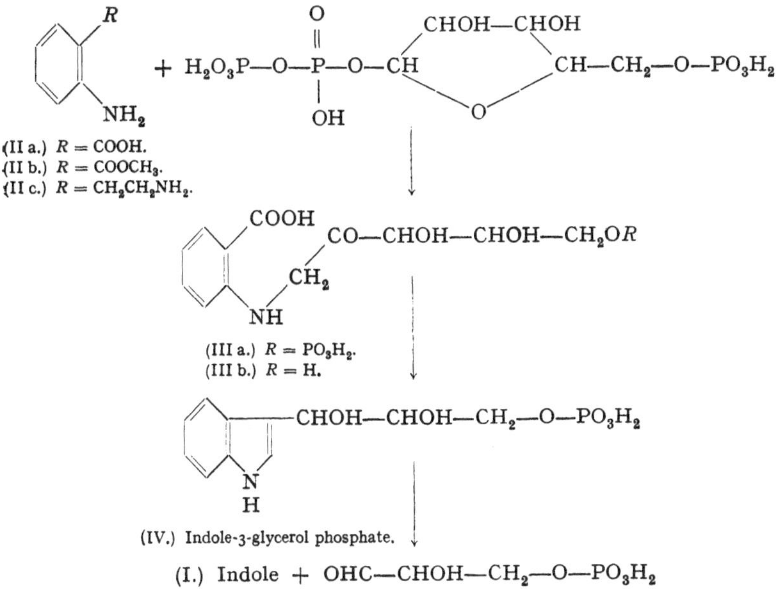
R
NH2
+ H2O3P—O—P—O—CH
O
OH
CHOH—CHOH
CH—CH2—O—PO3H2
O
COOH
CO—CHOH—CHOH—CH2OR
CH2
NH
CHOH—CHOH—CH2—O—PO3H2
N
H
(I.) Indole + OHC—CHOH—CH2—O—PO3H2

(II a.) $R$ = COOH.
(II b.) $R$ = $COOCH_3$.
(II c.) $R$ = $CH_2CH_2NH_2$.

(III a.) $R$ = $PO_3H_2$.
(III b.) $R$ = H.

(IV.) Indole-3-glycerol phosphate.

in *Saccharomyces* has also been presented (*222*). In *Neurospora*, however, the preliminary evidence indicates that indole is not a free intermediate in the formation of tryptophan from indole-3-glycerol phosphate (*334*). The presence of methyl anthranilate (IIb) in the same flower oils that contain indole (*62, 128, 301*) suggests that a pathway similar to that in bacteria may be operative in higher plants. *Escherichia coli* can produce indole by cleavage of tryptophan (*106, 119, 135, 196, 329, 330*); this possibility has not been explored in higher plants.

Two research groups have shown (*108, 204*) that indole can be utilized by plant extracts, presumably for the synthesis of tryptophan, although the quantitative aspects remain to be determined. One of these studies (*204*) indicated that some indoleacetic acid was also produced from indole, which may explain indole's weak, but genuine, activity in plant growth hormone bioassays (*293*). The conversion was not observed in the latter study, but seems a reasonable explanation of the hormonal action, since activity has otherwise not been found for any substance which is not an acid nor readily convertible to one.

Mann and Smithies (*191*) have shown that indole can be destroyed by the action of plant peroxidase.

## 2. Skatole (3-Methylindole).

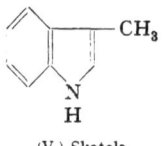

(V.) Skatole.

The position of skatole (V) as a genuine higher plant product is clouded by uncertainty as to whether its presence results secondarily from bacterial action.

DUNSTAN (59), who obtained crystalline skatole from steam distillates of Javanese filth-wood (*Celtis reticulosa* MIQ.) noted the early work on skatole formation by bacteria but he failed to state whether his material might have been decomposed. HERTER (127), noting GRESHOFF's earlier observation that only wounded branches had a skatole odor (113), investigated various parts of the plant and found skatole only in the old wood. In another isolation of crystalline skatole from wood (243), later attributed to a related species, *Celtis durandii* ENGL. (245), the odor was stated to be characteristic of decayed wood. A species of *Nectandra* (probably *N. globosa*) also contains skatole in its wood (238).

In order to determine where skatole was deposited in *Celtis* xylem, WEEHUIZEN developed two histochemical reagents for its localization (316). These both indicated it was only found in the medullary rays and wood parenchyma, the only parts of the wood containing appreciable protein. Nonetheless, WEEHUIZEN stated his belief that bacteria were not involved. This also seems likely from the studies of BOORSMA (35), who noted that the wood was especially rich in parenchymatous cells which were filled with crystals. He was able partially to purify a substance from the wood; this was not skatole, but he believed it was related to it. GRESHOFF (113) had no trouble isolating crystalline skatole from *C. reticulosa*. But when he investigated (111) two other plants with a clearly skatole odor, *Sterculia foetida* L. and an unspecified luminescent liverwort, he instead obtained crystals of m. p. 118°, which did not give indolic colors on strong acidification. He also notes that the putrefying blooms of *Balanophora* (111) have an intense skatolic odor, and implies that this may also be true of the rotten parts of *Solenostigma* (112). ·

BORZI investigated the putrefactive odor of *Visnea mocanera* L. flowers with a color reagent (37). He reported a reaction similar to indole, but on the basis of odor he thought the substance was more likely to be skatole. Both indole and skatole have been isolated from beet molasses (177), but since alkaline treatment was used, they may have been decomposition products of other compounds, notably hypaphorine (p. 260) which was isolated at the same time. A possible identification of skatole by paper chromatography has been made (176). It may be a photolytic product of indoleacetic acid (194); heating the latter compound also produces a skatole odor.

There is thus ample evidence that some precursor, if not skatole itself, is abundantly present in several plants. The aforementioned substance of m. p. 118° may even be this compound, but it does not appear to have been investigated further.

Very little is known of the metabolism of skatole. In bacteria, PROCTOR (228) reported *Pseudomonas* forms skatole from indoleacetic acid and then metabolized it via indoxyl (VI) and salicylic acid (VII) to catechol (VIII).

(V). Skatole. —→ (VI.) → (VII.) → (VIII.)

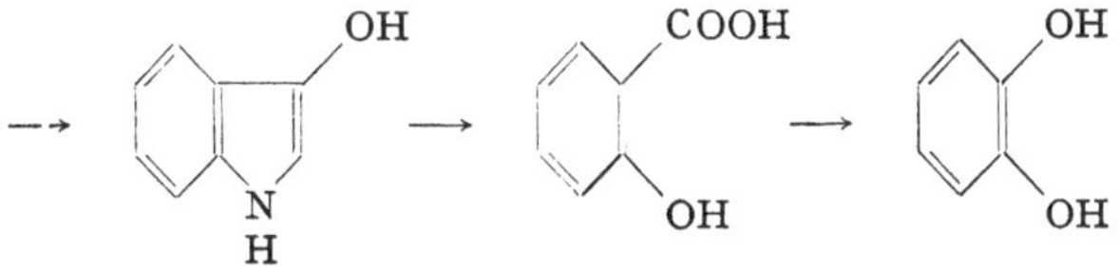

Indoleacetic acid is an unlikely precursor in living higher plants as it is too toxic to be present in any but minute amounts. It might arise in the cases above, after the death of the tissue, by bacterial action upon tryptophan. If, as WEEHUIZEN (316) believed, skatole formation in *Celtis* is not bacterial in origin, then some precursor other than indoleacetic acid must be present in relatively large amounts.

## II. Tryptophan and its Derivatives.

### 1. *L*-Tryptophan [2-Amino-3-(3-indole)-propionic Acid].

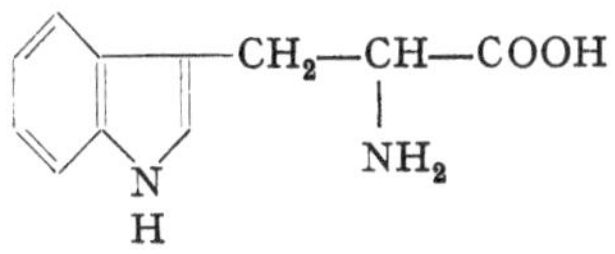

The amino acid tryptophan was first isolated from casein in 1903 (135), and it was soon recognized as the predominant indole derivative in plants. Because it is an essential animal nutrient, many assays of protein tryptophan in plant material have been reported and these have been assembled in tabular form (32). Despite the abundance of analytical data, only a few studies can be found which shed light on the role of tryptophan in the physiology of the plant, especially since many analyses of amino acid composition have used acid hydrolysis which destroys this amino acid.

Among the earliest studies are those of KRETZ (159) who carried out a series of histochemical tests, for the most part on wheat and beans. He used a tedious technique which was probably reasonably specific for protein tryptophan. This showed that the greatest concentration of the amino acid occurred in the shoot and root meristem. Not much protein tryptophan was found elsewhere except in storage tissues (potato tubers) and seeds. VIRTANEN and LAINE (309) were the first to consider changes in protein tryptophan during the life of a plant. In peas and red clover the proportion of tryptophan in protein was found to remain remarkably constant except shortly before flowering when it more than doubled, dropping back to normal values shortly afterwards. The authors suggested that this was probably related to the plant's need for hormonal indoleacetic acid at flowering time. Very similar results were obtained by McCoy and coworkers with oats (195). Not liable to simple interpretation were their assays of protein tryptophan in the major parts of their plants and in plants exposed to various photoperiods.

TSUI (302) also noted an increase in protein tryptophan before flowering of tomatoes, and the close relationship of total plant tryptophan to zinc nutrition. In short term experiments on bean seedlings HARDING (*121, 143*) noted no net synthesis of protein tryptophan. In maize, STEHSEL (*271*) found a steady increase of protein tryptophan in developing kernels but no clear relation to auxin levels was found. TEAS and NEWTON (*284*) investigated tryptophan differences in maize mutants.

Despite the fact that protein tryptophan should be readily available for metabolic processes by dynamic equilibrium with the uncombined amino acid (*307*), it might be expected that the concentration of free amino acid would be more responsive to metabolic demands, and indeed a number of investigators have found interesting concentration variations. Only two years after the isolation of tryptophan, SCHULZE and WINTERSTEIN (*249*) obtained a number of indications that the compound was present in uncombined form in *Vicia sativa* and *Lupinus albus* seedlings, although they were unable to isolate it. A decade later, VON LIPPMANN reported the probable isolation of a very small amount of tryptophan from the sap of bleached beet shoots (*177*). After these reports it was generally not recognized that appreciable free tryptophan exists in plant tissues until the development of paper chromatography and microbiological assays. Then CHRISTIANSEN and THIMANN (*48*) found tryptophan to be a conspicuous component of the free amino acids in the pea stem sections employed for auxin bioassays, and NASON (*206*) was able to trace free tryptophan during maize germination.

More detailed are the data of YAMAKI and NAKAMURA (*332*) who have demonstrated that the levels of free tryptophan and free indoleacetic acid are closely interrelated during the early development of the maize seedling. Likewise, NITSCH (*210, 215*), by a refinement of the microbiological assay for tryptophan, has shown that the development of strawberry fruits is closely controlled by the levels of available indoleacetic acid and that these levels vary similarly to that of the free tryptophan in the achenes, but less so to that in the receptacles (*212*). LUND (*184*) has reported similar experiments in styles and ovaries of tobacco but here the fall in free tryptophan approximately coincided with the rise in auxin levels, while the concentration of protein tryptophan stayed more or less constant on a weight basis. No change in free tryptophan was found during pollen tube germination. HENDERSON and BONNER (*125*) studied free and protein tryptophan in normal and crown gall tissue of sunflower and showed that less was present in the more metabolically active gall tissue. NITSCH and WETMORE (*215*), in agreement with the earlier histochemical studies of KRETZ (*159*), have found that the highest concentration of free tryptophan in seedling lupine tissue occurs in the meristem and in the cotyledons, where it reaches nearly 3 mg/g dry weight. Other studies showed that free tryptophan is found in the ovaries and pollen of squash, tobacco, and lily (*211*). Thus, there appears to be more than enough free tryptophan in plants to serve as the source of the indoleacetic acid formed by these tissues.

Surprisingly little is known about the metabolic role of tryptophan in higher plants. Most work has centered on the still unresolved question of its importance as a precursor of indoleacetic acid which is discussed below. Other aspects of its metabolism have attracted less attention and some have led to contradictory answers.

The oxidation of tryptophan by plant seedling extracts has been studied by WILTSHIRE (328). He found peas to be the most active source of nine plant preparations studied. In the pea system amines accelerated the tryptophan oxidation and this was traced to their formation of hydrogen peroxide in the presence of an amine oxidase (155), the peroxide then being utilized by a peroxidase to degrade tryptophan. Although the product was not fully characterized, its properties were compatible

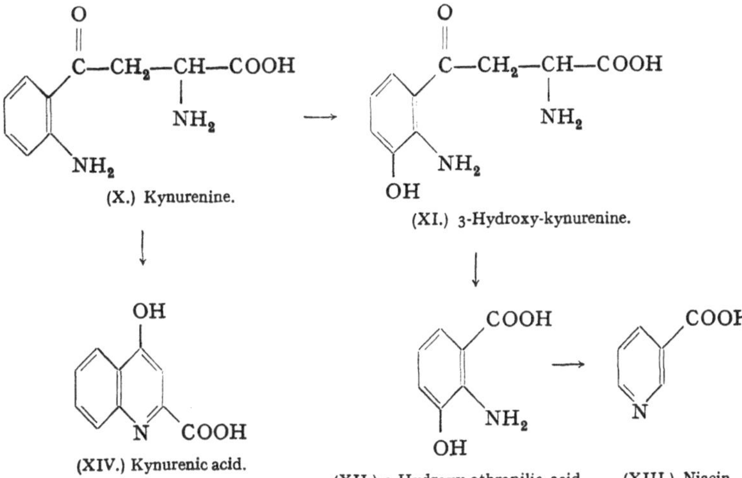

(X.) Kynurenine.

(XI.) 3-Hydroxy-kynurenine.

(XIV.) Kynurenic acid.

(XII.) 3-Hydroxy-athranilic acid.    (XIII.) Niacin.

with 3-hydroxykynurenine (XI). This oxidation thus probably runs via kynurenine (X) similar to the oxidative routes which occur in mammals and many bacteria (55, 106, 196, 270).

One might expect, therefore, that the 3-hydroxykynurenine formed could act in higher plants as a precursor of niacin (XIII) as is the case in these other organisms (107). But TERROINE, who had earlier shown (288) that embryos of *Phaseolus vulgaris* can synthesize a somewhat less than normal amount of niacin in in vitro cultures, states (285, 286) that the concentration of tryptophan in the medium, the age of the plant used, the nature of the inorganic nitrogen source, light, and vitamin $B_6$, were all without effect on the synthesis of niacin. TERROINE concluded (287) that there was no basis for believing tryptophan to be a niacin precursor in *Phaseolus*. It may be, however, that at the seedling stage this author used, niacin synthesis normally takes place predominantly in the cotyledons rather than in the embryos, or that in the relatively nitrogen-rich bean an ample supply of precursors nearer to niacin than tryptophan is available, and thus the amino acid fails to promote niacin formation.

Other investigators have reported positive results. GUSTAFSON (*115*), for instance, maintained intact leaves of cabbage, broccoli, and tomato plants for 48 hours with their petioles in solutions with and without tryptophan and showed a general trend towards more niacin in those leaves which received tryptophan. However, his results do not exclude the participation of bacteria. GALSTON (*83*) incubated pea epicotyls in solutions for 24 hours and, while he failed to obtain significant increases in the niacin content of the tissues with tryptophan, incubation with kynurenine showed a 25% increase in the niacin present, and 3-hydroxy-anthranilic acid more than doubled the niacin concentration. There is a possibility that bacteria may have contributed to this result, but the persistently negative results with tryptophan under the same conditions make it appear that they were not a major factor. Tryptophan did not promote niacin formation in three legumes (*255*).

ARONOFF (*7*) believes, that a pathway for niacin synthesis not involving tryptophan or 3-hydroxyanthranilic acid is present in plants.

NASON (*206*) is thus the only investigator to demonstrate unambiguously that tryptophan can stimulate niacin formation in plants. He grew maize embryos in sterile culture and showed a significant increase proportional to the tryptophan content of the medium, but the amount of niacin formed was less than one percent of the added tryptophan. It should be noted that maize seeds are relatively low in tryptophan (*271, 284*).

Further evidence that abundance of other precursors may be the reason that tryptophan itself fails to stimulate niacin formation, is provided by the work of SHANMUGA SUNDARAM et al. (*255, 256, 257*).

Although they did not obtain stimulation of niacin synthesis, the antimetabolites of pyridoxine, biotin and folic acid (all involved in niacin synthesis) strongly inhibited niacin formation, and each inhibition was competitive, being completely removed by the respective natural cofactor.

All experiments with tryptophan on intact plants are subject to the difficulty that commercial tryptophan may be contaminated with traces of indoleacetic acid, and in fact even heating the pure amino acid may result in the formation of considerable hormonal activity which could obscure the experimental results (*9, 100, 163*). Aware of this problem, and running appropriate controls, KULESCHA and GAUTHERET (*162, 164*) have made observations which at present are not easy to interpret. They report that, whereas niacin (XIII) and 3-hydroxyanthranilic acid (XII) are inactive in the *Avena* test and niacin fails to promote growth of Jerusalem artichoke callus, at 10 ppm kynurenine (X) shows both types of auxin activity, i. e. presumably indoleacetic acid is produced from it. Furthermore, kynurenic acid (XIV), which is inactive in the *Avena* test, shows a stronger growth promoting activity on the callus than does kynurenine. However, both substances produce the same amount of auxin in the tissue. These results would seem to indicate that these tissues, under the influence of the relatively high concentrations employed, may reverse the "usual" oxidative pathway and produce indoleacetic acid from kynurenine, perhaps via tryptophan. Tryptophan is active in the cultures at o.1 ppm so that a 1% conversion would explain the results obtained (*161, 162, 163*).

The biosynthesis of tryptophan in higher plants has been little studied. As noted earlier, many other organisms can form it from indole and serine (XV) (*55, 196, 283*).

$$\text{(I.) Indole} + \text{HOCH}_2\text{—CH—COOH} \quad \xrightarrow{\text{Pyridoxal phosphate}} \quad \text{(IX.) Tryptophan.}$$
$$\underset{\underset{\text{(XV.) Serine.}}{\overset{|}{\text{NH}_2}}}{}$$

(p. 255.)

An indication that a similar pathway may be operative here is provided by Tsui (*302*), who demonstrated that zinc-deficient tomato plants are low in tryptophan but rapidly form it after zinc is restored. Nason (*207*) found a similar situation in *Neurospora* and showed that zinc deficiency caused a reduced synthesis of the enzymes before tryptophan on the synthetic pathway, and that zinc was not a cofactor of these enzymes. Mudd and Zalik (*204*) have studied tomato preparations which utilized indole; there were indications that this utilization was increased by serine and that tryptophan was produced. Zinc-deficient plants were more active, in disagreement with Nason's results. Green-berg (*108*) has shown that indole disappearance in pea bud extracts is dependent on serine and pyridoxal phosphate; some tryptophan was produced but this was not followed quantitatively.

The above evidence is compatible with the belief that indole and serine link to form tryptophan as is true in some other organisms (but note *334*).

## 2. L-Abrine (Amino-N-methyltryptophan).

Ghatak and Kaul (*87, 88*) announced in 1932 the isolation of an indole derivative from *Abrus precatorius* L., which they named abrine. Hoshino (*136*) showed it to be *L*-tryptophan monomethylated on the amino group (XVI). The racemic form was synthesized at about the same time by Gordon and Jackson (*102*), who found it could be used by rats for growth in place of dietary tryptophan. The natural *L* form is nearly as efficient as *L*-tryptophan, but the *D*-isomer is believed ineffective (*101*). Its configuration was

confirmed by CAHILL and JACKSON (46). The compound isolation of this
is not reported from any other plant.

MILLER and ROBSON (200), who developed an improved synthesis, have
criticized the name "abrine", pointing out that it is easily confused with "abrin",
long the name of a poisonous protein mixture in the same plant (88). Their
alternative, d,l-α-methylamino-β-3-indolylpropionic acid, does not appear to have
sufficient brevity but has been occasionally used.

While nothing is known of abrine metabolism in plants, KYU-SUI (166)
concluded from the examination of the urine of dogs fed abrine that
the methyl group must be removed in the tissues. YOSHIDA and
FUKUYAMA (335) then found an enzyme in rabbit kidney and liver which
oxidatively demethylated abrine to tryptophan and formaldehyde.
MAGAKI has purified and characterized the same rabbit enzyme (187) and a
similar one from Escherichia coli (186). It may also be present in
Salmonella typhosa (5). Synthesis of the compound from amino-N,N-
dimethyltryptophan amino-N-oxide (XVII) has been shown to occur
under very mild non-enzymic conditions, as well as in mouse liver (76).
However, no enzymic formation of the N-oxide itself from amino-N,N-
dimethyltryptophan (XVIII) was noted, although it was readily
synthesized by means of hydrogen peroxide.

(XVII.) Amino-N,N-dimethyltryptophan amino-N-oxide.    (XVIII.) Amino-N,N-dimethyltryptophan.

### 3. L-Hypaphorine (Tryptophan Betaine).

(XIX.) Hypaphorine.

In 1890 GRESHOFF discovered (110) and
later isolated (112) an alkaloid in 3% yield
from the seeds of Erythrina hypaphorus BOERL.
which he named hypaphorine. Several studies
of its properties were made (112, 225, 321). The
stench of rotten wood from E. hypaphorus,
which had been traced to indole (113), was apparently due to hypaphorine,
since the compound decomposed readily to indole and trimethylamine
on heating. This observation led VAN ROMBURGH and BARGER (236) to
its synthesis based on the correct postulate that it was the trimethylated
betaine of tryptophan (XIX). The steric configuration was established
by CAHILL and JACKSON (46).

Some years later MARAÑON and SANTOS (192) carried out a second
isolation from E. variegata var. orientalis (L.) MERR.; this has since
been repeated several times (78, 229, 230). Another Erythrina species
was soon added (58, 79), and when FOLKERS and his coworkers examined

the genus systematically, fourteen more species were found to contain hypaphorine (*77, 78, 79, 188*). The percentage isolated from the seeds of various species is tabulated in two papers (*78, 79*), and runs as high as 5.8% in *E. acanthocarpa* E. MEY. FOLKERS noted that hypaphorine had been found in the seeds of every *Erythrina* species examined for it. Two more species have since been reported to contain the compound (*167, 168*).

One observation indicates that it may be more widely distributed, as VON LIPPMANN (*177*) isolated a substance from beet molasses which yielded indole and trimethylamine upon heating. Direct comparison was not possible, but analysis, melting point, and optical rotation support his conclusion that the substance was hypaphorine.

The first toxicological studies on animals were conducted by PLUGGE (*225*), who followed the distribution of hypaphorine between urine and the tissues. Pure hypaphorine was found relatively innocuous to a wide variety of animals, but oddly enough it was especially toxic to frogs. JACKSON (*142*) showed that hypaphorine could not replace tryptophan in rat diets.

### 4. Malonyl-tryptophan.

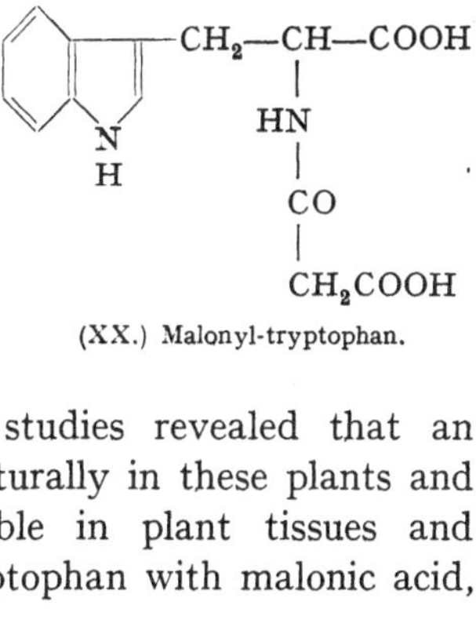

GOOD and ANDREAE (*91*) have investigated a substance which is formed when tryptophan is incubated in vitro with spinach leaves, and have determined its structure by degradation and synthesis to be the malonic acid peptide of tryptophan (XX). Tomatoes and peas were also shown to form the compound when tryptophan was added, and

(XX.) Malonyl-tryptophan.

more importantly, paper chromatographic studies revealed that an apparently identical compound is present naturally in these plants and in oats. Malonyl-tryptophan is quite stable in plant tissues and presumably is formed by conjugation of tryptophan with malonic acid, a plant product.

## III. Indole Bases and Some Related Compounds.

### 1. Gramine [3-(Dimethylaminomethyl)-indole].

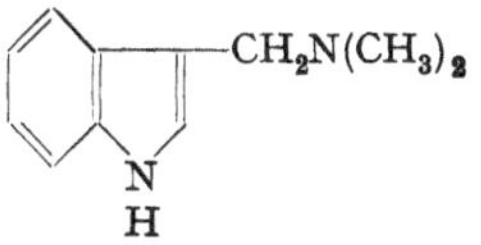

VON EULER and HELLSTRÖM (*65, 66*) discovered an indole alkaloid during an investigation of albino mutants of barley and isolated it in crystalline form. At first it was thought to occur only in the mutants (*67*), but later it was detected in normal barley strains and named gramine (*68*).

(XXI.) Gramine.

It represents about 0.1% of the dry weight of the seedlings; the amount is constant at first but decreases as the plant grows older, and finally disappears (*42*). Some indications were found that gramine content and nematode resistance were correlated. Before the structure was established, ORECHOFF and NORKINA (*219*) had examined *Arundo donax* L., an Asiatic reed (which was avoided by foraging camels) and isolated from it an alkaloid "donaxine". This turned out to be identical with gramine (*63, 64, 218*). No report of its presence in other plants has been made, but it seems a likely constituent of other grasses. *Arundo donax* L. also contains another low molecular weight alkaloid which may be an indole derivative (*185*).

The compound was first synthesized by WIELAND and HSING (*323*), who established that it was 3-dimethylaminomethyl-indole (XXI). It is noteworthy that gramine is simply synthesized in high yield from indole, formaldehyde, and dimethylamine by an application of the MANNICH reaction (*160*).

Gramine is the first indole alkaloid for which a biogenetic precursor has been unambiguously established. BOWDEN and MARION (*40*) fed tryptophan labeled with $C^{14}$ in the methylene group to barley seedlings and showed by degradation that gramine isolated from the plants had all its radioactivity in the same position. Later, LEETE and MARION (*173*) proved that the indole to methylene bond remains unbroken in this process. The mechanism of this surprising shortening of the tryptophan side-chain remains a mystery. The same workers have also demonstrated that *L*-methionine can supply the methyl groups; the *D*-isomer is less effective, and formate is a very poor methyl group source (*174*). Other studies with radioautographs indicated that after feeding tryptophan the gramine formation proceeds rapidly and probably occurs in the leaves (*41*).

## 2. Tryptamine [3-(2-Aminoethyl)-indole].

CH$_2$—CH$_2$NH$_2$

(XXII.) Tryptamine.

This base was isolated in 1% quantities from *Acacia floribunda* and *A. purinosa* by WHITE (*320*). He found the substance to be concentrated in the tops and in the flowers where it was associated with β-phenylethylamine, the latter apparently contributing to the perfume. Indications that other acacias contained tryptamine were noted. The only other isolation is recent, from the leaves of another legume, *Prosopsis juliflora*, the common mesquite (*73*).

Tryptamine has long been recognized as having adrenergic action in mammals, and has for some time been considered a possible intermediate in indoleacetic acid synthesis in plants. In support of this latter view is the work of SKOOG (*261*) who was able to show its slow auxin activity in oats, and other evidence is provided by GORDON and NIEVA (*97, 98*).

KENTEN and MANN (*155*) studied an amine oxidase from peas which took up oxygen in the presence of tryptamine; this enzyme was later highly purified (*189*). Unlike animal amine oxidases, it attacked both diamines and aromatic alkyl amines. The enzyme produces hydrogen peroxide, the aldehyde, and ammonia, and is inhibited by cyanide, as well as by carbonyl and copper reagents. The reaction with tryptamine was studied in detail (*49*), and indoleacetaldehyde definitely established as the major product. WERLE and ROEWER (*319*) have found a distinctly different enzyme in several plants. It is a monoamine oxidase which reacts with tryptamine and more strongly with phenyl-alkylamines but is not cyanide sensitive. WILDMAN et al. (*325*) have found no evidence for such an oxidation of tryptamine in other plants.

The presence of tryptamine in plants suggests the action of a tryptophan decarboxylase. At least two investigators have noted amino acid decarboxylases in plants (*216*, *318*), but did not test their enzymes on tryptophan. However, there were other indications that these enzymes, like the analogous decarboxylases in bacteria (*82*), are highly specific.

GALE (*82*) has tested over 1000 bacterial strains for the presence of a tryptophan decarboxylase with negative results. VON KAMIENSKI (*147–149*), in a detailed study of amines in higher plants and fungi, did not comment on the presence of tryptamine in over three hundred species he examined, and was unable to find decarboxylase activity on a number of compounds in several extracts tested. It thus appears that the enzyme is not widely distributed in plants, supporting the negative results with bioassays many workers have obtained with tryptophan. But recent work with radioactive tryptophan indicated the production of a small amount of tryptamine after incubation with watermelon slices (*56*).

### 3. Amino-N-methyltryptamine.

(XXIII.) Amino-N-methyltryptamine.  (XXIV.) Amino-N,N-dimethyltryptamine oxide.

YURACHEVSKII and STEPANOV (*339*) discovered a new alkaloid in the Chenopod *Gergensohnia diptera* BGE. which they named "dipterine". Later work (*337*) established its structure by comparison with the already synthesized amino-N-methyltryptamine (XXIII) (*137*). It was also found (*338*) in another member of the same family, *Arthrophytum leptocladum*, associated with an obviously closely related tricyclic indole alkaloid (*217*).

The name dipterine appears particularly unfortunate as it is reminescent of both Diptera and pterines, while the compound is not associated with either of these groups. The use of amino-N-methyltryptamine is preferable.

Although the substance has not been reported elsewhere, FISH and his colleagues (75) have shown that it and formaldehyde are readily produced non-enzymically from amino-N,N-dimethyltryptamine oxide (XXIV) in the presence of oxalic acid and ferric ions. Furthermore, a crude mouse liver homogenate oxidized amino-N-methyltryptamine to indoleacetic acid in good yield. As both precursor and product of these reactions occur in plants, the natural presence of such a pathway seems probable.

### 4. Amino-N,N-dimethyltryptamine and its N-Oxide.

-CH$_2$—CH$_2$N(CH$_3$)$_2$

(XXV.) Amino-N,N-dimethyltryptamine.

The remaining indole bases have been discovered in plants through their effects on animals. In this first example, the hallucinogenic effect of snuffs used by certain American Indian tribes has been traced to their content of indole bases such as amino-N,N-dimethyltryptamine (XXV). The compound has been identified in the seeds and pods of the legumes *Piptadenia peregrina* L. (BENTH.) and *P. macrocarpa* BENTH. by FISH and coworkers (74). Its N-oxide (XXIV) was also present, but perhaps was an artefact, since amino-N,N-dimethyltryptamine readily forms it. However, the N-oxide can also be obtained enzymically by a mouse liver preparation (75). The amine's presence in plants has been extended by isolation of amino-N,N-dimethyltryptamine from the leaves of a member of the Apocyanaceae, *Prestonia amazonicum*, by HOCHSTEIN and PARADIES (132).

No information on the plant metabolism or origin of this compound is available; presumably it is derived from tryptophan or tryptamine. Interestingly, FISH and coworkers (75) found mouse liver mitochondria to convert N,N-dimethyltryptamine actively to indoleacetic acid, apparently via N-methyltryptamine. Lower yields from the N-oxide using a non-enzymic acid solution of ferric ions were also obtained.

### 5. 5-Hydroxytryptamine [3-(2-Aminoethyl)-5-indolol].

5-Hydroxytryptamine (XXVI), also known as *serotonin* or *enteramine*, has attracted broad attention due to its action as an animal hormone. Exhaustive recent reviews on this subject are available (175, 220). In plants, the substance is apparently an active irritating principle of cowhage, *Mucuna pruriens* DC. (39), and the stinging nettle, *Urtica dioica* L. (51), where it occurs in approximately 0.02% quantities. It has also been isolated from mesquite, *Prosopis juliflora* (73). More surprising is the fact, arising from the discovery that monkeys fed bananas greatly increase their output of urinary 5-hydroxyindoleacetic acid (XXVIII) (1), that the banana is a relatively rich source of 5-hydroxytryptamine. The observation has been confirmed and extended to cotton fruits and skunk cabbage (*44*). WAALKES et al. (*311*) found from 28 to 65 µg/g of the substance in banana pulp and peel in company with several catechol amines. This would suggest that these compounds

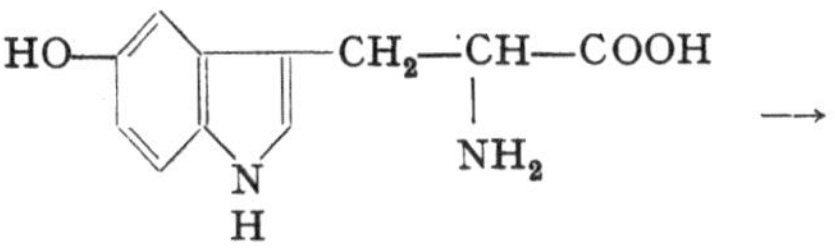

HO—⟨⟩—CH$_2$—CH—COOH $\longrightarrow$ HO—⟨⟩—CH$_2$—CH$_2$NH$_2$ $\longrightarrow$
                |
                NH$_2$

(XXIX.) 5-Hydroxytryptophan. 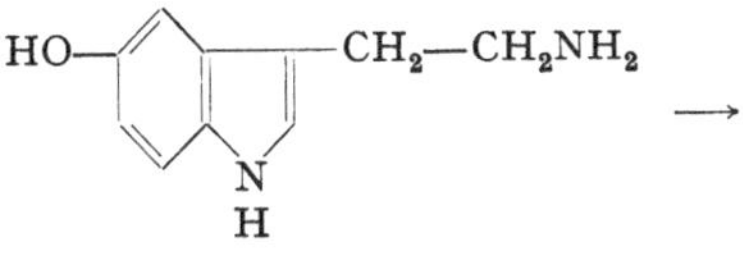        (XXVI.) 5-Hydroxytryptamine.

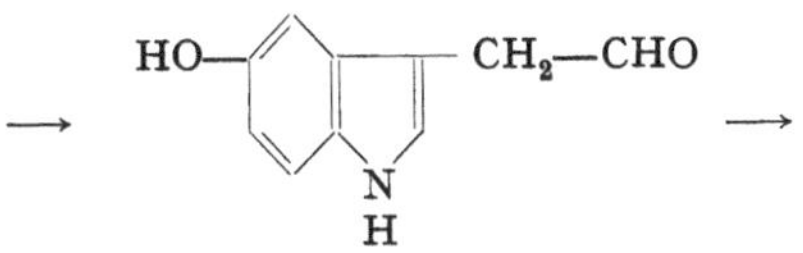

$\longrightarrow$ HO—⟨⟩—CH$_2$—CHO $\longrightarrow$ HO—⟨⟩—CH$_2$—COOH

(XXVII.) 5-Hydroxy-3-indole-acetaldehyde. 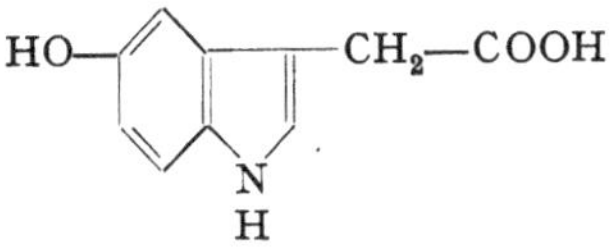    (XXVIII.) 5-Hydroxy-3-indoleacetic acid.

are involved in the melanizing process, and that 5,6-dihydroxyindoles are also present (*264*). However, these were not detected, nor was 5-hydroxyindoleacetic acid. The latter arises in monkeys via 5-hydroxy-indoleacetaldehyde (XXVII) by the action of an amine oxidase and an aldehyde oxidase.

Possibly, this enzyme occurs in the nettle, since COLLIER and CHESHER (*51*) described the disappearance of 5-hydroxytryptamine in nettle sting suspensions. However, later COLLIER (*50*) expressed the belief that this was not destruction but active adsorption, and suggested that 5-hydroxytryptamine was produced at some point in the nettle remote from the sting. The action of a plant amine oxidase has been shown upon 5-hydroxytryptamine, however (*49*).

The formation of 5-hydroxytryptamine takes place probably from 5-hydroxytryptophan (XXIX), as a specific decarboxylase for this compound is widely distributed in animals (*175, 220*). No definitive report of the hydroxyamino acid's presence in plants has yet been made, but DANNENBURG and LIVERMAN (*56*) have shown a compound with similar solubility and chromatographic properties to be produced in watermelon slices incubated with tryptophan. At least two *Chromobacterium* species produce 5-hydroxyindoles (*14*), and one does so via the conversion of tryptophan to 5-hydroxytryptophan (*201*).

## 6. 5-Methoxy-amino-
## N-methyl-tryptamine [3-(2-Methylaminoethyl)-5-methoxyindole].

A recent report of cases of "staggers" in sheep pastured largely on the grass *Phalaris arundinacea* L. has led to an investigation of this material (*327*) and the isolation and

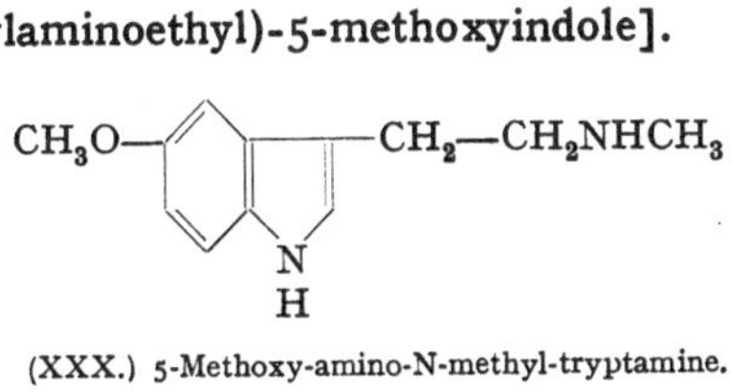

CH$_3$O—⟨⟩—CH$_2$—CH$_2$NHCH$_3$

(XXX.) 5-Methoxy-amino-N-methyl-tryptamine.

synthesis of the alkaloid (XXX). No fungi were present in the grass so that its position as a higher plant product is secure. Another indole was also noted.

### 7. Bufotenine [3-(2-Dimethylaminoethyl)-5-indolol] and its N-Oxide.

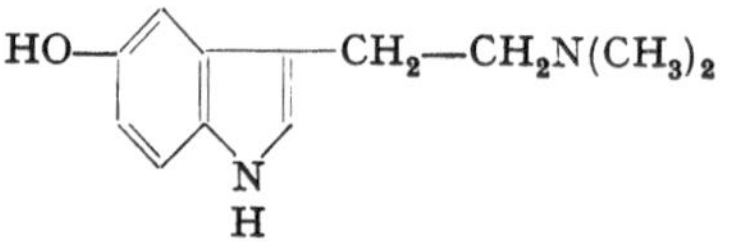

(XXXI.) Bufotenine. 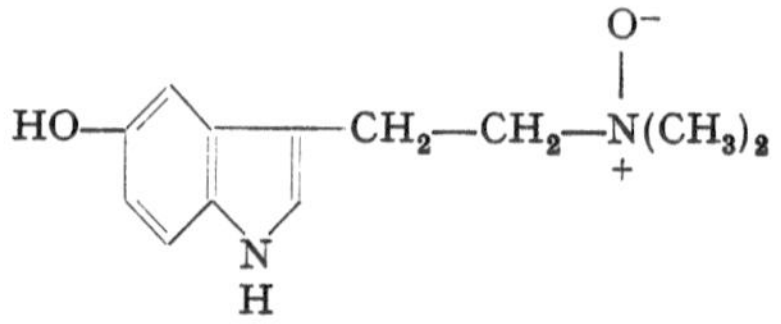                    (XXXII.) Bufotenine-N-oxide.

Bufotenine has been known as the adrenergic agent in toad poison since its isolation by WIELAND et al. (*322*) in 1934. These workers also established its structure (XXXI). Its isolation by STROMBERG (*278*) from the seeds of *Piptadenia peregrina* L. (BENTH.) in 1954 was the first indication that it was also a plant product. Bufotenine is physiologically active, being one of the effective ingredients of Cohoba snuff. The isolation has been amplified by FISH et al. (*74*), who detected bufotenine and its N-oxide (XXXII) in the seeds of this species and in *P. macrocarpa* BENTH. In this case, the N-oxide seems unlikely to be an artefact since bufotenine is more stable than amino-N,N-dimethyltryptamine. Another, unidentified 5-hydroxyindole derivative is present.

FISH et al. have suggested, purely on distributional grounds, that bufotenine is derived from amino-N,N-dimethyltryptamine (XXV, p. 264), but no experimental evidence of its metabolism is available. It is also produced by a fungus, *Amanita mappa* [WIELAND et al. (*324*)].

### 8. 5-Hydroxy-3-indoleacetic Acid.

The presence of this compound (XXVIII, p. 265) in higher plants seems very likely by analogy with the situation in animals, which produce it readily by oxidation of 5-hydroxytryptamine (*175, 220*). And in fact the existence of the compound in plants containing the amine has been briefly recorded (*303, 304*), although the identification was not carried to a definitive isolation (*73*). As noted above, COLLIER and CHESHER (*51*) have shown that nettle stings may contain an enzyme capable of producing 5-hydroxyindoleacetic acid from 5-hydroxytryptamine; as CLARKE and MANN's pea amine oxidase reacts with the amine (*49*), it seems a likely agent for the first step of the conversion.

Despite the close similarity to indoleacetic acid, tests indicate that the presence of the hydroxyl group has severely limited 5-hydroxyindoleacetic acid's potency as an auxin (*45, 293*), and it seems unlikely that it has any major action on plant growth.

# IV. Potential Precursors of 3-Indoleacetic Acid.

## 1. 3-Indoleacetaldehyde.

It should be stated at the outset that this compound (XXXIII) has never been isolated from plants. Nonetheless, there is persuasive evidence of its presence. Its identification was made possible by the meticulous work of LARSEN (*169*) who established that a neutral

(XXXIII.) 3-Indoleacetaldehyde.

substance found in peas, sunflowers and broadbeans could be converted in the presence of oxygen by the SCHARDINGER aldehyde-oxidizing enzyme to an auxin having much in common with indoleacetic acid. His studies, which were carried out on very small quantities of only partially purified material, also showed that the neutral compound could form a bisulfite addition complex and had a molecular weight of 158. LARSEN concluded that he was dealing with indoleacetaldehyde. HEMBERG (*123*) found much the same substance in potatoes. These reports were amplified by GORDON and NIEVA (*97, 98*); working with pineapple leaves and dandelion, they showed that the similar substance in these plants would react with dimedon, an aldehyde-specific reagent.

A reliable synthesis of the compound was first carried out by BROWN et al. (*43*), who found the substance to be unstable; they collaborated with BENTLEY and HOUSLEY (*23*) in establishing the biological characteristics of synthetic indole-acetaldehyde. The results were not incompatible with the data obtained with partially purified isolates, but unfortunately the natural and synthetic materials have never been compared directly. BENTLEY and HOUSLEY concluded that the observed biological activity of synthetic indoleacetaldehyde is likely due to the indoleacetic acid formed from it by plant tissues.

A number of recent workers have reported the presence of indoleacetaldehyde on paper chromatograms (*33, 49, 251, 317, 332*), but there is some inconsistency in the Rf values and color reactions. It should be noted that the bisulfite addition test, as mentioned earlier, is not specific for the aldehyde, since indole itself, and perhaps other indole derivatives, will form bisulfite complexes (*62, 128*), as do $\alpha$-keto acids and their esters (*314*). But the neutral substance's reactivity with SCHARDINGER enzyme and dimedon, and the ready oxidation of tryptamine in certain cases to a neutral substance make the presence of indoleacetaldehyde in plants seem very likely. A synthetic method just reported for producing stable indoleacetaldehyde will, it is to be hoped, settle this problem shortly (*105*).

The presence of an enzyme in oats which can convert natural indole-acetaldehyde to an acid auxin is well established (*34, 170*). The enzyme also oxidizes naphthalene acetaldehyde (*171*) but in this case a dismutation reaction was suggested as at least partially responsible for the conversion. A similar enzyme occurs in *Artemisia absinthia* (*8*) and in pineapple (*97, 98*). The most interesting aspect of this enzyme is its extreme sensitivity to ionizing radiation which has been shown by GORDON and is discussed in his review (*95*). X-rays interfere with auxin synthesis and this explains why doses of radiation too small to cause much direct indoleacetic acid inactivation, induce a large drop in auxin concentration. This has been observed after irradiation of the plant in both whole tissue and cell-free mung bean systems. Probably the same enzyme was studied by KENTEN (154), who used phenylacetaldehyde as a substrate instead of the then unavailable indoleacetaldehyde. He detected the enzyme in several plants and has provided the first detailed data on its properties. Its activity was also encountered in other studies (49).

## 2. 3-Indoleacetonitrile.

(XXXIV.) 3-Indoleacetonitrile.

Prompted by the frequent reports of neutral auxins in plants, HENBEST, JONES and co-workers (124) isolated this substance from cabbage and identified it as indoleacetonitrile (XXXIV). The unexpected nitrile nature of this compound and its remarkably high activity in the oat auxin bioassay have excited considerable interest in its function in growth regulation. The discoverers presented evidence for its occurrence in other Crucifers; they also noted that this compound may well be identical with some of the unidentified cabbage auxins reported earlier (133).

Unfortunately, indoleacetonitrile is not always clearly separable from other neutral indole compounds by paper chromatography, and some of the reports of its presence in extracts of plants other than Crucifers must be regarded with skepticism. Efforts to separate it have met with some success (30, 213, 214) and a specific test involving the microcrystalline appearance of its picrate has been suggested (253).

The first studies of its hormonal action (21, 23, 144) had the remarkable result that indoleacetonitrile was found to be a more potent auxin on oats than indoleacetic acid, but was virtually without activity on peas. This anomaly led THIMANN (291) to the conclusion that oats must hydrolyze the nitrile to the acid, and indeed, he was able to show that oat juice converted the compound to an acid auxin which was active on peas. MICHEL (198) has extended this observation to cabbage. Paper chromatography has confirmed that the auxin thus produced was indoleacetic acid (254, 276). The differential activity of indole-acetonitrile on oats and peas is thus explained; the stronger activity of the nitrile as compared to the acid would seem most likely attributable to an increased uptake by, or permeability in, the oat tissues.

The enzyme responsible for the nitrile → acid conversion has been found in several plants and partially purified (296). Its action is a true hydrolysis, which is not impeded by sulfhydryl or heavy metal reagents. Indoleacetamide, the expected intermediate, does not appear on chromatograms (*254, 276*), is not very active on oats (*144*), and is not attacked very rapidly by the enzyme (*296*); hence, it cannot be a free intermediate of the enzymatic hydrolysis. Thus, this enzyme, both in the nature of its substrate and of its reaction mechanism, seems to be unique.

For the degradation of 3-indoleacetonitrile to 3-indolecarboxylic acid cf. p. 280.

Although a number of nitriles occur in plants, the biogenesis of the nitrile group has not yet been established.

BONDE (*34*) showed that an acidic, water-soluble material in cabbage could be converted to a neutral auxin probably indoleacetonitrile; and HOUSLEY and BENTLEY (*138*) found that this conversion was accelerated by alkaline conditions and heat (see also *133*). The nature of the precursor remains unknown. JONES et al.'s initial suggestion (*144*) that indoleacetonitrile would arise by double dehydrogenation and decarboxylation of tryptophan seems implausible in view of the likely reactivity of the imino intermediate under aqueous conditions.

This reviewer would like to advance the hypothesis that indole-acetonitrile is produced from indolepyruvic acid via its oxime (XXXV). One oxime, oximino-succinic acid (XXXVI), is known to occur in plants, having been isolated by VIRTANEN and LAINE (*308, 310*) from sand-cultured peas infected with root nodules. The oxime of phenylpyruvic

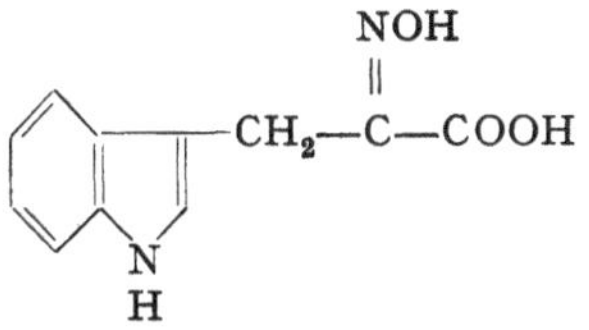

(XXXV.) Indole-3-pyruvic acid oxime. 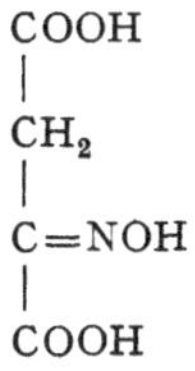        (XXXVI.) Oximino-succinic acid.

acid, a compound closely analogous to indolepyruvic acid, was prepared by BOUVEAULT and LOCQUIN (*38*) and shown by them to be converted to phenyl-acetonitrile by "acides ou alcalis aqueux . . . avec la plus grande facilité . . .". WATERS (*313, 314*) has extended this observation to the oximes of some other $\alpha$-ketoacids.

In fact, he was unable to recover the $\alpha$-ketoacid from its oxime in any substantial quantity, the nitrile always being the preponderant product. SHAW et al. (*258*) were unable to prepare the oxime of indoleglyoxylic acid, obtaining instead 3-cyanoindole. The oxime of indolepyruvic acid was prepared; however, its conversion to the nitrile was not attempted. The relatively mild conditions employed and the availability of the compound by synthesis make this seem like a promising pathway for biological investigations. DAKIN (*54*) has published a synthetic method by which amino acids can be easily converted to nitriles but no biological analog of his reagent seems likely.

### 3. 3-Indolepyruvic Acid.

$$\text{(XXXVII a.)} \qquad \overset{\displaystyle O}{\underset{\displaystyle \parallel}{}} \qquad -CH_2-C-COOH \qquad \longleftarrow \qquad \longrightarrow \qquad -CH=C-COOH \qquad \overset{\displaystyle OH}{\underset{\displaystyle |}{}} \qquad \text{(XXXVII b.)}$$

(XXXVII a.)                                                    (XXXVII b.)

This compound was one of the first indole derivatives shown to possess auxin activity (*158*), and was soon suggested as an intermediate in the biogenesis of indoleacetic acid (*290*), a proposal for which some tests were later made (*325*). BERGER and AVERY (*26*) noted evidence of indolic materials other than indoleacetic acid in a variety of maize which was reinvestigated by STOWE and THIMANN (*276*).

The latter workers found on paper chromatograms of maize extracts four areas, giving SALKOWSKI and EHRLICH color reactions, which were biologically active on oats. One of these was clearly indoleacetic acid, whereas another was an acidic material which darkened upon spraying with 2,4-dinitrophenylhydrazine reagent. This latter compound was compared with synthetic indolepyruvic acid (XXXVII) and found to display similar characteristics. The other two, presumeably indolic, spots remain to be identified.

Both natural and synthetic indolepyruvic acid decomposed into several compounds, including indoleacetic acid, when handled in aqueous media. This instability had been noted earlier (*325*), but the unsatisfactory nature of the then available syntheses (*25, 142, 276, 314, 330*) made a study of this behavior difficult.

After BENTLEY et al. (*22*) had developed an improved synthetic method, they came to the conclusion that indolepyruvic acid decomposed completely under the chromatographic conditions employed by STOWE and THIMANN (*276*). This conclusion rested largely on the belief that indolepyruvic acid reduces ammoniacal silver nitrate rapidly, a behavior not observed by the latter workers. The explanation for the discrepancy was found (*273*) in the fact that indolepyruvic acid, when synthesized, crystallized in the enol form (XXXVIIb) which is also indicated by the ultraviolet spectral maxima (*22*). The enol form reacts not only with ammoniacal silver nitrate, but also reduces certain dyes (*273*); however, it does not readily react with typical indole reagents, probably because of the conjugation of its double bond with the indole ring.

Upon dissolution in aqueous media, the enol of indolepyruvic acid tautomerizes to the ketonic form (XXXVII a), the rate and equilibrium depending on pH. The keto form has a typical indolic spectrum, reacts with indole and ketone reagents, but only slowly reduces ammoniacal silver nitrate. The spectral shift has since been studied under various conditions by KAPER and VELDSTRA (*150, 151*). A similar shift of tautomeric equilibrium had been noted earlier for *p*-hydroxyphenylpyruvic acid, a compound in whose enol form the double band is also conjugated to an aromatic ring (221). These results indicate that failure to obtain a rapid reaction with ammoniacal silver nitrate, after chromatography of indolepyruvic acid under alkaline conditions, is to be expected since the enol form is not then present.

Using the same chromatographic conditions, DANNENBURG and LIVERMAN (56) obtained a positive reaction with 1,2-diamino-4-nitrobenzene, a relatively specific ketone reagent and concluded that even in an oxygen atmosphere complete decomposition of indolepyruvic acid does not occur. In nitrogen, formation of only three spots was observed. Further evidence for the presence of both keto and enol forms is provided by the ingenious analysis of chromatographic and electrophoretic patterns of indolepyruvic acid decomposition provided in the several papers of SCHWARZ and BITANCOURT (31, 250, 251, 252).

All workers agree that slow decomposition of indolepyruvic acid to half a dozen or more substances does occur in aqueous media in the presence of oxygen. On the other hand, SHAW et al. (258) who have developed another improved synthetic method, report that pure crystalline material retains its identity indefinitely if kept dry in darkness.

These complications have made it impossible to estimate whether or not enzymes promote any formation of indoleacetic acid beyond that arising spontaneously.

Besides indoleacetic acid, the only other product of decomposition definitely identified is indolealdehyde (151, 258). Other compounds which have been suggested, but not yet unambigously established are, indolelactic acid (250), indoleglycolic acid (22, 151, 196 a), indoleacetaldehyde (250), indolenine-acet-aldehyde (250, 251), and indoleglyoxylic acid (196 a).

The problem is further complicated by the fact that under acid conditions of chromatography fewer spots have been obtained (22, 151). This has been interpreted as indicating greater stability of indolepyruvic acid under acid conditions, but since acid chromatography fails to separate acidic and neutral substances adequately (276) it may be that decomposition products are merely overlapping one another on the acid chromatograms. Another difficulty is that although the enol → keto change proceeds rapidly in base, the reverse keto → enol shift is not promoted by acid conditions (150, 221).

It should be emphasized that the instability of indolepyruvic acid does not exclude its participation in biological systems. SHAW and his colleagues (258) were able to recover as crystals 48% of the indolepyruvic acid they added to a bicarbonate solution after 24 hours.

The relative slowness of the decomposition is also shown by experiments which indicate that indolepyruvic acid can replace tryptophan in rat diets (25, 142), and in Salmonella (5) or Neurospora (15). It was used for indole formation by Escherichia coli under conditions where the un-inoculated medium gave no reaction (240). Furthermore, it produced nearly 10% indole from indolepyruvic acid dissolved in dilute aqueous ammonia (330); since no other indole compound of a number tested (119) gave rise to any indole, this observation is of particular significance.

Thus, there is no a priori reason to exclude the presence of indole-pyruvic acid in dry maize kernels, nor its transitory participation in an enzyme pathway. A conclusive isolation from higher plants is now desirable, but would seem difficult to realize for the compound itself. As KAPER (*150, 151*) has pointed out, the existence of such a complex and distinctive decomposition pattern on chromatograms is in itself considerable evidence for the presence of indolepyruvic acid in biological materials.

The suggestion by THIMANN (*290*) that indolepyruvic acid might be produced by oxidative deamination of tryptophan, is discussed by KAPER (*150, 151*) in explaining his results obtained with *Agrobacterium tumefaciens.* However, STOWE (*273*) has demonstrated that cell-free extracts of the same microorganism (and of two others) produce indole-pyruvic acid by a transamination from tryptophan which requires certain α-ketoacids and pyridoxal phosphate.

In the only available investigation concerning higher plants, MURAKAMI and HAYASHI (*205*) have noted active transamination of some amino acids in rice preparations, but only weak evidence for such a reaction with tryptophan. They concluded that transamination was an unlikely step on the usual pathway of indoleacetic acid formation in rice. However, the minute amounts of auxin normally formed in plants make such a conclusion lose its force.

### 4. Ascorbigen.

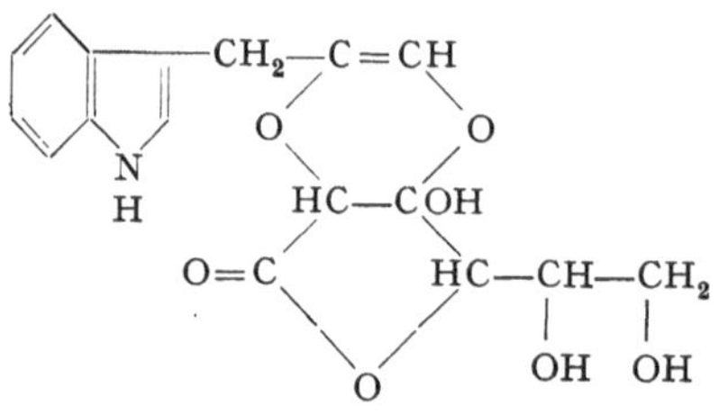

(XXXVIII a.) 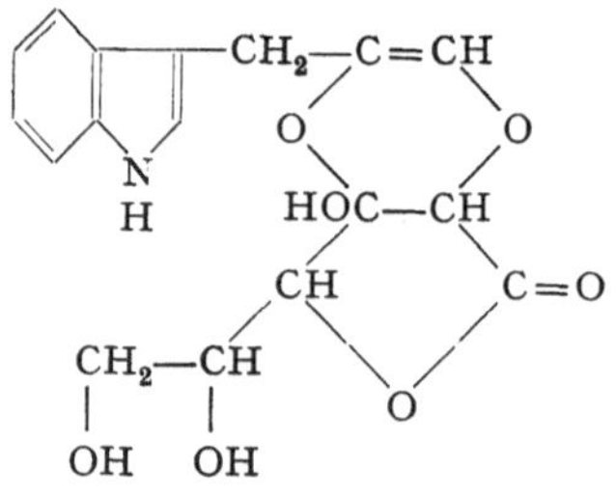                              (XXXVIII b.)

Several years' work by PROCHÁZKA (*227*) on ascorbigen, a substance in cabbage and related plants, which releases ascorbic acid on hydrolysis, has led to its isolation and to the discovery that the ascorbic acid is linked to an indole derivative. The compound is believed to have one of the two structures (XXXVIIIa) or (XXXVIIIb); these formulas differ only in the arrangement of the ether linkages. Upon chemical hydrolysis, ascorbigen releases ascorbic, indolecarboxylic and indole-acetic acids, and could also be a precursor of these compounds in tissue. Indeed, KUTÁČEK and colleagues (*165a*) showed its concentration to parallel that of indoleacetic acid in several plant parts. It increased after tryptophan injection.

## V. 3-Indoleacetic Acid.

The extensive work on the action of this plant growth hormone has been the subject of several reviews ($19$, $93$, $172$, $295$), and accordingly, this discussion will be limited to the most pertinent papers, in particular to those dealing with its isolation from, and its metabolism in, higher

(XXXIX.) 3-Indoleacetic acid.

plants. Investigations of the compound in fungi are reviewed by GRUEN ($114$), and indications of its presence in algae are discussed by BENTLEY ($20$).

The first search for indoleacetic acid in higher plants was carried out by HERTER ($127$), who did not find clear evidence for its presence. The earliest identification seems to have been that of VON LIPPMANN ($178$), who isolated the compound from sugar beet molasses. As he earlier had obtained tryptophan from the same source ($177$), the indoleacetic acid could have been an artefact, inasmuch as tryptophan has been shown to produce an auxin on heating ($9$, $100$, $163$).

The important function of indoleacetic acid as a plant growth hormone was revealed by KÖGL, HAAGEN-SMIT and ERXLEBEN ($156$), who isolated the substance from urine, by KÖGL and KOSTERMANS ($157$), who found it in yeast, and independently by THIMANN ($290$), who obtained it from a culture of the Phycomycete, *Rhizopus suinus*. Despite their demonstration of the strong growth promoting activity of the compound, it was some years before it was isolated from a higher plant. During the intervening time many did not consider it to be a natural plant hormone, hence the now outdated name, *"heteroauxin"*. Although the first isolation ($117$) of indoleacetic acid from the kernels of *Zea mays* is subject to the criticism that alkaline hydrolysis was used, later work by BERGER and AVERY ($26$), and its purification from the immature growing seed by HAAGEN-SMIT et al. ($116$) made its existence in higher plants incontrovertible.

Furthermore, the development of paper chromatography, first applied to plant auxins by YAMAKI ($331$), established that indoleacetic acid is widely distributed in higher plants. These reports have been summarized and tabulated by GORDON ($93$), LARSEN ($172$), and BENTLEY ($19$), and access to the literature is also provided by other workers ($6$, $30$, $152$, $153$, $213$, $276$, $277$, $317$).

In many other plants the compound has not been detected, but this could easily be due to a failure to separate it from other substances at the very low concentrations ($10^{-6}$ to $10^{-8}$ M) at which it normally occurs. Since, as far as is known, all higher plants are responsive to it in some degree, it is not unreasonable to suppose that indoleacetic acid is universally a part of their hormonal mechanisms.

However, there are some workers who believe that this is not necessarily the case, a viewpoint which has been summarized by BENTLEY (*19*).

In this connection, it should be realized that other compounds, not chemically related but having some growth-promoting action, such as phenylacetic acid and dihydrogibberellic acid, are also present in plants. Thus, the evidence from a bioassay, even if combined with chromatographic data, is not necessarily evidence of the presence of indole compounds—tests for chemical identification must also be applied. Equally serious is the problem caused by compounds present in larger, non-hormonal amounts which also may stimulate growth. For example, potassium (*267, 297, 298*), manganese (see *292*), sugars (*36, 267*), and various nitrogenous compounds, especially adenine (*86*), can promote auxin bioassays if they are not already supplied or otherwise taken into account, and pH is also a factor (*297*). Some substances may accentuate auxin-induced growth in trace amounts, as do cobalt (see *292*), and fatty acid esters (*274*). A related problem is the existence of synergists, compounds with little or no auxin activity themselves, but which can somehow enhance the potency of auxins in the bioassay (*305*). Various methods of auxin bioassay are detailed by LARSEN (*172*). It seems necessary to enumerate all these factors here, since they have frequently been neglected in bioassays which have been presented as evidence for the existence of indoleacetic acid and other indole auxins in higher plants.

The general topic of indoleacetic acid biogenesis has been reviewed frequently, most recently by GORDON (*93, 94*), who also has demonstrated the soluble nature and centrifugal properties of the enzyme system involved (*96*). Indoleacetic acid can arise from indoleacetaldehyde (XXXIII, p. 267), indoleacetonitrile (XXXIV, p. 268), indolepyruvic acid (XXXVII, p. 270), methyltryptamine (XXIII, p. 263), and dimethyltryptamine (XXV, p. 264); the evidence for each source has been discussed above.

Indoleacetic acid has frequently been encountered as a microbial product (*29, 80, 135*) and in fact its first isolation from any source was that by SALKOWSKI (*241*), who found the substance in bacterially decomposed protein. STOWE (*273*) traced this biogenesis in three bacteria to tryptophan which in cell-free extracts was vigorously transaminated to indolepyruvic acid; the latter then produced indoleacetic acid, probably for the most part spontaneously. KAPER (*150, 151*) has examined one of these microorganisms further, and was also unable to decide whether or not enzymes are involved in the final step of the process. However, MURAKAMI and HAYASHI (*205*), as was noted on p. 272, concluded that a similar pathway was unlikely to be important in rice.

There is much evidence, however, that tryptophan can be the ultimate source of indoleacetic acid in some plants. One argument dates from SKOOG's finding (*262*) that auxin is low in zinc-deficient plants, which TSUI (*302*) noted to parallel a reduction in the tryptophan content. The zinc requirement may act before indole formation, however, since MUDD and ZALIK (*204*) showed that zinc-deficient plants consume more indole than normal ones. Then there are the several papers cited earlier (*97, 98, 125, 184, 212, 326, 332*) indicating a relationship between tryptophan and auxin levels, and KULESCHA's data (*161, 164*) showing that tryptophan is utilized by plant tissue cultures dependent on auxin for growth. STEHSEL (*271*) found the activity of a tryptophan → auxin enzyme system to vary parallel with the free auxin measured in maize kernels. Finally, DANNENBURG and LIVERMAN's (*56*) radioactive tryptophan was converted in watermelon tissues to radioactive indoleacetic acid.

None of this evidence, however, excludes the possibility that some pathway not involving tryptophan does exist; in fact the existence of five immediate precursors of indoleacetic acid, directly convertible under physiological conditions, makes it likely that plants have evolved several normal routes to produce this vital compound.

Work by FAWCETT et al. (*69, 70*) indicates that indoleacetic acid may also arise by β-oxidation from long-chain indole acids. These compounds are not yet known to occur in higher plants, although tentative identifications of indolepropionic acid (*176, 196 b*) and indolebutyric acid (*33*) have been made; furthermore, indolepropionic acid can be produced from tryptophan by bacteria (*135, 329*).

Some synthetic indole acids with an even number of carbon atoms in the side-chain were found to be degraded to indoleacetic acid, while odd numbered carbon side-chains were shortened only as far as indolepropionic acid. Since the latter has moderate hormonal activity, all the higher indole-3-alkyl acids tested acted as auxins. This β-oxidation has also been noted by ANDREAE and GOOD (*3*) with indolebutyric acid.

Much work has been devoted to the production of active auxin from "bound" auxin, that is to say, the release of plant hormones, presumably indoleacetic acid or closely related compounds, from precursors of different solubility and in some cases of much higher molecular weight. The agents used for effectuating this release have ranged from proteolytic enzymes to heat treatments, and there has been considerable argument as to whether or not such methods are analogous to processes occurring in plants. In the case of protein "auxin precursors", for instance, SCHOCKEN (*248*) showed that some, and presumably all, tryptophan-containing proteins, including those of animal origin, will release auxin under conditions used by several earlier investigators. Further, GORDON and WILDMAN (*100*) found tryptophan to form auxin under several of these mild treatments (but see *9*). Yet there is some evidence, as is perhaps best shown in maize, for the formation of a non-proteinic complex (*27, 271*) from which sizeable amounts of indoleacetic acid are released upon hydrolysis.

STEHSEL, in a detailed study of this substance (*271*), which unfortunately is only readily available in abstract form (*272*), concluded that the complex had a molecular weight somewhat below 500. He believed that it released two molecules of indoleacetic acid during a non-oxidative hydrolysis which had a marked pH optimum, surprisingly at 7.5. His studies, however, indicate that the material was normally formed from indoleacetic acid, and thus was not a "precursor". He preferred the term "auxin complex". It should also be noted that studies of maize have shown marked variation in the auxin content of different strains (*10, 284*).

Other evidence for an auxin releasing substance of about twice the molecular weight of indoleacetic acid dates back over twenty years and has been summarized by SÖDING and RAADTS (*266*). Many other, often ill-defined, "auxin precursors" have been described and these are cited in the reviews (*19, 93*). It seems unlikely that the reality of the functions ascribed to these substances can be established until their chemical nature is determined and they can be studied under better defined conditions.

Conversely, GALSTON (*85*) has presented evidence that a coupling reaction takes place, i. e. free indoleacetic acid can be "bound" to higher molecular weight substances, in particular proteins, by a metabolic process in plant tissues.

Degradative means of auxin removal have been the object of much more study, initiated by TANG and BONNER's (*282*) discovery of a potent oxidase acting on indoleacetic acid. This prompted the suggestion that indolealdehyde (XL, p. 277) was formed (*312*) but conclusive evidence was for a long time lacking. RAY (*233*) has lately reviewed the decomposition of indoleacetic acid, hence it will suffice here to point out that several oxidizing enzymes exist, apparently varying slightly in action and end products. One fungal enzyme oxidizes indoleacetic acid to an unstable indolic intermediate (*232*), which then decomposes to several different compounds, at least one of which is oxindolic (*275*); indolealdehyde is not formed in appreciable amounts. Some other products have been proposed (*85*). In the most recent advance in the field of higher plant oxidases, STUTZ (*279*) has highly purified an enzyme from lupines, and has made the unexpected discovery that cytochrome c and cytochrome oxidase must be present for indolealdehyde to be produced in any quantity.

A related problem is that of photolysis of indoleacetic acid; this process may be important in plants as the initiating reaction in phototropic curvature. FISCHER (*72*) found that indolealdehyde is produced in in vitro photolytic reactions, but MAYR (*194*) as well as RAY and CURRY (*234*) have shown that such conditions actually lead to several products. MELCHIOR (*196a*) has extended these photolytic studies to include other indole derivatives. Earlier GALSTON (*84*) found that riboflavin can catalyze photolysis in vitro, an observation extended by FERRI (*71*) to include several chemically unrelated fluorescent compounds. On the other hand, some plant extracts protect indoleacetic acid from photolysis (*194*, cf. *196 a*). The relationship of these reactions to phototropism is detailed by THIMANN and CURRY (*294*). An understanding of the effects of shorter wavelengths on plants has been the object of studies on the sensitivity of indoleacetic acid to ionizing radiation (*99*) and to ultraviolet light (*53*).

There is no compelling reason for believing that indoleacetic acid is not the primary indole growth hormone in plants. All but two of the biologically active materials discussed elsewhere, with the possible exception of indole, can be readily converted to indoleacetic acid and their activity explained on this basis. The two exceptions, indoleglycolic acid (XLVI, p. 280) and 5-hydroxyindoleacetic acid (XXVII, p. 265), would require reduction of a hydroxyl group, and it seems more likely that they have genuine activity themselves (*293*). However, they are relatively weak growth factors in all tests so far reported, and their existence does not seriously disturb the central position of indoleacetic acid.

# VI. Products Formed from 3-Indoleacetic Acid.

## 1. 3-Indolealdehyde.

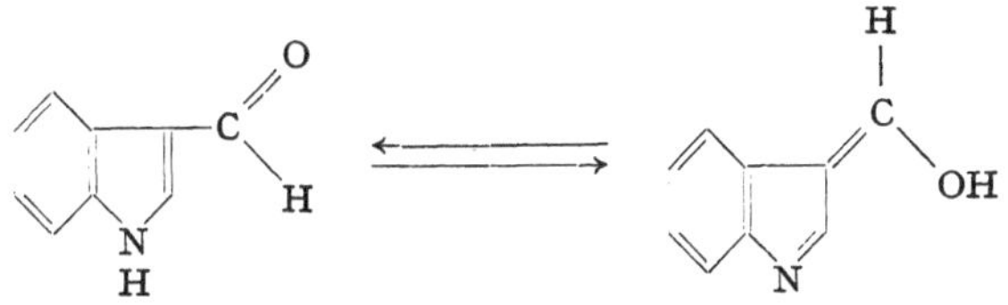

(XL a.)                    (XL b.)

As noted in the paragraphs above, indolealdehyde can be produced from indoleacetic acid both by the action of light (*72, 234*) and by plant oxidases, especially if cytochrome c and cytochrome oxidase are available (*279*). The aldehyde was first reported in plant material by VON DENFFER et al. (*57*) on chromatographic grounds, but a clear isolation by JONES and TAYLOR (*145*) from cabbage has now been made. Indole-aldehyde has also been identified in mammalian tissue (*202*). It is a remarkably stable aldehyde and does not spontaneously oxidize to the acid under normal conditions; this probably is due to its tautomerism to the hydroxymethylene-indolenine (XLb) form (*280*), the existence of which is indicated by the distinctive and non-indolic ultraviolet absorption spectrum (*232, 263, 275*). However, two laboratories (*69, 197*) have produced evidence that plant enzymes do exist which can carry out its oxidation to 3-indolecarboxylic acid (XLV, p. 280); the evidence for this will be discussed below.

## 2. Ethyl 3-Indoleacetate.

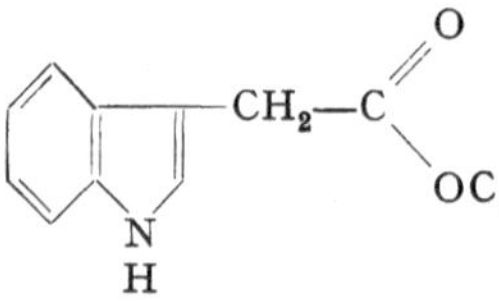

(XLI.) Ethyl 3-indoleacetate.

Evidence for a substance in maize with an exceptionally high parthenocarpy-stimulating action led REDEMANN et al. (*235*) to its isolation. The substance was identical with ethyl indoleacetate, but criticism was leveled at the method of isolation, viz. ethanol extraction of immature maize kernels (*124*). The presence of an esterase in the tissue could thus have catalyzed the reaction of the ethyl alcohol with endogenous indoleacetic acid—previously demonstrated to be present in this material (*116*). The extraction has, therefore, been repeated with ether and the criticism found to be valid. Accordingly, the original claim has been withdrawn (*81*). Another identification (*289*), this time in apples, has also been subject to criticism (*183*).

Nonetheless, several reports of materials with biological, chromatographic, and in some cases, color-reacting, properties similar to ethyl indoleacetate have been made (*30, 213, 214, 253*); and considering that an esterase capable of producing ethyl indoleacetate from indoleacetic acid does exist, the ester remains a likely constituent of higher plants. Ethyl indoleacetate has strong growth hormone activity; and SEELEY et al. (*254*) have presented chromatographic data indicating hydrolysis of synthetic methyl indoleacetate to indoleacetic acid in several plant tissues.

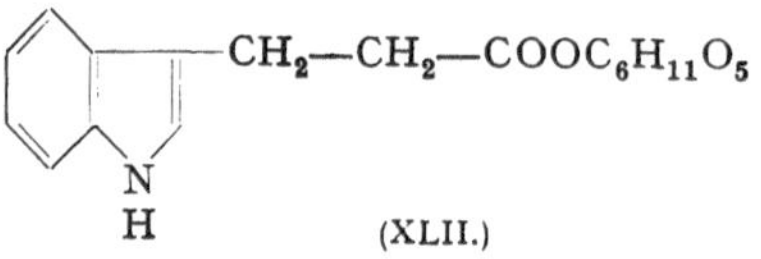

(XLII.)

A $\beta$-glucosidic ester (XLII) is produced from synthetic indolepropionic acid in cultures of *Bacillus megatherium* (*281*), and the occurrence of similar compounds remains a possibility in higher plants. Indeed, STEWARD, in an unpublished lecture, has reported evidence of an indoleacetyl arabinoside in coconut milk.

## 3. 3-Indoleacetamide.

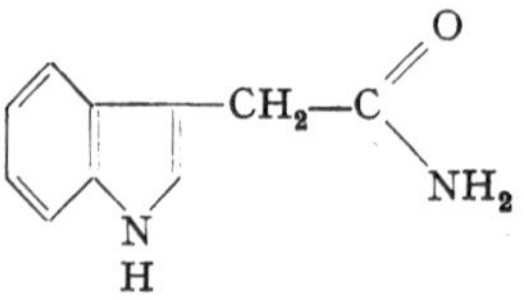

(XLIII.) 3-Indoleacetamide.

This substance, as noted earlier, does not seem to arise upon enzymatic hydrolysis of indoleacetonitrile (*254, 276, 296*). Instead, it was shown to be produced in many plant tissues after incubation with indoleacetic acid [GOOD et al. (*92*)]. The identification was made by chromatographic and color-forming properties, coupled with the demonstration that indoleacetic acid and ammonia were produced upon hydrolysis. Grasses form indoleacetamide in especially large amounts, but no trace of its natural occurrence could be found in any plant tissue which had not been treated with indoleacetic acid. ANDREAE and GOOD (3) showed that synthetic indolepropionic and indolebutyric acids are also weakly converted to their respective amides by pea tissue. The formation of an amide from benzoic acid, and probably from indolecarboxylic acid, was also indicated.

SEELEY et al. (254) found that the reverse (hydrolytic) reaction takes place when peas, maize, and tomatoes convert synthetic indole-acetamide to indoleacetic acid. This reaction cannot be especially favored in most tissues since the auxin activity of the applied compound is small (144, 254).

## 4. 3-Indoleacetylaspartic Acid.

Like indoleacetamide, this peptide of indoleacetic acid and aspartic acid (XLIV a) was demonstrated by ANDREAE and GOOD (2) to arise in tissues incubated with indoleacetic acid. They were led to this discovery by the observation that indoleacetic acid incubated with peas formed a substance which reacted with the SALKOWSKI reagent but was not affected by indoleacetic acid oxidase.

(XLIV a.) R = OH.
3-Indoleacetylaspartic acid.
(XLIV b.) R = NH$_2$.
3-Indoleacetylasparagine.

Hydrolysis of a partially purified preparation yielded indoleacetic and aspartic acids, and comparison with synthetic indoleacetylaspartic acid (90) showed identity. Indoleacetylasparagine (XLIV b) was clearly different. The auxin potency on peas was extremely small or non-existent, but on oats the peptide was as active as indoleacetic acid. Its formation in many other plants, especially onion and legumes, by indoleacetic acid was shown later (92). The failure of several investigators (56, 69, 254) to find this peptide can be attributed to their extraction methods. Synthetic indolepropionic and indolebutyric acids also formed the aspartic derivative (3).

ANDREAE and VAN YSSELSTEIN (4) have quantitatively followed the formation of indoleacetylaspartic acid in pea tissues at different indoleacetic acid concentrations and found it to parallel the indoleacetic acid uptake up to about 0.4 $\mu$M/g tissue, after which for the first time free indoleacetic acid accumulates in the stem sections. The process requires oxygen, and is relatively linear with time. If removed from the indoleacetic acid solution, the pea sections converted all the free indoleacetic acid in the tissues to indoleacetylaspartic acid which then remained stable throughout the experiment. Treatment with synthetic indolebutyric acid also led to the formation of indoleacetylaspartic acid, apparently from indoleacetic acid formed by $\beta$-oxidation (3). No report of the natural occurrence of this compound in plant tissues has yet been made, but the above studies indicate that it is surely a natural plant constituent.

## VII. Some other Indoles.

### 1. 3-Indolecarboxylic Acid.

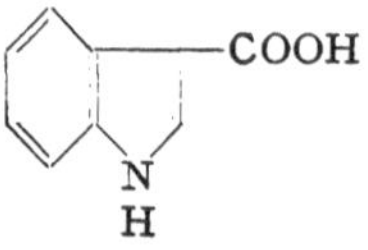

(XLV.)
3-Indolecarboxylic acid.

The presence of this acid in plants was suggested by chromatograms (72, 196 b) and has been substantiated by JONES and TAYLOR's isolation of the substance from cabbage (145). It can be produced enzymically from indolealdehyde as was shown by FAWCETT and his colleagues (69). MEYER (197) has investigated the conditions under which oat breis convert both indolealdehyde and indoleacetic acid to indolecarboxylic acid; the aldehyde → carboxylic acid step can also be carried out by the SCHARDINGER enzyme of milk. The compound has long been known to be formed by bacteria (29).

As in the detoxification of alkyl nitriles in animals, indolecarboxylic acid has been found to arise from indoleacetonitrile in plant tissues which do not form indoleacetic acid from the nitrile (254), and the indolecarboxylic acid so produced has been isolated (69, 254). The evidence indicated that indolealdehyde is the intermediate in this reaction, and FAWCETT et al. (69) have suggested that the indoleacetonitrile → → indolealdehyde reaction proceeds via either a cyanohydrin or a ketonitrile intermediate. Indole carboxylic acid is also produced from indoleacetonitrile by photolysis (336, 196 a).

The only other metabolic study of indolecarboxylic acid indicates that it can be converted to its amide by pea tissues (3), but since comparison with synthetic amide was not possible this must still be regarded as a tentative conclusion.

Other studies indicate that the heterocyclic ring of indolecarboxylic acid is broken and anthranilic acid produced in vitro by photolysis in the presence of riboflavin (197). It is also noteworthy that the synthetic compound is decarboxylated to indole under mild conditions (40), and can be synthesized from indole and carbon dioxide (280). 3-Indolecarboxylic acid does not act as a higher plant hormone (70), and in fact is an inhibitor, though only at relatively high concentrations (197). In an unpublished lecture HUSTEDE also indicated that it may be particularly significant as an algal hormone.

### 2. 3-Indoleglycolic Acid.

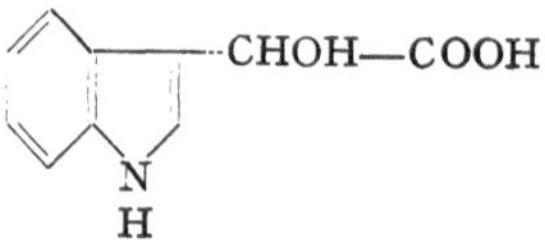

      

(XLVI.) 3-Indoleglycolic acid. 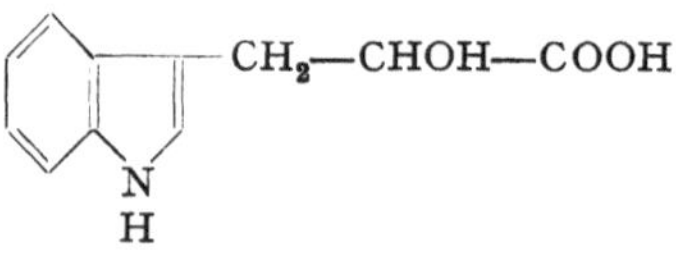        (XLVII.) Indolelactic acid.

A report of (XLVI) on chromatograms of three plant extracts has been made (72). It was based on a comparison with a synthetic sample whose preparation was not outlined and, unfortunately, the identification has been accepted by other workers (*56*, *196 a*, *196 b*). Two independent syntheses (*22*, *258*) have now shown that indoleglycolic acid is different from the natural product. Thus, this acid has not yet been found in plants. GREENBERG et al. (*109*) have established, however, that indoleglycolic acid is readily formed non-enzymatically from indole and glyoxylic acid. Since both compounds are known to occur in plant materials (*300*), the presence of the combined product seems reasonable.

BENTLEY and coworkers (*22*) [see also KAPER and VELDSTRA (*151*)], have stated that indoleglycolic acid is formed from indolepyruvic acid by spontaneous decomposition, but they fail to indicate a mechanism. Their evidence is based only on a comparison of chromatographic properties and color reactions, and the spot they implicate could equally well contain the keto form of indolepyruvic acid. There are indications that free indoleglycolic acid is unstable (*13*, *258*); so far only a salt has been prepared in stable form. The compound is only a weak auxin (*293*).

This is also true of the next higher analog, *indolelactic acid* (XLVII), a compound not yet detected in higher plants but produced by a fungus from tryptophan (*61*). It has been tentatively identified among the products of *Agrobacterium* acting on tryptophan (*151*), with its decarboxylation product, tryptophol.

### 3. Indican.

Although historically indican would rank as the most important of all the compounds considered, as it has served since antiquity as the source of indigo, nothing is known of its biogenesis.

(XLVIII.) Indican.

The compound is widely distributed in about twenty-seven genera of eleven plant families (*118*, *223*), but some of the reports have not been critically confirmed. Of these *Indigofera*, and particularly *I. tinctorum* L., were the most cultivated, although in more temperate zones woad (*Isatis tinctoria* L.), and *Polygonum tinctorium* AIT. have been important. The technological aspects are well described (*28*, *139*, *223*, *231*). Simple techniques for the detection of indican in plants are available (*208*, *237*).

Leaves are the richest source of indican which may reach 3.8% of the dry weight. The absolute amount is even more in the green plant as nearly half is lost on drying (*223*, *269*). In *Strobilanthus flaccidolius* NEES, shading the growing plant nearly tripled the yield obtained (*269*).

The enzyme responsible for the hydrolysis of indican has been named indimulsin (*122*) and was examined by BEIJERINCK (*17*). Other glucosidases, such as emulsin (*52*), also cleave indican, but are less active. Indimulsin is not very soluble (*299*), does not require oxygen, and appears to be on the chloroplasts (*17*). In the absence of any modern work, the specificity of indimulsin for indican might well be questioned.

Beijerinck has also established that the precursor of indigo in the woad is not indican, but a highly alkali-unstable compound. He at first thought the plant contained free indoxyl (*16*), but later (*18*) showed that this was not the case. A labile compound which he named "isatan" was present.

This has never been identified, but it was hydrolyzed by a probably particulate enzyme named "isatase" which was found only in woad. Isatase was specific, and did not attack indican. Microbes failed to decompose "isatan" readily. The indoxyl freed by isatase could be oxidized to indigo in the usual manner, but the enzyme itself was not oxidative. "Isatan" is thus yet another unidentified indole derivative which is present in relatively large amounts in a plant. (It should be noted that the name "isatan" has since become affixed to a clearly different bioxindole compound.)

A number of microbes, especially *Aerobacter* sp., and yeast also hydrolyze indican (*17, 120*). Two soil bacteria isolated by Gray (*104*) form indigo from indole; he did not believe indoxyl was an intermediate.

Miles et al. (*199*) have identified indigo as the product of a mutant of a fungus, *Schizophyllum commune*.

## References.

*1.* Anderson, J. A., M. R. Ziegler and D. Doeden: Banana Feeding and Urinary Excretion of 5-Hydroxyindoleacetic Acid. Science (Washington) **127**, 236 (1958).

*2.* Andreae, W. A. and N. E. Good: The Formation of Indoleacetylaspartic Acid in Pea Seedlings. Plant Physiol. **30**, 380 (1955).

*3.* — — Studies on 3-Indoleacetic Acid Metabolism. IV. Conjugation with Aspartic Acid and Ammonia as Processes in the Metabolism of Carboxylic Acids. Plant Physiol. **32**, 566 (1957).

*4.* Andreae, W. A. and M. W. H. van Ysselstein: Studies on 3-Indoleacetic Acid Metabolism. III. The Uptake of 3-Indoleacetic Acid by Pea Epicotyls and its Conversion to 3-Indoleacetylaspartic Acid. Plant Physiol. **31**, 235 (1956).

*5.* Arai, I.: Growth Factor for *Salmonella typhosa*. V. Mechanism of Growth Promoting Action of Tryptophan. (2). Additional Observations on the Inhibition of Growth of *Salmonella typhosa* by Several Indole Derivatives. J. pharmac. Soc. Japan **71**, 673 (1951) [Chem. Abstr. **45**, 10301 (1951)].

*6.* Armstrong, M. D., K. N. F. Shaw, M. J. Gortakowski and H. Singer: The Indole Acids of Human Urine. Paper Chromatography of Indole Acids. J. Biol. Chem. **232**, 17 (1958).

*7.* Aronoff, S.: Experiments on the Biogenesis of the Pyridine Ring in Higher Plants. Plant Physiol. **31**, 355 (1956).

*8.* Ashby, W. C.: Effects of Certain Acid Growth-regulating Substances and Their Corresponding Aldehydes on the Growth of Roots. Bot. Gaz. **112**, 237 (1951).

*9.* Avery, G. S., Jr. and J. Berger: Tryptophan and Phytohormone Precursors. Science (Washington) **98**, 513 (1943).

*10.* Avery G. S., Jr., J. Berger and B. Shalucha: Auxin Storage as Related to Endosperm Type in Maize. Bot. Gaz. **103**, 806 (1942).

*11.* BACCARINI, P.: Sopra la presenza di indolo nei fiori di alcune piante. Bull. soc. botan. Ital. **1910**, 96.

*12.* — Sulla presenza di indolo negli organi vegetativi di alcune piante. Bull. soc. botan. Ital. **1911**, 105.

*13.* BAKER, J. W.: Syntheses in the Indole Series. Part I. Synthesis of Indolyl-3-glyoxylic Acid and of *r*-3-Indolylglycine. J. Chem. Soc. (London) **1940**, 458.

*14.* BALLANTINE, J. A., C. B. BARRETT, R. J. S. BEER, S. EARDLEY, A. ROBERTSON, B. L. SHAW and T. H. SIMPSON: The Chemistry of Bacteria. Part VII. The Structure of Violacein. J. Chem. Soc. (London) **1958**, 755.

*15.* BARRATT, R. W. and W. OGATA: A Strain of Neurospora with an Alternative Requirement for Leucine or Aromatic Amino Acids. Amer. J. Bot. **41**, 763 (1954).

*16.* BEIJERINCK, M. W.: On the Formation of Indigo from the Woad *(Isatis tinctoria)*. Kon. Ned. Akad. Wetensch. Proc. **2**, 120 (1900).

*17.* — On Indigo-fermentation. Kon. Ned. Akad. Wetensch. Proc. **2**, 495 (1900).

*18.* — Further Researches on the Formation of Indigo from the Woad *(Isatis tinctoria)*. Kon. Ned. Akad. Wetensch. Proc. **3**, 101 (1901).

*19.* BENTLEY, J. A.: The Naturally-occurring Auxins and Inhibitors. Annu. Rev. Plant Physiol. **9**, 47 (1958).

*20.* — Role of Plant Hormones in Algal Metabolism and Ecology. Nature (London) **181**, 1499 (1958).

*21.* BENTLEY, J. A. and A. S. BICKLE: Studies on Plant Growth Hormones. II. Further Biological Properties of 3-Indolylacetonitrile. J. exp. Bot. **3**, 406 (1952).

*22.* BENTLEY, J. A., K. R. FARRAR, S. HOUSLEY, G. F. SMITH and W. C. TAYLOR: Some Chemical and Physiological Properties of 3-Indolylpyruvic Acid. Biochemic. J. **64**, 44 (1956).

*23.* BENTLEY, J. A. and S. HOUSLEY: Studies on Plant Growth Hormones. I. Biological Activities of 3-Indolylacetaldehyde and 3-Indolylacetonitrile. J. exp. Bot. **3**, 393 (1952).

*24.* BENTLEY, K. W.: Alkaloids of the Indole Group I and II. In: The Chemistry of Natural Products. The Alkaloids. Vol. I, p. 146. New York: Interscience Publ. 1957.

*25.* BERG, C. P., W. C. ROSE and C. S. MARVEL: Tryptophane and Growth. III. 3-Indolepropionic Acid and 3-Indolepyruvic Acid as Supplementing Agents in Diets Deficient in Tryptophan. J. Biol. Chem. **85**, 219 (1929/30).

*26.* BERGER, J. and G. S. AVERY, Jr.: Isolation of an Auxin Precursor and an Auxin from Maize. Amer. J. Bot. **31**, 199 (1944).

*27.* — — Chemical and Physiological Properties of Maize Auxin Precursor. Amer. J. Bot. **31**, 203 (1944).

*28.* BERGTHEIL, C.: The Fermentation of the Indigo-plant. Chem. Soc. (London) **85**, 870 (1904).

*29.* BERTHELOT, A.: Recherches sur le Proteus vulgaris. III B. Étude de la fonction indologène. Ann. Inst. Pasteur **28**, 849 (1914).

*30.* BITANCOURT, A. A.: Recherches physiologiques sur les auxines. Rev. gén. bot. **62**, 498 (1955).

*31.* BITANCOURT, A. A., K. SCHWARZ e A. P. NOGUERA: A Decomposição Espontânea de Alguns Derivados Indólicos. I. Métodos Experimentais. Arqu. Inst. Biol. (São Paulo) **24** (13), 169 (1957).

*32.* BLOCK, R. J. and K. W. WEISS: Amino Acid Handbook, Part II, p. 296. Springfield, Illinois: Thomas Publ. 1956.

*33.* BLOMMAERT, K. L. J.: Growth- and Inhibiting-substances in Relation to the Rest Period of the Potato Tuber. Nature (London) **174**, 970 (1954).

34. BONDE, E. K.: Auxins and Auxin Precursors in Acid and Nonacidic Fractions of Plant Extracts. Bot. Gaz. **115**, 1 (1953).

35. BOORSMA, W. G.: Über Aloëholz und andere Riechhölzer. Bull. Dépt. Agric. Indes Néerlandaises (Buitenzorg) **7** (Pharmacologie 3), 1 (1907).

36. BOOTH, A.: Non Hormonal Growth Promotion Shown by Aqueous Extracts. J. exp. Bot. **9**, 306 (1958).

37. BORZI, A.: Produzione d'indolo e impollinazione della *Visnea Mocanera* L. Atti Reale Accad. Naz. Lincei (Roma) [5] **13** I, 372 (1904).

38. BOUVEAULT, L. et R. LOCQUIN: Préparation des éthers et des acides α-cétoniques à l'aide des éthers α-oximidés (V). Bull. soc. chim. France [3] **31**, 1142 (1904).

39. BOWDEN, K., B. G. BROWN and J. E. BATTY: 5-Hydroxytryptamine: its Occurrence in Cowhage. Nature (London) **174**, 925 (1954).

40. BOWDEN, K. and L. MARION: The Biogenesis of Alkaloids. IV. The Formation of Gramine from Tryptophan in Barley. Canad. J. Chem. **29**, 1037 (1951).

41. — — The Biogenesis of Alkaloids. V. Radioautographs of Barley Leaves fed with Tryptophan-β-$C^{14}$. Canad. J. Chem. **29**, 1043 (1951).

42. BRANDT, K., H. v. EULER, H. HELLSTRÖM und N. LÖFGREN: Gramin und zwei Begleiter desselben in Laubblättern von Gerstensorten. Z. physiol. Chem. (Hoppe-Seyler) **235**, 37 (1935).

43. BROWN, J. B., H. B. HENBEST and E. R. H. JONES: 3-Indolylacetaldehyde and 3-Indolylacetone. J. Chem. Soc. (London) **1952**, 3172.

44. BULARD, C. et A. C. LEOPOLD: 5-Hydroxytryptamine chez les végétaux supérieurs. C. R. hebd. Séances Acad. Sci. **247**, 1382 (1958).

45. — — Mise en évidence de l'activité de l'acide 5-hydroxyindolacetique sur l'élongation des cellules végétales. C. R. hebd. Séances Acad. Sci. (sous presse).

46. CAHILL, W. M. and R. W. JACKSON: The Proof of Synthesis and the Configurational Relationships of Abrine. J. Biol. Chem. **126**, 29 (1938).

47. CERIGHELLI, M. R.: Sur l'indol des fleurs du Jasmin d'Espagne. C. R. hebd. Séances Acad. Sci. **179**, 1193 (1924).

48. CHRISTIANSEN, G. S. and K. V. THIMANN: The Metabolism of Stem Tissue During Growth and Its Inhibition. III. Nitrogen Metabolism. Arch. Biochemistry **28**, 117 (1950).

49. CLARKE, A. J. and P. J. G. MANN: The Oxidation of Tryptamine to 3-Indoleacetaldehyde by Plant Amine Oxidase. Biochemic. J. **65**, 763 (1957).

50. COLLIER, H. O. J.: The Occurrence of 5-Hydroxytryptamine in Nature. In: G. P. LEWIS, 5-Hydroxytryptamine, p. 5. London and New York: Pergamon Press. 1958.

51. COLLIER, H. O. J. and G. B. CHESHER: Identification of 5-Hydroxytryptamine in the Sting of the Nettle *(Urtica dioica)*. Brit. J. Pharmacol. **11**, 186 (1956).

52. CRAMER, F.: Über Einschlußverbindungen. IV. Die Hemmung der Glykosidspaltung durch Cyclodextrin. Liebigs Ann. Chem. **579**, 17 (1953).

53. CURRY, G. M., K. V. THIMANN and P. M. RAY: The Base Curvature Response of Avena Seedlings to the Ultraviolet. Physiol. Plantarum **9**, 429 (1956).

54. DAKIN, H. D.: The Oxidation of Amino Acids to Cyanides. Biochemic. J. **10**, 319 (1916).

55. DALGLIESH, C. E.: Metabolism of the Aromatic Amino Acids. Adv. Protein Chem. **10**, 31 (1955).

56. DANNENBURG, W. N. and J. L. LIVERMAN: Conversion of Tryptophan-2-$C^{14}$ to Indoleacetic Acid by Watermelon Tissue Slices. Plant Physiol. **32**, 263 (1957).

57. DENFFER, D. v., M. BEHRENS und A. FISCHER: Papierchromatographischer und papierelektrophoretischer Nachweis des β-Indoleacetonitrils und des β-Indolealdehyds in Extrakten aus Kohlpflanzen. Naturwiss. 39, 550 (1952).

58. DEULOFEU, V., E. HUG and P. MAZZOCCO: Studies on Argentine Plants. Part I. Hypaphorine from Erythrina crystagalli. J. Chem. Soc. (London) 1939, 1841.

59. DUNSTAN, W. R.: On the Occurrence of Skatole in the Vegetable Kingdom. Proc. Roy. Soc. (London) 46, 211 (1889).

60. DUNSTAN, W. R. and T. A. HENRY: Occurrence of Orthohydroxyacetophenone in the Volatile Oil of Chione glabra. J. Chem. Soc. (London) 75, 66 (1899).

61. EHRLICH, F. und K. A. JACOBSEN: Über die Umwandlung von Aminosäuren in Oxysäuren durch Schimmelpilze. Ber. dtsch. chem. Ges. 44, 888 (1911).

62. ELZE, F.: Über das Öl von Robinia pseudoacacia. Chem.-Ztg. 34, 814 (1910).

63. EULER, H. v. und H. ERDTMAN: Über Gramin aus schwedischen Gersten-sippen. Liebigs Ann. Chem. 520, 1 (1935).

64. EULER, H. v., H. ERDTMAN und H. HELLSTRÖM: Über das Alkaloid Gramin. Ber. dtsch. chem. Ges. 69, 743 (1936).

65. EULER, H. v. und H. HELLSTRÖM: Spektrometrische Messungen an Alkohol-extrakten der Laubblätter von Chlorophyllmutanten der Gerste. Z. physiol. Chem. (Hoppe-Seyler) 208, 43 (1932).

66. — — Über ein Indolderivat aus zwei chlorophyllmutierenden Gerstensippen. Z. physiol. Chem. (Hoppe-Seyler) 217, 23 (1933).

67. EULER, H. v., H. HELLSTRÖM und J. HAGEN: Über die aus Gerstenmutanten Albina 1 und 3 gewonnene Indolbase und ihre Umwandlung. Ark. Kemi, Mineral. Geol. 11 B (36), 1 (1934).

68. EULER, H. v., H. HELLSTRÖM und N. LÖFGREN: Zur chemischen Genetik chlorophyllmutierender Gerstensippen. Z. physiol. Chem. (Hoppe-Seyler) 234, 151 (1935).

69. FAWCETT, C. H., H. F. TAYLOR, R. L. WAIN and F. WIGHTMAN: The Metabolism of Certain Acids, Amides and Nitriles within Plant Tissues. Proc. Roy. Soc. (London) 148 B, 543 (1958).

70. FAWCETT, C. H., R. L. WAIN and F. WIGHTMAN: Beta-Oxidation of Omega-(3-Indolyl)alkanecarboxylic Acids in Plant Tissues. Nature (London) 181, 1387 (1958).

71. FERRI, M. G.: Fluorescence and Photoinactivation of Indoleacetic Acid Arch. Biochem. Biophys. 31, 127 (1951).

72. FISCHER, A.: Über die papierchromatographische und papierelektrophoretische Trennung von Indolderivaten. Planta 43, 288 (1954).

73. FISH, M. S.: Personal communication, 1958.

74. FISH, M. S., N. M. JOHNSON and E. C. HORNING: Piptadenia Alkaloids. Indole Bases of P. peregrina (L.) BENTH. and Related Species. J. Amer. Chem. Soc. 77, 5892 (1955).

75. FISH, M. S., N. M. JOHNSON, E. P. LAWRENCE and E. C. HORNING: Oxidative N-Dealkylation. Biochem. Biophys. Acta 18, 564 (1955).

76. FISH, M. S., C. C. SWEELEY, N. M. JOHNSON, E. P. LAWRENCE and E. C. HORNING: Chemical and Enzymic Rearrangements of N,N-Dimethyl Amino Acid Oxides. Biochim. Biophys. Acta 21, 196 (1956).

77. FOLKERS, K. and F. KONIUSZY: Erythrina Alkaloids. III. Isolation and Characterization of a New Alkaloid, Erythramine. J. Amer. Chem. Soc. 61, 1232 (1939).

78. — — Erythrina Alkaloids. VII. Isolation and Characterization of the New Alkaloids, Erythraline and Erythratine. J. Amer. Chem. Soc. 62, 436 (1940).

79. Folkers, K., J. Shavel, Jr. and F. Koniuszy: Erythrina Alkaloids. X. Isolation and Characterization of Erysonine and Other Liberated Alkaloids. J. Amer. Chem. Soc. **63**, 1544 (1941).

80. Frieber, W.: Beiträge zur Frage der Indolbildung und der Indolreaktionen sowie zur Kenntnis des Verhaltens indolnegativer Bakterien. Centralbl. Bakter. u. Parasitenk. **87**, 254 (1921).

81. Fukui, H. N., J. E. DeVries, S. H. Wittwer and H. M. Sell: Ethyl-3-Indoleacetate: an Artefact in Extracts of Immature Corn Kernels. Nature (London) **180**, 1205 (1957).

82. Gale, E. F.: The Bacterial Amino Acid Decarboxylases. Adv. Enzymology **6**, 1 (1946).

83. Galston, A. W.: Indoleacetic-Nicotinic Acid Interactions in the Etiolated Pea Plant. Plant Physiol. **24**, 577 (1949).

84. — Riboflavin-sensitized Photooxidation of Indoleacetic Acid and Related Compounds. Proc. Nat. Acad. Sci. (USA) **35**, 10 (1949).

85. — Some Metabolic Consequences of the Administration of Indoleacetic Acid to Plant Cells. In: R. L. Wain and F. Wightman, The Chemistry and Mode of Action of Plant Growth Substances, p. 219. London: Butterworths. 1956.

86. Galston, A. W. and M. E. Hand: Adenine as a Growth Factor for Etiolated Peas and its Relation to the Thermal Inactivation of Growth. Arch. Biochemistry **22**, 434 (1949).

87. Ghatak, N.: Chemical Examination of the Seeds of *Abrus precatorius*. III. Constitution of Abrine. Bull. Acad. Sci. United Provinces Agra and Oudh, Allahabad **3** (4), 295 (1934).

88. Ghatak, N. and R. Kaul: Chemical Examination of the Seeds of *Abrus precatorius* Linn. Part I. J. Indian Chem. Soc. **9**, 383 (1932).

89. Gibson, F. W. E., C. H. Doy and S. B. Segall: A Possible Intermediate in the Biosynthesis of Tryptophan: 1-Deoxy-1-N-o-Carboxyphenyl-Ribulose. Nature (London) **181**, 549 (1958).

90. Good, N. E.: The Synthesis of Indole-3-acetyl-*D,L*-aspartic Acid and Related Compounds. Canad. J. Chem. **34**, 1356 (1956).

91. Good, N. E. and W. A. Andreae: Malonyltryptophan in Higher Plants. Plant Physiol. **32**, 561 (1957).

92. Good, N. E., W. A. Andreae and M. W. H. van Ysselstein: Studies on 3-Indoleacetic Acid Metabolism. II. Some Products of the Metabolism of Exogenous Indoleacetic Acid in Plant Tissues. Plant Physiol. **31**, 231 (1956).

93. Gordon, S. A.: Occurrence, Formation and Inactivation of Auxins. Annu. Rev. Plant Physiol. **5**, 341 (1954).

94. — The Biogenesis of Natural Auxins. In: R. L. Wain and F. Wightman, The Chemistry and Mode of Action of Plant Growth Substances, p. 65. London: Butterworths. 1956.

95. — The Effects of Ionizing Radiation on Plants: Biochemical and Physiological Aspects. Quart. Rev. Biol. **32**, 3 (1957).

96. — Intracellular Localization of the Tryptophan-Indoleacetate Enzyme System. Plant Physiol. **33**, 23 (1958).

97. Gordon, S. A. and F. S. Nieva: The Biosynthesis of Auxin in the Vegetative Pineapple. I. Nature of the Active Auxin. Arch. Biochemistry **20**, 356 (1949).

98. — — The Biosynthesis of Auxin in the Vegetative Pineapple. II. The Precursors of Indoleacetic Acid. Arch. Biochemistry **20**, 367 (1949).

99. Gordon, S. A. and R. P. Weber: Studies on the Mechanism of Phytohormone Damage by Ionizing Radiation. I. The Radiosensitivity of Indoleacetic Acid. Plant Physiol. **30**, 200 (1955).

*100.* GORDON, S. A. and S. G. WILDMAN: The Conversion of Tryptophane to a Plant Growth Substance by Conditions of Mild Alkalinity. J. Biol. Chem. **147**, 389 (1943).

*101.* GORDON, W. G.: The Metabolism of N-Methylated Amino Acids. II. The Comparative Availability of *l*(—)-Tryptophane, *l*(+)- and *dl*-Amino-N-monomethyltryptophane for Growth. J. Biol. Chem. **129**, 309 (1939).

*102.* GORDON, W. G. and R. W. JACKSON: The Metabolism of Certain Monomethyl Tryptophanes. J. Biol. Chem. **110**, 151 (1935).

*103.* GOTS, J. S. and S. H. ROSS: The Accumulation of Indole-3-glycerol by Tryptophan Auxotrophs of *Escherichia coli.* Biochim. Biophys. Acta **24**, 429 (1957).

*104.* GRAY, P. H. H.: The Formation of Indigotin from Indol by Soil Bacteria. Proc. Roy. Soc. (London) **102** B, 263 (1927).

*105.* GRAY, R. A.: Preparation and Properties of 3-Indoleacetaldehyde. Arch. Biochem. Biophys. (in press).

*106.* GREENBERG, D. M.: Carbon Catabolism of Amino Acids. In: Chemical Pathways of Metabolism, Vol. II, p. 47. New York: Academic Press. 1954.

*107.* — Synthetic Processes Involving Amino Acids. In: Chemical Pathways of Metabolism, Vol. II, p. 113. New York: Academic Press. 1954.

*108.* GREENBERG, J. B.: Reactions of Possible Significance in the Synthesis of Indolic Auxins in Higher Plants. Thesis, Yale University, 1958.

*109.* GREENBERG, J. B., A. W. GALSTON, K. N. F. SHAW and M. D. ARMSTRONG: Formation and Auxin Activity of Indole-3-Glycolic Acid. Science (Washington) **125**, 992 (1957).

*110.* GRESHOFF, M.: Hoofdstuk. II. Eerste Bijdrage tot de Chemisch-Pharmacologische Kennis van Nederlandsch-Indische Leguminosen. 4. *Erythrina (Hypaphorus) subumbrans* HASSK. Mededeel. Lands Plantentuin, Buitenzorg **7**, 29 (1890).

*111.* — Onderzoek naar de Plantenstoffen: Sterculiaceae. Mededeel. Lands Plantentuin, Buitenzorg **25**, 36 (1898).

*112.* — Onderzoek naar de Plantenstoffen: Leguminosae, Erythrynia L. Mededeel. Lands Plantentuin, Buitenzorg **25**, 54 (1898).

*113.* — Onderzoek naar de Plantenstoffen. Urticaceae. Mededeel. Lands Plantentuin, Buitenzorg **25**, 175 (1898).

*114.* GRUEN, H. E.: Auxins and Fungi. Annu. Rev. Plant Physiol. **10** (1959), in press.

*115.* GUSTAFSON, F. G.: Tryptophane as an Intermediate in the Synthesis of Nicotinic Acid by Green Plants. Science (Washington) **110**, 279 (1949).

*116.* HAAGEN-SMIT, A. J., W. B. DANDLIKER, S. H. WITTWER and A. E. MURNEEK: Isolation of 3-Indoleacetic Acid from Immature Corn Kernels. Amer. J. Bot. **33**, 118 (1946).

*117.* HAAGEN-SMIT, A. J., W. D. LEECH and W. R. BERGREN: The Estimation, Isolation and Identification of Auxins in Plant Materials. Amer. J. Bot. **29**, 500 (1942).

*118.* HADDERS, M.: Systematische Verbreitung und Vorkommen der Indoxylglucoside. In: G. KLEIN, Handbuch der Pflanzenanalyse, Bd. III (2), S. 1062. Wien: Springer-Verlag. 1932.

*119.* HAPPOLD, F. C.: Tryptophanase-Tryptophan Reaction. Adv. Enzymology **10**, 51 (1950).

*120.* HARADA, T.: A New Bacterium in Urine which makes Lignin Red. V. The Relation between the Lignin Red Bacterium and A. I. Bacterium. J. Agric. Chem. Soc. Japan **23**, 96 (1949) [Chem. Abstr. **44**, 8419 (1950)].

*121.* HARDING, F.: Études sur le tryptophane. 1. Variation au cours de la germination. Arch. sci. physiol. 1, 193 (1947).
*122.* HAZEWINKEL, J. J.: Indican. Its Hydrolysis and the Enzyme Causing the Same. Kon. Ned. Akad. Wetensch. Proc. 2, 512 (1900).
*123.* HEMBERG, T.: Studies of Auxins and Growth-inhibiting Substances in the Potato Tuber and Their Significance with Regard to its Rest Period. Acta Horti Bergiani 14, 133 (1948).
*124.* HENBEST, H. B., E. R. H. JONES and G. F. SMITH: Isolation of a New Plant Growth Hormone, 3-Indolylacetonitrile. J. Chem. Soc. (London) 1953, 3796.
*125.* HENDERSON, J. H. M. and J. BONNER: Auxin Metabolism in Normal and Crown Gall Tissue of Sunflower. Amer. J. Bot. 39, 444 (1952).
*126.* HENDRICKSON, J. B.: Chemistry of Strychnine. In: R. H. F. MANSKE, The Alkaloids, Vol. VI. New York: Academic Press. 1959.
*127.* HERTER, C. A.: Note on the Occurrence of Skatol and Indol in the Wood of *Celtis reticulosa* (MIQUEL). J. Biol. Chem. 5, 489 (1909).
*128.* HESSE, A.: Über ätherisches Jasminblüthenöl. III. Ber. dtsch. chem. Ges. 32, 2611 (1899).
*129.* — Über ätherisches Jasminblüthenöl. IV. Ber. dtsch. chem. Ges. 33, 1585 (1900).
*130.* — Über ätherisches Jasminblüthenöl. VI. Ber. dtsch. chem. Ges. 34, 2916 (1901).
*131.* HESSE, A. und O. ZEITSCHEL: Über Orangenblüthenöl. II. J. prakt. Chem. 66, 481 (1902).
*132.* HOCHSTEIN, F. A. and A. M. PARADIES: Alkaloids of *Banisteria caapi* and *Prestonia amazonicum*. J. Amer. Chem. Soc. 79, 5735 (1957).
*133.* HOLLEY, R. W., F. P. BOYLE, H. K. DURFEE and A. D. HOLLEY: A Study of the Auxins in Cabbage Using Counter-Current Distribution. Arch. Biochemistry 32, 192 (1951).
*134.* HOOGEWERF, S. and H. TER MEULEN: Contribution to the Knowledge of Indican. Kon. Ned. Akad. Wetensch. Proc. 2, 520 (1900).
*135.* HOPKINS, F. G. and S. W. COLE: A Contribution to the Chemistry of Proteids. II. The Constitution of Tryptophan, and the Action of Bacteria upon It. J. Physiol. 29, 451 (1903).
*136.* HOSHINO, T.: Die Konstitution des Abrins. Liebigs Ann. Chem. 520, 31 (1935).
*137.* HOSHINO, T. und T. KOBAYASHI: Synthese des *d,l*-Eseräthols. Synthetische Versuche über Eserin. IV. Synthesen in der Indol-Gruppe. XIII. Liebigs Ann. Chem. 520, 11 (1935).
*138.* HOUSLEY, S. and J. A. BENTLEY: Studies in Plant Growth Hormones. IV. Chromatography of Hormones and Hormone Precursors in Cabbage. J. exp. Bot. 7, 219 (1956).
*139.* HURRY, J. B.: The Woad Plant and its Dye. London: Oxford Univ. Press. 1930.
*140.* HUSSONG, R. V. and S. QUAM: Relationship of the Consumption of Peppergrass by Cows to the Flavor and Indol Content of Butter. J. Dairy Sci. 26, 505 (1943).
*141.* INGLE, J. D.: Report of 31st Annual Meeting American Butter Institute, 1939. Dairy Sci. Abstracts 2, 256 (1940).
*142.* JACKSON, R. W.: Indole Derivatives in Connection with a Diet Deficient in Tryptophane. II. J. Biol. Chem. 84, 1 (1929).
*143.* JACQUOT, R. et F. HARDING: Les variations du tryptophane au cours de la germination de *Phaseolus multiflorus*. C. R. hebd. Séances Acad. Sci. 224, 1576 (1947).

*144.* JONES, E. R. H., H. B. HENBEST, G. F. SMITH and J. A. BENTLEY: 3-Indolyl-acetonitrile, a Naturally Occurring Plant Growth Hormone. Nature (London) **169**, 485 (1952).

*145.* JONES, E. R. H. and W. C. TAYLOR: Some Indole Constituents of Cabbage. Nature (London) **179**, 1138 (1957).

*146.* JULIAN, P. L., E. W. MEYER and H. C. PRINTY: The Chemistry of Indoles. In: R. C. ELDERFIELD, Heterocyclic Compounds, Vol. III, p. 1. New York: Wiley and Sons. 1952.

*147.* KAMIENSKI, E. S. v.: Untersuchungen über die flüchtigen Amine der Pflanzen. I. Methodik der Trennung und des Nachweises flüchtiger Amine. Planta **50**, 291 (1957/58).

*148.* — Untersuchungen über die flüchtigen Amine der Pflanzen. II. Die Amine von Blütenpflanzen und Moosen. Planta **50**, 315 (1957/58).

*149.* — Untersuchungen über die flüchtigen Amine der Pflanzen. III. Die Amine von Pilzen. Über den Weg der Aminbildung in Pflanzen. Planta **50**, 331 (1957/58).

*150.* KAPER, J. M.: Over de Omzetting van Tryptophaan door *Agrobacterium tumefaciens.* Proefschrift, Univ. Leiden, 1957.

*151.* KAPER, J. M. and H. VELDSTRA: On the Metabolism of Tryptophan by *Agrobacterium tumefaciens.* Biochim. Biophys. Acta **30**, 401 (1958).

*152.* KEFFORD, N. P.: The Growth Substances Separated from Plant Extracts by Chromatography. J. exp. Bot. **6**, 129 (1955).

*153.* — The Growth Substances Separated from Plant Extracts by Chromatography. II. The Coleoptile and Root Elongation Properties of the Growth Substances in Plant Extracts. J. exp. Bot. **6**, 245 (1955).

*154.* KENTEN, R. H.: The Oxidation of Phenylacetaldehyde by Plant Saps. Biochemic. J. **55**, 350 (1953).

*155.* KENTEN, R. H. and P. J. G. MANN: The Oxidation of Amines by Pea Seedlings. Biochemic. J. **50**, 360 (1952).

*156.* KÖGL, F., A. J. HAAGEN-SMIT und H. ERXLEBEN: Über ein neues Auxin („Hetero-auxin") aus Harn. Z. physiol. Chem. (Hoppe-Seyler) **228**, 90 (1934).

*157.* KÖGL, F. und D. G. F. R. KOSTERMANS: Hetero-auxin als Stoffwechselprodukt niederer pflanzlicher Organismen. Isolierung aus Hefe. Z. physiol. Chem. (Hoppe-Seyler) **228**, 113 (1934).

*158.* — — Über die Konstitutions-Spezifität des Hetero-auxins. Z. physiol. Chem. (Hoppe-Seyler) **235**, 201 (1935).

*159.* KRETZ, F.: Über den mikrochemischen Nachweis von Tryptophan in der Pflanze. Biochem. Z. **130**, 86 (1922).

*160.* KÜHN, H. und O. STEIN: Über Kondensationen von Indolen mit Aldehyden und sekundären Aminen, I. Mitt.: Eine neue Gramin-Synthese. Ber. dtsch. chem. Ges. **70**, 567 (1937).

*161.* KULESCHA, Z.: Recherches sur la transformation du tryptophane sous l'action des tissus de Topinambour. C. R. hebd. Séances Acad. Sci. **228**, 1304 (1949).

*162.* — Recherches sur l'élaboration de substances de croissance par les tissus végétaux. Thèse, Sorbonne, 1951.

*163.* KULESCHA Z. et R. J. GAUTHERET: Recherches sur l'action du tryptophane sur la prolifération des cultures de tissus de quelques végétaux. C. R. Séances Soc. Biol. **143**, 460 (1949).

*164.* — — Recherches sur l'action de la cynurénine sur les tissus de topinambour cultivés in vitro. C. R. Séances Soc. Biol. **145**, 245 (1951).

*165.* KUMMERT, E.: Über ätherisches Goldlackblütenöl. Chem.-Ztg. **35**, 667 (1911).

*165a.* Kutáček, M., M. Valenta und F. Icha: Untersuchungen über den Ascorbigen-gehalt von Kohlrabi während der Vegetation und den Zusammenhang zwischen Ascorbigen und Wachstum bei den Pflanzen der Familie *Brassicaceae.* Experientia **13**, 289 (1957).

*166.* Kyu-sui, C.: Über die im Tierkörper stattfindenden Veränderungen der an N substituierten Aminosäuren. I. Über die Entstehung von Kynurenin und Kynurensäure aus dem N-Methyltryptophan (Abrine) im Kaninchenorganismus. Z. physiol. Chem. (Hoppe-Seyler) **257**, 12 (1938/1939).

*167.* Lapière, C.: Les Alcaloïdes de L'*Erythrina tholloniana.* Bull. soc. chim. biol. (Paris) **31**, 862 (1949).

*168.* — Les Alcaloïdes de L'*Erythrina abyssinica.* J. Pharm. Belg. **6**, 71 (1951).

*169.* Larsen, P.: 3-Indole Acetaldehyde as a Growth Hormone in Higher Plants. Dansk. Bot. Ark. **11** (9), 1 (1944).

*170.* — Conversion of Indoleacetaldehyde to Indoleacetic Acid in Excised Coleoptiles and in Coleoptile Juice. Amer. J. Bot. **36**, 32 (1949).

*171.* — Enzymatic Conversion of Indoleacetaldehyde and Naphthaleneacet-aldehyde to Auxins. Plant Physiol. **26**, 697 (1951).

*172.* — Growth Substances in Higher Plants. In: K. Paech and M. V. Tracey, Modern Methods of Plant Analysis, Vol. III, p. 565. Berlin: Springer-Verlag. 1955.

*173.* Leete, E. and L. Marion: The Biogenesis of Alkaloids. IX. Further Investigations on the Formation of Gramine from Tryptophan. Canad. J. Chem. **31**, 1195 (1953).

*174.* — — The Biogenesis of Alkaloids. X. Origin of the N-Methyl Groups of the Alkaloids of Barley. Canad. J. Chem. **32**, 646 (1954).

*175.* Lewis, G. P.: 5-Hydroxytryptamine. London and New York: Pergamon Press. 1958.

*176.* Linser, H., H. Mayr und F. Maschek: Papierchromatographie von zell-streckend wirksamen Indolkörpern aus *Brassica*-Arten. Planta **44**, 103 (1954).

*177.* Lippmann, E. O. v.: Ein Vorkommen von Indol und Skatol. Ber. dtsch. chem. Ges. **49**, 106 (1916).

*178.* — Stickstoff-haltige Bestandteile von Rüben und Rübenprodukten. Ber. dtsch. chem. Ges. **57**, 256 (1924).

*179.* Louveau, C.: Produits de la fleur de Jasmin. Rev. marques parfum. et savon. **9**, 482 (1931).

*180.* — *Cheimonanthus fragrans.* Rev. marques parfum. et savon. **9**, 622 (1931).

*181.* Löwy, M.: Eine Reaktion auf Champignons. Chem.-Ztg. **33**, 1251 (1909) [Chem. Abstr. **4**, 622 (1910)].

*182.* — Der Champignon, eine indolbildende Pflanze. Chem.-Ztg. **34**, 340 (1910) [Chem. Abstr. **4**, 1755 (1910)].

*183.* Luckwill, L. C. and L. E. Powell, Jr.: Absence of Indoleacetic Acid in the Apple. Science (Washington) **123**, 225 (1956).

*184.* Lund, H. A.: The Biosynthesis of Indoleacetic Acid in the Styles and Ovaries of Tobacco Preliminary to the Setting of Fruit. Plant Physiol. **31**, 334 (1956).

*185.* Madinaveitia, J.: The Alkaloids of *Arundo donax* L. J. Chem. Soc. (London) **1937**, 1927.

*186.* Magaki, I.: Abrine Demethylase of *Escherichia coli*. Osaka Daigaku Igaku Zassi **7**, 359 (1955) [Chem. Abstr. **50**, 12147 (1956)].

*187.* — Abrine Demethylase of Rabbit Kidney. Osaka Daigaku Igaku Zassi **7**, 369 (1955) [Chem. Abstr. **50**, 12147 (1956)].

*188.* Major, R. J. and K. Folkers: Erythrina Alkaloid. U. S. Patent 2407713 Sept. 1946 [Chem. Abstr. **41**, 781 (1947)].

*189.* MANN, P. J. G.: Purification and Properties of the Amine Oxidase of Pea Seedlings. Biochemic. J. **59**, 609 (1955).

*190.* MANN, P. J. G. and W. R. SMITHIES: Plant Enzyme Reactions Leading to the Formation of Heterocyclic Compounds. 1. The Formation of Unsaturated Pyrrolidine and Piperidine Compounds. Biochemic. J. **61**, 89 (1955).

*191.* — — Plant Enzyme Reactions Leading to the Formation of Heterocyclic Compounds. 2. The Formation of Indole. Biochemic. J. **61**, 101 (1955).

*192.* MARAÑON, J. and J. K. SANTOS: Morphological and Chemical Studies on the Seeds of *Erythrina variegata* var. *orientalis* (L.) MERRILL. Philippine J. Sci. **48**, 563 (1932).

*193.* MARION, L.: The Indole Alkaloids. In: R. H. F. MANSKE and H. L. HOLMES, The Alkaloids, Vol. II, p. 371. New York: Academic Press. 1952.

*194.* MAYR, H. H.: Zur Photolyse von Indol-3-Essigsäure bei papierchromatographischen Arbeiten. Planta **46**, 512 (1956).

*195.* McCOY, T. A., T. H. SUBLETT and V. W. DOBBS: The Relation of the Amino Acid Composition to the Development of Oats. Plant Physiol. **28**, 77 (1953).

*196.* MEISTER, A.: Biochemistry of the Amino Acids. New York: Academic Press. 1957.

*196a.* MELCHIOR, G. H.: Über den Abbau von Indolderivaten. I. Photolyse durch ultraviolettes Licht. Planta **50**, 262 (1957/58).

*196b.* — Über den Abbau von Indolderivaten. II. Abbau durch ein Enzymsystem aus Weißkohl und ein Vergleich der Indolderivate in UV-bestrahlten und unbestrahlten Weißkohlpflanzen. Planta **50**, 557 (1957/58).

*197.* MEYER, J.: Die photolytischen Abbauprodukte der 3-Indolessigsäure und ihre physiologische Wirkung auf das Wachstum der *Avena*-Koleoptile. Z. Bot. **46**, 125 (1958).

*198.* MICHEL, B. E.: Growth Responses of Crucifers to Indoleacetic Acid and Indoleacetonitrile. Plant Physiol. **32**, 632 (1957).

*199.* MILES, P. G., H. LUND and J. R. RAPER: The Identification of Indigo as a Pigment Produced by a Mutant Culture of *Schizophyllum commune*. Arch. Biochem. Biophys. **62**, 1 (1956).

*200.* MILLER, E. J. and W. ROBSON: The Synthesis of r-α-Methylamino-β-3-indolylpropionic Acid. J. Chem. Soc. (London) **1938**, 1910.

*201.* MITOMA, C., H. WEISSBACH and S. UDENFRIEND: Formation of 5-Hydroxytryptophan from Tryptophan by *Chromobacterium violaceum*. Nature (London) **175**, 994 (1955).

*202.* MORTON, R. A. and N. I. FAHMY: Indole-3-aldehyde from Tissues. Nature (London) **182**, 939 (1958).

*203.* MOTHES, K.: Physiology of Alkaloids. Annu. Rev. Plant Physiol. **6**, 393 (1955).

*204.* MUDD, J. B. and S. ZALIK: The Metabolism of Indole by Tomato-plant Tissues and Extracts. Canad. J. Bot. **36**, 467 (1958).

*205.* MURAKAMI, Y. and T. HAYASHI: The Conversion of Tryptophan to Indoleacetic Acid by the Sap of Immature Kernels of Rice Plants. J. Agric. Chem. Soc. Japan **31** (7), 468 (1957).

*206.* NASON, A.: The Distribution and Biosynthesis of Niacin in Germinating Corn. Amer. J. Bot. **37**, 612 (1950).

*207.* — Effect of Zinc Deficiency on the Synthesis of Tryptophan by *Neurospora* Extracts. Science (Washington) **112**, 111 (1950).

*208.* NEGER, F. W.: Neue Methoden und Ergebnisse der Mikrochemie der Pflanzen. 1. Eine bequeme Reaktion zum Nachweis von Indigo in Pflanzen. Flora **116**, 323 (1923).

209. Nichols, R.: A Possible Source of Error in the Chemical Detection of Indolyl Acetic Acid in Plants. Nature (London) 181, 919 (1958).
210. Nitsch, J. P.: Le micro-dosage du L-tryptophane dans les plantes. C. R. Séances Soc. Biol. 145, 1809 (1951).
211. — Plant Hormones in the Development of Fruits. Quart. Rev. Biol. 27, 33 (1952).
212. — Free Auxins and Free Tryptophan in the Strawberry. Plant Physiol. 30, 33 (1955).
213. — Methods for the Investigation of Natural Auxins and Growth Inhibitors. In: R. L. Wain and F. Wightman, The Chemistry and Mode of Action of Plant Growth Substances, p. 3. London: Butterworths. 1956.
214. Nitsch, J. P. and C. Nitsch: The Separation of Natural Plant Growth Substances by Paper Chromatography. Beitr. Biol. Pflanzen 31, 387 (1955).
215. Nitsch, J. P. and R. H. Wetmore: The Microdetermination of "Free" L-Tryptophan in the Seedling of Lupinus albus. Science (Washington) 116, 256 (1952).
216. Okunuki, K.: Über ein neues Enzym Glutaminocarboxylase. Bot. Mag. (Tokyo) 51, 270 (1937).
217. Orechoff, A. P.: Chemistry of Alkaloids. Moscow: Akademia Nauk, S. S. S. R. 1955.
218. Orechoff, A. P. and S. S. Norkina: Alkaloids of Arundo donax L. J. Gen. Chem. (USSR) 7, 673 (1937).
219. Orechoff, A. P., S. S. Norkina und T. Maximowa: Über die Alkaloide von Arundo donax L. Ber. dtsch. chem. Ges. 68, 436 (1935).
220. Page, I. H.: Serotonin (5-Hydroxytryptamine); the Last Four Years. Physiol. Rev. 38, 277 (1958).
221. Painter, H. A. and S. S. Zilva: The Tautomeric Conversion of p-Hydroxyphenylpyruvic Acid. Biochemic. J. 41, 520 (1947).
222. Parks, L. W. and H. C. Douglas: N-Fructosyl Anthranilic Acid as a Possible Intermediate in the Synthesis of Indole by Saccharomyces. Biochim. Biophys. Acta 23, 207 (1957).
223. Perkin, A. G. and A. E. Everest: The Natural Organic Colouring Matters. London: Longmans Green and Co. 1918.
224. Platt, R. S., Jr. and K. V. Thimann: Interference in Salkowski Assay of Indoleacetic Acid. Science (Washington) 123, 105 (1956).
225. Plugge, P. C.: Tijdschrift voor inlandsche Geneeskundigen, Batavia 1, 933 (1893) [quoted by M. Greshoff, Mededeel. Lands Plantentuin, Buitenzorg 25, 61 (1898)].
226. — Onderzoek naar de Plantenstoffen. [Reported by W. G. Boorsma, Mededeel. Lands Plantentuin, Buitenzorg 31, 5 (1899).]
227. Procházka, Ž., V. Šanda and F. Šorm: On the Structure of Ascorbigen. Collect. Czech. Chem. Communs. 22, 654 (1957).
228. Proctor, M. H.: Bacterial Dissimilation of Indoleacetic Acid: A New Route of Breakdown of the Indole Nucleus. Nature (London) 181, 1345 (1958).
229. Rao, J. V.: Chemical Examination of Erythrina indica (white variety). Current Sci. (India) 14, 198 (1945).
230. Rao, P. S., C. V. Rao and T. R. Seshadri: Chemical Examination of Erythrina indica. Proc. Indian Acad. Sci. 7 A, 179 (1938).
231. Rawson, C.: The Cultivation and Manufacture of Indigo in Bengal. J. Soc. Chem. Ind. 18, 467 (1899).
232. Ray, P. M.: The Destruction of Indoleacetic Acid. II. Spectrophotometric Study of the Enzymatic Reaction. Arch. Biochem. Biophys. 64, 193 (1956).
233. — Destruction of Auxins. Annu. Rev. Plant Physiol. 9, 81 (1958).

234. RAY, P. M. and G. M. CURRY: Intermediates and Competing Reactions in the Photodestruction of Indoleacetic Acid. Nature (London) 181, 895 (1958).

235. REDEMANN, C. T., S. H. WITTWER and H. M. SELL: The Fruitsetting Factor from the Ethanol Extracts of Immature Corn Kernels. Arch. Biochem. Biophys. 32, 80 (1951).

236. ROMBURGH, P. VAN and G. BARGER: Preparation of the Betaine of Tryptophan and its Identity with the Alkaloid Hypaphorine. J. Chem. Soc. (London) 99, 2068 (1911).

237. ROSENTHALER, L.: Indoxylglucoside. In: G. KLEIN, Handbuch der Pflanzen-analyse, Bd. III (2), S. 1060. Wien: Springer-Verlag. 1932.

238. SACK, J.: I. Voorkomen van indol in bloemengeuren. II. Skatol in het hout van Nectandra Sp. Pharm. Weekblad Nederland 48, 307 (1911).

239. — Indol in Bloemengeuren. Pharm. Weekblad Nederland 48, 775 (1911).

240. SAITO, J.: Über den Einfluß der Konfiguration bei Indolbildung aus Indol-milchsäure durch Bakterien. Z. physiol. Chem. (Hoppe-Seyler) 214, 28 (1933).

241. SALKOWSKI, E.: Zur Kenntnis der Eiweißfäulnis, II: Die Skatolcarbonsäure, nach gemeinschaftlich mit H. SALKOWSKI in Münster i. W. angestellten Ver-suchen. Z. physiol. Chem. (Hoppe-Seyler) 9, 8 (1885).

242. SAXTON, J. E.: The Indole Alkaloids Excluding Harmine and Strychnine. Quart. Rev. Chem. Soc. (London) 10, 108 (1956).

243. Schimmel and Co.: Novelties. Semi-Annual Report, April/May 1903, p. 79.

244. — Neroli Oil (Oil of Bitter Orange-blossoms). Semi-Annual Report, Oct./Nov. 1903, p. 49.

245. — Semi-Annual Report, April 1914, p. 126.

246. — Bericht, Jubiläums-Ausgabe 1929, S. 70.

247. SCHLITTLER, E.: The Chemistry of Rauwolfia Alkaloids. In: R. E. WOODSON, Jr., H. W. YOUNGKEN, E. SCHLITTLER and J. A. SCHNEIDER, Rauwolfia: Botany, Pharmacognosy, Chemistry and Pharmacology, p. 50. Boston: Little, Brown Publ. 1957.

248. SCHOCKEN, V.: The Genesis of Auxin During the Decomposition of Proteins. Arch. Biochemistry 23, 198 (1949).

249. SCHULZE, E. und E. WINTERSTEIN: Über die aus den Keimpflanzen von Vicia sativa und Lupinus albus darstellbaren Monoaminosäuren. Z. physiol. Chem. (Hoppe-Seyler) 45, 38 (1905).

250. SCHWARZ, K.: Espectros U. V. dos Produtos da Decomposição Espontânea do Ácido Indolpirúvico. Arq. Inst. Biol. (São Paulo) 24 (6), 81 (1957).

251. SCHWARZ, K. e A. A. BITANCOURT: A Decomposição Espontânea de Alguns Derivados Indólicos. II. Ácido Indolpirúvico. Arq. Inst. Biol. (São Paulo) 24 (14), 183 (1957).

252. — — Paper Chromatography of Unstable Substances. Science (Washington) 126, 607 (1957).

253. SCHWARZ, K., R. DIERBERGER e A. A. BITANCOURT: Estudos sôbre o Câncer Vegetal. I. A Natureza Química das Auxinas de Alguns Tecidos Vegetais Normais e Tumorais. Arq. Inst. Biol. (São Paulo) 22, 93 (1955).

254. SEELEY, R. C., C. H. FAWCETT, R. L. WAIN and F. WIGHTMAN: Chromato-graphic Investigations on the Metabolism of Certain Indole Derivatives in Plant Tissues. In: R. L. WAIN and F. WIGHTMAN, Chemistry and Mode of Action of Plant Growth Substances, p. 234. London: Butterworths. 1956.

255. SHANMUGA SUNDARAM, E. R. B., G. RANGANATHAN and P. S. SARMA: Studies on the Interrelationship among Vitamins and Amino Acids. Influence of Desoxypyridoxine on the Biosynthesis of Nicotinic and Ascorbic Acid in Germinating Pulses. Current Sci. (India) 20, 122 (1951).

256. Shanmuga Sundaram, E. R. B. and P. S. Sarma: Studies on the Inter-relationships among Vitamins and Amino Acids. Part II. Influence of γ, 3 : 4 Ureylene Cyclohexyl Butyric Acid, Aminopterin and ω-Methyl Pantothenic Acid on the Biosynthesis of Nicotinic Acid in Germinating Green Gram (Phaseolus aureus). J. Sci. Ind. Res. (India) 13 B, 21 (1954).

257. Shanmuga Sundaram, E. R. B., M. O. Tirunarayanan and P. S. Sarma: Role of Biotin in the Conversion of Tryptophan to Nicotinic Acid. Current Sci. (India) 22, 211 (1953).

258. Shaw, K. N. F., A. McMillan, A. G. Gudmanson and M. D. Armstrong: Preparation and Properties of β-3-Indolyl Compounds Related to Tryptophan Metabolism. J. Organ. Chem. (USA) 23, 1171 (1958).

259. Shaw, K. N. F. and J. Trevarthen: Effect of Atmospheric Contaminants on Paper Chromatography of Urinary Indole and Phenol Acids. Nature (London) 182, 664 (1958).

260. — — Exogenous Sources of Urinary Phenol and Indole Acids. Nature (London) 182, 797 (1958).

261. Skoog, F.: A Deseeded Avena Test Method for Small Amounts of Auxin and Auxin Precursors. J. Gen. Physiol. 20, 311 (1937).

262. — Relationships Between Zinc and Auxins in the Growth of Higher Plants. Amer. J. Bot. 27, 939 (1940).

263. Smith, G. F.: Indoles. Part I. The Formylation of Indole and Some Reactions of 3-Formylindole. J. Chem. Soc. (London) 1954, 3842.

264. Sobotka, H., N. Barsel and J. D. Chanley: The Aminochromes. Fortschr. Chem. organ. Naturstoffe 14, 217 (1957).

265. Soden, H. v.: Über ätherische Öle, welche durch Extraktion frischer Blüten mit flüchtigen Lösungsmitteln gewonnen werden (ätherische Blütenextrakt-öle). II. J. prakt. Chem. 110, 273 (1925).

266. Söding, H. und E. Raadts: Über das Verhalten des Wuchsstoffes der Koleoptilenspitze gegen Säure und Lauge. Planta 43, 25 (1953/54).

267. Spear, I. and K. V. Thimann: The Effect of Onion Juice on the Growth Response to Auxin. Plant Physiol. 24, 587 (1949).

268. Spies, J. R. and D. C. Chambers: Chemical Determination of Tryptophan. Analyt. Chemistry 20, 30 (1948).

269. Spooner, R. C., H. L. Richardson, S. T. Tu, W. H. Yang and C. H. Wang: Indican Content of Szechwan Indigo and the Effect of Fertilizers. J. Chinese Chem. Soc. 10, 69 (1943).

270. Stanier, R. Y. and O. Hayaishi: The Bacterial Oxidation of Tryptophan: A Study in Comparative Biochemistry. Science (Washington) 114, 326 (1951).

271. Stehsel, M. L.: I. Interrelationships between Tryptophane, Auxin and Nicotinic Acid during Development of the Corn Kernel. II. Studies on the Nature of the Auxin Complex in the Corn Kernel. Thesis, Univ. of California, Berkeley, 1950.

272. Stehsel, M. L. and S. G. Wildman: Interrelations between Tryptophane, Auxin and Nicotinic Acid During Development of the Corn Kernel. Amer. J. Bot. 37, 682 (1950).

273. Stowe, B. B.: The Production of Indoleacetic Acid by Bacteria. Biochemic. J. 61, ix (1955).

274. — Growth Promotion in Pea Epicotyl Sections by Fatty Acid Esters. Science (Washington) 128, 421 (1958).

275. Stowe, B. B., P. M. Ray and K. V. Thimann: The Enzymatic Oxidation of Indoleacetic Acid. C. r. 8e Congrès Intern. Botan., Paris 1954, Sect. 11, 135 (suppl. vol. 1957).

276. STOWE, B. B. and K. V. THIMANN: The Paper Chromatography of Indole Compounds and Some Indole-containing Auxins of Plant Tissues. Arch. Biochem. Biophys. 51, 499 (1954).

277. STOWE, B. B., K. V. THIMANN and N. P. KEFFORD: Further Studies of Some Plant Indoles and Auxins by Paper Chromatography. Plant Physiol. 31, 162 (1956).

278. STROMBERG, V. L.: The Isolation of Bufotenine from *Piptadenia peregrina*. J. Amer. Chem. Soc. 76, 1707 (1954).

279. STUTZ, R. E.: Enzymatic Formation of Indole-3-carboxaldehyde from Indole-3-acetic Acid. Plant Physiol. 33, 207 (1958).

280. SUMPTER, W. C. and F. M. MILLER: Heterocyclic Compounds with Indole and Carbazole Systems. In: A. WEISSBERGER, Chemistry of Heterocyclic Compounds, Vol. VIII. New York: Interscience. 1954.

281. TABONE, J. et D. TABONE: Bio-estérification du glucose. V. Biosynthèse par *Bacillus megatherium* de l'ester β-glucosidique de l'acide indolpropionique. C. R. hebd. Séances Acad. Sci. 237, 943 (1953).

282. TANG, Y. W. and J. BONNER: The Enzymatic Inactivation of Indoleacetic Acid. I. Some Characteristics of the Enzyme Contained in Pea Seedlings. Arch. Biochemistry 13, 11 (1947).

283. TATUM, E. L. and D. SHEMIN: Mechanism of Tryptophan Synthesis in Neurospora. J. Biol. Chem. 209, 671 (1954).

284. TEAS, H. J. and A. C. NEWTON: Tryptophan, Niacin and Indoleacetic Acid in Several Endosperm Mutants and Standard Lines of Maize. Plant Physiol. 26, 494 (1951).

285. TERROINE, T.: Formation de l'acide nicotinique dans la germination; rôle du tryptophane. Arch. sci. physiol. 1, 445 (1947).

286. — Action rhizogène du tryptophane dans les phases initiales de la germination. Rev. gén. bot. 55, 249 (1948).

287. — Le cours de la synthèse nicotinique dans la germination. Existe-t-il un lien tryptophane-acide nicotinique? C. R. hebd. Séances Acad. Sci. 226, 511 (1948).

288. TERROINE, T. et J. DESVAUX-CHABROL: La synthèse de l'acide nicotinique au cours de la germination. Arch. sci. physiol. 1, 117 (1947).

289. TEUBNER, F. G.: Identification of the Auxin Present in Apple Endosperm. Science (Washington) 118, 418 (1953).

290. THIMANN, K. V.: On the Plant Growth Hormone Produced by *Rhizopus suinus*. J. Biol. Chem. 109, 279 (1935).

291. — Hydrolysis of Indoleacetonitrile in Plants. Arch. Biochem. Biophys. 44, 242 (1953).

292. — Studies on the Growth and Inhibition of Isolated Plant Parts. V. The Effects of Cobalt and Other Metals. Amer. J. Bot. 43, 241 (1956).

293. — Auxin Activity of Some Indole Derivatives. Plant Physiol. 33, 311 (1958).

294. THIMANN, K. V. and G. M. CURRY: Phototropism and Phototaxis. In: H. S. MASON and M. FLORKIN, Comparative Biochemistry, Vol. I. New York: Academic Press. 1959 (in press).

295. THIMANN, K. V. and A. C. LEOPOLD: Plant Growth Hormones. In: G. PINCUS and K. V. THIMANN, The Hormones, Vol. III, p. 1. New York: Academic Press. 1955.

296. THIMANN, K. V. and S. MAHADEVAN: Enzymatic Hydrolysis of Indoleacetonitrile. Nature (London) 181, 1466 (1958).

*297.* Thimann, K. V. and C. L. Schneider: The Role of Salts, Hydrogen-Ion Concentration and Agar in the Response of the Avena Coleoptile to Auxins. Amer. J. Bot. **25**, 270 (1938).

*298.* — — Differential Growth in Plant Tissues. Amer. J. Bot. **25**, 627 (1938).

*299.* Thomas, F., W. P. Bloxam and A. G. Perkin: Indican. Part III. Chem. Soc. (London) **95**, 824 (1909).

*300.* Towers, G. H. N., J. F. Thompson and F. C. Steward: The Detection of the Keto Acids of Plants. A Procedure Based on their Conversion to Amino Acids. J. Amer. Chem. Soc. **76**, 2392 (1954).

*301.* Tsuchihashi, R. and S. Tasaki: Essential Oil and Wax of Shuei Flowers (*Jasminium odoratissimum* L.). J. Chem. Ind. (Tokyo) **21**, 1117 (1918) [J. Soc. Chem. Ind. (London) **38** A, 117 (1919)].

*302.* Tsui, C.: The Role of Zinc in Auxin Synthesis in the Tomato Plant. Amer. J. Bot. **35**, 172 (1948).

*303.* Udenfriend, S., E. Titus and H. Weissbach: The Identification of 5-Hydroxy-3-indoleacetic Acid in Normal Urine and a Method for its Assay. J. Biol. Chem. **216**, 499 (1955).

*304.* Udenfriend, S., E. Titus, H. Weissbach and R. E. Peterson: Biogenesis and Metabolism of 5-Hydroxyindole Compounds. J. Biol. Chem. **219**, 335 (1956).

*305.* Veldstra, H.: The Relation of Chemical Structure to Biological Activity in Growth Substances. Annu. Rev. Plant. Physiol. **4**, 151 (1953).

*306.* Verschaffelt, E.: Une réaction permettant de déceler l'indol dans les parfums des fleurs. Rec. trav. bot. Néerl. **1**, 120 (1904).

*307.* Vickery, H. B., G. W. Pucher, R. Schoenheimer and D. Rittenberg: The Assimilation of Ammonia Nitrogen by the Tobacco Plant: a Preliminary Study with Isotopic Nitrogen. J. Biol. Chem. **135**, 531 (1940).

*308.* Virtanen, A. I., A. A. Arhimo, J. Sundman und L. Jännes: Vorkommen und Bedeutung der Oxalessigsäure in grünen Pflanzen. J. prakt. Chem. **162**, 71 (1943).

*309.* Virtanen, A. I. and T. Laine: Investigations on the Aminoacids of Plants. I. Tryptophan Content of Leguminous Plants at Different Stages of Growth. Biochemic. J. **30**, 1509 (1936).

*310.* — — Biological Fixation of Nitrogen. Nature (London) **142**, 165 (1938).

*311.* Waalkes, T. P., A. Sjoerdsma, C. R. Creveling, H. Weissbach and S. Udenfriend: Serotonin, Norepinephrine, and Related Compounds in Bananas. Science (Washington) **127**, 648 (1958).

*312.* Wagenknecht, A. C. and R. H. Burris: Indoleacetic Acid Inactivating Enzymes from Bean Roots and Pea Seedlings. Arch. Biochemistry **25**, 30 (1950).

*313.* Waters, K. L.: The Preparation of α-Alkoximino Acids and Their Derivatives. Thesis, Univ. of Maryland, 1945.

*314.* — The α-Keto Acids. Chem. Rev. **41**, 585 (1947).

*315.* Weehuizen, F.: Over Indol in Bloemen. Pharm. Weekbl. Nederland **45**, 1325 (1908).

*316.* — Über indoloide Düfte. Rec. trav. bot. Néerl. **8**, 97 (1911).

*317.* Weller, L. E., S. H. Wittwer and H. M. Sell: The Detection of 3-Indoleacetic Acid in Cauliflower Heads. Chromatographic Behavior of Some Indole Compounds. J. Amer. Chem. Soc. **76**, 629 (1954).

*318.* Werle, E. und A. Raub: Über Vorkommen, Bildung und Abbau biogener Amine bei Pflanzen unter besonderer Berücksichtigung des Histamins. Biochem. Z. **318**, 538 (1948).

*319.* WERLE, E. und F. ROEWER: Monaminoxydase in Pflanzen. Biochem. Z. **320**, 298 (1950).

*320.* WHITE, E. P.: Alkaloids of the Leguminosae. XIII. Isolation of Tryptamine from some *Acacia* Species. New Zealand J. Sci. Tech. **25** B, 157 (1944).

*321.* WICHMANN, A. W.: Kristallografische Aanteekeningen, Hypaphorin. Mededeel. Lands Plantentuin, Buitenzorg **25**, 197 (1898).

*322.* WIELAND, H., W. KONZ und H. MITTASCH: Die Konstitution von Bufotenin und Bufotenidin. Über Kröten-Giftstoffe. VII. Liebigs Ann. Chem. **513**, 1 (1934).

*323.* WIELAND, T. und C. Y. HSING: Synthese und Konstitution des Gramins. Liebigs Ann. Chem. **536**, 188 (1936).

*324.* WIELAND, T., W. MOTZEL und H. MERZ: Über das Vorkommen von Bufotenin im gelben Knollenblätterpilz. Liebigs Ann. Chem. **581**, 10 (1953).

*325.* WILDMAN, S. G., M. G. FERRI and J. BONNER: The Enzymatic Conversion of Tryptophane to Auxin by Spinach Leaves. Arch. Biochemistry **13**, 131 (1947).

*326.* WILDMAN, S. G. and R. M. MUIR: Observations on the Mechanism of Auxin Formation in Plant Tissues. Plant Physiol. **24**, 84 (1949).

*327.* WILKINSON, S.: 5-Methoxy-N-methyltryptamine: A New Indole Alkaloid from *Phalaris arundinacea* L. J. Chem. Soc. (London) **1958**, 2079.

*328.* WILTSHIRE, G. H.: The Oxidation of Tryptophan in Pea Seedling Tissues and Extracts. Biochemic. J. **55**, 408 (1953).

*329.* WOODS, D. D.: Indole Formation by *Bacterium coli*. I. The Breakdown of Tryptophan by Washed Suspensions of *Bacterium coli*. Biochemic. J. **29**, 640 (1935).

*330.* — Indole Formation by *Bacterium coli*. II. The Action of Washed Suspensions of *Bacterium coli* on Indole Derivatives. Biochemic. J. **29**, 649 (1935).

*331.* YAMAKI, T.: A New Method of Auxin Determination. Misc. Rep. Inst. Nat. Resources (Tokyo) **17/18**, 180 (1950).

*332.* YAMAKI, T. and K. NAKAMURA: Formation of Indoleacetic Acid in Maize Embryo. Sci. Papers Coll. Gen. Educ., Univ. Tokyo **2**, 81 (1952).

*333.* YANOFSKY, C.: The Enzymatic Conversion of Anthranilic Acid to Indole. J. Biol. Chem. **223**, 171 (1956).

*334.* YANOFSKY, C. and M. RACHMELER: The Exclusion of Free Indole as an Intermediate in the Biosynthesis of Tryptophan in *Neurospora crassa*. Biochim. Biophys. Acta **28**, 640 (1958).

*335.* YOSHIDA, T. and S. FUKUYAMA: Metabolic Changes in N-Substituted Amino Acids. VII. Demethylase. J. Biochem. Japan **34**, 429 (1941).

*336.* YOUSSEF, E. und O. KIERMEYER: Zur Photolyse von Indol-3-acetonitril. Planta **49**, 607 (1957).

*337.* YURASHEVSKII, N. K.: Alkaloids of *Girgensohnia diptera* BGE., Family Chenopodiaceae. II. J. Gen. Chem. (USSR) **10**, 1781 (1940).

*338.* — Alkaloids of *Arthrophytum leptocladum*, M. POP., Family Chenopodiaceae. J. Gen. Chem. (USSR) **11**, 157 (1941).

*339.* YURASHEVSKII, N. K. and S. I. STEPANOV: Alkaloids of *Girgensohnia diptera* BGE., Family Chenopodiaceae. J. Gen. Chem. (USSR) **9**, 2203 (1939).

*340.* ZIMMERMAN, P. W., A. E. HITCHCOCK and F. WILCOXON: Responses of Plants to Growth Substances Applied as Solutions and as Vapors. Contrib. Boyce Thompson Inst. **10**, 363 (1939).

*(Received, January 2, 1959.)*

# Some Biochemical Aspects of Disease in Plants.

By **A. E. Dimond**, New Haven, Connecticut.

## Contents.

## I. Introduction.

The purpose of this paper is to stimulate curiosity, not to satisfy it. The physiological and biochemical problems that confront the plant pathologist are legion. An improved understanding of the physiological processes involved in plant pathology inevitably leads to improvement in our ability to control plant diseases, which in the U. S. A. alone take a toll of some 3 billion dollars worth of crops annually.

Such studies improve our understanding of normal physiological functions. The history of human medicine contains many examples of how normal function came to be understood through a study of a pathological process. The function of insulin, determined through studies on diabetes, is an example. And now the role of the gibberellins in the growth of healthy plants has been discovered as a result of investigations begun years ago in Japan on the bakanae disease of rice seedlings.

The present paper, then, discusses a few cases where biochemical investigations have improved our understanding of plant pathology and plant physiology. Abnormal growth responses in plants resulting from the action of pathogenic bacteria and fungi are discussed, together with the role of enzymes and vivotoxins produced by pathogens in plant diseases. Our interest lies in how these cause disruption of normal processes in host tissue, such as dysfunction of the water-transporting system or blocking of normal biochemical pathways of the host plant. The topics discussed here are merely examples. A number of others could have been chosen equally well.

## II. Abnormal Growth in Plant Disease.

The growth and development of plants is often affected in plant disease. Sometimes the plant is merely stunted as a result of the commandeering of foods from the host by a pathogen. Frequently, however, growth effects are specific. Leaves may fall from the plant prematurely, owing to the development of an abscission layer at the base of the petiole (leafstalk), in response to products produced by the pathogen. The plant may flower prematurely or it may not flower at all. Some diseased plants are not deformed but are larger than normal. Other pathogens cause tumor formation, a result of abnormally rapid growth of localized cells. In other diseases, many growing points develop in a limited area, giving rise to the symptom known in some diseases as a "witches broom" and in others as "hairy root". These and many other symptoms result from abnormal growth in response to the formation of specific compounds by a pathogen or by a pathogen and host acting together. A few examples will be considered below.

### 1. Ethylene Production and its Effects.

Ethylene is a physiologically active compound in higher plants. Marigolds are highly sensitive to the compound, concentrations as low as 1 part per billion in air inducing epinasty, a downward curvature of leaves without wilting. Epinasty results from more rapid growth of the upper surface of the petiole than the lower surface. Other responses of higher plants to ethylene include accelerated ripening of fruits, the destruction of chlorophyll, dropping of leaves and the initiation of roots.

Ethylene has been found as a metabolic product of ripening fruit (*1, 33, 46*). It has also been detected in detached healthy rose and cotton leaves (*46*) and in healthy leaves, flowers and tubers of a variety of plants (*95*).

Injury increases ethylene production by living tissues. Thus, shredded rose and cherry leaves produce more ethylene than uninjured ones (*95*). HALL (*47*) has shown that about 20% more ethylene is produced by cotton leaves sprayed with two commercial defoliants, disodium 3,6-endoxo-hexahydrophthalate and the mixture sodium chlorate-sodium pentaborate. These defoliants are phytotoxic, and the excess ethylene apparently arises from the lesions caused by chemical treatment. The more ethylene produced, the more likely is defoliation to occur, other things being equal (47).

Leaf-spotting fungi cause from one to many lesions on a leaf surface, depending on the number of discrete infections that occur. The amount of ethylene arising from rose leaves infected with the black spot fungus, *Diplocarpon rosae*, and from cherry leaves infected with the cherry leafspot fungus, *Coccomyces hiemalis*, is considerably higher than of healthy, uninjured or even of healthy shredded rose and cherry leaves. Defoliation is a conspicuous symptom in these two diseases. No ethylene was detected from cultures of these two fungi, and the ethylene is, therefore, presumably a product of the injured host alone (95).

Some fungi, however, do produce ethylene in culture. *Penicillium digitatum*, which causes a decay of the stem end of citrus fruits is one of them (105).

Defoliation is not always or even frequently associated with leaf-spotting diseases. Leaf tissue probably varies among species of plants in the readiness with which it produces ethylene (95); and the increase in ethylene production that follows leaf injury presumably also varies, not only with the species of plant but also with the nature of the injury.

Thus, the rust fungi frequently produce leaf spots but defoliation seldom follows. In the rusts of hollyhock, mint, and ash, which are exceptional, defoliation occurs only when there are many lesions on a leaf. The powdery mildews are superficial obligate parasites and defoliation rarely is an attendant symptom. By contrast with the rusts and powdery mildews, a single lesion caused by the blackspot fungus on roses causes leaf fall. Some hosts, such as oak and bean, are rarely defoliated by leaf spotting diseases. Other hosts drop their leaves when attacked by one leaf disease, e. g., apple scab, but the leaves persist when attacked by other pathogens, e. g., fire blight of apple and cedar apple rust on apple.

Apparently, all possible combinations occur among the various leaf diseases on their respective hosts. The injured host may produce ethylene in response to the pathogen, as in rose black spot. The pathogen may produce ethylene in its own right. Leaves may be able to adhere to the plant, whether ethylene is produced or not.

In contrast with leaf-spotting fungi and bacteria, the vascular wilt diseases are caused by pathogens that live in the water-conducting tissues of the plant. If ethylene is produced, it can be translocated throughout the plant and produce other symptoms of ethylene injury. These are observed in some plant diseases. Epinasty is one of the first symptoms of Fusarium wilt of tomato (94). Later leaves turn yellow, though they do not fall. Occasionally adventitious roots develop in infected plants (32). Infected plants induce epinasty in healthy tomato plants confined with them, but healthy plants do not. The causal fungus, *Fusarium oxysporum* f. *lycopersici*, produces ethylene in culture (31). Ethylene, produced by the fungus and perhaps also by the tomato plant in response to the fungal invader, is apparently responsible for the epinasty observed in Fusarium-wilted tomato plants and for the yellowed leaves and the adventitious roots as well. Ethylene gas does not cause the abscission of leaves in the tomato plant (31).

In other diseases the symptoms suggest that ethylene is involved, but as yet its presence has not been demonstrated. Many examples could be given.

A few diseases in which defoliation is a conspicuous symptom are, angular leafspot of cotton, caused by *Xanthomonas malvacearum*, black rot of crucifers, caused by *X. campestris*, downy mildew of grapes, caused by *Plasmopara viticola*, and peach leaf curl, caused by *Taphrina deformans*. Diseases in which defoliation is preceded by yellowing of leaves and in which either epinasty or its equivalent occurs include Dutch elm disease, caused by *Ceratostomella ulmi*, Verticillium wilt of peppermint, Fusarium wilt of peas and cabbage yellows. In bacterial wilt of tomato, potato and other solanaceous crops, caused by *Pseudomonas solanacearum*, epinasty precedes wilting and adventitious roots develop on the stem over infected vessels of the tomato plant (92).

## 2. Gibberellin.

The bakanae or "foolish seedling" disease of rice illustrates how investigations of a pathological problem have led to the identification of a new series of naturally-occurring growth regulators and an improved understanding of how organized growth is regulated in plants. Only the highlights of this story will be mentioned here because this subject has been thoroughly reviewed by STODOLA (77) and by STOWE and YAMAKI (81).

The bakanae disease of rice is caused by the fungus *Gibberella fujikuroi*. Rice plants become infected at flowering or soon afterwards, so that rice grains carry the pathogen. In the Orient, losses from this disease may be very high. When seedlings are infected, they show no immediate symptoms of disease. However, about 30 days later, they are taller than healthy plants. Leaves are pale green and plants have the etiolated appearance of plants grown in dim light. The symptom of gigantism and etiolation is known as the "bakanae effect". Although the fungus attacks many other species of plants, including corn, cotton and sugar cane, only those strains of the fungus attacking rice produce the bakanae effect. When inoculated into other hosts, these strains produce the bakanae effect. Studies by KUROSAWA and later by SETO showed that the bakanae effect was obtained when rice seedlings were treated with fungus-free filtrates of culture media on which the pathogen had grown. This suggested the formation by the fungus of a compound capable of stimulating the growth of plants.

The active principle in culture filtrates was concentrated and found to be stable to heat and light, soluble in acetone, chloroform, ethanol, ether and water, capable of adsorption on charcoal and elution by acetone, alcohol or ether but not by water. The eluate produced the bakanae effect on a number of test plants.

Conditions favoring production by the fungus of this growth-stimulating substance were then thoroughly studied by the Japanese, by a team of investigators of Imperial Chemical Industries in England and by a group at the Northern Regional Research Laboratory in the U. S. A. As a result, the gibberellins are now produced by fermentation and are available in commercial quantities.

Simultaneously, studies on the chemical nature of the compounds responsible for the bakanae effect were begun by the Japanese, and prior to World War II, two crystalline compounds were isolated and called gibberellins A and B. A series of Japanese papers following this described the relation of these and other gibberellin-like compounds to one another, but unfortunately, these results were unavailable to western workers until after the end of the war.

The Japanese work then available stimulated the interest of both British and American authors in the gibberellins, and the isolation of gibberellic acid by both groups followed. Gibberellic acid (I) was converted to gibberene (III), which was identified as 1,7-dimethyl-fluorene, and its properties shown to coincide with a synthetic sample

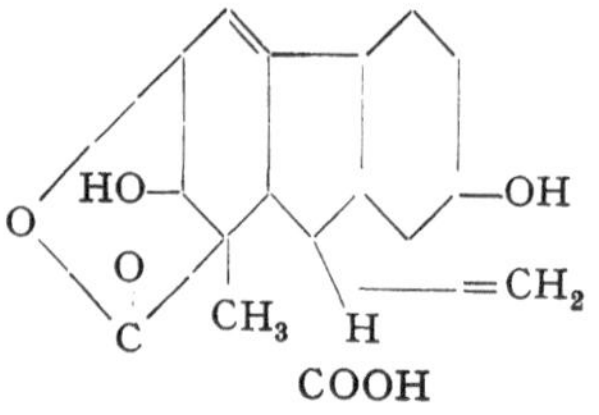

(I.) Gibberellic acid.

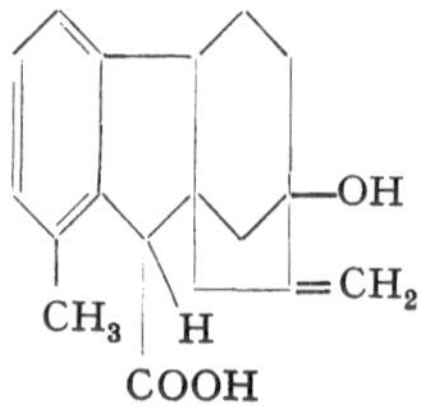

(II.) *allo*Gibberic acid.

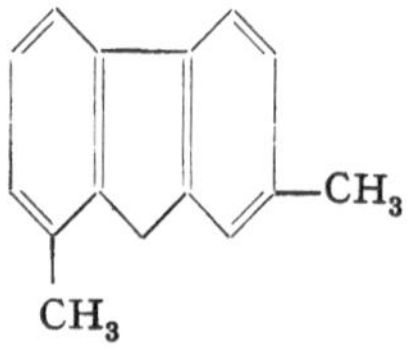

(III.) Gibberene.

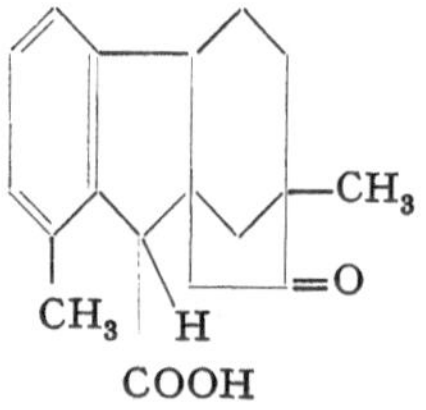

(IV.) Gibberic acid.

by Mulholland and Ward (66). Mulholland (65) subjected gibberellic acid to acid hydrolysis and obtained *allo*gibberic acid (II). Cross et al. (20–22) then proposed structures for gibberic (IV) and gibberellic (I) acids, which suggested that these compounds were unique in structure among naturally occurring plant products and that the growth effects in plants were produced by compounds having no chemical resemblance to the auxins.

A series of related compounds has been encountered by U. S. Department of Agriculture and by Japanese chemists, although the strains of the fungus and conditions of fermentation employed by the Imperial Chemical Industries workers were such that only gibberellic acid was produced. As a result the nomenclature of the compounds is somewhat fluid at the present time. A variety of gibberellin-like compounds is produced by different strains of the fungus.

The gibberellin-like compounds affect the growth of a very wide range of plants. The bakanae effect is no mere influence on hydration of cells; in fact gibberellic acid fails to affect the water uptake by pieces of potato tuber (*19*).

The dry weight of gibberellin-treated plants is sometimes higher than that of controls, although not invariably so (*17*). Effects of gibberellins observed on higher plants include increased cell elongation (*17*), increased pollen tube growth (*53*), promoted growth of genetic dwarf plants (*18*), altered leaf size (*48*), and retarded root growth (*17*). Treatment also affects cell division and induces earlier flowering (*60*), affects fruiting (*49*) and causes parthenocarpy (production of fruit without pollenation) at lower levels than auxins (*96*). Cloroplasts of treated plants contain less chlorophyll than normal plants do (*104*).

The physiological effects of gibberellin and of auxins are compared in *Table 1*.

Table 1. A Physiological Comparison of Gibberellin with Auxin.

| Effect | Gibberellin | Auxin |
|---|---|---|
| Root growth | No short term effect | Inhibited |
| Root initiation | Inhibited | Promoted |
| Epinasty* | No effect | Induced |
| Callus formation | No effect | Induced |
| Leaf and dwarf plant growth | Promoted | No effect |
| Parthenocarpy | Promoted | Promoted but less than by gibberellin |
| Hydration of cells | No effect | Promoted |
| Abscission** | No effect | Delayed |
| Dormancy | Broken | No effect |
| Apical dominance*** | Overcome | Promoted |
| Stem movement | Nonpolar | Polar† |

  * Epinasty: a downward curvature of petioles (leaf stalks) and leaves without wilting.
  ** Abscission: dropping of leaves.
  *** Apical dominance: dominance in growth of the terminal growing point over the lateral ones, giving the plant a growth habit similar to a fir tree in contrast with non-apically dominant growth, which habit is characteristic of maples.
  † Polar movement: movement in the stem in one direction only. Thus auxins move only from tip to base of the stem, whereas sugars show nonpolar movement and migrate either upwards or downwards in the stem with equal readiness.

Gibberellin occurs naturally in higher plants. Thus MacMillan and Suter (64) isolated pure gibberellin $A_1$ from developing seeds of runner bean. Gibberellin-like substances have been detected in extracts of wild cucumber (61) and in a variety of plants (67). The existence of genetic dwarfs in plants and their assumption of normal growth on application of gibberellin suggests their absence in the dwarfs and their presence in normal plants. Thus, the role of the gibberellins as naturally occurring growth-regulators and their occurrence generally in higher plants will probably follow eventually from the investigations begun with the bakanae disease.

### 3. Crown Gall Induction.

In the bakanae effect, growth is controlled and organized in a normal manner and the resulting tissues, if larger than usual, are still the result of an orderly growth process under the stimulus of a growth-promoting compound. In other diseases, growth is not organized and leads to abnormalities. The crown gall disease is such a case.

Caused by the bacterium *Agrobacterium tumefaciens*, the crown gall disease attacks many species of plants and produces galls on them in which the host cells undergo unlimited and uncontrolled growth. Only host cells potentially capable of further growth and cell division are affected. These burst into rapid and irreversible activity a few hours after the bacteria are present in tissues. Rapid growth will then continue even if the bacteria are eliminated. Here is a system permitting study of how growth is normally controlled and how growth can be released from control through the presence of factors introduced by the bacteria (9, 10, 12). Many aspects of this problem are still unsolved.

Cultures of *A. tumefaciens* vary in tumor-inducing ability. Some strains are avirulent and are unable to produce galls on any host. Other, highly virulent strains produce large, actively growing galls. Between these extremes are other strains that vary in virulence, and the size and rate of growth of galls resulting from them varies accordingly. Virulent bacteria have been attenuated by growing them on culture media containing glycine, these attenuated strains failing to produce galls ordinarily (84). However, in the presence of indoleacetic acid, attenuated bacteria induce galls in host tissues (13, 83). Avirulent cultures have been transformed to virulent ones by exposing them to material produced by virulent cultures. This transforming principle is a desoxyribonucleic acid. Only certain combinations of avirulent and virulent bacteria result in transformation of the avirulent strain. The transformed culture maintains its virulence permanently (55).

Host tissues may also vary in response to inoculation. For example, slices of about one carrot in ten inexplicably fail to respond to inoculation by gall formation (*25*). Two types of gall have long been noted in experimental studies and to some extent the type of gall depends upon the type of host tissue inoculated. One type of tumor is undifferentiated and contains rapidly dividing but undifferentiated cells. This type usually results when the host is inoculated toward the base of the stem. Another type of gall results when Turkish tobacco or Kalanchöe (a tropical shrub) is decapitated near the growing point and inoculated near the point of decapitation. The resulting gall is at first amorphous but later differentiates more and more, forming leaf- and bud-like structures. When placed in tissue culture, these galls continue to differentiate and can be grafted to growing plants. Such tissues eventually develop normally and recover from the abnormal growth (*7*).

How tissues are inoculated with crown gall bacteria alters the outcome of gall development. Unless host cells are potentially capable of growth, no gall will develop. On inoculated carrot slices, the cambium produces tumor tissue readily, the secondary phloem less so and the secondary xylem* still less so (*25*). Wounding of host tissue is essential for gall development, gall size being proportional to the volume of tissue affected by the wound at the time of inoculation (*71*). Gall induction occurs from the activity of the bacteria in intercellular spaces, however, and not from their presence in host cells (*71*). The size of a gall does not depend on the number of bacteria in the inoculum. Thus, a single bacterium introduced into host tissue can induce a gall, and the size of the gall depends on the depth of the wound (*51*).

Gall formation occurs in distinct stages. The *first phase* is a conditioning of host cells in response to wounding. Conditioning occurs whether crown gall bacteria are present or not. Thus, when potato, carrot or beet slices are incubated under moist conditions, the surface cells divide to form callus tissue, presumably in response to a wound substance, released from injured cells. Unless host cells are conditioned, bacteria are unable to induce galls. Also conditioning is temporary. Thus gall formation fails if crown gall bacteria are applied to the host surface as much as five days after it has been wounded (*71, 72*).

The relation between time of wounding of host tissue and responsiveness of cells to crown gall bacteria has been discussed by KLEIN (*57*) and by BRAUN (*9*).

Bacteria cause galls when applied to decapitated tomato stems o, 12 or 24 hours after wounding. Galls are smaller when applied 36 hours after wounding

---

* Xylem is the tissue functional in movement of water and mineral nutrients. Secondary xylem arises from cambium, as the stem increases in diameter. Phloem is the tissue in which sugars and other organic substances move in the plant. Secondary phloem also arises from division of cambial cells as the stem increases in girth.

and none develop if the bacteria are applied 48 hours after wounding. Moreover, beet discs which were washed 10 hours after slicing and then inoculated formed no galls, whereas discs in air formed both galls and callus tissues when inoculated 24 hours later. For washing times less than 10 hours, the size of gall developing on inoculation after washing was proportional to the time of washing (57). These time relations can be accounted for by the need for conditioned host cells (9, 14). Conditioning occurs slowly, attains a peak from two to three days after wounding and then declines as healing progresses (8). The process is not dependent upon temperature in the range from 25 to 32°, and occurs at about the time when cell division attending wound healing is most active (9).

The *second phase* is the transformation of normal host cells to tumefacient cells. If conditioning is inadequate, cells are not transformed, even if bacteria are present (8). Quite in contrast with the conditioning process, the transformation requires the presence of crown gall bacteria and is highly temperature-sensitive in the range from 25° to 32°. Thus, in tomatoes galls develop normally in inoculated plants incubated at 28°, develop poorly at 30° and not at all at 32° (72). In *Vinca rosea* the process is equally temperature-dependent, although the bacteria multiply in inoculated tissue at these temperatures (4). Virulent bacteria can be recovered from inoculated stems after 5 days at 32° but no galls develop (14). Even after one month at room temperature, virulent bacteria have been recovered but galls develop only when the tissues are wounded at this later time (57). The process of conversion of normal to tumefacient cells occurs in 36 to 48 hours (3).

The transformation process is caused by a tumor-inducing principle (3–5, 72). Antibiotics inhibit tumor formation when present during the early part of the transformation process (24). Thus, the tumor-inducing principle is produced by the crown gall bacteria. It is inactivated at 32° but not at 25°.

Large tumors form when bacteria act for period of 4 to 6 hours, totalling 30 hours at 25°, alternating with one hour exposure at 30°. By incubating Kalanchöe plants at temperatures ranging from 25° to 30° at 0.1° intervals, BRAUN (5) showed that measurable inactivation of the tumor-inducing principle is limited to a 2° interval and that no tumors were initiated above 29°. The activation energy obtained by applying the Arrhenius equation was estimated to be 80,000 cal. per mole, a value suggestive of the denaturation of a protein or of some similarly complex structure.

Host cells show a rise in desoxyribonucleic acid to a maximum 24 hours after inoculation with crown gall bacteria, a peak that is maintained for an additional 24 hours. Levels of desoxyribonucleic acid then decrease (56). Wounding of plants or inoculation of plants held at 30° with virulent bacteria results in no increase or only a slight increase of desoxyribonucleic acid during the period critical to transformation (57). Consequently, KLEIN (57) has suggested that the tumor-inducing principle is a polymerized desoxyribonucleic acid, produced by the crown gall bacteria in a wound sap medium.

Transformation of cells occurs gradually over a period involving a considerable number of hours, as little as one day if host cells are previously conditioned; and the rate of growth of gall tissue has been shown to depend upon the time allowed for the transformation process to occur. No galls are formed when host cells are exposed for but 24 hours to bacteria, but in the period from 24 to 60 hours exposure, the galls that result grow at a rate proportional to the time allowed for transformation (9, 14). The nature of the gall formed is also a result of the degree of transformation (12). Moreover, these galls, on subsequent transplanting to tissue culture or to a new host, maintain this growth rate as an enduring characteristic. The degree to which cells are transformed, then, controls the growth rate of the resulting tumors, and this can be controlled experimentally by eliminating the bacteria or the tumor-inducing principle from tissue after stated times of contact through appropriate heat treatment.

The *third phase* of tumor development involves the proliferation of transformed host cells in tumor development. This phase does not require the presence of bacteria. Thus, at temperatures as high as 47° for five days in a heat-tolerant host, the bacteria can be killed in gall tissue without serious damage to the host, yet the galls continue to grow as cell division continues (3, 4). Moreover, old crown galls in nature are often free of bacteria. Also, secondary tumors on sunflower and other hosts arise naturally at a point removed from the primary gall and these are frequently free of bacteria, yet host cells divide in them in an uncontrolled manner (16).

At this stage of gall development, growth-promoting factors in host tissue are functional (10, 73). Greater amounts of auxin have been reported in rapidly growing galls than in slow growing ones, in cultures of virulent strains than in avirulent ones. The ability of auxin to cause attenuated bacteria to be gall-forming has lent substance to this view (13, 83). Attenuated bacteria remain so after re-isolation from such galls. Avirulent strains of the bacteria are unaffected by auxin. When auxin is applied to cut ends of stems inoculated with attenuated bacteria and removed again after stated times, auxin does not promote tumors if removed before 36 hours. Tumor size is proportional to the time of exposure to auxin between 36 and 72 hours, but after 72 hours continued exposure to auxin has no further effect. When auxin is applied 15 days after inoculation of host tissues with attenuated bacteria, galls are formed (58). Thus, the transformed cell remains so for some time (57). The auxin at this stage must be present in the host in higher than regulatory amounts (9).

That auxin suppression in the plant prevents gall development is indicated from two lines of evidence. Application of maleic hydrazide, a well-known growth inhibitor, inhibits the development of crown gall tumors on carrot slices and on tomato plants (*86, 91*). Also, dosages of ionizing radiation that do not affect the pathogenicity of the crown gall bacteria prevent gall development in host tissues (*85*). Auxin synthesis in the plant is stopped by these same dosages of radiation (*41, 76*).

Crown gall tumor tissue contains chemical factors that influence both cell enlargement and cell division (*12, 15, 50*). When cells become transformed, they acquire the ability to produce these factors in sufficient quantity that they are no longer limiting (*12*).

Clearly, there are major biochemical problems to be solved in the crown gall problem. Among them are the nature of the materials present in the sap of wounded cells, the nature of the tumor-inducing principle and how it stimulates desoxyribonucleic acid formation for a limited time in host tissues, during which time cells are transformed to tumor cells. The implication of these problems for solving the riddle of unlimited growth wherever it occurs is obvious.

## III. Abnormal Metabolic Pathways in Disease.

Growth is an energy-consuming process and, when it has been altered by disease, metabolic processes are altered also. In crown gall, the host as well as the pathogen contributes to the pathological relationship. In other diseases the pathological process is clearly a one-sided affair. When a single metabolic process is disturbed, growth of the plant may be unaffected and symptoms of disease may be uniquely a result of a simple metabolic block. "Wildfire" of tobacco is an example of such a situation.

Tobacco wildfire causes lesions on leaves, where the causal bacterium, *Pseudomonas tabaci*, has entered. The lesions characteristically are surrounded by a chlorotic halo*. In contrast, the bacterium *P. angulatum*, cause of angular leafspot of tobacco, produces lesions lacking this halo. These two bacteria are indistinguishable from one another, even serologically, except in the symptoms they produce in the host and in one other respect (*2*).

In contrast with *P. angulatum*, *P. tabaci* produces in culture media a heat-stable toxic principle that causes haloing in tobacco leaves. Growth of the green alga *Chlorella vulgaris* is inhibited by this toxin.

---

\* Chlorotic halo: At the point of invasion of the bacteria a small spot of dead tissue appears. Surrounding this is a yellow (chlorotic) zone from which the green chlorophyll has disappeared. This yellow ring around the necrotic point of invasion is referred to as a chlorotic halo.

*References, pp. 316—321.*

However, the addition of liver extract to the culture medium in which the alga is grown overcomes the action of the toxin. Components of liver and yeast extract have no effect on the action of the toxin, except for *l*-methionine, which counteracts the growth-inhibiting effect of the toxin on *Chlorella*. Methionine sulfoximine, a known inhibitor of methionine, inhibits the growth of *Chlorella* and also produces haloing of tobacco leaves that is indistinguishable from that caused by the wildfire toxin (6).

Although the action of the pure toxin is reversed on *Chlorella* by methionine, the haloing in tobacco leaves is not. Neither can the action of methionine sulfoximine (V) in inducing chlorotic halos in tobacco leaves be reversed by methionine (VI). Mutants of *Chlorella* that are resistant to the action of methionine sulfoximine are also resistant to the wildfire toxin (*11*).

Chemical investigation has revealed the wildfire toxin as a derivative of a new α-amino acid. Acid hydrolysis of the toxin yields lactic acid and α,ε-diamino-β-hydroxypimelic acid to which the name tabtoxinine (IX) has been applied (*99, 100*). Tabtoxinine is inactive as a toxin, but is related to the α,ε-diaminopimelic acid isolated from diphtheria and tubercle bacilli (*102*). The wildfire toxin (VII) is considered to be the lactone of α-lactylamino-β-hydroxy-ε-aminopimelic acid. The toxin also undergoes inactivation in dilute alkali (*101*). .

The structural analogy between the wildfire toxin, tabtoxinine, methionine sulfoximine and methionine is evident from comparison of

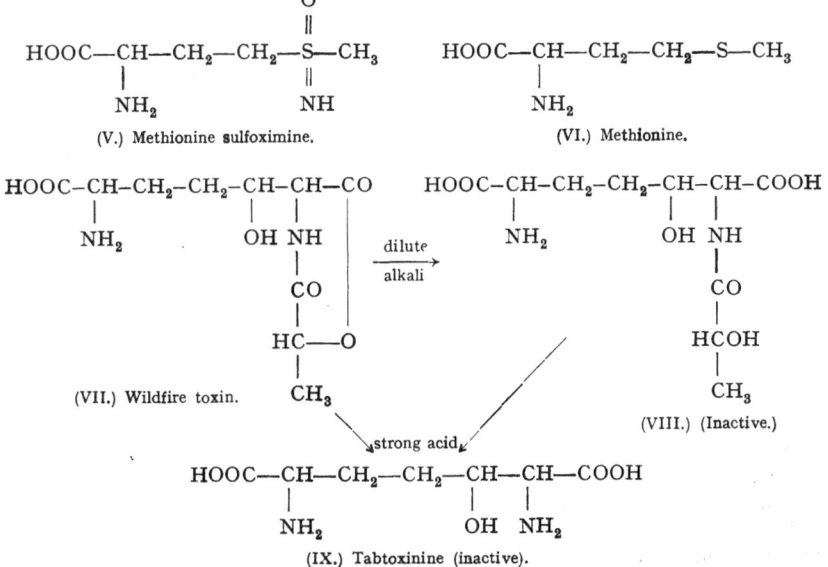

(V.) Methionine sulfoximine.

(VI.) Methionine.

(VII.) Wildfire toxin.

(VIII.) (Inactive.)

(IX.) Tabtoxinine (inactive).

the formulas (V), (VI), (VII), and (IX). Both methionine and methionine sulfoximine contain sulfur atoms and the analogy is straightforward. The wildfire toxin and tabtoxinine instead contain the group —CHOH—CH=. The biologically isosteric* nature of —S— and —C=C— is known to exist in fungitoxic molecules (52) and, quite evidently, is applicable here as well in a modified form.

*l*-Methionine, in counteracting the effect of the toxin, does not neutralize it, but rather reverses its action. The action of the toxin is based on a blocking of methionine utilization, rather than on inhibition of methionine synthesis (6). Methionine is a competitive inhibitor of the wildfire toxin (*11*).

## IV. Abnormal Water Economy.

Fungi and bacteria often produce compounds in culture that are phytotoxic, i. e., toxic to higher plants. More rarely they form compounds in the host that produce only a portion of the disease syndrome. Because the toxic components of culture filtrates on which pathogens have grown have so frequently been called "toxins" and assumed to damage the diseased host, DIMOND and WAGGONER (28) have proposed the term "vivotoxin" to describe compounds that injure the diseased plant and are produced by the pathogen during its course of attack of the host. In addition to vivotoxins, pathogens often liberate enzymes that attack living host tissue. Enzymes need no special name to designate them and are excluded from the term vivotoxin.

The Fusarium wilt of tomato serves to illustrate these concepts (26). Certain toxins have been postulated to have a role in this disease, but their role remains unproven. Also vivotoxins have been demonstrated in tomato plants attacked by *Fusarium*. Moreover, enzymes are now known to be responsible for certain phases of the attack of tomatoes by this pathogen. In many other plant diseases similar relations exist and Fusarium wilt of tomato will serve as a model to illustrate the role of vivotoxins and enzymes in plant disease.

Fusarium wilt of tomato is caused by a soil-dwelling fungus, *Fusarium oxysporum* f. *lycopersici*. This fungus invades the roots of tomato plants and promptly grows through root tissue toward the xylem, the tissue of root, stem and leaf having the function of water transport. Functional xylem cells are dead and lack a protoplast, and are merely water-filled

---

* Isosteric is used here in the chemical sense and refers to the action of one molecule in a manner similar to another having the same shape. In biological reactions a toxic or otherwise active molecule produces an effect similar to or identical with another molecule having the same shape and possessing the same terminally reactive groupings.

*References, pp. 316—321.*

capillary tubes, surrounded by living xylem parenchyma. Once the mycelium of *Fusarium* gains entry into xylem tubes, it remains there, growing up the root, into the stem and eventually into leaf petioles. In addition to the epinasty described earlier, the symptoms of this disease include a wilting of leaves that is at first mild and later severe. The oldest leaves wilt first and the progression of wilting is orderly in the plant, from oldest to youngest leaves, over a period of several weeks. An attendant symptom is a discoloring of xylem tissue, from brown to black, that is in marked contrast to the normal buff color of healthy vascular bundles. Severely wilted leaves become withered, dry and brittle. This is a consequence of acute water shortage (*63, 75*). Occasionally leaves of diseased plants show localized lesions and withering before the leaf tissue collapses from wilting, but this symptom is infrequent.

Toxins have been said to be the cause of wilting. When first proposed, this conveniently accounted for how diseased leaf tissue could be affected by a pathogen that does not invade the leaves beyond the petioles. Later workers frequently observed components in culture filtrates that were toxic to tomato cuttings. Most of this work, however, did not relate the appearance of symptoms in the diseased plant to the components in culture filtrates. GOTTLIEB (*44, 45*) displaced the tracheal fluid of Fusarium-wilted and healthy tomato plants and showed that a toxic material was present in diseased plants that increased the permeability of plant cells. This suggested a toxin was present, although its nature was not determined.

## 1. Lycomarasmin.

Examination of spent culture medium on which *Fusarium* had grown revealed the presence of a toxin, lycomarasmin, that has profound effects on the permeability of leaf cells in low concentration (*39*). Lycomarasmin was proposed as the toxin responsible for wilting of tomato plants attacked by *Fusarium* (*34*).

PLATTNER and CLAUSON-KAAS (*68, 69*) purified lycomarasmin and determined a number of its properties, including its empirical formula and molecular weight and found the hydrolytic products to be glycine, aspartic acid, pyruvic acid and ammonia. WOOLLEY (*97*) later investigated the nature of lycomarasmin and proposed the structure α-hydroxy-α-acetaminopropionyl-glycylasparagine (X, p. 312). This compound was synthesized and shown to coincide in its biological and chemical properties with lycomarasmin (*98*).

Lycomarasmin has never been detected in plants invaded by *Fusarium*, and while produced by the fungus in culture, no evidence has yet appeared to indicate that it is involved in damaging the diseased plant.

The amount of lycomarasmin necessary to damage leaf tissue has been carefully determined (*34, 39*). The rate at which lycomarasmin is produced in culture is insufficient to account for the rate at which infected plants succumb to wilting (*27, 74*). In culture, lycomarasmin production by

$$
\begin{array}{lll}
\text{H}_2\text{NOC—CH}_2 & & \text{CH}_3 \\
\quad\quad\quad | & & \quad | \\
\text{HOOC—CH—NH—CO—CH}_2\text{—NH—COH} \\
& & \quad | \\
& & \text{COOH}
\end{array}
$$

(X.) Lycomarasmin.

*Fusarium* apparently results from lysis of mycelium and increases as the culture ages (*27*). Lycomarasmin is absorbed by plant cuttings and becomes distributed in them uniformly (*39*). If it were the cause of wilting, symptoms should appear over the plant as a whole, rather than progressing in a regular fashion from oldest to youngest leaves. Lycomarasmin irreversibly increases the permeability of leaf cells (*34*), but even severely wilting tomato leaves from infected plants recover from wilting when floated on water (*30*).

Lycomarasmin chelates metals, and its toxicity to plant tissue is correlated with this property (*36, 87*). The unchelated molecule is not toxic and neither is the copper chelate (*87*). The iron complex is uniquely toxic (*68*). On mild heating or long standing at room temperature, the asparagine amide-N of lycomarasmin splits off, liberating ammonia. When this happens, the compound loses both chelating ability and toxicity (*29, 87*). If the iron chelate of lycomarasmin is involved in the disease, then Fusarium-infected plants should be less severely attacked if iron-deficient, but they are not (*75*). In fact, in acutely iron-deficient plants, pure lycomarasmin aggravates symptoms of iron deficiency, not of wilting or withering (*38*). Pure lycomarasmin, when mixed with another chelating agent, 8-quinolinol sulfate, is not toxic to tomato plants when iron is limiting, because the two chelating agents compete for what iron is available in the system, and the stronger chelating agent sequesters most of the metal. Also the iron complex of lycomarasmin can be rendered non-toxic in the presence of copper ion by virtue of the greater stability of the copper complex than of the iron complex. If lycomarasmin is present in *Fusarium*-infected tomato plants, the administration of 8-quinolinol sulfate or of copper sulfate should retard the rate of symptom development. When applied to the roots of different lots of infected plants, both 8-quinolinol sulfate and copper ion were detected in leaves, but neither affected the rate of symptom development (*87*). Thus lycomarasmin has not yet been detected in diseased plants and its action as a vivotoxin remains undetermined at present.

## 2. Fusaric Acid.

Fusaric acid is a second toxin produced by *F. oxysporum* f. *lycopersici* in culture and it has been investigated in connection with the Fusarium wilt of tomatoes (*40*). Fusaric acid was first described and characterized in connection with the bakanae disease of rice, when *Gibberella fujikuroi* was found to produce a compound that stunted growth of rice seedlings while it also produced gibberellin. The inhibiting compound was isolated in crystalline form and tentatively assigned two alternative structures by YABUTA, KAMBE and HAYASHI (*103*), one of them being correct. The structure of fusaric acid (XI) was proven by synthesis to be 5-*n*-butyl-picolinic acid (*70, 82*). *Gibberella fujikuroi* and *F. oxysporum* f. *lycopersici* also produce dehydrofusaric acid (*78, 80*), the structure of which has been established by ozonolysis to be 5-*n*-butyl-3'-enyl-picolinic acid (XII) (*79*). Fusaric acid is produced by a number of fungi in culture, including *Fusarium heterosporum*, *F. vasinfectum* and *F. orthoceros* in addition to those mentioned above.

Fusaric acid has been investigated as a wilting toxin in tomato wilt and in other wilt diseases caused by pathogens known to produce it (*35*). Conditions of fusaric acid production in culture are consistent with its acting as a vivotoxin in diseased plants. It is a chelating agent and produces necrotic lesions in interveinal areas of leaves similar to those caused by lycomarasmin (p. 312). Fusaric acid destroys the differential permeability of leaf cells (*37*). The compound has been detected in

$$HOOC-\text{(ring)}-CH_2-CH_2-CH_2-CH_3$$

$$HOOC-\text{(ring)}-CH_2-CH_2-CH=CH_2$$

(XI.) Fusaric acid.        (XII.) Dehydrofusaric acid.

Fusarium-infected but not in healthy plants (*54, 59*). Thus, fusaric acid has proven to be a vivotoxin in Fusarium wilt of tomato and functions in symptom development (*35*).

## 3. Pectic Enzymes.

How much of the syndrome of *Fusarium*-wilted plants is accounted for by the action of ethylene and of fusaric acid, respectively? Together, these will account for leaf-yellowing, epinasty and the necrotic lesions that occasionally appear on infected plants. They will account in part for the earlier flagging of leaves on a sunny day. But there are other symptoms, in fact the diagnostic symptoms, that have not been accounted for by the presence of these compounds in infected plants.

Leaves of infected plants become severely wilted before the plant succumbs. The vascular bundles are discolored as a consequence of disease. The stems and leaf petioles of infected plants are not only discolored; they transmit water very poorly, even under pressure. Dye solutions that readily pass through healthy stem tissue can hardly be forced through infected stems. The stem is apparently plugged and the plugs are stainable with ruthenium red, which has been interpreted as suggesting that the plugs are of pectic materials (*63*, *75*). However, the liquid content of the vascular bundles of infected stems has the same viscosity as that of water (*30*, *90*).

That reduced water flow through infected stems cannot result from the interference resulting from the presence of fungal mycelium, is apparent when the system is considered as a hydraulic one. By building a hydraulic model to scale and studying liquid flow through it at the same Reynolds number, the flow of water in plant stems was found to be laminar and not turbulent (*88*). Even if all xylem vessels contained mycelium, the fungus could contribute to, but not be a primary cause of, water shortage in infected plants. Leaves on infected plants become drought-hardened as a result of a gradual but ever-increasing water-shortage (*30*).

In culture, *Fusarium oxysporum* f. *lycopersici* produces two pectic enzymes, polygalacturonase and pectin methylesterase (*42*, *89*, *90*). Although some workers have reported the pectin-hydrolyzing enzyme produced by *Fusarium* to be a pectin depolymerase (instead of a polygalacturonase) (*42*, *43*, *93*), this distinction is no impediment to the interpretation of what happens in the diseased stem. Polygalacturonase has not been found in healthy tomato plants. Pectin methylesterase of fungal origin has been detected in diseased plants by virtue of its inactivation by surface active agents, whereas the pectin methylesterase in healthy plants is unaffected (*89*, *90*).

In the xylem vessels, that conduct water through stems and petioles, the cell wall is predominantly of cellulose and lignin. However, cell walls also contain pectins which are exposed where pits occur in these vessels, thus providing a substrate for the fungal pectic enzymes. Pectic enzymes, then, are freed by the fungus growing in these cells and, because they are present only in low concentration, their action on the substrate ceases before hydrolysis proceeds to the galacturonic acid stage. The polygalacturonic acid so liberated forms calcium salts in the presence of calcium ion, present in the transpiration stream, and this results in gel formation that plugs the stem (*26*, *89*, *90*).

The pectic enzymes have a second role in symptom production. By their action at vessel pits, these enzymes produce a mild maceration of cells through their action on the pectic middle lamella that cements cells together.

At this stage, another fungal enzyme enters the picture.

*F. oxysporum* f. *lycopersici* will grow on either salicin or tannic acid as a sole carbon source; therefore, it must produce $\beta$-glucosidase with which it liberates glucose from the glycoside. $\beta$-Glucosidase can be detected abundantly in culture filtrates. This enzyme is present in the tracheal fluid of diseased but not of healthy tomato stems. The initial vascular browning so characteristic of Fusarium-infected plants can be reproduced by allowing them to absorb gallic acid, pyrogallol, tannic acid or other polyhydric phenols (23). This discoloration and that in diseased plants first appears in living xylem parenchyma cells, rather than in the lumina of dead xylem vessels. Tomato stems are known to contain polyphenol oxidase (62).

Thus, the initial vascular discoloration of xylem parenchyma results, first from the macerating action of fungal pectic enzymes that pave the way for entry of fungal $\beta$-glucosidase into xylem parenchyma. Phenolic glucosides in these cells are then hydrolyzed and the resulting phenols become available as substrates for phenol oxidase, a normal component of the host cells. These are oxidized to quinones and, because the host cells are somewhat damaged by the action of both types of fungal enzymes, the quinones, instead of cycling in the polyphenol oxidase system as they normally do, are polymerized to dark-colored melanoid pigments. On accumulating, these pigments color the xylem parenchyma dark brown. As maceration of host cells proceeds, some of these pigments leave living host cells and become entrapped in the pectic gels that plug the water-conducting vessels. At this later stage the plugs in the xylem are visibly dark-colored.

Thus, symptoms of Fusarium wilt are a result of the action of several systems, some of fungal and some of host origin. The epinasty and yellowing of leaves result from ethylene, the lesions on leaves and to some extent the downward curvature of petioles results from fusaric acid, the plugging of stems and consequent wilting of leaves is caused by fungal pectic enzymes, and the fascular browning appears after the $\beta$-glucosidase of the fungus has entered host cells as a result of the action of the pectic enzyme (26).

# V. Conclusion.

In many other plant diseases biochemical and physiological relationships of pathogen to host are recognized and some progress has been made toward determining the exact nature of the processes involved. Soft rots of potatoes and fruits, dry rots of wood, and the rots of roots of living plants, all involve hydrolytic enzymes attacking pectins, cellulose or lignin. In such diseases as the rusts and powdery mildews there is a burst of respiration of host tissue as the pathogen begins to sporulate. Progress is being made toward explaining how respiration of host tissue is altered by the pathogen during its attack. Space does not permit a discussion of these aspects of the subject. Rather the discussion here has been deliberately limited to a few cases that are well investigated and in which pathogenesis is reasonably well understood.

The production of compounds by the pathogen that interfere with the normal functioning of the host may lead to discrete symptoms. When ethylene is produced either by the pathogen itself or as a result of the damage it causes in the host, epinasty and defoliation occur. Some pathogens produce vivotoxins in their hosts that interfere with respiration, with specific metabolic pathways or with the differential permeability of leaf cells. The interaction of host and pathogen in the case of crown gall, if complex, is of fundamental significance to an understanding of how growth is normally controlled in plant tissues.

A study of plant diseases has contributed to plant physiology in a straightforward manner. The discovery of the physiological action of the gibberellins and their role in normal growth is a direct result of investigations of the physiology and biochemistry involved in the bakanae disease of rice.

Although many plant pathologists have explored these relationships in the past, the exploration of biochemical aspects of this field has led to the discovery of many new types of molecules, and the field is still a fertile one.

### References.

*1.* Biale, J. B.: Postharvest Physiology and Biochemistry of Fruits. Annu. Rev. Plant Physiol. **1**, 183 (1950).

*2.* Braun, A. C.: A Comparative Study of *Bacterium tabacum* Wolf and Foster and *Bacterium angulatum* Fromme and Murray. Phytopathology **27**, 283 (1937).

*3.* — Studies on Tumor Inception in the Crown Gall Disease. Amer. J. Bot. **30**, 674 (1943).

*4.* — Thermal Studies on the Factors Responsible for Tumor Initiation in Crown Gall. Amer. J. Bot. **34**, 234 (1947).

*5.* — Thermal Inactivation Studies on the Tumor-Inducing Principle in Crown Gall. Phytopathology **40**, 3 (1950).

*6.* — The Mechanism of Action of a Bacterial Toxin on Plant Cells. Proc. Nat. Acad. Sci. (USA) **36**, 423 (1950).

*7.* — Recovery of Tumor Cells from Effects of the Tumor-Inducing Principle in Crown Gall. Science (Washington) **113**, 651 (1951).

*8.* — Conditioning of the Host Cells as a Factor in the Transformation Process in Crown Gall. Growth **16**, 65 (1952).

*9.* — Studies on the Origin of the Crown-Gall Tumor Cell. In: Abnormal and Pathological Plant Growth. Brookhaven Symposia in Biology, No. 6, p. 115, Brookhaven Natl. Lab. (1954).

*10.* — The Physiology of Plant Tumors. Annu. Rev. Plant Physiol. **5**, 133 (1954).

*11.* — A Study of the Mode of Action of the Wildfire Toxin. Phytopathology **45**, 659 (1955).

*12.* — A Physiological Study on the Nature of Autonomous Growth in Neoplastic Plant Cells. In: The Biological Action of Growth Substances, p. 132. Symposium XI, Soc. Exptl. Biol. Cambridge: Univ. Press. 1957.

13. BRAUN, A. C. and T. LASKARIS: Tumor Formation by Attenuated Crown-Gall Bacteria in the Presence of Growth-Promoting Substances. Proc. Nat. Acad. Sci. (USA) 28, 468 (1942).

14. BRAUN, A. C. and R. J. MANDLE: Studies on the Inactivation of the Tumor-Inducing Principle in Crown Gall. Growth 12, 255 (1948).

15. BRAUN, A. C. and U. NAF: A Non-Auxinic Growth-Promoting Factor Present in Crown Gall Tumor Tissue. Proc. Soc. exp. Biol. Med. 86, 212 (1954).

16. BRAUN, A. C. and P. R. WHITE: Bacteriological Sterility of Tissues Derived from Secondary Crown-Gall Tumors. Phytopathology 33, 85 (1943).

17. BRIAN, P. W., G. W. ELSON, H. G. HEMMING and M. RADLEY: The Plant-Growth-Promoting Properties of Gibberellic Acid, a Metabolic Product of the Fungus Gibberella fujikuroi. J. Sci. Food Agricult. 5, 602 (1954).

18. BRIAN, P. W. and H. G. HEMMING: The Effect of Gibberellic Acid on Shoot Growth of Pea Seedlings. Physiol. Plantarum 8, 669 (1955).

19. BRIAN, P. W., H. G. HEMMING and M. RADLEY: A Physiological Comparison of Gibberellic Acid with Some Auxins. Physiol. Plantarum 8, 899 (1955).

20. CROSS, B. E., J. F. GROVE, J. MACMILLAN and T. P. C. MULHOLLAND: Gibberellic Acid. IV. The Structures of Gibberic and alloGibberic Acids and Possible Structures for Gibberellic Acid. Chem. and Ind. 1956, 954.

21. — — — — Gibberellic Acid. VII. The Structure of Gibberic Acid. J. Chem. Soc. (London) 1958, 2520.

22. CROSS, B. E., J. F. GROVE, J. MACMILLAN, T. P. C. MULHOLLAND and N. SHEPPARD: The Structure of Gibberellic Acid. Proc. Chem. Soc. (London) 1958, 221.

23. DAVIS, D., P. E. WAGGONER and A. E. DIMOND: Conjugated Phenols in the Fusarium Wilt Syndrome. Nature (London) 172, 959 (1953).

24. DE ROPP, R. S.: Action of Streptomycin on Plant Tumours. Nature (London) 162, 459 (1948).

25. — Experimental Induction and Inhibition of Overgrowths in Plants. In: F. SKOOG, Plant Growth Substances, p. 381. Madison, Wis.: Univ. Wisconsin Press. 1951.

26. DIMOND, A. E.: Pathogenesis in the Wilt Diseases. Annu. Rev. Plant Physiol. 6, 329 (1955).

27. DIMOND, A. E. and P. E. WAGGONER: The Physiology of Lycomarasmin Production by Fusarium oxysporum f. lycopersici. Phytopathology 43, 195 (1953).

28. — — On the Nature and Role of Vivotoxins in Plant Disease. Phytopathology 43, 229 (1953).

29. — — Effect of Lycomarasmin Decomposition upon Estimates of its Production. Phytopathology 43, 319 (1953).

30. — — The Water Economy of Fusarium Wilted Tomato Plants. Phytopathology 43, 619 (1953).

31. — — The Cause of Epinastic Symptoms in Fusarium Wilt of Tomatoes. Phytopathology 43, 663 (1953).

32. FISHER, P. L.: Physiological Studies on the Pathogenicity of Fusarium lycopersici SACC. for the Tomato Plant. Maryland Agricult. Exp. Sta. Bull. No. 374, p. 261 (1935).

33. GANE, R.: Production of Ethylene by Some Ripening Fruits. Nature (London) 134, 1008 (1934).

34. GÄUMANN, E.: Some Problems of Pathological Wilting in Plants. Adv. Enzymology 11, 401 (1951).

35. — Fusaric Acid as a Wilt Toxin. Phytopathology 47, 342 (1957).

*36.* Gäumann, E. und E. Bachmann: Über den Einfluß des Lycomarasmins und seiner Schwermetallkomplexe auf die Wasserpermeabilität pflanzlicher Protoplasten. Phytopath. Z. **29**, 265 (1957).

*37.* — — Über den Einfluß der Ernährung auf die Schädigung der Wasserpermeabilität der Protoplasten durch Fusarinsäure. Phytopath. Z. **31**, 1 (1958).

*38.* Gäumann, E., E. Bachmann und R. Hütter: Über den Einfluß der Eisenernährung auf die Lycomarasmin-Empfindlichkeit der Tomatenpflanzen. Phytopath. Z. **30**, 87 (1957).

*39.* Gäumann, E. und O. Jaag: Die physiologischen Grundlagen des parasitogenen Welkens. I, II, III. Ber. Schweiz. bot. Ges. **57**, 3, 132, 227 (1947).

*40.* Gäumann, E., St. Naef-Roth und H. Kobel: Über Fusarinsäure, ein zweites Welketoxin des *Fusarium lycopersici* Sacc. Phytopath. Z. **20**, 1 (1953).

*41.* Gordon, S. A. and R. P. Weber: The Effect of X-Radiation on Indoleacetic Acid and Auxin Levels in the Plant. Amer. J. Bot. **37**, 678 (1950).

*42.* Gothoskar, S. S., R. P. Scheffer, J. C. Walker and M. A. Stahmann: The Role of Pectic Enzymes in Fusarium Wilt of Tomato. Phytopathology **43**, 535 (1953).

*43.* — — — — The Role of Enzymes in the Development of Fusarium Wilt of Tomato. Phytopathology **45**, 381 (1955).

*44.* Gottlieb, D.: The Presence of a Toxin in Tomato Wilt. Phytopathology **33**, 126 (1943).

*45.* — The Mechanism of Wilting Caused by *Fusarium bulbigenum* var. *lycopersici.* Phytopathology **34**, 41 (1944).

*46.* Hall, W. C.: Studies on the Origin of Ethylene from Plant Tissues. Bot. Gaz. **113**, 55 (1951/52).

*47.* — Evidence on the Auxin-Ethylene Balance Hypothesis of Foliar Abscission. Bot. Gaz. **113**, 310 (1951/52).

*48.* Hayashi, T. and Y. Murakami: The Biochemistry of Bakanae Fungus. 32. The Physiological Action of Gibberellin. J. Agricult. Chem. Soc. Japan **28**, 543 (1954).

*49.* Hayashi, T., Y. Takijima and Y. Murakami: The Biochemistry of Bakanae Fungus. 28. The Physiological Action of Gibberellin. J. Agricult. Chem. Soc. Japan **27**, 672 (1953).

*50.* Henderson, J. H. M. and J. Bonner: Auxin Metabolism in Normal and Crown Gall Tissue of Sunflower. Amer. J. Bot. **39**, 444 (1952).

*51.* Hildebrand, E. M.: A Micrurgical Study of Crown Gall Infection in Tomato. J. Agricult. Res. **65**, 45 (1942).

*52.* Horsfall, J. G. and S. Rich: Fungitoxicity of Sulfur-Bridged Compounds. Indian Phytopathology **6**, 1 (1953).

*53.* Kato, Y.: Responses of Plant Cells to Gibberellin. Bot. Gaz. **117**, 16 (1955/56).

*54.* Kern, H. und D. Kluepfel: Die Bildung von Fusarinsäure durch *Fusarium lycopersici* in vivo. Experientia **12**, 181 (1956).

*55.* Klein, D. T. and R. M. Klein: Quantitative Aspects of Transformation of Virulence in *Agrobacterium tumefaciens.* J. Bacteriol. **72**, 308 (1956).

*56.* Klein, R. M.: The Probable Chemical Nature of the Crown-Gall Tumor-Inducing Principle. Amer. J. Bot. **40**, 597 (1953).

*57.* — Mechanisms of Crown Gall Induction. In: Abnormal and Pathological Plant Growth. Brookhaven Symposia in Biology No. 6, Brookhaven Natl. Lab., p. 97 (1954).

*58.* Klein, R. M. and G. K. K. Link: Auxin as a Promoting Agent in the Transformation of Normal to Crown-Gall Tumor Cells. Proc. Nat. Acad. Sci. (USA) **38**, 1066 (1952).

59. LAKSHINARAYANAN, K. and D. SUBRAMANIAN: Is Fusaric Acid a Vivotoxin? Nature (London) 176, 697 (1955).

60. LANG, A.: Stem Elongation in a Rosette Plant, Induced by Gibberellic Acid. Naturwiss. 43, 257 (1956).

61. LANG, A., J. A. SANDOVAL and A. BEDRI: Induction of Bolting and Flowering in Hyoscyamus and Samolus by a Gibberellin-Like Material from a Seed Plant. Proc. Nat. Acad. Sci. (USA) 43, 960 (1957).

62. LINK, G. K. K., R. M. KLEIN and E. S. GUZMAN BARRON: Metabolism of Slices of the Tomato Stem. J. exp. Botany 3, 216 (1952).

63. LUDWIG, R. A.: Studies on the Physiology of Hadromycotic Wilting in the Tomato Plant. Macdonald College Tech. Bull. No. 20, p. 38, McGill Univ., Montreal (1952).

64. MACMILLAN, J. and P. J. SUTER: The Occurrence of Gibberellin A₁ in Higher Plants: Isolation from the Seed of Runner Bean (Phaseolus multiflorus). Naturwiss. 45, 46 (1958).

65. MULHOLLAND, T. P. C.: Gibberellic Acid. IX. The Structure of alloGibberic Acid. J. Chem. Soc. (London) 1958, 2693.

66. MULHOLLAND, T. P. C. and G. WARD: Gibberellic Acid. II. The Structure and Synthesis of Gibberene. J. Chem. Soc. (London) 1954, 4676.

67. PHINNEY, B. O., C. A. WEST, M. RITZEL and P. M. NEELY: Evidence for "Gibberellin-Like" Substances from Flowering Plants. Proc. Nat. Acad. Sci. (USA) 43, 398 (1957).

68. PLATTNER, PL. A. und N. CLAUSON-KAAS: Über ein Welke erzeugendes Stoffwechselprodukt von Fusarium lycopersici SACC. Helv. Chim. Acta 28, 188 (1945).

69. PLATTNER, PL. A., N. CLAUSON-KAAS, A. BOLLER und U. NAGER: Der hydrolytische Abbau des Lycomarasmins. Helv. Chim. Acta 31, 860 (1948).

70. PLATTNER, PL. A., W. KELLER und A. BOLLER: Konstitution und Synthese der Fusarinsäure. Synthese von 5-Äthyl- und 5-n-Hexyl-pyridin-2-carbonsäure. Helv. Chim. Acta 37, 1379 (1954).

71. RIKER, A. J.: Some Relations of the Crown Gall Organism to its Host Tissue. J. Agricult. Res. 25, 119 (1923).

72. — Studies on the Influence of Some Environmental Factors on the Development of Crown Gall. J. Agricult. Res. 32, 83 (1926).

73. RIKER, A. J. and J. E. THOMAS: The Interaction between Causative Agents in Diseased Growths. In: F. SKOOG, Plant Growth Substances, p. 405. Madison, Wisc.: Univ. Wisconsin Press. 1951.

74. SCHEFFER, R. P.: The Wilting Mechanism in Fusarium Wilt of Tomato. Thesis, Univ. Wisconsin, Madison (1952).

75. SCHEFFER, R. P. and J. C. WALKER: The Physiology of Fusarium Wilt of Tomato. Phytopathology 43, 116 (1953).

76. SKOOG, F.: The Effect of X-Irradiation on Auxin and Plant Growth. J. Cell. Comp. Physiol. 7, 227 (1935/36).

77. STODOLA, F. H.: Source Book on Gibberellin, 1828–1957. Northern Utilization Research and Development Div., U. S. Dept. Agric. Washington, D. C.: Govt. Printing Office, 556 pp. (1958).

78. STOLL, CH.: Über Stoffwechsel und biologisch wirksame Stoffe von Gibberella fujikuroi (SAW.) WOLL., dem Erreger der Bakanaëkrankheit. Phytopath. Z. 22, 233 (1954).

79. STOLL, CH. und J. RENZ: Über den Fusarinsäure- und Dehydrofusarinsäurestoffwechsel von Gibberella fujikuroi (SAW.) WOLL. Phytopath. Z. 29, 380 (1957).

80. Stoll, Ch., J. Renz und E. Gäumann: Über die Bildung von Fusarinsäure und Dehydrofusarinsäure durch das *Fusarium lycopersici* Sacc. in saprophytischer Kultur. Phytopath. Z. **29**, 388 (1957).

81. Stowe, B. B. and T. Yamaki: The History and Physiological Action of the Gibberellins. Annu. Rev. Plant Physiol. **8**, 181 (1957).

82. Tamari, K.: Biochemical Studies of the Bakanae Fungus. XIX. Synthesis of Fusarin. 1. J. Agricult. Chem. Soc. Japan **22**, 16 (1948) [Chem. Abstr. **46**, 5143 (1952)].

83. Thomas, J. E. and A. J. Riker: The Effects of Representative Plant Growth Substances upon Attenuated-Bacterial Crown Galls. Phytopathology **38**, 26 (1948).

84. Van Lanen, J. M., I. L. Baldwin and A. J. Riker: Attenuation of Cell Stimulating Bacteria by Specific Amino Acids. Science (Washington) **92**, 512 (1940).

85. Waggoner, P. E. and A. E. Dimond: Crown Gall Suppression by Ionizing Radiation. Amer. J. Bot. **39**, 679 (1952).

86. — — Crown Gall Suppression by Anti-Auxin. Science (Washington) **117**, 13 (1953).

87. — — Role of Chelation in Causing and Inhibiting the Toxicity of Lycomarasmin. Phytopathology **43**, 281 (1953).

88. — — Reduction in Water Flow by Mycelium in Vessels. Amer. J. Bot. **41**, 637 (1954).

89. — — Pectic Enzymes Produced by *Fusarium oxysporum* f. *lycopersici* in Infected Plants. Phytopathology **44**, 509 (1954).

90. — — Production and Role of Extracellular Pectic Enzymes of *Fusarium oxysporum* f. *lycopersici*. Phytopathology **45**, 79 (1955).

91. — — Assaying Effect of Growth Regulators upon Plant Tumors. Connecticut Agricult. Exp. Sta. Bull. No. 587, p. 14 (1955).

92. Walker, J. C.: Plant Pathology, 2nd Ed., p. 708. New York: McGraw Hill. 1957.

93. Walker, J. C. and M. A. Stahmann: Chemical Nature of Disease Resistance in Plants. Annu. Rev. Plant Physiol. **6**, 351 (1955).

94. Wellman, F. L.: Epinasty of Tomato, One of the Earliest Symptoms of Fusarium Wilt. Phytopathology **31**, 281 (1941).

95. Williamson C. E.: Ethylene, a Metabolic Product of Diseased or Injured Plants. Phytopathology **40**, 205 (1950).

96. Wittwer, S. H., M. J. Bukovac, H. M. Sell and L. E. Weller: Some Effects of Gibberellin on Flowering and Fruit Setting. Plant Physiol. **32**, 39 (1957).

97. Woolley, D. W.: Studies on the Structure of Lycomarasmin. J. Biol. Chem. **176**, 1291 (1948).

98. — Synthesis and Determination of Lycomarasmin Activity of some Derivatives of Aspartic Acid. J. Biol. Chem. **176**, 1299 (1948).

99. Woolley, D. W., R. B. Pringle and A. C. Braun: Isolation of the Phytopathogenic Toxin of *Pseudomonas tabaci*, an Antagonist of Methionine. J. Biol. Chem. **197**, 409 (1952).

100. Woolley, D. W., G. Schaffner and A. C. Braun: Isolation and Determination of the Structure of a new Amino Acid, Contained within the Toxin of *Pseudomonas tabaci*. J. Biol. Chem. **198**, 807 (1952).

101. — — — Studies on the Structure of the Phytopathogenic Toxin of *Pseudomonas tabaci*. J. Biol. Chem. **215**, 485 (1955).

102. Work, E.: The Isolation of $\alpha,\varepsilon$-Diaminopimelic Acid from *Corynebacterium diphtheriae* and *Mycobacterium tuberculosis*. Biochemic. J. **49**, 17 (1951).

*103.* YABUTA, T., K. KAMBE and T. HAYASHI: Biochemistry of Bakanae Fungus of Rice. I. J. Agricult. Chem. Soc. Japan **10**, 1059 (1934) [Chem. Abstr. **29**, 1132 (1935)].

*104.* YABUTA, T., Y. SUMIKI, K. FUKUNAGA and M. HORIUCHI: Biochemistry of Bakanae Fungus. 22. Chemical Composition of Rice Seedlings Treated with Gibberellin. J. Agricult. Chem. Soc. Japan **24**, 396 (1951).

*105.* YOUNG, R. E., H. K. PRATT and J. B. BIALE: Identification of Ethylene as a Volatile Product of the Fungus *Penicillium digitatum*. Plant Physiol. **26**, 304 (1951).

*(Received, December 12, 1958.)*

# The Chemical Structure of the Normal Human Hemoglobins.

By **W. A. Schroeder**, Pasadena, California.

With 14 Figures.

## Contents.

*References, pp. 371—378.*

*Acknowledgement.* It is a pleasure to acknowledge that during the preparation of this article the author had many stimulating discussions with Dr. Richard T. Jones, Dr. Jerome Vinograd, and Dr. Norman Weliky, and that some of the ideas expressed here arose from these discussions.

# I. Introduction.

The conspicuous red color of the blood has no doubt been a source of wonder, interest, and curiosity ever since it was first observed by man. It followed much later that hemoglobin, the cause of the red color, is neatly packaged in the red blood cells apart from the other constituents of the blood and that it serves as a carrier for oxygen in the body. Not only the oxygen-carrying function of hemoglobin has been the subject of much research but almost every chemical or physiological topic related to hemoglobin has received attention until, at the present time, the literature on hemoglobin is almost overwhelming in amount. No small part of this vast literature has come during the last ten years, since the discovery by Pauling, Itano, Singer, and Wells (99) of the first abnormal human hemoglobin which is now termed sickle-cell anemia hemoglobin or hemoglobin S. This discovery gave a new impetus to the study of hematological disorders for here, for the first time, was a pathological state unquestionably associated with a molecule that is slightly different in some way from the normal one. Because of this difference it is unable to function properly: a molecular disease had been detected. The detection of other abnormal hemoglobins soon followed so that today almost all the letters of the alphabet have been used to designate the various abnormal hemoglobins.

Parallel to the burgeoning interest in hemoglobin has been the development and improvement of analytical methods in protein chemistry. Fifteen years ago the determination of the structure of even a small peptide was no mean accomplishment: today, the amino acid sequence has been determined in insulin, corticotropin, glucagon, ribonuclease, and a host of smaller peptides. Now one can undertake the task of isolating and purifying a protein with the knowledge that methods are available that, potentially, can be used to determine its complete structure.

One can, then, hope to determine the amino acid sequence and the complete structure of hemoglobin.

If we proposed in an article such as this to discuss hemoglobin in detail from many aspects such as the physiological, the clinical, the genetic, the chemical, and so on, it would be necessary to extend its length many times. Rather, we wish to limit its scope to a detailed discussion of the chemical structure of the normal human hemoglobins. However, this limitation is an elastic one that will be stretched as needed to include those physiological, clinical, and other aspects that throw light on the main subject of discussion.

We must now decide which are the normal human hemoglobins.

Most of our attention will be given to the adult human hemoglobin (that is, hemoglobin A) which the majority of individuals possesses, but we cannot neglect the different hemoglobin (fetal hemoglobin or hemoglobin F) that has been known for almost a hundred years to be the major hemoglobin component in the circulatory system of the new-born human infant and that also must be considered a normal hemoglobin. In addition to these two hemoglobins, it has been reported that very early in fetal development there exists a third type, a "primitive" or "embryonic" hemoglobin. This we shall also discuss.

Those facets of the hemoglobin problem that we shall not discuss in detail have been covered in books and review articles of more or less recent publication. An excellent discussion of the state of knowledge in 1949 is to be found in the book of Lemberg and Legge entitled "Hematin Compounds and Bile Pigments" (86). Although this volume gives much basic information about the chemistry of hemoglobin, in general, mention of the type of hemoglobin from which the data were obtained is deliberately omitted. The reader should realize that many older investigations were made with animal hemoglobins and that the data may not be applicable to human hemoglobin. Unfortunately, in some original articles, the author has not seen fit to describe the species that was the source of the. hemoglobin. The year 1949 also saw the publication of the proceedings of a conference on hemoglobin that was held in Cambridge, England, in memory of Sir Joseph Barcroft (110). In 1957, a Conference on Hemoglobin was held in Washington, D. C. and its proceedings have recently been published (23). A Symposium on Molecular Heterogeneity of Hemoglobin was sponsored by the American Society of Biological Chemists in 1957 (32). Betke (15) has written an extensive discussion of the properties, differences, and clinical significance of fetal and adult human hemoglobin. Among the review articles may be mentioned those of Drabkin (31), Huisman (50, 51), Itano (67—70), Itano, Bergren, and Sturgeon (72), Pauling (97), Tuttle (128), and Wyman (134).

## II. Nomenclature.

Before proceeding further, let us describe the nomenclature of the compounds with which we have to deal. Hemoglobin is a conjugated protein in which the prosthetic group is the *heme* and the protein part is the *globin*. The heme has the same structure not only in human and animal hemoglobins but also in the cytochromes and in catalase. However, the behavior of the iron in the heme differs vastly in the three proteins. In the cytochromes, the heme iron is alternately oxidized and reduced as it acts as a link between atmospheric oxygen and cellular oxidation-reduction systems. In catalase, the heme iron is permanently in the ferric form, and the chromoproteid destroys hydrogen peroxide. In hemoglobin, normally, the iron is in the ferrous form, and because it can combine reversibly with oxygen, hemoglobin is able to fulfill its function as a transporter of oxygen; when it is in the ferric form, the ability to combine reversibly with oxygen (as well as carbon monoxide and other small molecules) is lost.

Because the iron in hemoglobin may be in the ferrous or ferric form and because small molecules or ions may combine with the iron, many derivatives have been prepared. The nomenclature in the literature, therefore, attempts by appropriate prefixes to the word "globin" to describe the state of oxidation of the iron as well as the nature of the combining group. *Table I* presents the names that have been applied to various derivatives. Sometimes they are confusing and unrelated to the chemical nature. The nomenclature may be extended to other derivatives as well as to denatured materials. The species that is the source of the hemoglobin can be readily indicated by the proper adjective, for example, sheep oxyhemoglobin. Actually, because we shall be most concerned with the protein moiety of hemoglobin, we shall have little reason to use these specific names but when it is necessary, the nomenclature of PAULING and CORYELL (98) will be used because it is most descriptive of the chemical entity. Consequently, the name "hemoglobin" will be used in a general sense when it is unimportant

Table I. Nomenclature for Derivatives of Undenatured Hemoglobin.

| Oxidation state of iron | Group attached to iron* | Nomenclature | | |
|---|---|---|---|---|
| | | LEMBERG and LEGGE [(86), there p. 209] | KEILIN (78) | PAULING and CORYELL (98) |
| Ferrous | none | Hemoglobin | Hemoglobin | Ferrohemoglobin |
| Ferrous | $O_2$ | Oxyhemoglobin | Oxyhemoglobin | Oxyhemoglobin |
| Ferrous | CO | Carboxyhemoglobin | Carboxyhemoglobin | Carbonmonoxy-hemoglobin |
| Ferric | none | Hemiglobin | Methemoglobin | Ferrihemoglobin |
| Ferric | $CN^-$ | Hemiglobin cyanide | Cyanmethemoglobin | Ferrihemoglobin cyanide |

* In addition to these more or less easily removable groups, the nitrogens of the tetrapyrrole structure and some group(s) from the globin, of course, are attached to the iron.

to the description of the data what may be the oxidation state of the iron or the nature of the attached group. In most instances, "hemoglobin" will actually be oxyhemoglobin because it is the compound that results whenever the coloring matter of the red blood cells is isolated in solution in the presence of atmospheric oxygen and no derivative is formed by the addition of chemicals.

In 1953, a group of investigators issued a statement concerning a system of nomenclature for abnormal human hemoglobins (8). According to this recognized system of nomenclature, English capital letters designate the various hemoglobins. Thus, normal adult human hemoglobin is hemoglobin A, fetal hemoglobin is hemoglobin F, and sickle-cell hemoglobin is hemoglobin S (in an earlier nomenclature it was hemoglobin b). No hemoglobin is designated as hemoglobin B, but other hemoglobins are named after the letters of the alphabet in the order of their discovery. This nomenclature will be used in any reference to the abnormal hemoglobins. In addition, this system suggested that possible sub-species of a given hemoglobin be designated as $A_1$, $A_2$, etc. or $F_1$, $F_2$, etc.

It is now necessary to adopt a nomenclature to differentiate the individual polypeptide chains of the molecule. A proposed system is discussed on p. 364.

## III. The Hemoglobin Molecule as a Whole.

Because crystallinity is ordinarily taken as an indicator of purity, it is not surprising that far more chemical investigations have been made with horse or dog hemoglobin, which crystallizes with ease than with human hemoglobin which is more difficult to crystallize. As a result, the literature often records only one or two determinations of a given constant for human hemoglobin instead of many that may be found for horse and other hemoglobins.

### 1. Iron Content.

An elementary analysis of a protein is of somewhat doubtful usefulness unless the percentage of one of the elements is small. In the case of hemoglobin, it is useful because a small but definite amount of iron is present. Modern analyses of the iron content of human hemoglobin have been infrequent. BERNHART and SKEGGS (14) report an iron content of 0.340% in crystallized adult hemoglobin. DRABKIN (33) states that his analytical value of 0.338% iron in crystallized adult hemoglobin was tentatively accepted by the Protein Commission of the International Union of Pure and Applied Chemistry in June 1955. The minimum molecular weight on the basis of a content of 0.338% iron is 16,520.

The calculated minimum molecular weight is, of course, greatly influenced by errors in the determination of the iron—thus 0.335% calculates to 16,700 and 0.340% to 16,400. The significance of these carefully determined percentages is open to question in view of the recently proved heterogeneity of human hemoglobin whether crystallized or uncrystallized (cf. p. 344). Although the results of the several investigators agree well, the agreement may simply mean that the iron content of all heme-containing components in the erythrocytes is essentially identical or that the samples were equally heterogeneous.

The binding capacity of hemoglobin for oxygen or carbon monoxide is one molecule per atom of iron [MORRISON and HISEY (91)]. It has been shown that all the iron is contained in the heme groups.

The iron content of human fetal hemoglobin apparently has not been determined. However, a comparison of its spectrum with that of adult hemoglobin shows that the heme to protein ratio is the same and this suggests that the iron content probably is the same.

## 2. The Molecular Weight.

It has long been known that the molecular weight of human hemoglobin normally is about four times the minimum molecular weight derived from the iron analysis, or 66,100 if the iron content is 0.338%. Osmotic pressure measurements with several derivatives were made by ADAIR (1) in 1924 and gave values of 60,000 to 68,000 with rather low precision ($\pm$ 10 to 15%). Molecular weights by ultracentrifugal methods fall generally within these limits: 59,400 (38); 63,000 (85, 122); 64,500 (40); and 66,500 (45). The molecular weight of fetal hemoglobin is said not to differ significantly from that of adult hemoglobin [JOPE and O'BRIEN (77); TAYLOR and SWARM (123)]. The normal molecular weights, then, are of the order of 66,000.

In contrast to the dissociation of some animal hemoglobins under some conditions [reviewed by WYMAN (134), there especially p. 434], definite evidence of reversible dissociation of human hemoglobin has come only recently. MOORE and REINER (90) observed the separation of adult human globin and hemoglobin into two components by electrophoresis and ultracentrifugation at low pH; the proportions varied with the pH. FIELD and O'BRIEN (38) suggest that MOORE and REINER's results may be explained on the basis of irreversible denaturation. FIELD and O'BRIEN observed a constant molecular weight for adult hemoglobin over the pH range of 6 to 11. Between pH 6 and 3.5, they found a reversible dissociation which was increased by dilution or by the presence of barbital. Dilution does not produce dissociation in the region of neutral pH. HASSERODT and VINOGRAD (44) have extended these observations to higher pH. Like FIELD and O'BRIEN, they found that the molecular weight is independent of concentration between pH 6 and 10, but unlike FIELD and O'BRIEN, they observed a decrease in molecular weight above pH 10, with complete and reversible dissociation to about half the normal molecular weight at pH 11.0 and denaturation at higher pH. It should be noted that FIELD and O'BRIEN (*38*) do not present specific data above pH 8. *Fig. 1*, taken from HASSERODT and VINOGRAD (*44*), is a combination of their data and those of FIELD and O'BRIEN. In this Figure, the sedimentation constant S (in SVEDBERG units), rather than the molecular weight, is plotted as a function of

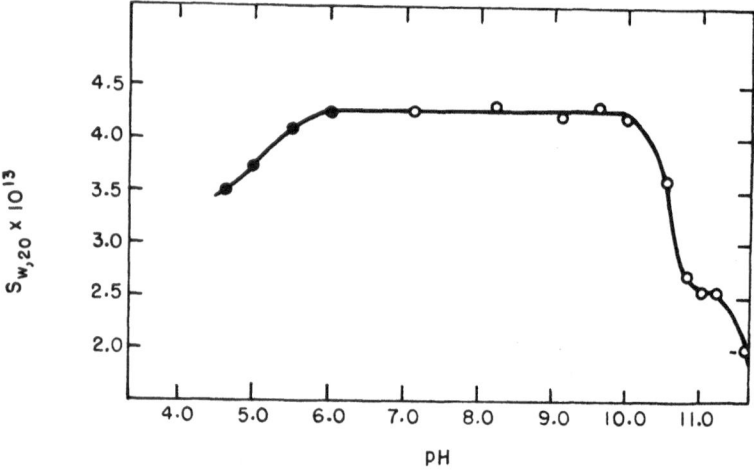

Fig. 1. The sedimentation constants of human carbonmonoxyhemoglobin as a function of pH.
[From: Proc. Nat. Acad. Sci. (U. S. A.) *45*, 12 (1959).]

pH: the sedimentation constant of 4.3 S is equivalent to a molecular weight of about 66,000 and of 2.5 S to about 36,000. On the basis of a less complete investigation, the sedimentation constant of fetal hemoglobin has been shown to behave in much the same way as a function of pH (*130*).

Unlike horse hemoglobin, adult human hemoglobin does not dissociate in 4 M-urea. [The significance of the dissociation of horse hemoglobin may be questioned in the light of the observation (*10, 19*) that 107 of 110 samples of horse hemoglobin showed the presence of two components.] However, GUTTER, SOBER, and PETERSON (*40*) observed that adult human hemoglobin dissociated to some extent in the presence of mercaptoethanol and to a greater extent when urea was also added. Inasmuch as adult human hemoglobin probably does not contain disulfide bonds, the observed dissociation in the presence of the reducing agent is without explanation at present.

*References, pp. 371—378.*

Rossi-Fanelli, Antonini, and Caputo (*107*) have reported that the sedimentation constant of adult human globin is 2.55 $S$ and that the molecular weight is 42,000. The same sedimentation constant has been observed in these laboratories (*43*) for adult human globin.

In a reversible equilibrium such as exists in the dissociation of hemoglobin at certain pH values, it is not possible to separate the particles of different molecular weight from each other by centrifugation despite the fact that they differ by a factor of about 2 in molecular weight. At high pH, the dissociation apparently is complete, and the molecular weight of 36,000 is somewhat higher than the expected 33,000 if dissociation were into equal parts. If the difference is really more than experimental error, it would suggest dissociation into unequal parts. The molecular weight of 42,000 for adult human globin [Rossi-Fanelli et al. (*107*)] seems very high although we need not postulate that the dissociation takes place in the same way when hemoglobin dissociates as it does when globin is prepared. The heme seems to serve an important function in holding the molecule together because the molecular weight changes greatly when it is removed. On the other hand, we can not exclude the possibility that the procedure for the removal of the heme alters some protein interactions that hold the molecule together. In a consideration of the structure of the molecule, much information will be gained if we can answer the questions: Is the dissociation into identical or non-identical parts? And, what role does the heme play in holding the molecule together?

A probable answer to the first question may now be given: the dissociation by change in pH very likely is into non-identical parts at an acidic pH (*71, 131*). The experiments that provided this conclusion required the use of other pertinent data for their evaluation, and further discussion of this topic will be postponed until p. 364. Likewise, the second question for which there is as yet no answer may more profitably be discussed later.

### 3. The Shape of the Molecule.

Although the environment will influence the shape of the hemoglobin molecule, only by X-ray diffraction of crystals can we hope to determine one exact shape. Bragg, Perutz, and collaborators have for many years been engaged in X-ray studies of horse ferrihemoglobin and they have also compared the general shape of a variety of hemoglobins including human. Bragg and Perutz conclude (*16*) that, with the possible exception of sheep fetal hemoglobin, many animal hemoglobins and human hemoglobin seem to have the same general external form. Especially important is the fact that, in all cases, the molecule has an axis of symmetry and therefore is composed of identical halves. Recently, Cullis, Dintzis, and Perutz (*28*) have described the molecule of horse ferrihemoglobin in these words: ". . . the external shape of the molecule . . . is seen to be an ellipsoid of dimensions 55 × 55 × 70 Å. There is a depression at the centre corresponding to a dimple or a pair of dimples at the surface of the molecule." Far more interesting is the conclusion of Perutz, Trotter, Howells, and Green (*100*) to the effect that a comparison of the Patterson projections of human and horse hemoglobin suggests a close relationship in the internal structure of the two proteins.

Zinsser and Tang's X-ray investigation of human fetal hemoglobin (*135*) was not sufficiently extensive to permit of any conclusions about the size and shape of the molecule.

## IV. The Prosthetic Group.

The linkage between the prosthetic group and the protein is easily disrupted in hemoglobin. This separation of the heme and globin is commonly done by a "modification of the method of Anson and Mirsky" (*9*). (Unfortunately, most authors do not mention the nature of the modification.)

Basically, the method of Anson and Mirsky involves the addition of a solution of hemoglobin to acidified acetone (hydrochloric acid or organic acids such as oxalic acid may be used) at 0° or below. Under these conditions, the globin precipitates and may easily be separated from the solution of heme.

We need devote little attention to the structure (I) of the heme which is well known from the painstaking work of Willstätter, Küster,

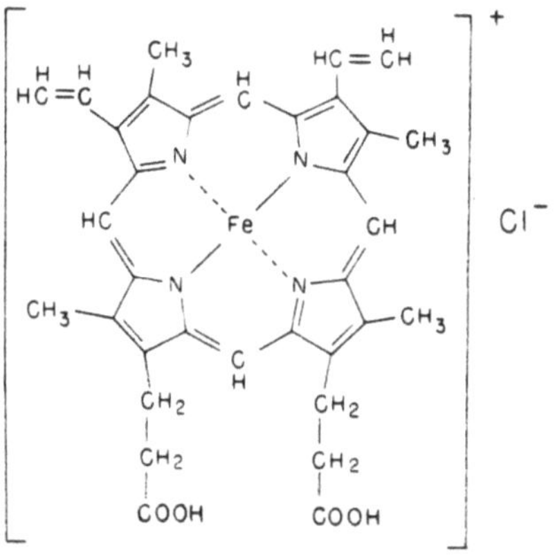

(I.) Hemin chloride.

H. Fischer, and others over a long period of time. The structure is shown as hemin chloride, the form in which it is usually isolated and in which it is most stable. Only one of the many resonance forms that are to be expected from so complex a molecule is depicted.

## V. The Linkage Between Heme and Globin.

The very lability of the heme-globin linkage has frustrated all attempts to determine directly its nature. In contrast, twenty years ago, THEORELL (*124, 125*) was able to show that the prosthetic group of cytochrome c was linked to the protein through cysteine residues by a thioether bond, and recently TUPPY and PALÉUS (*127*) have elucidated

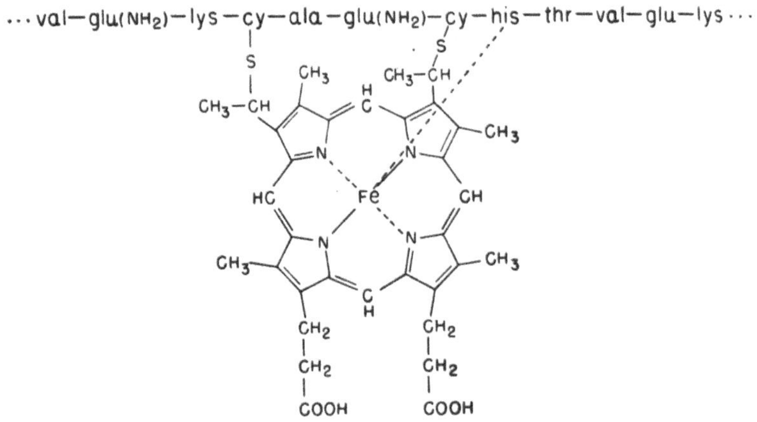

(II.) The heme-protein linkage in cytochrome c as proposed by THEORELL (*126*).

the sequence of amino acid residues in the immediate vicinity of the link, with the result shown in (II) as formulated by THEORELL (*126*). Any information about the heme-globin linkage, therefore, is based on indirect evidence.

The *abbreviations for amino acids and peptides* as used here and elsewhere in this article follow the common usage summarized by SANGER (*111*). The first few letters of each name designate the amino acid; thus, ala = alanine, arg = arginine, asp = aspartic acid, asp(NH$_2$) = asparagine, cys = cysteine, cy–S–S–cy = cystine, glu = glutamic acid, glu(NH$_2$) = glutamine, gly = glycine, his = histidine, hypro = hydroxyproline, ileu = isoleucine, leu = leucine, lys = lysine, met = methionine, orn = ornithine, phe = phenylalanine, ser = serine, thr = threonine, try = tryptophan, tyr = tyrosine, and val = valine. Thus, ala-gly-ileu denotes alanylglycylisoleucine whereas ala-(gly, ileu) denotes that alanine contains the free amino group but that the sequence of glycine and iso-leucine is undetermined.

The very lability of the link discounts the likelihood that it is covalent. Furthermore, it is not through the side-chains of the tetrapyrrole system. Thus, globin can be combined with modified hemes in which the vinyl side-chains have been reduced to ethyl groups or in which the propionic acid chains have been esterified: these modified hemoglobins still show reversible dissociation with oxygen. On the other hand, heme will combine with proteins other than globin but the products do not combine reversibly with oxygen. Therefore, some specialized area or structure must be present in the globin.

It is generally assumed that the link is through the iron atom. Presumably, six groups are in octahedral coordination around the iron atom and four of the positions are occupied by the nitrogens of the tetrapyrrole system in a plane. At least one of the two remaining positions must be occupied by some group from the globin. The large content of histidine in globin led Küster and Koppenhöfer (84) to suggest more than thirty years ago that the imidazole rings of the histidine residues are involved in the linkage between heme and globin. In support of this suggestion, Wyman and associates [reviewed in detail in (134)] have amassed a vast amount of differential titration and thermodynamic data, and Wyman concludes that two imidazole groups may be linked to the iron. It has long been known that when oxygen converts ferro-hemoglobin to oxyhemoglobin the acidity of the molecule is increased. Wyman interprets the data to mean that the entrance of oxygen alters the linkage of the imidazole groups so that the ionization of one is increased and the other decreased. Coryell and Pauling (25) have given a detailed description of the changes in bond type that would accompany oxygenation according to Wyman's mechanism. However, objections have been raised to this concept of the heme-globin linkage. Haurowitz and Hardin (46) discuss these objections in considerable detail. They point out among other arguments that the heme-globin linkage is labile to acid and stable to alkali but, on the other hand, that histidine itself shows no affinity for heme and that the imidazole derivatives that do combine are easily dissociated by alkali.

There is evidence in the work of Riggs (106) that the sulfhydryl groups are in some way a part of the linkage. The sigmoid shape of the oxygen dissociation curve means that there are interactions between the four hemes in the molecule. Indeed, Allen, Guthe, and Wyman (3) have concluded that the combining centers are inherently identical, that all the heme-globin linkages are the same, and that there is complete equivalence in mutual interactions of the hemes. Riggs (106) has treated hemoglobin with p-chloromercuribenzoate which probably is reasonably specific for the sulfhydryl group in proteins and has observed a large, though not complete, interference with the heme-heme interactions. Presumably, this interference would not have occurred had there been no interrelation of the sulfhydryl groups and the heme-globin linkage.

Not all of the experiments that have been described in broad outline above have been made with human hemoglobin. However, in view of the unique ability of the hemoglobins to combine reversibly with oxygen, we may suppose that the source of the hemoglobin is of little moment in such experiments. Yet, this conclusion may not be entirely sound because quantitatively the reaction of the various hemoglobins with oxygen is not identical. It is true that the *oxygen capacity* of the various hemoglobins is identical but on the other hand they differ in *oxygen affinity*. Clearly, the particular characteristic of the heme-globin link that confers upon hemoglobin its peculiar properties is subject to variation in degree. Conclusions that are reached with one kind of hemoglobin may not be applicable to another.

Related to the question of the heme-globin linkage is, of course, the spatial relationship of the two parts. The heme constitutes only about 4% by weight of the molecule and is, therefore, a relatively small part in bulk and weight. Is it attached to the surface of the globin or is it buried within the polypeptide chains? It is unlikely that the heme is buried far below the surface because it would then tend to be relatively inaccessible to oxygen but it may be present in some type of crevice in which it is reasonably accessible. GEORGE and LYSTER (39) have recently discussed the evidence for and against such a crevice configuration. The data of KON and DAVIDSON (79a) suggest that the heme groups of hemoglobin are shielded more from the external environment than the heme group of myoglobin. Again, all of the evidence is indirect and no certain conclusion can be reached. Although it is not possible to settle the question of the spatial relationships of the heme and globin, INGRAM, GIBSON, and PERUTZ (60) have used electron spin resonance to determine the orientation of the hemes in horse ferrihemoglobin, that is, to determine the plane of the tetrapyrrole ring system. They find that the heme groups are arranged in pairs and that those in each pair are related by the two-fold axis of symmetry. They stress that this method determines only the orientation of the hemes and not their position in the molecule. Human hemoglobin has not been studied thus.

## VI. The Normal Human Hemoglobins.

In the foregoing discussion, we have been able to consider, in general terms, the size of the hemoglobin molecule, the structure of the prosthetic group, and the linkage of the prosthetic group to the protein because the mammalian hemoglobins seem to have these features in common. Thus, although the exact structure of the heme group in human fetal hemoglobin does not appear to have been determined, it is unlikely that it differs from that of adult hemoglobin in view of the fact that the hemes of the latter and of sickle-cell hemoglobin are identical with each other (99) and with those of animal hemoglobins. We must now become more specific in our discussion because it is clear that the differences in the hemoglobins concern the globin part. First, however, let us consider the change in the type of hemoglobin in the individual human as a function of age.

## 1. The Change in the Type of Hemoglobin in the Individual Human as a Function of Age.

The tremendous change in conditions that occurs when the individual is born is also manifested in the change in the type of hemoglobin in his

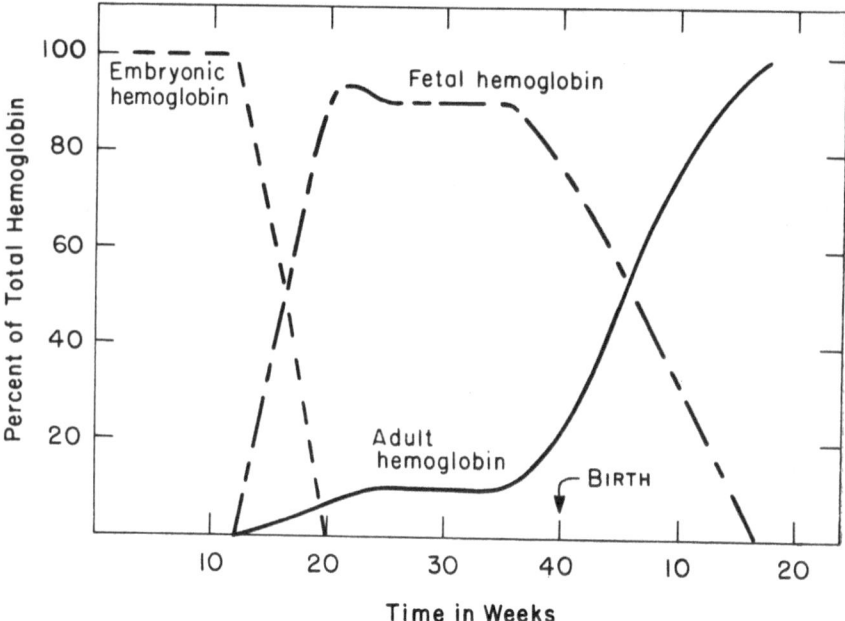

Fig. 2. Idealized representation of the change in the type of hemoglobin in the human individual as a function of age.

circulatory system. More than 90 years ago, Körber (8o) observed that the hemoglobin of the new-born human infant is denatured much more slowly by alkali than is adult human hemoglobin. The two hemoglobins have since been shown to differ in many other properties. Recently, it has been reported that another hemoglobin, an embryonic or primitive hemoglobin, precedes fetal hemoglobin and may be detected in very young fetuses [Drescher and Künzer (34); Allison (6); Halbrecht and Klibanski (41); Künzer (83); Halbrecht, Klibanski, Brzoza, and Lahav (42)]. If these reports can be substantiated, this hemoglobin presumably represents the third normal human hemoglobin.

On the basis of many reports in the literature, we may attempt to describe in an idealized way the change in the type of hemoglobin in the human individual as a function of age. This is portrayed in *Fig. 2* in which each of the hemoglobins is given as a percentage of the total hemoglobin during the pre-natal period and a short post-natal period. According to DRESCHER and KÜNZER (*34*), embryonic hemoglobin is present, but fetal hemoglobin is not detectable in fetuses 7 to 12 weeks in age. On the other hand, their data show only fetal hemoglobin in a 20-week-old fetus. Thus, as shown in Fig. 2, the embryonic hemoglobin must undergo a rapid decline and the fetal hemoglobin a similarly rapid increase between the 12th and 20th weeks of pre-natal life. WALKER and TURNBULL (*133*) report that adult hemoglobin makes its first appearance about the 13th week of pregnancy, increases to about 10% by 22 to 24 weeks, remains constant until about the 35th week, and then increases to an average of about 30% at term. The quantity they report at term is larger than the average of 20 to 25% that others typically report (*74, 79*) although, of course, individual variation is very great (0–30%). Thus, by the time that the embryonic hemoglobin has disappeared, some adult hemoglobin has already appeared along with the fetal hemoglobin and its percentage of the total slowly increases to term.

Birth unquestionably produces a marked change: the destruction of fetal hemoglobin begins, and adult hemoglobin takes its place. On the average, fetal hemoglobin has essentially disappeared from the infant's blood by the age of 16 weeks or 112 days (*74, 79*). This period of time is very similar to the average life-time of a red cell in the adult human, namely, approximately 120 days. On the other hand, the work of DRESCHER and KÜNZER (*34*) suggests that the destruction of embryonic hemoglobin and its replacement by fetal hemoglobin takes place in about half this period of time and that the two coexist during only a rather short period in embryonic life.

Recently developed methods show clearly that this picture of the sequence of hemoglobins in the individual is far too simple. "Adult" hemoglobin as ordinarily isolated certainly is not a "pure" substance nor does cord hemoglobin contain only fetal and adult hemoglobin. For a complete picture, we would have to follow some of the other components now known to be present. Of much greater interest, however, is the question:

## 2. Does an Embryonic Hemoglobin Actually Exist?

Only indirect evidence has so far been presented in the literature to support the claim of the existence of an human embryonic hemoglobin. DRESCHER and KÜNZER (*34*) deduced its presence by alkali denaturation

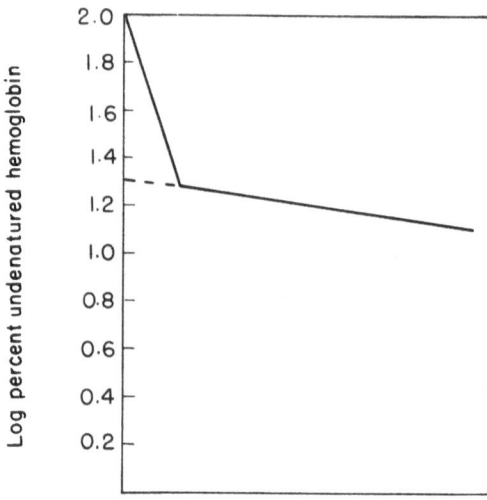

Fig. 3. Idealized results from the alkali denaturation of two hemoglobins that differ widely in resistance to alkali denaturation.

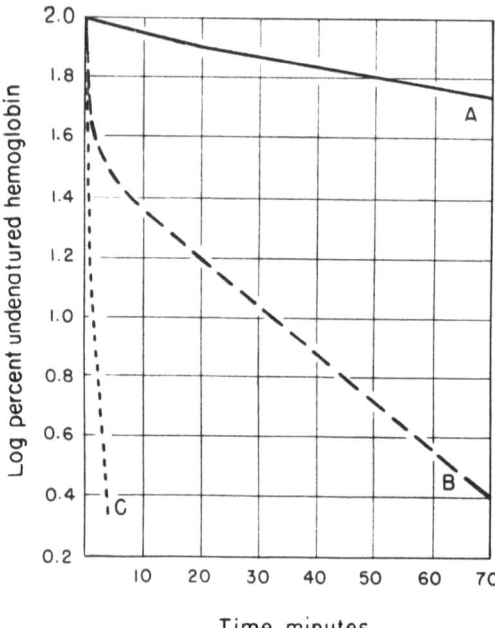

Fig. 4. Alkali denaturation of hemoglobins from a 20-week fetus *(A)*, a 9-week fetus *(B)*, and an adult *(C)*. [From: Klin. Wschr. *32*, 92 (1954).]

*References, pp. 371—378.*

studies. Because of the widely disparate rates of alkali denaturation of fetal and adult hemoglobins, the method has been widely applied to determine the amounts of "fetal" and "adult" hemoglobin in normal and pathological bloods. Recent discussions of the methodology have been given by JONXIS and HUISMAN *(75)* and CHERNOFF *(20)*. Because denaturation produces a change in the spectrum, the progress may be followed spectrophotometrically and the percentage of undenatured hemoglobin may be calculated at any given time. If two components that differ greatly in rate of denaturation are present, a result very much as shown in *Fig. 3* is observed. At pH values 12 to 13, the adult hemoglobin is completely denatured at room temperature in one to two minutes. If the portion of the curve of lesser slope is extrapolated to zero time, the percentage of the alkali-resistant fraction (in Fig. 3, about 20%) may be calculated. *Fig. 4* is taken from DRESCHER and KÜNZER's paper *(34)*. The presence of the embryonic hemoglobin is deduced from Curve *B* in which the rate of denaturation is intermediate between that of adult hemoglobin (Curve $C$) and fetal hemoglobin (Curve $A$). The statement that ,,...bei sehr jungen Feten im Alter von 7—12 Wochen kein fetales Hb nachweisbar ist" seems a straightforward conclusion. However, if we continue to examine Curve $B$ in the usual manner, by extrapolating the portion of lesser slope to zero time, the substance with the intermediate rate of denaturation constitutes only about 30% of the hemoglobin. The remainder has a rate of denaturation that approaches that of adult hemoglobin. Must we conclude that in very early fetal life still another hemoglobin which denatures rapidly is present or perhaps that adult hemoglobin is present? Yet, according to Curve $A$, both components must have disappeared by the 20th week of pregnancy. The evidence presented does not lead unambiguously to the interpretation of DRESCHER and KÜNZER.

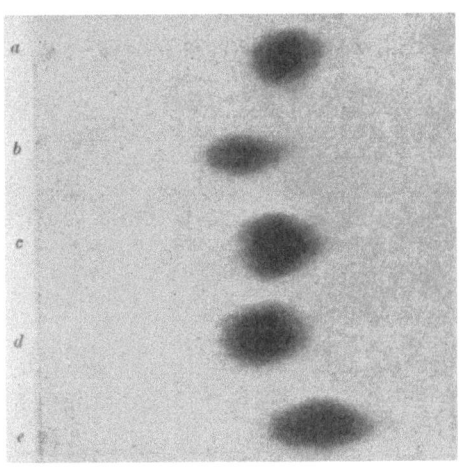

Fig. 5. Comparison of the paper electrophoretic behavior of hemoglobins from a normal adult (a and e) 20-week fetus (b), full-term cord blood (c), and mixture of b and c (d). The movement of the spots is toward the right. [From: Nature *178*, 794 (1956).]

Paper electrophoresis has been used by HALBRECHT and coworkers (*41, 42*) in their study of embryonic hemoglobin. They observed that the hemoglobin from a 10-week fetus moved more slowly than that from umbilical cord blood of the newborn or from adult blood. Indeed, the mobilities of material from 10- and 20-week fetuses were identical. *Fig. 5* is taken from the paper of HALBRECHT and KLIBANSKI (*41*). The results are rather curious if we consider the movement of Spot d in comparison with Spots b and c. Why, if Spot d contains the hemoglobins in Spots b and c, is its movement less than that of one of the component parts, namely, Spot c? Perhaps some unrecognized factor makes fetal hemoglobin move anomalously, and falsely suggests the presence of an embryonic hemoglobin. It is also of interest that in 20-week fetuses, HALBRECHT and co-workers find the embryonic hemoglobin but DRESCHER and KÜNZER do not. In a more extensive paper, HALBRECHT, KLIBANSKI, BRZOZA, and LAHAV (*42*) report the presence of the embryonic hemoglobin in 11 of 15 [9 of 15 are mentioned on p. 342 of Ref. (*42*)] fetuses that ranged in weight from 18 to 550 grams. On the basis of the relation of fetal weight to fetal age (*36*), these fetuses ranged from 12 to 23 weeks in age. Unfortunately, these authors do not report whether the fetuses that showed no embryonic hemoglobin were the older ones or whether there was random distribution throughout the sample.

Experiments that have been made in this Laboratory and that are still in progress may throw some light on the problem of the existence of an embryonic hemoglobin (*5*, *119*). In these experiments the

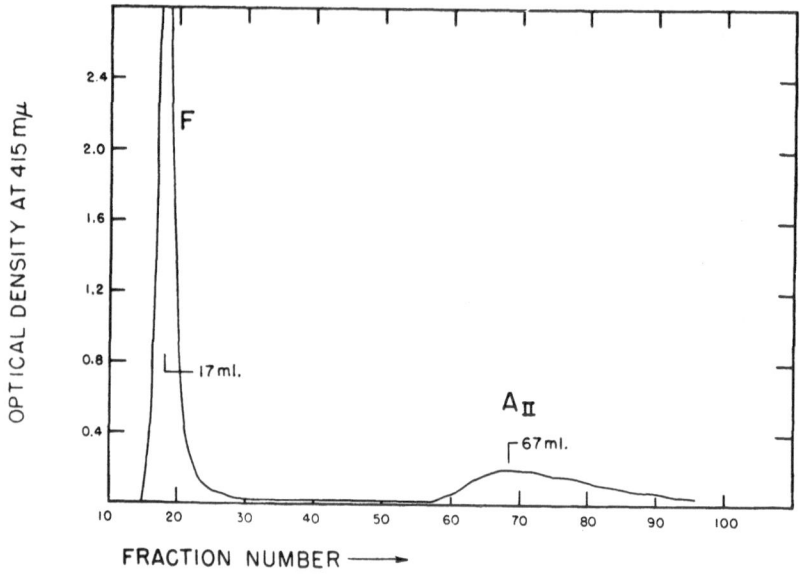

Fig. 6. Chromatogram of cord oxyhemoglobin on the ion-exchange resin IRC-50. In this and similar Figures optical density of the effluent from the column is plotted against the fraction number or effluent volume. [From: J. Amer. Chem. Soc. *80*, 1628 (1958).]

chromatographic method which was used easily and simply separates fetal and adult hemoglobin *(Fig. 6)*. In cord hemoglobin, components other than hemoglobins F and A are conspicuously present when Zone F is rechromatographed under proper conditions as shown in *Fig. 7*. The main component of cord hemoglobin is designated* as $F_{II}$ in this Figure and normally comprises 70 to 75% of the total hemoglobin. Zone $F_I$ accounts for about 10% of the total hemoglobin and it has been suggested (*5*) that it may be identical with the embryonic hemoglobin reported by others. It is unlikely, however, that this suggestion is correct: Zone $F_I$ is present to the extent of about 10% in all fetal samples (in age from 15 weeks to term) examined (*119*). Zone $F_I$ has a resistance to alkali denaturation that is of the same order as Zone $F_{II}$.

---

* A new nomenclature is not proposed here. It is a temporary nomenclature for reasons noted in Ref. (*5*).

Although the separation of hemoglobins A and F is incomplete by paper electrophoresis (Fig. 5), it may be made absurdly great by chromatography. Thus, it is to be expected that a separation of embryonic

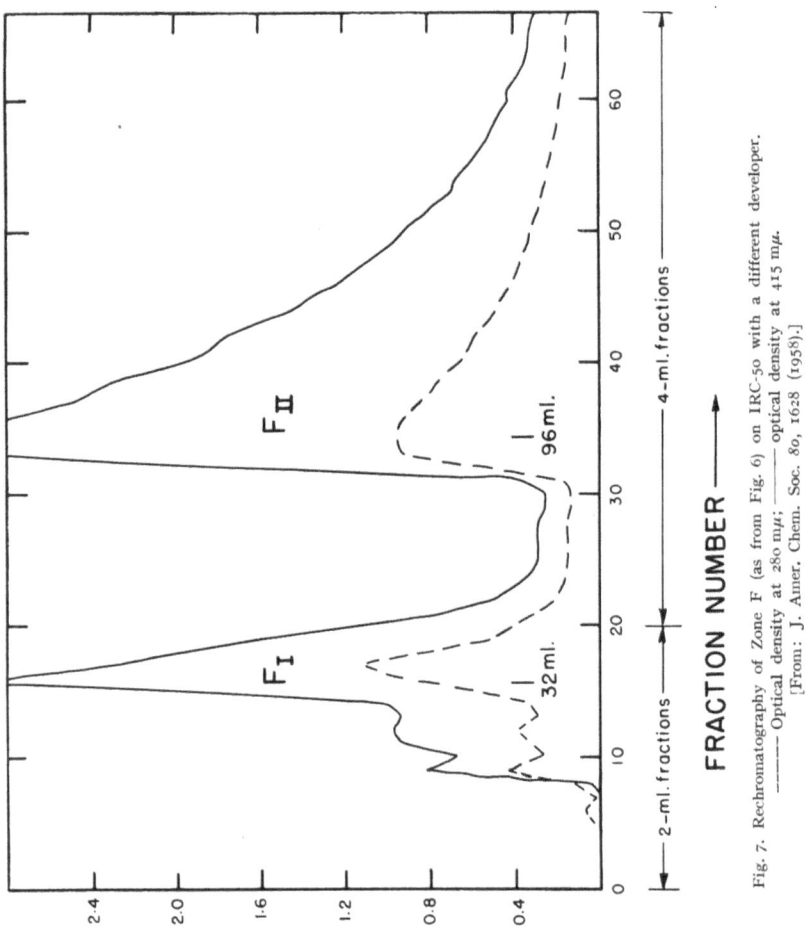

Fig. 7. Rechromatography of Zone F (as from Fig. 6) on IRC-50 with a different developer. —— Optical density at 280 mμ; ------ optical density at 415 mμ. [From: J. Amer. Chem. Soc. 80, 1628 (1958).]

hemoglobin and hemoglobin F of the order that is reported for paper electrophoresis (Fig. 5) could be greatly improved by chromatography. However, chromatographic studies of hemoglobin from fetuses in age from 15 weeks to term (119) have not demonstrated differences other than rather minor alterations in the proportions of the various components.

Zone $F_I$ is present to the extent of about 10% in all, Zone $F_{II}$ is roughly 70 to 75%, adult hemoglobin* is roughly 5 to 15%, and the remainder is composed of several minor components in only small amount (one of these constitutes about 2% in early samples and about 1% at term). Evidence for an embryonic hemoglobin has not been forthcoming. Chemical characterizations must be made to provide an unequivocal answer and these are now in progress.

## VII. The Purity of Hemoglobin in Hemoglobin Preparations.

The above comments on the components of cord blood hemoglobin point up a problem that is critical in any investigation of the structure of a protein: What is the purity of the protein sample? Related to this question are such questions as these: To what extent has the protein been altered by isolation? And, what is the stability of the isolated protein?

### 1. Isolation.

From the standpoint of isolation, hemoglobin is in many respects an ideal protein. Although the blood contains a complex mixture of proteins, the hemoglobin is essentially separately packaged in the red blood cells.

To isolate it, the cells are centrifuged and then washed repeatedly with 0.9% saline solution (which maintains the osmotic conditions of the blood and prevents the rupture of the cells) to remove the plasma proteins. The washed cells are then hemolyzed, for example, with distilled water and toluene. After high-speed centrifugation to remove cell remnants, the hemoglobin solution is readily separated from these insoluble parts. Recentrifugation removes the last traces of insoluble material and dialysis against distilled water takes out the inorganic ions and small molecules. Inasmuch as these operations may be (indeed, should be) carried out at 0 to 5°, the treatment to which the hemoglobin is subjected is short, simple, and mild compared to that often required in the isolation of other proteins.

### 2. Effect of the Isolation Procedure on the Purity of Hemoglobin Preparations.

Even as mild a procedure as that described above apparently is not without effect on the properties of hemoglobin. Thus, Riggs (106) notes that dialysis alters the oxygen dissociation curve. No doubt, this most important property of hemoglobin is adequately protected in the red cell even at the normal temperature of 37° but when the compound is in solution it is less stable even at temperatures around 0°. Definitive experiments on oxygen dissociation require the utmost care as Allen, Guthe, and Wyman (3) have also pointed out.

---

* This fraction has not been shown definitely to be adult hemoglobin and, indeed, seems to be composed of several components.

The importance of the effect of isolation on the properties of the protein is in reality relative and depends upon the property under consideration. For example, for anyone who is interested in the sequence of amino acid residues, an alteration in the oxygen dissociation curve is of little concern. Such an alteration is hardly likely to result in the cleavage of a peptide bond. Although the heme-globin link may well be disturbed by the isolation, our knowledge of the structure of hemoglobin is so slight at present that this effect has little direct bearing now on the determination of the structure. Its importance in the final determination of the structure cannot for a moment be denied.

Careless handling certainly can alter the properties of the hemoglobin molecule and even careful handling may produce a detectable alteration in some properties, but it appears that large alterations during the procedure of isolation can be kept to a minimum.

### 3. Heterogeneity of Hemoglobin Preparations.

As we have seen, the isolation of hemoglobin may produce a detectable effect on the molecule. What other evidence of homogeneity or heterogeneity exists? If hemolysates of normal adult red cells prepared as described above are examined electrophoretically or ultracentrifugally, there is little evidence of heterogeneity in the hemoglobin. Cord blood hemoglobin, however, has long been known to be heterogeneous through the evidence derived from alkali denaturation. Electrophoretically, the heterogeneity of cord hemoglobin is also detectable but ultracentrifugally it is not, because the molecular weights of the hemoglobins are very similar.

Evidence that normal adult hemoglobin is somewhat heterogeneous was indicated by alkali denaturation, immunological, and salting-out experiments [an excellent resume of work with each of these methods is presented in Refs. (20), (21), and (29)]. The incomplete denaturation of adult hemoglobin by alkali suggested that traces of fetal hemoglobin are present, and immunological methods likewise led to this conclusion. HUISMAN, JONXIS, and DOZY (54) have determined the amino acid composition of the alkali-resistant fraction of adult hemoglobin (0.3 to 0.4% of the total) and find that it is very similar to that of fetal hemoglobin. Salting-out procedures have indicated a considerable heterogeneity but the interpretation of the results has been criticized by ITANO [see discussion in (29)]. However, the application of the methods of starch block electrophoresis and chromatography has been necessary for a definite demonstration of the extent of the heterogeneity in hemolysates from normal adult blood and umbilical cord blood. Independently, KUNKEL and WALLENIUS (82) by starch block electrophoresis and MORRISON and COOK (93) by column chromatography detected the presence of minor components in hemolysates of adult red cells.

Kunkel and Wallenius carried out the starch block electrophoresis of normal adult hemoglobin at pH 8.6. Under these conditions the

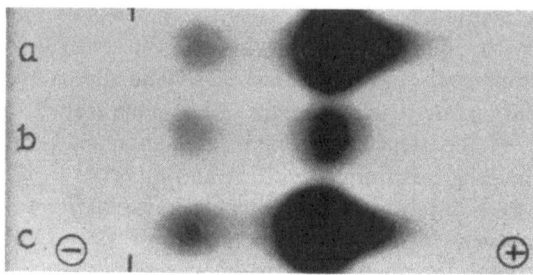

Fig. 8. Starch block electrophoresis of hemoglobin from adult red blood cells (a and c) as well as hemoglobin from hemoglobin E trait (heterozygous) cells (b). The site of application was at the lines. [From: Science *122*, 288 (1955).]

hemoglobins move anodically. As may be seen from *Fig. 8*, the main component is well separated from a slowly moving small spot but is poorly separated from a faster moving component which is apparent only because of the asymmetry of the main spot. The slowly moving

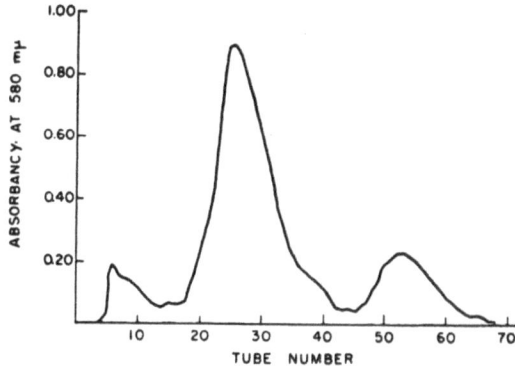

Fig. 9. Chromatography of normal adult oxyhemoglobin on IRC-50 at pH 6.3 with a gradient in sodium ion concentration. [From: Science *122*, 920 (1955).]

spot has been designated as hemoglobin $A_2$. Extensive investigation by Kunkel and co-workers (*81*) has shown that the amount in normal individuals is about 2.5% of the total hemoglobin; they have also studied many of its properties. Hemoglobin $A_2$ is not present in cord hemoglobin. These minor components are not artefacts of preparation or electrophoresis (*81*).

MORRISON and COOK (93) chromatographed normal adult hemoglobin on the carboxylic acid ion exchange resin IRC-50 and developed at pH 6.3 with a gradient in sodium ion concentration. In addition to the main component, both a more rapidly moving and a more slowly moving component were detected as *Fig. 9* shows. In the order of elution, the quantities were found to be about 10, 84, and 6% of the total hemoglobin. COOK and MORRISON (24) observed that the main component of cord hemoglobin behaved chromatographically as did the most rapidly moving

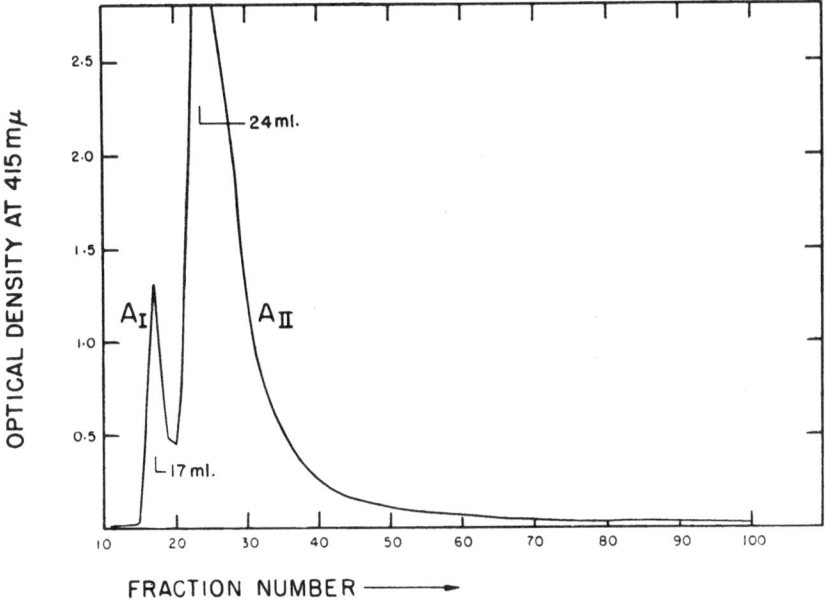

Fig. 10. Chromatogram of adult oxyhemoglobin on IRC-50. [From: J. Amer. Chem. Soc. *80*, 1628 (1958).]

component of adult hemoglobin but that it differed in other properties. Cord hemoglobin also contained components with chromatographic properties similar to the main component and the most slowly moving component of adult hemoglobin. PRINS and HUISMAN (102) by chromatography on IRC-50 under somewhat different conditions were unable to detect as many components in adult or cord hemoglobin as had MORRISON and COOK.

Amino acid analyses by STEIN and co-workers (121) have yielded very pertinent data. According to these analyses, the main component of adult hemoglobin as purified by starch block electrophoresis contains no isoleucine, whereas previously SCHROEDER, KAY, and WELLS (117), VAN DER SCHAAF and HUISMAN (129), and DUSTIN, SCHAPIRA, DREYFUS, and HESTERMANS-MEDARD (35) all reported the presence of small amounts of isoleucine, and only Rossi-Fanelli, Cavallini, and de Marco (*108*) failed to detect it. Thus, in most instances hemoglobin preparations must contain an impurity of rather high isoleucine content. This problem

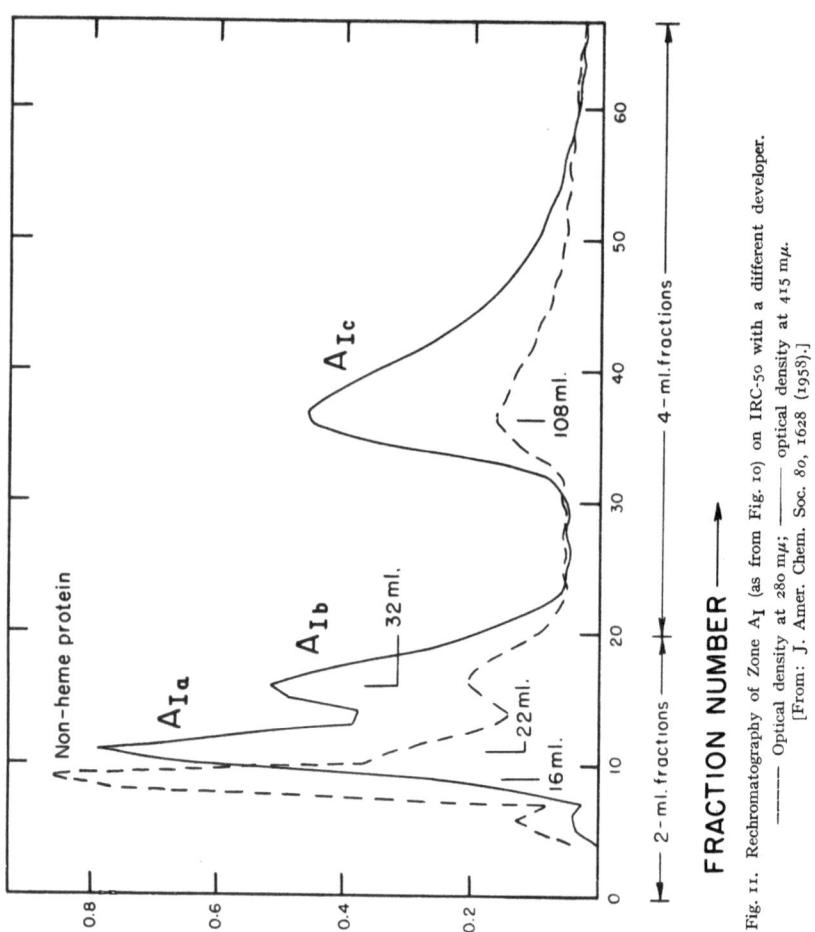

Fig. 11. Rechromatography of Zone A$_I$ (as from Fig. 10) on IRC-50 with a different developer. Optical density at 280 m$\mu$; —— optical density at 415 m$\mu$.
[From: J. Amer. Chem. Soc. 80, 1628 (1958).]

has received further attention in the chromatographic studies of hemoglobin by Allen, Schroeder, and Balog (*5*). They showed that crystallization was of no avail in purifying hemoglobin but rather that hemoglobin of crystallized preparations and hemolysates alike contained equal quantities of minor components and of isoleucine. If the minor components were removed by chromatography or electrophoresis, no isoleucine was detectable in the major component. We shall return to this observation when the amino acid composition of adult hemoglobin is discussed below.

ALLEN, SCHROEDER, and BALOG (5) have chromatographed hemoglobin on IRC-50 under conditions rather different from those of MORRISON and COOK (93) or of PRINS and HUISMAN (102). They, like PRINS and HUISMAN, observed only 2 zones on some chromatograms (*Fig. 10*, p. 343);

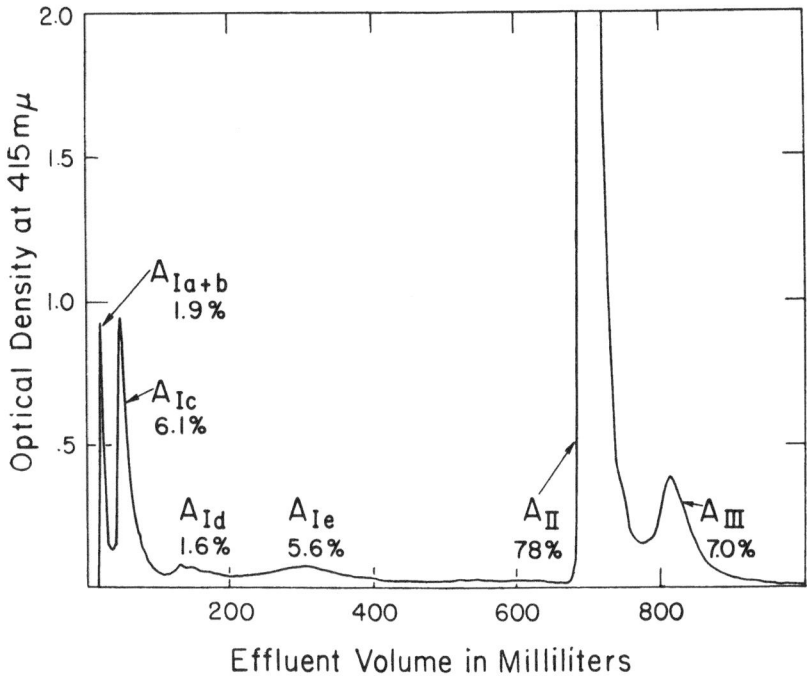

Fig. 12. Chromatogram of adult oxyhemoglobin on IRC-50 under improved conditions similar to those in Fig. 11 (22).

the smaller moved the more rapidly but the slowly moving zone of MORRISON and COOK was not evident. However, when Zone $A_I$ (Fig. 10) was rechromatographed under different conditions, one non-heme protein and three heme-containing proteins were detected *(Fig. 11)*. MORRISON (92) has mentioned that his most rapidly moving fraction (presumably equivalent to Zone $A_I$) can be separated into two components. Further use and improvement of the methods of ALLEN et al. (5) has led to the detection of a total of 6 or 7 components in addition to the main component (22). A typical result is presented in *Fig. 12*. The developer for this chromatogram is slightly stronger than that used for the chromatogram depicted in Fig. 11. As a result Zones $A_{Ia}$ and $A_{Ib}$

emerge together, but Zone $A_{Ic}$ is separate and Zones $A_{Id}$ and $A_{Ie}$ are now apparent. The movement of Zone $A_{II}$ is so slow at 6° with this developer that it and Zone $A_{III}$ are eluted from the column by warming to room temperature after the effluent volume has reached about 700 ml (Fig. 12). Zone $A_{III}$ may be identical with MORRISON and COOK's most slowly moving zone. Zone $A_{III}$ often breaks up into two zones but it is uncertain yet whether this is due to human variation or to some uncontrolled chromatographic variable.

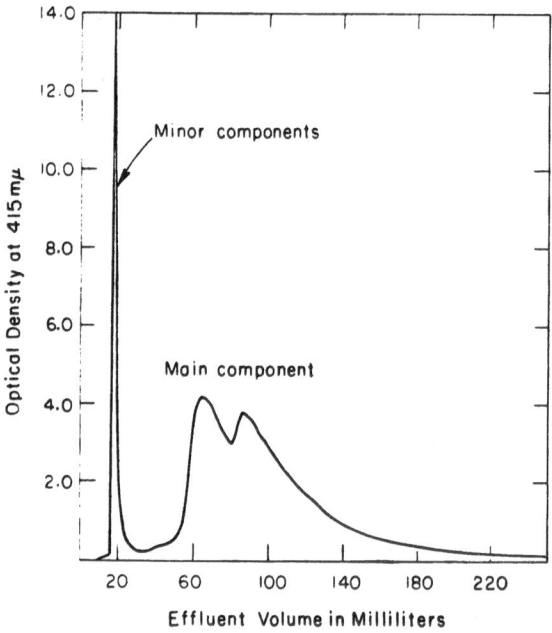

Fig. 13. Apparent chromatographic heterogeneity of the main zone of sickle-cell hemoglobin. [Unpublished experiments (73).]

One cannot with confidence say that the main zone of adult hemoglobin as isolated chromatographically under these improved conditions is entirely homogeneous. Indeed, some chromatograms suggest that it is not. The effect is best illustrated in chromatograms of sickle-cell anemia hemoglobin. Each of the more than twenty samples of sickle-cell hemoglobin exhibited the heterogeneity which is depicted in *Fig. 13 (73)*. This heterogeneity is readily apparent as the zone moves down the column unless an excessively large amount of hemoglobin (150 mg. on a 1 × 35-cm. column) is chromatographed in which case it seems to be homogeneous*. Under some conditions of chromatography, a similar

---

* Of the minor components in sickle-cell hemoglobin, some behave chromatographically like those in normal adult hemoglobin and others do not.

though somewhat less marked heterogeneity is observed with normal
hemoglobin: in this instance, it is apparent while the zone is on the
column but is less likely to be evident in the graph of the optical densities
of the fractions versus the fraction number.

HUISMAN, MARTIS, and DOZY (55) have described the chromatography
of human hemoglobin on a different ion exchanger, namely carboxymethylcellulose. They have used developers of low ionic strength coupled
with a gradient in pH. Although the main purpose of their experiments
was to separate the various normal and abnormal human hemoglobins,
the heterogeneity of normal adult hemoglobin shows up very prominently
on their chromatograms. Their results are essentially similar to those
of MORRISON and COOK (Fig. 9, p. 342) in that a minor zone both precedes
and follows the main component.

Methodologically, their procedures may be very useful. The capacity of
carboxymethylcellulose seems to be somewhat greater than that of IRC-50. In
addition, under the conditions used, the zones are well-shaped and the distribution
in the effluent is very symmetrical about the peak of the zone quite in contrast
to the skewed distribution in the effluent from IRC-50 columns (for example,
Figs. 6, 9, and 11, pp. 338, 342, 344).

The heterogeneity of cord hemoglobin has already been discussed
on p. 338.

## 4. Biological Significance of the Heterogeneity of Hemoglobin Preparations.

The minor components in hemoglobin preparations actually are present
in the red blood cell; they are not artefacts of preparation or of the
method of isolation (5, 81). Hence, they are likely to be of far more
interest to physiologists and geneticists than to those who are interested
in the chemical structure of hemoglobin. The latter can consider them
as "impurities" of which they are now happily aware and for whose
removal methods are available so that the study of the structure can
be pursued with a purer substance. The physiologist and geneticist,
however, ask what the function and source of the minor components
may be. One may speculate that they are under strict genetic control
as the major component(s) apparently is. Or one may suggest that they
mark the path of the history of the red cell from immaturity to destruction:
some may be incomplete hemoglobin in the process of synthesis, others
may be old hemoglobin from cells about to be destroyed, while the main
material represents the long period between these extremes. We may
ask further whether each cell contains all components or whether they
are segregated in individual cells. An insight into their function would

There is a greater variation in the number and amount of the minor components
in different samples of sickle-cell hemoglobin than is observed in different samples
of normal hemoglobin (73).

no doubt be gained if some of these questions were answered. In essence, these questions about function and source require some knowledge of structure on which the answers may be based and they will go unanswered until this knowledge is available. One significant piece of information is the fact that at least some of the minor components of adult hemoglobin contain isoleucine (5) and it is, therefore, unlikely that they mark part of the path of the synthesis and destruction of the main component which contains no isoleucine.

It is difficult to assess the significance of the apparent chromatographic heterogeneity that is shown by the main component of normal and sickle-cell hemoglobin. It is possible that we may be observing a dissociation of the molecule. Although the pH (about 7) at which the chromatograms are developed is not one at which the molecule dissociates in solution (see Fig. 1, p. 328), we cannot discount the possible influence of the ion exchange resin. Preliminary investigations suggest, however, that asymmetric dissociation is not responsible for this effect (115).

Although it has long been known that cord blood hemoglobin contained at least two components, conclusive demonstration that normal adult hemoglobin is heterogeneous has been made very recently. One must, therefore, realize that all but a few experiments have been made with a heterogeneous substance. This fact may explain some of the disagreements and discrepancies in the literature and it brings into question the meaning of carefully determined constants. Many authors have been distressingly lax in describing the details of the preparation of hemoglobin samples and in characterizing them. Because methods are now available for purification and characterization of hemoglobin preparations, an author can no longer be excused from the task of describing clearly the nature and purity of the sample with which he has worked.

Up to this point in this paper, we have intentionally refrained from using the terms "hemoglobin A", "hemoglobin F", etc., to any extent and have instead used "adult hemoglobin", "fetal hemoglobin", etc. We shall continue to do this because we wish to place upon the words "adult hemoglobin" and "fetal hemoglobin" the connotation of a material that has been isolated from the red cell but has not been rigorously purified. The terms "hemoglobin A" and "hemoglobin F" should be reserved for the main hemoglobin component of adult and cord blood.

## VIII. Chemical Investigations of Hemoglobin.

Hemoglobin has more and more been subjected to study by the methods of protein chemistry that have been devised in increasing numbers during recent years. The discovery of the first abnormal human hemoglobin was immediately followed by a determination of its amino acid composition by Moore and Stein's then newly developed method of starch chromatography. Its composition was compared with that of normal adult hemoglobin. Earlier, SANGER had applied his new method for N-terminal residues to several hemoglobins including adult and fetal human hemoglobin. We shall now consider the results of these and some other chemical investigations.

## 1. Amino Acid Composition.

### a) Adult Hemoglobin.

Because of the very evident heterogeneity of hemoglobin in most preparations, the significance of most of the published amino acid analyses may be called into question. On the other hand, the analyses have not been very successful in demonstrating a definite difference in the amino acid composition of the abnormal human hemoglobins, and it may be that the minor components and the main component are so similar in amino acid composition that the results are little influenced by the heterogeneity. Although no definitive analysis has been made on carefully purified hemoglobin, the published analyses should give some insight into the general amino acid composition of the molecule. We shall now consider only the results of complete analyses by modern methods.

References to older fragmentary analyses are given by SCHROEDER, KAY, and WELLS (*117*).

The analyses in *Table 2* are recorded as the nearest integral number of residues of each amino acid per 66,000 molecular weight and have been recalculated to a common basis from the results of the several authors. Somewhat unwarranted use has thus been made of the data of STEIN et al. (*121*). These authors clearly point out that they have made only a single complete analysis of a 22- and 70-hr. hydrolysate of each sample and hence that the results are not sufficiently extensive to permit the formulation of a complete amino acid composition. However, inasmuch as they present information on four samples of hemoglobin A from different sources, the values in Table 2 have been arrived at by averaging their values for each amino acid after discarding a few that seem unusually aberrant and after choosing the appropriate ones that reflect the influence of time of hydrolysis.

Before considering the results proper, let us first compare various features of the experimental work of the several authors such as preparation of hemoglobin, conditions of hydrolysis, methods of analysis, and replication of analyses.

Samples of hemoglobin prepared in several ways have been used for analysis. Thus, SCHROEDER, KAY, and WELLS (*117*), ALLEN and SCHROEDER (*4*), and ROSSI-FANELLI, CAVALLINI, and DE MARCO (*108*)

Table 2. Number of Amino Acid Residues per 66,000 Molecular Weight in Adult and Fetal Hemoglobin.

| Amino Acid | Adult Hemoglobin | | | | | | | Fetal Hemoglobin | | | |
|---|---|---|---|---|---|---|---|---|---|---|---|
| | SCHROEDER et al. (117) | DUSTIN et al. (35) | VAN DER SCHAAF et al. (129) | ROSSI-FANELLI et al. (108) | STEIN et al. (121) | ALLEN et al. (4) | No. of Residues per Molecule (?) | VAN DER SCHAAF et al. (129) | STEIN et al. (121) | No. of Residues per Molecule (?) | Residues in F Compared to A |
| Alanine | 76 | 78 | 76 | 73 | (70) | | 76 | 71 | 64 | 68 | —8 |
| Amide | 36 | — | 43 | 46 | — | | 26 | 43 | — | — | |
| Arginine | 13 | (10) | 13 | 13 | 12 | | 14 | 13 | 12 | 12 | ±1 |
| Aspartic acid | (55) | 55 | 53 | 50 | 49 | (57) | 52 | 53 | 49 | 52 | ±1 |
| Cystine/2 | — | — | 6 | 5 | 4—5 | | | 5 | 3—4 | — | |
| Glutamic acid | 32 | 33 | 32 | 33 | 30 | 32 | 32 | 34 | 32 | 34 | ±1 |
| Glycine | 43 | 42 | 40 | 38 | 39 | | 40 | 40 | 39 | 40 | ±1 |
| Histidine | 36 | 35 | 36 | 36 | 36 | | 36 | 32 | 33 | 32 | —4 |
| Isoleucine | 1 | 1 | 2 | 0 | 0 | | 0 | 9 | 7 | 8 | +8 |
| Leucine | 75 | 74 | 77 | 75 | 72 | | 74 | 76 | 69 | 74 | ±1 |
| Lysine | 44 | (40) | 48 | 48 | 42 | | 46 | 48 | 45 | 46 | ±1 |
| Methionine | 6 | 6 | 7 | 5 | 5 | | 6 | 9 | 7 | 8 | +2? |
| Phenylalanine | 31 | (34) | 32 | (38) | 29 | 30 | 30 | 32 | 29 | 30 | ±1 |
| Proline | (25) | (34) | 29 | 29 | 29 | | 30 | 25 | 21 | 24 | —6 |
| Serine | 32 | 33 | 35 | 32 | (29) | | 34 | 43 | 37 | 40 | +6 |
| Threonine | 32 | 32 | 34 | 33 | (30) | | 34 | 40 | 36 | 38 | +4 |
| Tryptophan | 5 | — | 4 | 7 | — | | 6 | 5 | — | ? | ? |
| Tyrosine | 11 | 13 | (16) | 11 | 11 | 12 | 12 | 13 | 9 | 12 | ±1 |
| Valine | 60 | (58) | 62 | 62 | 63 | | 62 | 54 | 53 | 54 | —8 |

analyzed crystallized hemoglobin; VAN DER SCHAAF and HUISMAN (*129*) used the hemoglobin directly from hemolysates; DUSTIN et al. (*35*) do not mention specifically whether crystallized material was used; and STEIN et al. (*121*) used an electrophoretically purified preparation. No doubt then the latter was the purest preparation analyzed.

Only VAN DER SCHAAF and HUISMAN (*129*) have made an extensive study of the effect of the duration of hydrolysis on the analytical results and have corrected for increased release or destruction as a function of time. STEIN et al. (*121*) used two times of hydrolysis but their data are too incomplete except for serine and threonine to permit a definite decision as to increase or loss. SCHROEDER, KAY, and WELLS (*117*) do not stress this point although ALLEN and SCHROEDER (*4*) were careful to determine the effect in their partial analysis. Some authors (*35, 108*) have not specifically described the exact conditions and time of hydrolysis.

Ion exchange chromatography has been the method of analysis by all authors except by SCHROEDER, KAY, and WELLS who used starch chromatography for what was the first complete amino acid analysis on a single sample of human hemoglobin.

The data of Table 2 are the average of rather different numbers of replicate determinations by the several authors: ten to fifteen (*117*), not given (*35*), ten (*129*), three (*108*), two on each of four samples (*121*), and sixteen (*4*).

Basically, the chromatographic procedures that were employed for these analyses all have about the same accuracy. For the most part, the recovery of quantities in a known amino acid mixture will be 100 ± 3% although the absolute quantity as well as the individual peculiarities of each amino acid will influence the accuracy and precision. Although we might expect the average of replicate determinations in different laboratories under ideal conditions to agree to better than ± 3%, yet we will probably not be far wrong if we decide that in the determinations by various authors a difference of 5 to 6% would not be exorbitantly great. Such a difference is one residue in fifteen to twenty. If we accept this criterion, then it is met, or only slightly exceeded, by glutamic acid, histidine, and leucine; and, if we omit the one most divergent value (in parentheses in Table 2), also by alanine, arginine, tyrosine, and valine. We shall now consider the discrepancies in the data for the other amino acids.

The determination of *amide* groups in the form of ammonia is somewhat of a by-product of amino acid analysis. Surprisingly enough, VAN DER SCHAAF and HUISMAN's value which was carefully determined by correcting for the increased amount of ammonia as a function of the time of hydrolysis is higher than that of SCHROEDER et al. who determined the amount at a single time of hydrolysis. DICKMAN and MONCRIEF (*30*) in an investigation of the amide groups alone report only 26 per molecule.

Of the determinations of *aspartic acid*, those of Allen and Schroeder (*4*) and Schroeder, Kay, and Wells (*117*) must be treated with some caution and probably should be ignored. The former was arrived at after applying a 15% correction and the latter is somewhat uncertain because of difficulties with recovery experiments. Van der Schaaf and Huisman's value is an extrapolated one because of the observed loss as a function of time of hydrolysis.

*Cystine*, or *cysteine*, is difficult to determine under any conditions. Probably the best value is that of Stein et al. (*121*) which was estimated as cysteic acid after oxidation of the protein. The cysteine, or sulfhydryl, content of hemoglobin is the subject of much controversy and will be discussed at length on p. 355.

There is no reason to suspect any of the determinations of *glycine* although they cover a somewhat greater range than expected.

Pure adult hemoglobin contains no *isoleucine*. This subject was discussed on p. 343.

The data for *lysine* show a rather wide range. Van der Schaaf and Huisman's value resulted after consideration of the hydrolytic losses whereas the other authors did not take this into account. The lysine content reported by Dustin et al. seems rather low.

*Methionine* is so difficult to determine with accuracy that the agreement of the reported amounts is surprising.

The careful determination of *phenylalanine* by ion exchange chromatography (*4*) agrees almost exactly with that obtained some years ago by starch chromatography (*117*). The data of Dustin et al. and Rossi-Fanelli et al. are unaccountably high.

Data of Rhinesmith, Schroeder, and Pauling (*104*) indicate that Schroeder, Kay, and Wells' result for *proline* is incorrect and that on correction it would agree with those of van der Schaaf et al., Rossi-Fanelli et al., and Stein et al.

Only van der Schaaf et al. and Stein et al. correct the data for hydrolytic destruction of *serine* and *threonine*. Even with correction, the values of Stein et al. are lowest of all. If we assume that the correction applied by van der Schaaf and Huisman is applicable to the others, the agreement is rather satisfactory.

*Tryptophan* is very difficult to determine accurately. Schroeder, Kay, and Wells on the basis of many control experiments showed that the tryptophan that survived basic hydrolysis could be accurately determined. Any value, thus, is a minimum value because loss during hydrolysis of the peptide bonds cannot be measured.

Van der Schaaf and Huisman's value for *tyrosine* is the only one that disagrees appreciably with the others.

With the exception of isoleucine which is absent in pure adult hemoglobin, the data of Stein et al. on what probably are the purest samples yet analyzed do not show striking deviations from the previously reported values.

With the knowledge that such use of the data must be done with caution, we have attempted to present a statement of the probable number of residues of each amino acid in the molecule. These numbers are all even because of the preponderant evidence (p. 363) that the molecule is made up of identical halves. Inasmuch as the isoelectric point is approximately at pH 7, Dickman and Moncrief's value (*30*) of 26 amide groups is more acceptable than the others because then the sum of aspartic and glutamic acid residues minus the amide groups essentially equals the sum of lysine and arginine residues.

### b) Fetal Hemoglobin.

Four analyses of fetal hemoglobin by ion exchange chromatography have been reported in the literature [Dustin, Shapira, Dreyfus, and Hestermans-Medard (*35*); Rossi-Fanelli, Cavallini, de Marco, and Trasatti (*109*); van der Schaaf and Huisman (*129*); and Stein, Kunkel, Cole, Spackman, and Moore (*121*)]. Dustin et al. and Rossi-Fanelli et al. analyzed crystallized material but because they present no evidence to show that adult hemoglobin had been removed by crystallization, these analyses will not be considered further. Van der Schaaf and Huisman first treated cord hemoglobin with alkali to denature the adult hemoglobin and then analyzed the alkali-resistant remainder. Their sample, however, must have been heterogeneous to the extent of about 10% because Zone $F_I$ (Fig. 7, p. 339) has essentially the same alkali resistance as does Zone $F_{II}$ (*88*) and, therefore, is not removed by such a procedure. Stein et al. analyzed fetal hemoglobin that had been isolated by starch block electrophoresis. Presumably, adult hemoglobin had been removed, but the extent of other heterogeneity cannot be estimated.

As in their analyses of adult hemoglobin, van der Schaaf and Huisman carefully studied the effect of varied times of hydrolysis upon the results whereas Stein et al., as they themselves acknowledge, have made only two determinations.

The data for fetal hemoglobin are presented in Table 2 (p. 350), again in terms of integral residues per 66,000 molecular weight. The agreement of the data is satisfactory in many instances although there are some very discrepant values. Although a suggested number of residues of each amino acid per molecule is again given, it is with greater reservation than in the case of adult hemoglobin and no doubt with some prejudice as to which set of data is to be assigned more weight. This information is presented mainly to aid in the comparison of the amino acid composition of the two proteins.

*c) Comparison of the Amino Acid Composition of Adult and Fetal Hemoglobin.*

The most striking difference in amino acid composition lies in the content of isoleucine. Whereas it is completely absent in adult hemoglobin, approximately 8 residues are to be found in fetal hemoglobin. Differences in content of other amino acids are also apparent, although the extent of the difference is doubtful at least to $\pm 2$ residues of that given in Table 2 (p. 350). Thus, in fetal hemoglobin as compared to adult hemoglobin, there is more isoleucine, methionine, serine, and threonine and less alanine, histidine, proline, and valine. These conclusions agree qualitatively with van der Schaaf and Huisman's (*129*) who also suggest (with less confidence) that more glutamic acid and less tyrosine may be present.

Unfortunately, the analyses for tryptophan in both proteins are few in number and probably not very reliable because of the difficulty in analyzing for this amino acid. A more satisfactory determination might be possible by the use of N-bromosuccinimide as suggested by the recent experiments of Patchornik, Lawson, and Witkop (*96*) who find this reagent to be very specific for tryptophyl bonds.

Some difference in tryptophan content is suggested by the spectra. The spectra of adult and fetal hemoglobin are essentially identical in positions of the absorption maxima and in extinction coefficients [Beaven, Hoch, and Holiday (*12*)] with the exception of the region around 290 m$\mu$ [Jope (*76*)]. Adult hemoglobin shows an inflection or a barely detectable maximum at 291.0 m$\mu$ whereas fetal hemoglobin shows a more definite maximum at 289.8 m$\mu$ (*76*). The definite maximum in the case of fetal hemoglobin is attributed to tryptophan and, presumably, reflects a difference in tryptophan content. The absorption due to tryptophan is not an isolated maximum in the spectrum but is super-imposed upon the rather broad absorption maximum at about 280 m$\mu$ which is due to phenylalanine and tyrosine and is so characteristic of all proteins. Jope suggests that fetal hemoglobin must have much less phenylalanine than adult hemoglobin, a conclusion that is not borne out by the amino acid analyses. If the phenylalanine and tyrosine contents are essentially identical in the two proteins, and this seems to be so, the difference in the spectrum presumably is due to tryptophan only. No one, apparently, has attempted to calculate to what extent the tryptophan contents must differ to produce the observed spectra.

## 2. Sulfhydryl Groups.

The determination of the cystine or cysteine content of a protein is as uncertain and difficult as the determination of tryptophan. Despite varied attacks on the problem by many investigators, the answers that have been obtained with the human hemoglobin are many and contradictory.

Any attempt to determine cysteine or cystine chemically in the hydrolysate of a protein is useless because of the lability of the compounds. Probably the best chemical method, therefore, is that of SCHRAM, MOORE, and BIGWOOD (113) in which the protein is first reacted with performic acid to oxidize cystine and/or cysteine (which, hence, cannot be differentiated) to cysteic acid. After hydrolysis, the cysteic acid is separated chromatographically on Dowex-2 and determined by the customary ninhydrin procedure. The method is not without its uncertainties. For example, under the conditions of the procedure, cysteine and cystine yield only about 90% of the theoretical amount of cysteic acid and the use of a correction factor is, therefore, required.

HOMMES, SANTEMA-DRINKWAARD, and HUISMAN (49) have applied this procedure to adult and fetal hemoglobin. The hemoglobins were prepared in the same manner as described above (pp. 351, 353). They conclude that adult hemoglobin contains eight half-cystine residues and fetal hemoglobin six. STEIN et al. (121) have likewise applied this method to their electrophoretically purified hemoglobins: the results do not yield integral numbers but suggest four to five half-cystine residues in adult hemoglobin and three to four in fetal hemoglobin. BROWN (18) by purely chemical methods has determined the sulfur distribution in adult globin. He reports about six half-cystine residues as well as five methionine residues (by difference from the total sulfur). STEIN et al. (121) suggest that discrepancy between their results and those of HOMMES et al. (49), as well as the somewhat erratic results that they themselves obtained, may be attributed to impurities that have been incompletely removed. This suggestion may well be valid: the reported isoleucine content of adult hemoglobin is now known to be due to "impurities" of high isoleucine content, and the possibility of "impurities" of high cysteine and/or cystine content cannot be ignored.

Such analyses cannot ascertain whether cysteine or cystine is present in the molecule and yet this factor is of vast import to its structure. As a result, many investigators have attempted to determine the number of sulfhydryl groups (hence, cysteine residues) by an investigation of the number of silver or mercury ions that each molecule of hemoglobin will bind. Inherent in this method is the assumption which is difficult to substantiate that the binding of these ions is done only by the free sulfhydryl groups. This subject has been investigated in detail by Ingbar and Kass (59); Benesch, Lardy, and Benesch (13); Ingram (61, 64); Hommes, Santema-Drinkwaard, and Huisman (49); Hommes, Dozy, and Huisman (48); Stein, Kunkel, Cole, Spackman, and Moore (121); Murayama (94, 95); and Allison and Cecil (7). An interesting discussion of this topic by several of the investigators involved has recently been published (23, there p. 238).

The binding of silver ion by adult hemoglobin differs in the native and denatured protein. According to Ingram (61) and Murayama (94), four reactive sulfhydryl groups are present in the native protein and eight in the denatured hemoglobin (13, 49, 61, 94). In considerable contrast, Stein et al. (121) could detect only four to five sulfhydryl groups in their electrophoretically purified samples in agreement with their analyses by the cysteic acid method.

Silver because it is univalent can combine with only one sulfhydryl group. Mercuric ion, on the other hand, can combine with two if the steric relationships are propitious. Indeed, Ingram (61) and Murayama (94) report that only six mercuric ions combine with denatured adult hemoglobin. Allison and Cecil (7) have titrated both with mercuric chloride and phenylmercuric hydroxide. The mercury in the latter must behave as a univalent ion and hence like silver. They find that only two of each ion combine with native protein and six with the denatured protein and conclude that silver ion must have bound to two sites other than sulfhydryl in the experiments of other workers. They conclude that there are only six instead of eight sulfhydryl groups in adult hemoglobin.

Cole, Stein, and Moore (22a) have examined the problem further with several preparations of hemoglobin and a variety of methods and conclude that about five cysteine residues are present.

Murayama, on the basis of titrations with both silver and mercury at 0° and 38°, has discussed extensively the possibilities of the steric relationships of the sulfhydryl groups to each other in hemoglobins A, F, S, and C.

Benesch, Lardy, and Benesch (13) concluded that no disulfide bonds exist in adult hemoglobin. Stein et al. (121) reach the same conclusion on the basis of the content of half-cystine by the cysteic acid method and of sulfhydryl by silver titration. The same conclusion derives from the data of Hommes et al. (49) but the numbers on which these conclusions are based are very different.

Similarly unsatisfactory experimental agreement is to be found in the data from fetal hemoglobin. Hommes et al. (49) found six half-cystine residues by the cysteic acid method and only four sulfhydryl groups by titration and are led to the conclusion [cf. also (48)] that a disulfide bond exists in the molecule. Six silver ions and three mercuric ions were bound in MURAYAMA's experiments (95) which are not designed to detect a disulfide bond. However, no evidence for a disulfide bond is present in the data of STEIN et al. (121) (three to four half-cystine residues by the cysteic acid method and an equal number by silver titration).

One cannot draw definite conclusions from the mass of conflicting data that has been briefly presented above. Much of the contradiction very likely stems from the minor components that were present in all but the samples of STEIN et al. (121) and from the difficultly justifiable assumptions that must be made as to the binding site of the silver or mercuric ions. Taken all in all, the evidence at the present time would suggest six sulfhydryl groups and no disulfide bonds in adult hemoglobin and perhaps four sulfhydryl groups in fetal hemoglobin. The evidence for or against a disulfide bond in fetal hemoglobin is less positive, although unpublished data to be discussed later (p. 367) indicate the presence of a relatively strong interchain bond of some sort.

### 3. N-Terminal Amino Acid Residues and Sequences.

Since the elaboration of SANGER's method for the determination of N-terminal amino acids*, the method has been widely used as the first step in the determination of the structure of a protein. Indeed, hemoglobin was one of the first proteins to which SANGER applied the method after his initial success in identifying the N-terminal residues of insulin. SANGER's method is based upon the reaction of dinitrofluorobenzene with the free amino groups of the protein to produce a yellow dinitrophenyl (DNP) protein. The bond between the DNP group and the amino nitrogen is usually more stable than the peptide bonds that link the amino acid residues so that, after complete hydrolysis, the hydrolysate will contain the N-terminal amino acid(s) which now has been tagged with the DNP group. A DNP-amino acid is no longer a dipolar ion and can, therefore, be extracted from the hydrolysate. Identification by chromatographic means is then easy. As one might expect, partial hydrolysis of the DNP-protein will lead to DNP-peptides, the determination of whose structure will identify the sequence or sequences in the vicinity of the N-termini. We shall now discuss what is known about the N-termini of adult and fetal hemoglobin.

### a) Adult Hemoglobin.

PORTER and SANGER (101) found five N-terminal valyl residues in adult hemoglobin. Since then, the value has been redetermined frequently, usually for the comparison of the N-terminal residues of normal with

---

* An N-terminal amino acid is defined as one with a free α-amino group and likewise a C-terminal amino acid as one with a free α-carboxyl group.

abnormal hemoglobins. Just as PORTER and SANGER, the following authors found five valyl residues in normal adult hemoglobin: HAVINGA (47); SCHAPIRA and DREYFUS (112); HUISMAN and DRINK-WAARD (53); and BROWN (17). MASRI and SINGER (87) reported that only four valyl residues were present but they do not make special comment about this finding. SCHRAMM, SCHNEIDER, and ANDERER (114) also found four valyl residues by a different technique; they used EDMAN's method in which the free amino group is first reacted with phenylisothiocyanate and the N-terminal amino acid is then split off as the phenylthiohydantoin. More recently, RHINESMITH, SCHROEDER, and PAULING (104) detected only 3.6 N-terminal valyl residues in adult human hemoglobin by SANGER's method. This result not only disagrees with most other determinations but is, of course, absurd because it is non-integral.

Because of this result, RHINESMITH, SCHROEDER, and PAULING undertook a detailed study of the quantitative aspects of SANGER's method. It must be realized that when a DNP-protein is hydrolyzed some destruction of the N-terminal DNP-amino acid occurs and that the extent of the destruction is largely determined by the structure of the DNP-amino acid itself. Most of the above authors have applied corrections of 20 to 35% to compensate for the hydrolytic destruction of DNP-valine when DNP-hemoglobin is hydrolyzed. Such correction factors are equivalent to 1 to 1.5 residues per molecule. RHINESMITH, SCHROEDER, and PAULING, on the basis of extensive control experiments, concluded that their correction was only 13%, an amount that was the sum of all operational losses (hydrolytic, extractive, and chromatographic). During these experiments, the N-terminal peptide, DNP-val-leu, was observed and identified even in 22-hr. hydrolysates. In further investigations (105), they studied the release of this peptide and its hydrolysis as well as the release of DNP-valine as a function of time of hydrolysis. The DNP-val-leu is released essentially quantitatively in an amount equivalent to two N-terminal residues per molecule within 15 minutes in refluxing 6 N hydrochloric acid and it then hydrolyzes at the rate expected of this peptide under these conditions. Consideration of these data led to the conclusion that four N-terminal residues are present in the molecule and that there exist two kinds of N-terminal sequences, one of which is val-leu. Several explanations were given for the non-integral number of end groups previously reported. RHINESMITH, SCHROEDER, and MARTIN (103) then were able to isolate two additional N-terminal peptides from partial hydrolysates of DNP-globin and to identify them as di-DNP-val-his and di-DNP-val-his-leu. The results were less quantitative than in the isolation of DNP-val-leu in that they totaled only about 1.3 end groups instead of 2 per molecule. It has not been possible to isolate longer N-terminal sequences.

Hence, there is now little doubt that adult hemoglobin has four N-terminal valyl residues per molecule. These four residues terminate sequences of two kinds and suggest that the molecule consists of identical halves, a conclusion supported by other evidence discussed on p. 363. The two polypeptide chains that terminate in the sequence val-leu have been named the $\alpha$ chains and those that terminate in val-his-leu the $\beta$ chains (*103*).

### b) Fetal Hemoglobin.

PORTER and SANGER (*101*) reported that fetal hemoglobin from a 30-week fetus contained 2.6 N-terminal valyl residues. HUISMAN and DRINKWAARD (*53*) and MASRI and SINGER (*87*) reported the presence of two N-terminal valyl residues; the former used the alkali-resistant fraction of cord hemoglobin and the latter corrected for the valyl residues that stemmed from the known content of adult hemoglobin in the sample. SCHROEDER and MATSUDA (*118*), who used chromatographically purified hemoglobin F from cord blood, substantiated the presence of two N-terminal valyl residues but, in addition, determined the existence of two N-terminal glycyl residues. DNP-Glycine is rather readily destroyed during hydrolysis and its presence was probably not detected by the earlier investigators because they used periods of hydrolysis from 16 to 48 hrs., during which the DNP-glycine is largely or completely destroyed. SCHROEDER and MATSUDA used a period of hydrolysis as short as 1 hour and succeeded in obtaining an almost quantitative yield of DNP-glycine. In addition, they determined that the N-terminal sequence of the valyl chains was val-leu. There is good evidence that the N-terminal sequence of the glycyl chains is gly-his-phe (*89*).

The presence of four instead of two chains in the molecule of hemoglobin F obviously requires a re-thinking of the possible similarities or differences in the structures of adult and fetal hemoglobin. Indeed, one may speculate that half of each molecule is identical inasmuch as both have two chains with the same N-terminal sequence, val-leu.

### 4. C-Terminal Amino Acid Residues.

A knowledge of the C-terminal residues of a protein is important in ascertaining whether or not branched polypeptide chains are present. If N-terminal and C-terminal residues are in equal number, unbranched chains probably exist (for example, interchain links through cystine residues do not affect this conclusion). If there are unequal numbers, branched chains may be suspected. Regardless of whether the determination is made for N- or C-terminal residues, it may be difficult to prove that a terminal residue has not gone undetected because of the presence of some blocking group such as a carbohydrate moiety.

Methods for the determination of C-terminal residues are far less satisfactory than those for N-terminal residues, and only one investigation of the C-terminal residues of hemoglobin has been published. HUISMAN and DOZY (52) have reacted several hemoglobins including adult and fetal hemoglobin with the enzyme carboxypeptidase which releases amino acids with a free α-carboxyl group if certain specificity requirements are met. They conclude that both adult and fetal hemoglobin contain one C-terminal tyrosyl and one C-terminal histidyl residue. ITANO (70) cites data that suggest that tyrosine is not C-terminal.

Another C-terminal method is the hydrazinolytic procedure of AKABORI and collaborators (2). In this method, the protein is heated with hydrazine to produce the hydrazides of all amino acids except the C-terminal amino acid which is released as the free amino acid. By taking advantage of the different chemical properties of the hydrazides and the free amino acid, a separation may be achieved without excessive difficulty and the free amino acid can be identified and estimated. Unsuccessful attempts (115) have been made in this Laboratory to apply the method, with modification, to adult hemoglobin. Several amino acids in greater or less amount were detected in the hydrazinolates. No trace of tyrosine was found in hydrazinolates of hemoglobin itself although 50% of the added amount was detectable if gly-tyr were mixed with hemoglobin prior to hydrazinolysis. Histidine (not characterized with certainty) was present to the extent of a fraction of a residue per molecule.

These results of C-terminal determinations are disappointingly inconclusive. It seems unlikely that tyrosine is C-terminal. One C-terminal histidyl residue per molecule is unreasonable from the evidence that there are two kinds of chains and that the molecule dissociates. Indeed, experiments to be discussed on p. 367 suggest that there are four individual chains in adult hemoglobin.

KAUFFMANN and BOETTCHER found 2 C-terminal histidins by hydrazinolysis and suggest the presence of branched chains (Added in Proof).

## 5. Investigation of Internal Sequences.

Although valuable preliminary information about the structure of a protein is obtained when the number and kind of N-terminal and C-terminal residues have been elucidated, yet the final structure determination requires much more: the complete amino acid sequence of the individual chains. The basic approaches to the determination of the complete sequence are not new: they have been tried with more or less success since the days of EMIL FISCHER. Ideally, one would hope to have a method by which amino acid residues were removed one at a time from one or the other terminus of the chain until the opposite terminus were reached. With care, EDMAN's method (p. 358) may sometimes achieve this very thing with short peptides (five or ten residues) but the method that would be successful for a long chain of 100 or more residues remains to be devised. The second approach is that which SANGER so brilliantly applied to the determination of the structure of insulin. Earlier use of this approach failed to a large extent because methodology was not sufficiently advanced. Thus, if a protein is partially hydrolyzed by one method, one mixture of peptides will be produced; if by another method, another mixture. Then, after the peptides from both hydrolysates have been individually isolated and identified, the pieces of the puzzle may be fitted together because of overlapping sequences.

The determination of the complete amino acid sequence of human hemoglobin is in its very early stages and our present knowledge has resulted more as a by-product of an investigation than from its primary object. The initial observation of PAULING, ITANO, SINGER, and WELLS (99) that hemoglobin A and hemoglobin S differed in isoelectric point showed that the number of charged groups in the two molecules differed although amino acid analyses were unable to detect the nature of the difference (117). INGRAM has now been able to show by a comparison of tryptic hydrolysates of hemoglobins A and S that one peptide only is different. These investigations are described in a series of papers by INGRAM and by HUNT and INGRAM (62, 63, 65, 66, 56, 57).

INGRAM first denatured adult hemoglobin at 90° for 4 minutes at pH 8 and then prepared the tryptic hydrolysate at pH 8 and 38° over a period of 90 minutes during which time the pH was kept constant by the addition of alkali. The precipitate which is produced by the heat denaturation does not dissolve completely during the tryptic hydrolysis and what remains was centrifuged off. The mixture of peptides in the supernatant solution was then separated by a combination of paper electrophoresis and paper chromatography. The mixture was spotted on paper and electrophoresed at pH 6.4 for several hours, and then after drying was chromatographed in a direction perpendicular to the electrophoresis with n-butanol-acetic acid-water as the developer. The resulting two-dimensional map is termed a "fingerprint" by INGRAM. In this way, 26 peptides were detected. One may ask how many peptides are expected. Trypsin is one of the most specific of the proteolytic enzymes and its action is essentially limited to the splitting of peptide bonds that involve the carboxyl groups of lysine and arginine. Adult hemoglobin contains about 46 lysyl residues and 14 arginyl residues (Table 2, p. 350) and hence a maximum of about 60 peptides could be present in a tryptic hydrolysate. Because only 26 peptides, roughly half of the maximum

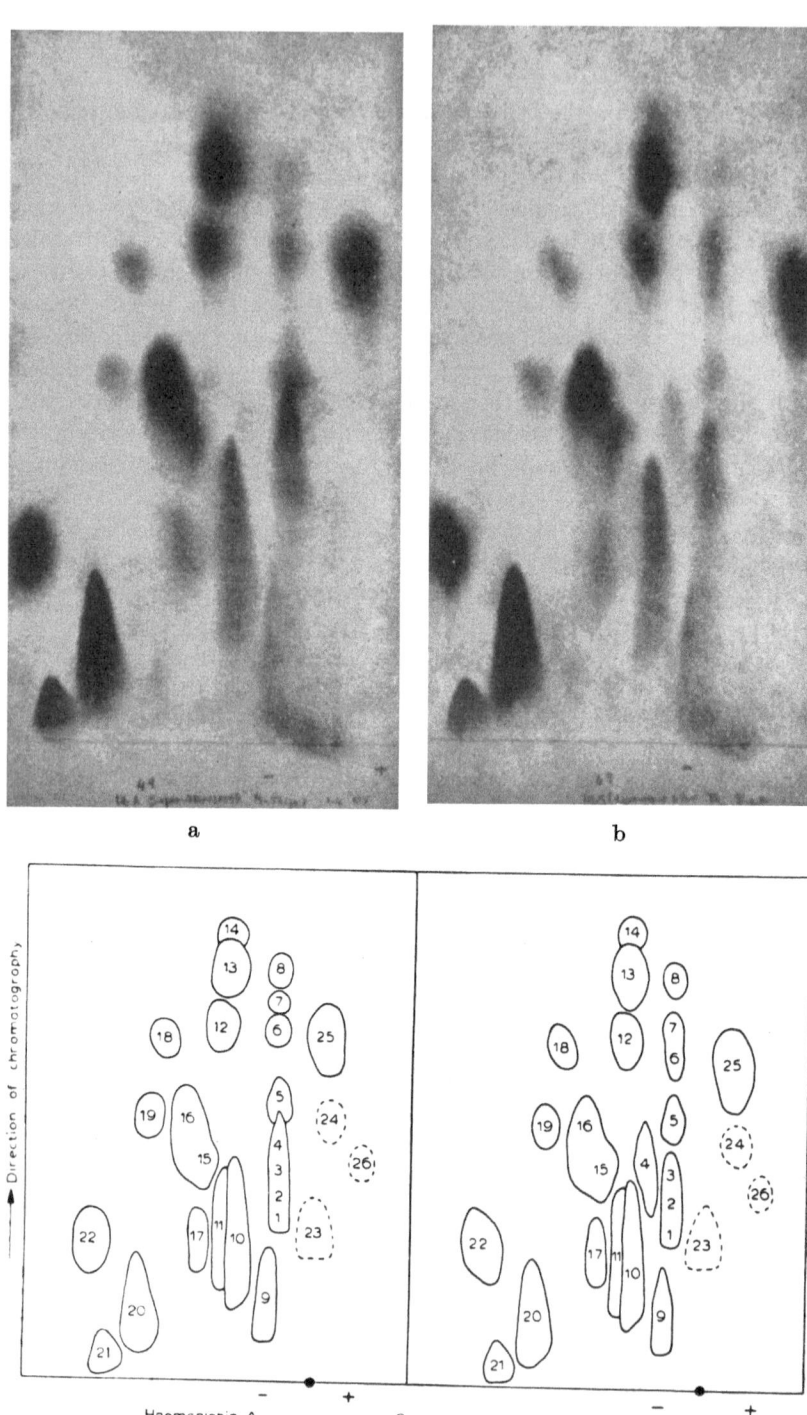

Fig. 14. Separation of peptides in tryptic hydrolysates of hemoglobins A and S by combined paper electrophoresis and paper chromatography: a is from hemoglobin A, b is from hemoglobin S, and c is a tracing of the two fingerprints. [From: Biochim. Biophys. Acta 28, 539 (1958).]

number, are found, this result suggests that identical half molecules are present, and this interpretation is substantiated by the N-terminal sequences and by the X-ray data. (Approximately 15 peptides would be indicative of identical quarter molecules.)

As already mentioned, when tryptic hydrolysates of hemoglobins A and S are thus compared, all of the peptide spots except one are coincidental. Fig. 14 compares the results from hemoglobins A and S. The differing spot is No. 4. The amino acid sequence of the peptides in the differing spots has been determined not only for hemoglobins A and S but also for hemoglobin C. These sequences are:

Hemoglobin A ... his-val-leu-leu-thr-pro-GLU-glu-lys ...
Hemoglobin S ... his-val-leu-leu-thr-pro-VAL-glu-lys ...
Hemoglobin C ... his-val-leu-leu-thr-pro-LYS-glu-lys ...

Added in Proof: In a recent reinvestigation of these peptides, Hunt and Ingram (66a) found that only one leucyl residue is present and that the above sequence is incorrect. The peptides are now formulated as:

Hemoglobin A ...val-his-leu-thr-pro-GLU-glu-lys...
Hemoglobin S ...val-his-leu-thr-pro-VAL-glu-lys...
Hemoglobin C ...val-his-leu-thr-pro-LYS-glu-lys...

This sequence leads Hunt and Ingram to suggest that peptide No. 4 arises from the N-terminus of the $\beta$ chains inasmuch as the peptide and the $\beta$ chains have same N-terminal sequence and because the -leu-thr- bond would be expected to hydrolyze easily in acid to give the di-DNP-val-his-leu that is observed in the partial acid hydrolysate of DNP-globin (103). This suggestion may, indeed, be correct because recent results in the writer's laboratory show that the N-terminal sequence in the $\beta$ chains is val-his-leu-thr-pro... and that of the $\alpha$ chains is val-leu-ser-pro-ala... (R. Shelton and W. A. Schroeder: unpublished).

The implications and ramifications of Ingram's work are many but they will not be discussed here. This sequence of amino acid residues is the only internal sequence that is known for adult hemoglobin. Brown (18) has isolated many cysteic acid peptides from hydrochloric acid and peptic hydrolysates of oxidized adult hemoglobin and has determined their amino acid composition. These peptides range in length from dipeptides to those with 14 residues. Essentially, however, only the amino acid composition and the N-terminal residues of some of these peptides have been determined.

Limited as this information is, nothing has been published about internal sequences in fetal hemoglobin.

Consideration of Ingram's results and of other data discussed above lead to the conclusions that the molecules of hemoglobins A and S are composed of identical halves, that only one amino acid residue in each half is exchanged, and that the exchanged residue is in one of the two kinds of chains. They do not tell us in which kind of chain the exchange has occurred, that is, whether the above sequence is in the $\alpha$ or $\beta$ chain. It has now been determined in this Laboratory (*131*, *132*) that the changed sequence is in the $\beta$ chains. These conclusions have been reached on the basis of the following ideas and experiments.

In the following discussion, this nomenclature will be used. The $\alpha$ chains by definition terminate in the sequence, val-leu, in hemoglobins A, F, and S; the $\beta$ chains in val-his-leu in hemoglobins A and S; and the $\gamma$-chains in glycine in hemoglobin F. Superscripts, as for example $\alpha^A$ or $\gamma^F$ designate the hemoglobin that is the source of the chain and subscripts, in the usual chemical sense, denote the number of each chain. Thus, hemoglobin A is represented by $\alpha_2^A \beta_2^A$.

As described on p. 327, a change in pH will dissociate the hemoglobin molecule. We may write the dissociation of hemoglobin A as

$$\alpha_2^A \beta_2^A \rightleftarrows \alpha_2^A + \beta_2^A \qquad \text{or} \qquad \alpha_2^A \beta_2^A \rightleftarrows 2\,\alpha^A \beta^A$$

depending upon whether it is asymmetrical or symmetrical and we may write similarly for hemoglobin S. If a mixture of hemoglobins A and S is taken to a dissociating pH and then returned to an undissociating pH, hybrid molecules (that is, molecules formed from parts of both hemoglobins) may be present. If one of the hemoglobins is radioactive, the nature of any hybridization should be determinable. Thus, if radioactive hemoglobin A $(\alpha_2^{A^*} \beta_2^{A^*})$ and non-radioactive hemoglobin S $(\alpha_2^S \beta_2^S)$ should hybridize through an asymmetrical dissociation and recombination, four possible compounds, $\alpha_2^{A^*} \beta_2^{A^*}$ and $\alpha_2^S \beta_2^S$ (the starting hemoglobins) and $\alpha_2^{A^*} \beta_2^S$ and $\alpha_2^S \beta_2^{A^*}$ (the hybrids) would be expected. The hybrids, depending upon which chain differs in the two hemoglobins, must act like either hemoglobin A or S. A symmetrical dissociation and recombination would give either the original molecules and no hybrids or the original molecules and the hybrid, $\alpha^{A^*} \beta^{A^*} \alpha^S \beta^S$, which is different from either hemoglobin A or S.

To test these possibilities, radioactive hemoglobin A was prepared by incubating blood rich in reticulocytes (immature red blood cells) with radioactive leucine (*132*). No radioactivity was present in hemoglobin S if radioactive hemoglobin A and non-radioactive hemoglobin S were mixed and separated chromatographically. But, hemoglobin S was radioactive after the mixture had been kept at pH 5 and 0° for 24 hours, returned to approximately pH 7 for chromatography, and resolved chromatographically. The radioactive hemoglobin S was diluted with non-radioactive hemoglobin S as a carrier and dinitrophenylated. After partial hydrolysis and isolation of the N-terminal peptides, the DNP-val-leu was radioactive and the di-DNP-val-his-leu was essentially non-radioactive. These results in conjunction with INGRAM's show that hybridization does occur and that the dissimilar chain in hemoglobins A and S is the $\beta$ chain and suggest that dissociation is asymmetrical. This latter conclusion is based mainly upon the fact that the molecular weight of hemoglobin is about half after dissociation and that there is no evidence of further dissociation within the sensitivity of the ultracentrifugal method. However, the possibility of some further dissociation, for example, thus

$$\alpha_2 \beta_2 \xrightarrow{\rightarrow} \alpha_2 + \beta_2 \xrightarrow{\rightarrow} 2\alpha + 2\beta$$

cannot be entirely discounted so that the hybridization could occur as a result of some small dissociation to the individual polypeptide chains rather than by the recombination of asymmetrical halves.

## 6. The Insoluble Residue from Tryptic Hydrolysates.

The portion of the denatured hemoglobin that is not dissolved by tryptic action has been termed the "core" (56). This is an unfortunate designation because of the connotations inherent in the word; it suggests that this portion may be very basic to the properties of hemoglobin (as indeed it may be) although there is no evidence to support this connotation. It may well be desirable to call it the "trypsin-resistant fraction" instead of "core" or "trypsin-resistant core" even if it leads to an alphabetical designation such as "TRF".

HUNT and INGRAM (56) have investigated this trypsin-resistant fraction which contains most of the heme of the protein. After removal of the heme, the fraction was digested with chymotrypsin and the products of hydrolysis were completely soluble. No difference was detected in the chymotryptic digests of the trypsin-resistant fractions of hemo-globins A and S. HUNT and INGRAM did not determine what percentage of the molecule this trypsin-resistant fraction amounts to although they note that it is "a considerable portion" [according to INGRAM (66) it amounts to about 30%]. They conclude that it is low in lysine and arginine but high in aromatic amino acids and in those with long non-polar side-chains. A quantitative estimation of the amount of the trypsin-resistant fraction has shown that it accounts for one-third of the weight of the molecule (11).

## 7. The Separation of the Polypeptide Chains of Hemoglobin.

The determination of the sequence of amino acids in hemoglobin A and F is complicated by the fact that two kinds of polypeptide chains are present in the molecules. Thus, though the peptides from a tryptic hydrolysate of whole hemoglobin may be separated by INGRAM's procedure or by column chromatography (*116*), one cannot know from these experiments alone whether a given peptide is derived from the $\alpha$ or $\beta$ chain. This same obstacle to structure determination was present in insulin and was solved by separating the two kinds of chains (SANGER). After oxidizing the interconnecting disulfide bonds in insulin to liberate the individual chains, they then could be separated by precipitation after appropriate alteration of the pH of the solution. Adult hemoglobin, at least, contains no disulfide bonds to connect the chains and a change in pH produces a reversible dissociation over a wide range of pH. Globin, however, has rather different properties. Not only is its molecular weight about half that of hemoglobin but it shows no evident tendency to reassociate. Its solubility properties also are vastly different.

The success of the hybridization experiments described on p. 364 has removed the necessity of separating the $\alpha$ and $\beta$ chains in order to determine the source of a peptide in a hydrolysate because by appropriate labelling it should be possible to tell whether a peptide originated in the $\alpha$ or $\beta$ chain. However, before the success of the hybridization experiments, globin was the subject of many experiments in this Laboratory in an attempt to separate the individual chains of adult and fetal hemoglobin (*120*). Globin for most of these experiments has been made essentially by the method of ANSON and MIRSKY (*9*), except that the heme was removed at —15° and all other operations were done at 0° or below. Sometimes, globin oxidized with performic acid has been used. It has not been possible to achieve any separation of the chains of adult hemoglobin by precipitation through change in pH. Moving boundary or paper electrophoresis has not produced unequivocal evidence of separation. Chromatography, likewise, has been unsuccessful. Success in separating the chains of fetal hemoglobin has come by the rather unexpected route of dialysis. Although dialysis as a method of separating the chains was found by chance, its success is understandable on the basis of the experiments of CRAIG and co-workers (*26, 27*). They have found that commercially available Visking cellulose tubing differs greatly in its permeability. This difference does not depend on the production lot of cellulose tubing but rather on the size of tubing. That which is designated as "20/32 inch inflated diameter" is the most permeable of the sizes that CRAIG has studied.

It was found that if globin from cord hemoglobin or from chromatographically purified hemoglobin F were dialyzed in 20/32 inch tubing at 0°–5° against distilled water with frequent changes of the dialyzing water, the material in the dialysate had N-terminal valyl residues almost exclusively (95–100%) whereas the material in the raffinate (that is, the solution in the dialyzing sac) had been enriched in N-terminal glycyl residues. Thus, the $\alpha^F$ chains have diffused through the membrane almost to the exclusion of the $\gamma^F$ chains. However, when the experiment was made with adult globin, it was not possible to find conditions under which there is any appreciable differential dialysis of the $\alpha^A$ and $\beta^A$ chains.

The conditions under which these dialyses were made differ vastly from those of CRAIG but his experiments did show that proteins of molecular weight as high as 45,000 diffused slowly through 20/32 inch tubing. One might expect, therefore, that some idea of the molecular weight of the hemoglobin components that diffuse through the membrane could be obtained by comparing with the diffusion rates of substances of similar molecular weight. We are here concerned with molecular weights of about one-half and one-quarter that of hemoglobin or of about 33,000 and 16,000–17,000. Experiments with the rates of diffusion of ribonuclease, lysozyme, chymotrypsin, and $\beta$-lactoglobulin (molecular weights approximately 14,000, 15,000, 21,000, and 35,000) suggest that both $\alpha^A$ and $\beta^A$ chains and the $\alpha^F$ chains pass through the membrane as particles of about 16,000–17,000 molecular weight but that the $\beta^F$ chains are essentially retarded because their molecular weight is around 35,000. If this conclusion is correct, we then must say that adult hemoglobin contains four individual chains unconnected by covalent bonds, whereas fetal hemoglobin has two such chains and two chains that are tightly connected in some way.

HOMMES, DOZY, and HUISMAN (48) conclude that fetal hemoglobin has a disulfide link on the basis of the difference between total half-cystine by analysis and sulfhydryl groups by titration. As confirming evidence they show that, after fetal globin has been reacted with thioglycolic acid to reduce the disulfide bond and then with iodoacetamide to combine with the sulfhydryl groups, the product contains two components by paper or moving boundary electrophoresis whereas only one was present before these reactions. Their results are interpreted on the basis of only two N-terminal residues of valine and hence only two chains in the molecule of fetal hemoglobin. It may be that they have observed a separation of the $\alpha^F$ and $\gamma^F$ chains.

INGRAM (66a) has recently successfully separated the $\alpha^A$ and $\beta^A$ chains chromatographically by the method that WILSON and SMITH (133a) devised for the separation of the chains of horse globin. In this procedure, globin is adsorbed on a column of Amberlite CG-50 (Type 2) from 11.7% formic acid and developed first with 2 M urea at pH 1.9 and then with a gradient between this solution and 8 M urea at pH 1.9. The $\alpha$ chains emerge first in the effluent. The separation of the $\alpha^F$ and $\gamma^F$ chains can also be achieved in this way (115).

# IX. Discussion and Conclusions.

In the above Sections, we have set forth, in greater or less detail, pertinent information about the chemical structure of human adult and fetal hemoglobins. A better understanding of both hemoglobins is certainly to be gained by considering the data not only individually for each kind of hemoglobin, but also collectively for both and in conjunction with the data from abnormal human hemoglobins and sometimes animal hemoglobins. We shall now try to draw together the information to give a picture of the present state of our knowledge about adult and fetal hemoglobin.

Although we do not wish to stress the point unduly, it should be clear that almost all investigations of the human hemoglobins have been made with more or less heterogeneous material. Because the influence of the heterogeneity will depend much upon the kind of experiment, the following discussion, for the most part, will make no attempt to assess the effect of heterogeneity on the results and conclusions.

## 1. The Hemoglobin Molecule as a Whole.

We have previously discussed the molecule as a whole, but we should now view it in a different light that takes into account the results of the chemical studies.

The molecule of 66,000 molecular weight contains four heme groups and four polypeptide chains. In both adult and fetal hemoglobin the four chains are of two kinds: the $\alpha^A$ and $\alpha^F$ chains are N-terminal in val-leu, the $\beta^A$ chains in val-his-leu, and the $\gamma^F$ chains in gly-his-phe. The $\alpha$ chains are identical in hemoglobins A and S (and C by inference) and possibly identical in hemoglobin F. Is each of the hemes associated with an individual chain? If so, portions of each kind of chain must have regions of identical or very similar structure if it is true as ALLEN, GUTHE, and WYMAN (3) conclude that the combining centers are inherently identical, that all heme-globin linkages are the same, and that there is complete equivalence in mutual interactions of the hemes. The heme-globin linkage very likely plays a part in holding the molecule together because the molecular weight of adult globin is roughly 33,000. Note now that when adult hemoglobin dissociates by change in pH, the dissociation seems to be asymmetric, $\alpha_2^A \beta_2^A \rightleftarrows \alpha_2^A + \beta_2^A$ (cf. p. 365). If, in globin, the parts result from asymmetric dissociation, then heme must have an interaction with both the $\alpha$ and the $\beta$ chains. One may then ask what is the effect of dissociation on the oxygen dissociation curve. WYMAN (134) states that pH does not alter the shape of the oxygen dissociation curve, and ALLEN, GUTHE, and WYMAN (3) later substantiated this statement for adult hemoglobin. However, the pH range of their experiments was only 6.4 to 8.9, a range in which hemoglobin does not dissociate appreciably. FERRY and GREEN (37) investigated a wider range of pH with horse hemoglobin but their data are incomplete below pH 6. It is doubtful, in any case, that data from horse hemoglobin would be applicable to human hemoglobin in view of the difference in their dissociation in strong urea solution. WYMAN's discussion (134) of heme-heme interaction and related problems is an excellent commentary in the light of information available in 1948. It may be suggested that a reconsideration of those and more recent data on the structure of the molecule as well as a study of oxygen dissociation at pH values below 6 and above 10 (necessarily at low temperature to prevent denaturation) might lead to a better understanding of the heme-globin linkage and the heme-heme interactions. An investigation of the dissociation of ferrohemoglobin as a function of pH might also lead to significant conclusions. In oxyhemoglobin, the six coordinate positions of the iron atom are occupied presumably by four nitrogens from the tetrapyrrole moiety, the oxygen molecule, and the link to globin. In ferrohemoglobin, on the other hand, oxygen is absent and the linkages to globin might be influenced in a way that is reflected in the dissociation. Although they have not studied the dissociation of ferrohemoglobin as a function of pH, HUTCHINSON and VINOGRAD (58) have compared the dissociation constants at pH 5 for several derivatives and have found the following:

Oxyhemoglobin ............... $14.6 \times 10^{-5}$ moles per liter
Carbonmonoxyhemoglobin....... $9 \times 10^{-5}$
Ferrihemoglobin .............. $22 \times 10^{-5}$
Ferrohemoglobin .............. $2 \times 10^{-5}$

Clearly, then the dissociation is much influenced by the state of oxidation of the iron in the heme and by the attached small groups. Ferrohemoglobin dissociates only to the extent of about one-seventh of the dissociation of oxyhemoglobin.

Further investigations of this type should lead to a much clearer understanding of the heme-globin linkage.

Although the heme-globin linkages must be responsible to some extent in holding the polypeptide chains together, other connections between chains must also be present. If they were not, the molecular weight of globin would be of the order of 66,000/4 whereas it is closer to 66,000/2. There is no evidence that adult hemoglobin contains disulfide bonds; with fetal hemoglobin, the evidence is conflicting. If no disulfide bonds connect the individual chains, the possibility of branched chains must be considered. It is unlikely that branched chains are present despite the fact that no satisfactory determination of the C-terminal residues has been achieved. We may cite the evidence that there are four

N-terminal residues but that the number of peptides in tryptic hydrolysates is that required by a molecule composed of identical halves. Thus, any branching would have to be very close to the C-terminus. A further evidence against branched chains is the dialysis of both $\alpha^A$ and $\beta^A$ chains at a rate similar to proteins with molecular weights of 17,000. These data apply to adult hemoglobin and to a certain extent to fetal hemoglobin. It seems probable that identical halves are present in fetal hemoglobin, but dialysis shows that $\alpha^F$ chains dialyze through a membrane and $\gamma^F$ chains do not. This difference in dialysis is almost an all-or-none effect and suggests that one kind of chain is only weakly linked together, if at all, and that the other kind is strongly linked. A disulfide bond would be a strong interchain link of the type required but evidence for its presence is, as remarked above, conflicting.

## 2. The Polypeptide Chains Themselves.

The rapid advances in recent years in chemical methods for studying the structure of proteins have so far been applied with good success to the hemoglobin molecule. As a result, we are better able to evaluate some of the problems still to be faced.

Although a definitive amino acid analysis of neither hemoglobin A nor hemoglobin F has yet been made, the available data do show the compositions with reasonable certainty and point out the differences in the two hemoglobins. If a more certain analysis is desired, analytical methods are not wanting nor are methods of purifying the proteins. It is of interest to point out in this connection that, if the $\alpha^A$ and $\alpha^F$ chains are identical*, the difference in amino acid composition between the two proteins resides in the $\beta$ chains. According to Table 2 (p. 350), 26 residues of certain amino acids are decreased in fetal hemoglobin relative to adult hemoglobin and 20 residues of others are increased. These numbers, of course, are subject to rather large error so that we may say that the total number of residues has not changed appreciably. Furthermore, the difference, except for histidine, is in uncharged amino acids, a fact which is also apparent in the similar electrophoretic behavior of the two proteins. The difference in the $\beta^A$ and $\gamma^F$ chains (if they are composed of approximately half of the 600 residues in the molecule) then, will be in somewhat less than 10% of the residues. One may wonder if there has simply been exchange of residues at certain sites.

---

* Added in Proof: Strong evidence for the identity of the $\alpha^A$ and $\alpha^F$ chains has now come from two sources. Hunt (55a) concludes this from a thorough study of the "fingerprints" of $\alpha^A$ and $\alpha^F$ chains. Independently, Jones, Schroeder, and Vinograd (unpublished) reached the same conclusion not only from "fingerprints" but also from the hybridization of hemoglobin F with other hemoglobins.

The presence of two kinds of chains in the molecules requires that some method be available for identifying the particular chain from which any peptide in a hydrolysate may arise. Ordinarily, the solution of the problem requires the physical separation of the chains. However, in the case of hemoglobin, the situation is unique: the physical separation of chains is not necessary because the hybridization of the hemoglobins gives a means for identifying the chain from which a peptide in a hydrolysate arises and, yet, the separation of the chains has actually been achieved (66 a). The problem of determining the sequence of amino acids in hemoglobin thus is reduced to polypeptide chains about the size of lysozyme or ribonuclease. Peptides from the individual chains may be sorted out by INGRAM's "fingerprint" method or by column chromatography. By labelling with particular radioactive amino acids and by proper colorimetric tests of the various spots on the "fingerprints" it should be possible to detect those peptides which contain tryptophan, tyrosine, cysteine, and arginine. In this way, the number of residues of such difficultly determinable amino acids as tryptophan and cysteine possibly can be estimated. It will also be possible to determine whether the trypsin-resistant fraction originates in one or both types of chain.

## 3. Final Remarks.

Although hemoglobin probably is as complex a protein as any upon which serious work to determine the complete structure has been begun, the progress already made shows that the molecule has a greater symmetry and simplicity than might have been expected. Many years of effort by many investigators will no doubt be required before the last amino acid residue is placed in the sequence and before the nature of the heme-globin linkage becomes apparent, but it is not too much to expect that this goal will be achieved.

### References.

*1.* ADAIR, G. S.: A Comparison of the Molecular Weights of the Proteins. Proc. Cambridge Phil. Soc. **1**, 75 (1924).

*2.* AKABORI, S., K. OHNO, T. IKENAKA, Y. OKADA, H. HANAFUSA, I. HARUNA, A. TSUGITA, K. SUGAE and T. MATSUSHIMA: Hydrazinolysis of Peptides and Proteins. II. Fundamental Studies on the Determination of the Carboxyl Ends of Proteins. Bull. Chem. Soc. Japan **29**, 507 (1956).

*3.* ALLEN, D. W., K. F. GUTHE and J. WYMAN, Jr.: Further Studies on the Oxygen Equilibrium of Hemoglobin. J. Biol. Chem. **187**, 393 (1950).

*4.* ALLEN, D. W. and W. A. SCHROEDER: A Comparison of the Phenylalanine Content of the Hemoglobin of Normal and Phenylketonuric Individuals: Determination by Ion Exchange Chromatography. J. Clin. Investigation **36**, 1343 (1957).

5. Allen, D. W., W. A. Schroeder and J. Balog: Observations on the Chromatographic Heterogeneity of Normal Adult and Fetal Human Hemoglobin: A Study of the Effects of Crystallization and Chromatography on the Heterogeneity and Isoleucine Content. J. Amer. Chem. Soc. 80, 1628 (1958).

6. Allison, A. C.: Notation for Hemoglobin Types and Genes Controlling Their Synthesis. Science (Washington) 122, 640 (1955).

7. Allison, A. C. and R. Cecil: The Thiol Groups of Normal Adult Human Haemoglobin. Biochemic. J. 69, 27 (1958).

8. Anonymous: Statement Concerning a System of Nomenclature for the Varieties of Human Hemoglobin. Blood 8, 386 (1953).

9. Anson, M. L. and A. E. Mirsky: Protein Coagulation and its Reversal. The Preparation of Insoluble Globin, Soluble Globin, and Heme. J. Gen. Physiol. 13, 469 (1929/30).

10. Bangham, A. D. and H. Lehmann: "Multiple" Haemoglobins in the Horse. Nature (London) 181, 267 (1958).

11. Barrett, H. W. and W. A. Schroeder: unpublished.

12. Beaven, G. H., H. Hoch and E. R. Holiday: The Haemoglobins of the Human Foetus and Infant. Electrophoretic and Spectroscopic Differentiation of Adult and Foetal Types. Biochemic. J. 49, 374 (1951).

13. Benesch, R. E., H. A. Lardy and R. Benesch: The Sulfhydryl Groups of Crystalline Proteins. I. Some Albumins, Enzymes, and Hemoglobins. J. Biol. Chem. 216, 663 (1955).

14. Bernhart, F. W. and L. Skeggs: The Iron Content of Crystalline Human Hemoglobin. J. Biol. Chem. 147, 19 (1943).

15. Betke, K.: Der menschliche rote Blutfarbstoff bei Fetus und reifem Organismus. Berlin: Springer-Verlag. 1954.

16. Bragg, W. L. and M. F. Perutz: The External Form of the Hemoglobin Molecule. II. Acta Crystallogr. 5, 323 (1952).

17. Brown, H.: The Free Amino Groups and Terminal Dipeptides of Human Adult Hemoglobin. Arch. Biochem. Biophys. 61, 241 (1956).

18. — The Sulfur Distribution and Some Cysteic Acid Peptides of Human Adult Hemoglobin. Arch. Biochem. Biophys. 67, 256 (1957).

19. Cabannes, R. et Ch. Serain: Étude électrophorétique des hémoglobines des Mammifères domestiques d'Algérie. C. R. Séances Soc. Biol. 149, 1193 (1955).

20. Chernoff, A. I.: The Alkali Denaturation Procedures. In: Conference on Hemoglobin. Nat. Acad. Sci., National Research Council, Washington, D. C., Publication No. 557, 1958, p. 172.

21. — Immunologic Aspects of the Human Hemoglobin. In: Conference on Hemoglobin. Nat. Acad. Sci., National Research Council, Washington, D. C., Publication No. 557, 1958, p. 179.

22. Clegg, M. and W. A. Schroeder: J. Amer. Chem. Soc. (1959).

22a. Cole, R. D., W. H. Stein and S. Moore: On the Cysteine Content of Human Hemoglobin. J. Biol. Chem. 233, 1359 (1958).

23. Conference on Hemoglobin. Nat. Acad. Sci., National Research Council, Washington, D. C., Publication No. 557, 1958.

24. Cook, J. L. and M. Morrison: Ion Exchange Chromatography of Human Hemoglobin. Federat. Proc. (Amer. Soc. exp. Biol.) 15, 235 (1956).

25. Coryell, C. D. and L. Pauling: A Structural Interpretation of the Acidity of Groups Associated with the Hemes of Hemoglobin and Hemoglobin Derivatives. J. Biol. Chem. 132, 769 (1940).

26. CRAIG, L. C., T. P. KING and A. STRACHER: Dialysis Studies. II. Some Experiments Dealing with the Problem of Selectivity. J. Amer. Chem. Soc. 79, 3729 (1957).

27. CRAIG, L. C., W. KONIGSBERG, A. STRACHER and T. P. KING: The Characterization of Lower Molecular Weight Proteins by Dialysis. In: A. NEUBERGER, Symposium on Protein Structure, p. 104. New York: Wiley and Sons. 1958.

28. CULLIS, A. F., H. M. DINTZIS and M. F. PERUTZ: X-Ray Analysis of Haemoglobin. In: Conference on Hemoglobin. Nat. Acad. Sci., National Research Council, Washington, D. C., Publication No. 557, 1958, p. 50.

29. DERRIEN, Y.: Studies on the Heterogeneity of Adult and Fetal Hemoglobins by Salting-out, Alkali Denaturation and Moving Boundary Electrophoresis. In: Conference on Hemoglobin. Nat. Acad. Sci., National Research Council, Washington, D. C., Publication No. 557, 1958, p. 183.

30. DICKMAN, S. R. and I. H. MONCRIEF: Primary Amide Groups of Human Hemoglobin. Proc. Soc. exp. Biol. Med. 77, 631 (1951).

31. DRABKIN, D. L.: Metabolism of the Hemin Chromoproteins. Physiol. Rev. 31, 345 (1951).

32. — (Chairman): Symposium on Molecular Heterogeneity of Hemoglobin. Federat. Proc. (Amer. Soc. exp. Biol.) 16, 740–773 (1957).

33. — Heredity and Environment in Structure of Hemoglobin. Federat. Proc. (Amer. Soc. exp. Biol.) 16, 740 (1957).

34. DRESCHER, H. und W. KÜNZER: Der Blutfarbstoff der menschlichen Feten. Klin. Wschr. 32, 92 (1954).

35. DUSTIN, J. P., G. SCHAPIRA, J. C. DREYFUS et O. HESTERMANS-MEDARD: La composition en acides aminés de l'hémoglobine foetale humaine. C. R. Séances Soc. Biol. 148, 1207 (1954).

36. EASTMAN, N. J.: Obstetrics. 10th ed., p. 168. New York: Appleton-Century-Crofts. 1950.

37. FERRY, R. M. and A. A. GREEN: Studies in the Chemistry of Hemoglobin. III. The Equilibrium between Oxygen and Hemoglobin and its Relation to Changing Hydrogen Ion Activity. J. Biol. Chem. 81, 175 (1929).

38. FIELD, E. O. and J. R. P. O'BRIEN: Dissociation of Human Haemoglobin at Low pH. Biochemic. J. 60, 656 (1955).

39. GEORGE, P. and R. L. J. LYSTER: A Survey of the Evidence for and against a Crevice Configuration for the Heme in Hemoglobin. In: Conference on Hemoglobin. Nat. Acad. Sci., National Research Council, Washington, D. C., Publication No. 557, 1958, p. 33.

40. GUTTER, F. J., H. A. SOBER and E. A. PETERSON: The Effect of Mercaptoethanol and Urea on the Molecular Weight of Hemoglobin. Arch. Biochem. Biophys. 62, 427 (1956).

41. HALBRECHT, I. and C. KLIBANSKI: Identification of a New Normal Embryonic Haemoglobin. Nature (London) 178, 794 (1956).

42. HALBRECHT, I., C. KLIBANSKI, H. BRZOZA and M. LAHAV: Hemoglobins and the Serum Protein Fractions in Early Embryonic Life. Amer. J. Clin. Pathol. 29, 340 (1958) [Chem. Abstr. 52, 13061 (1958)].

43. HASSERODT, U. and M. CLEGG: unpublished.

44. HASSERODT, U. and J. VINOGRAD: Dissociation of Human Carbonmonoxyhemoglobin at High pH. Proc. Nat. Acad. Sci. (USA) 45, 12 (1959).

45. HASSERODT, U., J. VINOGRAD and R. SRINIVASAN: The Molecular Weight of Human Hemoglobin. J. Amer. Chem. Soc. (in press).

46. Haurowitz, F. and R. L. Hardin: Respiratory Proteins. In: H. Neurath and K. Bailey, The Proteins, Vol. II A, p. 328. New York: Academic Press. 1954.

47. Havinga, E.: Comparison of the Phosphorus Content, Optical Rotation, Separation of Hemes and Globin, and Terminal Amino Acid Residues of Normal Adult Human Hemoglobin and Sickle Cell Anemia Hemoglobin. Proc. Nat. Acad. Sci. (USA) **39**, 59 (1953).

48. Hommes, F. A., A. Dozy and T. H. J. Huisman: Further Studies on the Cysteine-Cystine Content of the Foetal Human Haemoglobin. Biochemic. J. **68**, 309 (1958).

49. Hommes, F. A., J. Santema-Drinkwaard and T. H. J. Huisman: The Sulfhydryl Groups of Four Different Human Haemoglobins. Biochim. Biophys. Acta **20**, 564 (1956).

50. Huisman, T. H. J.: The Properties, Estimation Methods, Hematologic Features, and Some Other More General Aspects of Different Abnormal Human Hemoglobins. Clin. Chem. **3**, 371 (1957).

51. — Abnormal Hemoglobins. Clin. Chim. Acta **3**, 201 (1958).

52. Huisman, T. H. J. and A. Dozy: The Action of Carboxypeptidase on Different Human Haemoglobins. Biochim. Biophys. Acta **20**, 400 (1956).

53. Huisman, T. H. J. and J. Drinkwaard: The N-terminal Residues of Five Different Human Haemoglobins. Biochim. Biophys. Acta **18**, 588 (1955).

54. Huisman, T. H. J., J. H. P. Jonxis and A. Dozy: Is Foetal Haemoglobin Present in the Blood of Normal Human Adults? Biochim. Biophys. Acta **18**, 576 (1955).

55. Huisman, T. H. J., E. A. Martis and A. Dozy: Chromatography of Hemoglobin Types on Carboxymethylcellulose. J. Lab. Clin. Med. **52**, 312 (1958) [Chem. Abstr. **52**, 17361 (1958)].

55a. Hunt, J. A.: The Identity of the $\alpha$ Chains of Adult and Foetal Human Haemoglobins. Nature (London) (in press).

56. Hunt, J. A. and V. M. Ingram: Abnormal Human Haemoglobins. II. The Chymotryptic Digestion of the Trypsin-Resistant "Core" of Haemoglobins A and S. Biochim. Biophys. Acta **28**, 546 (1958).

57. — — Allelomorphism and the Chemical Differences of the Human Haemoglobins A, S, and C. Nature (London) **181**, 1062 (1958).

58. Hutchinson, W. D. and J. Vinograd: unpublished.

59. Ingbar, S. H. and E. H. Kass: Sulfhydryl Content of Normal Hemoglobin and Hemoglobin in Sickle-cell Anemia. Proc. Soc. exp. Biol. Med. **77**, 74 (1951).

60. Ingram, D. J. E., J. F. Gibson and M. F. Perutz: Electron Spin Resonance in Myoglobin and Haemoglobin. Orientation of the four Haem Groups in Haemoglobin. Nature (London) **178**, 905 (1956).

61. Ingram, V. M.: Sulfhydryl Groups in Haemoglobins. Biochemic. J. **59**, 653 (1955).

62. — A Specific Chemical Difference Between the Globins of Normal Human and Sickle-cell Anaemia Haemoglobin. Nature (London) **178**, 792 (1956).

63. — Gene Mutations in Human Haemoglobin: The Chemical Difference between Normal and Sickle Cell Haemoglobin. Nature (London) **180**, 326 (1957).

64. — The Sulfhydryl Groups of Sickle-cell Haemoglobin. Biochemic. J. **65**, 760 (1957).

65. — Abnormal Human Haemoglobins. I. The Comparison of Normal Human and Sickle-cell Haemoglobins by "Fingerprinting". Biochim. Biophys. Acta **28**, 539 (1958).

66. — The Chemical Difference between Normal Human and Sickle Cell Anaemia Haemoglobins. In: Conference on Hemoglobin. Nat. Acad. Sci., National Research Council, Washington, D. C., Publication No. 557, 1958, p. 233.

*66a.* INGRAM, V. M.: Private communication.
*67.* ITANO, H. A.: Human Hemoglobin. Science (Washington) **117**, 89 (1953).
*68.* — Clinical States Associated with Alteration of the Hemoglobin Molecule. Arch. Internal Med. **96**, 287 (1955) [Chem. Abstr. **50**, 2834 (1956)].
*69.* — The Hemoglobins. Annu. Rev. Biochem. **25**, 331 (1956).
*70.* — The Human Hemoglobins: Their Properties and Genetic Control. Adv. Protein Chem. **12**, 215 (1957).
*71.* — Asymmetric Dissociation and Hybridization of Hemoglobin Molecules. Correlation with Chemical and Genetic Subunits. Abstr., Meeting Amer. Chem. Soc., Sept. 1958.
*72.* ITANO, H. A., W. R. BERGREN and P. STURGEON: The Abnormal Human Hemoglobins. Medicine **35**, 121 (1956).
*73.* JONES, R. T. and W. A. SCHROEDER: unpublished.
*74.* JONXIS, J. H. P.: Foetal Haemoglobin and Rh Antagonisms. In: F. J. W. ROUGHTON and J. C. KENDREW, Haemoglobin, p. 261. London: Butterworths, and New York: Interscience Publ. 1949.
*75.* JONXIS, J. H. P. and T. H. J. HUISMAN: The Detection and Estimation of Fetal Hemoglobin by Means of the Alkali Denaturation Test. Blood **11**, 1009 (1956).
*76.* JOPE, E. M.: The Ultraviolet Spectral Absorption of Haemoglobins Inside and Outside the Red Blood Cell. In: F. J. W. ROUGHTON and J. C. KENDREW, Haemoglobin, p. 205. London: Butterworths, New York: Interscience. 1949.
*77.* JOPE, H. M. and J. R. P. O'BRIEN: Crystallization and Solubility Studies on Human Adult and Foetal Haemoglobins. In: F. J. W. ROUGHTON and J. C. KENDREW, Haemoglobin, p. 269. London: Butterworths, and New York: Interscience Publ. 1949.
*77a.* KAUFFMANN, T. und F.-P. BOETTCHER: Bestimmung der C-terminalen Aminosäuren von Menschen-, Pferde- und Rinderhämoglobin. Z. Naturf. **13b**, 467 (1958).
*78.* KEILIN, D.: A Comparative Study of Turacin and Haematin and its Bearing on Cytochrome. Proc. Roy. Soc. (London) **100** B, 129 (1926).
*79.* KLEINKNECHT, R.: Das Vorkommen der mittels Alkalidenaturierung unterscheidbaren Hämoglobintypen $Hb_1$, $Hb_2$ und $Hb_3$ im Säuglingsalter. Monatsschr. Kinderheilk. **101**, 360 (1953).
*79a.* KON, H. and N. DAVIDSON: Nuclear Magnetic Relaxation of Water Protons by Ferrihemoglobin and Ferrimyoglobin. J. Molecular Biol. (1959) (in press).
*80.* KÖRBER, E.: Über Differenzen des Blutfarbstoffes. Dissert., Dorpat, 1866.
*81.* KUNKEL, H. G., R. CEPPELLINI, O. MULLER-EBERHARD and J. WOLF: The Minor Basic Hemoglobin (Hb) Component in the Blood of Normal Individuals and Patients with Thalassemia. J. Clin. Investigation **36**, 1615 (1957).
*82.* KUNKEL, H. G. and G. WALLENIUS: New Hemoglobin in Normal Adult Blood. Science (Washington) **122**, 288 (1955).
*83.* KÜNZER, W.: Human Embryo Haemoglobins. Nature (London) **179**, 477 (1957).
*84.* KÜSTER, W. und G. F. KOPPENHÖFER: Über den Blutfarbstoff. Z. physiol. Chem. (Hoppe-Seyler) **170**, 106 (1927).
*85.* LAMM, O. and A. POLSON: The Determination of Diffusion Constants of Proteins by a Refractometric Method. Biochemic. J. **30**, 528 (1936).
*86.* LEMBERG, R. and J. W. LEGGE: Hematin Compounds and Bile Pigments: Their Constitution, Metabolism, and Function. New York: Interscience Publ. 1949.
*87.* MASRI, M. S. and K. SINGER: Studies on Abnormal Hemoglobins. XII. Terminal and Free Amino Groups of Various Types of Human Hemoglobins. Arch. Biochem. Biophys. **58**, 414 (1955).

88. Matsuda, G. and W. A. Schroeder: unpublished.
89. Matsuda, G., R. Shelton and W. A. Schroeder: unpublished.
90. Moore, D. H. and L. Reiner: Electrophoretic and Ultracentrifugal Analyses of Globin Components. J. Biol. Chem. 156, 411 (1944).
91. Morrison, D. B. and A. Hisey: The Carbon Monoxide Capacity, Iron, and Total Nitrogen of Dog Hemoglobin. J. Biol. Chem. 109, 233 (1935).
92. Morrison, M.: Discussion in: Conference on Hemoglobin. Nat. Acad. Sci., National Research Council, Washington, D. C., Publication No. 557, 1958, p. 166.
93. Morrison, M. and J. L. Cook: Chromatographic Fractionation of Normal Adult Oxyhemoglobin. Science (Washington) 122, 920 (1955).
94. Murayama, M.: Titratable Sulfhydryl Groups of Normal and Sickle-cell Hemoglobins at 0° and 38°. J. Biol. Chem. 228, 231 (1957).
95. — Titratable Sulfhydryl Groups of Hemoglobin C and Fetal Hemoglobin at 0° and 38°. J. Biol. Chem. 230, 163 (1958).
96. Patchornik, A., W. B. Lawson and B. Witkop: Selective Cleavage of Peptide Bonds. II. The Tryptophyl Peptide Bond and the Cleavage of Glucagon. J. Amer. Chem. Soc. 80, 4747 (1958).
97. Pauling, L.: Abnormality of Hemoglobin Molecules in Hereditary Hemolytic Anemias. Harvey Lect. 49, 216 (1953/54).
98. Pauling, L. and C. D. Coryell: The Magnetic Properties and Structure of Hemoglobin, Oxyhemoglobin and Carbonmonoxyhemoglobin. Proc. Nat. Acad. Sci. (USA) 22, 210 (1936).
99. Pauling, L., H. A. Itano, S. J. Singer and I. C. Wells: Sickle Cell Anemia, a Molecular Disease. Science (Washington) 110, 2865 (1949).
100. Perutz, M. F., I. F. Trotter, E. R. Howells and D. W. Green: An X-Ray Study of Reduced Human Hemoglobin. Acta Cristallogr. 8, 241 (1955).
101. Porter, R. R. and F. Sanger: The Free Amino Groups of Haemoglobins. Biochemic. J. 42, 287 (1948).
102. Prins, H. K. and T. H. J. Huisman: Chromatographic Behaviour of Haemoglobin E. Nature (London) 177, 840 (1956).
103. Rhinesmith, H. S., W. A. Schroeder and N. Martin: The N-Terminal Sequence of the β Chains of Normal Adult Human Hemoglobin. J. Amer. Chem. Soc. 80, 3358 (1958).
104. Rhinesmith, H. S., W. A. Schroeder and L. Pauling: The N-Terminal Amino Acid Residues of Normal Adult Human Hemoglobin: A Quantitative Study of Certain Aspects of Sanger's DNP-Method. J. Amer. Chem. Soc. 79, 609 (1957).
105. — — — A Quantitative Study of the Hydrolysis of Human Dinitrophenyl-(DNP)globin: The Number and Kind of Polypeptide Chains in Normal Adult Human Hemoglobin. J. Amer. Chem. Soc. 79, 4682 (1957).
106. Riggs, A. F.: Sulfhydryl Groups and the Interaction between the Hemes in Hemoglobin. J. Gen. Physiol. 36, 1 (1952).
107. Rossi-Fanelli, A., E. Antonini and A. Caputo: Pure Native Globin from Human Haemoglobin: Preparation and Some Physico-chemical Properties. Biochim. Biophys. Acta 28, 221 (1958).
108. Rossi-Fanelli, A., D. Cavallini and C. de Marco: Amino Acid Composition of Human Crystallized Myoglobin and Haemoglobin. Biochim. Biophys. Acta 17, 377 (1955).
109. Rossi-Fanelli, A., D. Cavallini, C. de Marco and F. Trasatti: Fetal Hb. I. Quantitative Analysis of the Amino Acids of Crystalline Human Fetal Hb and Some Technical Precautions. Boll. Soc. ital. Biol. sper. 31, 328 (1955) [Chem. Abstr. 49, 14982 (1955)].

*110.* ROUGHTON, F. J. W. and J. C. KENDREW (Editors): Haemoglobin. London: Butterworths, and New York: Interscience Publ. 1949.

*111.* SANGER, F.: The Arrangement of Amino Acids in Proteins. Adv. Protein Chem. **7**, 1 (1952).

*112.* SCHAPIRA, G. et J.-C. DREYFUS: Groupes N-terminaux de l'hémoglobine de la maladie de Cooley. C. R. Séances Soc. Biol. **148**, 895 (1954).

*113.* SCHRAM, E., S. MOORE and E. J. BIGWOOD: Chromatographic Determination of Cystine as Cysteic Acid. Biochemic. J. **57**, 33 (1954).

*114.* SCHRAMM, G., J. W. SCHNEIDER und A. ANDERER: Zur Bestimmung der Amino-Endgruppen verschiedener Hämoglobine und des Tabakmosaikvirus mit Phenylisothiocyanat. Z. Naturforsch. **11 b**, 12 (1956) [Chem. Abstr. **50**, 8789 (1956)].

*115.* SCHROEDER, W. A. and J. BALOG: unpublished.

*116.* SCHROEDER, W. A. and L. M. KAY: unpublished.

*117.* SCHROEDER, W. A., L. M. KAY and I. C. WELLS: Amino Acid Composition of Hemoglobins of Normal Negores and Sickle-cell Anemics. J. Biol. Chem. **187**, 221 (1950).

*118.* SCHROEDER, W. A. and G. MATSUDA: N-Terminal Residues of Human Fetal Hemoglobin. J. Amer. Chem. Soc. **80**, 1521 (1958).

*119.* SCHROEDER, W. A., G. MATSUDA and R. T. JONES: unpublished.

*120.* SCHROEDER, W. A., G. MATSUDA, N. MARTIN, L. M. KAY and M. CLEGG: unpublished.

*121.* STEIN, W. H., H. G. KUNKEL, R. D. COLE, D. H. SPACKMAN and S. MOORE: Observations on the Amino Acid Composition of Human Hemoglobins. Biochim. Biophys. Acta **24**, 640 (1957).

*122.* SVEDBERG, T. and K. O. PEDERSEN: The Ultracentrifuge. Oxford: Clarendon Press. 1940.

*123.* TAYLOR, J. F. and R. L. SWARM: Molecular Weight of Human Fetal Hemoglobin. Federat. Proc. (Amer. Soc. exp. Biol.) **8**, 259 (1949).

*124.* THEORELL, H.: Über die chemische Konstitution des Cytochroms c. Biochem. Z. **298**, 242 (1938).

*125.* — *l*-Cystin aus Porphyrin C. Enzymologia **6**, 88 (1939).

*126.* — Relations between Prosthetic Groups, Coenzymes and Enzymes. In: D. E. GREEN, Currents in Biochemical Research, p. 275. New York: Interscience Publ. 1956.

*127.* TUPPY, H. and S. PALÉUS: Study of a Peptic Degradation Product of Cytochrome c. I. Purification and Chemical Composition. Acta Chem. Scand. **9**, 353 (1955).

*128.* TUTTLE, A. H.: Human Hemoglobins. J. Chronic Diseases **6**, 528 (1957) [Chem. Abstr. **52**, 2226 (1958)].

*129.* VAN DER SCHAAF, P. C. and T. H. J. HUISMAN: The Amino Acid Composition of Human Adult and Foetal Carbonmonoxyhaemoglobin Estimated by Ion Exchange Chromatography. Biochim. Biophys. Acta **17**, 81 (1955).

*130.* VINOGRAD, J. and U. HASSERODT: unpublished.

*131.* VINOGRAD, J. and W. D. HUTCHINSON: $C^{14}$-Hybrids of Human Hemoglobins. I. Dissociation of Human Haemoglobin and the Isolation of $C^{14}$-Labelled Haemoglobin Hybrids. Nature (London) (submitted).

*132.* VINOGRAD, J. R., W. D. HUTCHINSON and W. A. SCHROEDER: $C^{14}$-Hybrids of Human Hemoglobins. II. The Identification of the Aberrant Chain in Human Hemoglobin S. J. Amer. Chem. Soc. (in press).

*133.* Walker, J. and E. P. N. Turnbull: Hemoglobin and Red Cells in the Human
   Fetus. III. Fetal and Adult Hemoglobin. Arch. Disease Childhood **30**, 111
   (1955) [Chem. Abstr. **49**, 11 125 (1955)].

*133a.* Wilson, S. and D. B. Smith: Separation of the Valyl-leucyl- and Valyl-
   glutamyl-polypeptide Components of Horse Globin by Column Chromatography
   and Fractional Precipitation. Can. J. Biochem. Physiol. **37**, 405 (1959).

*134.* Wyman, J., Jr.: Heme Proteins. Adv. Protein Chem. **4**, 407 (1948).

*135.* Zinsser, H. H. and Y.-C. Tang: X-Ray Observations on Single Crystals
   of Carbonmonoxyhemoglobin from Human Fetal Blood. Arch. Biochem.
   Biophys. **34**, 81 (1951).

*(Received, January 31, 1959.)*

# Paleobiochemistry and Organic Geochemistry.

By PHILIP H. ABELSON, Washington, D. C.

With 1 Figure.

**Contents.**

## Introduction.

Dynamic developments in petrochemical industry and in atomic energy portend a new attitude of technology and civilization toward the fossil fuels. In the past these fuels have been used largely as energy sources, being destroyed in the process. In the future, atomic energy will supply an increasing fraction of energy needs in the form of electric power and process heat. The fossil chemicals will gradually find more valuable utilization as chemical raw material, and hence greater interest will be focused on the exact nature of the organic compounds in rocks. These substances are also of philosophic interest, for there is little doubt that originally they were part of living entities.

Two methods of approach can be followed to obtain information about the nature of these organic substances. The first is a simple empirical one, in which the organic matter in a given geological formation is examined and some of the components are identified. In view of the large number of chemicals involved and the great variation in time of deposition, burial history, and environmental circumstances, such a procedure could lead to a proliferation of data.

A second approach is to attempt to understand the effects of possible variables so that a logical framework can be erected with which empirical findings can be correlated. In turn, such a structure would provide an improved mechanism for estimating the nature of substances that may be found in yet-unexplored localities. The variables to be considered are: the chemical composition of living matter, present and past; changes in chemical composition on the death of an organism; the effects of activities of anaerobic bacteria and of chemical reactions in natural environments.

This essay will describe some of the chemical and physiological common denominators of present life and show how they provide a basis for speculating on the composition of ancient life. The major components of living matter will be seen to be relatively few and probably to have been the same for the past 500 million years. After the death of an organism the remains are subject to enzymatic and other biological attack. In aerobic environments the organic matter is destroyed; in anaerobic circumstances only partial destruction takes place. After the cessation of biological alteration, chemical mechanisms become predominant. The reactions among the components of the sediments are one important feature. Another is the intrinsic instability of organic substances, almost all of which degrade spontaneously. Reaction rates are, in general, strongly dependent on temperature. This factor is considered here, therefore, together with typical effects of this variable. Finally, a few actual occurrences of organic substances in rocks will be described. It will be noted that some of the constituents of living creatures are sufficiently stable to remain unchanged after several hundreds of millions of years. Other substances have been found that appear to be only slightly altered from their original composition.

Virtually all the organic matter now present in the crust has been synthesized since the beginning of Cambrian times, about 500 million years ago. All the commercial coal deposits and all the world's petroleum reserves were formed during this interval. During Precambrian times sedimentary rocks were deposited that contained as much as 25% of reduced carbon, but most of this carbon is now graphitic. Studies of Precambrian rocks may one day yield much information about early life, but as of today little has been established.

In contrast, our knowledge of the past 500 million years is voluminous. In Cambrian times there appeared creatures that could secrete hard parts. Enormous numbers of shells, bones, and other hard parts preserved in sedimentary rocks provide the basis for a history of life's evolving patterns. Most of the present-day phyla were represented at that time. These fossils provide a number of different types of information. In the invertebrates the shells outline the dimensions of the creatures. The hinge lines of bivalves reveal mode of operation. Even more important are muscle scars on the shells, which show points of attachment of muscles and often permit a detailed reconstruction of the musculature of the animals. Similar remarks can be made about the vertebrates. From bones, size and musculature can be reconstructed, and, in addition, emergent holes from bones indicate something of the blood and nervous systems as well as of vision and hearing. Of the organic constituents of the soft parts of ancient animals little, in general, remains in situ, but, as we shall see, some of the original constituents of shells and bones are often preserved. In special cases accidents of infiltration and preservation result in outlining soft parts of the creatures so that details of anatomy can be at least partially inferred. Under special circumstances remarkable examples of preservation occur—insects in Baltic amber, and intact eye sockets of insects in the Eocene ($\sim$ 40-million-year-old) Green River shale of Wyoming, Utah, and Colorado.

In these reconstructions of past biology, physiology, and ecology, present-day observations of living creatures furnish models and analogies. Comparative biochemistry of modern life provides an even richer basis for speculation on the creatures of the past. In the next section some of the highlights of the chemistry of present forms will be presented, with particular emphasis on facets revealed by our increasing knowledge of the structure and function of proteins.

### Major Constituents of Living Matter.

The occurrence of carbohydrates, lipides, and proteins as major constituents of living things has been recognized for over a hundred years. As knowledge of biology has accumulated, there has been increasing awareness of the crucial importance of common denominators of life. In addition to common features of chemical composition, great similarities in chemical mechanisms and physiological functions have been found to exist.

In *Table 1*, p. 382 are shown some typical values of carbohydrates, lipides, and proteins in widely diverse phyla. These three major components are seen to account for about $90\%$ of the dry weight of these specimens. Most microorganisms, including photosynthetic algae which are ordinarily classified as plants, have similar compositions.

The essential activities of life are growth and reproduction. These processes and indeed life itself cannot proceed without energy, notably

Table 1. Major Constituents of Living Forms.

Data on animals from Florkin (22); data on algae from Milner (34).

| Living Form | % of Dry Weight | | |
| --- | --- | --- | --- |
| | Protein | Lipide | Carbohydrate |
| Sea urchin: | | | |
|    Unfertilized egg................ | 66.9 | 31.2 | 5.4 |
|    40-hour larva.................. | 60.6 | 17.4 | 3.4 |
| Oyster ....................... | 51.2 | 11.1 | 28.2 |
| Silkworm: | | | |
|    Larva........................ | 55.5 | 13.3 | 1.8 |
|    Imago ...................... | 63.4 | 24.3 | 6.5 |
| Alga (ash-free): | | | |
|    Chlorella (favorable environment) .. | 46.4 | 20.2 | 33.4 |
|    Chlorella (unfavorable environment) | 13.1 | 63.4 | 23.5 |

for the synthesis of chemicals. The production of useful energy in a biological system requires availability of chemicals that can react, a means of catalyzing the reactions, a mechanism for trapping resultant energy, and a means of transferring and releasing the energy when necessary.

All living systems utilize energy-rich adenosine triphosphate (ATP) as an important means of storing energy for quick release [Baldwin (6)]. Biosynthesis in all forms of life, including bacteria, seems to involve ATP. Analysis shows that ATP itself is produced in a limited number of ways. One of the schemes, utilizing glucose as a starting material, is known as the Meyerhof cycle. In a series of steps involving phosphorylation, glucose is converted into two molecules of pyruvate with the production of ATP. In an anaerobic environment the hydrogen is used to convert pyruvate to lactate. Under aerobic conditions the hydrogen may be used in a process that produces additional ATP.

Another scheme operating in aerobic environments is called the Krebs cycle. In this cycle acetic acid derived from pyruvate is oxidized after condensation with oxalacetate. Among the successive intermediates are $\alpha$-ketoglutarate, succinate, malate, and fumarate. The Krebs cycle is employed by such diverse forms as man, rat, pigeon, bacteria, yeast, molds, and algae, not only as a means of metabolism but also as a mechanism for the synthesis of aspartic acid, glutamic acid, proline, and arginine.

Most of the steps in the Meyerhof scheme and the Krebs cycle involve reactions that can proceed only slowly without the intervention of catalysts. Some of these reactions involve large activation energies. Keilin (30) has pointed out the contrast in the stability of succinate inside and outside biological systems. In the organism, succinate is destroyed virtually instantaneously, whereas in nature it has endured for more than 40 million years in Baltic amber.

The importance of proteins in their roles as enzymes makes the question of their composition of paramount interest. About 22 amino acids are the principal constituents of most proteins. Representative values of the amino acids in several different living forms are given in *Table 2*. Some notable similarities appear. About 21 amino acids are universally present and account for almost all the amino acids in all organisms. Among the dibasic amino acids, arginine and lysine are more abundant than histidine. Aspartic and glutamic acid are prominent constituents.

Table 2. Amino Acid Composition of Some Proteins
(calculated as g. of amino acid per 16.0 g. N).

| Amino acid | Rat, an average | Cod, meal | Squid | Lobster shell | Earth-worm | Tetra-hymena | Yeasts | Escherichia coli |
|---|---|---|---|---|---|---|---|---|
| Arginine | 4.5 | 5.7 | 12.6 | 6.6 | 5.7 | 5.3 | 5.4 | 8.5 |
| Histidine | 4.0 | 2.8 | 2.3 | 2.5 | 1.8 | 2.2 | 2.5 | 1.4 |
| Lysine | 9.0 | 9.5 | 10.0 | 2.9 | 5.6 | 5.9 | 6.6 | 9.4 |
| Tyrosine | 2.8 | 3.2 | 1.5 | 6.2 | 3.7 | 4.3 | 3.3 | 3.5 |
| Tryptophan | 1.8 | 0.8 | 0.3 | — | — | 0.9 | 1.1 | 1.9 |
| Phenylalanine | 7.5 | 4.6 | 2.1 | 7.1 | 4.8 | 5.1 | 3.0 | 5.0 |
| Cystine | 2.8 | 1.3 | — | — | 1.0 | 0.9 | 1.1 | 1.9 |
| Methionine | 2.1 | 2.9 | 1.1 | 0.3 | 3.0 | 1.7 | 1.3 | 4.6 |
| Serine | 8.6 | 4.9 | 3.2 | — | 3.7 | 3.9 | 4.2 | 5.9 |
| Threonine | 7.7 | 6.1 | 2.3 | 6.9 | 5.0 | 4.9 | 4.6 | 5.1 |
| Leucine | 10.2 | 10.2 | 5.3 | 7.2 | 7.8 | 8.4 | 6.0 | 9.5 |
| Isoleucine | 3.5 | 7.6 | 3.6 | 7.1 | 5.5 | 6.1 | 4.3 | 5.5 |
| Valine | 7.1 | 6.7 | 7.7 | 8.6 | 6.7 | 5.1 | 4.9 | 5.9 |
| Glutamic acid | 13.5 | 20.2 | 8.2 | 17.2 | 11.9 | 15.0 | 10.9 | 14.1 |
| Aspartic acid | 5.0 | 12.2 | 8.1 | 13.5 | 8.1 | 11.9 | 7.2 | 12.1 |
| Glycine | 3.8 | 5.8 | 5.9 | 10.3 | 4.7 | 5.1 | 4.0 | 5.4 |
| Alanine | 6.5 | 5.1 | — | 12.0 | 6.6 | 4.7 | 5.5 | 10.4 |
| Proline | 4.0 | 6.0 | 4.3 | 10.3 | — | 3.1 | 4.0 | 4.8 |
| Hydroxyproline | — | — | — | — | — | — | 0 | 0 |
| Diaminopimelic acid | — | — | — | — | — | — | — | 0.8 |
| Reference | MÜTING (35) | AGREN (5) | SUGIMURA (55) | DUCHÂTEAU (16) | BLOCK (11) | WU (73) | BLOCK (11) | ROBERTS (41) |

There are, however, some differences. Diaminopimelic acid is present in algae and in *Escherichia coli* [ROBERTS et al. (41)] but not in metazoans. Hydroxyproline is a component of collagen, which is an important constituent of mesodermal tissue of vertebrates. Hydroxyproline is not present in yeast.

Enzymes are so basic to the processes of life that it is difficult to even conceive of life without them. It is true that rather simple systems are capable of some of the catalytic activities of proteins. Iron in solution has one-millionth the activity of iron in catalase in splitting hydrogen peroxide. Some rare earths possess lipase activity. Recent studies in which proteins have been partially fragmented show that enzymatic activity may be retained by a small part of the molecule. Hence the question might well be raised: Is there any evidence about long-time trends in the amino acid content of proteins having definite biochemical functions?

Two recent studies are particularly valuable in providing evidence on the long-time stability of some biochemical systems. One of them is work by Sanger and Smith (44) on insulin, a protein instrumental in regulating the glucose content of blood. Sanger and Smith examined the amino acid content and the sequence of amino acids in insulin obtained from whale, sheep, cow, horse, and pig. In the insulin from all these species, the molecule consists of a total of 51 amino acid residues; moreover, 48 out of 51 of the constituent amino acids are identical and are arranged in an identical sequence. The odds that such a situation could occur by chance are negligible, and this finding is powerful evidence for the common ancestry of all five of these animals.

The other study was concerned with the amino acid content of collagen derived from eight widely different species of vertebrates. This substance is the principal protein in bone and is a component of other vertebral mesodermal tissue, including skin, tendon, and loose connective tissue. Eastoe (18, 19) analyzed the amino acids in collagens or only slightly modified collagens derived from sturgeon, cod, shark, lungfish, whale, ox, and man. Some of his results are shown in Table 3. The similarity in composition is striking. There are surely some real differences in amino acid content, but a change in only a relatively few amino acid residues would be required to arrive at an identical over-all composition for collagen from all these sources. Unfortunately, little is known about the exact amino acid sequence in these proteins. Nevertheless, the compositional pattern shown in Table 3 is intriguing and unusual. No other protein has been reported with a comparably large content of hydroxyproline. The large amount of glycine relative to other amino acids is surprising, and the ratio of arginine:histidine:lysine is also rather bizarre. The creatures are presumably descendants of a common ancestor, but the branching in the ancestral tree occurred at least 300 million years ago. In the meantime the various lines have existed under greatly differing environmental conditions and have consumed

Table 3. Amino Acid Composition of Vertebrate Collagens and Gelatins (calculated as g. of amino acid per 100 g. of dry ash-free protein).

| | Sturgeon swim-bladder collagen | Cod-bone gelatin | Shark-skin gelatin | Lungfish skin collagen | Whale-skin gelatin | Ox-bone collagen | Human bone collagen | Human tendon acid extract |
|---|---|---|---|---|---|---|---|---|
| Alanine ...... | 11.6 | 10.4 | 11.2 | 11.7 | 10.8 | 10.5 | 10.9 | 10.3 |
| Glycine ...... | 27.7 | 28.2 | 26.5 | 24.0 | 26.7 | 25.3 | 25.8 | 25.4 |
| Valine ....... | 2.31 | 2.32 | 2.71 | 2.56 | 2.64 | 2.65 | 2.97 | 3.10 |
| Leucine ...... | 2.55 | 3.26 | 3.32 | 3.37 | 3.56 | 3.93 | 3.60 | 3.57 |
| Isoleucine .... | 1.65 | 1.64 | 2.71 | 1.64 | 1.57 | 1.73 | 1.88 | 1.53 |
| Proline ...... | 12.8 | 12.4 | 13.9 | 14.8 | 16.2 | 14.7 | 15.3 | 15.2 |
| Phenylalanine | 2.53 | 2.04 | 2.43 | 2.60 | 2.32 | 2.88 | 2.49 | 2.46 |
| Tyrosine ..... | 0.46 | 0.63 | 0.26 | 0.19 | 0.72 | 0.56 | 0.86 | 0.69 |
| Serine ....... | 5.8 | 7.9 | 5.0 | 4.71 | 4.71 | 4.24 | 4.06 | 4.05 |
| Threonine .... | 3.79 | 3.22 | 3.24 | 3.18 | 3.12 | 2.52 | 2.35 | 2.30 |
| Methionine ... | 1.43 | 2.26 | 1.59 | 0.59 | 0.79 | 0.80 | 0.84 | 0.90 |
| Arginine ..... | 10.0 | 9.1 | 9.3 | 9.1 | 9.5 | 9.2 | 8.8 | 8.9 |
| Histidine..... | 0.83 | 1.24 | 1.26 | 0.80 | 0.96 | 0.96 | 0.96 | 0.87 |
| Lysine ....... | 3.46 | 3.66 | 3.76 | 3.63 | 4.14 | 4.11 | 4.40 | 3.29 |
| Aspartic acid. | 6.9 | 7.5 | 6.0 | 6.6 | 6.7 | 7.1 | 6.7 | 6.7 |
| Glutamic acid | 11.4 | 11.4 | 10.3 | 11.9 | 11.2 | 11.9 | 11.4 | 11.1 |
| Hydroxyproline | 11.8 | 8.3 | 10.9 | 9.8 | 12.8 | 14.1 | 14.1 | 12.6 |
| Hydroxylysine | 1.90 | 1.42 | 0.82 | 0.88 | 1.02 | 1.12 | 0.62 | 1.50 |
| Total ...... | 118.9 | 116.9 | 115.2 | 112.05 | 119.5 | 118.3 | 118.0 | 114.5 |

vastly different kinds of food. Nevertheless a compositional pattern has been maintained. These results, together with the study of SANGER and SMITH, indicate that the amino acid content of at least some of the proteins of animals may persist unchanged for hundreds of millions of years.

Another line of argument concerning the biochemistry of past creatures involves muscle. In all metazoans that have been studied one of the chief constituents is the protein myosin. This substance possesses enzymatic activity, and is capable of splitting energy-rich ATP and participating in the conversion of the resultant chemical energy into mechanical work. In many animals the adductor muscles terminate at the shell in what is called a scar. This is characteristically an indentation which outlines the shape of the muscle at its point of attachment. An example is the clam *Mercenaria mercenaria*. Muscle scars of this animal are clearly apparent, and specimens of shells formed 25 million years ago have scars identically comparable to those of today. The inference is strong that the older clams used the same biochemical system as present-day ones, and, in view of the comparable scars, perhaps the identically same proteins. One may speculate similarly about the muscle system of ancient bivalves. They have scars differing only in minor degree from those of today.

In summary, all available evidence from comparative biochemistry and paleontology points toward the long-time existence of many of the present-day biochemical systems. In dealing with organic chemicals in rocks our best initial assumption is to consider them as being originally derived from living creatures possessing biochemical functions virtually identical to those of creatures living today.

### Preservation of Organic Substances.

RUBEY (43) estimates that $65 \times 10^{20}$ g. of reduced carbon is now present in sedimentary rocks of all ages. In aerobic environments the inevitable fate of organic matter is destruction and ultimate conversion to such substances as carbon dioxide, water, and nitrogen. Under ordinary circumstances in nature this process is very rapid. Under special conditions, as in a dry environment (where bacterial action may be absent), the time required for oxidation is much longer. However, all organic substances combine with oxygen, many with relatively low activation energies. For instance, CONWAY and LIBBY (13) estimate a heat of activation of 25,000 cal/mole for the oxidation of alanine. This value implies a half-life of only about $10^4$ years at room temperature.

In the sedimentary rocks are strata, such as red beds, which were formed under aerobic conditions. They do not contain appreciable quantities of organic matter. In contrast, the great accumulations of organic matter are found where evidence indicates anaerobic environments. In fact, organic matter is in part their cause, for they occur in circumstances of high productivity of organic substance and limited aeration. Typical of such environments are bogs, swamps, and silled basins where the rate of aeration is low. In these circumstances any free oxygen is consumed. Although metazoans cannot exist in an environment devoid of free oxygen, many microorganisms can survive and grow and through their chemical activities alter their surroundings. Sulfate is reduced to hydrogen sulfide or sulfur; ferric iron is also reduced, and much of it ultimately appears as pyrite ($FeS_2$), a common constituent of sediments containing organic matter. The reducing nature of such environments is also indicated by observed values of $E_H$ ranging from $-0.2$ to $0.5$ volt. The chemical activities of anaerobes are numerous and varied. For example, in transformations of glucose, yeasts produce carbon dioxide and ethyl alcohol, while molds can produce a large number of organic acids. Other organisms convert glucose into carbon dioxide and methane. Potential rates of transformation of organic material are very great. One gram of yeast, for instance, can ferment many grams of glucose per day. Other microorganisms have comparable capacity to alter the nature of organic matter.

*References, pp. 400—403.*

Other reactions involve hydrogenation. Several examples of the reducing capacity of anaerobes could be cited, among them saturation of double bonds as in the conversion of the vinyl to an ethyl group in a porphyrin [TREIBS (65)] and reductive deamination of amino acids.

In natural environments many years are available for these transformations; the fact that much organic matter remains is the real puzzle. Part of the organic material is only a little altered. Evidence for this statement comes from chlorophyll. As chlorophyll is not produced by anaerobes, any remaining in the sediment must represent material originally synthesized in the presence of light. Relatively large amounts of chlorophyll or compounds derived from it are found in sediments.

Some substances appear to be more subject to attack than others. Only very small amounts of amino acids or unpolymerized carbohydrates have been found in sediments. In anaerobic environments the ratio of carbon to oxygen increases with time. These considerations suggest that the lipides are relatively immune.

In marine environments an important factor may be the limited amount of sulfate that sulfur bacteria use as a source of oxygen while consuming organic matter. Another limitation may be imposed by the thermodynamics of chemical reactions. If energy-producing reactions of the right type are not available, the organism cannot survive indefinitely in a moist natural environment where hydrolysis of peptide bonds proceeds at an appreciable rate.

A further limiting factor is accessibility. Insoluble keratinous proteins and chitins are attacked only very slowly. Organic material within a bone or shell may be protected from bacterial attack. Another protective mechanism seems to be the formation of large organic complexes which are relatively immune to immediate attack.

### Organic Complexes of High Molecular Weight.

In spite of the enormous potential chemical activity of bacteria, most of the organic matter deposited in anaerobic environments is not destroyed or even substantially altered by these agents. Instead, the original mixture of proteins, fats, and carbohydrates tends to appear in the form of a complex material of very high molecular weight called kerogen, which is not soluble in water, acids, or organic solvents.

In old sedimentary rocks most of the organic matter is present in the form of kerogen. HUNT and JAMIESON (29) have made an extensive study of the Frontier Shale, a 90-million-year-old formation in Wyoming, and have assessed the organic content to a depth of 900 feet and in an area of 800 square miles. Most of the organic matter was found to be in the form of kerogen amounting to 100 billion barrels, or 89% of the total organic matter. Second in quantity was asphalt, amounting to

9 billion barrels, or about 7% of the total. Asphalt is also a very complxe material, but it is soluble in organic solvents such as a mixture of 70% benzene, 15% acetone, and 15% methanol. A third fraction was hydrocarbon within shale, amounting to 2.5% of the total. Finally, there was oil which had migrated to sands within the shale. The actual recoverable oil in these sand reservoirs is 500 million barrels, or 0.3% of all the organic matter found in this section of the Frontier formation.

Examination of coals also indicates that most organic matter is present in the form of very large complex molecules. Both in shale and in coal these complexes have high adsorptive capacity and hence probably serve as a protecting and holding matrix for many smaller molecules. In view of the key roles of these large complexes it is important to try to find out how they are formed and how they can be studied. Owing to the complexity of the substances and the variable composition of original organic matter the problem is extremely difficult.

Kerogens of sediments appear to be related to humic acids of soils. Such evidence as is available indicates that reactions between carbohydrates like hemicellulose and proteins or amino acids may be an important factor in the formation of kerogen and humic acid. Hydrolysis of Recent sediments and soils yields hemicellulose or free sugars and amino acids. In the case of soils, humic acids have been isolated [OKUDA and HORI (37, 38)] which on hydrolysis yielded amino acids.

In a relatively short time after burial or deposition, however, most of the fixed nitrogen of soil and sediments is no longer in the form of protein but is firmly bound to organic chemicals that are insoluble in acid. STEVENSON (52) found that 76% of the fixed nitrogen in an Illinois farm soil was in non $\alpha$-amino compounds. TRASK (60) reported that about 69% of the fixed nitrogen in sediments from Lake Maracaibo was in a non-protein form.

A study by SOWDEN and PARKER (50) bears importantly on this problem. These workers treated soils with dinitrofluorobenzene to test for free amino groups. None were found, indicating both absence of free amino acids or peptides and absence of the free $\varepsilon$-amino group of lysine.

One way of considering the role of carbohydrates in the formation of kerogen is to study the fate of the remains of creatures not containing much carbohydrate. Vertebrates consist mainly of protein and fat with only small amounts of sugars. Organic remains of cadavers after long burial consist almost entirely of saturated fatty acids with no trace of kerogen present. DURAND and VIÈLES (17) studied a bird cadaver and found that 55% was adipocere and the remainder bones, skins, and membranes. The adipocere consisted of 80 to 90% free acids and the remainder glycerol and sterol. Earlier, WASMUND (70) reviewed case histories of examination of old cadavers and described their very high fatty acid content. Thus it is apparent that, in the absence of carbohydrate, protein and its attendant fixed nitrogen disappear.

The earlier discussion of the constituents of living matter including algae showed that the principal constituents are proteins, carbohydrates, and fats, with practically all fixed nitrogen in the form of protein. Hydrolysis of single-celled algae yields in the main simple soluble components, in complete contrast to the findings on the organic matter in sediments. There is, however, in such circumstances an important problem which has plagued analytical chemists [BLOCK and WEISS (11)]. When proteins are hydrolyzed in the presence of sugars considerable losses of some of the amino acids occur, through combination of the carbohydrates with amino acids.

In considering the problem of the formation of a substance like kerogen from proteins, fats, and carbohydrates it should be pointed out that only a nominal amount of cross-linking between proteins and carbohydrate polymers would be required to form a complex of enormous weight and size. The role of fats is not clear, but the unsaturated fatty acids might well be reactive. With time some amino acids and peptides appear to be lost, but the remaining mass apparently becomes even more tightly bound together.

### The Stability of Organic Substances.

All organic substances are thermally unstable and degrade at rates dependent on the properties of the individual substance and the temperature of incubation. At ordinary ambient temperatures these rates are often relatively low, and as a result some substances may persist for hundreds of millions of years. For instance, TREIBS (63) reported isolating porphyrins from formations as old as about 500 million years, and this work seems entirely correct. ABELSON (2) has reported identifying alanine, glycine, glutamic acid, leucines, and valine from a 450-million-year-old tribolite fossil and from a 300-million-year-old fish plate. Other examples of preservation of organic matter will be cited later.

The stability of organic substances can be estimated from laboratory experiments at moderately elevated temperatures. ABELSON (2) and CONWAY and LIBBY (13) have performed such experiments on aqueous solutions of alanine, and MONTGOMERY (personal communication) has studied the degradation of porphyrin. Results are shown in *Fig. 1*, p. 390. This illustration, published earlier, is reproduced here because it conveniently illustrates the sharp dependence of degradation on thermal conditions. ABELSON (4) has discussed the problem of stability tests and the degradation of other amino acids. He found that alanine, glycine, glutamic acid, isoleucine, leucine, and valine are of about equal stability, in good agreement with the observation of their occurrence in old fossils. Arginine, lysine, phenylalanine, tyrosine, serine, and threonine were shown to be less stable. None of these appear in old fossils, and only

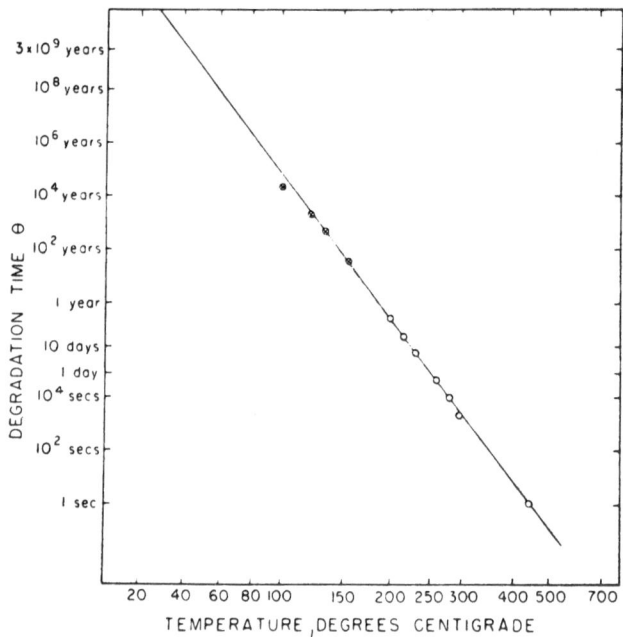

Fig. 1. The thermal degradation of alanine as shown by accelerated tests of aging. Shown is the time required for the decomposition of 63% of the original material. The upper four points were determined by the use of alanine tagged with $C^{14}$ by DWIGHT CONWAY at the University of Chicago. [From: Ann. New York Acad. Sci. 69, 276 (1957).]

small amounts of lysine and tyrosine are left in fossils after 25 million years.

Unfortunately, information on other substances is fragmentary. It is well established, however, that straight chain saturated hydrocarbons and aromatic hydrocarbons are more stable than porphyrins ($E \sim 58,000$ cal/mole). Fatty acids and sterols have been found in petroleum, and they seem sufficiently stable to endure for long periods of time. On the other hand, HANSON (personal communication) has stated that carotenoids are easily destroyed at temperatures of about 200°. ABELSON in some preliminary experiments has shown that aqueous solutions of purines and pyrimidines are not as stable as alanine.

These examples indicate that many substances of biological interest are capable of survival for long periods provided that temperatures are sufficiently low. Fortunately, in many areas in the world old sediments may be found that have not experienced temperatures much higher than 30°.

Similar remarks apply to kerogen. After the initial production of this material further changes occur only slowly if temperatures are sufficiently low. Kerogen from the 500-million-year-old Baltic Kolm does not differ markedly from that of comparatively recent shales. The Kolm shale has probably not been exposed to high temperatures. In other areas, however, where shales have experienced elevated temperatures, only graphite remains from the original organic material.

### Organic Substances in Soils and Rocks.

Some of the reported natural occurrences of organic chemicals will be described in the following pages. Emphasis will be placed on amino acids, carbohydrates, lipides and their degradation products, and porphyrins.

*Amino Acids in Soils.* The amino acid content of soils has been studied by many investigators over the past fifty years. SUZUKI (*56*) isolated alanine, leucine, valine, proline, and aspartic acid from hydrolysates of "humic acids". SCHREINER and SHOREY (*46*) isolated arginine and histidine from alkaline extracts of soils and detected the presence of lysine. Later discoveries were reviewed extensively by WAKSMAN (*69*) and by KOJIMA (*32*) and BREMNER (*12*).

The development of chromatographic techniques has made new and superior approaches to the problem feasible. STEVENSON (*52–54*) has applied these methods to an examination of some Illinois farm soils, obtaining quantitative analyses of hydrolysates of the constituent organic nitrogen substances. He identified and determined the amounts present of glycine, alanine, serine, threonine, valine, leucine, isoleucine, proline, phenylalanine, tyrosine, aspartic acid, glutamic acid, histidine, arginine, lysine, $\alpha,\varepsilon$-diaminopimelic acid, $\gamma$-aminobutyric acid, $\beta$-alanine, ornithine, $\alpha$-amino-n-butyric acid, methionine, methionine sulfoxide, methionine sulfone, cystine, and cysteic acid.

The first 15 of these amino acids are typical constituents of structural proteins. The presence of $\alpha,\varepsilon$-diaminopimelic acid is significant. This substance is not a constituent of woody plant material but is found only in microorganisms. Its occurrence in the hydrolysates is indicative that at least part of the amino acids were derived from bacterial proteins. STEVENSON (*53*) also identified chondrosamine in soil hydrolysates, and this is further evidence of a microbial contribution to the organic matter of soils. The $\beta$-alanine and $\gamma$-butyric acid occurring in the hydrolysates can be explained as products of the action of bacterial enzymes. Methionine sulfoxide and cysteic acid are almost certainly products of oxidation of methionine and cystine.

*Peats and Coals.* In view of the occurrence of amino acids in soils, their presence in peats is not surprising. Recently, two peats have been studied extensively [Swain, Blumentals, and Millers (57)]. The amino acids from hydrolysates of various zones in Cedar Creek peat from Minnesota and Dismal Swamp peat from Virginia were separated by paper chromatography. Among those measured were glycine, aspartic acid, glutamic acid, threonine, alanine, valine, and leucine.

A study [Shacklock and Drakeley (49)] using earlier techniques established the presence of diamino and monoamino acids in peat, lignite, and sub-bituminous coals. Monoamino, but no diamino, acids were found in bituminous and anthracite coals. The geological age of these latter specimens was not specified, but they were presumably Carboniferous. The apparent greater stability of monoamino acids is in accord with thermal stability studies quoted earlier.

Aspartic acid, glutamic acid, and glycine were found [Heijkenskjöld and Möllerberg (27)] in a hydrolysate of a British anthracite coal. The authors did not describe the formation from which the coal was collected, but state that it was about 250 million years old.

*Amino Acids in Sediments.* Another environment where amino acids may be found is in sediments and sedimentary rocks. Erdman, Marlett, and Hanson (20) have made a comparative study of the amino acid content of a typical Recent shallow-water, marine deposit and that of a similar sediment laid down approximately 30 million years ago. The Recent sample, collected on the inner continental shelf of the Gulf of Mexico, represents a narrow sedimentary layer at a depth of 120 cm.; this is below the zone of major bacterial activity. The age

Table 4. Comparison of a Recent and an Oligocene Marine Mud.

[From Erdman, Marlett, and Hanson (20).]

| Content | Recent | Oligocene |
|---|---|---|
| Carbonate carbon, % .. | 0.71 | 1.49 |
| Organic carbon, % .... | 0.53 | 0.27 |
| Organic nitrogen, % ... | 0.044 | 0.032 |
| Amino acids, $\mu$M/g .... | 3.0 | 0.51 |

Table 5. Principal Amino Acids in Recent and Oligocene Sediments.

(Arranged in order of decreasing abundance.)

[From Erdman, Marlett, and Hanson (20).]

| Recent | Oligocene |
|---|---|
| Valine | Alanine |
| Leucines | Glutamic acid |
| Alanine | Glycine |
| Glutamic acid | Proline |
| Aspartic acid | Leucines |
| Glycine | Aspartic acid |
| Proline | |
| Tyrosine | |
| Phenylalanine | |

of the sediment is probably not more than a few thousand years. The older sample was a section of a marine shale core cut from the Anahuac formation in Fort Bend County, Texas, at a depth of 5000 feet. Examination of these materials showed that the modern and ancient marine samples selected for study are quite similar with respect to concentrations of carbonate carbon, organic carbon, and fixed nitrogen. The concentrations of free or combined amino acids were lower in the Oligocene sample by a factor of 6. The Recent sediment contained 3.0 micromoles per gram. The nature of the amino acids present was also studied, with results shown in *Tables 4* and *5*.

*Amino Acids in Shells and Calcareous Tests.* Some of the most important occurrences of proteins and amino acids are in fossils. Within calcareous structures such as shells, the substances are protected from bacteria and often are not completely leached away by ground water. The amino acids do not react with the matrix, and other organic substances cannot come in contact with them. If the calcium carbonate of a fossil recrystallizes, soluble amino acids are, of course, lost. Many shells consist mainly of aragonite, which on recrystallizing changes to calcite, usually coarsely crystalline. Recrystallization of calcitic fossils can often be detected by noting the appearance of the specimen.

Recent shells and calcareous materials of biological origin have been examined [ABELSON (2)] as a guide to the understanding of occurrences of organic materials in fossils. It was noted that all specimens examined contained proteins in amounts corresponding to 6 to 40 micromoles of amino acids per gram. Forty other Recent arbitrarily chosen shells were surveyed; all contained protein. The amino acids appearing were those typically found in protein hydrolysates, with occasional specimens showing additional unidentified components.

ABELSON (3) has reported some observations bearing on the rate of hydrolysis of proteins in shells. The shell of the clam *Mya myarenaria* was employed to check the effect of relatively short exposures to geological environments. Recent specimens of this shell and specimens dated at 1000 years were compared. Solution of the Recent shell in dilute hydrochloric acid left a residue of filamentous light-colored protein. The 1000-year-old shell yielded protein which was amber-colored and had only relatively limited mechanical strength. Tests showed that protein content and amino acid content were identical in the two types of specimen. In another experiment, shells were dissolved in a mixture of dilute hydrochloric acid and trichloroacetic acid. Insoluble protein was removed from the mixture by centrifugation. The clear supernatant solution was examined for its content of free amino acids, peptides, or soluble protein. A negligible amount was found in each case, showing that the proteins of the 1000-year-old shell had not been broken into fragments of low molecular weight.

For studying older materials the clam *Mercenaria mercenaria* was a convenient object. This edible hard-shell clam, which lives today, is represented by fossil specimens dating back at least 25 million years. Specimens of Pleistocene age were tested which on geologic evidence are thought to be in the range 100,000 to 1 million years old. Comparison of Recent and older specimens revealed that marked changes had occurred in the fossil proteins. The material isolated by the usual protein precipitants was a black, tarry substance which yielded amino acids on hydrolysis. The clear supernatant solution contained peptides and free amino acids. The total amino acid content of the Pleistocene shell was only 18% of that found in Recent shells. When Miocene (25-million-year-old) shells were examined, amino acids were found, but no traces of proteins or peptides could be detected. These results are summarized in *Table 6*.

Table 6. Amino Acid Content of *Mercenaria mercenaria*.

| Age | Amino Acid Content, $\mu M/g$ | | |
|-----|------------------|------------------------------|---------------|
| | Bound protein | Soluble protein or peptide | Free protein |
| Recent ................. | 33.0 | 1.5 | < 0.35 |
| Pleistocene.............. | 2.1 | 2.25 | 1.0 |
| Miocene................. | 0 | 0 | 0.75 |

Initially most of the protein of the shell is evidently present in water-insoluble layers. For thousands of years only moderate changes occur, which do not affect the solubility of the protein. By the time 1 to 5% of the peptide bonds are broken (10 thousand to 100 thousand years), the protein fragments are much more soluble and some can be leached out of the shell. Some of the amino acids or peptides are probably entrapped in the aragonite structure. Ultimately, in the presence of water these peptide bonds are broken, leaving only free amino acids in the shell. The free amino acid content of the shell changed only moderately in the interval from 1 million to 25 million years.

ABELSON (2) studied the amino acid content of a series of 25-million-year-old shells from Chesapeake Bay, Maryland. All specimens examined were found to contain amino acids, including alanine, aspartic acid, glutamic acid, glycine, isoleucine, leucine (not resolved), proline and valine. The quantities isolated were about 1 part in 10,000 of the weight of the shells.

*Amino Acids in Bone.* Amino acids in fossil human bones have been studied by EZRA and COOK (21). They investigated a series of 20 fossil human and 3 fossil animal bones representing a time span from Recent to archaeologically very old. In bones exposed to burial for short periods most of the amino acids that are found in fresh bone are present in their normal amounts. In bones older than 5000 years, not all the amino acids are still present. Aspartic acid, glycine, and glutamic acid were detected in some of the oldest specimens, and only aspartic acid was found in a very ancient one. Other very old specimens contained no detectable amino acids.

In earlier work HEIZER and COOK (28) and COOK and HEIZER (14, 15) attempted to use the rate of disappearance of organic matter from bone as an auxiliary dating scheme calibrated by $C^{14}$ methods. This effort did not succeed very well. The difficulty lies in the fact that degradation of proteins is a function of other variables besides time. Variations in moisture, temperature, and pH of the environment and in oxygen tension could greatly affect the hydrolysis and ultimate leaching of proteins and their components. As an example of a very favorable environment for preservation may be cited the tar pits at La Brea in Los Angeles. There an asphalt matrix provided an essentially sterile, anaerobic environment in which the concentration of water was very low. The fossils have been preserved for 15,000 years (FLINT, personal communication). The total amino acid content of several specimens of bone was 10 to 12%. Studies of the peptide linkage showed that there were virtually no free amino acids and that the amino acids remained bound together. In such an environment proteins and peptides could be preserved for very long times. ABELSON (2) has examined a number of much older fossil bones.

Table 7. Amino Acid Content of Various Fossils.

| Name | Approximate age (years) | Formation | Amino acid content ($\mu M/g$) | Principal constituents |
|---|---|---|---|---|
| *Plesippus* (prehistoric horse) | Late Pliocene $5 \times 10^6$ | Hagerman Lake Beds, Idaho | 0.6 | Ala, gly |
| *Plesippus* (prehistoric horse, tooth) | Late Pliocene $5 \times 10^6$ | Hagerman Lake Beds, Idaho | 1.5 | Gly, ala, leu, val, glu |
| *Mesohippus* (prehistoric horse, tooth) | Oligocene $40 \times 10^6$ | White River, Nebraska | 0.31 | Ala, gly |
| *Mosasaurus* (dinosaur) | Cretaceous $100 \times 10^6$ | Pierre Shale, South Dakota | 1.8 | Ala, gly, glu, leu, val |
| *Anatosaurus* (dinosaur) | Cretaceous $100 \times 10^6$ | Lance, Lance Creek, Wyoming | 2.8 | Ala, gly, glu, leu, val, asp |
| *Stegosaurus* (dinosaur) | Jurassic $150 \times 10^6$ | Morrison, Como Bluff, Wyoming | 0.26 | Ala, gly, glu |
| *Dinichthys* (fish) | Devonian $360 \times 10^6$ | Ohio Black Shale | 3.0 | Gly, ala, glu, leu, val, asp |

Ala, alanine; asp, aspartic acid; glu, glutamic acid; gly, glycine; leu, leucine; val, valine.

(Specimens were furnished from the U. S. National Museum by C. LEWIS CAZIN and DAVID H. DUNKEL.)

In *Table 7*, p. 395 the amino acid contents of a variety of fossils are presented. Specimens were chosen from formations likely to have had a mild thermal history. The results show no trend with time, which is not too surprising, since different burial histories were involved. By means of radioactive tracers amino acids from fossils have been shown to be identical with present-day compounds.

The amino acids found in the fossils listed in Table 7 are those which laboratory tests have shown to be most stable. They are precisely those amino acids that could survive if a modern protein were to be stored in a fossil for similar periods of time.

*Lipides.* Fatty acids, waxes, and hydrocarbons have been found in soils, peats, brown coals, and petroleum. SCHREINER and SHOREY (*45*) isolated dihydroxystearic acid, $CH_3(CH_2)_7CHOHCHOH(CH_2)_7COOH$, from soils of low fertility. These included a Tennessee soil, classified as a Clarksville silt loam, containing 3.26% of organic matter. As much as 0.05 g. of the dihydroxy acid was found per kilo of soil. Subsequently, the same authors (*47*) identified α-hydroxystearic acid, $CH_3(CH_2)_{15}CHOHCOOH$, and lignoceric acid as constituents of a Maryland soil—the Elkton silt loam. In a review of the products obtained from peat, KIEBLER (*31*) lists additional examples of organic acids. SHABAROVA (*48*) isolated salts of oleic acid from a Black Sea ooze.

TANAKA and KUWATA (*59*) found palmitic ($C_{16}H_{32}O_2$), myristic ($C_{14}H_{28}O_2$), stearic ($C_{18}H_{36}O_2$) and arachidic ($C_{20}H_{40}O_2$) acid in petroleums from Japan, California, and Borneo.

Although many fatty acids are thermally stable enough to endure for long periods of time, only relatively small amounts have been isolated from ancient environments. Much of the organic acids which have been found appears to be the product of alteration of original straight-chain fatty acids. HANSON (*26*) suggests that unsaturated fatty acids can be converted into monocyclic and bicyclic naphthenes.

LOCHTE (*33*) reviewed current knowledge about petroleum acids and bases. On the basis of studies of mixtures of acids obtained from refinery products and in some cases from crude, a number of generalizations could be made, some of which are presented here:

(a) Phenols and carboxylic acids are found in all acid mixtures.

(b) The carboxylic acids consist of a mixture of the lower liquid and a few higher solid aliphatic acids and alicyclic acids which are known as naphthenic acids.

(c) Acids with less than 8 carbons are almost entirely aliphatic.

(d) Monocyclic acids start at $C_6$, range to $C_{20}$, and predominate between $C_9$ and $C_{13}$.

(e) Bicyclic acids start at $C_{12}$ and predominate above $C_{14}$.

WILLIAMS and RICHTER (72) isolated isovaleric, n-heptylic, n-octylic, and n-nonylic acids from a west Texas distillate. HANCOCK and LOCHTE (25) fractionated 70 liters of crude acids obtained from Signal Hill crude petroleum; they found a series of substances including acetic, propionic, isobutyric, n-butyric, isovaleric, and n-valeric acids.

Later investigations by LOCHTE and collaborators [QUEBEDEAUX et al. (39); NEY et al. (36)] resulted in identification of 2- and 3-methyl-pentanoic and n-hexanoic acids; 2-, 3-, 4-, and 5-methyl-hexanoic and h-heptanoic acids; n-octanoic; and n-nonanoic acids. In addition, a series of naphthenic acids were found.

Detailed explanation of the genesis of these various compounds must involve several sources. The n-butyric, n-hexanoic, and n-octanoic acids could come directly from hydrolysis of fats. Acetic and propionic acids and possibly others could be products of bacterial fermentation. Isovaleric, n-valeric, and 2- and 4-methyl hexanoic acids could conceivably arise from bacterial reductive deamination of amino acids. The origin of the other components is puzzling.

Because of its enormous economic interest, petroleum is the best-studied of the naturally occurring substances. ROSSINI et al. (42), who have conducted the most exhaustive investigation, have isolated more than 200 compounds from a Ponca City, Oklahoma, crude. The oil was produced from the uppermost part of the Wilcox sand formation of the Simpson Group, Middle Ordovician in age (450 million years) at a depth of about 3870 feet. The temperature in the well was 70°.

The largest class of components in the petroleum were straight-chain hydrocarbons. ROSSINI has pointed out that this is interesting on thermo-dynamic grounds, for the most stable hydrocarbons are the branched isomers. It is well known that there are 1858 possible isomers of $C_{14}H_3O$. Nevertheless, it was the relatively unstable n-tetradecane that was the largest $C^{14}$ component isolated. This fact points strongly to the genesis of these hydrocarbons from the straight-chain fatty acids of living organisms. There is a major question as to mechanism of this conversion, however. The conversion of the naturally occurring even-numbered carbon-containing fatty acids to odd-numbered straight-chain hydrocarbons could be explained as a chemical decarboxylation. The problem of genesis of even-numbered hydrocarbons is more puzzling. One interesting possible mechanism has been suggested by SHABAROVA (48), who postulates the reduction of the fatty acids by hydrogen and hydrogen sulfide. Substantial quantities of straight-chain hydrocarbons are present in petroleums from source rocks formed in every period of geologic time between the present and the Ordovician. There is little question that most of the Recent hydrocarbons were derived from fatty acids and possibly waxes. Young and old petroleum do not differ in major degree. Hence, it seems reasonable to take the position that the source materials and chemical processes involved were quite similar in the formation of young and old petroleum. Since large quantities of straight-chain hydrocarbons are produced from Ordovician rocks, one is led to infer that organisms were producing straight-chain fatty acids 450 million years ago.

*Carbohydrates.* Most of the annual production of carbohydrates is, of course, synthesized by plants, much of it in the form of polymerized sugars, notably cellulose. Since the tonnages involved are enormous it is not surprising that Recent peats, bogs, soils, and muds contain carbohydrates in various forms. WAKSMAN (68) listed some of the carbohydrates that have been identified in humus, including cellulose, mannan, galactan, levulan, and pentosan.

VALLENTYNE (66) has reviewed occurrences of sugars in lake waters and in lake seston. He and his co-workers identified sucrose, maltose, glucose, and fructose in seston. Free sugars were found in fresh-water sediments by VALLENTYNE and BIDWELL (67) and WHITTAKER and VALLENTYNE (71) in amounts of 10 to 3000 mg. per kilogram of organic matter present. Sugars identified included glucose, fructose, galactose, arabinose, ribose, xylose, sucrose, and maltose.

*Carbohydrates in Fossils.* Cellulose, which is a polymer of glucose, appears to be much more stable than the monomer. The reactive aldehyde group is involved in the formation of the polymer and is thus not available for other reactions. In any event, cellulose is the principal carbohydrate that has been isolated from fossils, though ABDERHALDEN and HEYNS (1) demonstrated the presence of chitin (a polymer of acetyl glucosamine) in a fossil coleopteron from the Middle Eocene (about 40 million years old).

GOTHAN (24) appears to have been the first investigator to isolate cellulose from an ancient environment—the Niederlauswitz (Miocene) formation in Germany. Others isolated and characterized it from a number of middle Tertiary environments. One of the more important investigations was that of STAUDINGER and JURISCH (51), who measured the specific viscosities of some Miocene preparations and found that the degree of polymerization had fallen from an original value of perhaps 2000 to 3000 to about 200—an indication that, on the average, original polymer molecules had been severed into about ten smaller units.

BARGHOORN (7) and BARGHOORN and SPACKMAN (10) also found relatively low degrees of polymerization in cellulose fractions isolated from early Tertiary lignites and fossil fruits. BARGHOORN (8, 9) considered that the degradation of cellulose was due to chemical hydrolysis rather than to bacterial decomposition.

*Porphyrins.* The occurrence of porphyrins in natural environments and in petroleums has been extensively investigated, owing in part to the ease with which they can be detected. Porphyrins related to hematin have been found, but most of the pigments isolated can be shown to be related to chlorophyll. When chlorophyll is exposed to geologic environments, magnesium is soon replaced by vanadium or nickel ions, giving rise to very stable complexes. The phytol moiety is lost, and the side-chains of the remaining tetrapyrrole are sometimes altered, for instance, by reduction. The complexes are hydrophobic and very soluble in petroleum, in which they are often found in relatively large quantities.

The major pioneering studies and perhaps the major contributions to the work on naturally occurring porphyrins were made by TREIBS (61–65). He devised useful procedures for isolating these pigments and found them in about a hundred specimens from a wide variety of geological environments, some as much as 500 million years old.

In a review of his work, TREIBS (65) reported finding metal porphyrins in 66 crude oils, 9 asphalts, 4 earth waxes, and 5 asphaltites. Seventy bituminous oil shales, 8 phosphorites, 1 guano, 7 cannel coals, and 17 coals contained both metal porphyrin complexes and green pigments. A Swiss oil shale was richest in porphyrin complex containing 4000 parts per million. The oldest occurrence was in "Eastern burning shales" of Cambrian age.

TREIBS was also aware of the importance of laboratory tests on the stability of porphyrins. He observed that, when pheophytin was heated to 250–320° in petroleum, phylloerythrin was formed. Further heating resulted in a decarboxylated form of desoxophylloerytherin.

The isolation of porphyrins from a series of geologic environments of ages dating back to the Cambrian is important to our views about the nature of past living things. These findings, added to earlier discussion of the ancient synthesis of fatty acids and proteins, help to build a fairly impressive structure of evidence indicating the long duration of important features of the present-day comparative biochemical plan.

*Precambrian Occurrences.* Discussion about ancient chemicals is necessarily speculative. Majority opinion agrees that life existed in Precambrian times. The most widely quoted reason for such belief is the profusion of phyla that appeared in Cambrian times. There is much reduced carbon in Precambrian sediments but very small quantities of extractable organic compounds. SWAIN (58) is one of few who have reported such substances, and the concentrations he found are quite low. There is marked contrast between the abundance of petroleum in Ordovician and that in earlier times. Failure to find petroleum in the older rocks has been ascribed to metamorphism. In view of the comparatively great stability of hydrocarbons this explanation would involve elevated temperatures and sizable depths of burial. It seems strange that the accidents of tectonics should subject all the older sediments to elevated temperatures with resultant destruction of virtually all organic material. A diligent search around the stable shields might uncover old undisturbed sediments containing quantities of extractable organic substances.

### The Perspective.

In the introduction to this survey the philosophic and the practical aspects of organic geochemistry were outlined. From our survey of the field it is apparent that much progress is being made in arriving at a fuller comprehension of the chemical nature of life, present and past. In contrast, our understanding of the constitution of the bulk of the chemicals of organic sediments is woefully poor. The reactions leading to the major components of the insoluble and relatively intractable kerogens are not known with any certainty. Despite some creditable efforts by RAUDSEPP (40) and FORSMAN and HUNT (23), no very good method yet exists for studying this material. It may be hoped that ultimately a gentle means will be found to break kerogen down into smaller entities that can be characterized. The discovery of such a method would open exciting new frontiers in the study of both the philosophic and the practical aspects of organic geochemistry.

### References.

1. ABDERHALDEN, E. und K. HEYNS: Nachweis von Chitin in Flügelresten von Coleopteren des oberen Mitteleocäns. Biochem. Z. **259**, 320 (1933).
2. ABELSON, P. H.: Annual Report of the Director of the Geophysical Laboratory, 1953–1954. Carnegie Inst. Wash. Year Book **53**, 97 (1954).
3. — Annual Report of the Director of the Geophysical Laboratory, 1954–1955. Carnegie Inst. Wash. Year Book **54**, 107 (1955).
4. — Some Aspects of Paleobiochemistry. Ann. New York Acad. Sci. **69**, Art. 2, 276 (1957).
5. ÅGREN, G.: Microbiological Determinations of Amino Acids in Foodstuffs. II. Acta Chem. Scand. **5**, 766 (1951).
6. BALDWIN, E.: An Introduction to Comparative Biochemistry, 164 pp. Cambridge, Engl.: Cambridge Univ. Press. 1949.
7. BARGHOORN, E. S.: Sodium Chlorite as an Aid in Paleobotanical and Anatomical Studies of Plant Tissues. Science (Washington) **107**, 480 (1948).
8. — Paleobotanical Studies of the Fishweir and Associated Deposits. In: F. JOHNSON, The Boylston Street Fishweir II. Papers Robert S. Peabody Found. Archaeol. **4**, 49 (1949).
9. — Degradation of Plant Remains in Organic Sediments. Botan. Museum Leaflets, Harvard Univ. **14**, No. 1, 1 (1949).
10. BARGHOORN, E. S. and W. SPACKMAN: Geological and Botanical Study of the Brandon Lignite, and its Significance in Coal Petrology. Econ. Geol. **45**, 344 (1950).
11. BLOCK, R. J. and K. W. WEISS: Amino Acid Handbook, 386 pp. Springfield, Ill.: Charles C. Thomas. 1956.
12. BREMNER, J. M.: Studies on Soil Organic Matter, I. J. Agric. Sci. **39**, 183 (1949).

*13.* CONWAY, D. and W. F. LIBBY: The Measurement of Very Slow Reaction Rates; Decarboxylation of Alanine. J. Amer. Chem. Soc. **80**, 1077 (1958).

*14.* COOK, S. F. and R. F. HEIZER: Archaeological Dating by Chemical Analysis of Bone. Southwestern J. Anthropol. **9**, 231 (1953).

*15.* — — The Present Status of Chemical Methods for Dating Prehistoric Bone. American Antiquity **18**, 354 (1953).

*16.* DUCHÂTEAU, G. et M. FLORKIN: Sur la composition de l'arthropodine et de la scléroprotéine cuticulaires de deux crustacés décapodes (*Homarus vulgaris* EDWARDS, *Callinectes sapidus* RATHBUN.). Physiol. Comparata Oecol. **3**, 365 (1954).

*17.* DURAND, J. F. et P. VIÈLES: Étude d'une adipocire d'un oiseau. Bull. soc. chim. biol. (Paris) **19**, 336 (1937).

*18.* EASTOE, J. E.: The Amino Acid Composition of Mammalian Collagen and Gelatin. Biochemic. J. **61**, 589 (1955).

*19.* — The Amino Acid Composition of Fish Collagen and Gelatin. Biochemic. J. **65**, 363 (1957).

*20.* ERDMAN, J. G., E. M. MARLETT and W. E. HANSON: Survival of Amino Acids in Marine Sediments. Science (Washington) **124**, 1026 (1956).

*21.* EZRA, H. C. and S. F. COOK: Amino Acids in Fossil Human Bone. Science (Washington) **126**, 80 (1957).

*22.* FLORKIN, M.: Biochemical Evolution, 157 pp. New York: Academic Press. 1949.

*23.* FORSMAN, J. P. and J. M. HUNT: Insoluble Organic Matter (Kerogen) in Sedimentary Rocks. Geochim. Cosmochim. Acta **15**, 170 (1958).

*24.* GOTHAN, W.: Neue Arten der Braunkohlenuntersuchung, IV. Braunkohle **21**, 400 (1922).

*25.* HANCOCK, K. and H. L. LOCHTE: Acidic Constituents of a California Straight-run Gasoline Distillate. J. Amer. Chem. Soc. **61**, 2448 (1939).

*26.* HANSON, W. E.: Some Chemical Aspects of Petroleum Genesis. In: P. H. ABELSON, Researches in Geochemistry, p. 104. New York: J. Wiley and Sons, Inc. 1959.

*27.* HEIJKENSKJÖLD, F. and H. MÖLLERBERG: Amino-acids in Anthracite. Nature (London) **181**, 334 (1958).

*28.* HEIZER, R. F. and S. F. COOK: Fluorine and other Chemical Tests of some North American Human and Fossil Bones. Amer. J. Physical Anthropol. **10**, 289 (1952).

*29.* HUNT, J. M. and G. W. JAMIESON: Oil and Organic Matter in Source Rocks of Petroleum. Bull. Amer. Assoc. Petrol. Geol. **40**, 477 (1956).

*30.* KEILIN, D.: Stability of Biological Materials and its Bearing upon the Problem of Anabiosis. Sci. Progr. **41**, 577 (1953).

*31.* KIEBLER, M. W.: The Action of Solvents on Coal. In: H. H. LOWRY, Chemistry of Coal Utilization, Vol. I, p. 677. New York: J. Wiley and Sons, Inc. 1945.

*32.* KOJIMA, R. T.: Soil Organic Nitrogen. I. Nature of the Organic Nitrogen in a Muck Soil from Geneva, New York. Soil Science **64**, 157 (1947).

*33.* LOCHTE, H. L.: Petroleum Acids and Bases. Ind. Eng. Chem. **44**, 2597 (1952).

*34.* MILNER, H. W.: The Chemical Composition of Algae. In: J. S. BURLEW, Algal Culture from Laboratory to Pilot Plant. Carnegie Inst. Wash. Publication No. 600, p. 285 (1953).

*35.* MÜTING, D. und V. WORTMANN: Zum Aminosäureaufbau der Serum- und Gewebeeiweißkörper gesunder Menschen und Tiere. Biochem. Z. **325**, 448 (1954).

36. NEY, W. O., W. W. CROUCH, C. E. RANNEFELD and H. L. LOCHTE: Petroleum Acids. VI. Naphthenic Acids from California Petroleum. J. Amer. Chem. Soc. 65, 770 (1943).

37. OKUDA, A. and S. HORI: Chromatographic Investigation of Amino Acids in Humic Acid and Alkaline Alcohol Lignin. Mem. Res. Inst. Food Sci., Kyoto Univ. No. 7, 1 (1954).

38. — — Identification of Amino Acids in Humic Acid. J. Sci. Soil Manure, Japan 26, 346 (1956).

39. QUEBEDEAUX, W. A., G. WASH, W. O. NEY, W. W. CROUCH and H. L. LOCHTE: Petroleum Acids. V. Aliphatic Acids from California Petroleum. J. Amer. Chem. Soc. 65, 767 (1943).

40. RAUDSEPP, KH. T.: A New Method of Investigation of the Chemical Structure of Combustible Minerals, and the Chemical Structure of the Estonian Shale-kukersite. Izvest. Akad. Nauk S. S. S. R., Otdel. Tekh. Nauk. No. 3, 130 (1954).

41. ROBERTS, R. B., P. H. ABELSON, D. B. COWIE, E. T. BOLTON and R. J. BRITTEN: Studies of Biosynthesis in Escherichia coli. Carnegie Inst. Wash. Publication No. 607. 1955. 521 pp.

42. ROSSINI, F. D., B. J. MAIR and A. J. STREIFF: Hydrocarbons from Petroleum. Amer. Chem. Soc. Monograph No. 121. New York: Reinhold Publ. Corp. 1953. 556 pp.

43. RUBEY, W. W.: Geologic History of Sea Water. Bull. Geol. Soc. Amer. 62, 1111 (1951).

44. SANGER, F. and L. F. SMITH: The Structure of Insulin. Endeavour 16, 48 (1957).

45. SCHREINER, O. and E. C. SHOREY: The Isolation of Dihydrocystearic Acid from Soils. J. Amer. Chem. Soc. 30, 1599 (1908).

46. — — The Presence of Arginine and Histidine in Soils. J. Biol. Chem. 8, 381 (1910).

47. — — Some Acid Constituents of Soil Humus. J. Amer. Chem. Soc. 32, 1674 (1910).

48. SHABAROVA, N. T.: The Organic Matter of Marine Sediments. Uspekhi Sovremonnoï Biol. 37, 203 (1954).

49. SHACKLOCK, C. W. and T. J. DRAKELEY: A Preliminary Investigation of the Nitrogenous Matter in Coal. J. Soc. Chem. Ind. (London) 46, 478 (1927).

50. SOWDEN, F. J. and D. I. PARKER: Amino Nitrogen of Soils and of Certain Fractions Isolated from them. Soil Sci. 76, 201 (1953).

51. STAUDINGER, H. und I. JURISCH: Über makromolekulare Verbindungen. 212. Mitt. Über den Polymerisationsgrad der Cellulose in Ligniten. Papier-Fabr. (tech.-wiss. Teil) 37, 181 (1939).

52. STEVENSON, F. J.: Ion Exchange Chromatography of the Amino Acids in Soil Hydrolysates. Soil Sci. Soc. Amer., Proc. 18, 373 (1954).

53. — Isolation and Identification of Some Amino Compounds in Soils. Soil Sci. Soc. Amer., Proc. 20, 201 (1956).

54. — Effect of Some Long-time Rotations on the Amino Acid Composition of the Soil. Soil Sci. Soc. Amer., Proc. 20, 204 (1956).

55. SUGIMURA, K., H. TAIRA, N. HOSHINO, H. EBISAWA and T. NAGAHARA: The Amino Content of Fish Muscle Protein. Bull. Japan. Soc. Sci. Fisheries 20, 520 (1954).

56. SUZUKI, S.: Studies on Humus Formation. III. Bull. Coll. Agr. Tokyo 7, 513 (1908).

57. SWAIN, F. M., A. BLUMENTALS and R. MILLERS: Stratigraphic Distribution of Amino Acids in Peats from Cedar Creek Bog, Minnesota and Dismal Swamp, Virginia. Limonology and Oceanography 4 (in press) (1959).

58. SWAIN, F. M., A. BLUMENTALS and N. PROKOPOVICH: Bituminous and other Organic Substances in pre-Cambrian of Minnesota. Bull. Amer. Assoc. Petrol. Geol. 42, 173 (1958).

59. TANAKA, J. and T. KUWATA: Higher Fatty Acids in Petroleum. J. Faculty Eng. Tokyo Imp. Univ. 17, 293 (1928).

60. TRASK, P. D.: Origin and Environment of Source Sediments of Petroleum. Houston, Texas: Gulf Publ. Co. 1932.

61. TREIBS, A.: Über das Vorkommen von Chlorophyllderivaten in einem Ölschiefer aus der oberen Trias. Liebigs Ann. Chem. 509, 103 (1934).

62. — Chlorophyll- und Häminderivate in bituminösen Gesteinen, Erdölen, Erdwachsen und Asphalten. Liebigs Ann. Chem. 510, 42 (1934).

63. — Chlorophyll- und Häminderivate in bituminösen Gesteinen, Erdölen, Kohlen, Phosphoriten. Liebigs Ann. Chem. 517, 172 (1935).

64. — Porphyrine in Kohlen. Liebigs Ann. Chem. 520, 144 (1935).

65. — Chlorophyll- und Häminderivate in organischen Mineralstoffen. Angew. Chem. 49, 682 (1936).

66. VALLENTYNE, J. R.: The Molecular Nature of Organic Matter in Lakes and Oceans, with Lesser Reference to Sewage and Terrestrial Soils. J. Fish. Res. Board Canada 14, 33 (1957).

67. VALLENTYNE, J. R. and R. G. S. BIDWELL: The Relation Between Free Sugars and Sedimentary Chlorophyll in Lake Muds. Ecology 37, 495 (1956).

68. WAKSMAN, S. A.: Humus: Origin, Chemical Composition, and Importance in Nature. Baltimore, Md.: Williams and Wilkins Co. 1936.

69. — Humus: Origin, Chemical Composition, and Importance in Nature, 2nd Ed. Baltimore, Md.: Williams and Wilkins Co. 1938.

70. WASMUND, E.: Die Bildung von anabituminösem Leichenwachs unter Wasser. Schr. Gebiet Brennstoff Geol. 10, 1 (1935).

71. WHITTAKER, J. R. and J. R. VALLENTYNE: On the Occurrence of Free Sugars in Lake Sediment Extracts. Limnology and Oceanography 2, 98 (1957).

72. WILLIAMS, M. and G. H. RICHTER: Acidic Constituents of a West Texas Pressure Distillate. J. Amer. Chem. Soc. 57, 1686 (1935).

73. WU, C. and J. F. HOGG: The Amino Acid Composition and Nitrogen Metabolism of *Tetrahymena geleii*. J. Biol. Chem. 198, 753 (1952).

(Received, April 2, 1959.)

# The Electron Gas Theory
# of the Color of Natural and Artificial Dyes:
# Applications and Extensions.

### By Hans Kuhn, Marburg a. d. Lahn.

#### With 30 Figures.

## Contents.

*Acknowledgement.* The writer is grateful to Dr. Walter Huber and Miss C. Harkort for help in the preparation of the manuscript and to Drs. E. W. Hughes and R. H. Eastman for critical reading of the text.

In a foregoing article (*71*) we have given a general discussion of some pertinent problems and principles and have treated the light absorption of symmetrical cyanine dyes and aza derivatives of cyanines in a simple manner. We shall present here along similar lines some important further applications of the theory.

For other approaches based on the free electron gas model cf. (*2–5*, *8–16*, *21*, *27–30*, *39–45*, *50*, *52*, *53*, *78*, *79*, *81*, *87–91*, *95*, *97–104*, *106*, *107*, *109*, *110*, *113*, *114*, *119*, *120*); and for reviews, cf. (*17*, *18*, *24*, *38*, *54*, *55*, *82*, *83*, *92*, *96*, *105*, *115*, *117*, *120 a*).

# I. Quantum Mechanical Determination of the Position of the Absorption Bands of Simple Dyes with an Unbranched, a Ring-Shaped, or a Branched Electron Gas.

## 1. A Simple Model of Vitamin $B_{12}$.

In vitamin $B_{12}$ we find a resonating chain indicated in (I) by heavy lines (*22*, *46*). By interchanging single and double bonds in this chain a limiting structure almost equivalent to (I) is obtained. Thus, we find here the conditions typical for symmetrical aza-cyanine type compounds

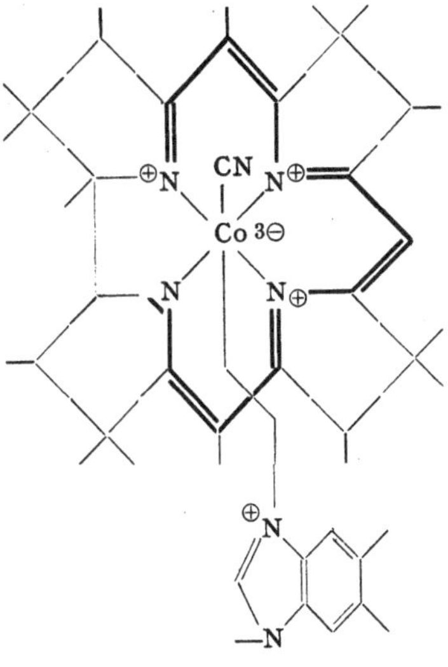

(I.) The chromophore of vitamin $B_{12}$.

that were considered in (*71*) by means of a simple theory: Each $\pi$ electron is described by a one-dimensional wave of DE BROGLIE wave length,

$$\Lambda = \frac{h}{m \cdot v} \qquad (1)$$

extending over the resonating chain of length $L$ [cf. (*71*), there equation (2)]; $h$ is PLANCK's constant ($h = 6.624 \times 10^{-27}$ erg sec.), $m$ is the mass of the electron ($m = 9.107 \times 10^{-28}$ g.), and $v$ its velocity.

According to *Fig. 1* the length $L$ is about $14\,l$ where $l = 1.39$ Å* is the bond length of the chain elements [cf. *(71)*, there p. 193]. There are seven $\pi$ electron pairs in the chain indicated in (I) by double bond symbols and by the bar at the N atom without charge sign. Thus, the electron states with wave functions with one to seven antinodes will be filled in the normal state of the molecule (Pauli principle). The wave functions for electrons in the highest occupied state and in the next higher state have 7 and 8 antinodes, respectively; the wave lengths $\Lambda$ equal $2\,L/7$ and $2\,L/8$ respectively (Fig. 1). The wave

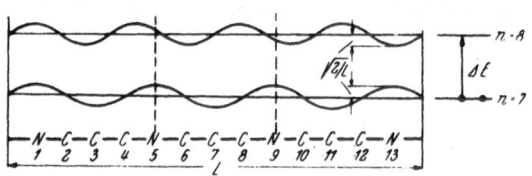

Fig. 1. Vitamin B$_{12}$ (I). — Highest occupied state ($n = 7$) and next higher state ($n = 8$). Wave functions and energy levels. In state $n = 7$ there are antinodes and thus charge accumulations at the positions 5 and 9 of the electronegative N atoms. This is not the case for state $n = 8$. Hence, the excitation energy of the aza-cyanine type compound (I) is higher than that of the corresponding cyanine.

length $\lambda_{max}$ of the absorption maximum is given by the excitation energy $\Delta E$ [cf. *(71)*, there p. 179], which is the difference in energy between the states 7 and 8:

$$\lambda_{max} = \frac{h\,c}{\Delta E} \quad (c = \text{velocity of light}). \tag{2}$$

We have considered in *(71*, there Section III. 1) the value of $\Delta E$ in the particularily simple case of the symmetrical cyanine type compounds where C instead of N is present at positions 5 and 9 (Fig. 1). By exchanging N for C at these positions $\Delta E$ is increased [cf. *(71*, there Section III. 2)]: In the highest occupied state the wave function has an antinode at positions 5 and 9, in the next state nearly a node. Thus, an electron in the highest occupied state shows an accumulation of the charge cloud at these positions, what an electron in the next state does not do. Hence, when N is introduced an electron in the highest occupied state will gain Coulomb energy, the energy level of this state will decrease, while the level of the next state will be far less affected. According to *(71*, there p. 197), the displacement of the energy level of a given state,

$$\varepsilon = - A\,\psi^2_{\text{position 5}} - A\,\psi^2_{\text{position 9}} \tag{3}$$

* For the sake of simplicity, the above value (which corresponds to the length of the CC one-and-a-half bond in benzene) will be used throughout this Chapter, and thus the difference between the lengths of CC and CN one-and-a-half bonds in resonating systems will be neglected.

where $\psi_{\text{position}\,5}$ and $\psi_{\text{position}\,9}$ are the respective values of the normalized wave function. $A$ is a constant characteristic of the hetero atom. According to (71, there p. 201), $A = 3.9 \times 10^{-20}$ erg cm. and $A = 6.0 \times 10^{-20}$ erg cm. for $-\overset{\oplus /\!/}{N}=$ and $-\overset{\oplus /\!/}{N}-R$ respectively, where $R$ is H or an alkyl group. In the present instance, due to the low electronegativity of the transition metal as compared to H or an alkyl, $A$ is clearly smaller than $6.0 \times 10^{-20}$ erg cm. but larger than $3.9 \times 10^{-20}$ erg cm., and the value $4.5 \times 10^{-20}$ erg cm. seems reasonable and shall be used here. For the highest occupied state we find from Fig. 1 and (71, there p. 197), $\psi_{\text{position}\,5} = \psi_{\text{position}\,9} = \sqrt{2/L}$, and by using equation (3), $\varepsilon = -4\,A/L$; similarly, for the next state $\varepsilon = -0.75\,A/L$. Thus, an increase in excitation energy, when N is exchanged for C at positions 5 and 9 of $3.25\,A/L = 3.25 \times 4.5 \times 10^{-20}/(14 \times 1.39 \times 10^{-8}) = 0.75 \times 10^{-12}$ erg is expected. The excitation energy of the cyanine type compound is found to be $2.80 \times 10^{-12}$ erg*. Hence, in the case of vitamin $B_{12}$, $\Delta E = 3.55 \times 10^{-12}$ erg, and $\lambda_{\text{max}} = 560$ m$\mu$**. Vitamin $B_{12}$ dissolves in water with red color and $\lambda_{\text{max}}$ in the visible region is located at 550 m$\mu$ (22), in good agreement with the calculated value. The absorption of vitamin $B_{12}$ in the ultraviolet region will be considered on p. 440.

## 2. A Simple Electron Gas Ring Model of the Phthalocyanines and Benzoporphyrines.

A phthalocyanine consists of a molecular skeleton and 21 $\pi$ electron pairs; 19 pairs are indicated in (II) by double bonds and 2 pairs by the bar at the nitrogen atoms in I and III. The benzene rings may be considered in a first approximation as separated resonance systems and thus an electron gas extending over a 16-membered ring and indicated in (III) can be considered as being responsible for the color. This gas consists of $21 - (4 \times 3) = 9$ $\pi$ electron pairs. Again, the state of each $\pi$ electron is described by a wave of DE BROGLIE wave length $\Lambda = h/(m\,v)$; an integer number of waves must extend over the

---

\* The absorption maximum of the imaginary molecule obtained from (I) by exchanging C for N at positions 5 and 9 may be assumed to coincide with the maximum of the symmetrical cyanine, $(CH_3)_2N-(CH=CH)_{\overset{\oplus}{j-1}}CH=N(CH_3)_2$, with $j = 6$, since in both molecules the resonating carbon chains are equally long and no unsaturated substituents are at the terminal N atoms. This compound has not yet been prepared, but since the lower members of the homologous series with $j = 2$, 3, and 4 show maxima at 309, 409, 511 m$\mu$, respectively (113), one may conclude that when $j = 6$, $\lambda_{\text{max}}$ will be located at $511 + 100 + 100 = 711$ m$\mu$.

\*\* If the $A$ value $3.9 \times 10^{-20}$ erg cm. (or $6.0 \times 10^{-20}$ erg cm.) is introduced instead of $4.5 \times 10^{-20}$ erg cm., which is certainly too low (or too high; cf. p. 407), the $\lambda_{\text{max}}$ value 580 m$\mu$ (or 520 m$\mu$) is obtained.

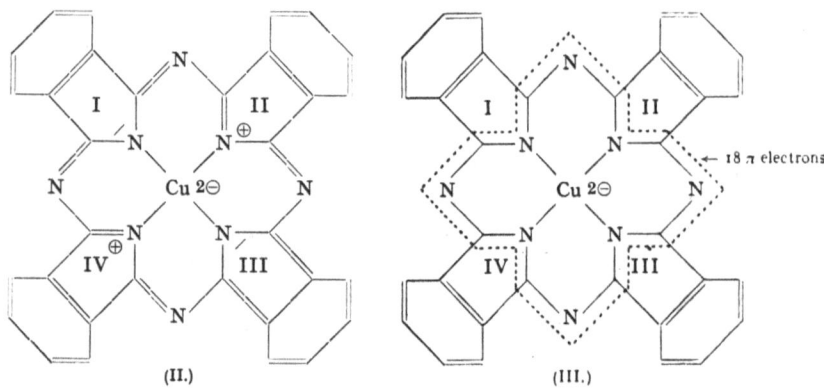

(II.) (III.)

circumference; thus, $\Lambda = L/n$, where $n = 0, 1, 2$, etc., and $L = 16\,l$ is the circumference of the ring. In the state $n = 0$ (lowest $\pi$ electron state) the wave has no nodes; the one-dimensional wave function $\psi$, and hence the cloud density along the circumference are constant *(Fig. 2)*. In the cases $n = 1, 2, 3, \ldots$, two electron-states are found for each $n$ (Fig. 2), viz. the states 1a, 2a, 3a, ... have antinodes but the states 1b, 2b, 3b nodes at the atoms II and IV. If we first neglect the difference in electronegativity between C and N, the same energy $\frac{m}{2}\,v^2 = \frac{h^2}{2\,m\,\Lambda^2} = \frac{h^2\,n^2}{2\,m\,L^2}$ must be attributed to both states with quantum number $n$, e. g., to states 4a and 4b. Due to the larger electronegativity of N, as compared to that of C, the energy level of state 4a is decreased by a certain value $-\varepsilon$, since the wave function has antinodes at each N atom. When the values of $\Lambda$ given in Section 1 (p. 407) are used

$$\varepsilon = -4 \times 4.5 \times 10^{-20} \times 2/(16 \times 1.39 \times 10^{-8}) - 4 \times 3.9 \times 10^{-20} \times$$
$$\times 2/(16 \times 1.39 \times 10^{-8}) = -3.02 \times 10^{-12} \text{ erg.}$$

In the case of 4b, $\varepsilon = 0$, since the wave function has nodes at each N atom. The energy levels of the eleven lowest states thus obtained are shown in Fig. 2; the nine $\pi$ electron pairs will occupy the nine lowest states. It is to be noted that the levels of states 5a and 5b coincide; in both states $\varepsilon = -1.51 \times 10^{-12}$ erg, i. e. $\varepsilon$ is half as large as in 4a. The first absorption band is caused by a transition of an electron from state 4b to states 5a, 5b, the next band by a transition from 4a to 5a, 5b. From the above it follows *(72)* that for 4b → 5a, 5b:

$$\Delta E = \frac{h^2}{2\,m\,L^2}\,(5^2 - 4^2) - 1.51 \times 10^{-12} \text{ erg}; \quad \lambda_{max} = 690 \text{ m}\mu,$$

and for 4a → 5a, 5b:

$$\Delta E = \frac{h^2}{2\,m\,L^2}\,(5^2 - 4^2) + 1.51 \times 10^{-12} \text{ erg}; \quad \lambda_{max} = 340 \text{ m}\mu.$$

*References, pp. 445—451.*

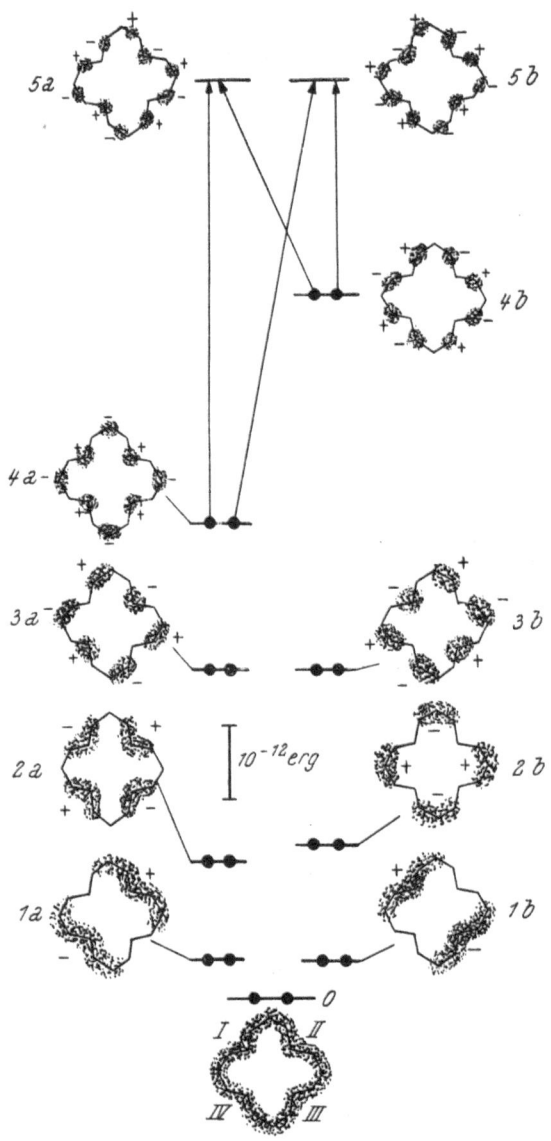

Fig. 2. Phthalocyanine (III). — Charge clouds and levels of the eleven most stable $\pi$ electron states. In states 4a and 4b we find the same number of cloud accumulations, but 4a is considerably lower than 4b since each cloud accumulation in 4a is at an electronegative N atom, in 4b at a C atom. In the normal state of the molecule the electron states O to 4b are filled and the visible band corresponds to a transition from 4b to 5a and 5b, the ultraviolet band to a transition from 4a to 5a and 5b.

Actually, the dye has only two strong bands in the visible and ultraviolet regions, and the calculated $\lambda_{max}$ position values are in good agreement with experimental data [unpublished work of H. NÖTHER, E. SCHNABEL and F. SCHÄFER cf. also (7)] as seen from *Fig. 3 a\**.

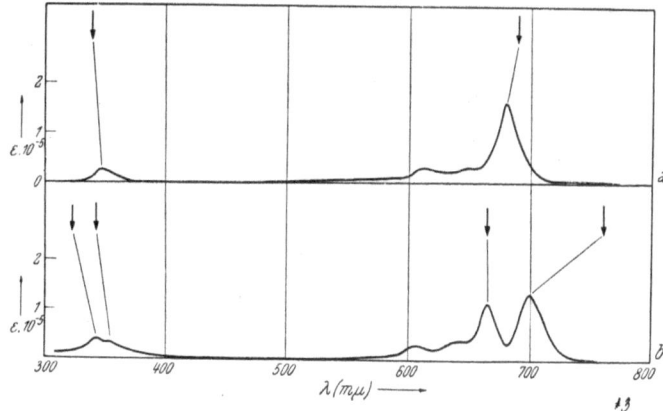

Fig. 3. Observed absorption spectra in chloronaphthalene (curves) and calculated positions of absorption maxima (arrows) of a) phthalocyanine (III); b) phthalocyanine free-base (IV).

In the phthalocyanine free-base (IV) the level of state 5 b is expected to be lower than that of state 5 a, since 5 b has antinodes and 5 a has nodes at the hydrogenated N atoms. Thus, a doubling of the two bands

(IV.)

considered before is expected and is actually observed [unpublished work of H. NÖTHER, E. SCHNABEL and F. SCHÄFER; cf. also (7)] *(Fig. 3 b)*. However, in each pair the spacing of the two bands is less than expected and this seems to be caused by the fact that the *A* values of the protonated

---

\* The smaller bands in (III) and (IV), which are not further considered here, might be due to a partial association of the dissolved dye molecules.

*References, pp. 445—451.*

and non-protonated N atoms are less different than assumed in the present calculation, as a consequence of the proximity of the protons of the H—N groups and the non-protonated N atoms.

In a benzporphine (V) the same situation is found as in a phthalocyanine, but the N atoms at the junctions between the pyrrol nuclei are replaced by CH groups. Correspondingly, the aza shift is less

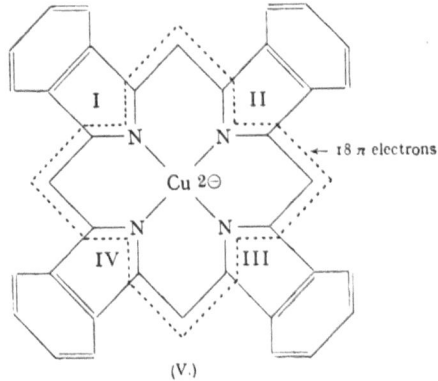

(V.)

pronounced in (V) than in (III). In (V) the levels 4a and 5a, 5b are lowered only about one-half as much as the corresponding levels in (III) by the aza type of disturbance. Thus, when proceeding from (III) to (V), a shift of the visible band to shorter wavelengths and a shift of the band in the ultraviolet region to longer wavelengths is expected. Such shifts are actually observed (7).

In the case of a porphyrine, which is obtained from (V) by replacing the benzene rings by double bonds, the situation is similar. However, while the $\pi$ electrons of the benzene rings in (V) form separate resonance systems, the $\pi$ electrons of the double bonds in a simple porphyrine have a larger effect on the $\pi$ electron system and may not be neglected. In the case of the porphyrines the simple ring model is less satisfactory than in the above cases but is still useful (73). In this case it is particularly important to consider the correlation of the $\pi$ electrons (p. 439 ff.), cf. PLATT (98) (100), GOUTERMAN (41 a), and (73). For earlier worte cf. SIMPSON (114), and (58, 61).

### 3. A Simple Model of Dyes with a Branched Electron Gas.

Let us now consider MICHLER's hydrol blue (VI) and acridine orange (VII). Both compounds possess a branched system of conjugated double bonds, and thus the $\pi$ electrons can be described as standing waves in the COULOMB field of a branched skeleton.

For MICHLER's hydrol blue we have assumed (71, there p. 174) for the sake of simplicity that the branches indicated by thin lines in (VI) can be neglected. Thus, the dye has been treated as an unbranched system of 6 pairs of $\pi$ electrons.

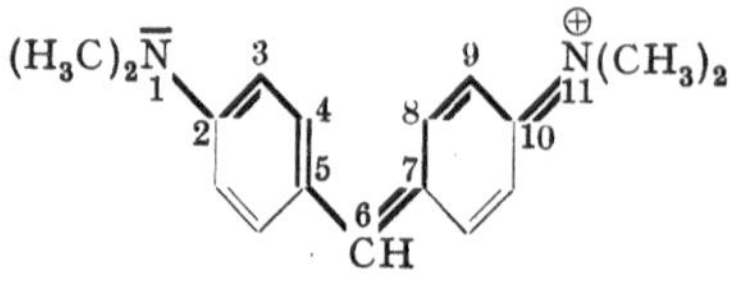

At this point we shall treat acridine orange in an analogous manner, viz. as a system of 7 pairs of $\pi$ electrons moving in a molecular skeleton branched as shown by heavy lines in (VII) as well as in *Fig. 4b (49)*.

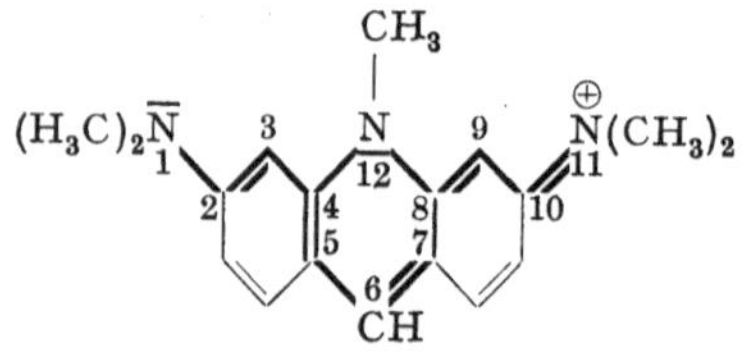

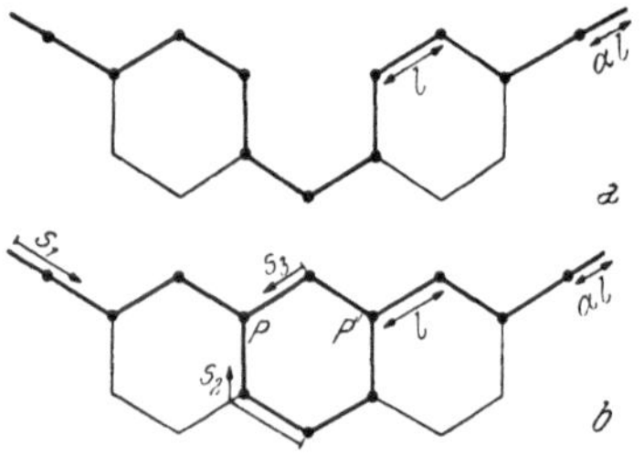

Fig. 4a. MICHLER's hydrol blue (VI). — It is assumed that the twelve $\pi$ electrons of the atoms located on the heavy line form an unbranched electron gas.

Fig. 4b. Acridine orange (VII). — It is assumed that the fourteen $\pi$ electrons of the atoms located on the heavy line form a branched electron gas which is extended along that line.

These $\pi$ electron pairs are represented in the formula by the 5 heavy double bond symbols and by the bars at the N atoms in positions 1 and 12.

The acridine orange molecule has the same dimethylamino end groups as MICHLER's hydrol blue. Hence we shall assume in both cases that the $\pi$ electron gas is extended by $\alpha l = 0.55 \, l = 0.77$ Å to either side of both terminal N atoms (*Fig. 4*)*. The problem then is, how to calculate the wave length $\lambda_{max \, VII}$ of the acridine orange maximum, i. e. the shift $\lambda_{max \, VII} - \lambda_{max \, VI}$ produced by the acridine bridge formation.

As in the case of an unbranched electron gas, one may neglect the component of the motion of the $\pi$ electrons across the line connecting the C and N atoms, along which the $\pi$ electron gas stretches, since this component gives a constant contribution to the energy of each $\pi$ electron state and thus has no effect on the distances between the energy levels of $\pi$ electrons; further we can assume as a model, that the $\pi$ electrons can be described by sine waves along each branch *(Fig. 5)*. These waves meet at the joints (points $P$ and $P'$ in Figs. 4b and 5). Let $\psi_{s_1}$ be the amplitude of the standing sine wave along branch $s_1$, $\psi_{s_2}$ and $\psi_{s_3}$ the amplitudes of the waves along $s_2$ and $s_3$ (Figs. 4 and 5). These three waves meet at $P$, where $\psi_{s_1}$

---

* The value, $\alpha = 0.55$ was found by introducing into equation (8) of (71) the values, $\lambda_{max} = 603$ m$\mu$ (observed $\lambda_{max}$ for MICHLER's hydrol blue) and $j = 5$.

must equal $\psi_{s_2}$ and $\psi_{s_3}$. Furthermore, a certain condition must hold for the slopes of these wave functions at point $P$, i. e. for the derivatives $\dfrac{d\psi_{s_1}}{ds_1}, \dfrac{d\psi_{s_2}}{ds_2}, \dfrac{d\psi_{s_3}}{ds_3}$. According to Chapter III (p. 433) it can be assumed with reasonable approximation

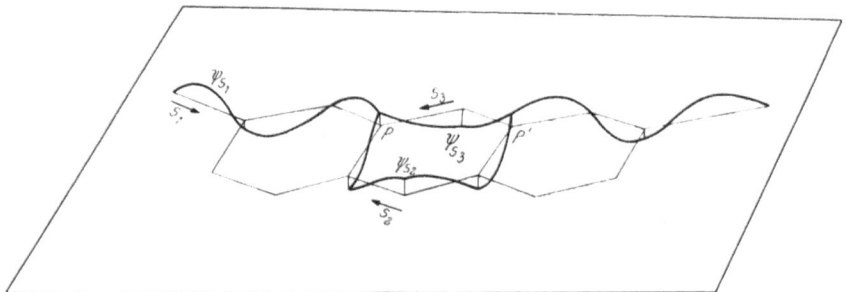

Fig. 5. Branching of electron gas in acridine orange (VII). — The $\pi$ electrons are described by sine waves along each branch.

that the sum of the slopes of the wave functions at a given branch point is zero, thus (60):

$$\left. \begin{array}{c} \dfrac{d\psi_{s_1}}{ds_1} + \dfrac{d\psi_{s_2}}{ds_2} + \dfrac{d\psi_{s_3}}{ds_3} = 0, \\[2mm] \psi_{s_1} = \psi_{s_2} = \psi_{s_3}, \end{array} \right\} \quad \text{at point } P. \qquad (4)$$

In this approximation the $\pi$ electron wave functions are given by the waves produced by vibrating a string which is branched similar to the molecular skeleton along which the electron gas stretches. In the

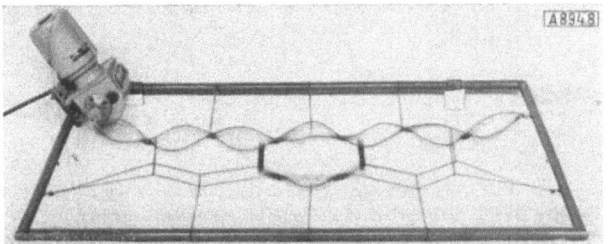

Fig. 6. Acridine orange (VII). — Standing transverse oscillation in a demonstration model consisting of a stretched string branched in a manner similar to the molecular skeleton. By varying slowly the rotational speed of a motor-driven excentric shown to the left of the Figure, the different standing oscillations of the string can be excited. [From: Angew. Chem. 71, 93 (1959).]

case of acridine orange a vibrating string suspended as shown in *Fig. 6* may be used as a model for the wave form of the $\pi$ electrons. Similar to the unbranched string considered in (71, there Fig. 4), the branched string may assume a number of standing transversal vibration states *(Fig. 7)*, and each of these stationary oscillations is obtained by exciting the string with a certain characteristic frequency. From the positions of maximum amplitude of the oscillation, the shape of the charge cloud can be found. The charge accumulations in the cloud will be at the positions of the antinode of the standing wave, as shown in Fig. 7 (49).

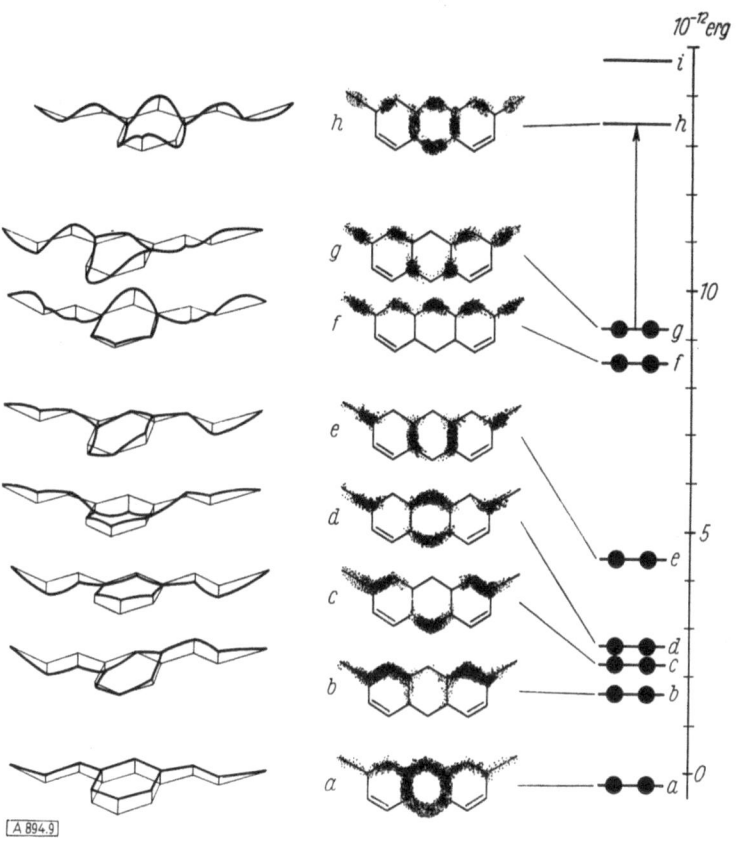

Fig. 7. Acridine orange (VII). — Charge clouds, standing electron waves and levels of the eight most stable π electron states. In the normal state of the molecule the electron states *a* to *g* are each occupied by two electrons. The band in the visible region corresponds to the jump of an electron, *g → h*. [From: Angew. Chem. 71, 93 (1959).]

According to the discussion in Sections 1 and 2 (p. 405 ff.) the energy of a given π electron can be regarded as the sum of two terms $(E + \varepsilon)$.
$E = \dfrac{m}{2}\,v^2 = \dfrac{m}{2}\left(\dfrac{h}{m\,\Lambda}\right)^2 = \dfrac{h^2}{2\,m\,\Lambda^2}$ is the kinetic energy and $\varepsilon = -A\,\psi^2$ is the perturbation energy due to the N atom at position 12 in (VII). $\psi$ is the value of the normalized wave function at position 12 and $A$, as mentioned before, equals $6.0 \times 10^{-20}$ erg cm. in the case of a —$NCH_3$— group (cf. 71, there p. 201). The values of $\Lambda$ and $\psi$ can easily be calculated as shown below. The energy levels thus obtained are given in Fig. 7.

In the normal state of the molecule the 7 pairs of $\pi$ electrons fill the 7 lowest energy levels, according to the PAULI principle. The excitation energy, and the position of the absorption band, are obtained in the same manner as in the cases considered above, and thus the value $\lambda_{max}$ VII $= 471$ m$\mu$ and the difference $\lambda_{max}$ VII $- \lambda_{max}$ VI $= -132$ m$\mu$ are obtained. The observed shift caused by the formation of the acridine bridge is, $\lambda_{max}$ VII $- \lambda_{max}$ VI $= 491 - 603 = -112$ m$\mu$ [(71), there p. 176], in agreement with the theory.

We shall give at this point some details of the theoretical treatment of acridine orange (49). On the basis of the symmetry one is allowed to consider only one half of this molecule, and we can distinguish two types of wave functions, viz. symmetrical [Equation (5)] and antisymmetrical ones [Equation (6)]:

$$\psi_{s_1} = B_1 \sin \frac{2\pi}{\Lambda} s_1,$$
$$\psi_{s_2} = B_2 \cos \frac{2\pi}{\Lambda} s_2,$$
$$\psi_{s_3} = B_3 \cos \frac{2\pi}{\Lambda} s_3. \tag{5}$$

$$\psi_{s_1} = B_1 \sin \frac{2\pi}{\Lambda} s_1,$$
$$\psi_{s_2} = B_2 \sin \frac{2\pi}{\Lambda} s_2,$$
$$\psi_{s_3} = B_3 \sin \frac{2\pi}{\Lambda} s_3. \tag{6}$$

$B_1$, $B_2$ and $B_3$ are constants. At joint $P$ (coordinates $s_1 = (3 + \alpha) l = 3.55\, l$; $s_2 = 2\, l$; $s_3 = l$) the conditions of Equation (4) (p. 413) must hold. For the case of a symmetrical wave function [Equations (5)]:

$$B_1 \frac{2\pi}{\Lambda} \cos \frac{2\pi}{\Lambda} 3.55\, l - B_2 \frac{2\pi}{\Lambda} \sin \frac{2\pi}{\Lambda} 2\, l - B_3 \frac{2\pi}{\Lambda} \sin \frac{2\pi}{\Lambda} l = 0,$$
$$B_1 \sin \frac{2\pi}{\Lambda} 3.55\, l = B_2 \cos \frac{2\pi}{\Lambda} 2\, l = B_3 \cos \frac{2\pi}{\Lambda} l. \tag{7}$$

By introducing the second expression into the first, the equation,

$$\operatorname{ctg} \frac{2\pi}{\Lambda} 3.55\, l - \operatorname{tg} \frac{2\pi}{\Lambda} 2\, l - \operatorname{tg} \frac{2\pi}{\Lambda} l = 0 \tag{8}$$

is obtained, which is fulfilled for the values: $\Lambda/l = 24.9928$ (state a); 7.4260 (state c); 5.6488 (state d); 3.6403 (state f); 2.9610 (state h), etc. For each of these values the values of the constants $B_1$, $B_2$ and $B_3$ can be determined from Equation (7) and from the normalization condition. In the special instance, $\Lambda/l = 2.9610$ (state h), the values $B_1 = 0.2524/\sqrt{l}$, $B_2 = -0.5298/\sqrt{l}$, and $B_3 = -0.4562/\sqrt{l}$ are obtained. In the case of antisymmetrical wave functions, the eigenvalues, $\Lambda/l = 8.9148$ (state b); 5.3292 (state e); 3.7003 (state g), etc. are determined in an analogous manner. The values $\Lambda = 3.7003\, l = 5.1434 \times 10^{-8}$ cm. (state g), and $\Lambda = 2.9610\, l = 4.1158 \times 10^{-8}$ cm. (state h) correspond to the highest occupied state and to the next one. The value of the normalized wave function at position 12 is zero in state g and equals $B_3 = -0.4562/\sqrt{l}$ in state h; thus $\varepsilon = 0$ in state g and $\varepsilon = -A\,B_3{}^2 = -6.0 \times 10^{-20}\,(0.4562)^2/(1.39 \times 10^{-8}) = -0.90 \times 10^{-12}$ erg in state h. Consequently, the excitation energy is $\Delta E = \dfrac{h^2}{2\,m}\left(\dfrac{1}{(4.1158 \times 10^{-8})^2} - \dfrac{1}{(5.1434 \times 10^{-8})^2}\right) + \varepsilon = 5.12 \times 10^{-12}$ erg $+ \varepsilon = 4.22 \times 10^{-12}$ erg, and hence $\lambda_{max} = \dfrac{h\,c}{\Delta E} = 471$ m$\mu$, as mentioned above.

Fig. 7 shows that the wave function of the highest occupied $\pi$ electron state g has a node at the positions 6 and 12, while that of the next state h has an antinode in both positions. Upon increasing the electronegativity of the atoms at 6 and 12, the energy level of state g will remain unaffected, while the level of state h will be lowered; thus a shift of the maximum towards the longer waves is to be expected [cf. (71), there Fig. 19].

On this basis one can qualitatively explain the spectral changes discussed in (71, there p. 177 ff.), such as the bathochromic shift observed upon exchanging the $NCH_3$ group at position 12 for the more electronegative O atom and the bathochromic shift that takes place on exchanging at position 6 the CH or $C-C_6H_4-COO^\ominus$ group for N.

A bathochromic shift could also be predicted when the N atom at 6 is substituted by the more electron attracting $NH^\oplus$ group, and such a shift has actually been observed. Thus, amethyst violet is violet in neutral solution (XIV, *Fig. 8*) but green in acids (XV, Fig. 8).

The calculated $\lambda_{max}$ values of dyes with the acridine type bridge are given in Fig. 8 (arrows), in good agreement with the observed values (bars)*.

These calculations (49) were based on the same assumptions as in the case of acridine orange. Again, the value $\alpha = 0.55$ was used for all compounds containing dimethylamino end groups, (VII)–(X), Fig. 8. The values $\alpha = 0.33$ and $\alpha = 0.36$, calculated from the observed $\lambda_{max}$ values of benzaurine and DOEBNER's violet were used in the case of dyes with oxygen end groups (XI)–(XIII) and amino end groups (XIV)–(XVII); Fig. 8. Thus, the influence of Br in eosine (XII, Fig. 8) and irisblue (XIII) has been neglected.

Let us now consider the details of the calculation for dye (X). Here $\Delta E = 5.12 \times 10^{-12}$ erg $+ \varepsilon$, as in the case of acridine orange, and according to (71, there Equations 10a and 10d), $\varepsilon = -9.9 \times 10^{-20}\,B_3{}^2 - 3.9 \times 10^{-20}\,B_2{}^2 = -9.9 \times 10^{-20} \times \dfrac{(0.4562)^2}{1.39 \times 10^{-8}} - 3.9 \times 10^{-20} \times \dfrac{(0.5289)^2}{1.39 \times 10^{-8}} = -2.27 \times 10^{-12}$ erg;

thus $\Delta E = 2.85 \times 10^{-12}$ erg and $\lambda_{max} = \dfrac{h\,c}{\Delta E} = 6.97 \times 10^{-7}$ cm. $= 697$ m$\mu$.

The branched $\pi$ electron gas model has been used to treat in the same manner the dyes mentioned in (71, there p. 178), and the concept of branching thus leads to a simple explanation of even such features as the deep color of the low-molecular weight WURSTER's blue and azulene (60). For further applications cf. (8, 44, 95, 106, 107, 109).

---

* For (XIV) cf. (112) (122); for (XV) cf. (122); for (XVI) and (XVII) cf. (84), and for the other cases cf. (71).

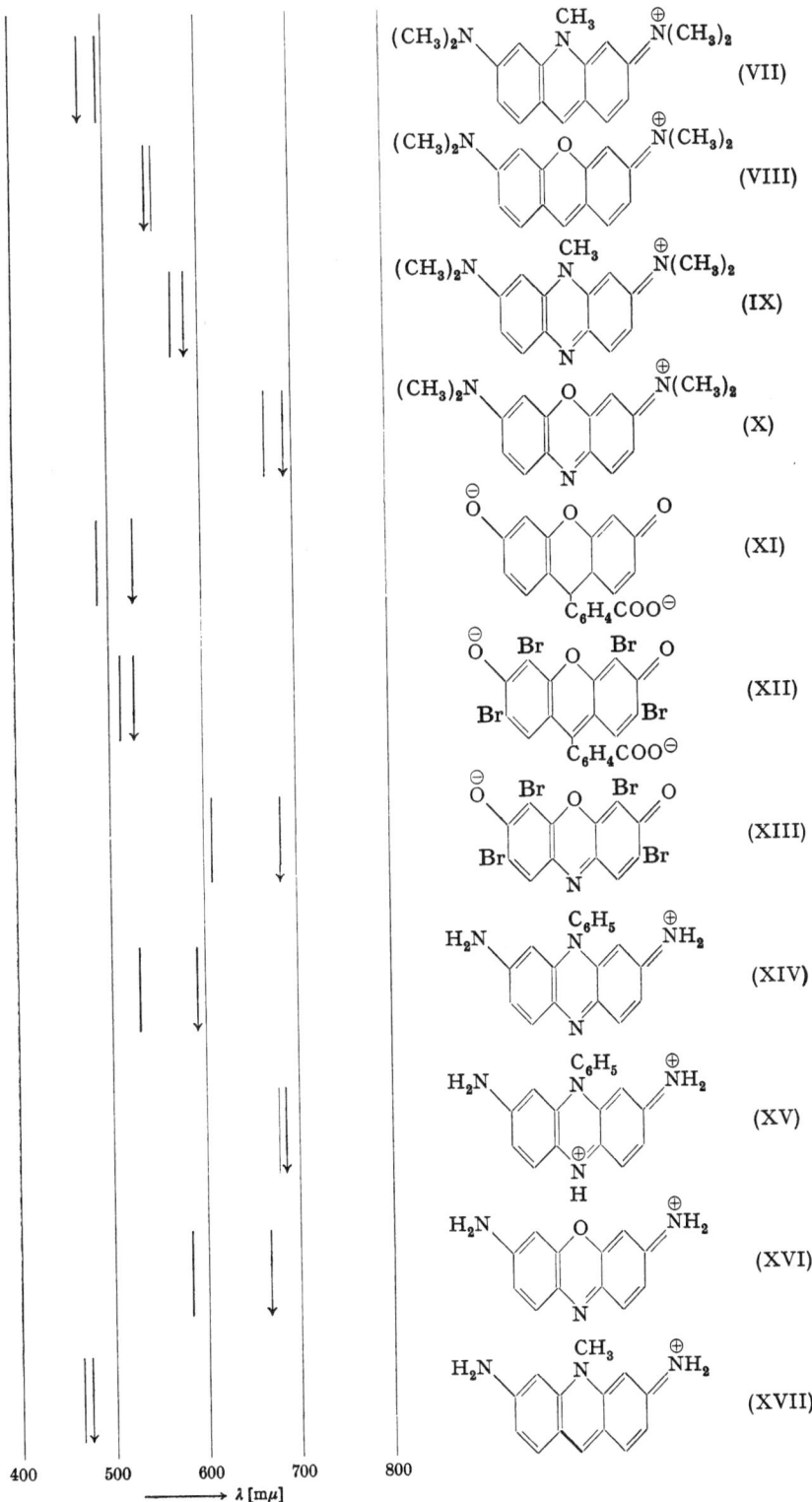

Fig. 8. Dyes with acridine type bridge: position of absorption maximum. — Arrow: calculated; bar: observed.

# II. A Refined One-dimensional Model
## Considering Irregularities of Potential along the Chain
## (Wave Shape Potential Model).

### 1. Qualitative Discussion of Some Quantum Mechanical Aspects of Polyenes and Symmetrical Cyanines (61).

In (71, there p. 170) we saw that a polyene absorbs at much shorter wavelengths than a symmetrical cyanine which contains a resonating

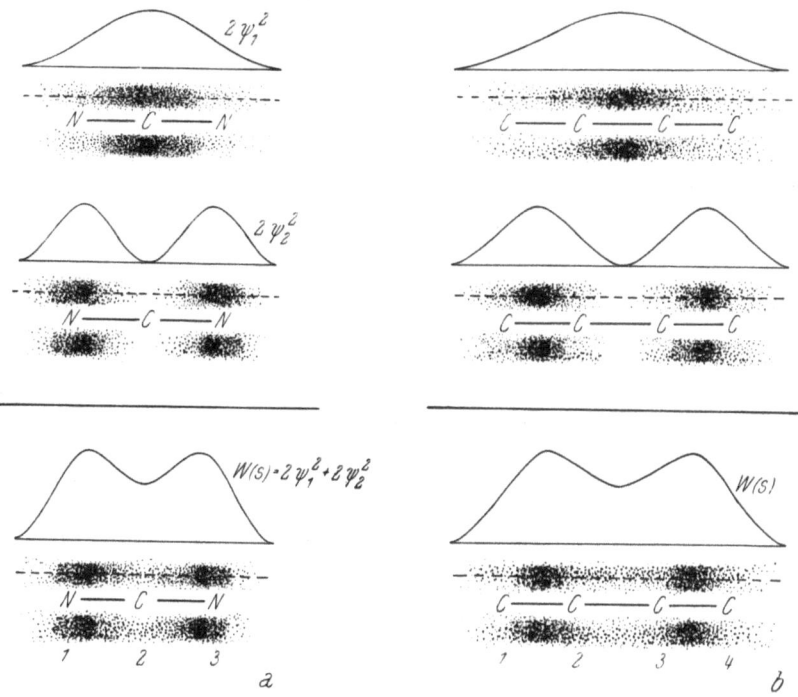

Fig. 9. Simple model: Density distribution along the dashed line and charge cloud accumulations of the two occupied π electron states. Over-all density distribution $W(s)$ and total charge cloud of the π electrons: a) amidinium ion (XVIII), and b) butadiene (XIX).

system of comparable length. Evidently, in the case of a polyene the excitation energy is much higher and, in contrast to the cyanines, is not determined by the chain-length and by the number of π electrons alone. In order to explain this surprising fact, let us consider the simplest

$$R_2\overline{\underset{1}{N}}-\underset{2}{CH}=\overset{\oplus}{\underset{3}{N}}R_2 \qquad\qquad H_2\underset{1}{C}=\underset{2}{CH}-\underset{3}{CH}=\underset{4}{CH_2}$$

(XVIII.)                            (XIX.)

representatives of the two series, viz. the amidinium ion (XVIII) and butadiene (XIX) which both contain four $\pi$ electrons.

When using the simple approximation method given in (*71*, there p. 190 ff.) we find that in the ground state of the molecule these electrons occupy the two lowest levels indicated in *Fig. 9* by the charge cloud accumulation and by the electron density distribution along the chain, e. g. along the dashed line. The total cloud and the over-all $\pi$ electron density distribution $W(s)$ along the chain are also given in this Figure. In the approximation mentioned the distribution in each single state is given simply by a sine square function; the density distribution $W(s)$ is then the sum of the two such functions and is a function with two maxima.

In the case of the amidinium ion the $\pi$ electron cloud extends over a chain of three charged atoms. The two maxima of the $\pi$ electron density distribution $W(s)$ are located, respectively, between the atoms 1–2 and atoms 2–3. The same $\pi$ electron density is found in the middle between atoms 1–2 and atoms 2–3; thus, the bonds between the two atom pairs mentioned must have identical properties. This conclusion (*61*) agrees with what follows from the resonance picture and represents a simple physical justification of that picture: According to the resonance theory (*93*), the molecule which can be symbolized by the two limiting structures,

$$R_2\overset{-}{\underset{1}{N}}-\underset{2}{CH}=\overset{\oplus}{\underset{3}{N}}R_2 \qquad\longleftrightarrow\qquad R_2\overset{\oplus}{\underset{1}{N}}=\underset{2}{CH}-\overset{-}{\underset{3}{N}}R_2$$

$$\text{(XVIII a.)} \qquad\qquad\qquad \text{(XVIII b.)}$$

is a resonance hybrid: the bond between atoms 1 and 2 is neither a single bond nor a double bond but represents an intermediate type. Since the same statement is valid for the 2–3 bond, the two bonds are equivalent.

In the case of butadiene (XIX) the $\pi$ electron cloud extends over a chain of four charged atoms. The two maxima of the $\pi$ electron density distribution $W(s)$ are between the atoms 1–2 and 3–4. Clearly, the $\pi$ electron density between atoms 2 and 3 is smaller than that between atoms 1–2 or 3–4; and by writing double bond symbols between 1–2 and 3–4, and a single bond between 2–3, the difference in the nature of these bonds is expressed (*61*). In contrast to (XVIII) it is impossible to write two equivalent limiting structures for butadiene.

This has an important corollary. Since the four C atoms are connected by equal $\sigma$ bonds, the bonds between the atoms 1–2, 2–3 and 3–4 would have the same length (about 1.5 Å), if the influence of the $\pi$ electrons could be neglected. However, the positive charges of the C atoms are within the COULOMB field of the negative electron cloud and they are attracted toward the accumulations of this cloud *(Fig. 10)*. Consequently, the "double bonds" between atoms 1–2 and 3–4 are shorter than the 2–3 "single bond", since the total $\pi$ electron charge density maxima are located between atoms 1–2 and 3–4. The length of the "double bonds" is 1.35 Å, that of the "single bond" is 1.47 Å *(I, III, I2I)*.

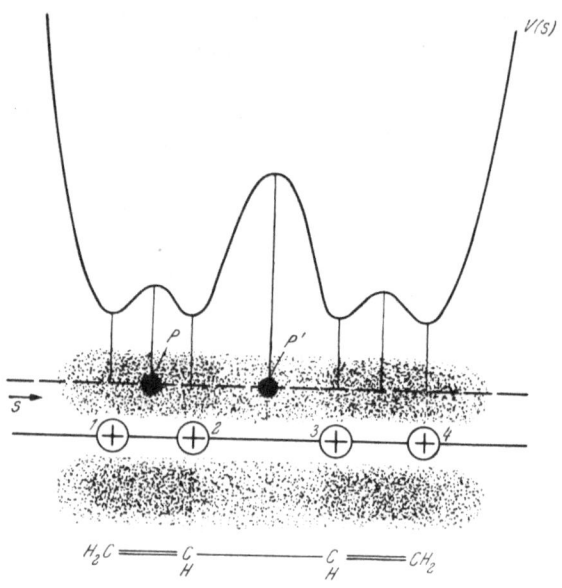

Fig. 10. Butadiene. — C atoms and cloud of $\pi$ electrons. Potential $V(s)$ of a $\pi$ electron, which is here assumed to move along the dashed line. Short and long bond distances are exaggerated, and the representation of the potential curve is schematic.

Let us now consider a single $\pi$ electron in the electrostatic field of the rest of the molecule, and assume first that it can move only along the zig-zag line of the carbon chain, say along the central line of the $\pi$ electron cloud above the molecular skeleton (dashed line in Fig. 10; coordinate *s*). This electron can be considered as being in the Coulomb field of the two nearest C atoms only, since the effect of the more distant carbons is practically neutralized by the other $\pi$ electrons *(64)*. The variation in the potential energy $V(s)$ of the electron along the chain is shown qualitatively in Fig. 10; its value is lowest in the proximity of an atom but increases beyond the ends of the chain; it is smaller at point $P$ between atoms 1–2 than at point $P'$ between atoms 2–3, since the distance of $P$ from atoms 1 and 2 is small, while that of $P'$ from atoms 2 and 3 is relatively large.

The light absorption corresponds to the jump of an electron from state $n = 2$ to state $n = 3$ *(Fig. 11a)*. In state $n = 2$ the electron cloud accumulations are in the two potential troughs, and thus the energy level of that state is particularly low. In state $n = 3$ there is a charge cloud accumulation at the potential peak, and thus the energy level of the state is particularly high. During the transition, $n = 2 \rightarrow n = 3$, negative charge has to be partially removed from the positively charged carbon nuclei, evidently giving rise to a positive addition to the expression for the excitation energy. In the corresponding transition in the amidinium ion such an unfavorable charge transfer does not occur.

An analogous situation obtains in a polyene molecule with $j$ conjugated double bonds *(Fig. 11 b)*. In this case the wave function in the highest

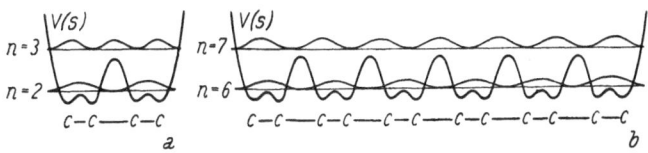

Fig. 11. Butadiene a) and dodecahexaene b). Variation of potential $V(s)$ of a $\pi$ electron along the carbon chain. — Density distribution of the excited electron before and after the transition, according to the simple model. In the state of the electron before the transition [state $n = 2$ in a), state $n = 6$ in b)] the charge is accumulated more heavily in the potential troughs, i. e. at the position of the positively charged C atoms than in the state after the transition [state $n = 3$ in a), state $n = 7$ in b)]. The state before the transition is thus particularly stable, the excitation energy particularly high. Short and long bond distances are exaggerated, and the representation of the potential curve is schematic. [From: Angew. Chem. 71, 93 (1959).]

occupied electron state has $j$ antinodes and $j$ charge cloud accumulations along the chain. There are $j$ potential troughs, and thus each cloud accumulation is precisely in a potential trough. This is, however, not the case in the next state ($j + 1$ charge cloud accumulations). Thus, in the first of these clouds the charge is more densely accumulated in the potential troughs, i. e. at the positions of the C atoms, than in the second one. We conclude that the excitation energy must be relatively high and that the polyene must absorb at particularly short wavelengths.

We have assumed that there are alternate shorter and longer bond distances between neighbouring atoms even in a polyene molecule, containing many conjugated double bonds. This assumption should now be justified. Let us first neglect the influence of the $\pi$ electrons on bond distances and consider a chain of equidistant C atoms *(Fig. 12a)*. If the $\pi$ electron states are described by sine wave functions, we find the over-all charge density distribution of the $\pi$ electrons in the ground state, $W(s)$, as given in Fig. 12a for $j = 6$. The maxima of $W(s)$ are approximately at the positions of the maxima of the density distribution of the highest occupied state, i. e. at the positions of the double bonds in the conventional structural formula. Therefore, this chain of equidistant C atoms cannot be a stable system, the C atoms are displaced towards the maxima of $W(s)$ as in the case of butadiene, and thus "pairs" of atoms are formed. According to Fig. 12a, the $\pi$ electron density in longer-chain polyenes is only a little larger at the maxima than at the minima; in butadiene, however, the corresponding density difference is appreciable (Fig. 9b, p. 418). Thus, it might seem at the first glance that in a polyene higher than butadiene the C atoms are less displaced and the character of "single" and "double" bonds becomes more and more alike, when the number of conjugated double bonds $j$ increases. However, this small displacement of the C atoms in long-chain polyenes changes the shape of the over-all electron cloud; the negative cloud tends to accumulate at the pairs of positive C atoms thus formed. In the larger COULOMB field of the cloud the atoms are further displaced and this increases the cloud accumulation at C atom pairs *(Fig. 12b)*. Thus, an equilibrium is reached

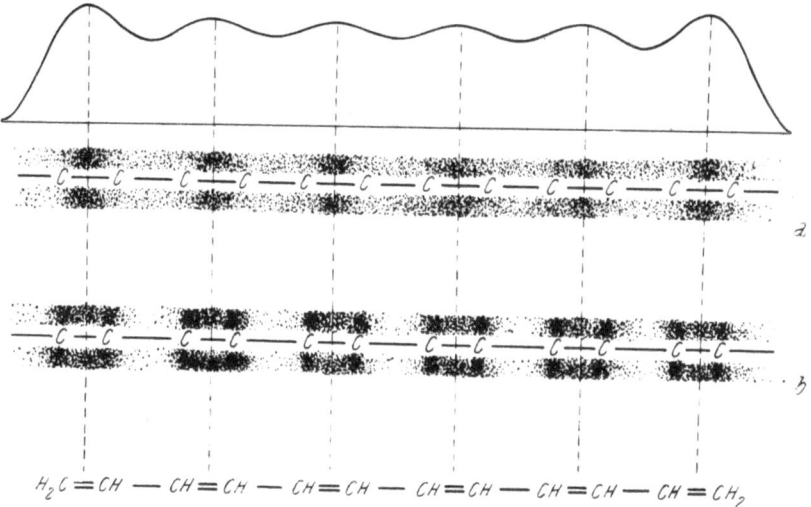

Fig. 12. Polyene with $j = 6$. — a) Chain of equidistant C atoms. Charge density distribution $W(s)$ and charge cloud of the $\pi$ electrons in the lowest state, according to the simple model. The C atoms are attracted to the maxima of $W(s)$ at the double bonds. Thus they are not in equilibrium position. b) Chain of C atoms in equilibrium position. Alternate short and long bonds. Charge cloud accumulation of the $\pi$ electrons. Each pair of positive C atoms attracts the negative $\pi$ electron cloud. Thus, the cloud is more heavily accumulated at the double bonds in the structural formula than in a). The C atoms joined by double bonds are held in the attractive COULOMB field of the cloud accumulations between them.

and it seems likely that an alternation of long and short bonds will be found as pronounced as in butadiene $(57$–$59)$*. This qualitative argument will be confirmed by the following quantitative treatment.

## 2. A Quantitative Treatment of the Light Absorption of Polyenes $(75)$.

The degrees of freedom of the $\pi$ electrons orthogonal to the zig-zag line of the carbon skeleton were disregarded above and shall now be considered. The average kinetic energy associated with these degrees

---

* Based on the same assumption is W. KUHN's model $(77)$ which replaces the polyene chain by a series of coupled linear oscillators, each oscillator corresponding to a double bond of the conjugated system. However, LENNARD-JONES $(80)$ by using a molecular orbital approach, found that the character of "single bonds" and "double bonds" should get more and more alike as we increase the length of the chain.

of freedom can again be neglected in a calculation of the distance between the energy levels of the $\pi$ electron states, since it can be considered as being equal in each state; hence, the $\pi$ electrons can be treated in a good approximation as electrons in a certain potential $V(s)$.

$V(s)$ is the average potential of an electron with coordinate $s$, i. e. of an electron placed on a surface, which is orthogonal to the zig-zag line of the carbon skeleton and which cuts this line in a point with coordinate $s$. The contribution

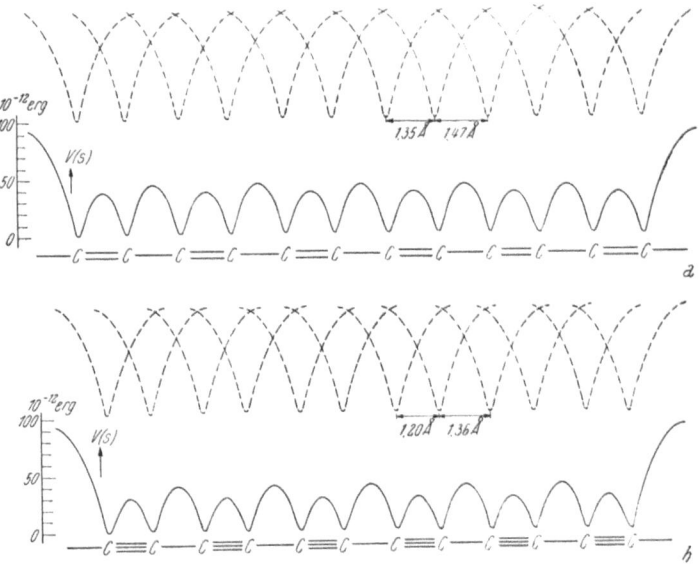

Fig. 13. Quantitative treatment of a) polyenes and b) polyacetylenes. — Potential $V(s)$ (full line) is obtained by superposing the contribution of each atom (dashed curves). Each dashed curve is equal in shape and size. The potential is lower at the positions in the middle of double or triple bonds than in the middle of single bonds, corresponding to the schematic representation of Fig. 11.

of a C atom to this average potential can be calculated approximately by using COULOMB's law, a reasonable value of the effective nuclear charge of C, and an appropriate distribution function of the $\pi$ electron cloud in the directions orthogonal to the carbon skeleton (48). The result is represented by any one of the dashed lines in *Fig. 13a*; and $V(s)$, the sum of the contributions of each C atom, is given for a particular case by the full line in the same Figure.

In a reasonable approximation $V(s)$ is given by the potential, considered before, of an electron moving along the central line of the $\pi$ electron cloud above the molecular skeleton (64); and our qualitative discussions are thus confirmed by a stricter treatment.

For a quantitative consideration one must find wave functions $\psi_n(s)$ and energy levels $E_n$ of an electron in the potential $V(s)$ of Fig. 13a,

i. e. one must find the eigenfunctions and eigenvalues of the SCHRÖDINGER equation,

$$\frac{d^2\psi_n}{ds^2} + \frac{8\pi^2 m}{h^2}(E_n - V(s))\,\psi_n = 0. \tag{9}$$

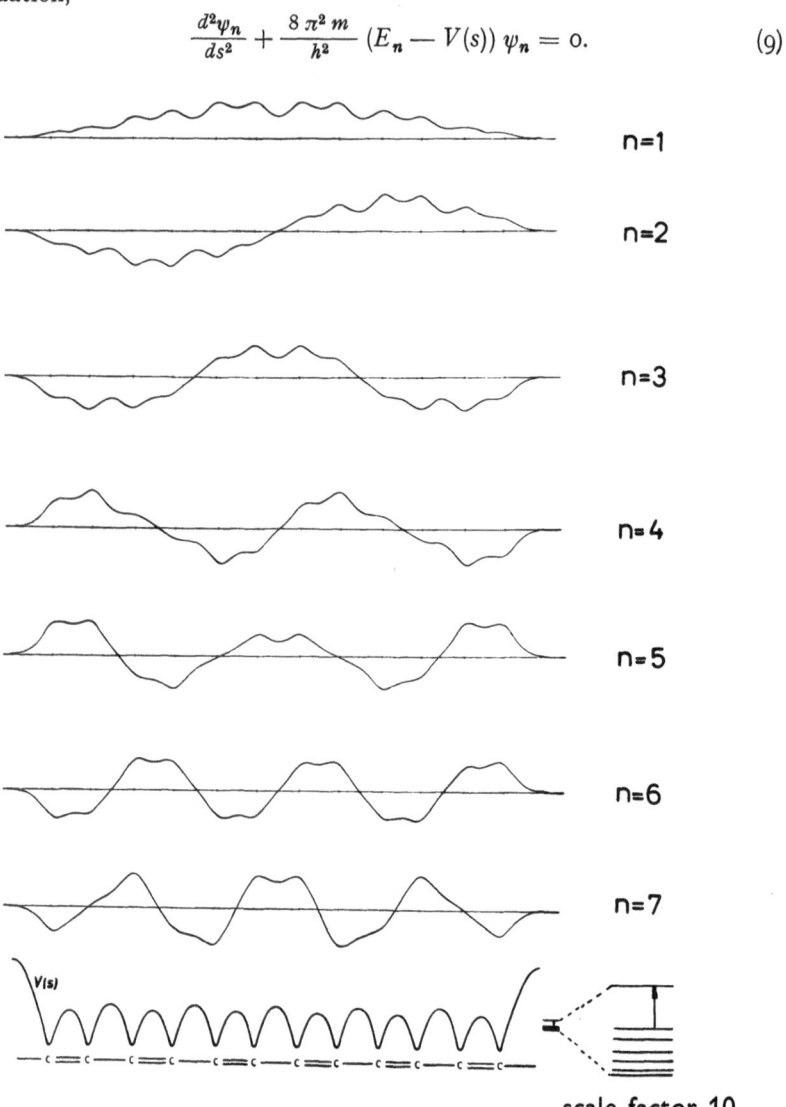

n=1

n=2

n=3

n=4

n=5

n=6

n=7

scale factor 10

Fig. 14. Polyene with $j = 6$: Strict treatment. — Potential $V(s)$ according to Fig. 13a, energy levels, and wave functions of the seven lowest states.

While it would be tedious to solve this problem numerically, it can be solved rapidly by using an electric analog computer (75). For $j = 6$ the result is shown in *Fig. 14*. Although each function has a very

complicated shape, it has the same approximate shape as the corresponding
sine wave function in the simple model. The large gap between the
sixth and the seventh energy levels should be noted.

The wave length $\lambda_{max}$ of the absorption maximum can be calculated
from $\Delta E = E_{j+1} - E_j$ in the usual manner, and the result is shown
in *Fig. 15a* for polyenes with $j = 2$ to 12 double bonds (full arrow),
compared with the observed data (bar) [cf. (71), there p. 171]. The good

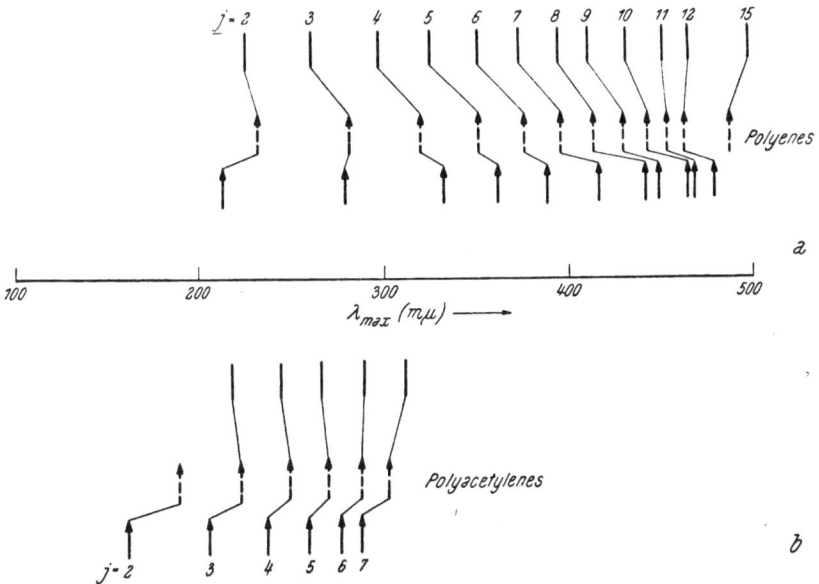

Fig. 15. Position of absorption maximum of a) polyenes and b) polyacetylenes. — Bar: observed; arrow: calculated; full arrow: strict treatment (potential according to Fig. 13); dashed arrow: simplified treatment [potential according to Fig. 17; values obtained from Equation (10) (p. 428) with $V_0 = 2.4$ eV (polyenes) and $V_0 = 3.4$ eV (polyacetylenes)].

agreement between theory and experiment is surprising, since no adjusted
parameter was introduced in the calculation.

This calculation is based on the assumption, that in a polyene all
"single bonds" have the same length, $r_s = 1.47$ Å, and all "double bonds"
have the same length, $r_d = 1.35$ Å (Fig. 13a). [These values are the
measured bond lengths of butadiene (*1, 111, 121*).]

Only few data are available for other polyenes and they are in very good
agreement with our assumed values. Thus, EICHHORN (*36*) found for $\beta$-ionylidene-
$\gamma$-crotonic acid ($j = 4$) the values, $\bar{r}_s = 1.47_3$ Å and $\bar{r}_d = 1.33_1$ Å; and according
to STURDIVANT and SLY (*118*) the corresponding values for 15,15'-dehydro-$\beta$-
carotene ($j = 11$, including the triple bond) are, 1.455 and 1.345 Å.

When the assumed values of the bond lengths in a polyene are even slightly
altered, the calculated $\lambda_{max}$ values change considerably. (For example, when the
above data are used for $j = 12$, $r_s = 1.47$ Å, and $r_d = 1.35$ Å, the value

$\lambda_{max} = 479$ m$\mu$ is found; however, if we set $r_s = 1.45$ Å, and $r_d = 1.36$ Å, $\lambda_{max} = 603$ m$\mu$ is obtained; the observed values lie between 461 and 483 m$\mu$.) Thus the good agreement of calculated and observed $\lambda_{max}$ values (Fig. 15a) represents an experimental justification of the basic assumption that in each polyene $r_s = 1.47$ Å and $r_d = 1.35$ Å.

This assumption concerning $r_s$ and $r_d$ can be justified theoretically in different ways (75). One is based on a consideration of the over-all $\pi$ electron density distribution $W(s)$ along the zig-zag chain in the ground state of the molecule, $W(s)$ being the sum of the squares of the normalized wave functions $\psi_n(s)$ of the occupied states, $W(s) = \sum\limits_{n=1}^{j} 2\,\psi_n^2$, where $j$ is the number of conjugated double bonds. $W(s)$ is shown in *Fig. 16a*

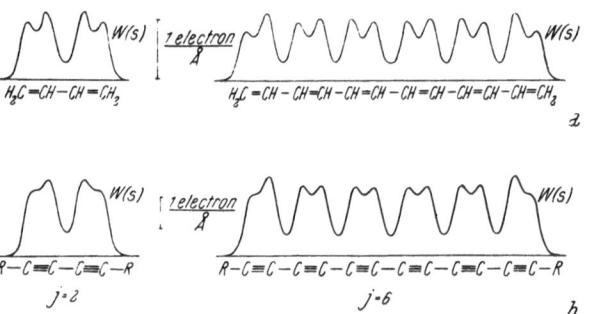

Fig. 16. Charge density distribution $W(s)$ of the $\pi$ electrons in the ground state, according to the strict treatment: a) polyenes, and b) polyacetylenes. — Note the large difference between the density at the single bonds and at the double or triple bonds. For $j = 6$ one finds the same pronounced difference between alternating short and long bonds ("double" or "triple bonds" and "single bonds") as in the case of $j = 2$

for $j = 2$ and $j = 6$. When we consider the points in the middle of the "single bond" of butadiene and of each "single bond" of a long polyene, we find identical values for $W$. The same statement is valid for points in the middle of double bonds. According to p. 420, the $\pi$ electron density in the neighbourhood of the middle of a C—C bond determines the length of this bond. Thus, each "single bond" or "double bond" in a long-chain polyene must have respectively the same length as the "single bond" or "double bond" in butadiene. This agrees with the basic assumption of our calculation and confirms the theory: the model is self-consistent.

No self-consistency is, however, obtained if it is postulated that the difference in character between "single bonds" and "double bonds" is less (or more) pronounced in a long-chain polyene than in butadiene. If one considers a polyene as a chain of equidistant C atoms and determines $V(s)$ and $W(s)$ in the usual manner, it is found that $W(s)$ assumes considerably larger values in the region of a double bond of a structural formula than in the region of a single bond; this confirms the result of the simple treatment (Fig. 12a, p. 422) and demonstrates that such a chain of equidistant C atoms cannot be a stable system.

*References, pp. 445—451.*

### 3. Treatment of the Light Absorption of Polyacetylenes.

In polyacetylenes [cf. (71), there p. 188] the difference in bond length between alternating long and short bonds is even more pronounced than in polyolefines: The $\pi$ electrons in the polyacetylenes have the same tendency to concentrate between successive pairs of C atoms as in polyenes; since there are twice as many $\pi$ electrons in a polyacetylene molecule than in the corresponding polyene, this tendency is more effective. A greater shift of the bands towards shorter wavelengths is both expected and observed.

The polyacetylenes can be treated quantitatively in a manner analogous to the polyenes. The potential $V(s)$ can easily be obtained by superposing the contributions of the C atoms (*Fig. 13b*, p. 423), and the contribution of a C atom is given by the curve valid for polyenes. The values 1.36 Å and 1.20 Å are used, respectively, for "single" and "triple" bonds, and these are the measured bond length of diacetylene (23, 94, 121). From the wave functions obtained with the electric analog computer the function $W(s)$ can be calculated, and in the manner just indicated the model is found to be self-consistent (*Fig. 16b*). The wavelength of the absorption maximum can be calculated as usual from the energies of the states $j$ and $j + 1$, where $j$ is the number of conjugated triple bonds. The result is given in *Fig. 15b* for $j = 2$ to $7$ (full arrows), in good agreement with the observed values (bars) [cf. (71), there p. 171].

### 4. Wave Shape Potential Model of Cyanines, Aza-cyanines and Some Other Dye Classes.

The method used above in Sections 2 and 3 can also be applied to cyanine and aza-cyanine type compounds as well as to more complicated structures such as unsymmetrical cyanines and compounds with hetero atoms placed at any position in the molecule. If the bond lengths in the resonating chain are known, the potential $V(s)$ is determined by superposing the individual contributions of the atoms. The potential trough contributed by a hetero atom (which can be obtained from considerations of nuclear charge and shielding effects) is similar in shape to the trough of a C atom considered before (Fig. 13, p. 423), but is lower because the electronegativity of the hetero atom is higher than that of carbon.

After the potential $V(s)$ has thus been determined, the wave functions and energy levels of the electron states can be obtained easily (by using an analog computer), even when the irregularities of the potential are of a complicated nature.

Since in general the bond lengths in the resonating chain are unknown, one must first make a guess about these lengths. Based on such a guess the probability distribution $W(s)$ of the $\pi$ electrons can then be calculated. Since the value of $W$

at a given bond determines the length of this bond, one can obtain values for bond lengths, that are more reliable than the values assumed first. Based on this new set of bond lengths, the calculation can be repeated, and repeated again, until self-consistency of the treatment is reached, i. e. until the bond lengths calculated from the resulting function $W(s)$ will agree with the values on which the calculation was based.

In the case of symmetrical cyanine type compounds the refined treatment leads to practically the same result as the simple treatment given in (71, there p. 190), and also in the case of aza-cyanines the results of the refined and of the simple treatment [cf. (71), there p. 194] are very similar. Thus in the particularly simple hypothetical molecule (XXIa, XXIb) (p. 431) the energy levels and wave functions resulting from these treatments (Figs. 18a and b, p. 430; 19a and b, p. 431; 20a and b, p. 432) are surprisingly close to each other (76).

### 5. A Simple Treatment of Polyenes, Polyacetylenes, and Unsymmetrical Cyanines (Sine Curve Potential Model).

A simplified treatment is useful even when a more precise treatment is available, since it shows the essential features of the

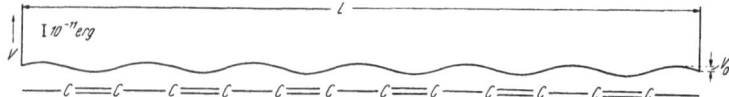

Fig. 17. Sine curve potential model of a polyene. — The potential energy of a $\pi$ electron as it moves along the zig-zag chain of the polyene (Fig. 13a) is replaced by the curve here indicated ($V$ has a sine curve variation of amplitude $V_0$ along the chain and rises sharply to infinity at the ends). Note the coincidence of the crests and troughs of the sine curve with the long and short bonds, respectively.

problem. Thus, let us suppose, instead of the potential of Fig. 13 (p. 423), a field which has a sine curve variation along the chain (amplitude $V_0$) *(Fig. 17)* (58, 59). The maxima of this curve correspond to the middle-points of the longer C—C bonds, and the minima, to those of the shorter ones. As in the case of the free electron gas model of (71), the potential energy is assumed to rise to infinity at both ends of the chain. The wave functions and energy levels of an electron in this type of potential can be calculated by using the analog computer, whereupon the excitation energy $\Delta E$ of a polyene or polyacetylene, resulting from this model, can be expressed in good approximation by Equation (10) [cf. (75)]*.

$$\Delta E = \frac{h^2}{8\,m\,L^2}\,(2\,j + 1) + 0.83\left(1 - \frac{1}{2\,j}\right)V_0 \text{ (polyenes, polyacetylenes). (10)}$$

* An earlier equation (58, 59) found by a simple approximation is identical with Equation (10), except that the factor 1 stands for 0.83.

*References, pp. 445—451.*

In the cases of polyenes and polyacetylenes we introduce the respective values, $V_0 = 2.4$ eV, and $V_0 = 3.4$ eV and obtain from Equation (10) the $\lambda_{max}$ values (Fig. 15, p. 425, dashed arrows) which have been found in good agreement with observed data as well as with the values obtained in Sections 2 and 3 (pp. 422 and 427).

The unsymmetrical cyanines can be treated in a manner similar to the treatment of polyenes. It must be considered however that these dyes are resonance hybrids between the conventional form and a structure derived by interchanging single and double bonds [see e. g., (XXa and XXb)]. Clearly, the difference in character between longer and shorter bonds is here less pronounced than in the polyene class.

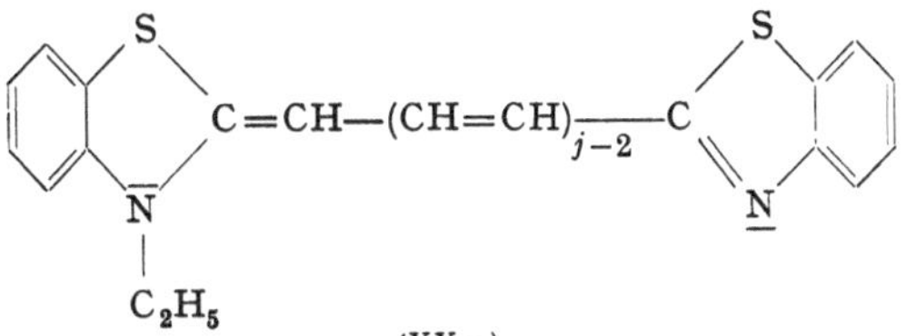

$$\text{C}=\text{CH}-(\text{CH}=\text{CH})_{\overline{j-2}}\text{C}$$

(XX a.)

$$\text{C}-\text{CH}=(\text{CH}-\text{CH})_{\overline{j-2}}\text{C}$$

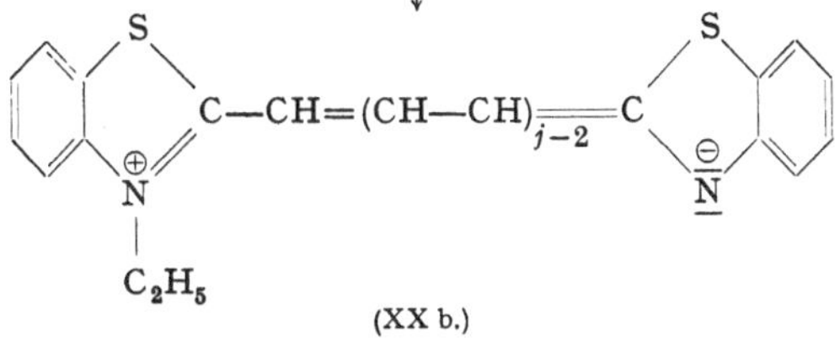

(XX b.)

The light absorption can again be treated (58, 59) by assuming a potential curve with sine wave variation, but the amplitude $V_0$ is here smaller than for polyenes. Equation (10) can be used but $j + 1$ should be substituted for $j$, since $\Delta E$ depends on the number of $\pi$ electron pairs in the electron gas ($j$ in polyenes and $j + 1$ in the cyanine series). Thus,

$$\Delta E = \frac{h^2}{8\,m\,L^2}(2\,j + 3) + 0.83\left(1 - \frac{1}{2\,j + 2}\right)V_0 \quad \text{(cyanines)}. \quad (11)$$

In the extreme case of symmetrical cyanines both limiting structures are equal and $V_0 = 0$. Then Equation (11) is reduced to Equation (6) in (71). In the whole intermediate range, viz. between $V_0 = 2.4$ eV and $V_0 = 0$ eV, the unsymmetrical cyanines are located.

The smaller the difference between the contribution of the two limiting structures of an unsymmetric cyanine, the smaller will be the difference between the nature of the bonds along the chain and the value of $V_0$. In compound (XXa, XXb), for example, structure (XXa) contributes considerably more to the normal state of the molecule than (XXb); thus, $V_0$ is relatively large. The value

$V_0 = 1$ eV provides the best agreement between observed $\lambda_{\mathrm{max}}$ values (71, there Fig. 1) and the values calculated from Equation (11).

## III. The Two-dimensional Electron Gas Model (65, 76).

The treatments discussed above were based on the assumption that the energy of a $\pi$ electron in a given state $n$ is the sum of two terms, viz. the energy $E_n$ associated with the coordinate $s$ along the molecular

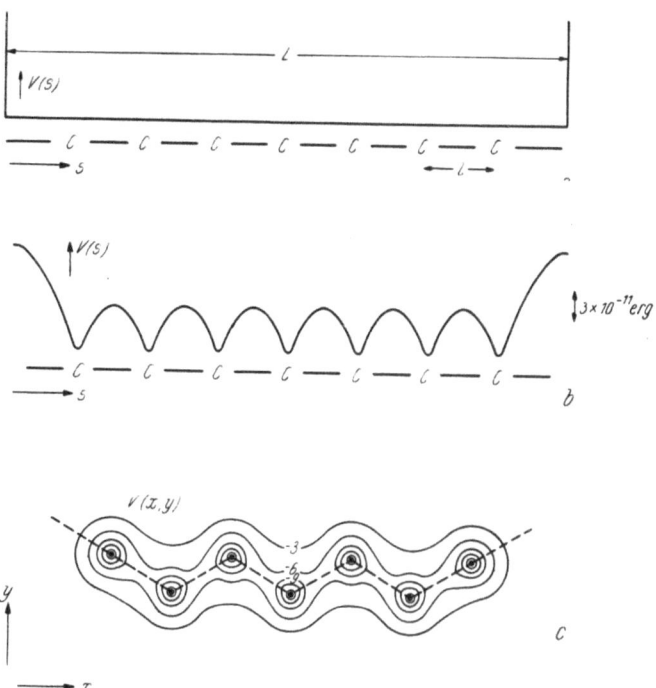

Fig. 18. Cyanine-like molecule (XXIa, XXIb). — Potential $V$, a) according to the simple free electron gas model; b) the one-dimensional wave shape potential model; and c) the two-dimensional electron gas model (distance between succeeding levels of contour lines $3 \times 10^{-11}$ erg).

chain, and the average kinetic energy associated with the degrees of freedom orthogonal to this chain. The latter term was supposed to be equal for each $\pi$ electron state and could be neglected; one-dimensional $\pi$ electron waves were considered. The reliability of this assumption and of the postulates on which the one-dimensional models of (71) and Chapters I and II (pp. 405 and 418) were based, shall now be proven by a next stage of refinement.

Let us consider, as a particularily simple case, the hypothetical ion (XXIa, XXIb) with the resonating structures,

References, pp. 445—451.

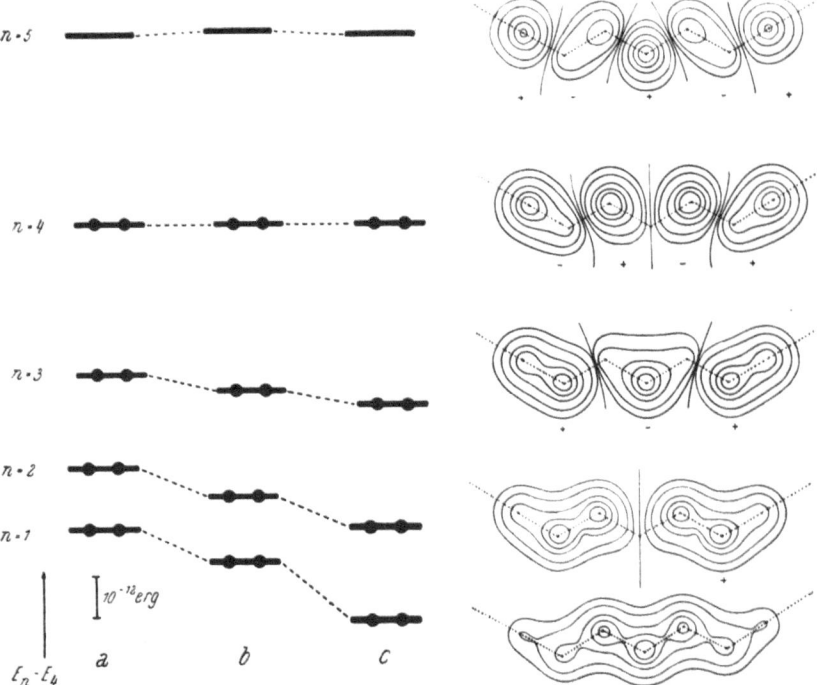

The plane of the centers of the atoms shall be chosen as the $xy$ plane of an $xyz$ coordinate system. As a reasonable approximation it is postulated

Fig. 19. Cyanine-like molecule (XXI a, XXI b). — a) One-dimensional free electron gas model; energy levels $E_n$ according to (71), there Equation (6), for $L = 8\, l = 11.1$ A; b) one-dimensional wave shape potential model; energy levels $E_n$ of an electron in the potential of Fig. 18 b, according to Section II. 4, p. 427; c) two-dimensional treatment. Wave functions $\psi_n(x, y)$ and energy levels $E_n$ of an electron in the potential trough of Fig. 18 c.

that the $\pi$ electrons are in a potential field $V(z) + V(x, y)$, where $V(z)$ is a certain (here not-specified) function of $z$. $V(x, y)$ is the potential of a $\pi$ electron averaged over $z$, and $V(x, y)$ is the sum of the contributions of each C atom. The contribution of a C atom can be obtained by nuclear charge and shielding considerations. A funnel-shaped potential surface is thus found; and the potential $V(x, y)$ of the molecule, given in *Fig. 18c*, was obtained by superposing the contributions of each C atom, assuming equal C—C bond distances of 1.39 Å and C—C—C bond angles of 120°. The $z$ part of the wave function of an electron in the potential $V(z) + V(x, y)$ can be separated from the $xy$ part, and then the problem of finding eigenfunctions $\psi_n(x, y)$ and eigenvalues $E_n$ of the SCHRÖDINGER equation,

$$\frac{\partial^2 \psi_n}{\partial x^2} + \frac{\partial^2 \psi_n}{\partial y^2} + \frac{8\pi^2 m}{h^2} [E_n - V(x, y)] \psi_n = 0 \qquad (12)$$

can be solved by using again an electric analog computer. The result is given in *Fig. 19c* for the five lowest states (76). The Figure also shows the energy levels found by using the simple free electron gas treatment of (71), assuming that the electron gas is extended by a bond length beyond each terminal C atom, as well as the levels given by the wave shape potential model of Chapter II (p. 427). The corresponding energy levels resulting from the three treatments are surprisingly close to each other. The contour line graph of each wave function is given in Fig. 19c; and *Fig. 20c* shows, for the case $n = 4$,

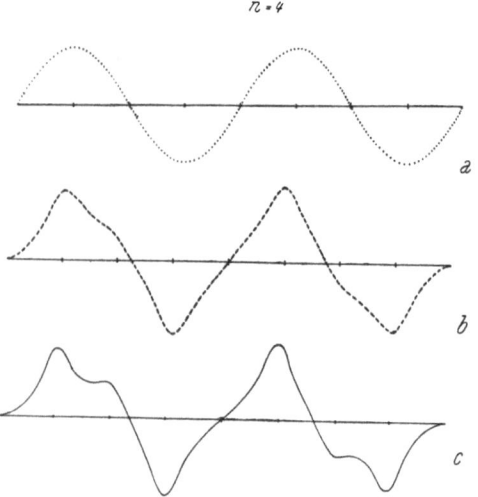

Fig. 20. Cyanine-like molecule (XXIa, XXIb). — Wave function of a $\pi$ electron in state $n = 4$. a) One-dimensional free electron gas model; sine wave; b) one-dimensional wave shape potential model; c) two-dimensional treatment. Course of the wave function of Fig. 19c along the zig-zag line of the carbon skeleton.

the course of the wave function along the zig-zag line of the carbon skeleton. Evidently, the sinusoidal wave function of the simple treatment *(Fig. 20a)* and the wave function given by the one-dimensional refined treatment *(Fig. 20b)* give approximately the course of the corresponding two-dimensional wave function along the chain (Fig. 20c).

A similar result is found in the case of other members of the homologous series of (XXI), i. e. in compounds, which differ from (XXI), by the number $j$ of double bonds in the limiting structures. The following values of $\lambda_{max}$ were found in this series (76):

| | $j = 1$ | $j = 2$ | $j = 3$ | $j = 4$ | $j = 5$ |
|---|---|---|---|---|---|
| Simple treatment .................... | 204 | 328 | 453 | 581 | 707 m$\mu$ |
| One-dimensional refined treatment....... | 221 | 320 | 442 | 585 | 736 |
| Two-dimensional treatment ............ | 275 | 335 | 450 | 550 | 685 |

References, pp. 445—451.

The result of the simple treatment (cf. *71*), according to which a bathochromic shift of $\sim$ 100 m$\mu$ must be expected when we add a double bond to the resonating chain, has thus been confirmed.

In the case of the cations obtained from these compounds by removing two $\pi$ electrons, the corresponding figures are (*76*):

| | $j = 1$ | $j = 2$ | $j = 3$ | $j = 4$ | $j = 5$ |
|---|---|---|---|---|---|
| Simple treatment | 340 | 460 | 580 | 710 | 830 m$\mu$ |
| One-dimensional refined treatment | 260 | 390 | 520 | 640 | 790 |
| Two-dimensional treatment | 230 | 360 | 490 | 620 | 750 |

Wassermann (*120b*) has investigated vitamin A acetate in benzene with trichloroacetic acid and he has obtained a compound with a strong absorption maximum at 650 m$\mu$ which is assumed to have the same chromophore as the above cation with $j = 4$. The value of $\lambda_{max}$ thus observed is in good agreement with the above value, $\lambda_{max} = 620$ m$\mu$, given by the two-dimensional treatment.

It must be postulated that the molecules considered here may assume a number of *cis-trans* isomeric forms, similar to the isomers of the polyenes investigated by Zechmeister and his school (*124–134, 34, 108*). The length $L$ in the simple treatment is considered to be equal in each of these isomers, and thus the energy values of all electron states are also supposed to be equal. The reliability of this assumption can be proved by applying the two-dimensional model to these stereoisomers. The potential trough of each isomer can be obtained by superposing the contributions of all C atoms as in the case of the all-*trans* forms discussed above; and the wave functions as well as the energy levels are determined by using the analog computer. Several *cis-trans* isomers have been treated in this manner and corresponding energy levels were found to be very close to each other (unpublished work by H. Martin). The assumption on which the one-dimensional model is based, has thus been vindicated.

The two-dimensional treatment can be applied to polyenes and polyacetylenes. Using for the bond distances the values given in Chapter II (p. 418) the position of the strong absorption bands were calculated and the results were found to be in good agreement with those of the one-dimensional treatment (Chapter II) and with observed data (*6a*). The assumptions, on which the one-dimensional wave shape potential model is based, were thus confirmed.

The two-dimensional model, applied to the case of molecules with branched electron gas (*68*; unpublished work by F. Schäfer and H. Martin) can be closely related to the branched electron gas model of Chapter I (p. 411). Each state of interest in the two-dimensional model corresponds to a state in the one-dimensional model. It is found that the one-dimensional wave gives approximately the course of the two-dimensional wave function along the zig-zag line of the carbon skeleton of each branch. The corresponding energy levels of the simple treatment and of the strict one are found to be close to each other. Thus, the postulates on which the branched electron gas model is based have been substantiated.

It must be mentioned, however, that the agreement between the branched electron gas model calculation and the strict treatment is not a very good one in the case of the lowest states. A more accurate prediction of the result of the strict treatment becomes available, when instead of Equation (4) (p. 413) a new branching condition is introduced (*66–69*) and thus a refinement of the one-dimensional branched model is obtained. In most instances this process causes only a slight change in the calculated values of $\lambda_{max}$.

The two-dimensional model may be applied to dye molecules of any complicated shape. The potential trough can be obtained in the usual manner by superposing the contributions of the atoms, and the analog computer calculation is rapid, even in very complicated instances. The necessary bond length data can be determined in a manner analogous to one-dimensional cases (Section II, 4, p. 427).

We hope that a predication of the light absorption characteristics of all important types of organic dyes will become possible in the near future.

## IV. Intensity and Structure of Absorption Bands.

### 1. Oscillator Strength of an Absorption Band.

The intensity of an absorption band can be measured by the oscillator strength $f$ defined by

$$f = 14.4 \times 10^{-20} \int_{\text{absorption band}} \varepsilon \, dv, \tag{13}$$

where $\varepsilon$ = molar decadic extinction coefficient in liter/(cm × mole); and $v$ = frequency of light in sec.$^{-1}$ *. The integral $\int \varepsilon \, dv$ is represented by the area between the curve and the $v$ axis. The $f$ values obtained by measuring this area in the case of some of the compounds considered before are indicated by the lengths of the bars in Fig. 21**. The $f$ values calculated as described below, using the one-dimensional electron gas treatment, are given by the lengths of the arrows (49, 62, 75). The approximate agreement of calculated and experimental values of $f$ is an excellent justification for the model, since the $f$ value of a band is a sensitive quantity which may differ greatly from one band of a molecule to another.

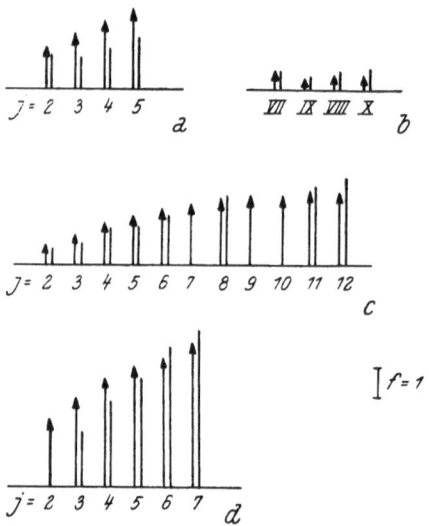

Fig. 21. Oscillator strength $f$ of absorption band. — Arrow: calculated. Bar: measured. $j$ = number of double or triple bonds in the resonating chain. a) Symmetrical cyanines [(71) dyes (IVa, IVb)]; b) acridine orange-like compounds (VII)–(X) (Fig. 8, p. 417); c) all-*trans* polyenes; and d) polyacetylenes. [From: J. Chem. Physics 29, 958 (1958).]

A certain discrepancy between calculated and measured values of $f$ is found in the case of symmetrical

---

* A solvent correction must be introduced in Equation (13) in a more refined treatment, cf. (86).

** The values in Fig. 21 are based on measurements by BROOKER (25) (cyanines), ZANKER (123), MICHAELIS and GRANICK (84), AUŠKĀPS (6) (acridine like compounds), BOHLMANN (19) and INHOFFEN and BOHLMANN (51) (polyenes, polyacetylenes).

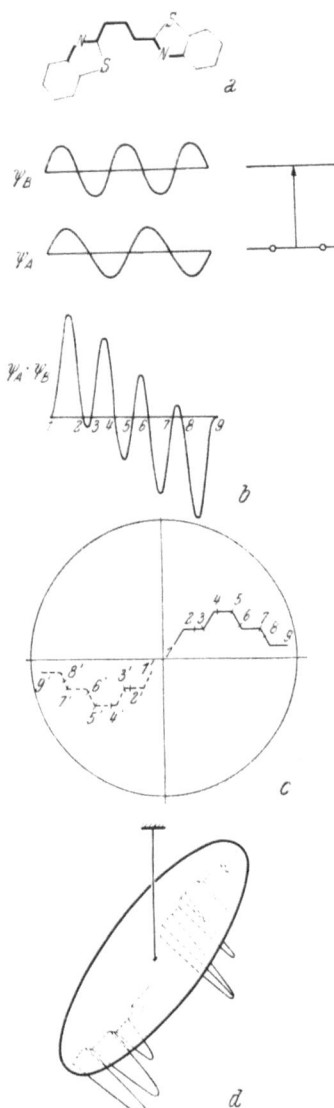

cyanines: the theoretical values given in Fig. 21 a, obtained by assuming all-*trans* configuration of the dye molecule, increase more rapidly with increasing chain-length than the measured values; thus, for $j = 3$, $f$ is expected to be larger by 30% than for $j = 2$; however, the measured values are equal in both instances. This discrepancy can be explained if the dye solution is considered to contain a mixture of *cis-trans* isomers, similar to those of the polyenes as studied by ZECHMEISTER et al. (*124–134, 34, 108*). For $j = 2$ only a single steric form must be postulated but five for $j = 3$.

If a statistical mixture of such stereoisomers is assumed, the calculated $f$ values for $j = 2$ and $j = 3$ will be equal, in agreement with the experiment (*62*). By chromatography ZECHMEISTER and PINCKARD (*131*) obtained experimental evidence for the presence of a stereoisomeric mixture in cyanine dyes. This is confirmed by the shape of the spectral band, which is narrow and has no fine structure, when $j = 2$, but is broader and shows a hump on the short wavelength side when $j > 2$ [cf. (*25*)]. This may indicate that the band is a superposition of the respective bands of several *cis-trans* isomers.

We shall show here (without giving any proof) how the oscillator strength $f$ can be calculated and how one may

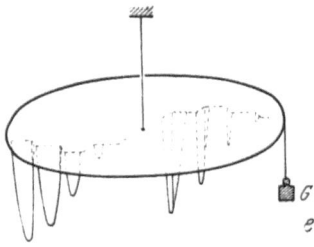

Fig. 22. Calculating device for predicting oscillator strength $f$ of absorption band. — a) Molecular skeleton and wave functions $\psi_A$ and $\psi_B$ of excited electron before and after the jump; b) product $\psi_A \psi_B$; c) molecular skeleton drawn on circular disk; d) disk c) with pasted areas of function $\psi_A \psi_B$ suspended at its center; and e) device d) equilibrated in horizontal position by weight $G$.

decide qualitatively whether or not a given transition is allowed or forbidden. The numerical calculation of $f$ is tedious; a simple and useful calculating device is shown in *Fig. 22* (cf. *62*).

It can be constructed by first plotting on a piece of cardboard the product $\psi_A \psi_B$ of the one-dimensional wave functions $\psi_A$ and $\psi_B$ which describe the states of the electron before and after light absorption against the distance along the molecular skeleton. For example, in the case of the cyanine (XXII) $\psi_A$ and $\psi_B$ are sine functions (Fig. 22a),

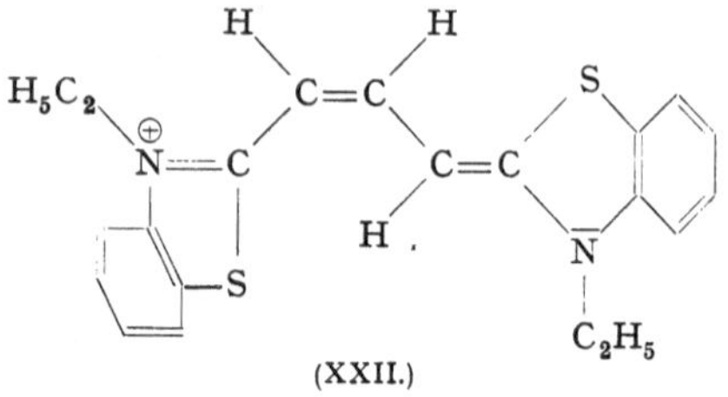

(XXII.)

and the function $\psi_A \psi_B$ (Fig. 22b) is easily obtainable. The area under the curve $\psi_A \psi_B$ is then cut out. The areas above the abscissa are attached in their correct positions (Fig. 22c) on a molecular skeleton drawn on a circular cardboard disc; thus, the area above points 1 and 2 in Fig. 22b is pasted between points 1 and 2 in Fig. 22c. The areas below the abscissa of Fig. 22b are attached to the disc at positions obtained from their position on the molecular skeleton by a 180° rotation about the center of the disc (Fig. 22c); thus, the area *below* the points 2 and 3 in Fig. 22b is attached between points 2' and 3' in Fig. 22c. The disc is then suspended at its center as shown in Fig. 22d.

By suitably adjusting the size and position of a weight $G$ suspended at the edge of the disc, the latter can be made to balance in a horizontal position (Fig. 22e). The weight $G$ is then a direct measure of the value of $f$ of the absorption band considered. Furthermore, a line drawn from the center of the disc to the point of application of the weight gives the orientation relative to the molecular skeleton in which the electric vector of light would preferentially bring about photoexcitation. Thus, the direction of polarization of the absorption band is obtained.

It can be shown that weight $G$ is proportional to the value of $\sqrt{X^2 + Y^2}$, where $X$ and $Y$, in the general case of a branched system, are given by the expressions

$$X = \sum_i \int_{s_i} \psi_{A,s_i} \, \psi_{B,s_i} \, x \, ds_i; \quad Y = \sum_i \int_{s_i} \psi_{A,s_i} \, \psi_{B,s_i} \, y \, ds_i, \tag{14}$$

in which $x$ and $y$ are the coordinates of a point of the $i^{th}$ branch in an $xy$ coordinate system located relative to the flat dye molecule so that the nuclei lie in the $xy$ plane. The integral is to be taken over the $i^{th}$ branch and the sum includes the contributions of each branch. The value of $f$ is then given by the expression,

$$f = 2 \, \frac{8 \, m \, \pi^2}{3 \, h^2} \, \Delta E \, (X^2 + Y^2), \tag{15}$$

where $\Delta E$ is the energy difference of the two states (excitation energy). The factor 2 is required in all cases where two electrons are in the state $A$.

On this basis it can be shown that the transition, state 3b → state 5b, in a phthalocyanine or a benzporphine (Fig. 2, p. 409) is forbidden and that the jump, 4b → 5b (lowest transition in the case of the phthalocyanine free-base) is allowed and corresponds to a band polarized in the direction of the axis II–IV (Figs. 2 and 23); the line connecting the non-protonated N atoms in the case of the free-base (IV, p. 410).

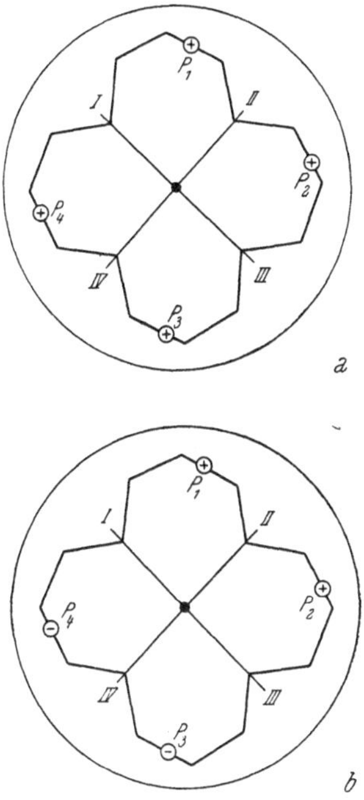

It can be seen from Fig. 2, that the wave function of state 3 b is symmetrical to the axis I–III and antisymmetrical to the axis II–IV, that is, when the function has the value $u$ at $P_1$ *(Fig. 23)* its value at $P_2$, $P_3$, $P_4$ is $-u$, $-u$, $u$, respectively. The wave function 5 b has the same symmetry as 3 b; thus, if the value at $P_1$ is $v$, the value at $P_2$, $P_3$, $P_4$ is $-v$, $-v$, $v$, respectively. Hence the product of the two wave functions at points $P_1$, $P_2$, $P_3$, $P_4$ has the value $u v$, $(-u)(-v)$, $(-u)(-v)$, $u v$, respectively, i. e. the same value $+ u v$ at all four points; thus, this product is a function which is symmetrical to both axes (Fig. 23 a); $G$ is zero and thus the transition 3 b → 5 b is forbidden.

In the case of the transition 4 b → 5 b the product of the corresponding wave functions is symmetrical to the axis II–IV and antisymmetrical to I–III, i. e. the function has the same absolute values at points $P_1$, $P_2$, $P_3$, $P_4$ but changes sign when we proceed from $P_1$ and $P_2$ to $P_3$ and $P_4$ (Fig. 23 b). The point of application of weight $G$ must be on the axis II–IV; hence the band is polarized in the direction of the latter.

Fig. 23. Phthalocyanine or benzporphine. — a) transition 3b → 5b (forbidden); and b) transition 4b → 5b (allowed, polarized in direction of axis II—IV).

The simple method discussed above is not successful in cases where two or more strong bands with identically directed transition moments must be expected. The band at the longer wave lengths is found to be weaker than expected or is not observed, the one at the shorter wave lengths is often found to be stronger.

Let us consider for example a *cis* form of a polyene with $j$ conjugated double bonds. Besides the first band considered in Section II. 2 (p. 422), which has a transition moment in the direction of the chain (transition $j → j + 1$), two further bands with a transition moment perpendicular to the chain are expected (transitions $j - 1 → j + 1$, $j → j + 2$)*.

---

\* In the case of the all-*trans* polyenes the $j$ values for both transitions are calculated to be zero and actually the corresponding bands do not occur, as it has been found by ZECHMEISTER and his school who has postulated and thoroughly investigated the *cis-trans* isomerism of the polyenes and who, in collaboration

However, as Dale (30) has recognized, only one of these bands, the one at the shorter wave lengths (transition $j \rightarrow j + 2$) is observed and corresponds to the cis-peak of Zechmeister and coworkers.

Referring to a polyene with $j$ conjugated double bonds Dale (29) has found the rule that the minor band or "overtone" band of band order $s$ ($s = 1$ for the main band, 2 for the first overtone, and so on) corresponds in wavelength location to the main band of a polyene with $j/s$ conjugated double bonds. If $j/s$ is not an integer, the corresponding wave length of the absorption maximum is obtained on the basis of an interpolation. Table 1 shows, for some examples which have been selected from a number of cases given by Dale (29), the good agreement between the observed $\lambda_{max}$ values of overtone bands and the values calculated by using Dale's rule.

Table 1. Overtone Bands of Polyenes.
(obs. = observed values; and calc. = calculated values by using Dale's rule).

| Compound | $j$ | Solvent | $s = 2$ $\lambda_{max}$ (m$\mu$) | | $s = 3$ $\lambda_{max}$ (m$\mu$) | | $s = 4$ $\lambda_{max}$ (m$\mu$) | | $s = 5$ $\lambda_{max}$ (m$\mu$) | |
|---|---|---|---|---|---|---|---|---|---|---|
| | | | calc. | obs. | calc. | obs. | calc. | obs. | calc. | obs. |
| Decapreno-$\varepsilon_1$-carotene ..... | 13 | cyclohexane | 378 | 392 | 318 | 321 | 280 | 280 | 252 | 246 |
| Lycopene ...... | 11 | hexane | 355 | 362 | 297 | 296 | 260 | 255 | 235 | 234 |
| 5,6-Dihydro-$\alpha$-carotene ..... | 9 | hexane | 324 | 329 | 270 | 266 | | | | |
| 1,1'-Dihydro-$\beta$-carotene ..... | 8 | ether | 309 | 312 | 258 | 254 | | | | |
| Tetradeca-heptaene..... | 7 | isooctane | 275 | 265 | | | | | | |
| Decapentaene .. | 5 | isooctane | 238 | 235 | | | | | | |

As a basis for this generalization Dale has postulated the selection rule that only the electrons of the outermost $\pi$ electron shell are excited by light; thus, according to Dale, the overtone bands correspond to the transitions $j \rightarrow j + 2$; $j \rightarrow j + 3$; ... $j \rightarrow j + s$.

Fig. 24 shows the energy levels of the $\pi$ electrons as obtained by the refined treatment of Section II. 2 (p. 422) in the cases $j = 2$ to $j = 12$ (unpublished work by F. Bär). The excitation energies for the transitions $2 \rightarrow 3$ in the case $j = 2$, $4 \rightarrow 6$ in the case $j = 4$, $6 \rightarrow 9$ in the case $j = 6$, $8 \rightarrow 12$ in the case $j = 8$, $10 \rightarrow 15$ in the case $j = 10$, and $12 \rightarrow 18$ in the case $j = 12$ (full arrows) are seen to be practically equal, as is expected from Dale's rule; similarly, the excitation energies of the transitions $3 \rightarrow 4$ for $j = 3$, $6 \rightarrow 8$ for $j = 6$, $9 \rightarrow 12$ for $j = 9$, $12 \rightarrow 16$ for $j = 12$ (dashed arrows) are practically equal, as is that of the transitions $4 \rightarrow 5$ for $j = 4$, $8 \rightarrow 10$ for $j = 8$, $12 \rightarrow 15$ for $j = 12$ (dotted arrows). As seen by interpolation of the data in Fig. 24, Dale's rule is found to be equally well fulfilled for non-integer values of $j$. Thus, these generalizations are substantiated by the

with Pauling et al. has given a qualitative explanation based on the classical picture of oscillating charges (cf. 124–134, 34, 108).

above treatment, but of course DALE's selection rule, which has been introduced as a postulate, is not explained.

An explanation of the situation is obtained by taking the inter-electronic interaction into account. A simple treatment is indicated

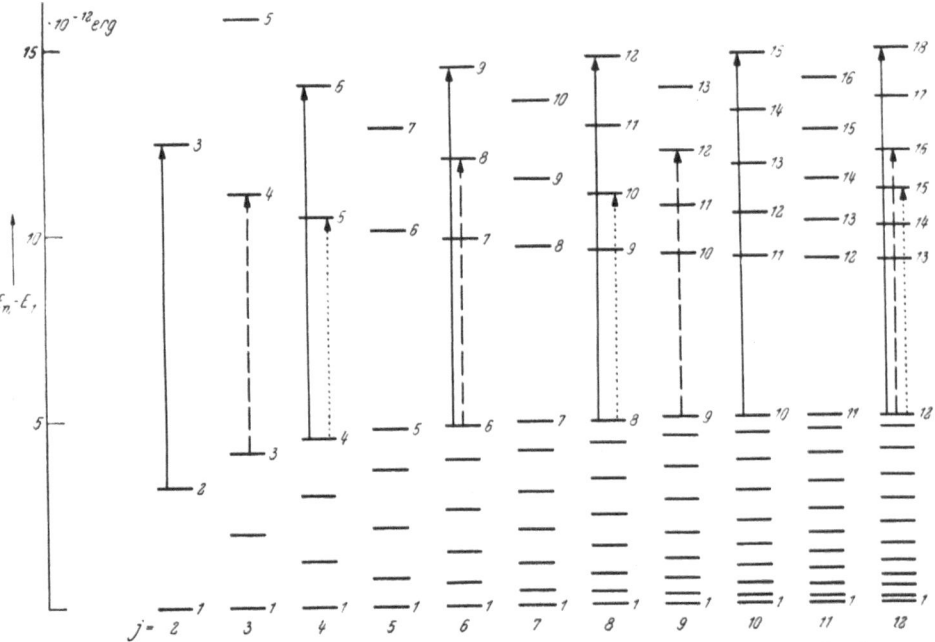

Fig. 24. Polyene with $j = 2$ to 12. Energy levels of $\pi$ electrons states. — As predicted from DALE's rule, the full arrows are practically equal in length, as are the dashed and dotted arrows, respectively.

below (73). Other methods to consider interelectronic interaction (by mixing configurations) have been developed by MOFFITT (85), DEWAR and LONGUET-HIGGINS (35a), PARISAR (91a), PLATT (98–100, 102), HAM and RUEDENBERG (44).

Let us consider a molecule which is irradiated with light whose frequency is in the region of a given absorption band (band 1) of the molecule. In terms of the classical theory of light absorption the electron responsible for this band (electron 1) will be excited by the action of the electric field of the light wave, or more precisely, by the component of the electric vector in the direction of the transition moment of this band (direction 1). The treatment given before is based on the assumption that the field acting on electron 1 within the molecule is identical with the external field of the light wave. It must, however, be considered that the other electrons in the molecule as a group will behave as a dielectric body which will be polarized in the field of the light wave, i. e. charge separation will occur (Fig. 25).

If the polarization of this dielectric body in the direction 1 is in phase with the alternant field acting on it (case of Fig. 25), the field acting on electron 1 will be smaller than the external field, since the latter will be diminished by the field due to charge separation; thus, the absorption is reduced and the $f$ value of the band is lowered by the field effect of the other electrons in the molecule. If the polarization of the dielectric body is opposite in phase to the alternant field acting on it, the field, acting on electron 1 will be larger than the external field and the $f$ value of band 1 will be enlarged by the field effect.

The polarization of a dielectric body is small unless the frequency of the incident light is not much different from the frequency range of an absorption band (band 2) with transition moment in the direction 1 considered. Let us assume that band 2 is present and that the influence of the other absorption bands of the body may be neglected. Then we find the first-mentioned case (in-phase oscillation of the dielectric body with the field) if the frequency of the incident light (frequency of band 1) is smaller than the frequency of band 2, and the latter case (opposite-phase oscillation of the body with the field) if it is larger.

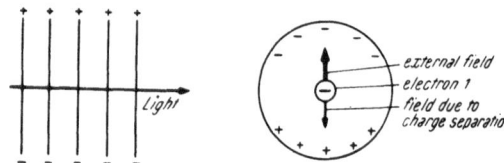

Fig. 25. Electron 1 (responsible for light absorption) in the alternant electric field of the light wave (external field) and in the field due to electron displacement in the rest of the molecule, induced by the light (field due to charge separation).

Thus, if band 2 is at higher frequencies than band 1, the field acting on electron 1 is diminished by the influence of the other electrons in the molecule, and it can even be compensated if the frequencies of the two transitions 1 and 2 are sufficiently close; band 1 will then not appear. However, if band 2 is at lower frequencies than band 1, the field acting on electron 1 will be enhanced and band 1 enlarged.

Thus, in the case considered above (transitions $j \rightarrow j + 2$ at higher frequencies, $j - 1 \rightarrow j + 2$ at lower frequencies; cf. Fig. 24) the band corresponding to the transition $j \rightarrow j + 2$ should be stronger than first expected; the band corresponding to the transition $j - 1 \rightarrow j + 1$ should be weaker and should not appear according to a quantitative treatment based on the above considerations. Dale's selection rule can thus be explained, and the $f$ value of an overtone band can be calculated; the result obtained in the case of 15,15'-*cis*-$\beta$-carotene was found to be in good agreement with the experiment (74).

Similar to a polyene with *cis* configuration, vitamin $B_{12}$ should show a *cis*-peak which should be particularly pronounced due to the large lateral extension of the electron gas in this case. An $f$ value, twice as large as that of the visible band (cf. Section I. 1, p. 405) is expected; actually the substance has a strong band in the ultraviolet ($\lambda_{max} = 360$ m$\mu$) whose $f$ value is twice the $f$ value of the band in the visible region (22).

The field effect considered above is important for the understanding of the complicated spectra of chlorophyll, bacterio-chlorophyll and the porphyrines, which have been successfully attacked in the way considered here (73, 74) or by using the method of configuration interaction (98, 100, 41 a).

## 2. Structure of Absorption Bands.

In many dyes the absorption band has no fine structure but in other instances (polyacetylenes) a pronounced structure is observed (*Fig. 28 a*, p. 443). Hence, the problem of predicting the shape of a band will be discussed at this point. Let us first consider a photo-transition in a diatomic molecule (*116*), and assume that the molecule does not dissociate upon

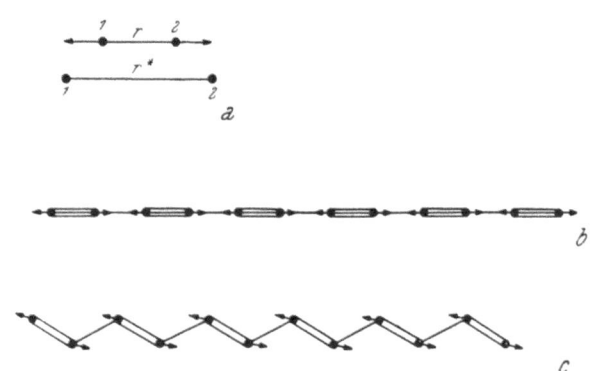

Fig. 26. Vibration excited by electron jump. — a) Diatomic molecule. Equilibrium bond distances $r$ in ground state und $r^*$ in excited state; b) polyacetylene with $j = 6$; and c) all-*trans* polyene with $j = 6$.

excitation. We shall designate the equilibrium distance between the two nuclei by $r$ in the normal state and by $r^*$ in the excited state.

Classically, the distance between the two nuclei before and immediately after the transition is $r$, but then the atoms start oscillating about their new equilibrium positions in the excited molecule with amplitude $r^* - r$ and with the frequency,

$$\nu_0 = \frac{1}{2\pi} \sqrt{\frac{k}{m}}, \tag{16}$$

where $k$ = force constant of the bond in the excited molecule, and $m$ = reduced mass, i. e. $m = m_1 m_2/(m_1 + m_2)$, where $m_1$ and $m_2$ are the masses of nuclei 1 and 2 *(Fig. 26a)*. In the case $r^* \simeq r$ the vibrational energy is small but it increases with increasing difference between $r^*$ and $r$.

According to quantum mechanics, however, the vibrational energy of the electronically excited molecule can assume any one of the discrete values $(v + 1/2)\, h\, \nu_0$, $(v = 0; 1; 2; 3; \ldots)$; thus, the distance between successive vibrational states is $h\nu_0$ *(Fig. 27)*. A photo-transition from the normal state into any one of these states is possible, but most probable is the transition into that state in which the vibrational energy equals or nearly equals the vibrational energy found by the above classical picture (FRANCK-CONDON principle). Thus, if $r^* \simeq r$, the transition into state $v = 0$ is most probable, and the larger the difference between $r^*$ and $r$, the larger is the quantum number $v$ of the state which is excited with the highest probability *(Fig. 27)*.

The situation in a complicated molecule is closely analogous to the one described here. Let us first discuss a polyacetylene (75). According to Section II. 3 (p. 427), the antinodes and thus the cloud accumulations

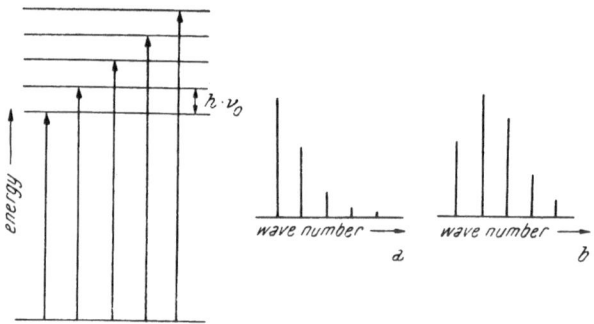

Fig. 27. Vibrational structure of absorption band. Level of normal state and vibrational levels of excited state. Absorption peaks corresponding to the transitions marked by arrows. Successive peaks in distance $v_0/c$; case a) $r^* \simeq r$; and case b) $r^*$ notably different from $r$. [$r$ and $r^*$ equilibrium bond distances in normal and excited state (cf. Fig. 26a).]

of the four $\pi$ electrons in the highest occupied state are at the triple bonds, and this is the actual reason for the presence of alternating short and long bonds. If the molecule is excited, the shape of the cloud of the promoted electron is changed, and thus one less electron has the tendency to participate in the formation of alternating bonds. Correspondingly, the difference between long and short bonds is less pronounced in the photo-excited state than in the normal state. By a quantitative treatment according to p. 427, it was found for $j = 6$, that the equilibrium triple bond distances are longer by 0.02 Å in the excited state than in the normal one, and that the equilibrium single bond distances are shorter by 0.02 Å. Thus, the normal vibration shown in Fig. 26b is excited, i. e. each triple bond is periodically stretched and compressed, and each single bond accordingly compressed and stretched. The frequency $v_0$ of this vibration and thus the distance between successive vibrational peaks of the electronic band can easily be calculated; and from the above value of the difference in equilibrium bond length between normal state and electronically excited state the intensity of the different vibrational peaks can be calculated. Thus, a distribution similar to a) in *Fig. 27* is found. The result of the calculation is given in Fig. 28 b, in excellent agreement with the experiment (*20*).

In the preceding Sections the value of $\lambda_{max} = h\,c/\varDelta E$ ($\varDelta E$ being the excitation energy if all nuclei are fixed to their position in the ground state) was identified with the wavelength of the highest vibrational peak of the band. Strictly speaking, $h\,c/\varDelta E$ is the wavelength of the center of gravity of the band. This has been considered in Fig. 28 b, and it can be noted that a somewhat better agreement of the theory with the observation was then obtained than had been found before (compare Fig. 28 with Fig. 15 b, p. 425).

A polyene constitutes a case similar to that of a polyacetylene. The normal vibration of Fig. 26 c is excited, i. e. each double bond is periodically stretched and compressed, and each single bond accordingly compressed and stretched. The change in bond length on exciting the molecule is larger for polyenes than for polyacetylenes, since the ratio of the number of $\pi$ electrons in state $j$ (highest occupied electron state in the ground state of the molecule) before the jump, to the number

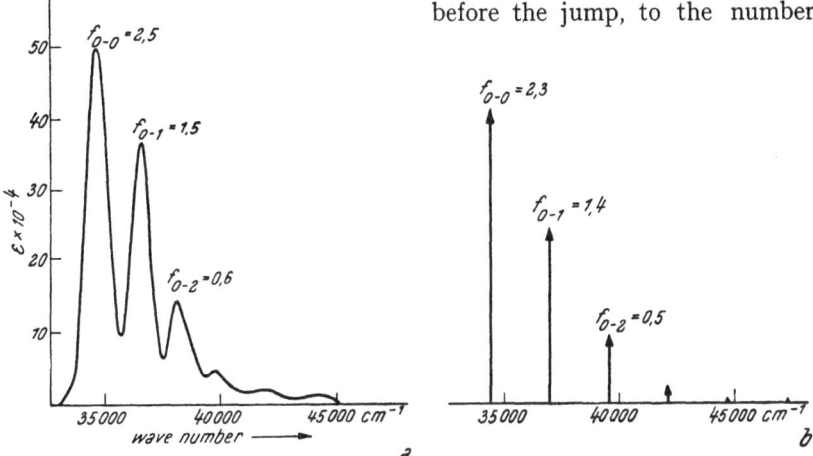

Fig. 28. Polyacetylene with $j = 6$. Vibrational structure of strong absorption band. — a) Observed absorption curve (*20*) and measured values of oscillator strength of the different vibrational peaks; and b) predicted position and oscillator strength of vibrational peaks. [From: Angew. Chem. 71, 93 (1959).]

of electrons in state $j$ after the jump is $4 : 3$ in polyacetylenes but $2 : 1$ in polyenes. Thus, in a polyene we must expect a structure of the absorption band similar to b) in Fig. 27, instead of a). The result of the quantitative treatment (*75*) is in good agreement with the observation. The qualitative interpretation of the vibrational structure of the polyene bands was first given by PLATT (*99*) who has noted that the pronounced vibrational structure of the band confirms experimentally the fact that even in a long polyene a marked alternation of long and short bonds is present.

The state considered here obtained by the transition $j \to j + 1$ may be described according to the resonance theory as a hybrid between limiting structures, e. g. (XXIIIa, XXIIIb), where the reference structure (XXIIIa) contributes about 50%.

$$C=C-C=C-C=C-C=C-C=C-C=C \qquad \text{(XXIIIa.)}$$

$$\overset{\pm}{C}-C=C-C=C-C=C-C=C-C=C-\overset{\mp}{C} \qquad \text{(XXIIIb.)}$$

$$\overset{\pm}{C}-C=C-C=C-\overset{\mp}{C}-\overset{\pm}{C}-C=C-C=C-\overset{\pm}{C} \qquad \text{(XXIIIc.)}$$

$$\overset{\pm}{C}-C=C-\overset{\mp}{C}-\overset{\mp}{C}-C=C-\overset{\pm}{C}-\overset{\pm}{C}-C=C-\overset{\mp}{C} \qquad \text{(XXIIId.)}$$

In the case of the first overtone band (transition $j \to j + 2$) DALE (30) has shown, on the basis of his selection rule (cf. p. 438) and by using the free electron model in its simplest form, that the electron density in the single bond in the middle of the conjugated chain in (XXIIIa) does not change during the transition, since the wave function of the promoted electron has a node at that bond before and after the transition. According to DALE the migration of charge from the double bonds to the single bonds is most pronounced in the middle of each half molecule; thus, (XXIIIc) instead of (XXIIIb), contributes to the excited state of the molecule, besides (XXIIIa). In the excited state, corresponding to the transition $j \to j + 3$, the limiting structure (XXIIId) contributes to this state besides (XXIIIa). From the approximate vibrational pattern of the different transitions thus obtained DALE has drawn some conclusions concerning the vibrational structure of the polyenyne spectra (30; for connected papers cf. 31–33).

According to Figs. 26b and c (p. 441), the atoms in a polyacetylene or in an all-*trans* polyene hydrocarbon vibrate along the chain. These atoms do not hit solvent molecules and their vibration is practically unhindered. Hence, a sharp vibrational structure is observed. However, in the case of most symmetrical cyanine dyes no vibrational structure is obtained. Nevertheless the band width is about as large as the total band width in a polyacetylene, even when only a single steric form is present [e. g., (71), there p. 172, dye (IV); $j = 2$].

By using the refined method described in Section II. 4 (p. 427), it was found in such a case that the distance between the central C atom and the two adjacent C atoms are appreciably increased upon excitation (unpublished results by H. MARTIN). The central C atom starts oscillating in a direction perpendicular to the chain, and the vibrational energy will be dissipated immediately by the solvent molecules. Vibration along the chain cannot develop and the absence of a fine structure in the spectral band can thus be explained.

In the few instances of symmetrical cyanines where a vibrational structure has been observed (25), the resonating chain is shielded by side groups, which could prevent the vibrational energy from dissipating by denying access to solvent molecules.

According to the simple treatment discussed in Section I. 3 (p. 411), the visible band of azulene is caused by the jump of an electron from state a to b (Fig. 29). Evidently, the density of the $\pi$ electrons in the bond between the two rings is strongly increased on excitation, while that in all other bonds remains nearly constant. Thus, the equilibrium length, of the bond between the rings decreases and the C atoms linked by the bond start vibrating against each other. A pronounced vibrational structure of the visible band of azulene is expected, since the vibrating atoms are perfectly shielded by the residual parts of the molecule.

The frequency of the vibration, $\nu_0$, is given approximately by Equation (16) (p. 441), where $k$ (force constant of a one-and-a-half C—C bond) is $7.6 \times 10^5$ dyne cm.$^{-1}$ and $m_1 = m_2 = 2\,m$ (mass of a C atom) is $2.0 \times 10^{-23}$ g. Hence, $\nu_0 = (1/2\,\pi)\,\sqrt{7.6 \times 10^5/1.0 \times 10^{-23}} = 4.4 \times 10^{13}$ sec.$^{-1}$.

The distance between successive vibrational peaks is thus expected to be $\nu_0/c = 1500$ cm.$^{-1}$. The observed band of azulene is

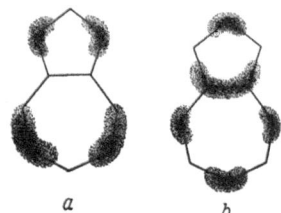

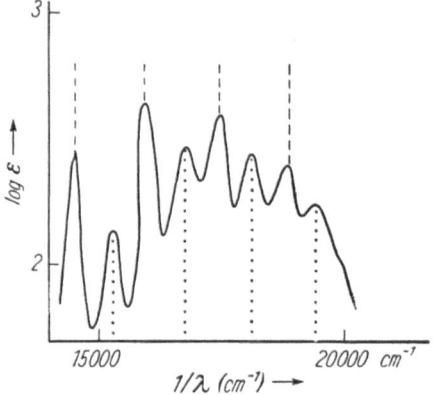

Fig. 29. Azulene: Charge cloud accumulations of excited electron a) before transition; and b) after transition. — Note the increase of $\pi$ electron density in the bond between the rings as a consequence of the $\pi$ electron transition. [From: Angew. Chem. 71, 93 (1959).]

Fig. 30. Azulene [in a mixture of 1 part of methanol and 4 parts of ethanol at — 170° C (26)]. Absorption band in the visible region. The two series of peaks are marked by dashed and dotted lines, respectively. [From: Angew. Chem. 71, 93 (1959).]

shown in *Fig. 30 (26)*. It is evident, considering the height of the peaks, that this band results from a superposition of two series of peaks, and that in each series the distance between adjacent peaks is 1400 to 1500 cm.$^{-1}$, in good agreement with the theory. In order to explain the presence of the two series of peaks, it seems necessary to postulate the existence of two isomeric forms of azulene, which probably correspond to two different steric forms of the seven-membered ring.

### References.

1. ALLEN, P. W. and L. E. SUTTON: Tables of Interatomic Distances and Molecular Configurations Obtained by Electron Diffraction in the Gas Phase. Acta Crystallogr. 3, 46 (1950).
2. ARAKI, G. and S. HUZINAGA: Theory of Absorption Spectra of Unsymmetrical Cyanines. J. Chem. Physics 22, 1141 (1954).
3. ARAKI, G. and T. MURAI: Molecular Structure and Absorption Spectra of Carotenoids. Progr. Theor. Phys. (Kyoto) 8, 639 (1952).
4. — — Theory of Absorption Spectra of Carotenoids According to Tomonaga-Gas Model of $\pi$-Electrons. J. Chem. Physics 20, 1661 (1952).
5. — — Theory of Absorption Spectra of Catacondensed Hydrocarbons According to Platt's Circular Model of Free Electrons. J. Chem. Physics 22, 954 (1954).

6. Auškāps, J.: Quantitative Untersuchungen über die Absorptionsspektren organischer Farbstoffe. Acta Univ. Latviensis (Chem. Ser.) 1, 279 (1930) [Chem. Zbl. 1931 I, 2588].

6 a. Bär, F., W. Huber, H. Martin, G. Handschig and H. Kuhn: Nature of the Free Electron Model. The Case of the Polyenes and Polyacetylenes. J. Chem. Physics (in print).

7. Barrett, P. A., R. P. Linstead, F. G. Rundall and G. A. P. Tuey: Phthalocyanines and Related Compounds. Part XIX. Tetrabenzporphin, Tetrabenzmonoazaporphin and their Metallic Derivatives. J. Chem. Soc. (London) 1940, 1079.

8. Barriol, J. et S. Nikitine: Théorie du modèle métallique comportant des ramifications. Application au naphtalène. J. phys. radium 15, 426 (1954).

9. Barrow, G. M.: Chemical Bond: A One-Dimensional Square Well-Type Model. J. Chem. Physics 26, 558 (1957).

10. Basu, S.: Free Electron Network Model for Cyanines and Diphenyl Polyenes. J. Chem. Physics 22, 1270 (1954).

11. — Nitrogen Electronegativity Correction in Free Electron Network Theory. J. Chem. Physics 22, 1625 (1954).

12. — Free Electron Model for Tropolone. J. Chem. Physics 22, 1776 (1954).

13. — Free Electron Treatment of the Orientation of Substituents in Aromatic Molecules. J. Chem. Physics 22, 1952 (1954).

14. Bayliss, N. S.: A "Metallic" Model for the Spectra of Conjugated Polyenes. J. Chem. Physics 16, 287 (1948).

15. — The Potential Energy in Conjugated Polyenes and the Effective Nuclear Charge of the Carbon Atom. J. Chem. Physics 17, 1353 (1949).

16. — Conjugated Compounds. II. Simple Potential-Energy Functions, Absorption Spectra, and Ionization in Linear Polyenes. Austral. J. Sci. Res. A 3, 109 (1950).

17. — Spectroscopy. Annu. Rev. physical Chem. 3, 229 (1952).

18. — The Free-Electron Approximation for Conjugated Compounds. Quart. Rev. Chem. Soc. (London) 6, 319 (1952).

19. Bohlmann, F.: Konstitution und Lichtabsorption, VI. Mitt.: Zur Deutung von Polyacetylen-Spektren, sowie Darstellung von Bis-tert.-butyl-decapentain-(1,3,5,7,9). Chem. Ber. 86, 63 (1953).

20. — Polyacetylene, IV. Mitt.: Darstellung von Di-tert.-butyl-polyacetylenen. Chem. Ber. 86, 657 (1953).

21. Bolton, H. C.: The Electrical Polarizability of Conjugated Molecules: The Use of Free-Electron Orbitals. Trans. Faraday Soc. 50, 1265 (1954).

22. Bonnett, R., J. R. Cannon, A. W. Johnson and A. Todd: Chemistry of the Vitamin $B_{12}$ Group. Part IV. The Isolation of Crystalline Nucleotide-free Degradation Products. J. Chem. Soc. (London) 1957, 1148.

23. Brockway, L. O.: The Electron-Diffraction Investigation of the Molecular Structure of Cyanogen and Diacetylene (with a Note on Chlorine Dioxide). Proc. Nat. Acad. Sci. (USA) 19, 868 (1933).

24. Brooker, L. G. S. and W. T. Simpson: Spectroscopy. Annu. Rev. physical Chem. 2, 121 (1951).

25. Brooker, L. G. S. and P. W. Vittum: A Century of Progress in the Synthesis of Dyes for Photography. J. Photogr. Sci. 5, 71 (1957).

26. Clar, E.: Aromatic Hydrocarbons. Part LVIII. The Structure of Azulene. J. Chem. Soc. (London) 1950, 1823.

27. Coulson, C. A.: Free-Electron Wave Functions for Conjugated Molecules. Proc. Phys. Soc. (London) A 66, 652 (1953).

28. — Note on the Applicability of the Free-Electron Network Model to Metals. Proc. Phys. Soc. (London) A 67, 608 (1954).

29. DALE, J.: Empirical Relationships of the Minor Bands in the Absorption Spectra of Polyenes. Acta Chem. Scand. 8, 1235 (1954).

30. — The Free-Electron Model, "Overtone" Bands, and Vibrational Structure in Absorption Spectra of Polyenes and Polyenynes. Acta Chem. Scand. 11, 265 (1957).

31. — Infrared Absorption Spectra of ortho- and para-Linked Polyphenyls. Acta Chem. Scand. 11, 640 (1957).

32. — Ultraviolett Absorption Spectra of ortho- and para-Linked Polyphenyls. Acta Chem. Scand. 11, 650 (1957).

33. — Ultraviolet Absorption Spectra of Chain Molecules Consisting of Alternating Benzene Rings and Ethylenic Bonds. Acta Chem. Scand. 11, 971 (1957).

34. DALE, J. and L. ZECHMEISTER: On the Stereochemistry of Azines: Cinnamalazine and Phenylpentadienalazine. J. Amer. Chem. Soc. 75, 2379 (1953).

35. DEWAR, M. J. S.: The Electronic Theory of Organic Chemistry, pp. 311–312. Oxford: Clarendon Press. 1949.

35 a. — and H. C. LONGUET-HIGGINS: The Electronic Spectra of Aromatic Molecules: I Benzenoid Hydrocarbons. Proc. Phys. Soc. (London) A 67, 795 (1954).

36. EICHHORN, E. L.: Thesis, Univ. Amsterdam, 1956.

37. FIERZ-DAVID, H. E.: Künstliche organische Farbstoffe. Berlin: Springer. 1926.

38. FÖRSTER, TH.: Molecular Electronic Spectroscopy. Annu. Rev. physical Chem. 8, 331 (1957).

39. FROST, A. A.: Delta Potential Function Model for Electronic Energies in Molecules. J. Chem. Physics 22, 1613 (1954).

40. — Delta Function Model. I. Electronic Energies of Hydrogen-Like Atoms and Diatomic Molecules. J. Chem. Physics 25, 1150 (1956).

41. FROST, A. A. and B. MUSULIN: A Mnemonic Device for Molecular Orbital Energies. J. Chem. Physics 21, 572 (1953).

41 a. GOUTERMAN, M.: Study of the Effects of Substitution on the Absorption Spectra of Porphin. J. Chem. Physics 30, 1139 (1959).

42. GRIFFITH, J. S.: Note on the Generalized Free-Electron Model of Conjugated Polycyclic Hydrocarbons. J. Chem. Physics 21, 174 (1953).

43. — A Free-Electron Theory of Conjugated Molecules. Part I. Polycyclic Hydrocarbons. Trans. Faraday Soc. 49, 345 (1953).

44. HAM, N. S. and K. RUEDENBERG: Electronic Interaction in the Free-Electron Network Model for Conjugated Systems. I. Theory. J. Chem. Physics 25, 1 (1956).

45. — — Electronic Interaction in the Free-Electron Network Model for Conjugated Systems. II. Spectra of Aromatic Hydrocarbons. J. Chem. Physics 25, 13 (1956).

46. HODGKIN, D. C.: X-ray Analysis and the Structure of Vitamin $B_{12}$. Fortschr. Chem. organ. Naturstoffe 15, 167 (1958).

47. HORNIG, J. F., WALTER HUBER and H. KUHN: Nature of the Free Electron Approximation: The Simple Example of the $H_2^+$ Ion. J. Chem. Physics 25, 1296 (1956).

48. HUBER, WALTER, J. F. HORNIG und H. KUHN: Über den Potentialverlauf entlang der Molekülkette im verfeinerten eindimensionalen Elektronengasmodell. Untersuchungen am Beispiel des Wasserstoffmolekülions. Z. physik. Chem. Neue Folge 9, 1 (1956).

49. HUBER, WERNHARD, H. KUHN und WALTER HUBER: Elektronengasmodell zur quantitativen Deutung der Lichtabsorption von organischen Farbstoffen. II. Teil C. Farbstoffe vom Acridintypus. Helv. Chim. Acta 36, 1597 (1953).

50. HUZINAGA, S. and T. HASINO: Electronic Energy Levels of Polyene Chains. Progr. Theor. Phys. (Kyoto) 18, 649 (1957).

51. INHOFFEN, H. H., F. BOHLMANN, J. H. ALDAG, S. BORK und G. LEIBNER: Synthesen in der Carotinoid-Reihe, XXI. Kondensation von Carotinoidketonen und -aldehyden mit Diacetylen; zugleich eine weitere Synthese des β-Carotins. Liebigs Ann. Chem. 573, 1 (1951) (s. insbes. S. 8).

52. JAFFÉ, H. H.: Free Electron Wave Functions as Approximations to MO Wave Functions for Conjugated Molecules. J. Chem. Physics 20, 1646 (1952).

53. — The Use of Free Electron Model Wave Functions in the Derivation and Representation of LCAO MO Wave Functions of Conjugated Molecules. J. Chem. Physics 21, 1287 (1953).

54. KAUZMANN, W.: Quantum Chemistry. An Introduction. New York: Academic Press. 1957.

55. KOTANI, M., Y. MIZUNO, K. KAYAMA and H. YOSHIZUMI: Quantum Theory of Electronic Structure of Molecules. Annu. Rev. physical Chem. 9, 245 (1958).

56. KUHN, H.: Elektronengasmodell zur quantitativen Deutung der Lichtabsorption von organischen Farbstoffen. I. Helv. Chim. Acta 31, 1441 (1948).

57. — Free Electron Model for Absorption Spectra of Organic Dyes. J. Chem. Physics 16, 840 (1948).

58. — A Quantum-Mechanical Theory of Light Absorption of Organic Dyes and Similar Compounds. J. Chem. Physics 17, 1198 (1949).

59. — Theoretische Deutung der Lichtabsorption organischer Farbstoffe. Z. Elektrochem. 53, 165 (1949).

60. — Quantenmechanische Behandlung von Farbstoffen mit verzweigtem Elektronengas. Helv. Chim. Acta 32, 2247 (1949).

61. — Lichtabsorption organischer Farbstoffe. Chimia 4, 203 (1950).

62. — Elektronengasmodell zur quantitativen Deutung der Lichtabsorption von organischen Farbstoffen. II. Teil A. Ermittlung der Intensität von Absorptionsbanden. Helv. Chim. Acta 34, 1308 (1951).

63. — Elektronengasmodell zur quantitativen Deutung der Lichtabsorption von organischen Farbstoffen. II. Teil B. Störung des Elektronengases durch Heteroatome. Helv. Chim. Acta 34, 2371 (1951).

64. — Chemische Bindung und Zustände von Elektronen in Molekülen. Experientia 9, 41 (1953).

65. — Lichtabsorption organischer Farbstoffe. (Neuere Ergebnisse der Elektronengasmethode.) Chimia 9, 237 (1955).

66. — Note on the Branching Condition in the One-Dimensional Free Electron Gas Model. J. Chem. Physics 22, 2098 (1954).

67. — Verfeinertes eindimensionales Elektronengasmodell. Verzweigungsbedingung und Orthogonalitätsrelation. Z. Naturforsch. 9 a, 989 (1954).

68. — Physical Basis of the Free-Electron Gas Model of Branched Molecules. J. Chem. Physics 25, 293 (1956).

69. — Die Verzweigungsbedingung in der Elektronengasmethode. Z. Elektrochem. 58, 219 (1954).

70. — Zweidimensionales Elektronengasmodell organischer Farbstoffe. Angew. Chem. 69, 239 (1957).

71. — The Electron Gas Theory of the Color of Natural and Artifical Dyes: Problems and Principles. Fortschr. Chem. organ. Naturstoffe 16, 169 (1958).

72. — Neuere Untersuchungen über das Elektronengasmodell organischer Farbstoffe. Angew. Chem. 71, 93 (1959).

*73.* KUHN, H. und WALTER HUBER: Elektronengasmodell organischer Farbstoffe. Feldeffekt als Ursache von Intensitätsanomalien bei Absorptionsbanden. Helv. Chim. Acta **42**, 363 (1959).

*74.* — — Lichtabsorption der Porphine und *cis*-Polyene. Angew. Chem. **71**, 140 (1959).

*75.* KUHN, H., WALTER HUBER et F. BÄR: Modèle de l'électron libre amélioré à une dimension. Position et structure des bandes d'absorption des poly-ynes et polyènes. Longueur des liaisons. Calcul des fonctions d'onde moléculaires, p. 179. Paris: Centre National Recherche Sci. 1958.

*76.* KUHN, H., WALTER HUBER, G. HANDSCHIG, H. MARTIN, F. SCHÄFER and F. BÄR: Nature of the Free Electron Model. The Simple Case of the Symmetric Polymethines. J. Chem. Physics (in print).

*77.* KUHN, W.: Über das Absorptionsspektrum der Polyene. Helv. Chim. Acta **31**, 1780 (1948).

*78.* LABHART, H.: FE Theory Including an Elastic $\sigma$ Skeleton. I. Spectra and Bond Lengths in Long Polyenes. J. Chem. Physics **27**, 957 (1957).

*79.* — FE Theory Including an Elastic $\sigma$ Skeleton. II. Changes of Molecule Dimensions due to the Optical Excitation. J. Chem. Physics **27**, 963 (1957).

*80.* LENNARD-JONES, J. E.: The Electronic Structure of Some Polyenes and Aromatic Molecules. I. The Nature of the Links by the Method of Molecular Orbitals. Proc. Roy. Soc. (London) A **158**, 280 (1937).

*81.* LICHTEN, W.: The Free-Electron Theory and the Virial Theorem. J. Chem. Physics **22**, 1278 (1954).

*82.* LONGUET-HIGGINS, H. C.: Recent Developments in Molecular Orbital Theory. Adv. Chem. Physics **1**, 239 (1958).

*83.* LONGUET-HIGGINS, H. C. and G. W. WHELAND: Theories of Valence. Annu. Rev. physical Chem. **1**, 133 (1950).

*84.* MICHAELIS, L. and S. GRANICK: Metachromasy of Basic Dyestuffs. J. Amer. Chem. Soc. **67**, 1212 (1945).

*85.* MOFFITT, W.: Configurational Interaction in Simple Molecular Orbital Theory. J. Chem. Physics **22**, 1820 (1954).

*86.* MULLIKEN, R. S. and C. A. RIEKE: Molecular Electronic Spectra, Dispersion and Polarization. The Theoretical Interpretation and Computation of Oscillator Strengths and Intensities. Rep. Progr. Physics **8**, 231 (1941).

*87.* NIKITINE, S. et S. G. EL KOMOSS: Étude du modèle métallique à trois dimensions. J. chim. phys. **51**, 129 (1954).

*88.* — — Calcul du spectre d'absorption de quelques colorants dans l'approximation du modèle métallique tenant compte des ramifications des chaines métalliques. J. phys. radium **15**, 536 (1954).

*89.* OLSZEWSKI, S.: Remarks on the Theory of Absorption Spectra of Symmetrical Cyanine Dyes and Polyenes. J. Chem. Physics **26**, 1205 (1957).

*90.* OOSHIKA, Y.: A Semi-empirical Theory of the Conjugated Systems. I. General Formulation. J. Phys. Soc. Japan **12**, 1238 (1957).

*91.* — A Semi-empirical Theory of the Conjugated Systems. II. Bond Alternation in Conjugated Chains. J. Phys. Soc. Japan **12**, 1246 (1957).

*91 a.* PARISER, R.: Theory of the Electronic Spectra and Structure of the Polyacenes and of Alternant Hydrocarbons. J. Chem. Physics **24**, 250 (1956).

*92.* PARR, R. G. and F. O. ELLISON: The Quantum Theory of Valence. Annu. Rev. physical Chem. **6**, 171 (1955).

*93.* PAULING, L.: The Nature of the Chemical Bond and the Structure of Molecules and Crystals. Ithaca, N. Y.: Cornell Univ. Press. 1945.

94. PAULING, L., H. D. SPRINGALL and K. J. PALMER: The Electron Diffraction Investigation of Methylacetylene, Dimethylacetylene, Dimethyldiacetylene, Methyl Cyanide, Diacetylene, and Cyanogen. J. Amer. Chem. Soc. **61**, 927 (1939).

95. PERKAMPUS, H. H.: Die Berechnung der Lichtabsorption der Acene mit Hilfe des verzweigten Elektronengasmodells von H. Kuhn. Z. Naturforsch. **7 a**, 594 (1952).

96. PITZER, K. S.: Quantum Chemistry. New York: Prentice-Hall. 1953.

97. PLATT, J. R.: Classification of Spectra of Cata-Condensed Hydrocarbons. J. Chem. Physics **17**, 484 (1949).

98. — Electronic Structure and Excitation of Polyenes and Porphyrins. In: A. HOLLAENDER, Radiation Biology, Vol. III, p. 71. New York: McGraw-Hill. 1956.

99. — Wavelength Formulas and Configuration Interaction in Brooker Dyes and Chain Molecules. J. Chem. Physics **25**, 80 (1956).

100. — Molecular Orbital Predictions of Organic Spectra. J. Chem. Physics **18**, 1168 (1950).

101. — Isoconjugate Spectra and Variconjugate Sequences. J. Chem. Physics **19**, 101 (1951).

102. — Spectroscopic Moment: A Parameter of Substituent Groups Determining Aromatic Ultraviolet Intensities. J. Chem. Physics **19**, 263 (1951).

103. — Free-Electron Network Model for Conjugated Systems. III. A Demonstration Model Showing Bond Order and "Free Valence" in Conjugated Hydrocarbons. J. Chem. Physics **21**, 1597 (1953).

104. — The Box Model and Electron Densities in Conjugated Systems. J. Chem. Physics **22**, 1448 (1954).

105. POPLE, J. A.: The Molecular-Orbital and Equivalent-Orbital Approach to Molecular Structure. Quart. Revs. Chem. Soc. (London) **11**, 273 (1957).

106. RUEDENBERG, K.: Free-Electron Network Model for Conjugated Systems. V. Energies and Electron Distributions in the FE MO Model and in the LCAO MO Model. J. Chem. Physics **22**, 1878 (1954).

107. RUEDENBERG, K. and C. W. SCHERR: Free-Electron Network Model for Conjugated Systems. I. Theory. J. Chem. Physics **21**, 1565 (1953).

108. SANDOVAL, A. and L. ZECHMEISTER: Some Spectroscopic Changes Connected with the Stereoisomerization of Diphenylbutadiene. J. Amer. Chem. Soc. **69**, 553 (1947).

109. SCHERR, C. W.: Free-Electron Network Model for Conjugated Systems. IV. J. Chem. Physics **21**, 1413 (1953).

110. — Free-Electron Network Model for Conjugated Systems. II. Numerical Calculations. J. Chem. Physics **21**, 1582 (1953).

111. SCHOMAKER, V. and L. PAULING: The Electron Diffraction Investigation of the Structure of Benzene, Pyridine, Pyrazine, Butadiene-1,3, Cyclopentadiene, Furan, Pyrrole, and Thiophene. J. Amer. Chem. Soc. **61**, 1769 (1939).

112. SHEPPARD, S. E. and A. L. GEDDES: Effect of Solvents upon the Absorption Spectra of Dyes. IV. Water as Solvent: A Common Pattern. J. Amer. Chem. Soc. **66**, 1995 (1944).

113. SIMPSON, W. T.: Electronic States of Organic Molecules. J. Chem. Physics **16**, 1124 (1948).

114. — On the Theory of the $\pi$-Electron System in Porphines. J. Chem. Physics **17**, 1218 (1949).

115. SPONER, H.: Electronic Spectroscopy. Annu. Rev. physical Chem. **6**, 193 (1955).

*116.* SPONER, H. and E. TELLER: Electronic Spectra of Polyatomic Molecules. Rev. Mod. Physics **13**, 75 (1941).

*117.* STAAB, H. A.: Einführung in die theoretische organische Chemie. Weinheim: Verlag Chemie. 1959.

*118.* STURDIVANT, J. H. and W. G. SLY: Private communication. Cf. W. G. SLY: A Preliminary Report on the Crystal-structure Determination of 15,15'-Dehydro-$\beta$-carotene. Acta Crystallogr. **8**, 115 (1955).

*119.* TAKIZAWA, E. I. et N. IMAI: Note sur la corrélation des électrons $\pi$ en colorants organiques. I. et II. Mem. Fac. Engin. Nagoya Univ. **4**, 216 (1952); **5**, 59 (1953).

*120.* WALSH, A. D.: Far Ultra-Violet Spectra, Ionisation Potentials, and Their Significance in Chemistry. Quart. Revs. Chem. Soc. (London) **2**, 73 (1948).

*120 a.* — Theory of Molecular Structure and Spectra. Annu. Rev. physical Chem. **5**, 163 (1954).

*120 b.* WASSERMANN, A.: Mode of Proton Addition to Conjugated Double Bonds. J. Chem. Soc. (London) **1959,** 979. For related papers by the same author cf. J. Chem. Soc. (London) **1954,** 4329; **1955,** 581; **1958,** 1014, 3228; **1959,** 983, 986.

*121.* WIERL, R.: Elektronenbeugung und Molekülbau. II. Ann. Physik [5] **13**, 453 (1932).

*122.* WIZINGER, R.: Mono- and Polyatomic Chromophores. Mededel. Vlaamse Chem. Ver. **19**, 65 (1957) (see esp. p. 94).

*123.* ZANKER, V.: Über den Nachweis definierter reversibler Assoziate („reversible Polymerisate") des Acridinorange durch Absorptions- und Fluoreszenzmessungen in wäßriger Lösung. Z. physik. Chem. A **199**, 225 (1952).

*124.* ZECHMEISTER, L.: *cis-trans* Isomerization and Stereochemistry of Carotenoids and Diphenylpolyenes. Chem. Revs. **34**, 267 (1944).

*125.* — Some Stereochemical Aspects of Polyenes. Experientia **10**, 1 (1954).

*126.* — Some in vitro Conversions of Naturally Occurring Carotenoids. Fortschr. Chem. organ. Naturstoffe **15**, 31 (1958).

*127.* ZECHMEISTER, L. and A. L. LeROSEN: Contribution to the Stereochemistry of Diphenylpolyenes. Science (Washington) **95**, 587 (1942).

*128.* — — Stereoisomeric Diphenyloctatetraenes. J. Amer. Chem. Soc. **64**, 2755 (1942).

*129.* ZECHMEISTER, L., A. L. LeROSEN, W. A. SCHROEDER, A. POLGÁR and L. PAULING: Spectral Characteristics and Configuration of Some Stereoisomeric Carotenoids Including Prolycopene and Pro-$\gamma$-carotene. J. Amer. Chem. Soc. **65**, 1940 (1943).

*130.* ZECHMEISTER, L. and E. F. MAGOON: Spectral Maxima of Stereoisomeric Polyenes. Chem. and Ind. **1957,** 431.

*131.* ZECHMEISTER, L. and J. H. PINCKARD: On Stereoisomerism in the Cyanine Dye Series. Experientia **9**, 16 (1953).

*132.* ZECHMEISTER, L. and A. POLGÁR: *cis-trans* Isomerization and Spectral Characteristics of Carotenoids and Some Related Compounds. J. Amer. Chem. Soc. **65**, 1522 (1943).

*133.* — — *cis-trans* Isomerization and *cis*-Peak Effect in the $\alpha$-Carotene Set and in Some Other Stereoisomeric Sets. J. Amer. Chem. Soc. **66**, 137 (1944).

*134.* ZECHMEISTER, L. and W. A. SCHROEDER: On the Occurrence of Stereoisomeric Carotenoids in Nature. Science (Washington) **94**, 609 (1941).

*(Received, May 1, 1958 and, in part, April 7, 1959.)*

# Namenverzeichnis. Index of Names. Index des Auteurs.

# Sachverzeichnis. Index of Subjects. Index des Matières.

## Fortschritte der Chemie organischer Naturstoffe. Progress in the Chemistry of Organic Natural Products. Progrès dans la chimie des substances organiques naturelles. Herausgegeben von L. Zechmeister, California Institute of Technology, Pasadena, California, U. S. A.

*Bisher erschienen:*

**Erster Band:** Mit 41 Abbildungen im Text. VI, 371 Seiten. Gr.-8⁰. 1938.
Ganzleinen S 348.—, DM 72.25, $ 17.20, sfr. 74.—

**Zweiter Band:** Mit 24 Abbildungen im Text. VII, 366 Seiten. Gr.-8⁰. 1939.
Ganzleinen S 348.—, DM 72.25, $ 17.20, sfr. 74.—

**Dritter Band:** Mit 10 Abbildungen im Text. VI, 252 Seiten. Gr.-8⁰. 1939.
Ganzleinen S 264.—, DM 55.45, $ 13.20, sfr. 56.80

**Vierter Band:** Mit 47 Abbildungen im Text. VIII, 499 Seiten. Gr.-8⁰. 1945.
Ganzleinen S 474.—, DM 99.10, $ 23.60, sfr. 101.50

**Fünfter Band:** Mit 34 Abbildungen. VIII, 417 Seiten. Gr.-8⁰. 1948.
Ganzleinen S 305.—, DM 50.40, $ 12.—, sfr. 52.20

*Über den Inhalt der fünf Bände erteilt der Verlag bereitwilligst Auskunft.*

**Sechster Band:** Mit 32 Abbildungen. VIII, 392 Seiten. Gr.-8⁰. 1950.
Ganzleinen S 338.—, DM 55.80, $ 13.30, sfr. 57.80

*Inhalt:* **Deuel, H. J. jr.,** and **S. M. Greenberg.** Some Biochemical and Nutritional Aspects in Fat Chemistry. — **Lederer, E.** Odeurs et parfums des animaux. — **Hoffmann-Ostenhof, O.** Vorkommen und biochemisches Verhalten der Chinone. — **Reti, L.** Cactus Alkaloids and Some Related Compounds. — **Bonner, J.** Plant Proteins. — **Dhéré, Ch.** Progrès récents en spectrochimie de fluorescence des produits biologiques.

**Siebenter Band:** Mit 12 Abbildungen. VII, 330 Seiten. Gr.-8⁰. 1950.
Ganzleinen S 325.—, DM 53.70, $ 12.80, sfr. 55.50

*Inhalt:* **Jeger, O.** Über die Konstitution der Triterpene. — **Heusser, H.** Konstitution, Konfiguration und Synthese digitaloider Aglykone und Glykoside. — **Niemann, C.** Thyroxine and Related Compounds. — **Cook, A. H.** Penicillin and its Place in Science. — **Stoll, A.,** and **B. Becker.** Sennosides A and B, the Active Principles of Senna. — **Williams, J. W.** Some Recent Developments in the Chemistry of Antibodies.

**Achter Band:** Mit 47 Abbildungen. XI, 400 Seiten. Gr.-8⁰. 1951.
Ganzleinen S 427.—, DM 70.50, $ 16.80, sfr. 72.20

*Inhalt:* **Frey-Wyssling, A.,** and **K. Mühlethaler.** The Fine Structure of Cellulose. — **Stacey, M.,** and **C. R. Ricketts.** Bacterial Dextrans. — **Leloir, L. F.** Sugar Phosphates. — **Kenner, G. W.** The Chemistry of Nucleotides. — **Schinz, H.** Die Veilchenriechstoffe. — **Asahina, Y.** Neuere Entwicklungen auf dem Gebiete der Flechtenstoffe. — **Galinovsky, F.** Lupinen-Alkaloide und verwandte Verbindungen. — **Paller, M.** Brechwurzel-Alkaloide. — **Corey, R. B.** X-Ray Diffraction Studies of Crystalline Amino Acids and Peptides. — **Zechmeister, L.,** and **M. Rohdewald.** Some Aspects of Enzyme Chromatography.

**Neunter Band:** Mit 20 Abbildungen. XI, 535 Seiten. Gr.-8⁰. 1952.
Ganzleinen S 498.—, DM 82.50, $ 19.60, sfr. 84.50

*Inhalt:* **Inhoffen, H. H.,** und **H. Siemer.** Synthetische Chemie der Carotinoide. — **Baxter, J. G.** Synthesis and Properties of Vitamin A and Some Related Compounds. — **Meunier, P.** Les Antivitamines. — **Stoll, A.** Recent Investigations on Ergot Alkaloids. — **Tomita, M.** Die Alkaloide der Menispermaceae-Pflanzen. — **Dean, F. M.** Naturally Occurring Coumarins. — **Borsook, H.** The Biosynthesis of Proteins and Peptides, including Isotopic Tracer Studies. — **Kalckar, H. M.** The Enzymes of Nucleoside Metabolism. — **McNutt, W. S.** Nucleosides and Nucleotides as Growth Substances for Microorganisms. — **Campbell, D. H.,** and **N. Bulman.** Some Current Concepts of the Chemical Nature of Antigens and Antibodies.

*Weitere Bände siehe nächste Seite!*

*Fortsetzung von vorhergehender Seite*

**Zehnter Band:** Mit 19 Abbildungen. IX, 529 Seiten. Gr.-8°. 1953.
Ganzleinen S 498.—, DM 83.—, $ 19.80, sfr. 85.—

*Inhalt:* **Alder, K.,** und **Marianne Schumacher.** Anwendungen der Dien-Synthese für die Erforschung von Naturstoffen. — **Mark, H.** Physical Chemistry of Rubbers. — **Asselineau, J.,** et **E. Lederer.** Chimie des lipides bactériens. — **Rosenkranz, G.,** and **F. Sondheimer.** Syntheses of Cortisone. — **Chatterjee, A.** Rauwolfia Alkaloids. — **Feinstein, L.,** and **M. Jacobson.** Insecticides Occurring in Higher Plants.

**Elfter Band:** Mit 67 Abbildungen. VIII, 457 Seiten. Gr.-8°. 1954.
Ganzleinen S 448.—, DM 74.80, $ 18.—, sfr. 77.40

*Inhalt:* **Peat, S.** Starch: Its Constitution, Enzymic Synthesis and Degradation. — **Freudenberg, K.** Neuere Ergebnisse auf dem Gebiete des Lignins und der Verholzung. — **Inhoffen, H. H.,** und **K. Brückner.** Probleme und neuere Ergebnisse in der Vitamin-D-Chemie. — **Schmid, H.** Natürlich vorkommende Chromone. — **Pauling, L.,** and **R. B. Corey.** The Configuration of Polypeptide Chains in Proteins. — **Schroeder, W. A.** Column Chromatography in the Study of the Structure of Peptides and Proteins. — **Lemberg, R.** Porphyrins in Nature. — **Albert, A.** The Pteridines.

**Zwölfter Band:** Mit 15 Abbildungen. X, 550 Seiten. Gr.-8°. 1955.
Ganzleinen S 497.—, DM 82.80, $ 19.80, sfr. 85.10

*Inhalt:* **Haagen-Smit, A. J.** Sesquiterpenes and Diterpenes. — **Jones, E. R. H.,** and **T. G. Halsall.** Tetracyclic Triterpenes. — **Tschesche, R.** Neuere Vorstellungen auf dem Gebiete der Biosynthese der Steroide und verwandter Naturstoffe. — **Haxo, F. T.** Some Biochemical Aspects of Fungal Carotenoids. — **Warren, F. L.** The Pyrrolizidine Alkaloids. — **Thompson, E. O. P.,** and **A. R. Thompson.** Paper Chromatography in the Study of the Structure of Peptides and Proteins. — **Roche, J.,** et **R. Michel.** Acides aminés iodés et iodoprotéines. — **Slotta, K.** Chemistry and Biochemistry of Snake Venoms. — **Beadle, G. W.** Gene Structure and Gene Action.

**Dreizehnter Band:** Mit 48 Abbildungen. XII, 624 Seiten. Gr.-8°. 1956.
Ganzleinen S 645.—, DM 107.50, $ 25.60, sfr. 110.10

*Inhalt :* **Cole, A. R. H.** Infrared Spectra of Natural Products. — **Schmidt, O. Th.** Gallotannine und Ellagen-Gerbstoffe. — **Tamm, Ch.** Neuere Ergebnisse auf dem Gebiete der glykosidischen Herzgifte: Grundlagen und die Aglykone. — **Nozoe, T.** Natural Tropolones and Some Related Troponoids. — **Price, J. R.** Alkaloids Related to Anthranilic Acid. — **Chatterjee, A.,** S. C. Pakrashi and G. Werner. Recent Developments in the Chemistry and Pharmacology of Rauwolfia Alkaloids. — **Graßmann, W.,** und **E. Wünsch.** Synthese von Peptiden.

**Vierzehnter Band:** Mit 38 Abbildungen. VIII, 377 Seiten. Gr.-8°. 1957.
Ganzleinen S 450.—, DM 75.—, $ 17.85, sfr. 76.80

*Inhalt :* **Bohlmann, F.,** und **H. J. Mannhardt.** Acetylenverbindungen im Pflanzenreich. — **Tamm, Ch.** Neuere Ergebnisse auf dem Gebiete der glykosidischen Herzgifte: Zucker und Glykoside. — **Brockmann, H.** Photodynamisch wirksame Pflanzenfarbstoffe. — **Birch, A. J.** Biosynthetic Relations of Some Natural Phenolic and Enolic Compounds. — **Sobotka, H.,** N. Barsel and J. D. Chanley. The Aminochromes. — **Morton, R. A.,** and **G. A. J. Pitt.** Visual Pigments. — **Brown, H.** The Carbon Cycle in Nature.

**Fünfzehnter Band:** Mit 81 Abbildungen. VI, 244 Seiten. Gr.-8°. 1958.
Ganzleinen S 246.—, DM 41.—, $ 9.75, sfr. 42.—

*Inhalt:* **Schlubach, H. H.** Der Kohlenhydratstoffwechsel der Gräser. — **Zechmeister, L.** Some in vitro Conversions of Naturally Occurring Carotenoids. — **Hartwell, J. L.,** and **A. W. Schrecker.** The Chemistry of Podophyllum. — **Hodgkin, Dorothy Crowfoot.** X-ray Analysis and the Structure of Vitamin $B_{12}$.

**Sechzehnter Band:** Mit 27 Abbildungen. VI, 226 Seiten. Gr.-8°. 1958.
Ganzleinen S 240.—, DM 40.—, $ 9.50, sfr. 41.—

*Inhalt:* **Freudenberg, K.,** und **K. Weinges.** Catechine, andere Hydroxy-flavane und Hydroxy-flavene. — **Wiesner, K.,** and **Z. Valenta.** Recent Progress in the Chemistry of the Aconite-Garrya Alkaloids. — **Tamelen, E. E. van.** Structural Chemistry of Actinomycetes Antibiotics. — **Bonner, J.** Protein Synthesis in Plants. — **Kuhn, H.** The Electron Gas Theory of the Color of Natural and Artificial Dyes: Problems and Principles.